FACILITY DESIGN AND MANAGEMENT HANDBOOK

Eric Teicholz Editor-in-Chief
President
Graphic Systems, Inc.

McGRAW-HILL
New York San Francisco Washington, D.C. Auckland Bogotá
Caracas Lisbon London Madrid Mexico City Milan
Montreal New Delhi San Juan Singapore
Sydney Tokyo Toronto

Library of Congress Cataloging-in-Publication Data

Facility design and management handbook / Eric Teicholz, editor-in-chief.
p. cm.
Includes index.
ISBN 0-07-135394-1
1. Facility management. 2. Factories—Design and construction. 3. Production planning.
I. Teicholz, Eric.

TS177.F326 2001
658.2—dc21 00-050002

McGraw-Hill

3 4 5 6 7 8 9 0 DOC/DOC 0 6 5 4 3 2 1

P/N 135395-X
PART OF
ISBN 0-07-135394-1

The sponsoring editor for this book was Wendy Lochner and the production supervisor was Pamela A. Pelton. It was set in Times Roman by Progressive Information Technologies.

Printed and bound by R. R. Donnelley & Sons Company.

This book is printed on recycled, acid-free paper containing a minimum of 50% recycled de-inked fiber.

Dedicated first and foremost to my wife Judy and son Adam, who are a constant inspiration to me. Secondly, to my colleagues and co-authors, who made the book possible.

ABOUT THE EDITOR-IN-CHIEF

Eric Teicholz is president and founder of Graphic Systems, Inc., of Cambridge, Massachusetts, which specializes in all aspects of facilities management, including automation consulting, systems integration, market research, and training. He has specialized in facilities design and management for 25 years. A contributing editor of *Facilities Design & Management,* Mr. Teicholz lectures internationally, and has published hundreds of articles and nine books. He received his postgraduate degree in architecture from Harvard, where he has taught in the Graduate School of Design and served as Associate Director of the Laboratory for Computer Graphics and Spatial Analysis. He has written four other books for McGraw-Hill, *Computer-Aided Facility Management, CIM Handbook, Δ/E Systems Update,* and *CAD/CAM Handbook.*

CONTENTS

PREFACE

". . . as prefaces, though seldom read, are continually written, no doubt for the behoof of that so richly and so disinterestedly endowed personage Posterity (who will come into an immense fortune), I add my legacy to the general remembrance."

Charles Dickens, The Pickwick Papers, *1837*

RATIONALE AND THE CHALLENGE

This is the tenth book that I have written or edited. All of them have been concerned with technology and most have dealt with facility management. The first challenge for this book came from Wendy Lochner, my Editor at McGraw-Hill. She tried persistently to get me to expand the scope of the book beyond FM technology as an end in itself. These discussions led to the plan for a general handbook on facility management dealing with its various functions, the relationship between them, and an exploration of how technology pervades most of the diverse functions that comprise FM.

The next challenge was that most of my professional experience is associated with the technology side of FM, so I decided early to seek out the best authors possible to cover some aspects of technology as well as other innovative components of facility management. FM means so many different things to different people. FM tasks change in scope and importance based on the size of an organization, public vs. corporate ownership, leased vs. owned space, centralized vs. distributed management, single vs. multiple buildings and so forth.

The next hurdle had to do with the collaboration between chapter authors. Individual chapter authors, in some cases, addressed subjects that overlapped. Authors needed to communicate both to insure minimal overlap as well as to build on content in other chapters. This was accomplished by developing a collaborative project Extranet web site where it was possible to store documents, post various versions of chapters, biographical and contact information as well as host news, support email and so forth.

This effort was only partially successful. Although the software functioned well, most authors did not avail themselves of the software's collaborative aspects–something I thought held great potential for enabling authors to discuss ideas on-line. Additionally, most of the authors did their writing at home, using relatively slow modem connections to the book site. The site evolved into a relatively simple repository of documents. Even this function was extremely beneficial since many of the authors posted their own documents on the site as well as using it to communicate with other authors, the editor, and the publisher.

FM is changing at a rapid pace, not only in terms of technology (e.g., effectively using the Internet) but in terms of organizational issues associated with how buildings are designed, constructed and managed. The production period of the book itself has been one further challenge. The manuscript for this book was submitted to the publisher in July of 2000, with a scheduled publication date of March 2001. This means that the selected topics had to be ones that would be likely to remain relevant eight months beyond the time the manuscript was completed.

I hope that the resultant handbook provides a context for facility managers to understand how innovative organizational and technical practices can assist them in the practice of their profession.

FORMAT

The book attempts to organize the text into functional areas that follow the life cycle of a project: from planning, to design and analysis, to implementation. An introduction to facility management was added at the front and a technology section added at the back. In some cases, a subject matter might apply to more than one functional area. In such cases, the primary function area was used as the repository for the chapter. The 31 chapters that comprise this book therefore include five functional areas:

Section I: *INTRODUCTION* contains a single chapter, a general introduction to facility management theories, concepts and organizational models.

Section II: *PLANNING* contains six chapters, dealing with Benchmarking, Strategic Planning, Business Transformation, Financial Management, Customer Service, and Disaster Recovery Planning.

Section III: *ANALYSIS AND DESIGN* contains nine chapters, including Alternative Workplaces, Facilities Conditions Assessment, Thinking Globally–The Competitive Edge, Sustainable Design, Smart Buildings/Intelligent Buildings, Lighting, Ergonomics and Workplaces: Managing the New Healthcare Real Estate Portfolio (case study), Implementing Technology at Rocketdyne (case study).

Section IV: *IMPLEMENTATION* contains eight chapters, including Project Management and Implementation, Real Estate Portfolio Management, Supporting the Mission of the Organization: An Approach to Portfolio Management (case study), The Design and Construction Process, Space and Asset Management, Operations and Maintenance, Energy Management, Security.

Section V: *TECHNOLOGY* contains seven chapters including Overview and Current State of FM Technology, Integrating the Internet into Facilities Management, The Birth and Development of a Collaborative Extranet for the A/E/C Industry, Technology at the Architect of the Capitol (case study), Michigan State University Facility Management Master's Level Certificate Program (case study), Space Management at the University of Minnesota (case study), Teaching Technology at the University of New South Wales (case study).

A detailed summary of each chapter associated with a particular section appears at the beginning of that section.

Eric Teicholz

ACKNOWLEDGMENTS

I received a great deal of help and support in designing and producing this book. First, there was the technical support provided by Brian Giuffrida and his colleagues at Framework Technologies in setting up and initially hosting their collaborative project web software. When Graphic Systems took over the hosting of the software, they then trained Cara Rodgers from my staff to take over all management of the software. I am extremely grateful to Cara for maintaining that role throughout the writing and production stage of the manuscript. She communicated with authors, taught them how to use the site, and handled most of the administrative tasks associated with production.

Bob Boes, a personal friend of mine, helped formulate content and recommend authors for some chapters. I am also indebted to Francoisé Szigeti, another author, for assistance in recruiting international authors and in helping determine the organization of the book. Likewise, BOMI provided support and assistance in defining content.

Two other staff from Graphic Systems deserve considerable credit for production assistance. First is Deana DiMenna, who had primary responsibility for formatting and printing the final document, and second is Deb Theodore, a professional writer as well as GSI's comptroller, who edited a number of book chapters.

Finally, I would like to thank my colleagues and co-authors who spent many hours discussing subjects, content, format and administrative aspects of the book and many more hours writing and editing their chapters.

P • A • R • T • 1

INTRODUCTION

CHAPTER 1 FACILITY MANAGEMENT—AN INTRODUCTION

1.1 Introduction

1.2 FM Principles and Theories

The underpinnings of the discipline of facilities management (FM) are found in the intersection of work, workers, and workplace. This section traces the evolution of the theories of facilities management and shows how FM deals with a wide variety of information and traditional professions, but in a unique way—as it affects what people do and where they do it. The role of FM in influencing the workplace has a direct affect on performance and productivity. Several key studies of the impact of workplace on performance are summarized and discussed. Also included in this section is a discussion of FM clients, both internal and external; the role of FM in most organizations; how FM relates to and can support organizational business strategy; and approaches to managing the FM function.

1.3 FM Organizational Models

The roles and responsibilities of FM organizations vary depending on the nature of the organization they serve. This section examines differences across broad categories of organizations and how those distinctions influence the way in which FM identifies, procures, and uses resources to plan, deliver, and manage workspace. The discussion focuses on differences between public and private sector organizations; single versus multiple locations; owned versus leased space; large versus small facilities; and in-house versus outsourced or entrepreneurial organizations.

1.4 FM Orientation by Company/Space Type

The nature of the business in which an organization is engaged will influence the roles, responsibilities, and organizational "home" for the FM function. Just as medicine is different from software engineering, so too are facilities that support these activities. These factors, in turn, influence the nature of the FM organization and the emphasis placed on particular activities and services. This section discusses five models of FM and the responsibilities, organization type, and industry in which they are most commonly found.

1.5 Placement of FM within Corporate Structures

Because organizational structure is unique and specific to particular organizations, the FM function is not commonly found in the same place in every organization. In some circumstances, FM may not be used to identify facilities' roles and responsibilities. This section examines the historical evolution of FM as a corporate function, the relationship of FM duties to other corporate units, and several of the more common reporting and communication channels for FM.

1.6 FM Strategic Plan and Mission as a Reflection of the Corporate Vision and Mission

The broader corporate or organizational vision and mission are the foundation for the direction in which the organization is headed. This section discusses elements, methods, and outcomes of strategic planning, and the ways in which FM needs to structure its strategies, plans, and activities to support, enable, and reflect the broader corporate strategy.

1.7 Characteristics of a Successful Facilities Manager

Effective FM requires a broad and diverse set of skills and knowledge. Several organizations and researchers have examined and profiled FM professionals. This section presents the results of those examinations and discusses the background, education, skills, and experience of successful facilities managers.

1.8 Summary

CHAPTER 1
FACILITY MANAGEMENT—AN INTRODUCTION

Timothy Springer
President
Hero, Inc.
E-mail: hero_inc@ameritech.net

1.1 INTRODUCTION

Facilities management (FM) is a multidisciplinary or *transdisciplinary* profession drawing on theories and principles of engineering, architecture, design, accounting, finance, management, and behavioral science. These disciplines each have a rich history of theory, research, and practice. Facilities management, as a new discipline, builds on this foundation to create a new set of theories and practices.

1.2 FM PRINCIPLES AND THEORIES

In its original definition of the roles and responsibilities of facilities managers, the International Facilities Management Association (IFMA) identified 41 responsibilities under eight headings. According to the IFMA, the scope of Facilities Management covers real estate, planning, budgeting, space management, interior planning, interior installation, architecture and engineering services, and building maintenance and operations (see Table 1.1).[1] In a study of how successful companies manage their facilities, Wilson (1985) reduced that list to five: real estate, long-range planning, building projects, building administration, and office support.[2]

During the latter half of the 1970s, Herman Miller, Inc. established the Facilities Management Institute (FMI). This organization helped establish the new profession of FM and gave birth to IFMA. In viewing the underlying model for facilities management as a profession, FMI developed a three-element model of people, process, and place.

1.2.1 FM as People, Process, and Place

The diagram (Figure 1.1) of three interlocking circles represents the important role facilities and facilities management play in integrating employees, work processes, and workplaces into a coherent, productive, holistic system. FM serves to coordinate the interface between *what* people do and *where* they do it. Thus, FM touches on elements of human resources, process engineering,

TABLE 1.1 IFMA Classification of FM Roles and Responsibilities

Maintenance Operations	**Architectural/Engineering Services**
Furniture maintenance	Code compliance
Finishes maintenance	Construction management
Preventive maintenance	Building systems
Breakdown maintenance	Architectural design
Exterior maintenance	
Custodial/housekeeping	**Real Estate**
Landscape maintenance	Building leases
	Site selection
Administrative Services	Acquisition/disposal
Corporate artwork	Building purchases
Mail services	Property appraisals
Shipping/receiving	Subleasing
Records retention	
Security	**Facility Planning**
Telecommunications	Operational plans
Copy services	Emergency plans
	Strategic plans
Space Management	Energy planning
Space inventory	
Space policies	**Financial Planning**
Space allocation	Operational budgets
Forecasting needs	Capital budgets
Furniture purchase	Major financing
Furniture specifications	
Furniture inventory	**Health and Safety**
Interior plans	Ergonomics
Furniture moves	Energy management
Major redesign	Indoor air quality
Trash/solid waste	Recycling program
Hazardous materials	Emissions

Source: Facility Management Practices, Research Report #16, International Facility Management Association, copyright 1996. BOMI © 1997, 2nd qtr. pp. 1–9.

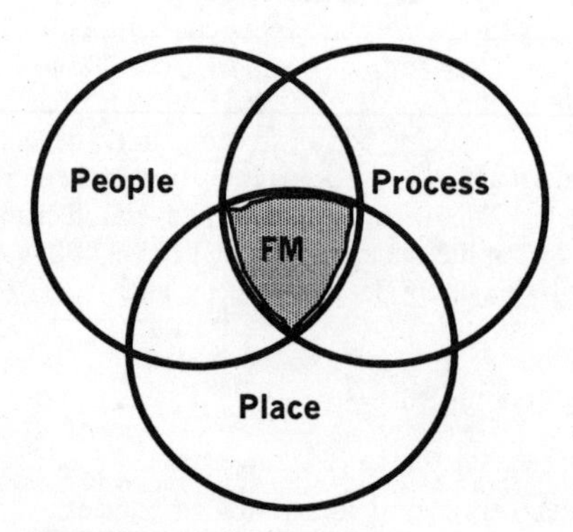

FIGURE 1.1 FM Components.

ergonomics, architecture, and interior design. Consequently, critical components of FM include planning and maintaining, and providing the assets, both large and small, that support the endeavor of people at work.

1.2.2 The Impact of Environment on Productivity

The role of the physical work environment on performance and productivity has been the subject of many studies over the past 15 years. In a compilation of over 40 studies of productivity improvement, changes in performance and productivity were shown for everything from reduction in glare (3–5%) to total facility redesign (18%).[3] Included in this compilation are three key studies that examine multiple factors in the environment. These studies demonstrate with great consistency that the physical place of work has a significant impact on people's performance. For example, adequate space was shown to have a positive impact in each of the three studies, while noise or lack of speech privacy had a negative effect.

Table 1.2 summarizes the results of the three key studies. The plus and minus signs next to the variables in the Aetna study (#3) reflect the weights given those variables. For example, space adequacy and ease of circulation and order are positive influences, whereas storage, speech privacy, and ability to concentrate—or the lack of each of these—are negative influences.

In Table 1.2, the first column shows performance mandates for buildings taken from the American Institute of Architects (AIA) *Building Systems Integration Handbook.*[4] These mandates are elements that a building must provide. They serve as useful categories for looking at variables in each of the three studies listed:

- The first study (Springer, 1982) was conducted at a major insurance company. The research examined the impact of the physical work environment on the performance and productivity of computer terminal operators. In a simulation laboratory that painstakingly recreated all work conditions, the study found that an optimized working environment yields a 10–15% improvement in

TABLE 1.2 Environment and Performance: 10 Years' Data

BSIH[1]	Springer[2]	BOSTI[3]	Aetna[4]
Performance mandates:			
Spatial	Floor space	Floor area, enclosure	(+) Space adequacy
	Aisle width	Layout	(+) Ease of circulation
	Storage	Storage—personal	(−) Storage
	Work surface	Relocation frequency	(+) Order
		Work surface width	
Acoustical	Noise	Noise	(−) Speech privacy
		Ease of communication	(−) Ability to concentrate
Visual	Lighting for VDT	Glare	(+) Lighting for VDT
	Glare		(−) Glare
			(+) Appearance
Thermal		Temperature fluctuation	(−) Temperature
Air quality			(−) Air quality

Sources:
1. Rush, R. (Ed.). 1986. *The Building Systems Integration Handbook.* John Wiley & Sons.
2. Springer, T. J. July 1979. *CRT Furniture Requirements.* State Farm Insurance Companies.
2a. Springer, T. J. 1982. *Visual Display Terminal Workstations: A Comparative Evaluation of Alternatives.* State Farm Insurance Companies.
3. Brill, M., et al. 1984. *Using Office Design to Increase Productivity.* Workplace Design and Productivity, Inc.
4. Sullivan, C. July 1989. Employee productivity and satisfaction as a result of the home office renovation. The Aetna Casualty and Surety Company.

performance when compared to the traditional workplace. Improvements included changes to the lighting, spatial configuration, furniture and support equipment, and seating.[5]

- The second study was conducted by the Buffalo Organization for Social and Technological Innovation (BOSTI, 1984). This research looked at the workspace as a collection of 18 facets, including physical enclosure, esthetics, privacy, furniture, status, communication, and thermal control. The research involved some 6,000 workers in 80 organizations. The results suggest improvements in the workplace have a reasonable result of an improvement in productivity of ~5%. The upper limit of benefits reported was 15%, while the average reported was 12%.[6]
- The third study (Aetna, 1989) was conducted by another large insurer. This research looked at the full range of environmental elements, including space, acoustics, visual quality, thermal comfort, and air quality. A comparison of traditional workplaces with a new workplace where environmental elements, such as space, light, adjustments, and configuration were optimized, yielded an improvement in performance of 10–15%. In other words, the new "improved" environments allowed people to be 10–15% more productive when compared with traditional work environments.[7]

Clearly, the consistency of these results over time shows that design and management of the physical work environment is a critical contributor to the performance and financial well-being of organizations. From seemingly minor elements, such as reducing glare, to major elements, such as the design or reconfiguration of entire buildings, the physical facility can either impede or enhance worker performance. The research shows that when done well, investments in facilities can yield returns on investments that contribute to the business bottom line. The workplace is a tool that when appropriately configured and designed to fit the people who work there and the work that they do, can yield direct and measurable impact on the performance and productivity of the entire organization. Thus, FM and the role it plays in planning, design, and management of workplaces is key to business success.

More recently, Mike Brill and his colleagues at BOSTI have begun to release the results of research conducted since 1994 that examined the impact of workplace on individual performance, team performance, and job satisfaction. Preliminary results are consistent with earlier results. A rough rank order of workplace characteristics that have the greatest effect on performance includes the following:

1. Acoustic privacy
2. Support for spontaneous collaboration and impromptu meetings
3. Support for face-to-face meetings, one-on-one, in individual's workspace
4. Support for administrative services, office chores
5. Sufficient storage—a place to put your "stuff," much of which is paper
6. Group space that is distraction-free
7. Taking a break
8. Dedicated project rooms
9. Appropriate ergonomics and physical comfort
10. Accommodating technology[8]

1.2.3 Defining the Client: Internal and External Customers

One of the difficulties facilities managers face is defining exactly who their clients are. Different client groups want different things from a facility. Internal clients, those who are part of the organization, have different criteria and needs than external clients, those who are not part of the "host" organization.

From a purely financial perspective, the company or organization that occupies a facility is the primary client of FM. Yet, even that distinction is not as simple as it sounds. Companies are owned by shareholders, run by executives, managed by managers, and staffed by employees. Each may have a different set of expectations and requirements for the facility.

In general, FM clients are distinguished as internal or external. Internal clients or customers include all the people who use the facility. This group includes employees, managers, and executives of the organization; contract employees in areas such as maintenance or security; customers of the business itself; and those people who come to the facility to conduct business with those who work there. Increasingly, the challenge facilities managers face in satisfying internal customers is the demand for quick response and quality service.

It is important to understand that frequency of request for service may not be an indicator of the importance of the service provided. For example, on a day-to-day basis facilities managers will spend considerable time responding to requests and the needs of employees and staff. They may be asked to adjust the heat, replace light bulbs, fix a squeaky door, or clear snow from the parking lot. However, the few times a year that the facilities manager reports to senior management may determine the size of staff and budgets; approval of projects, both major and minor; and what services are kept in-house or outsourced.

Facilities managers must serve the needs of employees and managers of the organization in a way that enhances their ability to perform. In addition, facilities managers must meet these goals while simultaneously keeping an eye on the efficiency and effectiveness of the building itself and the expenditure of resources, both human and capital. Financial accountability while maintaining quality work environments is a key challenge facing facility managers.

External customers of FM include the general public, investors and shareholders, and government agencies. The general public has concerns over appearance, safety, and environmental responsibility. Consequently, the facilities manager must ensure that his or her organization is a good corporate citizen. This could mean taking extra care in maintaining an attractive and appropriate facade for the buildings, ensuring the grounds are well kept, providing space for community activities, minimizing noise and disruption during construction projects, or finding environmentally friendly ways of dealing with waste.

Investors and shareholders want a good return on their facility investment while expecting the facility to represent the organization. The buildings an organization owns and occupies are the physical manifestation of that organization. It is the job of the facility manager to insure the buildings reflect the corporate culture while maintaining and operating them in the most cost efficient manner. The facility manager can meet these requirements by implementing cost savings practices that can include energy efficient building systems and products, as well as using durable and attractive, low maintenance materials.

Government agencies monitor compliance with standards, regulations, and statutes. To serve these customers, the facility manager must be familiar with laws, codes, and regulations. Familiarity with the requirements imposed by such regulations as fire safety codes and the Americans with Disabilities Act (ADA) will help the facilities manager ensure their buildings comply with government regulations.

1.2.4 Customer Service and Quality

Consumers of services (i.e., customers) have become more sophisticated, informed, and demanding. With experience comes knowledge of what is needed and what is possible. With this knowledge comes the ability to better express and explain customer requirements. As customers become better at expressing what they need and want, the quality and value of products and services must improve. Accompanying this expectation for improved value and quality is the need for continuous improvement in service and process delivery.

A large and rapidly growing body of knowledge has evolved around the topic of customer service and quality improvement. Total quality management (TQM) and continuous quality improvement (CQI) are but two of the many terms and acronyms associated with the national and global drive toward quality improvement. Starting with a definition of quality, the link between quality and customer service becomes apparent:

> We have found that like all important ideas, quality is very simple. So simple, in fact, that it is difficult for people to understand. (*Roger Hale, CEO, Tennat Company*[9])

What is quality? The TQM/CQI literature talks about quality as generally meeting or exceeding customer expectations. It is not more and better of everything, but *does* involve three dimensions: input, process, and output.

In the case of FM as customer service, *input* involves understanding the customer's needs and expectations, *process* is taking action to meet or exceed those needs and expectations, and *output* is following up to determine if the action was appropriate and if the customer is satisfied. This becomes a continual process of communication, action, and follow-up, always striving to meet and exceed customer expectations while operating effectively and efficiently.

1.2.5 FM's Role as Corporate Administrative Support

A major trend in business today is movement toward flatter organizational structures with fewer middle managers. Most workers are being given more responsibility and more authority to carry out work. This trend toward fewer managers is coupled with a reduction in clerical workforce. The remaining administrative staff are fewer, higher skilled, higher quality, serving more people or serving only the most senior executives. The workforce loss at the top and bottom of organizations puts added burden on the professionals in the middle. So, people in all types of jobs, including facilities managers, are truly challenged. Jobs are more complex. Work is less broken up into pieces, less routine, more varied and more interesting. Consequently, training and new skills are needed to prepare people for dealing with the broader responsibilities that accompany this new way of working.

Administrative support staff, of which facilities units are often a part, need to act more like generalists, professionals, and problem solvers.[10] As generalists, facilities managers need to know something about the broad range of issues that affect facilities. For example, the facility manager may not be an expert computer aided design (CAD) operator, but she needs to know what to expect from a CAD system and how to read CAD drawings. Similarly, a facility manager may need to know enough about building systems like heating ventilation and air conditioning (HVAC) to know when to call the specialist. Like the physician who is a general practitioner, the facility manager needs to be able to diagnose the problem and call in, and manage, the specialist. This problem identification role of the generalist is different from the specialized knowledge needed to fix the problem. Problem solving for facility managers in this new model of work involves collecting and synthesizing information, working with other units in the organization, and identifying the needed resources and course of action that will alleviate the problem. It is a longer view than the traditional reactive mode of "crisis management" that has characterized facility and building management in the past. Thus, the technical, human, and management skills required of facilities managers in the future have increased the professional nature of the job. This "professionalization" of the job is supported by cross-training, creation of teams, and continuing transfer of routine tasks from people to technology. The move toward headcount reduction and the increased professionalizing of the remaining support staff offer some advantages to the facilities organization. These include:

- More specialists on staff, including electrical, mechanical, and HVAC engineers
- Outsourcing certain functions to specialty contractors, such as housekeeping
- More training for FM staff both internally and through external education programs
- Task and outcome specific job assessment related to responsibilities and competencies (measurement and rewards can be tied directly to specific abilities and expectations rather than using general performance assessment and review)

- Matching staff competencies with job requirements
- Increased decentralization and local autonomy of FM[11,12]

Disadvantages of this trend in organizations for facilities units include:

- Larger span of control. This means people have more for which they are responsible and accountable.
- Greater impact when key people are absent. When units are expected to do more with less, there are increased demands on everyone. When someone is ill or absent, the strain on the remaining organization increases.
- Reduced ability to handle the unexpected. With lean organizations, everyone is expected to handle a full workload. Consequently, when the unexpected arises, it places an additional burden on time, energy, and human resources.
- Increased need to plan for changes. By maximizing the available resources, any changes that are expected must be planned for well in advance to avoid overtaxing the unit's capacity.

1.2.6 Relationship of FM to Corporate Business Strategy and Planning

Businesses regularly engage in a process of strategy formation and planning. Whether this is formally called strategic planning (see also Chapter 3) or not, it involves the identification of a direction for the business that aligns the organization's goals and resources with changes in market conditions and opportunities. For example, a retail organization may identify a goal for increasing profits by a certain amount. It could then outline a plan to accomplish this goal through a combination of increased sales, reduced headcount, and closing certain locations and opening new stores to serve new markets. Corporate strategic planning will have a direct impact on facilities management. Strategic planning requires that organizations make decisions about the direction of the company, the business it will be in, the resources required, assets to be managed or disposed of, and how market share will be captured. When facilities planning is linked to corporate strategic planning, the overall FM function can operate from a strategic reference point. Because real estate and buildings represent a major investment for any organization, the role of FM in acquiring, managing, and disposing of these assets can be critical to the success of the business. Costs of acquisition, operation, maintenance, and repair, as well as revenues from leases and sales, are just some of the ways facilities represent assets to an organization. Thus, facilities and the assets they represent can contribute to the strategic plan.[13]

Because the purpose of strategic planning is to articulate appropriate responses to current and future business conditions, the process is founded on the need for change. Facilities managers must be prepared to accommodate changes that emerge from the strategic planning process. For example, changes in interest rates can have a major impact on such FM decisions as leasing versus buying a building. Similarly, businesses may need to grow or shrink their workforce based on the demand for their product or service. Critical among changes in organizations is the need to either increase, decrease, or change the complement of space and its uses. Staffing assignments will change as well. For example, many organizations today are looking for ways to increase sales and expand service. They are exploring ways of increasing the time their salesforce spends calling on customers. One result of increasing time salespeople spend with customers is a reduction in the time they spend in the office. Consequently, some of these organizations have implemented space use plans, such as shared offices, "hoteling" workspaces, or much smaller workstations, each of which can significantly reduce the amount of office space required. These space reduction strategies have led to a surplus of space and the need for facilities organizations to dispose of extra space and consolidate real estate holdings. This is just one example of how implementing a strategic plan will involve the FM function as it is manifested in the physical workplace.

1.2.7 FM as Proactive versus Reactive Management

By its very nature, FM is often the practice of reacting to the crisis of the moment. Whether it is responding to calls from employees who are too hot or too cold, dealing with a burst water main, or organizing a crew to clean up after an event, the facilities manager can easily find himself reacting to one issue after another. However, reactive management seldom leaves time for planning, anticipation of problems, and dealing with issues before they become problems. Proactive management, or management that is based on planning and anticipation of crises, is generally the preferred mode of operating. Proactive management allows the facilities manager to avoid crises and make better use of time, staff, and resources.

Management Styles. The structure, culture, and politics of an organization strongly affect how facilities are planned, delivered, and managed. Organizations vary widely in terms of management styles and change over time. Management that is adequate when a business is young and growing may not be sufficient as it becomes larger and more stable. Several models have been offered to explain management style in facilities organizations. These models conform to the state of the organization, thus, they are called models of stability, control, and flexibility.

Stability. In stable environments, facilities tend not to be at the forefront of management's thinking. As the name implies, stability is characterized by organizations where change is relatively slow and the need to address facilities' issues is handled on a case-by-case basis. Depending on the size of the organization, facilities may not be part of the formal job duties of any one manager or executive. In small companies, facilities are considered only to the extent that they influence, either positively or negatively, the ability of the company to grow and prosper. In large organizations, facilities may be of interest to certain key executives who make decisions for the entire organization based on their personal experience and opinions. Because change is slow and the need to address problems is an infrequent event in large stable organizations, there is a tendency to react to problems as they arise and not incorporate facility considerations in any planning that occurs.

This style of managing facilities is associated with several costs. First, there is the cost of inappropriate action; that is, the expense of fixing the resulting consequences of those actions. Second is the hidden cost of forcing people to work in environments that are not well suited to their needs, thereby accommodating problems until they become sufficiently critical to warrant attention. In light of the research presented earlier regarding the potential benefit of well-designed environments on performance and productivity, this can amount to a significant "hidden" cost.

Control. The second approach to managing facilities relies on more formal control, recognizing the importance of physical assets to the company or organization, and placing responsibility for managing facilities with an organizational unit specifically and formally charged with maintaining the investment. Procedures and policies are often formalized and institutionalized in such an organizational unit, and outcomes are measured in terms of cost avoidance. For example, by instituting a recycling program, a company can save money on landfill charges while reselling scrap and generating revenue. While planning may involve facility considerations, there is not a clear understanding of the importance of facilities in the larger business plan. Similarly, operations and strategy are not clearly defined as distinct. As a consequence, the importance of facilities decisions may not be considered in longer-term goal setting and decision making. Instead, control and management are emphasized and bureaucracy grows to serve this need. Facilities management is more technical than bureaucratic and concerned primarily with cost containment, not cost avoidance; therefore, an approach that emphasizes control will limit benefits.

Flexibility. The last approach to FM reflects recent trends in general management. This model is characterized by a more flexible approach involving decentralized responsibilities across the organization. This model coincides with the movement in organizations toward matrix management and beyond to flexible or organic forms of organizational structure. Matrix management is an approach to organizations that strives to combat the traditional "column" or pyramid approach to responsibility. In a columnar or pyramidal organization, lines of communication and chain of command are clearly defined and formalized. People are members of well-defined units and traditionally communicate only within their prescribed unit or column. Information flows up and down within clearly defined channels. By contrast, matrix organizations use a grid to represent communication and organizational connections. A person in a matrix organization may communicate with other people throughout the company who are working together in cross-functional teams to solve complex problems. Who one reports to is more a function of the problem to be solved rather than of traditional hierarchies of "boss" or subordinate. With matrix organizations, there is a resulting emphasis on cross-functional teams, outcome orientation, and decentralization of responsibility. While matrix management is looked to as a better model, it can be difficult to implement and has been problematic in some places. This discussion is not intended to imply that matrix management is better, and, in fact, it may not be appropriate in every situation.

In flexible or "organic" organizations, executive management tends to provide strategic direction and guidelines, but allows for local control and flexibility in implementation. Skills reside in communities of practice. Ecologies of complementary skills and operations coordinate and collaborate to serve customers and markets. Here FM is understood as an important management function. It is proactive and planning based. It relies on information as important input to decision making, and it includes consideration of clients' needs and quality service in addition to technical concerns over efficiency and cost effectiveness.[14,15]

1.2.8 FM as Profit Center, FM as Overhead

In economic terms (see also Chapter 5), the question of profit versus overhead is addressed as one of benefit versus cost, or the benefit-to-cost ratio (BCR). Traditionally, items such as buildings, furnishings, and other elements for which facilities managers have responsibility are treated as expenses or cost items. This is due, in part, to accounting conventions dictated by the tax code. Matters of depreciation and other accounting principles determine how assets are treated on the corporate books and, in turn, how management views those assets. Increasingly, however, businesses are looking for ways to minimize costs and maximize benefits. For facilities, this presents both a challenge and an opportunity. The application of such economic analyses as life-cycle costing (LCC), benefit-to-cost ratio (BCR), savings-to-investment ratio (SIR), and activity-based costing (ABC) enable facilities managers to show how appropriate expenditures on facilities can be viewed as investments on which one can expect a return rather than as fixed or sunk costs.[16,17]

Many examples can be found to illustrate this point. Improvement in a single aspect of operations, for instance, changes in energy systems, can yield direct and measurable benefits to the corporate bottom line, and companies such as Prudential Insurance Company, Pennsylvania Power and Light, and Georgia Power have realized millions of dollars in cost savings by wisely investing capital for energy conservation and improved energy efficiency.[18]

The Pennsylvania Power and Light (PP&L) example is particularly noteworthy. The original project involved replacing the lighting system in a drafting area in order to reduce lighting and maintenance costs. Prior to modifying the system, PP&L was spending $12,745 annually to light the office. The new lighting system cost $48,882. With the new system, energy efficiency increased 75% and maintenance costs fell to $3,106 annually—a savings of $9,639. The energy savings (see also Chapters 13 and 23) resulted in a direct payback period of slightly more than five years. The benefit did not stop there, however. The change in lighting system not only yielded energy savings, it also affected the performance of the people working in the office. In the area affected by the lighting change, groups of people worked on plans for large clients of the utility company. This was extremely detailed and visually taxing work. The new lighting system not only was more energy efficient, but provided more and better quality of light. Consequently, the people working in this area were better able to see the details in their work and were able to work faster and with fewer errors. This resulted in a 13.2% improvement in the performance of the people working in the office. The economic benefit of this performance improvement in terms of improved productivity was $414,110 annually.

This example illustrates several critical points. First, people are the most expensive element of most workplaces. In offices and other "white-collar" work environments, personnel costs amount to 85–90% of costs over the life of a building. Thus, in the PP&L example, while the energy savings alone would justify the expenditure for the new lighting system, the impact of the modest performance improvements resulting from relatively small investments in building systems yielded enormous benefit and profit to the organization's bottom line.[19] This impact, large benefit from small improvements in the major cost centers, is called leveraging. As shown in the performance and productivity research described earlier, investment in facilities can be leveraged to yield significant improvements in overall organizational performance.

Facilities management practices from operations to "softer" matters, such as improved communication and workflow resulting from effective space planning, can all have direct and indirect impact on the bottom line and profitability of an organization.

1.2.9 Churn and Change: The Basic Driver of FM Workload

The term churn has been coined to describe the pace of change of physical workplaces within an organization. *Churn rate* has been defined as "the volume of occupant moves within a facility or complex and is expressed as a percentage of the number of total offices occupied during a given year."[20] In other words, a churn rate of 33% means that one-third of an organization's people and/or workplaces are moved in a given year. Churn, or relocation, is driven by several factors. The most common factor driving churn is reorganization and relocation associated with the addition or loss of staff, loss or gain of leased space, installation of new equipment, reorganization of functional units, or changes in work processes. Strategic decisions can lead to downsizing, consolidation of units, acquisition or disposition of properties, and other major changes that affect an organization's churn rate.

Understandably, when changes in organizational strategy, function or structure occur, the physical work environment is expected to change as well. Churn becomes one of the basic drivers of FM activities and workload. Move planning and management are time intensive, critical activities performed or managed by facilities organizations. The success of the move plan and implementation can determine the success of the organization's change. If everything goes smoothly and is completed on time and within budget, the organization can implement its planned change without interruption. However, if plans go awry and deadlines are missed and costs run over, the ability of the organization to successfully function and to implement change is challenged. Unfortunately for such a critical function, successful management of churn goes largely unnoticed outside FM. If churn is not successfully managed, however, many people will know it and the facilities manager will be held accountable.

1.3 FM ORGANIZATIONAL MODELS

Every organization is unique and organizations change over time. Organizations differ widely in how well they identify, procure and use resources necessary to plan, provide and manage physical workplaces. Since FM units are reflections of the organizations of which they are a part, there is not a typical FM organization. Facilities managers share a common broad range of responsibilities—however the emphasis and application varies widely across different organizations and organization types.

To simplify the many different types of organizational cultures, structures and types, the following discussion will focus on broad distinctions.

1.3.1 Public versus Private Sector

Public Sector FM. Few FM managers are challenged to handle as diverse a type of facility with as limited resources as the public sector facility manager.[21] These individuals work at the municipal, state, and federal government level, as well as for public universities and educational institutions. While many people don't think of public university and educational institutions as public sector organizations (like government), they are more similar than different since they operate as publicly funded, not-for-profit organizations. Public sector facilities managers must operate within large bureaucracies. Consequently, nearly every action is subject to multiple reviews by diverse groups and must comply with a host of regulations and standards. An example of the challenge of the public sector is found in government procurement policies. Regulations and guidelines for procurement are thick with detail and obtuse in meaning. Attention to these details is intended to insure fairness to all in the procurement and acquisition of products and services. While the concept of fairness may be served, this process does not always result in the best or most sensible choice based on business criteria. For example, the lowest bid on a project may not provide the most complete or highest quality solution. Consequently, cost avoidance at procurement could result in higher cost of ownership, operation, or maintenance over the life of the product or service. Change in the public sector is difficult due to regulations and the laborious process of modifying or changing those regulations. Competition for scarce resources offers little incentive for cross-unit cooperation. On the positive side, public sector facilities management organizations often deal with large, complex projects and issues in an effective manner. They are often more stable, better organized, and have clearly written procedures, making the job of managing the process more effective.

Recent initiatives to streamline the federal government have given rise to innovative and creative approaches to developing and managing public sector workplaces. The General Services Administration, the largest "landlord" in the world, has recently published the results of several years' research and inquiry into the requirements for a comprehensive approach to developing and managing workplaces. The result is a model that integrates people, space, and technology to improve productivity, employee satisfaction, and the effectiveness of the organization.[22]

Private Sector FM. In contrast to the public sector, private sector organizations are increasingly focused on reducing layers of management, "leaning" the structure, and maximizing the efficiency and effectiveness of the organization. As a result, facilities managers in the private sector must respond to more flexible organizations engaged in more rapid change. Accompanying the increased flexibility and pace of change is growing pressure to reduce costs and contribute to the overall profitability of the company. Unlike the public sector, where the emphasis is on staying within an allocated budget, in the private sector the challenge is to avoid costs and to reduce expenditures. Simultaneously, the private sector facilities manager must focus on providing high quality service to his or her customers both internal and external. Decisions in private sector organizations do not always have to be fair, but they almost always have to make good business sense.

1.3.2 Single versus Multiple Locations

An important determinant of how a facilities unit is structured and organized is the number and location of the facilities it manages. More and more, organizations are going through changes in attempts to streamline their operations and put more of their resources closer to their customers. Consequently, it is increasingly common to find individuals and parts of business units dispersed geographically. This geographical disbursement is often accompanied by a decentralized organizational structure. These dispersed parts of organizations are often electronically networked. Changes in where and how work gets done has led to changes that affect FM.

Deconstruction into smaller, more autonomous, more experimental units tends to erode the size, power, and roles of an organization's center—their headquarters. The size of the staff is often reduced and the capacity of the central office is diminished, especially in long-range planning, management information systems, and real estate and FM. Increasingly, some of the capabilities lost by erosion of center of the organization are outsourced and some people move from headquarters to consulting.

In such organizations, the role of the facility manager may change to something more like a coach/advisor (*"If you do it like this, you'll work better."*).[23] This environment is often found in organizations that have adopted the matrix or flexible styles of management. Facilities managers find themselves part of several cross-functional teams. Consequently, they must understand the contribution of facilities to the various teams and projects on which they are working. In this context, they can best serve the role of advisor and coach, pointing out where the facility can help or FM can facilitate improvements in function and output.

Multi-Site FM. In practice, multi-site FM is most often found in two different settings. First, there are large organizations with multiple locations within the same city or region. This type of organization is characterized by a headquarters staff that centralizes administration and management, but decentralizes operational control to the individual sites. However, complete decentralization is rare in this situation.

The other form of multi-site FM involves widely dispersed organizations with locations in multiple cities, states, regions, or countries. These may or may not report to an intermediate line of management at the regional or division level. Headquarters staff is reduced and provides policy, monitoring, and oversight functions. The FM department at headquarters is almost all staff (as opposed to line management) having no direct responsibility for any general administrative services.[24] In this model, administration and operations are dispersed to the individual sites. The small size of the FM organization and the multiple duties of FM staff, usually result in reliance on outsourcing and consultants in order to minimize undue costs. Large national and international firms tend to use this model.

Single-Site FM. Single-site FM may involve one building or multiple buildings; however, the model remains essentially the same. With a single site, the tendency is to use a small staff with multiple responsibilities. Control of facilities involves administration of projects, leases, contracts, budgets, and procurement. Management responsibilities rest in a centralized department. The size and complexity of the organization will determine the degree to which internal staff or outsourcing and consultants are used. For example, a small, lean organization may choose to hire an outside security service. Conversely, a large complex organization with highly sensitive on-site product development may choose to trust security to in-house employees.

1.3.3 Owned versus Leased Space

The decision to own or lease space is influenced by a number of factors. Economic considerations include considerations that will maximize benefit to the organization. Leasing can increase financial leverage by enabling a firm to use more assets than it could under a secured loan for purchase. Leasing may also offer greater flexibility since it may be easier to cancel a lease than to liquidate a purchased asset. Some organizations may find tax advantages, such as depreciation to offset taxable income, makes owning a building more attractive than leasing it.

The role of the facility manager changes depending on whether he or she is managing an owned building, a leased building, or serving as a landlord. In the case of leasing, either as tenant or landlord, FM responsibilities are usually spelled out in the lease. The facilities manager should be familiar with the basics of lease administration. Leasing space can be a long-term strategy for flexible management of space needs. Leasing also can be used as an effective, albeit often costly, short-term solution for space problems. Serving as a landlord brings additional considerations for FM staff. Either case involves considerations of strategy and organizational direction. Increasingly, lease administration and negotiation are thorny legal areas fraught with liabilities and legalities that require the advice of expert counsel.[25] In the case of owned facilities, questions of acquisition, maintenance, repair, and disposal are addressed over the life of the facility. Facilities staff in owned buildings tend to be larger and accommodate a broader range of services either through in-house skills and personnel or by contracting with outside sources.

In a study of FM practices, the majority of buildings of the facilities surveyed are owned by the company occupying them. The most likely types of buildings to be owned were those requiring heavy investments in specialized construction that made it difficult to adapt them to other uses. Examples include computing centers, research laboratories, education and training centers, and health care facilities. Those building types most commonly leased include storage, warehouse, and office space. The type of industry and use to which the building was put influenced whether it was owned or leased. Industries more likely to lease office and storage space include hospitality, energy, and mining. Those most likely to own space include government, education, research, retail, and chemical industries.[26]

1.3.4 Large versus Small Inventories

As is true of almost all organizations, with increased size comes complexity and the need to increase specialization and depth. Just as an individual homeowner can be a jack-of-all trades, the manager of a large apartment complex will require the skills of many people like electricians, plumbers, and other specialists. FM units reflect the size and complexity of their parent organization. Larger organizations, whether single or multiple site, require more in the way of facilities support. Often, responsibilities that might not constitute a full-time job in a smaller organization take on a magnitude of scope and responsibility in large organizations that requires several people to perform. It is common for certain responsibilities, such as real estate and building project work, to separate from building administration as an organization and its facility inventory grows. These functions then become the purview of departments separate from the facilities unit. As a result, FM may not have input into or control over some longer-term strategic decisions, such as real estate acquisition, disposal, or building construction. Also, in the largest organizations, the administrative functions separate from what might be called office services or operations.[27] Facilities managers still provide input to these functions, but the administrative responsibility will often reside with another unit. This apparent reduction in scope of responsibility is counterbalanced by a depth of functioning within the organization. For example, in an organization of 500 to 1,000 people, the need to track furniture inventory and manage moves may be part of a job for one person. However, in an organization of 10,000 people or more this one responsibility may be a full-time job for several people within the facilities organization.

Conversely, smaller organizations tend to rely on generalists who handle multiple responsibilities. The size of the organization may not require a degree of specialization, but the breadth of scope of services provided by the facilities organization, if a formal definition exists, will normally be broader.

1.3.5 In-House Staff versus Entrepreneurial Organization

One of the greatest changes in recent years to the role of the facilities manager is the issue of appropriate allocation of functions within the organization. The central question behind the issue of allocation of function is "Should this be done with in-house employees or contracted to an outside firm?" Driven by the demands on business to control costs, many functions are being outsourced. This is true for facilities as well as other functions. Contract employees are becoming common in many units within business organizations. The challenge to facilities managers is to identify those functions that are most efficiently and cost effectively managed within the organization and also those activities that are reasonably outsourced. Some facilities organizations are finding it constructive to adopt an entrepreneurial model where the cost of providing services is bid on by both in-house and outsourced units. This approach forces the internal units to stay efficient and competitive. It also helps make the case within the organization for the value represented by an efficient and effective FM unit that knows the organization better than any outside contractor could. In some instances, facilities managers have been able to demonstrate not only the cost savings to the organization by using in-house services, but they have been able to negotiate more attractive contracts with outside sources when it has been appropriate.

1.4 FM ORIENTATION BY COMPANY/SPACE TYPE

In addition to variation by number of sites, size of the organization, lease versus owned facilities, and the other distinctions discussed previously, FM, as it is practiced, may vary significantly depending upon the nature of the business of which it is a part. Wilson (1985) identified five basic functions and an equal number of different models of FM and characterized them by the responsibility, organization type, and industry. The five functions are:

1. Real estate: includes purchasing and leasing property
2. Long-range planning: a strategic function that includes gathering information on long-term space needs and formulating budgets and future plans
3. Building projects: handling capital funded projects, such as designing a new building, refurbishing an existing building, or major repair work
4. Building administration: handling the building structure, envelope, and building systems. This also includes building-related services such as cleaning, security, etc.
5. Office support services: operational support of the business activity, such as fleet, travel, photocopying, purchasing, etc.

The related models of FM are:

- Real estate based
- Office support based
- Integrated building based
- Integrated, organization based

Real estate based models include units responsible for real estate and building projects work. Responsibilities end with the building envelope. This model predominates among retailers, commercial insurance, and dispersed manufacturing organizations.

Office support based models include units responsible for building management and office support services. Generalist skills predominate. Real estate and building projects will be outsourced, and long-range planning is done by ad hoc committees. This model tends to find favor among small, growing firms; leased facilities of manufacturing firms; and larger facilities of "administrative" organizations like insurance companies.

Integrated, building based models involve all-inclusive facilities units. Responsibilities extend from the boiler room to the boardroom, from real estate to office support services. This model is a multidisciplinary group usually headed by a manager or executive who has a good knowledge of the company's mainstream work and is often found in financial firms where the building is owner occupied. It is also found in some more stable manufacturing firms.

The integrated organization based model includes everything but office support services. This model emphasizes space planning more than the other models. Design skills predominate. This model is found in large organizations with a strong corporate culture and multiple owned sites that are recognized as a powerful asset base to the company.[28]

1.5 PLACEMENT OF FM WITHIN CORPORATE STRUCTURES

Organizational structure is most often unique and specific to a particular organization. Consequently, there is no one place where FM is commonly found in organizations. In fact, FM may not even be called FM in some organizations. FM units can be found in many different places within organization structures. It is rare to find organizations in which FM units are at the executive board level. More commonly, facilities units report through one of several channels—real estate, property management, administrative systems, personnel and administration, finance, or general services. This list is neither comprehensive nor exhaustive, but represents some of the more common lines of reporting where FM can be found.

1.5.1 Communication and Reporting Channels

There is always risk involved when communicating with other people. Messages pass through many different channels and can, therefore, be subject to inaccurate interpretations. Many perceptual filters can potentially break down the original message. For this reason, reporting channels should remain as concise as possible in order to lessen the chance for error. In communicating a specific task, and all of the procedures necessary for completing the task, the manager or supervisor must ensure that the message sent is the same message that was received.

Reporting channels within the organizational structure should also be concise. A manager in charge of the task specialist should report his or her review directly to his or her supervisor or to a top-level person directly related to the status of the subordinate performing the task. This diminishes the chances that other persons along the reporting channel have, with personal motives, to alter or filter a report or review on that employee.

The channels of communication within a corporate structure must remain open and candid, but only among those in the direct channel. The channel should be as concise as management will allow. A prolonged and tangled channel suffers many different influences on the message in transit.

The implementation of task specialization benefits most from the shortest communication channel. One leader exposing one subordinate to the uses and methods of the task, as well as appraising his successes and hardships in the procedure, allows for open and unaltered communication. Thus, any problems are transferred from one sender to the other without any unnecessary complications.

1.5.2 Related Functions Performed by Other Departments

Using the broadest definition of FM—the forty-one responsibilities under eight broad headings from the IFMA shown in Table 1.1—it is rare to find an organization where the formal FM unit is directly responsible for all of these elements. While there are no universal generalities and each organization is unique, there are some common responsibilities that are usually delegated to units other than the facilities unit. It is common to find a situation in which responsibilities are distributed among the FM unit, sister departments such as real estate, and remote departments such as purchasing. For example, real property management is often the purview of the real estate function and not directly a FM responsibility. Similarly, security can often be outsourced, or part of the responsibilities of a sister unit report through the same executive channel. Telecommunications has become a sophisticated specialty in itself and is often the responsibility of a separate department.

1.5.3 Historical Placement and Evolution of FM

In her study of facility organizations, Wilson (1985) described three ways that FM units evolve:

- *Downward:* from the building acquisition function. In this type of FM unit, the role of the property manager tends to be primary. Staff usually has real estate, design, surveying, or building construction backgrounds.
- *Upwards:* from office services. FM becomes a distinct unit only when the company grows to a certain size and complexity to fully occupy one large or several buildings and thereby warrant a distinct FM unit.
- *Outwards:* from a team set up to manage the development of a new, custom designed building. Once the building project is complete and the facility is occupied, the team stays in place to manage the facility.

1.6 FM STRATEGIC PLAN AND MISSION AS A REFLECTION OF THE CORPORATE VISION AND MISSION

The strategic plan developed for facilities is a critical resource control tool available to the FM department. It is also a corporate endorsement of facilities priorities and policies. The elements of a strategic plan for facilities flow from the organizational mission, goals, objectives, and strategic plan. Using the organizational strategic plan as the basis for facilities strategies enables FM to establish clear parameters for facilities actions. The FM plan comprises a strategy derived from analysis of the facilities actions that will be required to support and carry out the corporate mission, goals, and objectives. The FM strategic plan enables FM professionals to develop probable scenarios, based on corporate planning, in order to anticipate facilities actions that will be required in conjunction with corporate planning.

A methodology termed multiple scenario analysis will provide an action framework for the facilities professional to develop several probable scenarios for any set of major actions for which they may potentially be called. This scenario development allows FM to plan for what is most likely to happen, what may happen, and what will probably not happen in the following year. This will enable scheduling of facilities priorities and activities accordingly.

Step 1: The Mission Statement

The first step in development of a strategic plan for FM is to develop a mission statement and policies with corresponding goals and objectives for the FM department.

Defining a mission statement for the FM department may appear difficult, but the appropriate content of that statement will evolve from the organization's mission, goals, and objectives. For example, if the corporation is downsizing, the mission statement of FM will require language to reflect that the mission of FM includes dealing purposefully and effectively to support downsizing actions. FM statements for a company that is divesting should focus on the successful sale, divestiture, and redeployment of existing facilities resources.

Step 2: Goals and Objectives

Goals and objectives are quantitative and qualitative statements that evolve from the mission statement. Facilities actions are turned into numbers and action statements when developing goals and objectives. Policies, goals, and objectives also set functional parameters and operating standards for the FM department.

Goals are quantitative, issues stated in numbers. Specific goals may include facilities upgrade, completion of a series of moves, and so on. Objectives are qualitative; for example, completion of a standards program, determination of functional procedures which eliminate staff downtime and disruption during change projects, and reduction of multiple vendors down to a package of three or four from whom most furniture purchases are made.

Goals and objectives will be unique to the FM department that develops them. They depend somewhat upon the management personnel in place, the capabilities and aspirations of the people, and the general background of the staff. It is important that goals, objectives, and policies be put in writing. That way, no one assumes that particular goals are desired or any unwritten policies are in place.

Step 3: Time Horizon

The facilities strategic plan must cover three time horizons:

- *Long Range:* A scenario covering two to five years out

- *Intermediate Range:* A scenario covering six months to two years in the future
- *Short Range:* A scenario from the present to six months in the future

The FM function is most effectively operated with a current year action plan, a two-year facilities strategy, and a five-year long-range plan. A planning horizon that seeks to project out more than five years is unfeasible because of rapid changes in business conditions. Major business changes that are occurring in all types of businesses potentially preclude the extrapolation of any of today's planning into the future. For example, IBM traditionally employed a strategic planning scenario of seven years. The time frame was based on very practical reasoning. It takes two years to develop a product and five years to market it. Thus, the company operated on a seven-year cycle.

Careful analysis of organizational cycles will provide solid information about the appropriate FM planning horizon. The budget cycles of most organizations have evolved as a result of organization rhythms. A facility strategy that coincides with the budgeting process is probably the most feasible answer for an appropriate planning horizon.

Short-term deployment of resources such as short-term space forecasts and furnishing purchases are reflected in the current year operating plan. Potential longer-term deployment of resources occurs in the medium and longer range strategic plan scenarios.

Step 4: Assessing the Organization's External Milieu

There may be external events that will affect FM operations that are not anticipated by corporate planning. For example, corporate strategic planners may be unaware of planned legislation that will demand an increase in the indoor air quality of the office environment. They may also be unaware of rising furniture costs and potential material shortages. Corporate planners or economists are not likely to track increasing utility costs and co-generation legislation.

All of these events and more must be tracked by the FM organization. If current information about any of these items is not available to use in preparation of the first strategic plan, it must be obtained through a process referred to as situation analysis. Completing a situational analysis in order to proceed with strategic planning involves capturing information about each external element or factor that can have a significant influence, negative or positive, on the FM department in the coming year. It is important that the FM department clearly understand what these factors are, along with the potential impact they can have on FM functions.

Tracking external information may be new to many FM organizations, but, in many instances, no one else in the organization will be tracking these items. Corporate planners will expect that the FM department is tracking information that is facilities specific.

Another important part of external analysis is to trace the major changes that have occurred in the organization's specific industry and within the organization over the past five years. It is also important to note where these major change patterns coincided with FM developments and priorities and where they did not. This information will provide a historical basis for projecting the future. Even though growth may no longer be a constant, other factors that influence corporate policies and decisions may still be in place. Some factors may be forecast with a fair degree of confidence, while other predictions may not flow as easily from current external events.

Step 5: Key Variables

The next step in the process is to determine the key variables that will make or break the success of the FM function. A good example of this would be electrical power needs. The facilities budget must include adequate funds to upgrade the electrical service to several of the organization's buildings. If it does not, installation of computer workstations will overload electrical systems in the facility. Allocation of a budget to upgrade these electrical requirements is a key variable, as the lack of it will cause critical electrical service problems to a significant portion of organizational staff.

Each key variable should be listed in order of priority, with respect to its impact on either the FM department or the organization or both. Some companies use a weighting system to determine the impact of these variables.

Step 6: Strategic Alternatives

Once key variables have been thoroughly explored, development of alternative strategies can begin. The purpose of development of more than one strategy is to be able to choose one that is maximally feasible or to choose elements of several strategies to combine into the most plausible scenario. There are several guidelines that will prove useful in strategy development:

- *Develop Multiple Scenarios:* It is important that more than one strategy (scenario) be developed. If only one strategy is presented to staff or senior management, there is no area for discussion; the scenario will appear set and inflexible. An even number of scenarios is recommended, avoiding a tendency to choose the one "in the middle." More than five scenarios will probably result in two scenarios that are very similar with only minor differences.
- *Make Each Scenario as Different from the Other as Possible:* The value of multiple scenarios is that it spurs thinking about the differences between the various paths and about the value of any of those differences. Multiple scenarios, which are quite different, present the opportunity to combine the best aspects of each scenario into a more ideal situation. The willingness of FM to present several scenarios with differing solutions also points to a certain amount of flexibility in the organization's facilities professionals to solve problems in the organization's best interest, rather than only relying on conventional paradigms.
- *Make Each Scenario Realistic:* Avoid presenting (as some planners do) one scenario which is totally unfeasible. Most rationalize doing so for reasons of opening negotiations with a scenario that can easily be thrown out. Many corporate managers will regard this tactic as game playing.
- *Enhance Each Scenario with Assumptions:* Present the external factors that led to the assumptions underlying each scenario during articulation of each scenario.
- *Present Each Scenario as an Executive Summary:* Avoid lengthy verbiage in scenario presentation. Highlight the key points and distinct differences between scenarios. Attach any necessary elaboration as an addendum. Present the scenarios in a format in which they can be easily compared.
- *Name, but Avoid Labeling the Scenarios:* Naming scenarios allows people to easily identify and recall them, but calling one scenario the best and one the worst will taint the responses. If the scenarios are valued in this way, they are not all as plausible. Similarly, select names that do not imply value (e.g., "doomsday" may have negative connotations).

The appropriateness of the final selected strategy should be evaluated against several criteria:

- Internal consistency of the strategy
- Consistency with the environment
- Flexibility within the strategy
- Appropriateness in light of available resources
- Satisfactory degree of risk
- Appropriate time horizon
- Workability

Use of a facilities strategy which corporate management can review, evaluate, and follow will bring FM efforts into agreement with other corporate activities. This process and action will make a significant contribution to improving the effectiveness of FM at the corporate level.

1.7 CHARACTERISTICS OF A SUCCESSFUL FACILITIES MANAGER

1.7.1 Attitudes and Values

Because FM is so diverse, the successful facilities manager must be willing to learn because he or she will never be knowledgeable in everything. Consequently, the facilities manager must be able to recognize when expert advice is needed and where to go to get it. Also, he or she must be able to deal well with the pressures of change and be willing to take certain risks. The successful facility manager should be accessible, approachable, not bothered by interruptions, and able to handle exceptions quickly and easily. In a word, the successful facilities manager must be flexible.[29,30]

1.7.2 Goals

The primary goal of the facility manager is promotion of his or her department as a concerned, efficient, effective service function. The facility manager must be a persuasive advocate for the FM unit. He or she should be ambitious and strive for promotions. This implies a highly motivated individual who also strives to assist others in realizing their goals and aspirations.

1.7.3 Communications and Interpersonal Skills

Perhaps the most important skills a facilities manager must possess or develop are communication skills and the ability to deal with people. The ability to get the message across to various constituents, from the boardroom to the boilerroom, is key to realizing the goals of FM, so oral and written communication are essential skills for FM. In addition, increasingly, the ability to make sophisticated economic arguments is becoming a part of the FM function as well. While staff and specialists may do the actual calculations, the facilities manager must be able to make sense of the numbers and, of equal importance, he or she must be able to explain them to others. In addition, facilities managers must be able to use information and information technology effectively to communicate with their own units as well as internal and external customers of FM services.

While FM is increasingly a technology-rich practice, the ability to interact with a wide variety of people is essential to success as a facility manager. The facilities manager must know and understand both the formal and informal channels of authority in their organization. His or her relationship with other managers and executives in the organization can be key to the success of the FM unit. In addition, the manager must be able to say no diplomatically since it is often impossible to satisfy the needs of everyone. Negotiating skills are increasingly important as competition for resources increases. The ability to give and receive feedback and collect and use information from others is an important part of the job. Listening, as well as speaking, is vitally important. Finally, a genuine interest in the welfare of all customers—staff, employees, management, executives, clients, and customers alike—is at the heart of FM.

1.7.4 Time Management and Multi-Tasking Skills

The demands of the FM function require a facilities manager to be an effective time manager as well. While planning and proactive management is preferred, there will always be a reactive element to the job. The ability to handle multiple demands, to arrange priorities and separate the demands in terms of their importance is the mark of a successful facilities manager.

1.7.5 Understanding and Accepting Diversity

People are different. In fact, no two people are exactly alike. In order to provide effective, efficient service to a workforce, the facility manager must recognize this fact and accept the wide variety of people who work in most organizations today. North America, and the United States in particular, has a history of diversity in its population. The "melting pot" includes differences in gender, ethnicity, physical abilities and characteristics, age, size, and almost any other distinguishing characteristic one can imagine. Increasingly, this diversity in the population is manifesting itself in the workforce. Considerations such as the Americans with Disabilities Act (ADA) require organizations to accommodate people with diverse characteristics. The successful facilities manager understands the value of diversity and strives to accommodate the variety of requirements of the entire population.

1.7.6 Characteristics of Good Facilities Managers

Cotts and Lee (1992) observed the following characteristics of good facilities managers:

1. Technically competent
2. Capable of good oral and written communication
3. Comfortable with reaction
4. Service oriented
5. Cost conscious
6. Outgoing, perhaps even political
7. Decisive
8. Slightly legalistic
9. Capable of concurrent problem solving
10. Comfortable with and capable of quantitative measurement
11. Action oriented
12. Able to deal well with people[31]

In Table 1.3, Becker (1988) presents a profile of the facilities manager.[32]

TABLE 1.3 Facilities Manager Profile

Experience and qualifications	Skill with people
• Minimum 'O' levels or equivalent • Management of multi-site complex, preferably hi-tech	• Able to influence; listen • Give and receive feedback • Negotiating skills
Mental abilities	**Ability to get things done**
• Good written and verbal communication • Able to understand and monitor budget process • Innovative; able to think on feet • Able to explain decisions; use information effectively	• Anticipates problems; forward planning • Manages time effectively • Delegates responsibility while remaining accountable
Attitudes and values	**Motivations and interests**
• Willing to learn; flexible • React positively to change; approachable • Can cope with pressure • Risk-taking	• Sets high standards; goal oriented • Self-motivated; determined to succeed and help others do so • Ambitious, seeks senior management role

Source: Becker, F. 1988. *The Changing Facilities Organization.* Haverhill, England: Project Office Furniture.

TABLE 1.4 IFMA Profiles

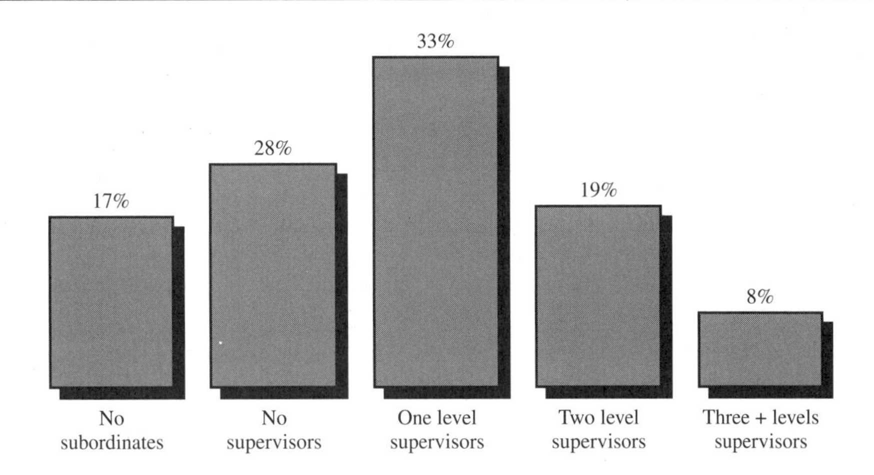

Data profiles by management levels

	Specialists (nonsupervisory)		Unit supervisor		Section head		Manager		Director	
Demographics										
Average work experience		17 yrs		19 yrs		22 yrs		24 yrs		25 yrs
Average FM experience		8 yrs		9 yrs		11 yrs		12 yrs		14 yrs
Average age		40		42		44		46		47
Percent female		47%		37%		22%		14%		13%
Percent college degree		71%		61%		65%		72%		76%
Median base salary		$41,670		$46,750		$55,000		$65,000		$71,000
Total compensation		$42,000		$48,000		$57,500		$68,800		$74,300
Primary job functions										
Multi-function	4	1%	425	41%	915	64%	655	81%	312	87%
Administrative	83	11%	85	8%	82	6%	22	3%	11	3%
Arch/energy	141	19%	83	8%	86	6%	28	3%	2	1%
Bldg/energy	24	3%	46	4%	31	2%	15	2%	2	1%
Operations/maint	59	8%	154	15%	182	13%	59	7%	23	6%
Real estate	41	5%	19	2%	32	2%	8	1%	6	2%
Space management	389	52%	213	21%	113	8%	25	3%	2	1%
Other	10	1%	2	—	0	0%	0	0%	1	—
Sectors										
Services	407	54%	585	57%	844	59%	447	55%	162	45%
Mfg/production	227	30%	289	28%	366	25%	189	23%	70	20%
Gov/education	114	15%	152	15%	231	16%	177	22%	126	35%
Total sample	748	17%	1,026	23%	1,441	33%	813	19%	358	8%

Source: Profiles '94 Salary Report, Research Report #12, International Facility Management Association.

1.7.7 Typical Backgrounds

With so many different types of people engaging in the duties and responsibilities that characterize FM, in truth there is no "typical" facilities professional. However, there are certain characteristics that when averaged across the profession provide a profile of who are facilities professionals (see Table 1.4).

In a survey of over 2,000 facilities professionals, the following profile emerged:

- Forty-three years of age, with twenty years of full-time work experience, about nine of those years directly involved in FM.
- Has worked for the past ten years with a service sector employer and has held current position for four years.
- Responsibilities include facilities totaling 500,000 square feet, half of which are offices.
- Has two levels of supervisors reporting directly, with a total staff of twelve plus contracted labor.
- Job entails multiple management responsibilities with subordinates assigned to specialty work.
- Prior to FM, held management jobs in other departments.
- Holds a bachelor's degree in business and total compensation in 1988 was $50,000.[33]

Across the 2,350 individuals surveyed, the average age of facilities professionals ranged from 39 to 47 years. The older the individual, the more likely they are to have the experience necessary to hold a higher level position in an organization. Average work experience ranged from 16 to 25 years and experience in FM covered 6 to 12 years' span. Of this group, 66–77% completed college and 21% held master's degrees. A large number of this group held business degrees (43%). Other majors included engineering (18%), liberal arts (13%), and architecture (12%) or interior design (8%). The vast majority (74%) were male. The majority (53%) came from other, non-facilities management or supervisory positions. The next largest group came to FM through construction or engineering (20%). Median (50th percentile) salaries ranged from $36,000 for non-supervisory specialist to $64,300 for director level. The largest group (section head) had a median salary of $48,200.[34]

1.8 SUMMARY

As one embarks on the new century and what the future holds for FM, the only constant appears to be change. Rapid progress in technology, approaches to work, business organization, and organization behavior will continue to present challenges to FM professionals. Trends that are present and will continue include moves toward reductions in corporate staffs through downsizing and decentralization, outsourcing of functions and services, and emphasis on cost effectiveness and competitiveness. This suggests that the successful FM professional in tomorrow's workplace will be someone who is flexible, able to think quickly, has a grasp of business and finance issues and how they relate to their unit, can communicate ideas effectively, and is able to manage time and multiple demands easily. FM as a profession will continue to grow and the professionalism of the practice will increase as organizations like the BOMI, BOMA, International Society of Facilities Executives (ISFE), Association of Physical Plant Administrators (APPA), American Institute of Safety and Plant Engineers (AISPE), and International Facility Management Association (IFMA) continue to grow and offer ways of recognition for professionals.

ENDNOTES

1. International Facility Management Association. 1986. *Demographics and trends: The IFMA report #2*. Houston, Texas: International Facility Management Association.

2. Wilson, S. 1985. *Premises of excellence: How successful companies manage their offices.* London: Building Use Studies.
3. Springer, T. J. 1986, 1990. *Improving productivity in the workplace: Reports from the field.* Geneva, IL: HERO, Inc.
4. Rush, R. (Ed.). 1986. *The building systems integration handbook.* NY: John Wiley & Sons.
5. Springer, T. J. 1982. VDT workstations: A comparative evaluation of alternatives. *Applied Ergonomics*, 13 (3): 211–212.
6. Brill, M., Margulis, S., Konar, E., and BOSTI. 1984, 1985. *Using office design to increase productivity.* Buffalo, NY: Workplace Design and Productivity, Vols. 1 and 2.
7. Sullivan, C. 1989. Aetna.
8. Brill, M., Keable, E., and Fabiniak, J. February 2000. The myth of open-plan. *Facilities design and management*, pp. 36–38.
9. Hale, R. 1992. Quest for quality. Tennat Company.
10. Brill, M. 1993. *Now offices, no offices, new offices . . . wild times in the world of office work.* Buffalo, NY: Teknion, pp. 10–11.
11. Becker, F. 1988. *The changing facilities organization.* Haverhill, England: Project Office Furniture, pp. 24–25.
12. Brill, M. Ibid. p. 11.
13. Steelcase. 1989. *Facilities management process.* Grand Rapids, MI: Steelcase Inc., p. 1.
14. Steele, F. 1986. *Making and managing high quality workplaces.* NY: Teachers College Press, pp. 21–22.
15. Becker, F. 1990. Ibid. pp. 11–13.
16. Ruegg, R., and Marshall, H. 1990. *Building economics: Theory and practice.* NY: Van Nostrand Reinhold.
17. Hope, J., and Hope, T. 1997. *Competing in the third wave: The ten key management issues for the information age.* Boston: Harvard Business School Press.
18. Springer, T. J. 1986. Ibid.
19. Ibid. pp. 53–54.
20. International Facilities Management Association. 1988. IFMA Facilities Practices 1988. *Research Report #4.* Houston, TX: International Facilities Management Association.
21. Building Research Board. 1990. Committing to the Cost of Ownership. Washington, DC: National Academy Press, p. xi.
22. The integrated workplace—A comprehensive approach to developing workspace. Washington, DC: U.S. General Services Administration. Office of Government-wide Policy, Office of Real Property.
23. Brill, M. 1993. *Now offices, no offices, new offices . . . wild times in the world of office work.* Buffalo, NY: Teknion, pp. 8, 34.
24. Cotts, D., and Lee, M. 1992. *The facility management handbook.* NY: Amacom, p. 27.
25. Ibid. pp. 107–108.
26. International Facilities Management Association. 1994. *Benchmarks II Research Report #13.* Houston, TX: International Facilities Management Association, pp. 10–11.
27. Wilson, S., Strelitz, Z., and O'Neill, J. 1985. *Premises of excellence: How successful companies manage their offices.* London: Herman Miller and Building Use Studies, p. 19.
28. Ibid. pp. 22–23.
29. Becker, F. 1990. Ibid. p. 57.
30. Cotts, D., and Lee, M. 1992. Ibid. p. 40.
31. Ibid. p. 41.
32. Becker, F. 1990. Ibid. p. 57.
33. International Facilities Management Association. 1994. Profiles 1994: IFMA Survey of Facility Management Professionals and Compensation. *Research Report #12.* Houston, TX: International Facilities Management Association, p. 8.
34. Ibid. pp. 9–13.

P • A • R • T • 2

PLANNING

CHAPTER 2 BENCHMARKING

2.1 Introduction

This introductory section provides the basis for benchmarking as a process that can help the organization and support the quality customer service and expense issues facility departments must successfully address.

2.2 What is Benchmarking?

This section provides definitions and a review of benchmarking from a business and facility management perspective where the science and art factors of benchmarking are discussed. Also included is a general discussion of misconceptions of what benchmarking is and should do and what financial and performance information many organizations currently do or do not capture.

2.3 Benchmarking Survey

Facility management benchmarking and survey processes are reviewed, including how organizations use and misuse benchmarking information and reports and how these should be used.

2.4 Apples to Apples

This section reviews the objectives of benchmarking and benchmarking processes, such as getting started and annualizing the effort. "Apples to Apples" comparison issues, benchmarking definitions, and the benchmarking of functional area costs, costs per full time equivalent (FTE), consumption, and benchmarking processes, procedures, quality, productivity, and performance are discussed.

2.5 Reports

The process to obtain benchmarking information, validating the information, and using "benchmarking clubs" and consultants are reviewed. Examples of reports are provided with special emphasis on quartile reporting and looking at efficiency and effectiveness of these types of reports. The development and management of dependable benchmarking databases are included.

2.6 Benchmarking Maturity

This section reviews the different levels of maturity which businesses and facility organizations progress through as they become more aware and confident in their benchmarking processes, the results from their benchmarking reports, and how the application of benchmarking influences the quality of services being provided to their customers.

2.7 Lessons Learned

Many organizations find benchmarking a difficult task, as the benchmarking results are not necessarily an indication of success or failure. These lessons learned provide a guide to assist the facility organization with leadership and proactive opportunities to ensure the success of an on-going benchmarking program.

2.8 Conclusion

Benchmarking, if properly used and managed, can provide a great benefit to you, your staff, team, customers, and your organization. Your successful benchmarking program will continue to be an on-going part of your department operations and service metric program.

CHAPTER 3 STRATEGIC PLANNING

3.1 Introduction

The chapter on strategic planning starts with a brief overview, which recognizes that the value of effective strategic planning is that it guides in building an essential knowledge base. This knowledge base enables its users to correctly identify the situation at hand and develop recommendations that are appropriate, specific, and viable—with favorable implications for an organization's future. All at once a process, a way of thinking, and an exercise in making a map for reaching a goal, this chapter views effective strategic planning as starting with the question, "What do we want to achieve, and why?"

Strategic planning is then examined including the three fundamental elements that underlie the strategic planning process. There is an acknowledgment that without the three elements in place, the chances of success of strategic planning activities are greatly reduced and the chances for optimal performance almost nonexistent.

A walk-through of the process follows some fundamental truths about strategic planning with brief explanations of each. The reader is then taken through an exploration of the key elements of a methodology including design of the process; collection, analysis, and interpretation of data; and recommendations followed by a word on strategic communications. Included, where relevant, are case studies of organizations that have launched strategic planning projects addressing a number of key planning activities, with examples of lessons learned and successful outcomes.

3.2 Primacy of Process: The Strategic Planning Model

This section of the chapter bases itself on the fundamental assumption that process is more important than outcome and the fact that inherent in the approach to problem solving is the requirement for understanding the client's desired outcomes. In addition, creating a map to guide the project requires understanding not only of more than the key elements of a range of project processes, but also an understanding of how to manage them.

With models, it explains that strategic planning resides in the plan phase of projects, but also that the information it makes use of and its influence extend to the design, construct, and manage phases as well. In addition, it explains how an organization's information output needs to be captured and used to affect or evaluate the performance of a strategic plan.

From there the chapter lays out some technical dimensions to the process, specifically:

- "There": Identify and agree upon a mutually desired outcome
- "Here": Identify the conditions that describe the company now
- What primary disciplines are involved?
- The process is both linear and cyclical
- Data are both quantitative and qualitative
- How do you make the strategic planning process value-centered?
- Observations on the process: human dimensions
- Consider the culture in designing the process
- Garner sustained support from senior leadership
- Enlist internal leadership to facilitate the process
- Determine the extent of desired community involvement
- Map significant project issues and constraints early
- The right problem, the right solution: finding both

3.3 Process: Data Collection and Analysis

This section of the chapter moves from the creation of the foundation for the strategic planning process into the specific activities involved in data collection and analysis. Key chapter sections and their primary arguments are:

- *Capture and Organize Data:* Stating that the best strategic planners never rely on a single method to capture information, but use a variety of tools over the course of the project, the key to methods employed is that they will vary by organization. In addition, the type of information gathered depends upon the problem being solved and the level of detail required, as well as the availability of information and sources.
- Examples of tools for data collection given in this section include surveys, interviews, workshops, Web-based information gathering, pilot projects, and post-occupancy evaluations.
- *Analyze and Interpret Data:.* The quality of the analysis and interpretation hinges on the value of the collected information and the questions asked.
- *Create and Test Alternatives:* Creating and testing alternative scenarios allows a more focused and finite search for information by attempting to defend a hypothesis or scenario and by building a business case of supportive data.
- *Develop Strategic Recommendations:* Linked to the overall project goals and to each specific business goal, the final recommendations and the way they are presented will depend on the audience and whether the desired result is a decision or simply information about the approved strategy. Typical tools or formats presented include final reports, electronic versions, workshops and presentations.

3.4 Project Deliverables

Explained under project deliverables is how the strategic delivery of project data and analysis is crucial not only to the clear and effective communication of recommendations to the client, but also to the successful implementation of the plan. The series of project deliverables—strategic programs, lease audits, cost structure evaluations, implementation frameworks—is described, with the rationale for each. This discussion includes:

- Analysis of the strategic environment—the demand profile
- Analysis of the current framework—the supply profile
- Recommendations for a new framework

The conclusion is made that only after all corporate real estate and facility assets have been analyzed, the client's business challenge understood, and the operational and business requirements determined, can a strategic framework be established.

Several steps leading to the development of a holistic strategy are then explained in detail, including:

- Implementation plan
- Evaluation and refinement of strategy
- Strategy for continuous improvement

Finally, a brief discussion about deliverables and the balance necessary between the time it takes to develop deliverables, the needs of the client, and the speed of change in the client's business environment.

3.5 Strategic Communication and Change Integration

Change management and strategic communications are explained as working in tandem to address the process for change, while recognizing that most organizations face the same external forces, pushing them in the same directions. Change management and strategic communications concern themselves with the crucial "how" when moving from a to b. Some key considerations addressed in this section include:

- Addressing the variables in your audience
- Building trust in a time of change
- Selecting the medium—communications collateral
- The basic steps to strategic communications planning

Closing out this section is a discussion on the systematic documentation of process and progress, which explains why it is crucial to be able to illustrate how the plan evolves.

3.6 The Last Word

The chapter closes with a final word on strategic planning and how it should not be undertaken lightly.

CHAPTER 4 BUSINESS TRANSFORMATION AND FACILITY MANAGEMENT

Downsizing and outsourcing activities in recent years have exhausted the energy of most facility managers. The demand to satisfy customers and to cut operating costs is their daily agenda. In order to meet the needs of the agile corporation, facility managers must organize themselves to face these challenges. This chapter introduces the concept of business transformation and the processes required for the facility management (FM) profession.

4.1 Introduction

"Change" overwhelms every organization. Everyone tries to predict the future and the future organization. Reengineering and reinventing became the key activities of any corporation. How about the facility management organizations?

4.2 The Changing Business Environment

Acquisitions and mergers dominate today's business scene. All of them strive for the leadership role in achieving maximum return on investment for their shareholders. Such approaches influence greatly the design and servicing of the workplace.

4.3 Transformation in the Business of FM

Corporations are removing their non-core operations. Various companies sold off their real estate holdings. Such corporate strategy mandates changes for in-house real estate or facilities departments.

4.4 The Business Transformation Process

In order to change, one must understand the fundamentals of business transformation. This section deals with the five components of transformation: vision, integration, communication, transformation, and measurement.

4.5 Three Key Issues in FM Transformation

This section deals with the three key elements in a successful transformation: the mission critical issues related to the business operations, the core competency issues of human resources, and the workspace design issues related to the management changes.

4.6 Three Obstacles in Transformation

People are reluctant to change. This section deals with the three elements that stifle the process of change: the lack of time, the stronghold of established corporate culture, and weak leadership.

4.7 Three Companies in Transformation

This section provides examples of three facility organizations in different industry sectors and how they deal with their transformation processes in relationship to their corporate strategy.

4.8 Conclusion

A summary of the transformation process and the importance of the human capital in an organization are provided here.

CHAPTER 5 FINANCIAL MANAGEMENT FOR FACILITY MANAGERS

All business decisions have a financial ingredient. Financial analysis and management is a key skill set that all facility managers must have. It is important for facility managers to push facility services and projects to the forefront of their organization's agendas by using accepted financial analyses. This chapter provides an overview, details, and examples of how fundamental financial analysis tools can be applied to everyday facility management tasks.

5.1 Fundamental Principles of Financial Management for FM

This section details the goals of basic finance and accounting principles and traces the evolution of common financial analytical tools. It covers the fundamental data and metrics used in financial analysis and provides a basic cost benefit review and examples related to FM. This section also details financial metrics such as NPV (net present value) and RONA (return on net assets).

5.2 Facility Asset Management (FAM) Financial Model

This section reviews the fundamental process of asset management. It reviews basic requirements for asset inventories and condition assessments, and provides details of developing operational budgets based on physical asset requirements. A review of different budget approaches and a checklist is included, along with sample calculations for making "repair vs. replace" decisions which link expense and capital budgeting for optimal life-cycle leveraging.

An overview of typical financial performance measurements along with an adaptation of the Balanced Scorecard for FMs is included. A review of financial forecasting is also included.

5.3 Life-Cycle Analysis

This section contains a review of basic LCC (life-cycle costing) techniques along with full-cost accounting and LCA (life-cycle assessment) review. The implications of commissioning on the life-cycle costs of a facility are also discussed.

5.4 On-Going Facility Finances

This section reviews the work tasks required in keeping facility asset inventories updated, dealing with various contract pricing issues, and using financial calculators. The importance of accurate base data and integration with computer-integrated facility management (CIFM) systems is also covered. Financial forecasting using simulation and system dynamics is detailed along with a case study.

Web sites for financial management along with a glossary of terms, acronyms, and a reference section are included.

CHAPTER 6 ULTIMATE CUSTOMER SERVICE

6.1 Introduction

To the successful facility manager, the customer is one of the pillars of the organization—identifying customers and understanding their requirements is one of the primary functions of the service delivery chain. Unfortunately, many facilities management (FM) organizations fail to realize that the customer actually drives the process, the quality measures, and the very definition of quality. However quality is defined, if the customer's definition of quality is not included, it will do no good. Consequently, that definition must be used in the structuring of service delivery to ensure the meeting and exceeding of the customers' expectations. Involving the customer in the service delivery process is a sure-fire, easy way to build them into the routine of the structure. Finally, measuring customer satisfaction with service and personnel is the critical step that allows FM organizations to come full circle in the ultimate customer service.

6.2 Identifying Customer Groups

Do you know who your customers are? You might be surprised by how many you really have and where to locate them.

6.3 Determining Customer Perceptions

How does an FM organization find out its customers' perceptions and expectations? Learning how to discern and then talk with customers about the difference between perceptions in service and expectations of service are important skills that are distinguished in this section.

6.4 Expectations and Perceptions

This section expands the concept of needs assessment to include facts regarding customer expectations. It also describes the importance of perceptions and how to explore them within an organization. Finally, the section provides a checklist for customer evaluation criteria.

6.5 Defining Quality FM Services and Structuring Service Delivery Accordingly

Without defining *quality*, where do we head? Goals are the only way to move forward, and the only way to build a service delivery structure or improve an existing one. How do you go about defining quality?

6.6 Involving Customers in the Service Delivery Process

Provision of services is an interactive process from the beginning! Be certain to structure your processes so that the customers' needs and desires are an integral part of the whole.

6.7 Assessing Customer Satisfaction with the Service Delivery Structure and Performance of FM Personnel

A healthy, effective customer service relationship begins with listening to your customers and knowing how they feel about your service. Measure, measure, measure! Competition can be fierce in today's business world and satisfaction assessment is vital to your organization's welfare.

6.8 Linking Customer Feedback to the Continuous Service Delivery Improvement Process

Finally, linking customer feedback to service delivery efforts will strengthen the FM organization. There is no better purpose for gathering customer satisfaction data if not to put it to use in your service goals and objectives, and this final step creates a continuous "loop" between an organization's service and customers.

7.1 Introduction

Disaster recovery planning, business continuity, contingency plan, crisis management—whatever you call it, planning to effectively meet emergencies is an important part of the survival process for any business.

7.2 Business Impact Analysis

Defining the cost of business interruption in dollars per day gets the attention and support from the top administrators of a company and makes it possible to have the complete organization's support in formulating the plan.

7.3 The Plan

An overview of risk assessment and response will clearly define the disaster action plan. This section recommends procedures to create the recovery plan.

7.4 The Team

When a serious business interruption occurs, a team of critical personnel is required to evaluate the situation and to decide on the course of action to be followed to recover. This section defines this team.

7.5 What to Do?

The first step in response to any disaster is to stop the damage and stabilize the loss site. This section defines that step for building structure, industrial equipment, electronics, and a variety of information storage media including paper, magnetic media, optical media, and film media.

7.6 Conclusion

CHAPTER 2
BENCHMARKING

Edmond P. Rondeau, AIA, CFM
Director
Global Operations and Learning for
International Development Research Council (IDRC)
E-mail: ed.rondeau@idrc.org

2.1 INTRODUCTION

Benchmarking provides a methodology that can help organizations improve a wide range of facility-related business processes, procedures, cost, consumption, and quality-related service areas. By measuring and validating expense, consumption, quality, and service practice programs, benchmarking can provide the facility professional with a guide to optimize facility management (FM) practices. It can also provide convincing new evidence with which to gain and retain the support of customers and senior management, and to more effectively promote benchmarking within a quality-focused facility enterprise.

2.2 WHAT IS BENCHMARKING?

Benchmarking is discovering the specific practices responsible for high performance, understanding how a facility best practices work, and adapting and applying these practices to your organization. While benchmarking has been described as partly science and partly art, it has also been described by one corporation executive as ". . . the continuous process of measuring products, services, and practices against the toughest competitors of those companies recognized as industry leaders."[1] This applied business concept of benchmarking will be used throughout this chapter.

The International Facility Management Association (IFMA) based in Houston, Texas, has developed a method for facility benchmarking that you may find useful to review in developing a benchmark for current FM services. The IFMA (www.ifma.org) periodically sponsors benchmarking research projects and the results are published in benchmarking reports that are used by facility professionals throughout North America. The following benchmarking reports are available through the IFMA Library Service: *Benchmarks II, Research Report #13*, and *Benchmarks III, Research Report #18*. The IFMA Library Service also sells a number of books and publications on benchmarking.

The International Development Research Council (IDRC) based in the Atlanta, Georgia area, has also developed and published a benchmarking system. The IDRC (www.idrc.org) has jointly developed with the Institute of Management Accountants (IMA), a Research Bulletin (July 1997) called *The Accounting Classification of Workpoint Costs*. This publication provides cost classification guidelines for facility and corporate real estate professionals and accountants. It can help to determine appropriate real estate/facility costs and the assignment of these costs to specific accounts associated with the use of technology, real estate, support, facilities, and other assets and liabilities associated with individual workers, business units, and other elements of the organization.

The Building Owners and Managers Association (BOMA) based in Washington, DC, publishes an annual benchmarking report known as the *BOMA Exchange Report.* The information found in this BOMA (www.boma.org) report is focused on cost per square foot information for a number of cost categories from the perspective of a landlord or building owner.

Another nonprofit organization that has developed a benchmarking methodology is the American Productivity and Quality Center (APQC), which is based in Houston, Texas. This organization's benchmarking process and related information should be reviewed by facility professionals as it defines and uses benchmarking from a business perspective. The APQC (www.apqc.org) defines "benchmarking as the process of identifying, learning, and adapting outstanding practices and processes from any organization, anywhere in the world, to help an organization improve its performance. Benchmarking gathers the tacit knowledge—the know-how, judgments, and enablers—that explicit knowledge often misses."[2] APQC also has a *Code of Ethics for Benchmarking* that you may choose to review and consider adopting.

"Why should we benchmark our facility service?" is a question often asked. Here are some good reasons why you, your staff, your customer, your management, and your service provider organizations can benefit from benchmarking:

- Benchmarking reduces the need to reinvent the wheel (why invest the time and costs when someone else may have done it already—and often better, cheaper, and faster?)
- Benchmarking can help increase the pace of change and restructuring by:

  Using tested and proven practices

  Convincing skeptics who can see that it works

  Overcoming inertia and complacency by creating a sense of urgency when benchmarking gaps are identified
- Benchmarking can lead to "outside the box" ideas by looking for ways to improve from outside of your industry
- Benchmarking often forces organizations to examine present processes, which often leads to improvement in and of itself
- Benchmarking can initiate a process where implementation is more likely because of the involvement of process owners

2.3 BENCHMARKING SURVEY

Benchmarking should be included as part of your "quality measurement and management" and performance management/scorecard programs within your FM services programs. When you look at site/facility issues, you will see that there are many opportunities to review and investigate all aspects of the life of facility development and management. Figure 2.1 depicts an overview of the site/facility cycle from strategic planning, to real estate requirements, plan development, and operations of the life of the organization's development, whether an owned or leased property. Any portion of the site/facility cycle can be benchmarked and/or included in a benchmarking survey as your peer organizations will have had experience with similar facility operations.

One way to review quality is to benchmark survey your FM performance against other organizations' FM departments identified as "best in class." This kind of survey provides a basis for comparing current services and performance with those of other organizations. You might first wish to look at the quality process issues shown in Figure 2.2 that you have developed or are in the process of developing. This quality process has many similarities to the benchmarking process shown in Fig-

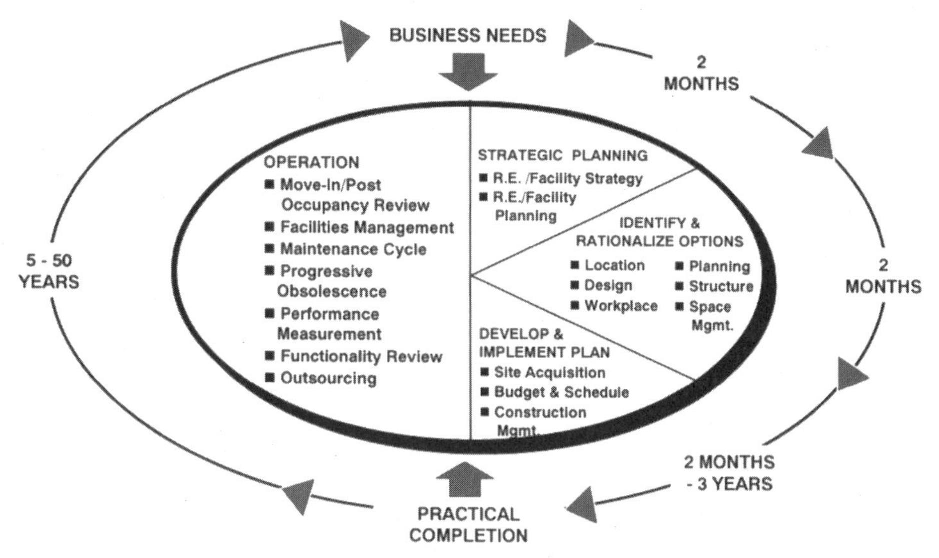

FIGURE 2.1 The Site/Facility Cycle.

ure 2.4, including identifying, developing, setting a baseline, and monitoring the facility-related services you are providing for your customers.

When looking at what area or areas to benchmark, moreover, you might wish to look at the needs analysis issues shown in Figure 2.3. These include investigating, evaluating, formulating, and integrating the facility service delivery areas you and your staff and service contractors/partners are providing for your customers. This needs analysis process is designed to consider looking at all ser-

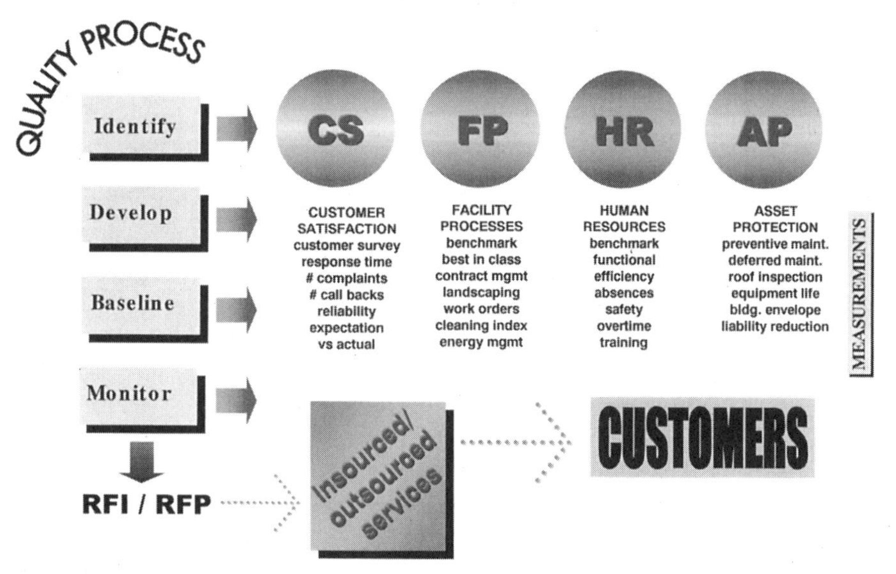

FIGURE 2.2 Quality Process Issues. (***Source:*** *Alex Lam, President, The OCB Network.*)

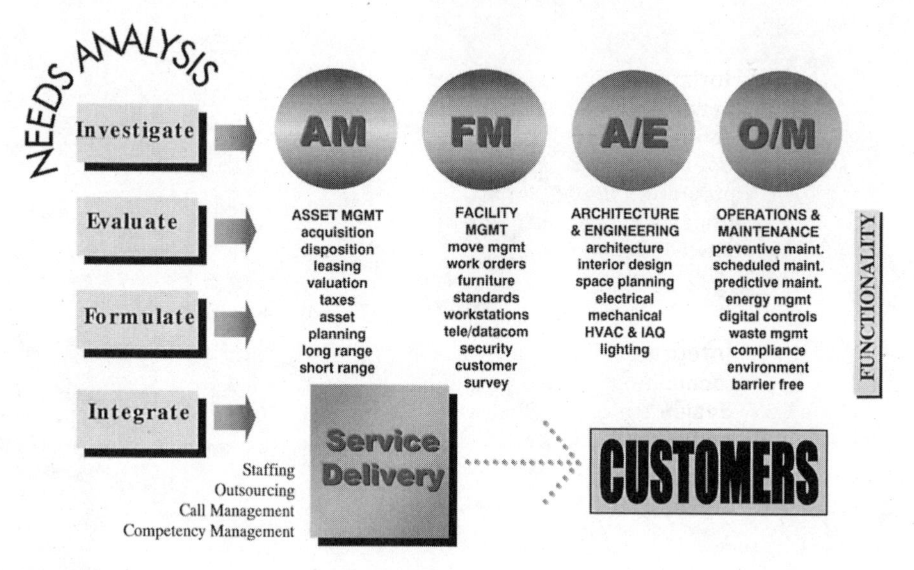

FIGURE 2.3 Needs Analysis for Facility Services. (***Source:*** *Alex Lam, President, The OCB Network.*)

vices across an integrated approach, as shown in Figure 2.5, vs. looking at services as silos of services shown in Figure 2.6.

After your review of the quality and needs analysis processes, your benchmarking survey could begin by addressing some or all of the following questions:

- What should our customers expect and by when?
- What are the relative costs and benefits for providing these services?

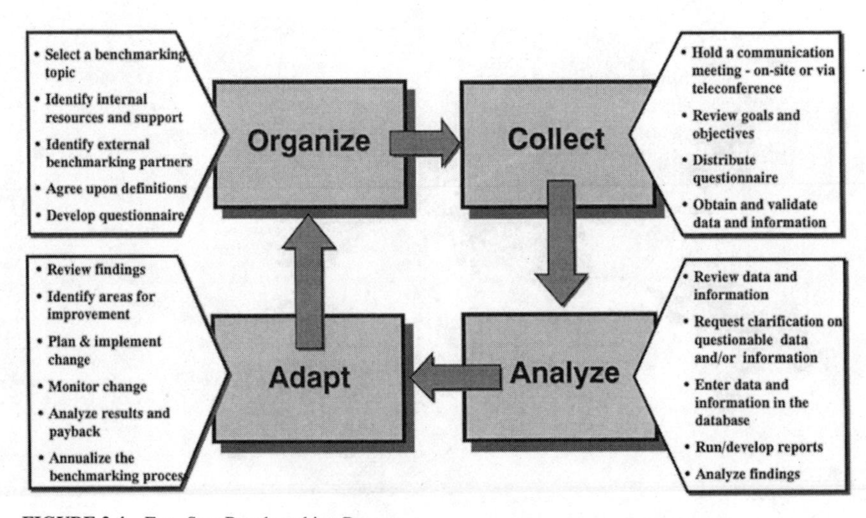

FIGURE 2.4 Four-Step Benchmarking Process.

Horizontal Processes
(between functional areas)
Identification of opportunities for consolidation and/or integration of personnel, processes, materials, etc. between functional areas.

Integrated Functions
(combining support functions)
Evaluation of management systems and other processes that apply to all functional areas

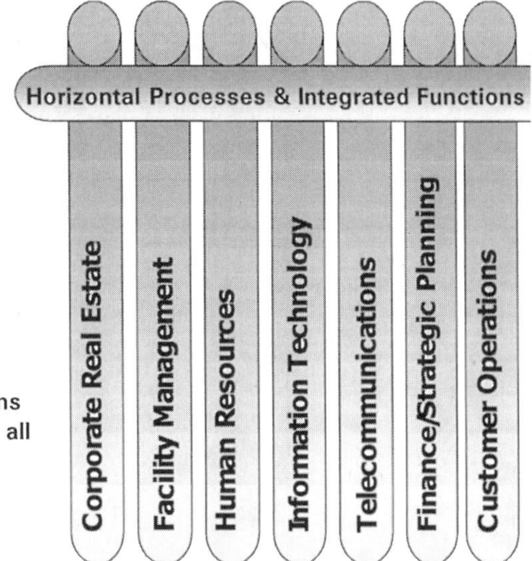

FIGURE 2.5 Horizontal Processes and Integrating Functions.

- What does senior management expect?
- What services, processes, and/or costs should be benchmarked?
- Can comparable benchmarking partners be found?
- What are the benchmarking definitions needed for the items/areas/services to be benchmarked?
- What geographical area(s) will be included in the benchmarking survey? By city, county, state or province, region, country, or global areas?
- What industry and building type(s) should be included in the benchmarking survey?

Vertical Processes
(within functional areas)
Detailed analysis of each functional area, with the intent of identifying best practices in facility management and strategies that can be applied to improve performance and reduce cost.

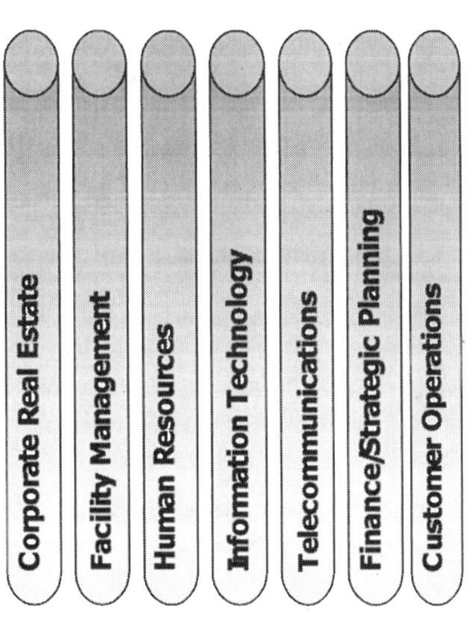

FIGURE 2.6 Functional Area Silos and Vertical Processes.

- If looking at cost(s), how old are the facilities being benchmarked?
- What are the hours of facility operation?
- How many days of the week are the facilities open for operations?
- What activity is going on inside/outside the facilities?
- What is the gross and net square footage of each facility?
- How many FTEs receive facility services in each facility?
- How old are the benchmark facilities?
- How many floors in the facilities?
- What utilities are used at each facility?

  Electricity

  Natural gas

  Heating oil

  Water

  Other utility source
- What specific and general activities are being performed in each facility?
- How large is the site in acres?
- What are the upper quartile, median, and the lower quartile costs as compared to your benchmark costs?
- What benchmarking should be performed annually?

You should check with your legal counsel if any information requested in the benchmarking gathering process may be considered as confidential; for instance, before contacting a direct competitor to be a partner. If you, your management, and/or your legal counsel are uncomfortable sharing confidential information, consider selecting another partner, or developing and signing a security/nondisclosure document such as a *Benchmarking Confidentiality Agreement.* This agreement would formally define the responsibilities of each partner and protects the information shared; this should satisfy the attorneys from both organizations. Do not ask competitors for sensitive data or cause the benchmarking partner to feel they must provide data to keep the process going.

You and your benchmarking partners should consider retaining an ethical third party to assemble and "blind" competitive and noncompetitive data, with inputs from legal counsel in direct competitor sharing, to ensure that data/information reports are not organization specific. Any information obtained from a benchmarking partner should be treated as internal, privileged communications. If "confidential" or proprietary material is to be exchanged, then a specific agreement, as discussed previously, should be executed to indicate the specific material to be protected, the duration of the period of protection, the conditions for permitting access to the material, and the specific handling requirements that are necessary for that material.

The benchmarking survey process should be managed and organized as a project and managed by a benchmarking project manager. This person is assigned to organize and lead the benchmarking team; to help develop the areas to be benchmarked and the questionnaire; to develop and manage the survey budget and personnel resource requirements; to help select and work with the benchmark survey partners; to review the findings, assist in the development of on-going improvement programs, and provide leadership in this process on a yearly basis.

The project manager should focus initial benchmarking efforts on critical business issues that have high payoff and are aligned with your organization's values and strategy. Consider concentrating benchmarking energies on those areas that would be most likely to improve long-term organizational capabilities.

If you are considering benchmarking for the first time, begin by selecting relatively "simple" areas to benchmark. This could include areas where you know your organization has up-to-date, detailed, accurate, and verifiable information. Use your first benchmarking survey as a learning experience and build upon that experience. It is important that you and your benchmarking team stay focused on the survey objectives and goals, and ensure that you manage your customers' and management's expectations on what the survey process can do for your organization.

If you are the benchmarking project manager, you must provide leadership to ensure that the process receives management and business unit(s) support. The survey process, expectations, objectives, and on-going change requirements must be explained to the survey team. Where necessary, the project manager must provide training to members of the team. The project manager must use a benchmarking process that provides a specific beginning, development, analysis, and implementation program. The benchmarking process may include the following four steps (see Figure 2.4):

1. *Organize:* Put the pieces of the benchmarking process together by:
   - Selecting benchmarking topics that are relevant and can be measured by you, your team, and your partners
   - Identifying internal resources, time requirements, and support from senior management and business units
   - Identifying external benchmarking partners who have similar facilities and have "best in class" reputations
   - Agreeing upon definitions to ensure an "apples to apples" acquisition and analysis of information
   - Developing interview guide(s) and questionnaire(s) that are relevant, simple, effective, can be shared in advance, and can be completed in a timely manner by your company and by your partners
2. *Collect:* Obtain, review, and validate the actual benchmarking information by:
   - Holding a communication meeting—on-site or via teleconference with benchmarking partner(s)
   - Reviewing benchmarking goals and objectives
   - Distributing the benchmarking questionnaire with time completion requirements
   - Obtaining benchmarking data and information for the questionnaire and validating benchmarking data and information
3. *Analyze:* Provide a detailed review and question-specific benchmarked material, analyze specific material, enter data into the benchmarking database, run sorted reports, and review report findings by:
   - Reviewing data and information by asking if the data and information make logical sense, identifying missing data/information items, and asking if this provides meaningful help
   - Requesting clarification on questionable data and information if you identify possible errors, omissions, or misinterpretations, especially in the application of benchmarking definitions
   - Entering data and information in the database and spot checking to ensure that the data/information was entered correctly
   - Running/developing reports and asking if the resulting reports make logical sense and provide meaningful help
   - Analyzing findings by looking for meaningful areas for improvement gaps in service and comparing findings against the upper quartile, median, and lower quartile against your facility benchmarking data
4. *Adapt:* Look at the findings and identify how these could be implemented to improve service, costs, staffing, processes, procedures, etc.; develop a plan for change; report on the progress of the change; and review change over time with benchmarking as an annual review process by:
   - Reviewing findings with your partners, staff, customers, and senior management
   - Identifying areas for improvement based on a critical review of the findings and identified gaps in service, quality, customer satisfaction, etc.

- Planning and implementing change by working with the benchmarking team, staff, and management to use the findings to decide what changes in processes, procedures, cost, and/or quality management will be used
- Monitoring change to ensure that the concepts and details of the implementation program are being used and are providing some or all of the improved service that was sought or expected
- Analyzing results and payback to provide the benchmarking team, staff, the customer, and management with on-going reports of progress whether from:

  Improved customer service

  Improved quality

  Improved expense tracking

  Improved response time

  Improved staff performance

  Improved use of resources

  Improved contractor/supplier performance

  Improved use of furnishing inventory

  Improved up-time on mechanical and electrical equipment

  Improved use of facility space

  Reduced churn

  Reduced response time

  Reduced operating costs, consumption, etc.

- Annualizing the benchmarking process requires that you and your staff make benchmarking an on-going part of your department activities and measurement process.

When considering a benchmarking survey, it is important to understand how much you know about benchmarking, the information you believe is available, the time it will take to obtain the information, and the results you or your customers expect to achieve. Benchmarking cost items is where most organizations start in their benchmarking survey programs. Consider choosing a small number of facility expense areas (two or three) that total the largest annual facility expenses where you have access to accurate yearly information and on a by-facility basis. The three most common expense areas that most organizations initially choose to cost benchmark are:

1. Maintenance and operations
2. Cleaning
3. Utilities

These facility expense items often comprise 60–80% of yearly facility expenses. By selecting these items, you and your customer should focus not only on cost, but also on quality and customer satisfaction for the level of service you are providing for a specific expense.

When your benchmarking survey process proceeds to an on-site visit, the following are recommended:

- Provide your meeting agenda in advance for comment by your partners
- All benchmarking partners should be professional, honest, courteous, and prompt
- Introduce all partners and attendees and explain why they are present
- Adhere to the agenda
- Use language that is universal, not your business or industry jargon or buzz words
- Ensure that all parties have agreed on and provide information based on common definitions
- Be sure that no party is sharing proprietary information unless prior approval has been obtained by all parties, from the proper authority

- Share information about your own process, and, if asked, consider sharing study results
- Ask pertinent questions and validate the information received
- Offer to facilitate a future reciprocal visit
- Conclude meetings and visits on schedule
- Thank your benchmarking partners for sharing their information, processes, etc.

While many benchmarking programs focus initially on cost per square foot, you should consider including the cost per FTE[3] (full-time equivalent) person, and the amount of product or service purchased, also known as the amount consumed. These additional benchmarking areas can provide you, the facility professional, your management, and your customers with other important comparisons that are often more informative than just looking at cost. For benchmarking to be successful, it should also look at the processes and procedures that drive cost, efficiency, and effectiveness. Benchmarking can seldom be successful without also looking at these areas.

2.4 APPLES TO APPLES

Consistent definitions are very important for a successful benchmarking survey program. All parties to the benchmarking program must agree on the definitions to be used to insure an "apples to apples" comparison, and all parties must be able to provide the benchmark information based on the agreed upon definitions.

For facility requirements, the benchmarking process could address common areas with definitions that may include processes, procedures, and/or cost areas for the following:

- Real estate, asset, and site management services
- Maintenance and operations
  - Mechanical
  - Electrical
  - Plumbing
  - Exterior building maintenance
  - Facility operations
  - Sub-contracts
- Landscaping and grounds maintenance
- Parking lot/parking garage services
- Utilities
- Cleaning
- Design services
- Construction services
- Moves and relocations
- Mail services
- Receiving and shipping services
- Reception services
- Telephone services
- Information services
- Salaries and overhead service costs
- Management fees
- Etc.

TABLE 2.1 Benchmark by Industry Type

Aerospace	Government
Brewery	Health care
Chemical	Hotel
Commercial retail	Insurance
Computer office equipment	Media
Computer services	Military
Distributor	Oil exploration and refining
Electric generation	Pharmaceutical
Electronics	Printing
Engineering	Property management
Financial/banking services	Telecommunications
Food and beverage	Textile manufacturing
Gas distribution	Tobacco

It is also important that benchmarking definitions include by-industry and by-building categories. This insures that you, your team, and your partners understand what types of buildings will be included in the benchmarking survey and what industries will be included in the survey. You also need to ensure that you benchmark like industry types; you may choose to use the list shown in Table 2.1 to ensure that the team and partners agree on the industry to be included in the benchmarking survey. For example, you could use the listing of building types shown in Table 2.2 to establish what building types will be included/benchmarked at a pharmaceutical (Table 2.1) site, such as research buildings, manufacturing buildings, office buildings, warehouse buildings, etc.

Sorting by building and industry type also results in an "apples to apples" benchmarking program that provides "like compared to like." You can also look across industry lines to see if there are "best practices" from one industry that might be applicable in another industry. For example, there are benchmarking surveys where the partners have specifically sought to cross industry lines to visit with organizations that are known for their quality and efficient processes for a core service, such as distribution. Distribution for some organizations is key to their success and a benchmarking program may seek to understand how they achieve their excellence in customer service and how these unique capabilities can be transferred to other industries.

TABLE 2.2 Benchmark by Building Type

Bank/financial	Mental hospital
Church	Municipal
Colleges	Nursing home
Elementary school	Office
Factory	Parking garage
Gas station	Research
Golf club house	Restaurant
Grocery	Retail
Hotel	Secondary school
Judicial	Sports center
Laboratory	Telecommunications
Law court	Theatre
Manufacturing	Warehouse

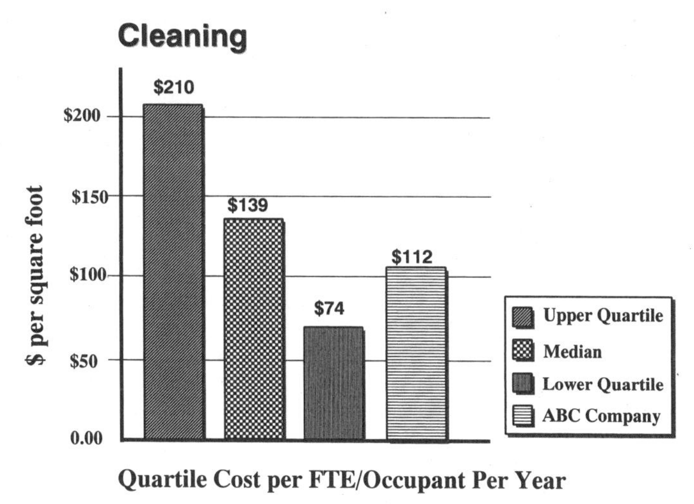

FIGURE 2.7 Benchmarking Report by Cost Per FTE.

2.5 REPORTS

Given the resources that are required to develop and maintain on-going benchmarking programs, you should not undertake benchmarking unless you understand what you are trying to measure, identify, or implement. Your resulting benchmarking report(s) must provide information to you, your management, and customers in a format that is understandable, defendable, reasonable, and provides opportunities to improve services, processes, and procedures with areas to reduce facility-related expenses.

If you are benchmarking costs per square foot or cost per FTE, you should develop benchmarking reports on a yearly basis. More importantly, you should benchmark these issues on a quartile basis, as shown in Figure 2.7 and Figure 2.8, not on a purely cost-to-cost comparison basis, which

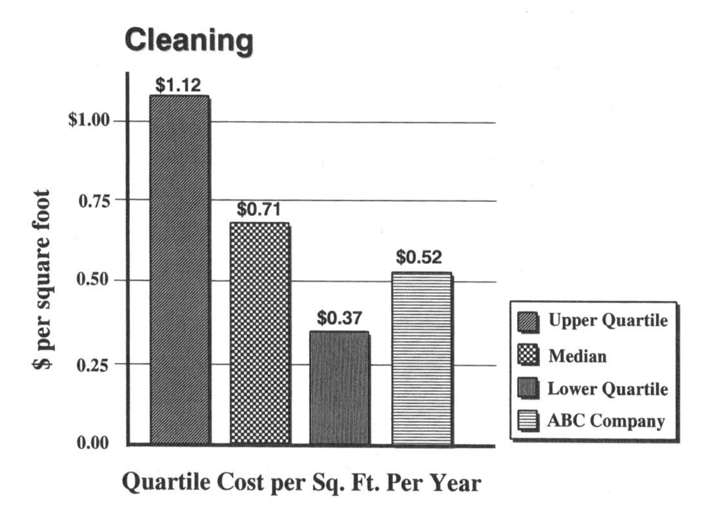

FIGURE 2.8 Benchmarking Report by Cost Per Square Foot.

some "benchmarking clubs" provide to their members. Benchmarking by quartiles provides the opportunity to review your costs against a range of costs through the use of an upper quartile, median quartile, and lower quartile cost against your specific cost. This provides you with the opportunity to look at your cost against a range of costs.

You may find that your management and/or customers expect you to benchmark purely on a cost basis. Cost benchmarking does not address the quality, efficiency, effectiveness, or customer satisfaction issues that are part of a total benchmarking program. Your program should include benchmarking processes, procedures, and best practices. While cost is important, your benchmarking research and investigation of current processes, procedures, and best practices could provide the opportunity to make substantial reduction in costs. A change in processes could result in reduced staff, reduced staff time, reduced direct and indirect costs, and/or improved service(s) to your customers, as shown in Table 2.3.

Remember that your benchmarking reports may show that you are providing excellent service to your customers, at a reasonable and competitive price, and that your peers may regard your facility operations as "best in class." Regardless of the results, you must market the results of your benchmarking programs and let your customers and management know how well your facility operations are performing against internal and external benchmarks.

You should consider using the benchmarking process that was discussed previously and as shown in Figure 2.4. Using this process requires that you develop a program to organize, collect, analyze, and adapt the benchmark information into a positive improvement program for your customers and your organization.

Implicit in the benchmarking process is that you and your staff will be able to access and provide accurate information to your benchmarking partners. Unfortunately, many organizations have not effectively or efficiently captured historically accurate facility information. Some organizations cannot readily report on how many facilities they own or lease, or how much they are paying in lease payments and operating expenses. Without this basic information, benchmarking becomes guessing at best, which can be very dangerous for you, your staff, your customers, your management, and your benchmarking partners. Because the information cannot be validated, benchmarking reports will be inaccurate, and reports can provide the wrong conclusions.

A complete benchmarking program cannot be done from your desk. To successfully benchmark and identify comparable benchmarking partners, you must know what you have in the way of assets, processes, procedures, and expenses and can obtain access to this information. You must also know how to look at processes and expenses and be able to **validate** any and all information that you provide and receive from your benchmarking partners. This also means that you might need to travel to their site and ask the right questions to validate their information. Sometimes the hardest validation requires that you are able to identify the information that *is not* there. If you cannot validate the information provided, then you run the risk of not being sure that you are getting accurate and/or an "apples to apples" benchmark.

Process benchmarking is often the most time consuming and the hardest to do because it requires funds to pay for the study, personnel time, time to become intimately involved with your own processes, and then time and resources to evaluate your partner's processes. The resulting benchmarking report could identify:

- The current benchmarked processes
- The bottlenecks in each process
- The benefits of the processes reviewed
- The recommended processes
- Possible benefits to the recommended new process

  Improved service to customer(s)

  Reduced time to complete the process

  Reduced number of process steps

  Reduced costs

TABLE 2.3 Benchmarking Report of Recommendations

Bottlenecks	Benchmark recommendations
1. *Absence of a Defined Process:* Without a defined lease approval process, the process can take place a different way for each lease. Without a defined process it would be difficult to identify bottlenecks, benchmark the process, or recommend improvements.	1. *Use Flow Chart and Benchmarking:* These two items will help to define the actual process, identify the steps and time involved, and provide the basis for developing an ideal process.
2. *Lease Document Diversity:* There is no pattern to the way the lease document and its contents are formatted. It appears the document is being started from scratch each time and the document is difficult to administer during the life of the lease.	2. *Develop a Model Lease:* The format and language of no two leases are alike. The model lease would provide a specific format in specific language which would be the same or similar for each lease. The model would protect the interests of the tenant and landlord and would include a term sheet to document all business issues and the body of the lease that would include all legal issues. With a model lease, it is realistically probable that once the business terms were agreed upon, the lease document for a deal may only go through one iteration with no need to review the legal points of the lease. This could be the greatest foundation for productivity increase.
3. *Extensive Queue Time and Too Many Iterations:* The amount of time the lease document is in transit or is waiting to be worked on by the landlord or tenant is queue time. The queue time often is a major factor in the total time it takes to process a lease. The number of times the lease document is revised, iterations, by the originator of the lease document often delays the process as a legal or business item revision may not be acceptable to the other party and the next iteration of the lease must address these issues.	3. *Standardize, Identify Queue Time, Track Process, and Audit Results:* Develop a standard process for lease approval, reduce steps, and examine and attack queue times. Use the benchmarking process to identify the current process and develop and implement a streamlined process to reduce the time in queue, reduce the number of steps and iterations to no more than two iterations by using a model lease. This would include reducing the number of outside groups and persons who are not necessarily contributing to the process but are creating bottlenecks.
4. *Transmittal of the Lease and Lease Iterations:* Document may be misplaced, delayed, or lost in transit. Time for mail or express delivery must be included.	4. *Implement an Electronic Network and Adopt an Electronic Document Log:* Benchmark receiving, changing, red lining, and sending the lease manuscript electronically. The *actual* time, cost, and savings from using this communications process in an actual test remains to be completed. Exhibit "H," "Electronic Transfer of Lease Documents," in the study discusses time, cost, and potential savings. The transmission of lease iterations via the electronic network can save mailing/express time and expense. The document log will provide an audit trail and systematic process to be able to track the status of each iteration and benchmark the queue time and process against the established benchmark.
5. *Authority Outside of Corporate Real Estate:* Corporate real estate must often wait for others within the corporation to make decisions and to sign the lease document.	5. *Empower Corporate Real Estate:* Focus on signatures and changes in signature authority and process. If the business terms are within previously agreed upon approvals from senior management and the model lease is used, corporate real estate will be empowered to sign the lease without the time delay for further approvals. More time could be available to negotiate the business terms and construction/build-out and move-in can occur on time or sooner.

continued on the following page

TABLE 2.3 Benchmarking Report of Recommendations *(Continued)*

Bottlenecks	Benchmark recommendations
6. *Lack of Resources, Staff, Legal, Accounting, Etc. to Make a Decision:* Timely response to lease document review may not be possible with workload and/or the perceived importance of the lease review is different within each organizational unit.	6. *Identify Client Expectations and Communicate Resources:* Ensure that the client's expectations are identified and the lack of resources, staff, priorities, etc., from corporate real estate or other internal departments are addressed. Communicate to the in-house client what the lease approval process will entail and require in the way of time, meeting, information and planning requirements, and personnel resources.
7. *Participating Units Independently Set Priority of Their Portion of the Work Flow:* Other departments, including legal, accounting, facilities, telecommunications, etc., may have other corporate business issues which they choose or are directed to prioritize above lease document review.	7. *Establish a Flow Chart, a Communications Process for Setting Priorities, and a Log to Monitor Process Time:* With other internal departments being required to complete other projects or work, their priorities often may not mesh with corporate real estate time or process requirements. A communication process should be established to set levels of priority for each lease, the steps within the process, and a log which is attached to the lease as it makes the approval rounds to provide a means of tracking and auditing the steps and process time against an established benchmark.
8. *Adversarial Negotiating Tactics:* Divergent internal and external requirements by the landlord and tenant during this phase may inhibit the quick resolution of initial or new business or legal issues.	8. *Use Detailed RFP and Letter of Intent, Convergent Internal and External Requirements by the Landlord and Tenant, and a Model Lease:* With quality time spent in the front end of the lease process to develop a quality and detailed RFP and Letter of Intent, 95% of the business and associated legal issues would be addressed in the Business Terms and the Model Lease. The win-win attitude by both parties to the lease should be part of the negotiating strategy and continue throughout the life of the lease.
9. *Word Processing: Grammar and Verification of Document Changes:* Lease format, grammar, and legal language used is provided by an attorney who has written the lease for review by another attorney. Review comments are made manually in no consistent format. Changes made in the next iteration of the lease document are often "redlined" manually and a change may be inadvertently not "redlined." Verification of document changes must be made manually by reading and comparing the previous iteration with the latest iteration.	9. *Use Computer Software to Identify Closed vs. Open Issues:* With today's computer software, the landlord or tenant can reduce the time to identify, handle, review, send, and receive iterations electronically. Grammar simplification, intent, and verification of document changes is available via software.
10. *Readability:* Lease document or the intent is often not written in plain English and is difficult to understand by those administering the lease. Paragraphs may be a page or more in length and items within the paragraph may have many interrelated requirements which are hard to follow by the lease administrator. The education to understand the document is often far above the level of the person administering the lease.	10. *Use a Model Lease, Software, and an Established Readability Index:* A Model Lease written in plain English and developed within a readability index of 40 for those who will administer the lease will provide a very "readable," manageable, and understandable document. Highly involved, lengthy paragraphs and wordy sentences with many interrelated caveats are unacceptable. A clear, concise, easily read, understandable, and manageable lease document is the goal which would greatly decrease the time to review and process the document at any point in the life of the lease.

Reduced staff
Reduced bureaucracy
More productive staff
Trained or re-trained staff
Clarified steps and hand-off points

- The steps or requirements needed to adopt the new process
- The expected benefits to the organization
- The analysis of the new process over time
- An on-going review of the new process

Many facility organizations do not have a detailed understanding of the processes and procedures being used for the services they provide for their customers. Benchmarking provides the opportunity to ask the questions and seek the reasons why current processes are in place and could be changed. A resulting process benchmarking report for the real estate lease approval process could provide information similar to that shown in Table 2.3.

2.6 BENCHMARKING MATURITY

Benchmarking can reveal striking differences in the financial returns generated from benchmarking projects. Some benchmarking professionals have reported many times higher financial benefits than did others. The same organizations also reported higher non-financial benefits, such as improvements in customer satisfaction and cycle time in the benchmarked process.

High, average, and low-return benchmarking results have led to the strong presence of certain underlying conditions that facilitate benchmarking. That presence has been identified as "maturity" for companies that actively participate as study partners. Widely recognized best practice companies are often flooded with benchmarking requests and companies that partner more effectively are those that have greater benchmarking maturity.

Mature organizations have people who are not awed by process, metrics, and benchmarking. They understand quality, process management, benchmarking, and to some extent, reengineering, which has been deployed throughout their organization.

How "mature" is a given organization in the benchmarking process? Maturity refers to the relative presence of certain conditions that tend to promote, facilitate, and optimize a process, procedure, and/or the consistent capturing of cost-related information. An organization can be relatively mature in some areas and less so in others. By gauging one's own degree of maturity in a given process, it is generally possible to get a fix on:

- How the organization can better facilitate the process by improving specific underlying conditions
- Why the process or procedure may not be meeting customer and/or senior management expectations (if this is the case)
- What steps may boost the organization's maturity level prior to gaining a commensurate level of benchmarking experience

Conditions/attributes that you may choose to select to facilitate the benchmarking process include:

- The organization's general awareness of benchmarking concepts
- Support of benchmarking from senior management
- The organization's documentation of its own best practices, processes, and associated cost information

- An environment that supports the sharing of best practices internally
- An environment that supports sharing externally with other organizations

Together these provide a self-test that generally reveals an organization's relative degree of benchmarking maturity.

2.7 LESSONS LEARNED

All processes and efforts for change will experience some degree of success and failure. The following benchmarking lessons should be reviewed before beginning your benchmarking survey program:

- Use benchmarking to create a sense of urgency or use the findings as a compelling reason for change. Competitive or best-in-class benchmarking can create a sense of urgency as well as demonstrate the value of looking outside for ideas and comparison.
- Limit the number of major projects to fit the resources that will be available for implementation. In the initial zeal and excitement that can come from imagining the improvements and gains made possible by the transfer of best practices, there may be tendency within your organization to forget that there is only so much an organization can invest and a finite amount of change that can be supported at any one time. It is demoralizing to find best practices only to discover that the investment dollars have already been spent for this year or that a facility does not have the extra staff needed to handle another implementation.
- As you get into the benchmarking process, do not let measurement get in the way. Your internal scoreboards may contain many inconsistencies in data collection and be open to interpretation about local causes for the differences in performance. Rather than spend time debating "who's best" and why the measures are not consistent, focus on those areas where dramatic differences in performance point to a real underlying process difference and not just an artifact of measurement. Overanalysis is an easy pitfall to fall into, especially when management by political action is the historical norm. Comparative measures can improve over time, but only if you start to act on them.
- You should consider, where necessary, a reward system that encourages sharing and information transfer. Real internal transfer is a people-to-people process and usually requires personal sharing or enlightened self-interest. Your management can help by promoting, recognizing, and rewarding people who model sharing behavior, as well as those who adopt best practices. It helps to design approaches that reward for organizational improvement as well as individual contributions of time, talent, and expertise.
- It is necessary to constantly reinforce the need for people at all levels to take responsibility for voluntarily participating in the benchmarking activity of sharing and leveraging knowledge. You and your team can ask regularly what people are learning from others and how they have shared with management ideas they think important. Be sure to support benchmarking and improvement teams that invest or "give up" resources to make this sharing happen, especially if they do not directly benefit.
- Use technology as a catalyst to support networking and the internal search for best practices, but do not rely on it as a solution. People are no longer tied to just their local organizational structure for information and knowledge. They can step out and learn elsewhere how to improve faster than those tied to smaller information sources. Use a combination of new information technology tools such as e-mail, "best practices databases," internal directories, and group software to support employees seeking knowledge and collaboration across the organization. However, do not let your organization rely on technology tools for solutions. As has been shown, although people have telephones, they do not automatically "reach out" to someone they do not know.

- As a benchmarking leader, you will need to consistently and constantly spread the message of sharing and leveraging knowledge for all services within the organization. The most effective transfer processes are those that are demand-driven rather than push approaches. You can help create and support demand, but cannot make it happen. You and your management can encourage collaboration across boundaries of vertical and horizontal structure, time, and function. Some ways to do this are to:

  Share and publicize success stories

  Provide infrastructure and support

  Change the reward system to remove barriers

2.8 CONCLUSION

In summary, you should strongly consider beginning an on-going benchmarking program that will take advantage of the tremendous untapped reservoir of knowledge in your own backyard. Whether you call this internal benchmarking, transfer of best practices, or knowledge management, it can lead to reduced costs and increased revenue, speed, and customer satisfaction.

Remember that it is your responsibility to provide the atmosphere conducive to benchmarking success. Your knowledge of your organization, staff, customers, and management will be tested during your benchmarking survey programs. It will require that you, your staff, and benchmarking team work at encouraging the sharing of information, the transfer of knowledge, and the adoption of relevant "best practice" processes and procedures.

ENDNOTES

1. David T. Kerns, CEO, Xerox Corporation, "Competitive Benchmarking, What It Is and What It Can Do For You." Quality Office, Xerox internal publication, Stamford, CT, 1987.
2. Source: American Productivity and Productivity Center, "Benchmarking Terms." APQC's training course "Applying Benchmarking Skills in Your Organization," Houston, TX.
3. FTE calculation must include all organization personnel, including full-time employees, contract, and temporary employees who receive facility-related services.

CHAPTER 3
STRATEGIC PLANNING

Loree Goffigon[1]
Vice President and Strategic Planner
Gensler
E-mail: loree_goffigon@gensler.com

3.1 INTRODUCTION

In a world of increasingly complex variables, the requirement to solve the right problem takes precedence over solving the problem right. Since the work we do has important implications for the futures of the organizations with which we work, we are obliged to get it right from the outset. Fortunately, we are helped in this effort by a number of decision-making frameworks at our disposal to support organizational direction and align internal strategies with external initiatives.

The framework that concerns us here is strategic planning. Because organizations are systematic by nature, they function at their maximum effectiveness when all actions are interrelated and aligned around a common purpose. This, of course, requires that you understand thoroughly the nature and business of your client's organization. The value of effective strategic planning is that it guides you in building this essential knowledge base, enabling you to identify correctly the situation and develop recommendations that are appropriate, specific, and viable, with favorable implications for the company's future.

Strategic planning is a process, a way of thinking, an exercise in making a map for reaching your goal. Because it defines the ultimate objectives before it develops an action plan, effective planning starts with the question, "What do we want to achieve, and why?"

Three fundamental elements underlie the strategic planning process: context, vision, and optimization.

Context is equivalent to purpose. It informs the entire project process and must be understood before you can design a plan to support a business strategy. The context for real estate and facilities strategic planning might include such parameters and drivers as the organization's overall business strategy and value proposition, the technology and human resources strategy, the existing real estate and facilities infrastructure, and even competitor benchmarks. All this information provides a platform for understanding and formulating what needs to be accomplished. The context sets the stage for linking goals with opportunities and constraints so that the desired end and the path for getting there are plausible and attainable.

Building a common vision for what is to be accomplished may be one of the most elusive requirements of strategic planning. Yet its articulation—the framing of the question—is the important first step in the planning process because it directs all forward motion. Internal alignment around the desired future state is critical. First, it underlies the development of decision criteria, and, second, it greatly increases your ability to execute the plan once it has been designed. As planning moves into implementation, alignment around this vision ensures that change initiatives are all presented with one voice—a benefit that cannot be overemphasized.

Optimization comes into play when you think of the organization as a system of interlocking parts. It hinges on the belief that the whole of the organization is greater than the sum of its parts. All parts drive the whole and the system functions optimally when all parts are in tune. It is precisely the interrelated nature of organizational resources and needs that makes strategic planning such a compelling process. Without the three elements—context, vision, and optimization—the chances of success are greatly reduced and the chances for optimal performance almost nonexistent. But when they are in place, even though the way is not black and white, clear-cut right or wrong, the planning process may impact future directions and actions in significant ways:

- Strategic planning is both a tool and a mindset for reaching a future as yet unknown. The shared understanding of context, vision, and resources provides a foundation for the development of strategies and actions that are viable and sustainable within a climate of change.
- Strategic planning always understands that every decision has a cost, in terms of both the investment made and the opportunity seized. But it assumes that the best understanding of the problem leads to the best chance of making more informed less risky (and costly) decisions. Needs assessment and analysis help clarify what is and identify opportunities for what could be so that the financial investment is qualified with an understanding of what is at stake.
- Strategic planning makes information *explicit* that has heretofore been *implicit*. The newly explicit information enables informed and realistic responses to immediate challenges such as work practices, real estate requirements, hardware and software needs, communications, and training. Analysis of the right data enables you to forge flexible long-range plans. Strategy development and resource allocations become more specific and readily defensible.
- Strategic planning establishes an internal direction that ideally leads to increased productivity and optimization of costs. Applied to workplace issues, it creates a comprehensive, measurable process linking technology, work practices, and human factors to develop the tools that will support client needs, increase efficiency, control costs, and create a more effective workplace strategy.
- A comprehensive strategic plan focuses on establishing and effectively communicating desired performance objectives. Great benefits follow from a plan whose clear mission and comprehensive strategy create opportunities for employee self-assessment and mid-course correction. Its post-implementation evaluations at set periods ensure that the desired outcomes are reached within the context of the stated goals.

The following discussion of strategic planning walks the reader through the process. It explores the key elements of a methodology, including design of the process; collection, analysis, and interpretation of data; and recommendations. A word on strategic communications will focus on project approach, activities, and tools for creating a common vocabulary to discuss the components of strategic planning. And, finally, case studies of organizations that have launched strategic planning projects will address a number of key planning activities, with examples of lessons learned and successful outcomes.

3.2 PRIMACY OF PROCESS: THE STRATEGIC PLANNING MODEL

> Process is more important than outcome. When the outcome drives the process we will only ever go to where we've already been. If process drives outcome we may not know where we're going, but we will know we want to be there. *(Bruce Mau)*

The success of any effort, if Bruce Mau is right, is not focused on the final product, but on the way we arrive at it. The process is somewhat like navigating by the stars. You know which star you want to follow when you set out from Miami in a sailboat. You never arrive at the star. But if you have prepared well, it leads you to landfall on the coast of France, which is what you had in mind when you set out.

The process of strategic planning works something like that. Inherent in your approach to problem solving is the requirement for understanding your client's desired outcomes, pinpointing their stars. Mapping an appropriate course to guide the project requires that you understand more than the key elements of a range of project processes. You must also know how to manage them.

3.2.1 The Strategic Planning Model

The process of strategic planning ideally occurs early, though it may function as an intervention at any point in an organization's life cycle. As illustrated in Figure 3.1, strategic planning clearly resides in the plan phase, but both the information it makes use of and its influence extend to the design, construct, and manage phases as well.

Throughout its life cycle, an organization constantly generates information that may be captured from any quadrant of the process at any time. This steady stream of information, sometimes called "exhaust" from the organization's "engine," merely disappears if it is not captured and used to affect or evaluate the performance of a strategic plan.

Early in the information gathering, it is important to document the assumptions on which the data are based. Then, throughout the process, it becomes easy to pinpoint those pieces of data that also need to be updated as assumptions change.

3.2.2 Observations on the Process: Technical Dimensions

The goal of the strategic planning process is to get "there" from "here." This requires that you first understand "there," the desired end result. And you must understand "here," the existing conditions within the client company. What is their business? How do they really make money? Do life cycle costs align with the operations and real estate strategies?

1. *"There": Identify and Agree Upon a Mutually Desired Outcome:* "What are we really attempting to achieve?" is the question. If you decide to undertake a strategic planning process, it is critical that you and everyone involved be crystal clear about why and what you hope to gain. Project proponents must carefully assess, understand, and clarify the underlying purpose, function, and desired outcomes. All participating parties must buy in on those outcomes because to a large extent they will drive the design of your strategic planning process.

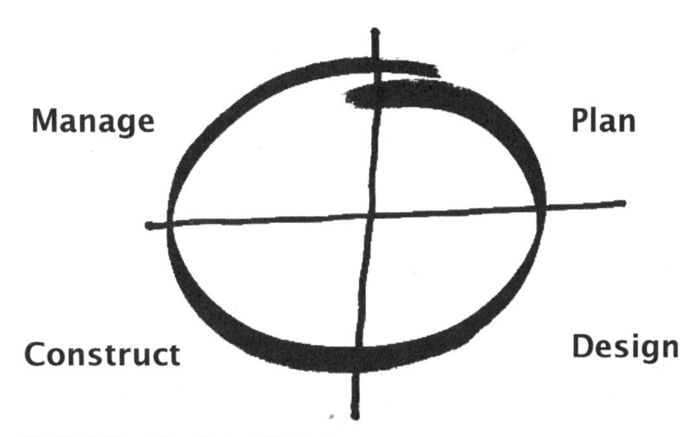

FIGURE 3.1 Client/Project Life Cycle.

Independent of industry, the answer to "the question" will determine the direction of the process. The question to be answered may have to do with capital tied up in non-performing assets, or the cost of churn, or the cycle time required to reach a business decision. Excessive vacancy or the inability to know where empty cubicles are located could reduce the reaction time it takes for a company to respond to a business condition. Finding the answer to this question is the whole point of the strategic planning process.

2. *"Here": Identify the Conditions That Describe the Company Now:* The design of the process helps you to find the rational basis for what you already know and to discover what you do not know. This information clarifies what is important in the client's present state, your point of departure, and informs your direction as you embark from "here" toward "there."

You need to know the pressure points in the business because, whatever they are, the process, information, and analysis will respond to these points. Pressure on the business might have to do with specific geography or configuration revealed in corporate history. If the organization has just undergone a major merger or acquisition, facilities may be redundant and geographically dispersed. A number of questions will help you arrive at a good understanding of the forces acting on the business:

- *Pressure Points:*

  What are the issues driving change in your industry? Your organization?

  What are the internal and external factors impacting change?

  What are the challenges and opportunities facing your industry?

  What is the one thing that could fundamentally change your business?

  In a world of constant change, how do you address change?

  How volatile is your business?

  How well do your business plan and operations plans support these changes?

- *Priorities and Opportunities:*

  What are the priorities of your organization?

  Are decisions cost-driven or revenue-driven?

  What is the level of your comfort with risk?

  How do you typically seize opportunities in your business?

- *Organizational Structure:*

  Does your organizational structure support the way products and services are delivered?

  How can your organization better support employees in terms of work process, systems, and environments?

- *Technology:*

  What is the role of technology in your organization?

  What anticipated technology changes will impact systems or work processes?

- *Work Environment:*

  What is the role of the work environment in your organization?

  How effectively is space used?

  How does it compare with others in your industry?

  What do you want your space to say about you?

3. *What Primary Disciplines are Involved?* Just as the cognitive map of the client affects the framing of the strategic question, the disciplines involved will affect the format of the answer. Engineering data will require engineering analysis, while economic data will require economic analysis. It does not follow, however, that the solution will be engineering or economic in content. Keys to success are often counter-intuitive, found outside the context of the information being analyzed. Engineering data, for example, may lead to an economic solution, while the proper response to economic data may lie in engineering.

4. *The Process is Both Linear and Cyclical:* The strategic planning process is linear inasmuch as each phase builds upon the last. Goals and objectives of the strategic plan determine the type of data collected, which leads to particular analysis, scenario development, strategic recommendations, and implementation.

The process is also cyclical because business goals determine the data to be collected. These data in turn affect the goals. When data appear relevant to the problem but irrelevant to any specified goal, re-examine the goals. When a void appears in the data related to a stated goal, re-examine or expand the data. This cyclical process is an extremely important characteristic of the process, for it allows the information to build on itself as the project proceeds.

5. *Data are Both Quantitative and Qualitative:* The type and quantity of information you collect about an organization will vary with each assignment. But it all falls basically into two categories: quantitative and qualitative.

Quantitative information is statistical in nature. It may take the form of headcounts, square footages, growth rates, market share, and revenues. There is a tendency to gravitate toward quantitative information because it is readily available in most businesses, which usually measure and monitor success statistically.

Qualitative information is more intuitive than statistical, and more important in the strategic planning process than quantity because it provides insight beyond the numbers into the soul of an organization. This type of information, which may include corporate culture, image statements, management philosophy, new ways of working, and company goals, can be found in annual reports or embedded in company newsletters or a letter from the President. These data are invaluable for opening a view into the corporate culture, the essence, or "heart," of the organization. A strategic plan may respond to all the statistical data available, but if it fails to address the culture—the heart and soul of the organization—it will not succeed.

6. *How Do You Make the Strategic Planning Process Value-Centered?* If Kevin Kelly is correct when he writes in *New Rules for the New Economy* that, "Wealth in the new regime flows directly from innovation, not optimization," then innovation is fundamental to the design of a successful process. Innovation, however, is an outcome rather than an item on your checklist ("Oh, let's innovate today!"). A plan for communication, as shown in Section 3.5 of this chapter, should be on your checklist. Regular and purposeful communication between your clients and yourself becomes a tool whose naturally occurring result is the possibility of innovation. Unforeseen discoveries will emerge as you go, and these will become crucial in the work of reaching your clients' goals. In such a scenario, it is acceptable (and probably desirable, as uncomfortable as that may be) not to be sure how the process will turn out. You are allowed to recognize that the focus is on the journey, on finding the new, reassured by Kelly's observation that "wealth is not gained by perfecting the known, but by imperfectly seizing the unknown."

7. *Be Sensitive to the Human Dimensions of the Project:* More is involved in the whole process than the technical aspects of setting the course and collecting data. The process gains a further dimension of complexity through the need for identifying and working with a large variety of people who are stakeholders in the planning, execution, and outcome.

8. *Consider the Culture in Designing the Process:* The values held by the corporate culture will shape answers to the questions you must ask in designing the process. Experience shows that when the process is co-created, with extensive buy-in, strong internal support, and community outreach, the strategic plan is more effective. Any organization whose culture places a high value on broad participation usually develops a strategic plan with higher impact than one that does not, because its plan is swept forward by the weight of the body of people sitting behind it. So you need to ask:

- *Leadership and Work Style:*

  Is the culture driven by senior leadership? How much regard is given to other voices?

  Does leadership ever, or periodically, involve a core group of employees who represent the organization's various business units?

  Have previous planning processes led employees to expect a number of task forces or working groups to meet on specific topics throughout the strategic planning process?

  Would your best plan involve a hybrid of some of the elements described above?

- *Attitudes toward Learning and Change:*

  Do your employees understand what is driving change in the organization?

  Do your corporate culture and values support change and planning?

  What is the attitude toward learning?

  How is learning celebrated?

- *Communication:*

  What level of communication do you need to maintain before, during, and after a strategic planning process?

  How does the organization access and manage information?

  Is there a system to support the exchange of information?

  How do you address rapid change in information?

  How important is face-to-face communication in your organization?

  How do people typically communicate?

  How can you best assist your people to do their best work?

9. *Garner Sustained Support from Senior Leadership:* Without support and unconditional buy-in from senior leadership, most strategic planning efforts are destined to languish in a report on a library shelf. This is not rocket science. Their engagement and support are essential to creating the momentum and securing the resources to prepare and implement the strategic plan. Their commitment must hold firm throughout the life cycle of the project—from inception to implementation and evaluation.

The designation of a person to serve as primary contact is often a function of the realm that initiated the process. If the client contact is the Chief Financial Officer, the question and its analysis may well be financial in nature. If the client contact is the Chief Operations Officer or Chief Information Officer, the analysis as well as the measures of performance may be operational or technological.

10. *Enlist Internal Leadership to Facilitate the Process:* A team of "project champions" is key. These are leaders who have the passion and commitment to steer the ship and wave the "Strategic Planning is Good" banner. They have the dedication to see the project through to implementation and beyond. They are the keepers of the flame, people who will guide the process and keep the content well-focused. They facilitate the dialogue among participants, ensure that essential reviews and approvals from senior leadership are in hand before moving on to the next step, and take steps to alter course should that prove necessary. Most strategic planning efforts are extraordinarily dynamic and require ongoing refinement and monitoring, so champions correspondingly need the ability to manage frequent change and to tolerate and handle ambiguity. (Not everyone does this well.)

11. *Determine the Extent of Desired Community Involvement:* An appropriate level of community involvement is one of the most important elements of any successful strategic planning effort. This is true, whether the project is contained within the micro-community of the business, or it is so vast that it affects the macro-community of the geo-political area where the business is located. In either case, the questions to be asked and the steps taken differ chiefly only in scale. The proponents must ask for whom they are creating this plan. Who ultimately will be held responsible and accountable for executing it? Whom do you intend to benefit from this effort? All these questions have one and the same answer: the employees—secretarial/administrative staff, sales force, senior leadership, executive management. And, on a larger scale, the beneficiaries include the macro-community.

You will need to know answers to these questions:

- Who are the external and internal stakeholders in the process? How much do they want to play—and how much do you want them to play? Will their input be genuinely factored into the solution?
- Does the corporate culture require a simple or extensive community involvement program?
- Is it appropriate to the culture to solicit constituent feedback through a series of "open houses" or "town hall meetings" at critical points in the process?

12. *Map Significant Project Issues and Constraints Early:* Many well-intentioned strategic planning efforts have been blindsided in mid-process by a single impediment or a series of key attitudes or issues recognized too late. A brief but pointed audit of the organization's vital signs will help you check the enterprise's pulse and uncover potential obstacles. If a negative attitude toward strategic planning lurks in the organization, this exercise will enable proponents to factor in that knowledge as they design their strategy. It will also clarify appropriate levels of scope, depth, breadth, senior participation, duration, and community outreach for the project. Answers to questions like these will be useful:

- What has been the general experience with developing a strategic plan in the past? Was it successful? Why or why not? How deeply did the strategic planning effort penetrate the organization, and how extensive was the outreach effort in the corporate community?
- What issues do the organization's senior leadership and shareholders have with respect to a strategic plan? What are the hot buttons?
- Are there any leadership or shareholder personalities that may influence the conduct or implementation of the strategic plan? What are their needs? How should they be involved in the design and ongoing review process to contribute positively to the process?
- Are there other organizational constraints to be considered? Schedule? Budget? Control?
- What is the communication/decision-making process?

The decision-making process will determine the format and the form in which the data are collected, analyzed, and documented. Centralized decision making requires one format, decentralized another. A hierarchical review process requires that you reiterate the information as approval moves up the ladder, concluding with an executive summary presentation to senior management.

3.2.3 The Right Problem, the Right Solution: Finding Both

Identifying the right problem and the right solution means not merely treating a symptom, but first analyzing the driving business issues facing your client. When your client increases the company's salesforce and hires you to determine how much more space to lease, the process may well lead to a solution not evident at first. Let us say that in listening to the business issues, you discover that the salesforce will spend an average of 25% of their time in the office and 75% at job sites. You see that the need is not for more real estate so much as for more efficient use of the space the client already has. The process has led you to identify the right problem, and enabled the right solution to suggest itself.

3.2.4 Case Study #1: Paragon Biomedical, Inc., National Strategy

Having identified a unique niche in the Contact Research Organizations (CRO) market, Paragon Biomedical, Inc., by 2000 had grown into a $15 million a year company. Since 1991, Paragon grew an average of 10% faster per year than the rest of the CRO industry. However, as the customer base grew, the increase in workload outpaced the development of the company's processes and systems.

The question was how to align the work environment with business goals and the evolving culture of the firm. Paragon had successfully employed the "virtual" firm model (a nationwide network of associates working from home) in the past, but now workplace was the missing link. So, the firm commissioned Gensler to assist in developing a national strategy for company workplaces that would link people, processes, and technology.

To start with, Gensler helped Paragon define and document their key requirements through a series of meetings, focus groups, individual interviews, and site visits in Boston, Chicago, and Irvine, California. In this process, Paragon identified the following needs: access to each other and to information (nationwide), support for technology and data management, and education for software applications and for systems and processes. The common issues driving the nationwide real estate selection process were, in order of importance: culture and image, functionality, and cost.

With a fundamental understanding of Paragon's need for an interactive, community-specific, functional environment, the consultant-proposed real estate scenario combines regional satellite centers. Their virtual network of individual offices will enhance the overall communication and strengthen Paragon's ability to share and leverage company-wide resources. This plan will create and maintain a sense of community and culture, foster client involvement and interaction, and support mentoring and learning through a network of spaces. It will provide Paragon with the functionality to create maximum scalability for future growth and the ability to leverage resources to the fullest extent in the twenty-first century.

3.2.5 Case Study #2: Arthur Andersen, Los Angeles, California

Founded in 1913, Arthur Andersen had always been known as a global multidisciplinary professional service organization that would go "beyond the numbers" by providing quality services including assurance, business and financial consulting, and tax and legal services to improve the business performance of their clients. As part of a lease re-negotiation effort, Arthur Andersen in Los Angeles asked Gensler to design a work environment that would optimize the use and flexibility of space while breaking down the traditional hierarchical office space structure. Educating staff and obtaining their support for the new work environment was a critical concern of the firm's executive council.

The parallel agendas of change management and design provided the consultant with a unique opportunity to run the full course from workplace consultation to design implementation. Gensler responded by canvassing the staff for ideas and work preferences. This dialogue continued for the duration of the project and employee feedback was consistently incorporated into the plan. Indispensable solutions in the redesign increased employee mobility, centralized company information, enhanced client service, and emphasized a multidisciplinary team approach to the delivery of service.

During initial occupancy, several training sessions during the lunch hour familiarized employees with their new facilities, relating the space back to the goals that they themselves had agreed to:

- *Encouragement of Communications, Chance Collaboration, and Employee Interaction:* On each floor, universal workstations are grouped in "neighborhoods" to define the expansive floorplate. They are strategically aligned with teaming spaces, training rooms, a service center (to centralize office support functions such as copying, faxing, printing, collating, etc.), conference rooms and coffee bars, all of which wrap around the "town center." Each "town center," which provides hoteling lockers, kiosks, "plug and play" work counters, private telephone rooms, and a pantry, is designed to serve as a social hub—encouraging communications, chance collaboration, and employee interaction.

- *Implementation of New Technologies to Increase Employee Efficiency and Productivity and Serve as the Connective Tissue of the Redesign:* New technology support includes fully wiring the organization for both Intranet and Internet capabilities. It also includes the implementation of voice and video conferencing, faxing capabilities, and identifying an official ISG-office liaison. Whether making a hoteling reservation, accessing a portable computer from a client site, or utilizing state-of-the-art multimedia centers, advancing technology has made this alternative office environment a practical reality. Employees now have a choice of how and where to work on any given day—a flexibility that fosters their productivity and job satisfaction.

This design project implemented dramatic change in the workplace and initially evoked skepticism among Arthur Andersen employees. It was crucial, therefore, to introduce a workplace training program that would guide the partners, managers, and staff on how to maximize the benefits of the new environment. Along with technical assistance and counseling, a complementary "user guide" handbook aided in implementing the skills and techniques important to a successful transition.

3.3 PROCESS: DATA COLLECTION AND ANALYSIS

We know that goals and objectives, set early in the project, determine the scope of work and become the measures for performance. They are the touchstones that ensure continued alignment of the strategic plan with the original intent. Goals and objectives emerge during initial research and are confirmed in subsequent interviews and work sessions. Visioning sessions, even including an occasion like the project kick-off, are opportunities to discuss the question, "To be ultimately successful in supporting the business plan or mission of the company, what must this strategic plan do or be?" Answers to that question will lead to explorations of what the project *might become*, an outline for accomplishing the project, and the growth of consensus and mutual understanding in relation to it.

When the groundwork for the strategic planning process has been laid, it is time to attack the specifics. Each phase has its own peculiar need for information, and various methods of gathering the information are appropriate to each one.

3.3.1 Capture and Organize Data

The best strategic planners never rely on a single method to capture information, but use a variety of tools over the course of the project. The methods you employ will vary as a function of the way the organization typically captures and conveys information. Based on the profile of the company and the issues being addressed, data may come from real estate, finance, engineering, facilities management (FM), human resources, operations, even legal. Many of these information sources provide revealing insight into how the company looks at itself and how it ultimately measures success. Factors such as schedule, availability, client type, level of detail, and business drivers also determine which tools are the most appropriate for each engagement. Figure 3.2 shows "Step One" in the process of data collection and analysis.

The type of information you gather depends upon the problem being solved and the level of detail required, as well as the availability of information and sources. Basic quantitative and qualitative information typically includes such items as project goals, head counts, growth rates and density targets, as well as existing facility information, such as space standards and real estate data. By the end of the project, your techniques for data gathering will have varied from high-tech surveys, to face-to-face interviews, to interactive work sessions with end-users.

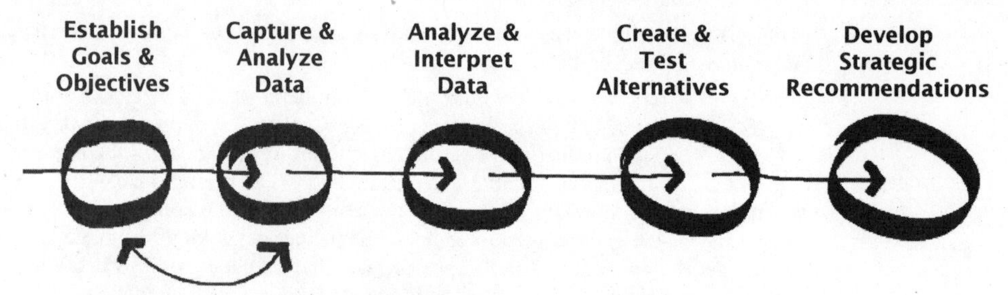

FIGURE 3.2 Linear Yet Cyclical, Step One.

The first, or primary, data collection occurs early in the process and focuses on gathering existing, readily accessible information. Secondary data collection occurs in one-on-one interviews with the users, group interviews, work sessions, or facility walk-throughs. The third cycle occurs when alternatives are formed. A critical step is the work session in which the project team verifies and evaluates all the information collected to date.

Examples of tools for data collection are:

- *Surveys:* Surveys or questionnaires are among the most common forms of gathering quantitative data. Because they are one-way communication, they are particularly time-effective in polling a wide population or a large number of departments or groups. They are also effective in collecting data about individual use of the workplace from a cross-section of staff in order to understand what goes on there or to build statistical support for a trend hypothesis.
- *Interviews:* Interviewing is a far more personal (and accurate) method of gathering data. It is one of the most common ways to gather qualitative data because, being two-way, it allows exploration of complex ideas. Interviews confirm survey results and often replace surveys as the primary means of gathering data for high-level strategic planning with executives. Early interviews with individual senior managers are commonly followed by a work session to confirm the collected information and build consensus.
- *Workshops:* Workshops are essentially group interviews. They are particularly effective in highly collaborative organizations since all participants feel that they have an individual and collective "say" about the strategic plan. Workshops may be the primary means of gathering information and confirming the interpretation of data gathered by surveys or interviews. The author has found that incorporating a series of workshops throughout the process builds consensus, avoids hidden agendas, and leads to a better strategic plan.
- *Web-Based Gathering:* Technology now enables gathering information quickly and accurately through the means of a company Intranet or ProjectNet (Internet). This works especially well in high tech companies that are accustomed to communicating by way of computer. The one drawback of this method is that often only computer-proficient staff respond.
- *Pilot Projects:* Pilot projects are most effective in testing and refining recommendations before making a larger financial commitment to implementation throughout the firm. Recommendations that lend themselves to pilot projects include space standards, teaming configurations, workplace strategies, new technologies, or information-sharing strategies. A bonus of a pilot done effectively is that the test group may become advocates for the merits and advantages of the proposal.
- *Post-Occupancy Evaluations:* Strategic plans are not static recommendations. They are dynamic, living strategies, continually refined to respond to business change. Post-occupancy evaluations (POEs) provide a simple way to capture information about how well the strategy supports the original goals of the project. The first POEs should be done approximately three months after occupancy and others later, whenever the business assumptions begin to change. These evaluations allow the strategist to anticipate when the facility strategic plan should be modified.

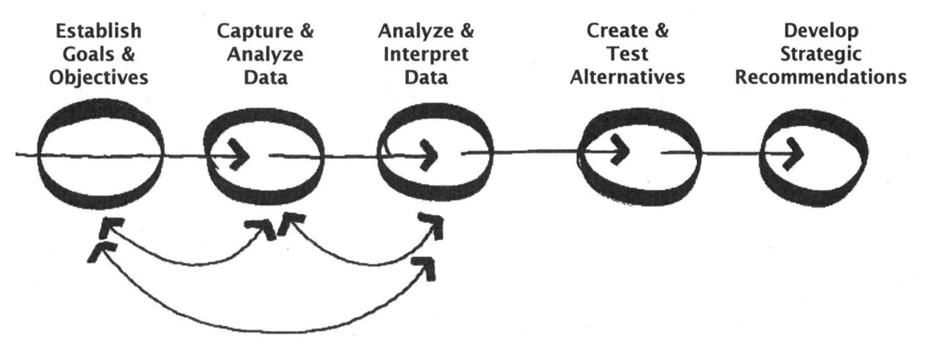

FIGURE 3.3 Linear Yet Cyclical, Step Two.

3.3.2 Analyze and Interpret Data

The value of strategic planning depends on filtering, interpreting, and transforming information into meaningful knowledge that can provide strategic direction. The value of the collected information is only as good as the questions asked and the quality of the analysis and interpretation. Figure 3.3 shows "Step Two" in the process of data collection and analysis.

Aspects of analysis and interpretation are important to understand, especially in relation to goals. Objectivity is key. Bias with respect to findings will merely confirm preconceptions, or perhaps even a hypothesis that is often embedded in the goals outlined early in the strategic planning process. If the team knows how to recognize bias in interpretation, the process is designed to advance opportunity and reduce preconception. The process at this juncture will also reveal voids, conflicts, or "soft spots" in the data that need to be pursued further. In addition, the data can also point the team in a new and unexpected direction, and it is exactly while exploring a new direction that innovation and new discoveries may occur.

With the advent of databases and the ability to track vast quantities of information, data collection sometimes falls prey to the phenomenon of "information glog," an overwhelming amount of information whose implications are exceedingly unclear. Paradoxically, innovation and discovery often exist on the periphery of such a problem, and the relevancy of data may not become apparent until the problem is explored further.

3.3.3 Create and Test Alternatives

As you create and test alternative scenarios, your search for information becomes more focused and finite. Here, since you are actually attempting to defend a hypothesis or scenario, you are building a business case of supportive data. Information now fills in the gaps to create an arguable defense for the scenario, and you will often articulate the data in defense of the alternative. Figure 3.4 shows "Step Three" in the process of data collection and analysis.

Business goals to be supported may include such things as growth ranges, business or market opportunities, potential legislative changes, or even mergers and acquisitions. Real estate goals may include renovation, remodel, expansion, consolidation, or decentralization strategies. These proposed outcomes are weighted against various criteria—timing, financial impacts, exiting, and support of each project goal. Typically, the result comes down to three or four radically different alternatives, and other scenarios become derivations or hybrids of those alternatives. One "remain as existing" scenario, kept as a baseline for comparison, enables you to assess the advantages and disadvantages of each alternative.

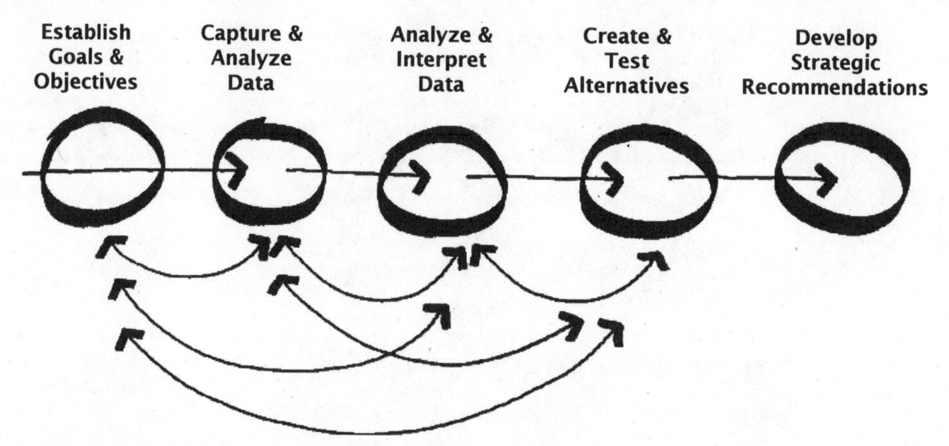

FIGURE 3.4 Linear Yet Cyclical, Step Three.

3.3.4 Develop Strategic Recommendations

At this point in the process, the plan is fine-tuned and presented to senior management for approval. The data may be presented either as *summary data,* used in an executive summary and distilled to its essence, or as *tracking data,* used as a measure and diagnosis of performance in monitoring the success of the plan. The final recommendations must be linked not only to the overall project goals, but also to each specific business goal. Ways of telling the story vary depending upon the audience and whether the desired result is a decision or simply information about the approved strategy. Figure 3.5 shows "Step Four" in the process of data collection and analysis.

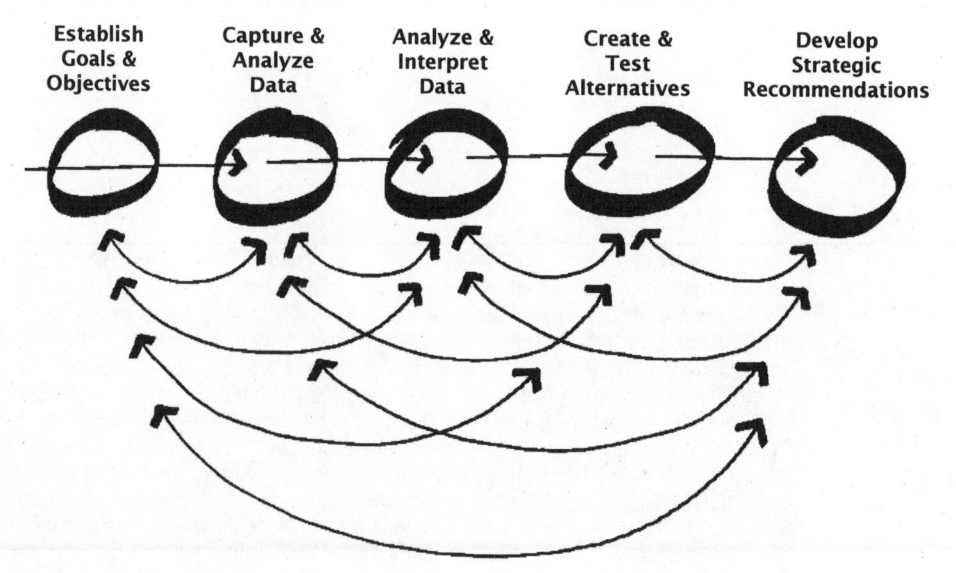

FIGURE 3.5 Linear Yet Cyclical, Step Four.

The following are some typical tools or formats:

- *Final Reports:* The best final reports are executive summaries that include goals, process, participants, business assumptions, major findings (i.e., head counts and real estate issues), alternatives explored, final recommendation, and next steps to implement. A second report containing all the detailed information and analysis may be provided as a back-up document.
- *Electronic Versions:* A slide presentation is an effective tool to document and communicate the final recommendation. When presenting several alternatives, the electronic version may not be the ideal tool because it is sometimes difficult to respond to unanticipated questions by clicking back and forth between slides. A hard copy will enable you to field unexpected questions more easily.
- *Workshops:* If you have been working with the senior management team throughout the process, your presentation of the final alternatives as a workshop or group work session is particularly effective. In such a setting, the pros and cons of each alternative can be easily debated and the best one selected or a hybrid recommendation created.
- *Presentations:* Formal presentations are typically reserved for meetings with senior management to solicit support or financial backing. These "big picture" presentations should include only the executive summaries, outlining the *why* and *how* of the plan.

3.4 PROJECT DELIVERABLES

The strategic delivery of project data and analysis is crucial not only to the clear and effective communication of recommendations to the client, but also to the successful implementation of the plan. The series of project deliverables—strategic programs, lease audits, cost structure evaluations, implementation frameworks—is described in the following sections, with the rationale for each.

3.4.1 Deliverable 1: Analysis of the Strategic Environment—The Demand Profile

The first step in any strategic planning process should involve an evaluation of business challenges, a study of the client's strategic response to these challenges, and an analysis of a client's operational requirements. In evaluating the business challenge, the consultant must determine and appreciate the strategic issues that are facing the client's business. In so doing, you will gain insights into market characteristics and their impact on the corporation's strategies and eventual performance. The external drivers that you evaluate will point to opportunities to help achieve better business results through an improved corporate real estate and FM strategy.

In studying the client's strategic response to the business challenges, you must undertake an in-depth review of corporate strategy, business unit strategies, and existing CRE/FM strategies. Do these strategies align or misalign with key corporate and business unit strategies? The extent to which they do either provides insights into the weaknesses and strengths of the current strategies.

Having completed these evaluations, you are now ready to determine the client's operational requirements and analyze these in the context of the existing strategies' weaknesses and strengths. This collective set of analyses can be combined into a quantitative and qualitative program of requirements addressing macro-level needs and requirements for each major business group, and quantifying the demand for space required in both the short- and long-term.

The product of these understandings is the *demand profile*, a summary that can be accurately documented and subsequently refined. At this point, the consultant should establish a statement of project vision and confirm this statement with the client. The accompanying dialogue allows all stakeholders to arrive at consensus on project opportunities and encourages buy-in and acceptance of the recommended plan or strategy. Most importantly, this step enables the client to avoid costly changes to the strategic plan at a later time in the process. This collective analysis of the strategic environment in which your client operates defines the context in which your real estate and facility recommendations will be implemented.

3.4.2 Deliverable 2: Analysis of the Current Framework—The Supply Profile

The full meaning of the demand profile becomes clear when it is analyzed in relation to the *supply profile*, whose accuracy depends on a series of quantitative and qualitative evaluations. On the quantitative side, the purpose of the supply profile is to collect and analyze a series of components for all real property assets:

1. Existing facility and real estate data including area, occupancy costs, lease commitments, and ownership statistics
2. Current types of use, with assessments of highest and best use
3. Suitability as workplace/facility
4. Macro-building condition

Assessments of current real estate and facility assets may be grouped into four primary categories:

- *Real Estate Assessment:* Includes building name, location, ownership information or lease terms/options, building square footage information, occupancy information, construction type, and zoning classifications.
- *Technical Assessment*: Focuses on mechanical/electrical/plumbing and life safety systems.
- *Architectural Assessment*: Focuses on the overall condition of the building exterior and interior. This assessment should include a review of optimal floor and building capacity, an evaluation of existing utilization, and an evaluation of suitability for existing/proposed uses.
- *Site Assessment*: Focuses on evaluating existing site conditions, site utilization, relationship to adjacent sites and their uses, zoning criteria, and planning policies.

When these data have been collected and analyzed, a financial analysis model may be developed for both the individual and aggregate analyses of the real estate and facility assets. To ensure buy-in, it is critical to develop this model in conjunction with the client.

On the qualitative side, the three categories of practice within the corporate infrastructure should be evaluated: human resources, information technology, and corporate real estate practices and policies. Each should be evaluated individually and in conjunction with the others. The purpose is to gain an understanding of the infrastructure disciplines within the client's organization and to develop appropriate performance criteria for purposes of evaluation. This analysis will allow you to identify opportunities to integrate the infrastructure with workplace/facilities so that work processes and the requirements of the departments in the overall business support each other more fully.

The classification of data and evaluations into these two distinct deliverables, demand and supply, ensures clarity of purpose and discipline of analysis. And, since organizations typically know less about their real estate and facilities than they think they do, your collection, reconciliation, alignment, and confirmation of real estate and facility data can be one of the most time consuming and resource intensive portions of the project. Yet, the effort is essential because the validity and effectiveness of your recommendations depends upon their basis on solid data.

Finally, performance measures already in place must be identified before you can begin to develop scenarios. Further criteria for evaluation, developed in conjunction with the client, may include projected costs, schedule, and risks associated with each strategy. These performance criteria provide a basis for systematically measuring the success of the project, adjusting the strategy to change as the client's business changes, and monitoring success of the strategy in such a way that they call attention to areas where improvements are needed and additional opportunities might be found.

3.4.3 Deliverable 3: Recommendations for a New Framework

Only after all corporate real estate and facility assets have been analyzed, the client's business challenge understood, and the operational and business requirements determined, could you establish a new strategic framework. Several steps lead to the development of a holistic strategy that results in real business value to the client.

First, develop and document a variety of possible scenarios and viable strategic alternatives to set before the client. Each will reflect all inputs collected in the previous activities and be vertically integrated with business unit objectives and corporate strategies. To assess the feasibility and cost/benefit implications of each alternative, each will factor in multiple variables, including the current and future space needs of operational functions, building condition and the cost of improvement, strategies for exit and reuse of existing facilities, risk factors of implementation, and schedule of implementation. Along with each of several strong strategic responses, a financial analysis is essential to help determine which is the most economically feasible. To maintain buy-in with the client's financial executive leadership, it is important to keep this group informed as you conduct the financial analysis.

In the final analysis, two to three strategies should emerge as the most likely to be economically successful. These two to three should then be weighted with regard to a number of factors: economic outcomes, ease of implementation, likely performance results, and impact on the client's vision for their company and the project. Out of this weighting, one strategy should surface as the best option.

A written document articulating the recommended strategy becomes the map that sets forth both the overall strategic direction and specific action steps. This deliverable serves as the basis for any future real estate decisions and implementation strategy. Its parts should summarize objectives and requirements, current real estate with evaluation, alternative strategy scenarios with economic and qualitative evaluations and overall assessments of each, and, finally, the recommended strategic plan showing both short and long-term strategies, the recommended timetable, strategy goals, and performance yardsticks.

Important to this deliverable is a "communication plan." At this point in the process it is essential to manage the perceptual impact of the change that will result as the strategy is implemented. The communication plan will detail for stakeholders the changes that will be occurring and how concerns can be addressed (see Section 3.5).

3.4.4 Deliverable 4: Implementation Plan

Once the recommended strategy has been developed and articulated, a strong implementation plan is necessary for its success. Any strategy will offer several possible options for implementation. First, brainstorm and document the possible implementation plans for the recommended strategy. Each option should contain a tactical planning schedule that includes phasing, move sequencing, potential costs, and potential risks to successful implementation. Based on these criteria, select the appropriate implementation plan.

Next, articulate the chosen implementation plan to show why this option was selected over the others. Include an analysis of the required phasing, sequencing, and financing options. This deliverable is a necessary foundation for the development of a detailed funding and improvement plan.

3.4.5 Deliverable 5: Evaluation and Refinement of Strategy

Shortly after implementation of the recommended strategy has begun, it is important to evaluate the initial impact of the strategy. This will help to maintain executive support for the implementation of the strategy and buy-in to the strategy. Further, it will enable modification of the plan to address changes in the client's business environment.

First, evaluate the early outcomes of the recommended strategy against the performance measures that were developed and articulated in an earlier deliverable.

Second, develop and articulate a recommendation for refining the strategy. The recommendation should be based on two findings: aspects of the strategy that are proving less successful than was anticipated and on changes in the outside business environment. Depending on the amount of change that has occurred, it may be important to revisit some of the earlier deliverables, such as the operational and programmatic data about the client or their corporate strategy. This deliverable should refine the implementation strategy to reflect your findings with regard to it. In the end, this step shapes the recommended strategy so that it can yield the best results possible.

3.4.6 Deliverable 6: Strategy for Continuous Improvement

Change is inevitable with every strategy. Since the business environment changes and evolves rapidly, your client must have a plan for continuous improvement. This will articulate processes both for monitoring the performance of the recommended strategy and for adjusting the strategy as required by ongoing internal and external change.

3.4.7 The Essential Deliverables

The six deliverables we have just described are essential in developing a successful strategy. To omit even one could have serious consequences in the outcomes resulting from the strategic plan. However, a balance must be struck between the time it takes to develop these deliverables, the needs of the client, and the speed of change in the client's business environment. Without such balance, the deliverables could be out-of-date before they are implemented. But this plan is almost self-revising, designed to accommodate revision again at the end. Speed is indisputably essential, but not if it diminishes or shortchanges the efficacy of the strategy and its implementation.

3.4.8 Case Study #3: Deloitte & Touche, Los Angeles, California

Due to rapid expansion and lease expirations in its Los Angeles basin locations, Deloitte & Touche, a quality provider of professional services, was forced to address the issue of future space requirements. A productive partnership between Gensler and a real estate firm designed an accommodation strategy to revamp D&T's entire real estate approach on the macro level, while continuing to address standard workplace issues.

The consultant's approach to this project rested on the theme *"building strategy from the inside out."* After thoroughly assessing D&T's existing real estate conditions, including head count projections, downtown and satellite space per square foot, and floorplate efficiency, Gensler and Studley began the real estate visioning process. Leadership, knowledge sharing, relationship building, and cultural clarity were identified as prevailing themes to guide emerging internal elements such as communication, collaboration/concentration, training, flexibility/mobility, information flow, quality of life, and community/culture. Alternative work programs were high on the list of priorities, offering employees ways to balance their work life with options such as flex-time, job sharing, building location, touchdown centers, and building amenities.

The end result included a centralized downtown location with satellites instead of branch offices. This offered great flexibility in terms of real estate. Many options were available regarding lease term, space design, and phased implementation. Methods of controlling space included purchase, lease, use of existing buildings, or building new ones. Associated deal structures could take new advantage of growth, availability of capital, and termination rights. Qualitative criteria included a variety of amenities, enhanced security, parking, touchdown support for floorplates, and phased implementation.

The rapid growth of the firm, increasing professional specialization, and geographically dispersed offices demanded a higher level of communication to ensure that all employees were adequately informed about firm direction, current projects, and upcoming changes.

3.5 STRATEGIC COMMUNICATIONS AND CHANGE INTEGRATION

Increasingly, clients expect consultants to provide services that reach out into the organization to help support internal change initiatives linked to real estate strategies. Representing the "people side" through change management and strategic communications has as much to do with human behavior and culture as with space and technology. Change management and strategic communications work in tandem to address the *process* for change, while recognizing that most organizations face the same external forces, pushing them in the same directions. Change management and strategic communications concern themselves with the crucial "how" when moving from a to b.

Driven by where an organization wants to be, change can be related to any number of variables:

- Leadership
- Work process
- People
- Physical space
- Technology
- Or combinations thereof

Understanding that none of these drivers can be addressed in isolation, the process for change must be designed with all of them in mind. Moreover, the process must be based on the fundamental assumption that the link between them all is people.

3.5.1 Addressing the Variables in Your Audience

Simple rule: understand the audience as best you can before tailoring messages. Communications in all its forms is about message delivery and reception—involving both senders and receivers; both play a role in controlling message delivery. All too often in creating a message, senders do not incorporate their intended audiences into its construction and subsequently limit the chances for communication to take place. In other words, they send out messages in a linear format not considering the variables in the receiver that can affect how the message is "heard." Incorporating those variables is fundamental to every strategic communications program and makes the process of message delivery and receipt less arbitrary.

The audience variables that affect message delivery include the receiver's needs, interests, and accessibility. The path runs something like the communications model shown in Figure 3.6.

In this model, the receiver's significant role in the construction and delivery of the sender's message increases the likelihood of message reception. The importance of the content of the message, its accessibility to the receiver, and medium of delivery are built into the process. When planning a strategic communications program as part of a change initiative, these variables become increasingly important to ensure that the change leader's messages are delivered effectively.

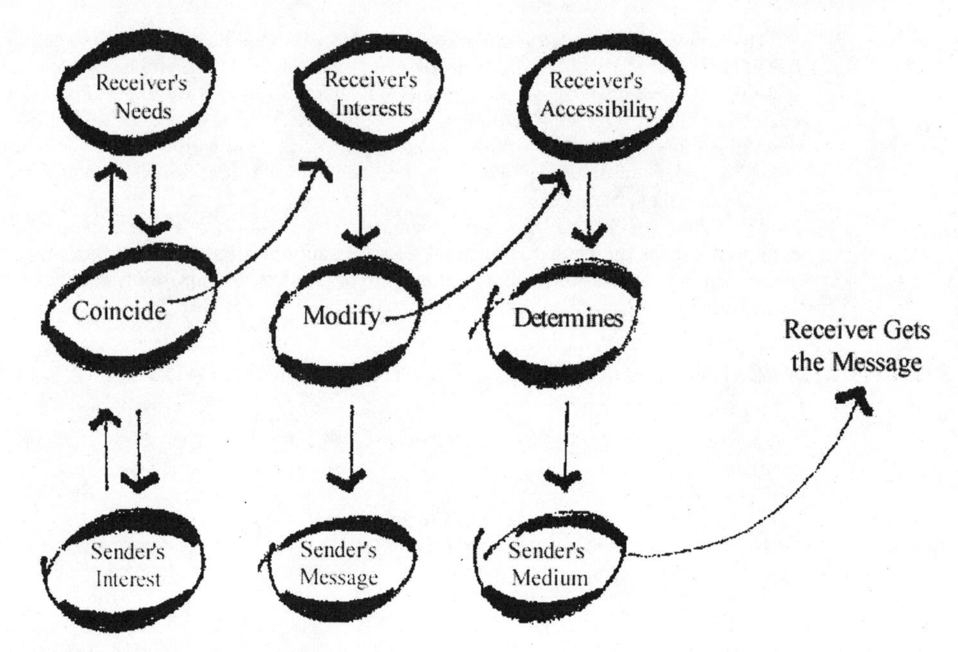

FIGURE 3.6 Receiver-Responsive Communications Model, in Which the Receiver Arbitrates the Success of the Communication.

3.5.2 Building Trust in a Time of Change

An underlying assumption to strategic communications is that the audiences with whom you are communicating are integral to the success of the changes being implemented. Well-crafted, strategic communications programs build credibility for change leaders. Information, delivered in a timely manner, furthers understanding of the situation at hand and lessens the chances that the target audience will create their own interpretations of the situation—allowing the change leaders to maintain a planned proactive communications position rather than an unpredictable reactive one.

By this, it is not meant that the change leaders are altering or "spinning" the truth. Strategic communications is not about creating false information to quell the disgruntled masses, so to speak. Rather, it is about creating and delivering messages in such a way as to increase the chances that audiences will receive them and being prepared for potential feedback—in all its forms. Assuming that audiences in any change situation are not sophisticated and will not understand the truth could prove to be very costly to change leaders in terms of both credibility and potential backlash.

3.5.3 Selecting the Medium—Communications Collateral

What medium will best reach your audience? When planning a strategic communications program, change leaders must determine how messages are to be delivered. Assuming that the audience needs and is interested in getting information about the changes at hand and the change leaders are interested in delivering information to meet organizational objectives, the receiver's accessibility needs to be determined and the medium for the message planned accordingly.

The variables here include addressing the cultural norms for the organization, and what "stands out." For example, an organization heavily dependent upon e-mail may or may not benefit from the use of a different medium in order for messages not to get lost. In the attempt to stand out, however, it is important to stay within the realm of vehicles which the target audience will use. It is also important to note that within the target audience, there will be a variety of learning styles: auditory learners, who focus on the spoken word; visual learners, who learn best from images; and kinesthetic learners, who learn best through a physical experience or, learning by doing. Ideally, any medium for your message will incorporate all three learning styles in its delivery.

Other considerations for message mediums include ensuring that the medium:

- Clearly identifies its target audience, who understand the message is intended for them
- Shows respect for the audience through the quality and content of the message, uses clear and tone-appropriate language, and is of a quality that prevents them from interpreting a cost savings measure or too casual language as a lack of respect
- Is not so unique as to "outshine" or impede the message, making it less likely to be received
- Takes into account the audience's past experiences and current situations, as both will affect how messages are interpreted.

Ideally, change leaders will build an evaluation stage into a strategic communications program in order to measure receipt of the message and effectiveness of the medium. The value of measuring program effectiveness will be to enhance future programs and further organizational learning.

3.5.4 The Basic Steps to Strategic Communications Planning

Boiled down to the essentials, the steps to successful planning for strategic communications are:

1. Isolate the issue
2. Determine the organizational objective to be achieved through communicating
3. Define the target audience
4. Develop the strategy by determining:
   - How the message should affect the defined audience
   - How the campaign will meet the organization's objectives
5. Select the media by determining what will work best for the audience in terms of credibility: web site, brochures, one-on-one messaging, direct mail, speeches, advertising, newsletters, personal letters, and the like
6. Determine the timetable
7. Calculate the budget
8. Design the evaluation process by determining:
   - How to measure success
   - How to generate audience feedback to ensure messages have been received as intended

3.5.5 Systematic Documentation of Process and Progress

It is crucial to be able to illustrate how the plan evolves. You will want to show the process graphically: assessment and discovery discussions fuel the visioning, which drives the preparation of strategic directions, which are supported by specific tactical steps for implementation. The story must live on the surface of the strategic planning process, visible to everyone, in order to accomplish the following:

- Illustrate that the planning process is credible, rational, well-directed, and achieving the outcome agreed upon at its inception

- Provide a graphic report for latecomers or people not involved in the process to see at a glance both the status of the project and the strategic implementation steps emerging from the findings of the assessment, discovery, and visioning phases
- Serve as a viable process record; a "how we did it" guide for future strategic planning efforts

3.5.6 Case Study #4: AT&T

AT&T's Global Real Estate (GRE) division retained Gensler to deliver a change management program that would assist in implementing a module for an integrated sales center. The model includes a new shared work environment and introduces hoteling, nondedicated workspace for employees frequently out of the office. Drivers of the sales integration strategy were a desire for increased sharing of tacit information and a sense of community, as well as a business decision to control real estate costs for investment in more strategic business areas. In most cases, the shift to a shared environment takes place concurrently with office consolidations or relocations.

Gensler's approach to the change management program for this project was three-fold. Due to the scope and schedule of the project (15 sites, 1500 employees, and a six-month time frame, including pre- and post-occupancy activity), the consultant designed a master process or template that provided a framework, but allowed for customization at each site as necessary. The elements of the template included:

1. *Strategic Communications Plan:* The need to inform both the management and employees of each office of decisions that have been made gave rise to a planned process, showing how needs would be met and issues addressed. The timing of information delivery is critical. If people learn what they need to know, when they need to know it, then they will neither dismiss it because it is too early to be relevant, nor panic because it is too late to be effective.

Sample materials include:

- Communiques to management and employees
- Information boards on-site
- User Guide for new systems
- Reference Guide for new software system
- Quick Start Guide for first day in the new space
- Post-occupancy surveys

2. *Facilitation and Coaching:* In order to understand issues specific to each site, the consultant conducts site visits and senior leadership meetings at each site. The first interaction with the client's change managers, these visits drive the customization of the template. Meetings typically involve introducing the change management program and coaching senior management regarding potential issues and key messages. In addition, focus group sessions with key employees from each business line discuss "on the ground" issues—again, to ensure that all issues are addressed before move-in and that key messages are delivered. Post move-in, focus groups or town hall meetings assess ongoing issues, including the effectiveness of the space itself.

3. *Orientation and Training:* Finally, the consultant delivers an orientation/training module to introduce employees to the new work environment and to the technical aspects of the hoteling software used to track space in the shared environment. These sessions also include a strong emphasis on the cultural shift and the change of mindset that may be required for adapting to the new environment.

3.6 THE LAST WORD

Strategic planning is a highly evolved and intensely collaborative process. It is demanding in its requirements of time, commitment, resources, and physical and emotional energy. For these reasons, it should not be undertaken lightly. The costs are significant and, when the effort is ill-defined and poorly executed, usually not recoverable. Strategic planning means that before taking a step forward, you must take a step back. From this vantage-point, a clear, encompassing perspective begins to take shape. This perspective leads to an understanding of the questions that every organization can ask to its benefit: Where are you going? Why? And which route must you take to get there?

ENDNOTE

1. Many people had a hand in writing this chapter. Their names, with the Gensler office where they work, follow: Francisco Acoba and Charles Elliott, Washington, DC; David Gensler, London; Jim Oswald and Darren Roberts, Los Angeles; Janet Pogue and Chris Nims, Denver.

CHAPTER 4
BUSINESS TRANSFORMATION AND FACILITY MANAGEMENT

Alex K. Lam, MTS MRAIC
President
The OCB Network
E-mail: alex@ocbonline.com

4.1 INTRODUCTION

> Systems are only dear to those who cannot take the whole truth into their hands and who want to catch it by the tail. A system is like the tail of truth, but truth is like a lizard. It will leave its tail in your hands and then escape you. It knows that within a very short time it will grow another. (*Ivan Turgenev*[1])

Leo Tolstoy received this correspondence from his friend and fellow writer Ivan Turgenev about the systematization of truth during their time. It seems that people tend to be engrossed in the systematizing efforts and miss the real issue—truth. The analogy of catching a lizard's tail is particularly interesting. It reflects what businesses are doing at the close of the twentieth century. No one knows what the future is like. It is as agile as a lizard. Everyone wants to lay hold of it, but all they get is a fragment—its tail. It keeps running away and it keeps growing new tails.

Facility managers are like those who are trying to shape their organization to these tails. Most companies handle their business transformation no differently. Business transformation is not new. Businesses have been transforming their way of doing business throughout history; it is only in the last decade or so that it has become more aggressive. The rate of change is more rapid and the need to change is more urgent.

Businesses are here because they provide a product or a service to their consumers. Whether you are selling groceries, books, furniture, or just providing a delivery service, your business must change with the needs of the customers. In the past, customers and competition were the two key elements that influenced companies to change. The formula for change was straightforward: how to have more customers and how to beat your competitor.

Today these forces are more complicated. Technology has brought the world to your doorsteps. Competition is no longer the person who opens a store across from you. Competition is someone who enters into cyberspace before you. Customers are more sophisticated and better informed. They also have more choices than ever and are no longer limited by space and time.

In this truly techno-info world, businesses are reengineering and reinventing themselves in order to stay alive. New ways of doing business and the accelerated mode of delivery require a new breed of workers. This, in turn, requires a new kind of workplace, which, in turn, requires a new kind of facility management (FM) organization. Few, if any books, on FM talk about business transformation—the closest to it is on *organization* or *outsourcing*. What most FM organizations do is just keep up with the changes and adapt to the business environment as best they can. Businesses are constantly bombarded with the corporate pressures of downsizing, cutting costs, and labor problems while still providing world-class service to their customers.

Different companies take different approaches to business transformation. It is the purpose of this chapter to present the guiding principles of business transformation based on the author's own experience and observation. The model is the cumulative results of the failures and successes of a particular company's transformational experience.

4.2 THE CHANGING BUSINESS ENVIRONMENT

In August 1999, the three major banks in Japan announced that they had formed a new alliance to be the largest banking group in the world. This alliance of Dai-Ichi Kangyo Bank Ltd., Industrial Bank of Japan Ltd., and Fuji Bank Ltd. has total assets that exceed $1.2 trillion.[2] This is an example of a trend today. In order to compete and be a leader in the industry, companies are joining forces to form new entities which would spell superiority in finance, technology, research, products, and service. Take a look at the world today. Mega-companies are being formed practically every day[3] (see Table 4.1).

How would this affect the facility manager? What does the merger mean to the facility manager? What would happen to the facilities and the office space? Will there be a reconfiguration of workspace? How will facility managers prepare for this?

Martha A. O'Mara of the Institute for Corporate Real Estate suggested that because of the changes in the corporate form, FM groups need to reconsider how they would provide services to their client. Organizations are no longer hierarchical structures, but more and more relational in a hybrid organizational structure. In other words, the traditional corporate boundaries are now being altered between customers and suppliers and between employers and employees.[4] Relationships may now span cities and countries. FM groups must find new and innovative ways to serve their customers.

Changes in the business world, whether for growth, consolidation, or minimization, have great impact on FM (see Figure 4.1). These changes bring about physical changes. Workspace needs to be provided or modified to fit the new organization. Michael L. Joroff of the Massachusetts Institute of Technology suggested that this new environment requires "an integrated spatial-technological solution that provided both real-time information via computer-controlled sensors and opportunities to 'feel' the actual production process."[5] The need to grow or to downsize keeps the facility man-

TABLE 4.1 Top Ten Mega Acquisitions

Buyer	Target	Value (Billions USD)
America Online	Time Warner	160.0
Vodafone AirTouch	Manesmann	148.6
MCI WorldCom	Sprint	128.9
Exxon	Mobil	85.2
Bell Atlantic	GTE	85.0
SBC Communications	Ameritech	80.6
Vodafone	AirTouch	74.4
Pfizer	Warner-Lambert	73.7
British Petroleum	Amoco	61.7
AT&T	MediOne Group	61.0

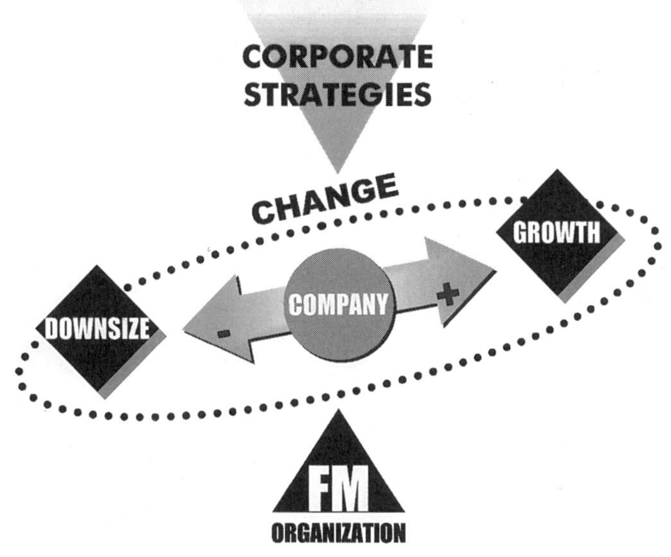

FIGURE 4.1 Corporate Changes and Facility Management.

ager busy. The FM organization must be able to meet these new requirements and practices when the workplace needs to be reconfigured and people need to be reassigned or redirected.

Many companies went through restructuring and reengineering in the last decade. There were external activities involving mergers and acquisitions jockeying into superior positions in the industry. Still other companies maintain their existing operation with some re-alignment and restructuring of business units. Most companies shed their noncore businesses and outsource. FM is noncore and, henceforth, outsourced. Outsourcing can put a great disruption in the facility operation. It is important, therefore, for organizations to transform themselves to anticipate these changes.

4.3 TRANSFORMATION IN THE BUSINESS OF FM

In May 1999, Royal Bank of Canada, Canada's largest financial institution, announced that they would sell off their downtown landmark headquarters in Toronto together with an extensive portfolio of other properties. It was a total package of thirty-three buildings across Canada.[6] In September 1999, Oxford Properties Group together with the Ontario Municipal Employees Retirement System (OMERS) and GE Capital Canada Inc. bought the Royal Bank package for Can.$827 million, completing one of the most extensive Canadian real estate transactions.[7]

A month later, Scotia Bank announced the selling off of a large real estate portfolio of eight to ten buildings. At the same time, Canadian Imperial Bank of Commerce, the most aggressive bank of Canada's big five, said that they were reviewing their holdings. A few years back, they established a separate real estate entity called CIBC Development Corp. What is going on?

The Globe and Mail, Canada's national newspaper, quoted Andrew Lennox, senior vice-president of real estate at Scotia Bank as saying, "We do not belong in the real estate business, per se . . . we are not developers. What we are is an institution that uses real estate."

For the same reason, Bell Canada, the largest telecommunication company in Canada sold off their commercial real estate holdings to a developer in Toronto for over Can.$750 million in 1998. The first group that started this trend was Canadian Pacific Ltd., which sold off their real estate for Can.$932 million in September 1996.

These large corporations all have their own corporate in-house real estate or FM groups. Many of them self-perform all the operations and maintenance work and only contract out work when they do not have in-house expertise. What will happen to these in-house groups in a merger or acquisition? How do they retain their tribal knowledge and expertise within the organization? How do they deal with the loss of productivity or efficiency during the transition? How do they maintain employee moral during these turbulent times? How do they maintain customer satisfaction?

4.4 THE BUSINESS TRANSFORMATION PROCESS

Outsourcing and downsizing are not business transformation—they are merely one outcome of transformation and are definitely not "the" outcome. Massive layoffs as a result of such decisions seemed to be in vogue in the 1990s and struck terror in employees' hearts. Harold Geneen, the former chairman of ITT, describes downsizing as a "medieval medical practice of bleeding patients to make them get better" and "nothing is more destructive of a company's culture than these mindless mass layoffs . . . whatever the former culture was . . . it is shattered and replaced by a new culture of fear, anger, and looking out for number one. People won't speak up, won't innovate, won't take chances."[8]

Initiating change and reducing the workforce should not be heralded to be the salvation for the bottom-line, although the bottom-line is part of the equation. In order to achieve success, change must be introduced with care and planning.

There are five components to the business transformation process: vision, integration, communication, transformation, and measurement, and each is described in the following sections.

1. *Vision:* Vision has no meaning if nobody understands it, if it is not shared, and if there is no way to get there. Vision should be the overarching governance of all corporate strategies. It must tell the employees where the corporation wants to go and what the corporation is going to look like. The change agent must understand the vision and how this vision was developed. In turn, this understanding must be shared with the employees.

In order to get there, there must be an analysis of the "what is" situation and the "what should be" scenario (see Figure 4.2). In order to do a "what is" evaluation, historical data is required. Every company must keep and maintain a good database of critical information about its operation, resources, and liabilities. Computerized FM programs provide such recordkeeping. Unfortunately, a lot of companies do not keep accurate records of their facilities, people information, and their business processes. By measuring the gap between the "what is" and the "what should be" environment, a how-to-get-there strategy can then be formulated.

Some people use process-mapping techniques to identify these differences. The process map will identify redundancy and disconnects. Disconnects are the "missing links" of the business process where there is no accountability and no one accepts responsibility. Process mapping provides a bird's eye view of the business operation. This is the first step of any business transformation.

2. *Integration:* The purpose of integration is to ensure the diverse components of the different groups are sharing the same culture, vision, and operations. When companies are acquired or business units are being re-aligned, there should be a transformation process to unify the parts.

Figure 4.3 shows four business units as A, B, C, and D. These letters may represent the newly acquired companies or the four newly established departments or divisions within the company. Each one can be very different from the others in terms of business culture, practice, or tradition. Due to the sheer speed of change, most companies did not allow time to design and execute a proper integration process. The result is confusion among the worker as well as the customers.

This is particularly true in FM. Downsizing and consolidation bring together various parts of the department with different standards and different ways of doing the same thing. Work processes tend to be different because different people are running the business. The people may be from different localities with strong local influences and culture.

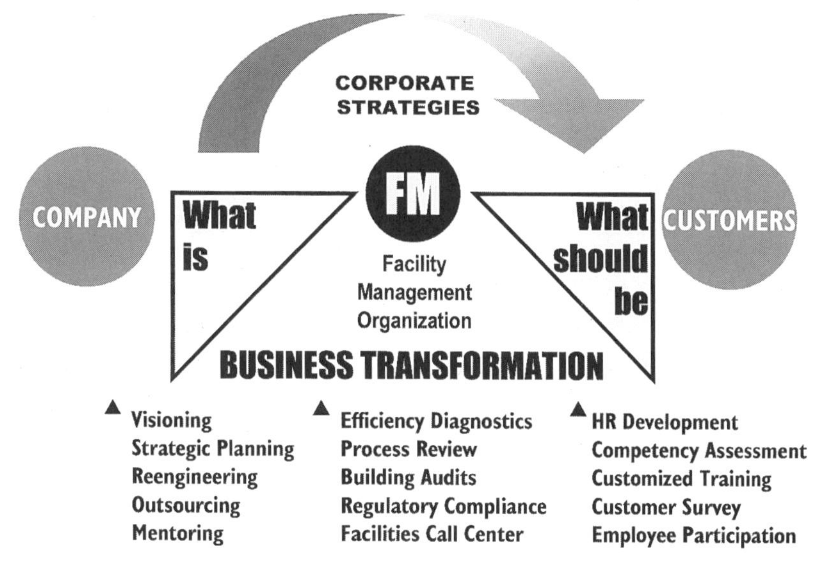

FIGURE 4.2 Gap Analysis in FM.

No matter how many groups there may be, a well-implemented transformation process should result in a common new corporate identity and culture. In other words, no more four separate voices, but one voice—one *brand.*

3. *Communication:* Poor communication yields disastrous results. Communication from the top down must be well presented so there is understanding and comprehension of the big picture.

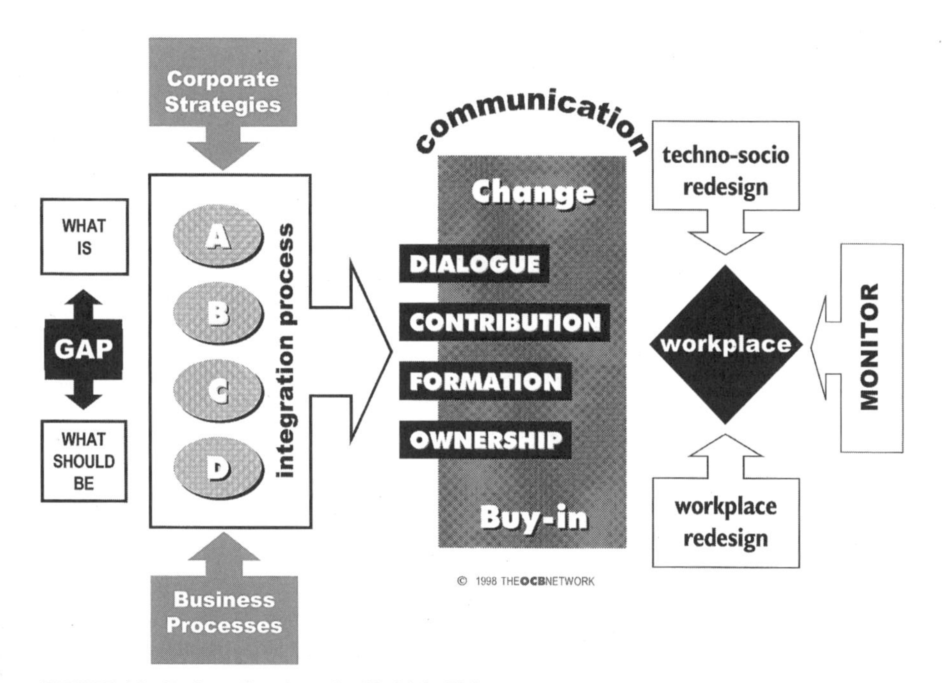

FIGURE 4.3 Business Transformation Model for FM.

Vision must be made clear and easy to understand. The best buy-in for any change is not some global mission slogan uttered by some charismatic change-agent. There must be a clear communication from the leaders about the vision and the reason.

Employees are our best assets. This much-abused slogan will backfire when action does not match vision. Over the years, companies invested millions of dollars in the training of their employees. This wealth of knowledge (or competencies) resides with the employees and should not be swallowed up by the black hole of downsizing. When employees walked out the door, knowledge went out with them.

Communication must be two-way. Unfortunately, FM organizational changes tend to be top down. Very seldom do the employees have an opportunity to input their concerns or suggestions.

An effective transformation process should include employee participation. The vision and mission must be explained clearly so that each one of them understands the direction of the organization. Employees must be allowed to participate by giving their input to the process. Many employees have profound knowledge and experience in their particular field of expertise, whether it is technical or customer oriented. Let the employees contribute to the transformation process; better still, provide a forum or exercise for them to come up with ideas to reinvent themselves.

There is no need to sell *change* to the employees. When their contribution becomes the formation of a process, they will institute change themselves because they now own the process. The shared-vision is part of their contribution. There is no need for *buy-ins* because the employees have invented it and now they have the ownership.

4. *Transformation:* Transformation is change. There are two basic components to change—the socio-technical aspect and the workplace redesign aspect. Socio-technical redesign involves the behavioral aspect of change—to change the human aspect of the workplace. This involves personal change, as well as relational changes. Some people refer to this as the soft skills.

Personal change involves one's own attitude and behavior toward the new paradigm. Relational change refers to dealing with attitudes toward others. This may mean a complete new approach to managerial approaches, such as team environment, self-directed or self-managed teams, or competency-based organization. In a competency-based organization, the evaluation system of the workforce may be changed to accommodate a new set of rules and measurement factors.

Such changes may be implemented in stages. When Bell Canada's real estate department went through socio-technical redesign, it was introduced in several stages. First, the employees went through a series of training sessions on dealing with soft skills. Courses such as conflict resolution, communication, and customer service, were part of the curriculum. Second, the employees were empowered to manage their own area of operation in total autonomy. Employee suggestions were taken seriously and rewarded monetarily. The final stage introduced self-directed work teams when the team reached maturity. In this case, pilot projects on a small scale were tested to ensure success before the program was rolled out to the entire organization.

By the end of the transformation period, the company increased productivity and employees were energized. Facility managers went from an old paradigm of operating at 20,000 square feet per manager to over 1 million square feet per manager. Staff was reduced or redeployed to other important tasks.

5. *Measurement:* Success cannot be realized unless it is monitored and measured. An evaluation system must be instituted to tell you whether you have arrived or not. The result may call for celebration or redesign. Transformation is an on-going process and has to be maintained, analyzed, and enhanced. If you do not measure, you do not know when you are successful. If you do not measure, you will not be able to improve. Hands-on type professionals tend to shy away from measurements because they are action oriented and want to get the job done.

Measurement metrics may fall into these three categories:

- Performance measurement related to own operation

  Gross square feet served per facility employee

  Operating cost per square foot

  Operating cost per company employee

Absenteeism, or better, presenteeism of own staff
Number of complaints per month
Safety records

- Performance measurement related to service

Timeliness of service delivery
Quality index of service provided
Cost of service
Number of repeat calls
Customer satisfaction index

- Performance measurement related to design

Rentable square-foot per employee
Cost of construction per square foot
Cost of construction per employee
Construction actual versus budget estimates
Churn rate

Measurement can be applied to practically anything. Most measurement deals with performance-related tasks, like the previous examples. However, few people measure the effectiveness of the organization by measuring the competencies. There are three key activities in determining core competencies: "First, it must offer significant added value. Second, it must be unique and offer lasting competitive advantage. Third, it must support multiple end products or services."[9] Core competency issues will be discussed later in this chapter in more detail. Core and supporting competencies must be measured and should be measured against a benchmark. Benchmarking with industry or outside organizations may not be as accurate as benchmarking against one's own organization. A team of subject-matter experts within the organization can establish this benchmark. An effective measurement is to compare year over year within the same group.

4.5 THREE KEY ISSUES IN FM TRANSFORMATION

4.5.1 Mission Critical Issues

A key factor that many people neglect is to ask the simple question: What is the one thing in our company that, when it fails, will bring the entire company to a halt? This one thing may be the computer center or it may be just a simple component within a complex system. A very good example is the "millennium bug." One embedded chip can bring the entire operation to a standstill.

There was a situation in Montreal a few years back during a commercial power failure in the city. A prominent university was within the area of blackout. The standby power of the university came on-line, as planned, to supply essential power to the university's computer system. The university was pleased that it did not lose any data. However, the refrigeration unit in the medical research lab was not linked to the emergency power. Therefore, all of the medical research specimens died in the process. For the university, the computer system where they stored all the students' records was their mission critical item, but for the medical department, the refrigeration unit should have been on their mission critical list.

In some cases, it may be some very small element within a large complex system that can cause complete failure. Another power situation in Toronto revealed that the small $14.99 fan supplying fresh air to the diesel room for the standby power unit became a mission critical item. It was not connected to the emergency power supply and caused the failure of the diesel engine.

Ask the question: What is that one thing which when it fails will bring the business to a halt? It may be the UPS in a data center; the telecommunication lines in a call center; some relays in an energy management system; or even the chain of command in an organization that can block the decision-making process in a major business transaction. It may be the supply line in a manufacturing plant or even your supplier of a key component in an assembly line. Whatever it may be, organizations must process map the critical functions and think through where this critical link may be.

The role of a facility manager is to ensure the company is housed in a safe and productive environment for conducting business. The critical factor can be manifold: security, essential power supply, or air-conditioning.

Whenever we look at any business transformation, we must identify all the mission critical elements. This should be a systematic approach and every department should be canvassed.

4.5.2 Core Competency Issues

Companies invest millions of dollars training their employees over the course of their employment. It is so sad to see many companies fall into the trap of letting their knowledge assets walk out the door through downsizing exercises. Many downsizing exercises include a competency assessment program to ensure they have all the necessary information about individuals who will fit the new corporate profile and remain in the company. The projection of such an image regarding competency assessment makes people afraid. They view this as an excuse for the company to get rid of them. The end result is either a poor reception or inaccurate results. However, a competency-based approach to human resources planning and development is an important step in building organizations with a competitive advantage through its people. The key value of competency assessment is to allow the corporation better resource allocation and diversification.[10]

Competency can be defined as any skills, knowledge, behavior, or other personal characteristic that is essential to perform a job and differentiate average from superior performance. Based on competency research by R. E. Boyatzis, David Dubois defines competency models as, "those competencies that are required for satisfactory or exemplary job performance within the context of a person's job roles, responsibilities, and relationships in an organization and its internal and external environments."[11]

The goal of building a competency model is to identify what it takes to deliver excellent performance in the target job or jobs. It is to provide a framework for recruitment and selection, performance management and reward, training and development, as well as career planning. Furthermore, the goal is also to design a practical tool that will fit and reflect the culture and future needs of the organization. Therefore, each organization must develop its own functional model to suit its specific business requirements.

The methodology is based on a set of pre-determined values of competencies compared to an individual's achievements at those levels and can be very subjective in nature. Therefore, companies should use an expert panel to determine the competency level an employee has to eliminate biases and errors in judgment. A fully developed process is complicated and hard to use, so most people become frustrated with it.

The first step in competency management is to define the objective of the exercise. What does the organization want to achieve and why is the organization doing it? The FM organization must partner with their human resources team to set the boundaries for the facility competencies so that they will not only align and support the corporate values and strategies, but also help formulate the facilities operations. Ask the question: "In what way would the FM team contribute to the corporation's success?" during the entire process for checks and balances.

The competency management process follows the following sequence:

- Establish an expert panel
- Appoint in-house subject matter experts
- Invite outside subject matter experts

- Develop competency models
- Simplify and establish job categories
- Determine the competency levels for each category

The first step in the competency management process is to establish the "expert panel." The expert panel consists of professionals in the field who are recognized by their peers as possessing mastery of the skills. Sometimes it may include professionals from outside the company, as well as customers who can contribute a more balanced view of the service delivery requirements of the company. Subject matter experts guide the whole process to ensure the final result is realistic and supports the corporate goals. Competencies for facility managers can be primarily divided into two skill categories: technical and behavioral.

In 1992, the International Facility Management Association (IFMA) embarked on a detailed study on competencies for facility professionals as the first step in their certification process. This is the first, and perhaps the only, comprehensive document on competencies for FM. In the study, IFMA defined competency as simply *able to do* with respect to the ability to perform duties and responsibilities to the standards of the profession.[12] The International Development Research Council (IDRC) established also their five core competency areas for corporate real estate professionals when they prepared their document for their own certification.[13] BOMI Institute, in their facilities management administrator program, established their curriculum based also on technical competencies, as shown in Figure 4.4. These are all technical competencies required for the facility practice.

Technical competency should be established by the expert panel and in accordance with the needs and requirements of the company. For FM organization, technical competency may be under the following key functional areas:

- Functional business management
- Realty management
- Facilities design and planning
- Facilities project management
- Facilities operations

FM Core Competencies

IDRC **BCCR designation** Board Certified Corporate Real Estate	**IFMA** **CFM designation** Certified Facility Manager	**BOMI** **FMA designation** Facility Management Administrator
Strategic Planning for Real Estate Professionals	Real Estate	Design, Operation, and Maintenance of Building Systems
Transactions	Operation & Maintenance	Fundamentals of Facilities Management
Site Selection & Location Analysis	Human & Environment Factors	Technologies for Facilities Management
Management of Facilities	Finance	Facilities Planning & Project Management
Project Execution	Communication	Real Estate Investment and Finance
	Facility Function	Environmental Health and Safety Issues
	Planning & Project Management	Ethics is Good Business
	Quality Assessment & Innovation	

IDRC - International Development Research Council
IFMA - International Facility Management Association
BOMI - BOMI Institute

FIGURE 4.4 FM Core Competencies.

- Facilities maintenance
- Financial management
- Environmental management
- Business continuity planning

Behavioral competencies are sometimes referred to as "soft skills" or "emotional competencies."[14] They relate to the management of people and business. Often, they contain values that the corporation embraces as their culture. Some companies embrace in their culture a behavioral list of over thirty-nine areas. Following is a simplified list of key skills for facility managers:

- Listening and comprehension
- Strategic business sense
- Achievement orientation
- Personal influence
- Customer focus
- Coaching
- Teamwork
- Leadership
- Critical thinking
- Personal development
- Life-long learning

John Micklethwait and Adrian Wooldridge, in their provocative book *The Witch Doctors,* summarized the definition of core competencies by Gary Hamel and C. K. Prahalad as "the skills and capabilities, codified and uncodified, that give a company its unique flavor and that cannot be easily imitated by a rival. These collections of knowledge constitute its expertise, which is what gives companies their competitive edge."[15]

4.5.3 MacroErgonomic Issues

Workplace redesign follows the human side of change. The movement of personnel, the new team culture, and the new work processes all have to be supported by the physical workplace. Not only are socio-technical changes necessary, the physical environment to support work shares equal importance. Changes that translate into new work activities must be supported by workplace redesign to foster productivity; such is macroergonomics.

"MacroErgonomics (ME) is an emerging field in which the overall physical work environment is aligned with the business objectives of the organization."[16] Workplace design deals with the effectiveness of the physical work environment that fosters the corporate team and entrepreneurial culture. The end product generally would include some kind of new workplace configuration using universal planning and workstation design. Flex hours and telecommuting were also a factor in the workplace redesign, granting the employees more flexibility in approaching work.

The new generations of workers that are coming into the workforce render today's workplace non-geographically relevant.[17] The workplace is not even territorial relevant. The Gen-X and Gen-*i* (the Internet generation) workers come with a completely different value system toward their concept of work. The traditional office and even the open office concept of modular system officing ideas will not necessarily satisfy their needs. The way of working is now governed by technology and individual's goals and motivation. To respond to this trend, many companies have adopted various officing strategies from telecommuting to a teaming environment, from office landscape to teamscape concepts. To provide for such a new work environment must require a thorough understanding of how people work.

Transformation in the physical workplace can yield measurable improvements in productivity, employee satisfaction, and realty cost savings. Facility managers must take this into consideration when designing a transformation process. Socio-redesign deals with the way people work. It means a change in value, attitude, and habits. Techno-redesign deals with the way technology as a tool is changing the work performance, and workplace design falls into this category. Workplace must be seen as a tool to enable the workers to perform work more effectively. It is very similar to upgrading to a faster computer, except the facility process is more involved.

The socio-techno redesign of the work must be matched by workplace redesign. Such a redesign may result in a more free-flow environment where the space is designed with opportunities for the workers to engage in spontaneous and impromptu discussions, brain-storming, and even problem-solving encounters. Considering the cost of space and furniture is only a small fraction of the operating cost of a company, there is no excuse to sacrifice an opportunity to provide an exciting work environment to motivate creativity, innovation, and interaction.

Take the office of the ad agency of Chiat/Day in Venice Beach, California, designed by the architect Clive Wilkinson. Creativity is the lifeline of their business. Their approach to officing is to allow the workspace to energize creativity. "Chiat/Day doesn't build new offices: It reinvents itself every time it moves."[18] They see their company not as an office with workers, but a community of many different teams of people who need to interact in various and many ways. The concept of an urban community life reflects their thinking in space design. "What makes a ghetto, and what makes a neighborhood? What's the right mix of public space, play space, and private space? These were the questions that we were asking ourselves," explains Laurie Coots, chief marketing officer of Chiat/Day.[19]

The Nortel Brampton Centre in Canada, designed by Hellmuth, Obata, and Kassabaum (HOK) is another example where a similar concept transforms the way people work in a traditional manufacturing company. The 600,000 square-foot factory (circa 1963) resurrected in 1995 as a city inside a building, based on a city plan with main streets, colonnades, and plazas. The idea is to encourage people to interact spontaneously and since the entire office is all on one floor, they call such encounters *horizontal happenstances*.[20] The redesigned office space can be a powerful force in motivating people in the new agile work process. Physical workplace redesign must be part of the business transformation formula.

4.6 THREE OBSTACLES IN TRANSFORMATION

Change is not easy. Clinical psychologist Harvey Robbins co-authored *Why Change Doesn't Work* with Michael Finley and claims that in 40,000 years the rules of change have not changed. "People generally will embrace initiatives provided the change has positive meaning for them . . . people believe what they see. Actions do speak louder than words. . . . Change is an act of the imagination. Until the imagination is engaged, no important change can occur."[21]

Business transformation involves change. In order for it to be successful, the process has to be well-planned and well-communicated, supported by provable results. In addition, change agents must understand what drives the people to change. Until a proper understanding is formed of the forces that prevent people from changing, no plan is workable. Some of the obstacles encountered by change agents are described in the following sections.

4.6.1 Time

People are too busy to think. Downsizing has placed additional burdens upon the workers. Many work long hours and find there is no time to do anything inventive. Even if people are willing to change or they see a need to change, time prohibits them from doing it. People spend all their time just putting out fires and have little time left to think and plan for the future.

Rosabeth Moss Kanter of Harvard Business School observed that "midlevel managers at downsizing firms complain about having to do many more mundane chores because of cuts in support staff. They report working more hours but getting less substance work done. They have less time and energy to invent the future."[22]

This is particularly true for facility managers. Downsizing has hit them hard. Experienced people are early retired or laid off. The less experienced ones are burdened by extra workload in an extremely active workplace. They are running between satisfying the customers' demands and meeting their operational requirements. The continued drive to reduce costs further robs them of their time.

4.6.2 Culture

The corporate culture is a strong influence on how people behave in the workplace, and more importantly, how people think in relationship to innovation and creativity. Companies that have a long history tend to have a well-embedded culture. This is a paradigm that they manage very well with, as this paradigm represents the rules, guiding principles, standards, and protocols.[23] Companies need to explore new ways or introduce new ideas so that strong culture becomes a restriction.

The FM people have a strong value system. Their work is technical and they enjoy their work and are dedicated to look after their facilities. Their intention of doing a good job becomes their stumbling block. Because they think technically, it is difficult for them to get out of their box and think differently. Their mentality is "don't bother me with change . . . let me do my job" and they view management activities as time-wasting exercises.

Business transformation must take this into account and allow such employees to free their minds and think out of the box. It requires a paradigm shift—old paradigms will not work in the new environment—a new paradigm is required. To manage a business is to manage within a paradigm, but to change, the change agents must facilitate between paradigms.

Values tend to be developed and established early in life. Kurt Hanks calls it a "mindset map."[24] A mindset map is a mental model of how one views the world. This is developed and molded throughout one's growing up period by both positive and negative influences. This will in turn order the way one interprets life (see Figure 4.5).

Individuals within the same generation tend to have the same value toward the way they view life. The wider this value is (value dispersion), the more different they are from each other in accepting and approaching life (generation gap). This governs the way they work in society and in business. For example, people who are born after World War II would have a very different view of money than the Gen-X people.

Technology makes this generation relatedness supersede time and space. Teenagers in Toronto, Miami, and Kiev have more in common with each other than they do with their parents.[25] Dave Arnott suggests that "people who join an organization with a particular value set will have the same value set for their entire tenure."[26] In any corporation today, it is not unusual to find a value dispersion of four generations—such dispersion of values stifles change.

4.6.3 Leadership

A leader leads. Many great ideas and purposes failed because of poor leadership. Either the leader is too dictatorial and causes a lot of harm to the troops or is so weak that the troops run in circles. Stephen Covey in his paper, *Three Roles of the Leader in the New Paradigm,* identifies that the leader of the future should have "the humility to accept principles and the courage to align with them, which takes great personal sacrifice. Out of this humility, courage, and sacrifice comes the person of integrity."[27] Servant leadership style is rare in the FM profession.

In business transformation, leadership becomes extremely important. Any change activity will bring about confusion, mistrust, and doubt unless the leader can provide a clear direction and gain the confidence of the people. No clear direction as to where the business is going generally restricts

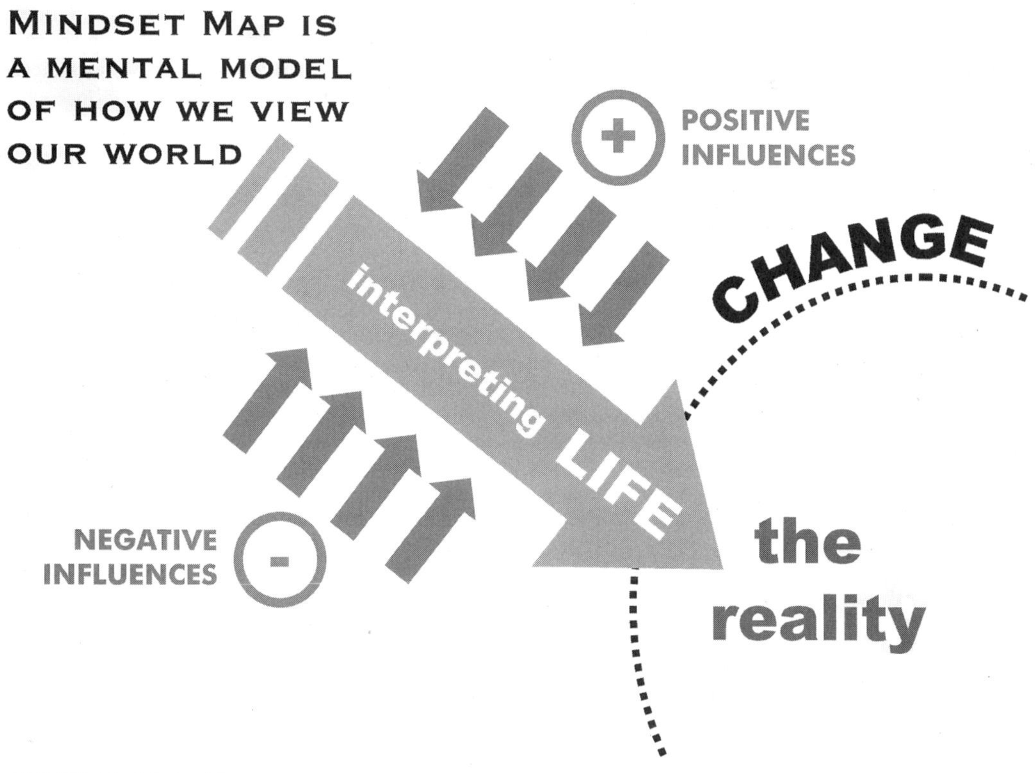

FIGURE 4.5 Mindset Map.

the effectiveness of change. Changing direction in mid-course worsens the situation. Changing direction too often spells disaster. The credibility of the leader is also a strong factor in achieving success. Honesty, genuineness, and integrity are the three important ingredients of a change leader. The change leader is the conductor of a great symphony—not understanding the instruments and the quality and capacity of the player will never produce good music. The sound that comes out will be painful to the ear and so will business transformation with a weak leader. Reengineering and restructuring demand managers to be observant leaders. In this chaotic time, the manager must possess leadership skills. Poor leadership with no direction guarantees failure in any business transformation.

4.7 THREE COMPANIES IN TRANSFORMATION

Canada Post was the first major corporation in Canada to go through outsourcing in their transformation process in 1994. They owned 3,200 properties across Canada. More than half (1,800 properties) were small rural properties or offices which represented only 10% of their total space and 5% of their occupancy cost. The local postmaster managed most of these offices. The head office provided training, standards, and policies to ensure uniformity in application.

In 1993, they began looking into reducing costs and human resources. Cost and customer service improvement were the reasons to proceed to the outsourcing scenario. They divided the territories into east and west and outsourced it to two vendors in a 5-year contract. Their employees went from 150 to 50, with one-third of them going over to the vendors.

Nicholas Sirenko, Manager—Asset Management, Real Estate Department of Canada Post reflected, "Relationship with your vendors plays such an important role in ensuring success in an outsourcing scenario. This relationship is not static but is an evolving one. Out of such a relationship you build trust and flexibility. Business agreements such as this are living agreements. It should allow you to make changes along the way. As the outsourcing industry becomes more and more mature in Canada, which means that there are more and more players on the scene, and they build up more and more expertise and clients. The benefit at the end is not so much cost reduction, but the ability to take advantage of the best practices and innovation coming from the outsourcers."[28]

Celestica, on the other hand, is in a growth mode. In 1999, the $5 billion computer hardware manufacturing company continues to grow by acquisition to maintain leadership in the industry. Their target for year-end 2000 is to be a $10 billion company. They now have operations in the United Kingdom, the United States, Canada, Ireland, the Republic of China, Hong Kong, Thailand, Malaysia, Brazil, and Mexico, and are now planning for European expansion.

The FM organization in Celestica is a traditional organization in spite of the rapid growth in their international ventures. The staff looks after the Toronto main office and manufacturing plant and warehouse. In order to meet the rising need of the expanding company, they did not increase their staff, nor did they alter the existing arrangement. Instead, they established a new staff function within the head office to oversee the expansion program and establish standards and procedures for the new acquisitions.

In their transformation process, they realized that they were no longer a North American company, but a global one participating fully in the global economy. Right at the start they outsourced all new activities while leaving the existing facility staff intact serving the home office. They maintained flexibility in their approach. In a global operation, they respected the different cultures, languages, and local business ethics and practices.

Jaan Meri, Director of Corporate Facilities and Real Estate, Celestica, remarked, "Whatever we do will be different next year. Fast-pace companies have to focus on growth."[29] To ensure brand consciousness, they provide a facilities council meeting to share best practices and experience twice a year. They control quality and service by adopting standard practices and promoting ISO 14000 certification.

In 1992, the author embarked on a five-year transformation process for the facilities department in Bell Canada, the largest telecommunications company in Canada. The objective was to align the department to the volatile business climate and prepare each individual employee within the department for the reality of the work environment of the future. Bell Canada at that time continued to lose their market share in the telecommunication business similar to the early break up of AT&T in the United States. Competitors continuously appeared on the scene, taking a piece of the action from the 100-year-old establishment.

Bell Canada went through a series of transformations. After analyzing their existing condition, they introduced a series of training on emotional competencies, teaching both the managers and technicians the behavioral skills. The second step was to form functional teams and break down the hierarchical organization. Self-directed work teams were formed and empowered. The next step was to redesign the entire organization into a subsidiary wholly owned by Bell Canada. Competency assessment was introduced to recruit the right people into the new organization. People in the new organization had to have the profile for the new entrepreneurial culture. The new company was launched in 1995.

Throughout the entire process, which spanned three years, the employees were fully informed of the direction and the outcome of the exercise. They understood the why's and the how's of the entire process. Communication was paramount to ensure a truly shared-vision and came in the form of newsletters, Intranet messages, posters, slogans, and celebrations. At every milestone, the employees were given the opportunity to participate and reinvent their future. The entire process resembles the transformation model of Figure 4.2.

Communication and training are two enablers to position the new direction with the employees. Throughout the entire process, people were informed of each step of the transformation. Partnered with this, they were provided training to have a better understanding and acquire new skills. With each training session, the mystery of the new work environment began to unfold before them.

Because of the diversity in the group, people perceived things differently. Do not expect people to comprehend the entire process in one sitting—understanding takes time.

Training is important at times of change. Unfortunately, most companies cut the training budget first at times of restraint. James Kouzes, associated with the Tom Peters Group, stressed in his book *Credibility* that training should be the top priority for any winning company. He cited the example of the Ford Motor Company. Training was the backbone of the Ford Motor Company's effort in the 1980s to restore its competitive position after years of continuous erosion of its market share.[30] Indeed, today Ford has built one of the best training centers in the country—The Fairlane Training and Development Center in Dearborn, Michigan.

4.8 CONCLUSION

The new economy of the new millennium brings in rapid change. Managing change becomes the critical corporate strategy in order to survive in the new global competition. Everyone tries to use a crystal ball to predict the future and jockey their organization into position to be on the leading edge. Books, seminars, and workshops abound on the subject. Words like competitive advantage, customer-driven, service-oriented, core competencies, outsourcing, reengineering, and global marketplace have now become commonplace in the workplace. The process of turning these terms into reality creates tremendous havoc in the workplace and in people's lives.

FM organizations are having a hard time coping with ever-changing corporate directions. The incredible movement of personnel at all levels creates a tremendous challenge to the facility manager. The desire to satisfy and the threat of extinction via outsourcing coexist in today's facility organization. As long as the corporate organization is there, facility organization must be organized to serve it.

Business transformation must bring these elements together. The specific transformation model must be strategic in the core of its design, aligning with the strategies and values of the company. The tactical implementation has to be achievable and realistic. People must be the focus of the outcome. Without the people, it is an empty organization. Transformation requires strong leadership, and yet this leadership must also be sensitive to the guiding force of the followers. Thomas Petzinger, Jr. of *The Wall Street Journal* wrote, "Organizations themselves are becoming leaderful. . . . This institutional leadership emerges from the interplay of the few and the many, the top and the bottom, the individually powerful and the collectively powerful."[31]

Good effective organization yields high productivity for the workers. As our clients are getting more and more sophisticated, we must also position ourselves to anticipate their needs and requests. The productivity of a facility manager depends on how this person is provided with the tools and moral support to perform the tasks. Business transformation is a supporting exercise.

People are human beings and they are not just numbers in an organization. Charles Handy, Britain's foremost business guru, saw his predictions in *The Age of Unreason* became reality in the 1990s. He now sees that we are living in a paradox of trying to attain efficiency and economic growth and neglecting the individual man and woman who made it happen. "We were not destined to be empty raincoats, nameless numbers on a payroll, role occupants, the raw material of economics or sociology, statistics in some government report. If that is to be the price, then economic progress is an empty raincoat. There must be more to life than to be a cog in someone else's great machine, hurtling God knows where."[32]

If we are to make sense of the future, we must take up the challenge. We must make a difference. We must be the people who will fill that empty raincoat.

ENDNOTES

1. Houston, J. 1998. *Reforming theology for spiritual renewal*. Unpublished lecture in audio form given at the "Treasures in the Attic" Conference, Regent College, Vancouver.
2. Kageyama, Y. 1999. Japanese banks to create global giant, *The Globe and Mail* (August 21).

3. Evans, M. 2000. Merger mates have eyes on each other's assets. *The Globe and Mail.* (January 11).
4. O'Mara, M. A. 1999. *Strategy and place*. New York: The Free Press.
5. Horgen, T. H., Joroff, M. L., Porter, W. L., and Schon, D. A. 1999. *Excellence by design—Transforming workplace and work practice.* New York: John Wiley & Sons.
6. Howlett, K. 1999. Lineup forms for real estate portfolio. *The Globe and Mail.* (May 28).
7. Wong, T. 1999. Oxford seals Royal Bank deal. *The Globe and Mail.* (September 23).
8. Geneen, H. 1997. *The synergy myth*. New York: St. Martin's Press.
9. Campbell, A., and Luchs, K. S. 1997. *Core competency-based strategy*. London: International Thomson Business Press.
10. Ibid.
11. Dubois, D. 1993. *Competency-based performance improvement—A strategy for organizational change*. Amherst: HRD Press.
12. IFMA. 1992. *Competencies for facility management professionals.* Houston: International Facility Management Association.
13. BCCR. 1997. *BCCR Candidate Handbook.* Norcross: The Board on Certification for Corporate Real Estate (BCCR).
14. Goleman, D. 1995. *Emotional intelligence*. New York: Bantam Books.
15. Micklethwait, J., and Wooldridge, A. 1996. *The witch doctors—Making sense of management gurus.* New York: Random House.
16. O'Neill, M. 1998. *Ergonomic design for organizational effectiveness.* Boca Raton: Lewis Publishers.
17. Becker, F., and Steele, F. 1995. *Workplace by design—Mapping the high-performance workspace*. San Francisco: Jossey-Bass Publishers.
18. Ransdell, E. Work different. *Fast Company*. (June).
19. Ibid.
20. Chadderdon, L. 1998. Nortel switches cities. *Fast Company*. (August).
21. Robbins, H., and Finley, M. 1996. *Why change doesn't work*. Princeton: Peterson's Pacesetter Books.
22. Kanter, R. M. 1995. *World class.* New York: Simon & Schuster.
23. Barker, J. A. 1993. *Paradigms—The business of discovering the future.* New York: HarperBusiness.
24. Hanks, K. 1994. *The change navigator—Preparing a new kind of leader for an uncharted tomorrow*. Menlo Park: Crisp Publications.
25. Adams, M. 1997. *Sex in the snow—Canadian social values at the end of the millennium.* Toronto: Viking.
26. Arnott, D. 2000. *Corporate cults.* New York: AMACOM.
27. Hesselbein, F. 1996. Goldsmith, M., and Beckhard, R. (Eds.). *The leader of the future.* San Francisco: Jossey-Bass Publishers.
28. Interview with Nicholas Sirenko, Manager, Asset Management, Real Estate Department, Canada Post Corporation.
29. Interview with Jaan Meri, Director of Corporate Facilities and Real Estate, Celestica.
30. Kouzes, J. M., and Posner, B. Z. 1993. *Credibility*. San Francisco: Jossey-Bass Publishers.
31. Petzinger, Jr., T. 1999. *The new pioneer—The men and women who are transforming the workplace and marketplace.* New York: Simon & Schuster.
32. Handy, C. 1995. *The empty raincoat—Making sense of the future*. Great Britain: Arrow Books Limited.

CHAPTER 5
FINANCIAL MANAGEMENT FOR FACILITY MANAGERS

Fred Klammt
Principal
Aptek Associates
E-mail: fredk@jps.net

5.1 FUNDAMENTAL PRINCIPLES OF FINANCIAL MANAGEMENT FOR FM

Financing is all about creating value. It is about making smart decisions and leveraging assets to allow an organization's workers to be productive without disruption. Facility management's (FM's) overall goal is to take care of the physical assets of the organization, avoid disruption to ongoing business operations, and leverage (extend the life of) assets. Leveraging involves financial management skills and, in the facility world, focuses on two main areas: capital project evaluations and operating budgets.

Within these areas a wide array of financial analysis and decision-making principles and tools can be applied to FM. Facility renovation projects are always competing for capital dollars with other departments' projects. Facility managers constantly evaluate the best buy between competing alternatives. Difficult repair or replace decisions need to be made before a costly repair is authorized. The return on investment for a new database or computer-integrated facility management (CIFM) system needs to be detailed. Many other facility applications require financial analysis and management skills.

Most facility managers report to the operating and financial side of their organizations and to the chief financial office (CFO). It is the responsibility of the CFO to increase cash flow as much as possible while reducing risk. If an asset is not continuously producing cash flow or adding value, the CFO must take action. Underperforming assets can contribute to poor overall performance and make an organization vulnerable.

It is therefore essential that the FM function be aligned with the overall organizational goals and with the CFO's business objectives. Financial analysis must be done with a proper perspective. Although the financial side takes priority in many cases, there are other considerations and priorities, such as operational, safety, and political issues that may have higher priority than financial returns. It makes absolutely no sense to do full-cost accounting and life cycle analysis for a project on a building whose lease expires in the next two years.

An important concept in all financial analysis is the concept of the time value of money. This means that today's liquid cash can be put to work earning income and will be more valuable in the future than it is today. This is the basic concept of economics that drives today's financial markets.

Two other areas that facility managers need to be aware of are:

- Financial analysis and management must always be kept in the proper perspective: the overall mission of the company
- What is the overall financial view of their companies, such as:

  What was the company's stock price yesterday, last week, last year?

  Why was the stock price up or down?

  What was the key driver of corporate profits last year?

  What will be the key driver in the next decade?

  Who is the company's most important customer?

  Who is the company's largest supplier?

  What is the company's average productivity rate, lost time rate?

5.1.1 How Financial Analysis Fits Into FM's Organizational Role

How financial analysis fits into facility manager's organizational role is a key question that is unique for each company and department. FM is an internal service and a liability on the bottom line. As such, it must create value and promote its financial contribution to the overall organization. Financial analysis provides a standardized tool that helps CFOs and executive management to decide where to put money. It is important that facility managers follow accepted financial practices so that others can easily understand their analysis and make fair comparisons to other projects. It is also important that they market their services and push their agendas to the forefront by using accepted financial analysis.[1]

This section deals with generally accepted financial principles from engineering and accounting disciplines. Each organization will have its own customized financial practices and facility managers must be adept at using their internal financial analysis tools.

5.1.2 The Goals of Finance and Accounting Principles

The goals of finance and accounting principles are to provide standardized tools that all companies and finance managers can use, enabling comparisons between different departments and companies. Success in decision making is enhanced by understanding the financial impacts of various alternatives.

While most companies have unique and confidential financial goals, there are some financial goals that are common to all for-profit companies:

- Lowest operating costs
- Highest revenue
- Maximize returns and profits
- Reduce risks
- Conserve capital
- Shortest time to market
- Minimum turnover

Facility managers must know and understand their company's overall business plan and financial goals. They need to know how these assets are currently performing and their role in the overall company's goals. They must leverage their company's physical assets in pursuit of, and to add value to, these goals.

5.1.3 The Fundamental Financial Analysis Tools

The fundamental financial analysis tools that facility managers use to perform project evaluations and budgeting have evolved over the past ten years and now are focusing on return on net assets (RONA). Here is a listing of these financial tools in chronological order:

1. Simple payback (SPB)
2. Life-cycle costing (LCC)
3. Return on investment (ROI)
4. Activity based costing (ABC)
5. Total cost of ownership (TCO)
6. Return on net assets (RONA)

Each financial tool provides value and insight in a specific application. The history of these financial tools has evolved toward higher and higher stakeholders. From simple engineering analysis, such as payback periods, to the total cost of ownership—which encompasses more than life-cycle costs—we have now arrived at a financial indicator that is used in most corporate boardrooms.

Every organization has its own unique internal financial analysis methods. In this section, we will focus on seven financial tools that are commonly used by many organizations involved in FM:

- Cost benefit analysis
- Payback periods
- Risk and return
- Internal rate of return
- Present value analysis
- Return on net asset (RONA)
- Off-balance sheet assets

The next section of this chapter will deal with life-cycle costing.

5.1.3.1 Essential Data. Before we get started, there are some basic data that are needed for all financial analysis:

Discount Rates. Discount rates are essential to financial analysis. Other terms used are interest rate, cost of capital, and hurdle rate. A simple discount rate is the company's cost to raise capital. This is the weighted average on all sources of debt and equity financing used by the company. Many companies use a higher discount rate to reflect various risks. It is important for facility managers to know what their current organization's discount rate is and to use it in all financial analyses.

Expense Budgets. Budgeting consists of two basic areas: expense and capital. Expense budgets deal with ongoing operating expenditures and focus on the labor and materials to keep a facility in good repair. There are always costs beyond a facility manager's control which will impact budgets and financial performance. It is important to plan for these. All budgets have some built-in contingencies. Sometimes these contingencies are determined as a percentage of the total expense budget—generally 2–5%. Many times they are listed as "miscellaneous" or "emergencies" line items.

Common expense items for all facilities include:

- *Services*: custodial, landscaping, security, and food services
- *General Repair*: carpentry, plumbing, and roof
- *Technical Repair*: electrical, HVAC, fire, and mechanical

Capital Budget. Capital budgets focus on replacing aging or underperforming assets, and on renovations and tenant improvements (TIs) to accommodate corporate churn and customer requirements. In some organizations, facility managers have a specific window of time to submit next year's capital budget requests. Sometimes these requests are nicknamed "blue sky," or "wish lists." After the facility manager submits capital requests, internal financial analysts lump all the capital requests from all departments together and rank the projects by financial returns and fund those within the company's overall capital budget. Other companies will fund capital projects throughout the year based on need and fund availability.

Project Evaluations. Project evaluations are important to the discipline of FM. When is a project to be funded and when is it not? How do you decide? One of the main decision factors in giving a project go-ahead is financial analysis. Facility management needs to highlight the answers to these questions in their project submittals:

- What benefits does the project bring?
- Will the project enhance business operations? How?
- What is the payback of the project?
- Is this a customer requirement? What is the customer's benefit?
- What is the risk? . . . of doing nothing? . . . of alternative projects?
- Is there an eminent safety, environmental, or health risk?

Before submitting their capital budget requests, facility managers need to perform an internal analysis and push the project's best benefits to the forefront. When financial analysts perform their funding analysis, they are usually not aware of certain hidden risks or needs that are difficult to put into financial calculations. It is the facility manager's responsibility to call attention to these.

5.1.4 Cost Benefit Analysis

Facility managers typically provide cost benefit analyses for all substantial *capital* expenditures. For *expense* budgets, they need to provide cost benefit analyses whenever an increase in budget is requested. For example, if a facility manager wants to increase the frequency of preventive maintenance on electrical systems or on roof inspections, an analysis of the benefits of a more reliable electrical system or longer roof life would show the financial returns that can be expected and make it easier to obtain expense budget funding.

It is somewhat more difficult to calculate cost benefits for expenses used in providing an internal service that does not generate revenues, than to calculate cost benefits for capita. However, calculating cost benefits of avoiding disruption to operations and extending the life of physical assets is straightforward and needs to be done. Facility managers need to determine the benefit and value these improvements provide their organization. While it is impractical to calculate the return on investment for facility maintenance, it is easy to calculate the risk of lost business productivity lost for each minute of electrical system disruptions and downtime for the electrical and back-up systems of a building.

If an overall facility operations and maintenance budget represents 1% of the organization's overall labor and operating budget, it is ludicrous to cut corners within a 1% budget that can affect 90%+ of the company's productivity and on-going operations. The facility manager must have the skills and marketing capability to put forward the cost benefit of their expense budgets and capital improvements and the associated risks to the overall business of not funding these initiatives.

Manufacturing operations take no shortcuts when it comes to their budgets and potential impact on production run-time. Having a production line stop due to poor maintenance of the electrical system is not tolerated in the manufacturing industry. Mean time between failure (MTBF) and reliability centered maintenance (RCM), along with numerous other programs are essential to reducing the risk of production stoppage.[2] Similarly, facility managers must have supporting data and benefits of minimizing electrical failures, providing good indoor air quality (IAQ), and having properly operating physical assets.

a. Determine all benefits of the project
b. Determine the life of the project or asset
c. Determine all front-end costs—include everything
d. Calculate all on-going (monthly, quarterly, annual) costs
e. Determine and add up all the on-going benefits
f. Calculate the simple payback as follows:

$$\frac{\text{Start-up costs (c)} + [\text{on-going costs (d)} \times \text{life cycle (b)}]}{\text{Start-up benefits (a)} + [\text{on-going benefits (e)} \times \text{life cycle (b)}]}$$

g. Financial results: SPB, ROI, IRR, NPV

FIGURE 5.1 Generic Financial Return Calculation Process.

Here is a generic process for calculating return on investment (ROI). This methodology will be used in several subsequent areas of this chapter. In calculating ROI, the discipline of process is most important. If using financial calculators, see section 5.1.6, Risk and Return, within this chapter.

Let's take the example of a CIFM (computer integrated facility management) software package and calculate the return on investment for such a project following the generic process shown in Figure 5.1.

a. Benefits of the Project

- Reduced labor—improved productivity
- Improved accuracy of data
- Faster retrieval of needed data

b. Life of the Project

In this example, the life of the project is not dependent on the life of the software, but on the life of the CIFM concept, which could be as long as the building or company exists. A realistic CIFM life may be twenty years.

c. Front-End Costs

- Hardware and software; prerequisite needs (workstation, OS, teledata/power connects)
- Purchase, maintenance, training

d. On-Going Costs

- Meetings, training, data entry, consulting
- Supplies, equipment, space, utilities, maintenance
- Upgrades, debugging/troubleshooting, compiling

e. On-Going Benefits

- Timeliness of data
- Space savings
- Budgeting savings–construction, space planning

f. Simple Payback Calculation

$$\frac{\text{Start-up costs (c)} + [\text{on-going costs (d)} \times \text{life cycle (b)}]}{\text{Start-up benefits (a)} + [\text{on-going benefits (e)} \times \text{life cycle (b)}]}$$

In a hypothetical case, this calculation may look like this:

$$\frac{\text{80k (c)} + [\text{20k (d)} \times \text{20 (b)}]}{\text{100k (a)} + [\text{50k (e)} \times \text{20 (b)}]} = \text{480k/1,100k} = \text{0.436 years}$$

g. Financial Results

Convert the calculation in "f" to your own internal financial methodology or use one of the financial tools discussed in the following section.

In this example, the costs of $480k to start and operate the CIFM system will be recovered within 3–4 months by the benefits the CIFM system will provide the organization. It is important for the facility manager in this CIFM example to document and accurately enumerate the benefits the CIFM system will provide to the organization. For example, in many companies, "time to market" is a critical competitive advantage. A CIFM system may contribute (e.g., by shortening moving cycles) to this timing and, therefore, this benefit needs to be included.

5.1.5 Payback Period

Payback period is a simple but powerful analytic tool that is commonly used to assess a project's return on investment—how quickly will the initial investment be recovered? There are several variations of payback period analysis. Simple payback analysis is most commonly used due to its ease of calculations. The definition of payback period is "the number of years needed to recover the initial investment."

When comparing two or more projects, the project with the shorter payback period is preferred. Cash inflow (benefits gained), rather than net earnings (as used in rate of return calculations) are used in payback analysis. A more robust payback analysis involves:

- *Upside:* Paybacks are easy to calculate and have applicability to most facility capital projects.
- *Downside:* Payback analysis does not account for the value of time, and it does not measure profitability: *what happens to cash flows after the project payback has occurred?*

Figure 5.2 illustrates an example using payback period analysis. In this example, it is important to note that there are many other possible calculations, such as lamp disposal costs, lumen depreciation curves, lamp cleaning, and possible socket replacement costs, etc. There are also other benefits, such as improved safety, productivity, etc.

5.1.6 Risk and Return

Let's face it, life has risks. Business decisions and operational management involves assessing the possible risks and reducing them to maximize returns. Minimizing risk is becoming more important in all business activities. Technology and business processes change quickly, unpredictably, and with great uncertainty. Within this environment it is difficult to minimize risk because future factors are continually evolving and changing. Risk management is fast becoming the preferred decision-making tool. What is the risk of not doing something, or the risk of implementing a solution?

The ultimate goal in risk management is the creation of a single overall picture of the uncertainty facing an organization. Risks are interconnected and interdependent. How do we describe and define them? The "Risk Spectrum" identifies "global risks" and four types of "organizational" risks: financial/market, political/regulatory, legal liability, and operational.

The key to risk management is to look at integrated, or strategic, aspects of financial/market, political/regulatory, legal liability, and operational risks together, rather than separately. The biggest problem facing us, however, is how to measure all these risks in terms of their potential likelihood, their possible consequences, how they correlate with each other, and how they will affect the facility operation. Risk assessors use engineering estimates for property exposures, leading to maximum foreseeable loss (MFL) and probable maximum loss (PML). Actuarial projections are employed for expected loss levels where sufficient loss data are available. Scenario analyses and Monte Carlo simulations are used when data are thin, especially to answer "how high is up?" questions. Probabilistic and quantitative risk assessments are used for toxicity estimates for drugs and chemicals,

Sample project: Retrofit warehouse fluorescent lighting system to HPS system

Existing lighting system:

100—F40's, 2 × 4 fixtures with 3 lamps per fixture;
5000 run h/yr
$0.10/KW, $5.00/KW demand.
Lamps have 15,000 h life

Proposed lighting system:

50–70/Watt HPS fixtures;
Lamps have 24,000 h life

Energy savings calculation:

Existing system current cost
100 fixtures × 3 lamps/fixture × (0.040 + 10% ballast) KW/lamp = 13.2 KW
13.2 KW × 5000 h/yr = 66,000 KWH
Proposed system future cost
50 fixtures × (0.070 + 10% ballast) KW/lamp = 3.85 KW
3.85 KW × 5000 h/yr = 19,250 KWH
Difference
13.2 KW − 3.85 KW = 9.35 KW × $5.00/KW × 12 mo/yr = $561/yr
66,000 KWH − 19,250 KWH = 46,750 KWH × $0.10/KWH = $4,675.00/yr
Total energy savings: $561 + $4,675 = $5,236

Maintenance and material cost savings:

Existing system maintenance and material cost
15,000 lamp life/5000 run h = 3 yr replacement cycle (0.33/yr)
Assume 0.20 h per lamp change, labor cost is $40/h
100 lamps × 0.2 h/lamp × 0.33/yr × $40/h = $264/yr labor
100 lamps × $1/lamp × 0.33/yr = $33/yr material
Proposed system maintenance and material cost
24,000 lamp life/5000 run h = 4.8 yr replacement cycle (0.21/yr)
Assume 0.5 h per lamp change, labor cost is $40/h
50 lamps × 0.5 h/lamp × 0.21/yr × $40/h = $210/yr labor
50 lamps × $50/lamp × 0.21/yr = $525/yr material
Difference
Maintenance costs = 264 − 210 = $54/yr
Material costs = 33 − 525 = <$492>/yr
Total maintenance and material savings = $54 + <$492> = <$438>

Simple payback calculation:

Cost for retrofit is $25,500.00
(50 fixtures × ($250 materials + $160 labor)) + $5000 fluorescent fixture removal costs
Annual Savings: $5,236 energy + <$438> labor & materials = $4,798.00
SPB = 25,500/4,798 = 5.3 years

FIGURE 5.2 Example Using Payback Analysis. A lighting retrofit has a very straightforward simple payback analysis. There are two major components of this example: energy and labor + material savings. Each must be calculated separately and must be based on the same assumptions. This example follows an engineering type calculation and does not take into account benefits of improved visual perception (i.e., faster parts picking in warehouse), safety improvements, etc.

and to support policy decisions. For political risks, facility managers rely on qualitative analyses of "internal experts."

Organizations need a combination of all tools so they can deliver sensible and practical assessments of their risks to their stakeholders. The first stakeholder is the Board, now required to take responsibility for risk in many jurisdictions. Other stakeholders are employees, customers, suppliers, regulators, investors, financial analysts, lenders, and, finally, the public and the communities where facility operations occur. Each stakeholder group has a different perspective on risks, possible outcomes, and desired responses. Each constituency requires a different description. The challenge is to develop a common "language of risk" that will enable us to communicate effectively with these various stakeholders.

Facility managers can reduce the risk of physical asset failure by establishing strategic planning processes and perform due diligence on workscope performance and supplier issues. Here are some other risk issues:

1. *Confidentiality:* Owners must take steps to minimize this risk by identifying beforehand what information is not to be made available to internal and outsourcer's employees. Contract provisions may include restrictions of moving employees to competitors.

2. *Intellectual Information:* Accessibility to corporate intellectual property is a high-risk issue. Basic protective measures must be included in all operations. Copyrights, trademarks, patents, and non-disclosure agreements must be used and actively enforced.

3. *Inappropriate Workscope:* There is an assumption made that an outsourced workscope is not part of a company's core competency. Later, it is discovered that part of the core competency was indeed outsourced, thereby compromising strategic contributions. It is therefore extremely important that each function be closely scrutinized before outsourcing takes place and not to place critical competitive competencies with the outsourcer. Another solution would be to bring the outsourcer into the decision-making process and work with a supplier that has this depth of experience.

4. *Service Levels:* This is a sizeable risk when outsourcing. Too often, customer service levels and quick responsiveness decreases over time. All too often technical specification contracts allow for lower service level—which is why performance-based contracts are used to specify service levels. These contracts can also provide cost incentives and penalties for specific measurable service-level criteria—specific levels that must be clearly specified and monitored.

Financial risk is the focus of this chapter and the tools presented are useful in determining which projects and expenses present the least financial risks.

5.1.7 Internal Rate of Return

Many people are familiar with return on investment (ROI). The more common financial indicator used by companies internally is internal rate of return (IRR). IRR is used mostly in capital budgeting and is a measure of the rate of profitability.

The technical definition of IRR is the discount rate that yields a new present value of zero (see next section). Stated another way, IRR makes the present value of cash flows equal to the initial investment. If the IRR of a specific project exceeds a specified cutoff rate, the project is added to the recommended list.

$$\text{rate of return} = \text{net income/investment}$$

IRR is used to select the most financially beneficial investment for the company. Each year a company must decide which projects to fund. From a financial perspective, the IRR is used to rank projects based on financial returns to the company. IRR varies between companies. There is no one algorithm that is used by all companies. Each company has its own variables and unique versions. The correct form and methods to use in calculating IRR depends on the organization's finance people. Every organization has its own rules, methods, and formats. It is important for facility managers to understand and follow their customized internal financial analysis methods.

- *Upside:* In financial circles, IRR is the most popular and often used analysis tool. It is relatively straightforward and allows for equitable comparisons between capital investments. It includes the time value of money and cash flow concepts.
- *Downside:* The IRR method may give unrealistic rates of return—they may be too high! A 50% IRR is an unrealistically high rate of return—the IRR calculation may be suspect.

5.1.8 Present Value Analysis

Of all the financial tools available to the facility manager, present value analysis, commonly referred to as net present value (NPV), is the most useful. The basic definition is:

$$\text{NPV} = \text{present value of future cash flows} - \text{net investment}$$

NPV deals with the time value of money. The basic concept is "The present value of a cash flow is what it is worth in today's dollars. Present value incorporates the time-value principle by discounting future dollars (computing their present value) using the appropriate discount rate (interest rate). In investment analysis, this discount rate is the cost of capital."[3] Figure 5.3 gives an example of determining whether to replace a roof or increase the preventive maintenance budget using the NPV formula.

Many facility staff shy away from applying financial accounting tools to their daily jobs, monthly reports, and annual summaries, due to their complexity and time requirements. The usual reason for this is that financial accounting is beyond their expertise and too complex. By learning how to do these financial calculations and taking the time to do them on a regular basis, facility managers can reduce the breakdowns and reactive nature of their operations by obtaining proper funding and recognition of the added value their operation provides to the organization. Financial accounting transcends all areas of business, and FM is no exception. By learning and applying some simple financial tools, a great return for a small investment of time can be gained.

- *Upside:* There are three distinct advantages to present value analysis: 1) It uses cash flows rather than net earnings (i.e., payback analysis); 2) It recognizes the time value of money; and 3) Value increases when funding only projects with positive net present values.
- *Downside:* Present value analysis presumes that one can make detailed predictions about future cash flows. Cost of labor, materials, and overhead are hard to predict the farther out in time one goes. The discount rate will also change over the life of a project.

5.1.9 Return on Net Asset (RONA)

Return on net asset (RONA) is defined as:

$$\frac{\text{Change in net assets}}{\text{Total net assets}}$$

This represents the change in value of an organization's assets over the past year. If it is growing, then the company is using its assets better than before; if it is declining, it may represent a change in asset leveraging strategy. As with most financial ratios, there are no quick and simple answers to one ratio—It must be compared with other ratios.

What is important is the fact that RONA is a key ratio within many organization's composite financial indexes. And, it is the most important financial health indicator of physical assets; hence, its importance to facility managers.

To optimize RONA (the ratio of dollars per asset), it is possible to reduce operating cost per asset or to decrease the amount of assets owned without affecting returns. This requires providing only what is needed at the right time for each asset. The FAM (Facility Asset Management) (see next section) process can be used to optimize RONA.

Problem statement: Do we increase roofing PM budget or get a new roof?

Roofing facts

Installed 1986
Replacement cost = $1 million
Rated life = 17 yr
Discount rate = 6%

Operational cost calculation

Current operating costs:	Future operating costs:
PM = $3,000	PM = $10,000
Repairs = $400	Repairs = $200
Energy = $2,500	Energy = $2,500
EHS = $800	EHS = $0
TOTAL = $6,700	TOTAL = $12,700

Difference between future − current operating costs = $6,000 per year

Capital cost calculation

Current replacement cost:
Expected life at current PM = 17 yrs
Therefore replacement year 'A' would be 2003
NPV of 'A' is $746,200
Future replacement cost:
Assume that increased PM budget will increase life of roof by 10 years
Therefore replacement year 'B' would be 2013
NPV of 'B' is $458,100
Difference between future − current replacement cost = $288,100

Summary

	Operational $	Current NPV $
Current PM	$6,700	$746,200
Increased PM	$12,700	$458,100
Difference	$6000 × 10 yr = $60K	$288,100

Conclusion

For an increased preventive maintenance budget of $60,000 over ten years, the capital budget of $1 million for roof replacement is delayed by 10 years saving $288,100 of capital value at 6% discount rate.

FIGURE 5.3 Example of Roof Repair or Replacement.

As more and more companies have off-balance sheet assets, it will be harder to compare the performance of assets between different companies. Internet-based companies (those with significant investments in electronic commerce) usually have a very high RONA.

5.1.10 Off-Balance Sheet Assets

Many businesses are investing more in bits and bytes than in bricks and mortar. Financial pundits claim that the current accounting and financial systems are increasingly irrelevant due to the growth of intangible assets: ideas, brands, ways of working, franchises, and the influence of the Internet on the economy. Intangible assets are replacing physical assets. They claim that managers cannot decide whether a project is worthwhile, or even how to assess the value of a project. These pundits maintain that a new knowledge-based financial and accounting system needs to be established. The old accounting system is based on antiquated principles that rely on double-entry bookkeeping that is increasingly irrelevant today. Companies' net asset value is increasingly relying more on intangible assets than on physical assets.

Intangible assets can be grouped into four main areas:

- Product innovation
- Company brand value
- Structural assets
- Monopolies

Facility managers are concerned with structural assets that focus on better, smarter, different ways of doing business (not bricks and mortar). Physical assets are increasingly limited in their usage and have conflicting ways to be used. Intangible assets (such as databases and software) can be used in multiple ways in multiple locations by multiple customers for multiple transactions—thereby increasing substantially their value over the limited uses of physical assets. Value is created through transactions that have no physical interaction. Software and Internet transactions add value based only on the existence of a communication link and an authorization code. Perhaps FM in the future will be more involved with the infrastructure management of data and communication centers. Storefronts will still need to be taken care of by facility managers.

Some of the solutions currently being considered include:[4]

- Satellite accounts (capitalize intangibles)
- Normalized earnings (past + future)
- Knowledge asset valuations
- Percentage revenue from new products

These solutions do not present any one acceptable financial measurement system, but together they all point at the amount of future value a company has based on its current knowledge base. Bricks and mortar, as well as the management of facilities, are not going away soon and there needs to be an awareness that, increasingly, more value is being created through telecommunications and software transactions that have no physical assets in play.

5.2 FACILITY ASSET MANAGEMENT (FAM) FINANCIAL MODEL

It is the responsibility of the facility manager to take care of the physical assets of the organization—facility managers are the stewards of an organization's assets. Effective management of physical assets distinguishes the FM from other professional managers. One of the best processes to manage these assets and add value to the organization is through the FAM process.

5.2.1 What Is FAM and How to Get Started

FAM is a combination of engineering, finance, and business management principles that track the performance of an organization's physical assets. The goal of FAM is to increase the RONA (see previous section).

The simplest way to review a company's assets is to look at a summary of replacement values of physical assets within a summary database. This database would ideally contain the following information:

- Asset description
- Date asset was put into service
- Current condition of asset

- Current replacement value of asset
- Depreciation to date
- Major repairs to asset
- Performance index

*5.2.1.1 **Asset Inventories.*** In order to get started in the FAM process, an inventory of all assets must be done. This is a simple, yet difficult task for most facility managers. Simple, because it involves merely the counting and assessment of physical things; complex, because it reaches across all organizational boundaries and can be easily forgotten about. Facility managers are usually too busy with their daily operations and capital projects and assume that catching up on the changing physical inventories can be done at the end of the fiscal year or whenever it is needed. It is time consuming to inventory all physical assets every few years. Most companies struggle to maintain a continually accurate physical asset database. The asset inventory is the foundation upon which all operations are built, and it is critical to any financial and operation management: "*If you don't know what you have . . . how do you know what to do?*" The key to successfully maintaining an accurate asset inventory lies in the process.

By installing a simple process that monitors the dynamic changes to asset inventories as daily work and projects are done, facility managers can be assured of a continually accurate asset inventory. The first question of what to inventory depends on the organization's need for detail and available resources. The answer lies somewhere between "*door knobs to buildings.*" While inventorying every doorknob may improve the detail accuracy, it adds little value and becomes very cumbersome and costly to update. Inventorying only the number of buildings or square feet occupied is too general and does not provide useable details to make decisions about major operational assets. The next question is how to update the asset inventory database.

Before plunging into inventorying all assets, it is important to have an update process in place. This is especially true for large facilities. During the process of inventorying assets, the assets will change. Work done on the electrical system or on an exterior landscaping project may change the number of E-panels or trees that were inventoried. A simple process to update the asset inventory may include tying asset codes into each work order or supplier invoice and integrating/linking the two databases.

*5.2.1.2 **Condition Assessment.*** Condition assessments (CAs) are an important and integral part of asset inventories (see also Chapter 9). CA must be a continuous process based on accurate asset inventories and sound engineering principles. CA is essential in forecasting repairs, capital renewal budgets, and replacement needs. A standardized process must be used to make valid comparisons and appropriate investments. Individual bias, corporate politics, changing assumptions, etc., have no place in a good CA process. The CA must be strictly based on a technical and engineering evaluation of the current state of an asset.

The purpose of a CA process is to provide a long-term, time-phased program that accurately forecasts work needed to maintain assets in a productive and reliable state. A uniform inspection process must be established. Keeping an updated asset inventory and CA of those assets in a database is essential to good FM practice. There are standard engineering and economic equations that can provide facility managers with long-term management systems for their physical assets.

*5.2.1.3 **Leveraging Assets.*** Asset value is defined as: the present value (price) of all future cash flows associated with that particular asset. This is usually related to stocks and bonds, but is also applicable to physical assets. Extending the life of the physical assets of a company is one of the important responsibilities of a facility manager.

5.2.2 Maintenance and Operations (O&M) Budgeting

In developing a budget for new projects, capital renewal, and facility operations, it is important to understand the needs and service levels of the overall organization. An on-going facility budget is like a roadmap for the future of the FM department—what will be needed to operate and maintain physical assets to support the business mission? Other departments' budgets affect the facility budget. One example: Information services (IS) will need to adjust its budget based on facility plans for retrofits, reconfigurations, additions, or conversions of building areas that will have different network and telecommunication needs. Likewise, the facility department may need to adjust its budget based on IS plans for increasing data centers, networking, and other technology improvements.

The highest costs in FM are the operational and maintenance costs over the lifespan of the physical assets. Many companies refer to these budgets as their operations and maintenance (O&M) budgets.

It is common for the O&M costs of a facility to be more than ten times higher than the original construction costs over the total lifespan of the physical assets. Therefore, it is essential (from the start) that the proper preventive maintenance and repairs are performed to assure effective operational and financial performance of physical assets.

Budgets are essential to controlling a company's costs. Adherence to budgets requires planning, on-going financial discipline, and constant adjustments. In deciding what the appropriate annual operating budget for a facility is, there are several approaches. Table 5.1 gives a comparison of these budgeting methods. These methods are based on typical organizations' progression:

1. *Zero-Based:* When a building or asset is first operated, there is little history available and few benchmarks available for the specific operations incurred. The first year may well become the baseline year to learn about the asset's operating costs.

2. *Historical-Based:* Once an asset has been in operation for at least a year, a history is established. It is commonplace to budget for next year's asset operations based on the same or ±5–10% of last year's budget.

3. *Benchmarked—Unit Costing:* To encourage continuous improvement in operating costs and to remain competitive within their industry, many organizations are looking to benchmarking their costs and setting their budgets accordingly. Benchmarked budgeting is an important first step in pointing out areas for improvements. Underperforming areas are quickly identified and investigation and improvements can be started quickly. The disadvantage is the difficulty of comparing one company's unique internal customers and operations with another.

One particular process in use (especially for capital projects) is unit costing. While unit costing can yield competitive bids and can be useful in standardized systems and projects, it can also be misleading when specialized custom projects and services are needed. R. S. Means has historical and benchmarked data commonly used by FM and suppliers to budget this way.[5]

4. *Activity-Based:* Based on activity-based costing (ABC) and activity-based management (ABM) general accounting principles, this activity-based budgeting focuses on each activity within a process and allocates the actual incurred costs for each activity physically being done. There is no "smearing" of overhead and indirect costs across the entire company's operations. ABC is an extremely long and cumbersome process, but the payoff can be tremendous. Companies that have gone through ABC have raised their competitiveness to a higher level.

TABLE 5.1 Comparison of Operations and Maintenance Budgeting Methods

Budget method	Accuracy	Fairness	Most used
Zero-based	N/A	N/A	First-year
Historical-based	Very small	Small	Most common
Benchmarked	Questionable	Minimal	To get competitive
Activity-based	High	Good	Control labor costs
Asset-based	Very high	Excellent	Leverage assets

5. *Asset-Based:* Based on the discussion in the previous section on FAM, asset-based budgeting—when used in conjunction with ABC—is the most accurate and best financial operational budgeting method. It is based on the current condition history of all physical assets that the FM is responsible for and leaves little room for error. It does require adequate resources and continual updating to provide value. Most facility budgets are grossly underfunded when FAM budget costs are totaled.

Sound budgeting leads to better management. By requiring all departments to plan and adhere to a budget, financial discipline is instilled throughout the company. Senior management has an accurate indicator of the company's costs and can maintain cash flow. Without this financial discipline, companies would be held hostage to large fluctuations of operating costs.[6]

Here is a checklist for FM budget performance:

- Is it aligned with the business mission?
- Does it accommodate other departments' plans?
- Does it take into account regional COLA?
- What will the asset's overall condition be?
- Does it have adequate contingencies for unforeseen emergencies?
- Does it include last year's carry over?

One budgeting process that has been successfully used by many companies is to base their expense budgets on specific work programs that start from the asset inventory, specify service levels, and cost out each work task in individual line items. Here are the steps in setting up a simple work program and budget:

1. *Physical Inventory:* See the previous section on asset inventories.
2. *Service Levels:* Decide what level of service customers need. Do they want a low-end (Kmart, Yugo) type facility? Do they want a high-end (Nordie, Lexus) type facility?
3. *Performance Frequencies:* Decide what the minimal technical service frequencies need to be and add the customer requirements. There is a direct relation between service levels of step #2 and how often a service is performed.
4. *Financial Factors:* How long does each service take? What materials and equipment are needed? What is the cost per hour for each service? Verify that internal external hourly rates are equitable.
5. *Crunching Numbers:* The arithmetic is simple, but the sheer reiteration can be overwhelming. Most facilities can support anywhere between 50 to 100 work program line items. Each item must be calculated with its own unique inventory units and service level frequency. Most calculations follow this formula:

$$[\text{inventory item}] \times [\text{performance frequencies/year}] \times [\text{hours to perform} + \text{materials}] \times [\text{unit labor} + \text{material costs}]$$

A sampling of a simple work program and budget may look similar to the illustration in Table 5.2. It lists the inventory of physical assets, service levels, and performance frequencies, along with expected costs per unit of labor and material. There is also a column for major services—such as overhauling a chiller, or major repairs anticipated for an HVAC rooftop unit that may occur once every five years or so.

5.2.2.1 Financial Impact of Facility Disruptions. What is the cost of one hour of downtime for a company's accounts receivable department? Manufacturing production lines know exactly what one minute of downtime costs their operations, and they avoid it at all costs. Facility managers need

TABLE 5.2 Sample of Typical Work Program and Budget

Line item #	Budget descriptions	Work tasks	Inventory units	Unit of measure	Annual freq.	Hours/ service	Major $/serv	Material $/serv	$ per hour	Total annual cost
2.0	Site inspects	Inspect buildings	54	Sites	52	2.0		1	35	$199,368
3.0	ME PM	OH doors	178	Doors	2	1.5		25	75	$48,950
3.1		Garage equip+	318	Units	2	3.0		55	42	$115,116
3.2		Air compress.	18	Units	1	2.0		15	65	$2,610
4.0	ME service calls	Requests	514	Units	30%	3.0		20	46	$24,364
4.1		Kitchen equip	46	Units	20%	6.0	20	50	40	$33,028
5.0	Plumbing systems	Repairs	638	Fixtures	20%	2.0		60	45	$19,778
6.0	HVAC PM	Pkg. units	77	Units	4	3.0	200	60	53	$269,192
6.1		Chillers	2	Chillers	12	10.0	300	183	68	$48,872

to know what the downtime costs of their facilities are if the electrical or HVAC systems fail and take appropriate steps to ensure their reliable operation.

Reliability centered maintenance (RCM) is the discipline of assuring reliability of physical systems. The first step in RCM is to identify and prioritize the physical assets (see FAM process) according to their impact on the organization's overall business operations. A simple start would be to identify the obvious high priority (e.g., electrical) and low priority (e.g., furniture) assets that could impact business operations. The next step in RCM is to identify all plausible ways in which the equipment could perform below expectations. Once the likely failure modes and their effects have been identified, a cost effective condition-based maintenance solution is planned and carried out. This may be as simple as adjusting the time schedules for programmed preventive maintenance, or as complex as acquiring sophisticated diagnostic equipment and scheduling facility downtime to perform the condition testing.

Another way of assessing the financial impact of disruptions is to look at industry benchmarks for various industries' downtimes. Consider the example in the following section.

5.2.2.2 Capital and Expense Linkage—Repair versus Replace. One of the most difficult decisions facing facility managers is when to replace a physical asset. Is it best to replace it just before it breaks? Is it best to replace it before the next major repair? Or is it best to replace it now? Consider these two scenarios:

1. After thousands of dollars of repairs have been made to an asset (which is halfway through its useful life) another major failure occurs. Is it more prudent to replace the asset or to make another major repair?
2. A critical physical asset has reached the end of its rated life, but shows no sign of deterioration and has an excellent performance record. Is it replaced or not?

All capital assets depreciate and eventually become outdated, obsolete, or nonoperational. Primary assets that are critical to a key business operation need to be replaced on a preset condition or run-time. Assets that are critical to human comfort and general business operations need to be replaced when they are approaching the end of their useful life, and before major repair expenditures are made. There comes a time when an asset's next failure is the trigger to replace it rather than to repair it. So how does one decide whether to repair or replace?

There are many questions that need to be answered:

- Is there a safety or health risk?
- Is the asset critical to a key business operation?
- Has the asset reached the end of its useful life?
- Will the expense budget support the next major repair?
- Will the capital budget support replacement costs?

A simple example is the calculation of repairing or replacing a roof, as shown in Figure 5.3. Roof replacement involves the expenditure of capital funds, whereas preventive maintenance involves expenditure of expense funds. When facility managers request additional funding for increasing preventive maintenance schedules on the roof to lengthen roof life (defer capital expenditures), they need to provide the financial comparisons so that the organization's financial decision makers have a proper analysis at hand. When this financial comparison is provided, the likelihood of obtaining the appropriate funding increases.

5.2.3 Financial Performance Measurements

Key performance indicators (KPIs) are an important measurement tool for businesses. FM departments must have their own KPIs that are aligned with the core business KPIs. Here is a listing of some typical FM KPIs (over a unit of time):

- Operational hours
- Areas of responsibility (AOR) (total sq. ft., number of customers, business volume, etc.)
- Response times
- Rework
- Value added
- Number and performance of suppliers
- Employee/supplier employee turnover
- Innovations—new processes
- Employee satisfaction
- Customer satisfaction
- Plan versus actual on contracts
- Number of items on punchlists

Financial performance cannot be the sole determinant of performance of facility operations. CFOs may focus on financial indicators, but they too (along with FM) realize that there are more important drivers behind the financial numbers. Performance results are what counts. A properly executed measurement system reflects the strategy of the organization, and avoids the costly mistake of guiding an organization in directions that are not aligned with its strategy. This is what most of the old measurement systems do. The importance of measurement comes down to the saying: "*You can't improve what you can't measure.*"

The accepted way of measuring is to develop metrics (measurement ratios) based on a business strategic plan which provides key business drivers and criteria for metrics that managers would most like to monitor. Processes are then designed to collect information relevant to these metrics and reduce it to numerical form for storage, display, and analysis. Decision makers examine the outcomes of various measured processes and track the results to guide the company and provide feedback. Usually these outcomes are in the form of a written report or are viewed on an Intranet. Because these reports consist of mainly financial indicators, today's measurement systems focus organizations on past performance and encourage a short-term view of the company's business strategy. There is a better way.

5.2.3.1 Balanced Scorecard (BSC). The best method for evaluating business performance is the balanced scorecard (BSC). BSC is not a fad, it is an important tool in managing any project, operation, or budget. A properly constructed BSC gives you the high altitude view of an operation and can drill down to pinpoint and improve problem areas quickly. Since initial publication in the Harvard Business Review of January 1993, the concept of the BSC has been interpreted in many different ways. While some people have chosen to view the BSC simply as a focused set of financial and nonfinancial measures, the proper use of a BSC is much more.[7]

A new approach to management measurement was developed in the 1990s by Robert Kaplan (Harvard Business School) and David Norton (Renaissance Solutions, Inc.). Recognizing some of the weaknesses and vagueness of previous measurement approaches, the BSC provides a clear prescription of what a company should measure in order to "balance" the financial perspective with other vital operating metrics. The basic premise of the BSC is that you need more than one or two key indicators of performance and it needs to be more than just financial data. The BSC focuses on four main areas:

- Learning and growing
- Business processes
- Customer
- Financial

It can be seen that these four perspectives relate directly to four Malcolm Baldridge Quality Criteria categories: human resources (Category 5); business process management (Category 6); customer focus (Category 3); and business results (Category 7). These four areas account for over 70% of the total possible scoring in this Quality Award Criteria. The BSC translates vision and strategy into a tool that effectively communicates strategic intent, and motivates and tracks performance against the established goals.

5.2.3.2 Designing and Using a BSC. "Building a balanced scorecard seems simple, but is deceptively challenging." This quote summarizes the frustrations of one executive who had embarked upon the development of a BSC with disappointing results. The design of a BSC should not be underestimated. There are two essential ingredients to the successful design of a BSC:

1. An architect who has a framework, philosophy, and a methodology for designing and developing the new management system.
2. A client who will be totally engaged and assume ultimate ownership of the project, understanding that he or she must live with the results long after the architect leaves.

One way to start using a BSC approach is to reengineer FM measurement systems. Define what learning and growing in each of the four areas really means to the organization or project strategy, then define what succeeds in each of those areas and measure it. There are many other approaches to BSC that involve more detailed steps.

Just having a great BSC does not do anything by itself. Facility managers need to take action based on what the BSC tells them. Measurement without action has no value. The whole purpose of the BSC is to prompt a manager about where and when to take action. BSC enables a manager to react like a pilot—making continual adjustments to regain a proper flight pattern toward the goal. Most BSCs have some sort of an automatic rating system applied to them. This shows the progress (or lack of) in each of the key areas. Most companies that use BSC roll-up their BSCs throughout the entire organization. Individuals, small project teams, and departments are usually where BSCs get started. The data is then summarized on a department or division-wide basis and so forth. The final summary "report card" that the senior executives see is the final product. If specific interventions are needed in one or more areas, it is easy to drill down through the organization, pinpoint the root cause problem area, and take corrective actions.

A properly managed BSC will show you the progress of implementing a strategic plan. If the goal is to improve certain areas and the BSC does not show improvement occurring, then the strategic and/or logistic plans need to be revised to overcome a roadblock, or to redirect action plans. A good BSC should tell the story of the business strategy.

Three criteria help determine if the performance measures within the BSC do, in fact, tell the story of a strategy:

1. Cause and Effect Relationships: Every measure selected for a BSC should be part of a chain of cause and effect relationships that represent the strategy.

2. Performance Drivers: Measures common to most companies within an industry are known as "lag indicators." Examples include market share or customer retention. The drivers of performance ("lead indicators") tend to be unique because they reflect what is different about the strategy. A good BSC should have a mix of lead and lag indicators.

3. Linked to Financials: With the proliferation of change programs underway in most organizations today, it is easy to become preoccupied with a goal such as quality, customer satisfaction, or innovation. While these goals are frequently strategic, they also must translate into measures that are ultimately linked to financial indicators.

5.2.3.3 ***Applying BSC to FM.*** The four main areas of the BSC can be applied here. Here is one example of measuring the value of outsourcing using a BSC. You need to adapt these areas to the specific overall business strategy that led you to the outsourcing contract in the first place.

1. Learning and Growing: Metrics here include training hours per year for the facility manager and supplier's workers. More importantly, measure the value and skills/performance improvement of the workers. Turnover in employees is an applicable indirect metric.

2. Business Processes: This is the number of processes reengineered and/or automated, data-streaming accuracy, integration of processes into business strategy, number of new processes introduced, etc.

3. Customer: CSI 5 Customer Satisfaction Index. This is survey results and comparisons to baseline, number of complaints, customer retention, response time, average speed to answer (ASA), etc.

4. Financial: This is the most common: $/sq. ft., total cost, $/employee, etc.

By looking at all of the indicators simultaneously a more holistic picture of the overall health of the operation can be obtained. The BSC has been selected by many government agencies as an appropriate general structure for a strategic measurement and management system. Government offices, warehouses, bases, and facilities have much the same support services and user needs as a private company's. There are degrees of efficiency within operating and support processes that can be measured, benchmarked, and improved. Here is a quote out of a recent U.S. government manual on measurement:

> A performance measurement system such as the balanced scorecard allows an agency to align its strategic activities to the strategic plan. It permits—often for the first time—real deployment and implementation of the strategy on a continuous basis. With it, an agency can get feedback needed to guide the planning efforts. Without it, an agency is "flying blind."

Companies that have been using the BSC for some time and have successfully integrated the tool into their management processes see significant financial and operational results. The CEO of a large insurance company estimates that by using their BSC as a catalyst for MIS improvements, he was able to increase product profitability by nearly $30 million over a two-year period. Others describe the intangible benefits gained from being able to "understand the drivers of business success" and "educate the organization, deciding how, when, and where to raise the bar." By developing a BSC that truly tells the story of the FM business strategy, a foundation will be set for a management system that is capable of driving dramatic improvements in performance.

5.2.4 Forecasting Future Financial Requirements

Every year facility managers are asked to prepare next year's budget and submit their budget adjustments for long-term financial projections. Occasionally they are asked to submit 3–5 year budget plans for both expense and capital expenditures.

The ultimate objective of all service businesses is to maximize profits and customer satisfaction. This goal must be in the forefront of all facility strategies. The problem is that no one variable can completely determine either of these outcomes—in fact there can be hundreds of variables that influence the outcome.

Uncertainty about the future causes a decline in the value of money; in other words, risk increases with time (see section on risk). No one can predict with certainty either the future value of money or future facility operational needs. However, there are tools that can be used to predict some of the future risks and forecast what the most likely facility budget needs to be.

There are four distinct characteristics about services that impact financial forecasting:[8]

- Intangible
- Perishable
- Heterogeneous output
- Simultaneous production and consumption

Accurately forecasting services (budgets), such as FM, requires answers to some difficult questions:

- Who are the future customers? What will they require?
- What is the capacity/utilization of the current service system?
- What are the equipment requirements?
- What will the future tasks be? New—current—abandoned?
- How will resources be assigned to these tasks?
- What are future reliability needs?

Some of the answers to these questions can be answered by the long-term business strategy of the company. Another answer comes from performance measurements (see previous section). Historical performance measurements provide the best foundation for future forecasting. All services require an understanding of customers' needs. Facility managers need to know what the future churn rates, response times, and other customer expectations will be, in addition to having a comprehensive database of physical asset conditions.

5.2.4.1 Leading Edge Forecasting. With recent advances in computers and software systems, simulation tools are now available to help overcome challenges in planning and managing service systems. Simulation is particularly helpful in performing "what-if" scenario analysis for forecasting future system performance. Simulation helps achieve the greatest improvement in service system performance in the shortest amount of time.

The problem with current forecasting technologies is that they are limited to two-dimensional spreadsheet analyses. They neglect the dynamics of a system that involves multiple variables that must be tracked in order to reflect the reality of the system. The processing performed by spreadsheets (no matter how complex the macros are) are static in nature and do not take variability of performance variables into account. The results from spreadsheet analysis for systems are therefore flawed. The concept of a "moving time clock" cannot be introduced into spreadsheet analysis. Hence, the need for a better analytical tool: simulation.

Simulation combines the friendliness of a spreadsheet, the methodology of flow diagramming, the visual benefits of a CAFM/CIFM system, and the intuition of facility managers to help make better business decisions. It reduces financial resources by avoiding inappropriate investments. It is more accurate and complex than all other financial analytical tools. Simulation moves a manager from "As-is" toward "To-be."

Simulation is the only technology that truly allows one to forecast the future by taking into consideration all of a system's randomness and variability. Simulation allows one to advance the time clock and fast-forward to the see the future responses to the current or proposed system. Without simulation, a manager is left to predicting the future based on past results. *It is tantamount to driving a car looking out the rear window.* Simulation performance measures are both quantitative and time-based and measure the efficiency and effectiveness of a system configuration and operating logic.

Simulation can also be used to facilitate resource-capacity investments. Simulations can be used for:

1. *Decision Support:* Causes of problems and consequences of solutions
2. *Scenario Planning:* Explore alternative futures in real-time
3. *Customer Education:* Investigate value and uses of complex products and services
4. *Training:* Build skills and understanding in an interactive and risk-free environment

Another use for simulation is to predict and fund the "right" amount of capacity to deliver and manage the business effectively. The objective here is to achieve efficiency and service-level targets without overspending. This application is particularly useful to large FM organizations that have vast, spread-out resources and critical service delivery targets to meet. The use of simulation as an operational decision support tool can assist in looking forward to the impact a current decision will have on future asset and business operations. A proper FM simulation model can provide the means to anticipate problems and risks and to alleviate them before they occur.

Table 5.3 illustrates a comparison of traditional financial strategic forecasting tools and a comparison to simulation tools advantages.

All too often, right after a new facility, new project, or new service process is started, there are significant problems identified that compromise the ability of facility managers to meet their business objectives. By simulating the project or process before it is implemented, problems can be resolved and correction actions taken without affecting the physical environment, the customer, or the business operation.

The reader is referred to various excellent publications on simulation for further information.[9]

TABLE 5.3 Traditional and Simulation Forecasting Comparison

From traditional tools	To business simulation
1. Forecasting from historical data	1. Forecasting based on system dynamics
2. Static, hard-coded results	2. Individualized results that can be experienced
3. Assumptions are inflexible and hidden	3. Greater realistic control of analysis
4. Navigation and pace preset	4. Navigation and pace driven
5. Generic responses with no business model	5. Robust business models including customer and competitor behaviors
6. Learn by review + diagnostics of past	6. Learn by doing, analyzing future system response

5.3 LIFE-CYCLE ANALYSIS

The concept of life-cycle analysis is the most important part of financial management for all businesses—including FM. Life-cycle thinking means taking a holistic view of a product or service—from the raw material extraction through production, distribution, usage, and final disposal. Life-cycle thinking is referred to in many different labels. This section will deal with:

- Life-cycle costing (LCC)
- Life-cycle assessment (LCA)
- Full-cost accounting
- Commissioning

From an overall business perspective, life-cycle analysis can provide a sustainable competitive advantage. While mergers and acquisitions and, increasingly, technology speed and bandwidths may drive companies to new heights, life-cycle analysis is what keeps businesses around for the long term. Short-term thinking, from quarter to quarter, may be suitable to the retail industry; but long-term thinking is what sustains businesses through the rough spots of down cycles and where life-cycle analysis adds value.

No business area has a longer life cycle than real estate and buildings. Some buildings are built to last over 150 years, whereas other buildings have a designed life of only 30 years. The commercial life cycle of buildings is different than their useable life: shopping centers cycle in 11 years, warehouses in 25 years, with most commercial building somewhere in-between.[10] Besides being used for analysis in design alternatives, life-cycle thinking is also an important part of facility operations.

Perhaps the single most important program a facility manager can undertake to reduce life-cycle costs is a robust preventive maintenance program. In section 5.2.2.2, Capital and Expense Linkage—Repair vs. Replace, the example of roofing repair or replace analysis, a form of life-cycle analysis, was used to look at the impact an increased preventive maintenance program would have on roof life. Prudent design decisions need to be made in the earliest stages of facility and project designs since they will irreversibly affect the next 15–30 years of a facility's operation.

5.3.1 Life-Cycle Costing (LCC)

This section deals with optimizing the life-cycle costing (LCC) of facility and real estate projects. The scope of this section is to deal within the closed system of the facility itself and not with the issues outside of the facility. Those issues are dealt with in the next section under life-cycle assessment (LCA). An easy definition of LCC is: "The accounting of cost in relationship to activities performed in developing a product or service from creation to retirement."

According to ASTM, the definition of LCC is:

> ASTM E833-89: (LCC is . . .) A technique of cost evaluation that sums over a given study period the cost of initial investment less resale value, replacements, operations, energy usage, and maintenance and repair of an investment decision. The costs are expressed in either lump sum present value terms or in equivalent uniform annual values.

There are many ways to reduce life-cycle costs. Figure 5.4 shows that the maximum impact on life-cycle costs occurs in the first conceptual preliminary design phase of a project. Once the design and construction of a project are completed, it gets more difficult to impact future life-cycle costs. However, there are steps a facility manager can take throughout the life of the facility to ensure that life-cycle costs do not increase.

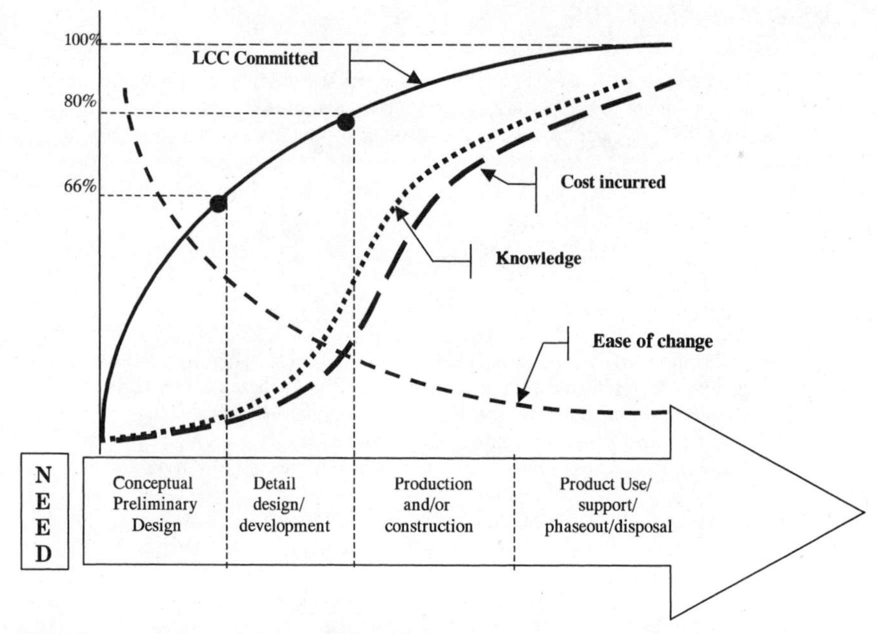

FIGURE 5.4 Cumulative Life Cycle—Phases of LCC.

The preliminary design phase deals with evaluation of possible alternatives. These possibilities are usually limited to the team of senior executives with regard to their vision, needs, and financial constraints. It is at this point that the senior facility manager needs to bring forth various operational scenarios and alternative design strategies that can be optimized through robust design and provide input to the LCC analysis.

The discipline of LCC analysis usually follows these steps:

1. Define need for analysis
2. Establish analysis approach
3. Select a model to facilitate evaluation process
4. Generate appropriate information
5. Evaluate alternatives
6. Recommend solution

The facility manager can have the most influence in steps 4 and 5 with regard to providing accurate information on long-term operational costs and introducing feasible and proven design and construction alternatives. Many of these alternatives come from the hard lessons learned during the daily operation of facilities.

The relationships of various facility design and operational phases for a facility are shown in Figure 5.4.[11]

During the conceptual preliminary design, facility managers need to ask questions about the future facility:

- What is the expected physical life of the building?
- What is the expected life of the components and systems within the building?
- What will be needed along the life cycle of the building to realize its full life?

These and other questions are critical in establishing a long-term FM strategy that provides proper funding and responsiveness to building and customer needs and reduces the risks of not attaining the entire building's life. The benefits of this approach are:

- Continued value of asset
- Minimal long-term annual maintenance costs
- Minimal breakdowns and operational disruptions
- Resale value maximized
- Visual and aesthetics appeal increased

Most new construction and retrofits deal with a sizeable initial capital cost. Once initial investments are known, a comparison based on future operational costs and benefits can be made. Similar assumptions are made for all alternatives so that equitable comparisons are possible and the life cycle of physical assets can be optimized.

5.3.2 Full-Cost Accounting (FCA)

Full-cost accounting (FCA) refers to accounting methods that allocate environmental costs, both direct and indirect, to a product, product line, service, process, or work activity. It looks beyond the specific closed system within an organization and addresses the company's total cost of ownership.

FCA goes beyond LCC and includes the impact the financial investment has outside its immediate occurrence. For example, if you were installing a new chiller, the LCC would provide the financial information only from the standpoint of the installed cost to when the unit is salvaged. FCA looks at the total cost of the raw materials used to manufacture the chiller, the real transportation costs to get the chiller to your site, the cost of acquiring and disposing all the chemicals used by the chiller, the real energy costs to power the chiller, the cost of caring for the chiller, etc. In other words, the entire life cycle of the chiller must be considered—beyond its facility use and disposal.

FCA offers a framework to aid decision makers with short- and long-term program planning and it can help identify measures for streamlining and improving operations. Unlike other common methods of accounting that record only current outlays of cash, FCA takes into account all of the monetary cost of resources used or committed to facility projects which may differ from cash outlays. Many companies and communities nationwide are already using FCA as a way to make programs sustainable in the long term, and provide the best service for the least cost.

5.3.3 Life-Cycle Assessment (LCA)

An even more comprehensive and holistic view than life-cycle costing (LCC) or full-cost accounting analyses is life-cycle assessment (LCA). LCA involves enlarging the boundary of the system within which the facility or project is viewed. LCC deals with a specific closed system limited to the immediate project and its location. FCA limits itself to the system or project in question. It does not take into account the project's impact on its surrounding community, on the resources required to make and deliver the materials and equipment used at the building site, nor on its environmental impact. LCA includes the design implications for the environment and the overall industrial ecology. The overall goal of LCA is to make more environmentally sound decisions regarding business operations—including facility design and operations. LCA came to the forefront with its inclusion in the ISO14001 series of worldwide standards. It embraces a concept called "design for the environment" (DfE), which addresses the design of products and processes (including services) in ways that can eliminate or minimize the creation of environmental damages. DfE lowers the overall waste disposal costs, reduces the expenses of regulatory compliance, and reduces total life-cycle costs. LCA also includes the concept of *industrial ecology,* which studies how materials and energies flow, how they are transformed within broad industrial and consumer activities, and what

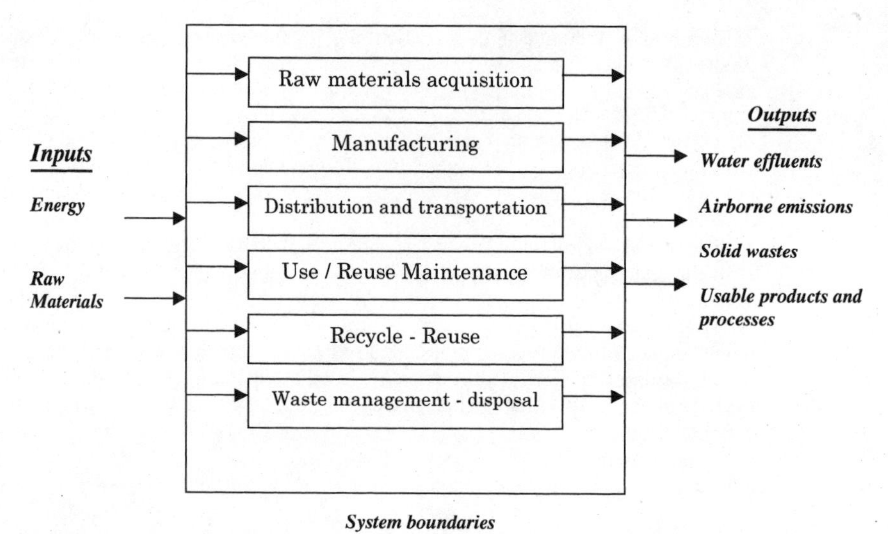

FIGURE 5.5 Life-Cycle Assessment Process.

effects they have on the environment. Industrial ecology includes factors such as political, social, regulatory, and economical issues. It seeks to place these systems within the framework of regional and global issues rather than just the specific localized issue that other analyses focus on.

LCA encourages businesses to understand the full range of environmental costs and to integrate these into basic decision making. Figure 5.5 illustrates this process.

The basic steps in doing LCA are:

1. Define goals and scope of assessment
2. Perform life-cycle inventory analysis
3. Assess life-cycle impact
4. Evaluate and implement improvements

LCA studies are time and resource intensive. Data gathering is complex and expensive. The reader is referred to the extensive literature that is available on ISO14001 and 14040/1 for additional details on this methodology. As more industries and businesses embrace a more holistic financial analysis, LCA will be pushed into the forefront of financial management tools. Facility managers need to become aware of their role in LCA analysis and look outside their immediate property boundaries to obtain a true picture of the resources their facilities are using.

5.3.4 Commissioning

Commissioning a building is a time-proven process for turning over from construction to operation. The definition of commissioning is "delivering owner functional intent." As building systems get more complex and LCC, FCA, and LCA analysis become a necessity, real estate directors and facility managers are turning their attention to developing a sound commissioning process for their new projects.

The need for commissioning and providing a building operations and maintenance manual was first recognized more than thirty years ago in Europe and specifically in the United Kingdom. As the built environment aged, it became vitally important to plan for maintenance which would prevent structures from deteriorating so far that they were no longer useful or even safe. Millions of dollars were allocated annually to this task. In the meantime, construction continued and new buildings were being completed at a record rate. It was determined that planning and implementing preventive maintenance programs from the first day offered both practical and economic advantages. Commissioning recognizes these advantages by including the start-up and maintenance costs along with the construction costs of a building. Today, maintenance is a multi-billion dollar business far outpacing the original construction costs, and the process of commissioning can substantially reduce lifetime maintenance costs.

The actual commissioning process should include all aspects of the building: its systems along with the site, structure, systems, and finishing requirements. How will the landscaping be maintained after it is fully grown? How will the exterior envelope be sealed from the elements—what if effervescence sets in? Are there architectural elements that will have a negative impact on the occupants—how can they be mitigated? Is the HVAC air balance within specifications and what is required to maintain high indoor air quality (IAQ)? What will the basic cleaning requirements be for the interior finishes—what chemicals should be avoided? How heavy will the traffic flow be within each area and what damage might occur? Is there enough data available to determine priorities when faced with multiple challenges? Are training classes available? What kind of support can be expected from the manufacturers and local contractors? All these questions can best be answered by both the design and construction and maintenance personnel, rather than any single party.

To get off on the right foot, the start-up and commissioning of a building needs to include the following:

1. At least one complete "full-load" test of all building systems
2. All operating manuals for substantial systems
3. Specification and cut sheets for equipment and special materials
4. Training by the installing contractors directly to the future operators of that system
5. Videos of the training or, at the very least, easy to understand, nonthreatening manuals
6. From the construction group—all warranty documents, evidence of code compliance, occupancy permits, etc.
7. From the material suppliers—all warranty documents, recommended maintenance procedures, content information, etc.
8. Release of mechanics and other liens
9. Maintenance schedule and procedural specifications for systems and finishes
10. Signed contracts for specialized first-year maintenance
11. A database of all equipment suppliers, local vendors, manufacturers, and replacement part order forms
12. A database of all material suppliers, local vendors, manufacturers, and ordering specifications

By focusing on one aspect of the building commissioning process, it becomes readily apparent how this can positively impact a project and its overall occupancy. From a technical systems standpoint, there are four major systems to look at when it comes to commissioning a building: electrical/lighting; HVAC; roofing; and environmental, health, and safety (EHS). These four areas also, coincidentally, are the four major issues within the ongoing operation and maintenance of any building. Each of these sections requires its own special attention, so we can only give a cursory view in this article. Here is a short list of some things to check for during building commissioning for each system:

- Electrical/lighting

  Proper grounding, sequencing, phase balancing

  All lamps and fixtures operative

Controls and adjustments work properly

Attic stock of special lamps and ballasts provided

Fixture manufacturer data and warranty information received

- HVAC

The air balance report shows design air volume delivered to each zone and space

All thermostats are properly located and calibrated

High IAQ

All chiller and boiler units are functioning and have been broken in properly

Maintenance crews have been trained on all system requirements

Water treatments for all systems are in place

EMS and remote controls calibrated and working properly

- Roofing

Are all "foreign" objects removed from the roof?

Do all roof drains work properly—best to check with flowing water.

Is access to roof appropriately restricted?

Are warranty documents provided and understood and authorized contractors in place?

Are all RTUs, exhaust fans, skylights, access hatches, screens, etc., properly secured?

- Environmental, health, and safety (EHS)

Are all safety issues resolved—hand rails, eyewashes, etc.?

Are hazardous materials identified and MSDs in place?

Waste management and recycling program in place.

Is everyone trained on the EHS issues?

Have ADA accommodations been provided for and communicated?

Whew—what a short list! The real list is ten times longer. And, if you do all of these items properly on a 1 million+ square foot building, you will spend hundreds of hours. The HVAC section alone may take this long.

Reality is usually filled with a last minute rush of getting the building ready for occupation and these commissioning items are "necessarily" skipped over or delayed. But at what cost? Or maybe a simple checklist is crossed off instead of meticulously verified. This all means that you are already working with deferred (ignored) maintenance before your first occupants enter the building. Maintenance costs will increase, not decrease, over time as a result of poor commissioning. The upcoming thousands of hours that you will be spending responding to these issues (that could have been resolved in a few hundred hours) is the first example of the "pay for it now or pay for it later" syndrome.

5.3.4.1 Commissioning's Potential Financial Benefits. Looking at just one example of what needs to be considered, it is easy to see why commissioning is so important and has been embraced more recently in the United States. Commissioning can ensure that:

- The performance of new equipment has been tested and verified in the presence of the staff who will be responsible for their future operations.
- The intended benefits of the new equipment, energy saving features, etc., are communicated to and understood by the operators, users, and managers.
- Routine maintenance and calibration is started on queue and not deferred. It is clearly understood what is to be done and when.

- Attic stock of replacement components and materials, or contacts of suppliers where they may be purchased, is provided.
- All manuals, cut sheets, specifications, as-built drawings, maintenance and warranty information are conveyed to the staff that will be operating the building.
- Budgets correctly reflect the expenditures involved and provide a reliable source for determining historic data.
- Deferred maintenance expenses, including premature replacement costs, are kept to a minimum.
- Successful design is recognized and costly problems are avoided in the future.
- Feedback on successes and failures are given to the design team.

In general, the old "turn key" approach to building start-up is slowly going out of vogue. Building owners and occupants demand a more proactive approach along with a properly trained staff that knows how to keep the building systems operating smoothly and the physical appearance from deteriorating. Maximum comfort, attractive surroundings, and significant savings are not conflicting points, but a common goal.

Getting the flywheel of building operations and asset management off to a smooth start is what commissioning is all about. Once deferred maintenance or other programs come into play, the flywheel's momentum is changed. As the added weight slows the process down, the energy (money) needed to bring the momentum back up to an acceptable level becomes more and more frightening. Building commissioning is the grease that can keep everything moving at a steady rate, support continuing smooth operations, and reduce future costs.

5.4 ON-GOING FACILITY FINANCES

Not only can capital and major repair projects benefit from LCC and ROI analyses, the daily operations of the facility can also benefit from financial analyses. Conversely, daily FM operations need to accurately track and provide good data for major repair and capital project financial analyses.

Three major areas that impact financial analysis and are important for FMs in keeping accurate data are:

1. Asset inventories
2. Computer-integrated facility management (CIFM)
3. Contracts

5.4.1 Updating Asset Inventories

As discussed in section 5.2.1.1, asset inventories, a fundamental part of ongoing facility management is maintaining an accurate inventory of physical assets. Keeping the asset database updated is fundamental to conducting valid condition assessments and essential in providing data for all financial budgeting and analysis.

The physical workplace is a dynamic, ever-changing world. Physical assets are continually being switched around, replaced, removed, or eliminated. Some of these changes are minor, others are major—most changes are insidious. Changes to physical assets occur during renovation projects and during major repairs. Facility managers need to track these asset inventory changes on a continual basis. Consider these two on-going changes:

- When a $100' \times 100'$ turf area in the landscaping is converted to a patio or parking area—that's a 10,000 sq. ft. reduction in turf maintenance and a 10,000 sq. ft. increase in sweeping or pavement care inventory.

- One of the most insidious inventory changes occur when electricians add duplex outlets or other minor electrical changes for customers—these small changes can add up to electrical phase imbalances and potentially overload circuitry.

While this change may not seem significant, it is the constant repetition of these small changes that end up as major changes to your asset inventory. Unless the facility manager keeps track of all these small changes, the accuracy of the physical assets database is reduced and eventually will be so inaccurate that it is rendered useless and would require an overall asset inventory count to be retaken—a timely and expensive process.

As stated in asset inventory, a process to update the physical assets needs to be in place. For small facilities, this is relatively straightforward. For large conglomerates, this is a time-consuming process, but even more important. This updating process needs to be kept going throughout the life of the facility to ensure proper replacement cycles and reliable operation. The updating process can include simple changes, such as including physical asset coding on work orders and contractor invoices. It may involve bar coding and scanning technology.

5.4.2 Contract Pricing Issues

Every facility requires some contract labor to keep it operational. Whether the function is entirely outsourced, or only one general contractor is used for special, occasional work; supplier management and contract issues require some knowledge of pricing.

There comes a time in every contract negotiation or outsourcing initiative when pricing and financial issues are considered. Sometimes special financial analysts are brought in—internal support analysts, auditors, or third party external analysts. The overall intent of contract financial analysis needs to be clearly stated up front: What is the goal or purpose? Is this analysis being performed before, during, or after a contract or outsourcing service is provided? If before, then the intent may be to compare internal to external costs and will influence the outsourcing initiative. If during an ongoing contract, then the intent may be to compare cost savings and benefits, or to verify contract performance. If after, then the intent may be to focus on the contract renewal pricing and workscopes, or to perform a post mortem of lessons learned.

5.4.2.1 Getting Started. The basic steps in any facility financial analysis are to collect internal and external costs, total them, compare them, and project them along their life-cycle future costs. As discussed earlier in this chapter, in the initial collection stage it is important to pull together ALL costs that are associated with a specific work activity or service. There are many hidden and unassociated costs that are impacted by facility operations and these may need to be included as well.

The next step is to decide: *For what time period will data be reviewed?* Will it be for the previous fiscal year (most common)? Will it be for the last six months or three months? Whatever time period is chosen, internal and external operational data collection need to follow these minimum criteria:

- Substantially stable operation over entire period
- Complete financial data and details available
- CAFM–CIFM data accurately captured over entire period
- Consistent operation throughout the period (no work stoppages, etc.)

Once the time decision is made, the workscope boundaries need to be specified. Financial analysts may collect the above data for each specific trade or work activity. This is usually very cumbersome and difficult. Most organizations do not track detailed costs for each separate work activity. Costs are usually lumped into general categories such as HVAC, electrical, carpentry, workstations, general administration, and so forth. If the categories cannot be compared fairly to external services, then it will be difficult to have a fair and equitable outsourcing analysis. Some companies use their CIFM systems to provide this detail and embark on a 3–6 month time period in which they collect all the data and required details.

Next, for the selected time period the following are added together for the specified workscope:

- Internal

  Labor costs (include indirect percentage, usually ranging between 120 to 200%)

  Material and equipment costs (purchases, leases, rentals)

  Stores inventory (carrying costs)

  Travel (fleet costs if applicable)

  Warranty costs

  Training

  Administrative

  Insurance

- External

  Supplier labor costs

  Material and equipment

  Mark-ups and profit margins

  Subcontractor costs

  Bonding, insurance

- Support

  Occupancy costs ($/sq. ft.)

  CIFM maintenance

  Purchasing, procurement

  Administrative

The reason to break down the costs in detail is to allow for easier comparisons now and in the future. It is difficult to go back and break down lump costs unless they were rolled up initially. For example, labor costs may be comparatively similar, but support cost comparisons may be way out of line and warrant further investigation.

5.4.2.2 Crank and Grind. Once financial analysts have prepared this data collection then it is time to crank and grind. First, they look at the current proposals from outsourcers or what the current outsourcer's performance has been, and then look at comparative internal performance. Comparative analysis can include straight-line analysis; that is, subtracting one from the other over time, or could involve NPV (net present value) and RONA (return on net asset) conversions.

5.4.2.3 The Downside. For those in the FM operations business, there is one major downside to financial analysis: it deals strictly with the financial risk. It does not take into account health and safety risks or operational risks. Seldom is any consideration given to market and competitive conditions or to customer satisfaction. Most financial analysts do not know the difference between an HVAC and a UPS systems nor what is the right amount of preventive maintenance for either system at different usage and aging patterns. It is the facility manager's responsibility to make sure that these issues are not drowned out by the financial background noise. Internal costs are sometimes misleading. They do not take into account market conditions and are hamstrung by internal overhead and administrative loading. External costs can also be misleading: a supplier may not be billing the true cost of providing the service (loss leaders, etc.)

5.4.3 Internet, Software, Financial Calculators

The ever-changing Internet contains a wealth of information on financial software and analytical tools and can assist in doing financial analyses. Appendix A in this chapter lists some Internet sites (as of January 2000) that facility managers may find useful regarding financial analysis and financial calculators. There are also many financial calculators (sometimes called business calculators) on the Internet. Among the most popular are Hewlett Packard and Texas Instrument financial calculators that can perform many financial functions. Some common financial and mathematical functions performed by financial calculators are:

- TVM (loans, savings, and leasing)
- Amortization
- Cash flow analysis
- Cash flow functions/No. of functions
- Print table with HP 82240B
- Bond price and yield
- Depreciation methods
- Interest rate conversions
- Currency conversions and unit conversions
- Statistical analysis
- Std deviation/mean/weighted mean
- Forecasting, correlation coefficient
- +, −, ×, /, %, 1/x, +/−, 1n, X, e^x, n!, y^x, SUM x, SUM x^2, SUM y, SUM y^2, SUM xy
- LOG, 10x, PI, x

Financial calculators are essential to performing simple calculations. Rather than resorting to financial interest tables to perform discounting and ROI analysis, facility managers need to know how to use the basic functions of a financial calculator. Financial calculators are beneficial, especially when compared to laborious manual calculations and look-up tables—but beware of their shortcomings and use them judiciously. A word of caution on using financial calculators: do not believe their results unless you use these calculators on an on-going basis or know the algorithms and approximate answers to the financial question. A financial calculator will most likely give you the wrong answer if you are unfamiliar with its operation. This is because *most financial calculators have multiple memories that may or may not be cleared, have different ways of entering various interest and discount rates, and different time periods.*

5.4.4 Integrating with CIFM Systems

Financial analysis within FM needs accurate input data. These data are usually supplied by a CIFM system. CIFM systems are a great resource for historical data tracking, ratio analyses, and comparative data analyses. Unfortunately, there are no stand-alone CIFM systems that have fully-integrated financial analysis capabilities. Several CIFM systems can provide basic lease management, asset value, and repair cost data summaries. Repair or replace decisions, ROI, RONA calculations, and other financial analyses are usually done with third-party software modules or with Excel™ macros.

CIFM systems are essential to providing the basic data needed to perform financial analysis and forecasting:

- Year asset was installed
- Operating costs

- Energy, EHS issues
- Asset condition
- Repair and trend history
- Major repair details
- Warranty information
- Replacement cost
- Depreciation

Advantages of using CIFM systems to aid in financial analyses include:

- Improved budget analysis and planning
- Improved forecasting
- Faster access and retrieval
- Faster turnaround time
- Better decisions
- Reduced operating costs
- Improved capital outlays
- Less labor needed

CIFM systems provide quantitative benefits to the facility manager. These are some of the areas that need to be considered when doing LCC analysis of CIFM systems:

- Longer equipment life
- Less time to drawing updates
- Greater space utilization
- Fewer duplicate asset purchases
- Improved disaster recovery
- Less corrective maintenance
- Effective parts inventory

The advantages of using third-party accounting and financial software packages are that in-house expertise is not required to perform the calculations. A CIFM system can also provide a wealth of additional information that is more important than the financial analysis:

- EHS and legal issues
- Safety risks
- Worn-out equipment
- Technology obsolescence

When looking for CIFM compatibility with financial analysis tools, consider these issues:

- Ease of use
- System interfaces
- Flexibility
- User interfaces
- Other departments' interface

CIFM systems are a great resource for historical data tracking, ratio analysis, and comparative data analysis; however, they do little in assisting with financial forecasting.

5.4.5 Using Simulation for Financial Forecasting

Using historical data to predict what will happen in the future is like driving your car by looking out the back window—past events and performance do not tell you what is coming in the future. Sophisticated forecasting tools have been available to large industries for many years. These tools are now starting to be used by facility managers and real estate directors to make more intelligent forecasts of future system performance and reduce future risks.

One cannot use a static model (CIFM) to study a dynamic problem. By modeling current and future business scenarios, facility managers can eliminate, or at least mitigate, much of the uncertainty that inevitably happens when facilitating changes. It will save capital budgets by avoiding inappropriate investments. When used effectively, it can also improve the on-going operational bottom line.

5.4.5.1 Simulation Overview. People create and run simulations in their minds every day: planning investments, coordinating transportation, analyzing relationships, and making business decisions. For most of us, the term "simulation" conjures up images of academies predicting the economy, the military playing war games, airlines teaching pilots, MBA's modeling businesses, engineers modeling power plants, and video game enthusiasts playing SimCity™. As useful as these simulation tools can be in all these situations, only a few people have the resources, knowledge, and tools to use simulation. There are currently several business simulation models that have been developed for the real estate and facility professional. FMs can also use several of the simulation software platforms that are available to develop simulations for their own processes and predict future behaviors of their processes, thereby reducing risks to their business operations.

5.4.5.2 System Dynamics. At the heart of these simulation products is a sophisticated methodology for developing continuous simulations, developed in the 1950s and 1960s at MIT by Professor Jay W. Forrester (inventor of computer memory), called System Dynamics. More recently, Peter Senge revitalized systems thinking concepts in his books on *The Fifth Discipline*. Systems thinking involves changing paradigms about the way the world works, the way corporations work, and the human role in each. Senge and his colleagues teach managers to look for interrelationships among system elements; to avoid placing blame in favor of finding the true, long-term solution to a problem. System dynamics takes this one step further.

System dynamics is less well-known than system thinking. Its purpose is to give managers a tool to understand the complex systems that they are charged with controlling. The methodology uses computer simulation models to relate the structure of a system to its behavior over time. It is based on defining the changing relationships between entities using formulas. Continuous simulation is best used for analyzing behaviors that cannot be easily viewed as discrete events. With this simulation software, the ability to use this complex theory is now put in the hands of managers, marketers, and mere mortal real estate and facility managers.

Simulation allows you to run hypothetical "what-if" scenarios on a computer model without disturbing the actual system. It allows understanding of randomness and variability in a business environment. Simulation allows you to forecast by looking forward and analyzing what is coming. It allows you to contain future risks and capture the dynamics of your business environment in a flexible, breakable, changeable model using multiple "what-if" scenarios.

Simulation models can also be used to manage change and can result in dramatic improvements in business performance and profitability. Each scenario offers its own ROI alternative. Alternatives that do not match business objectives can be eliminated and tradeoffs in implementation strategies and their effect on customers and profits can be explored.

Simulation combines a familiarity with spread sheets, the methodology of flow diagramming and process mapping, and the visual benefits of graphical outputs—all to help make improved business decisions and financial forecasting. Business process reengineering (BPR) unleashed the power of information technology by enabling organizations to break the rules and create new ways of working.

TABLE 5.4 Traditional Tools vs. Business Simulation

From traditional tools	To business simulation
1. Forecasting from historical data	1. Forecasting based on system dynamics
2. Static, hard-coded results	2. Individualized results that can be experienced
3. Assumptions are inflexible and hidden	3. Greater realistic control of analysis
4. Navigation and pace preset	4. Navigation and pace driven
5. Generic responses with no business model	5. Robust business models including customer and competitor behaviors
6. Learn by review + diagnostics of past	6. Learn by doing, analyzing future system response

Business process simulation enables organizations to reengineer more effectively. It allows them to think "outside the box" and determine new ways to run a business. Table 5.4 lists the traditional methods and the corresponding business simulation objectives to move towards. Business scenarios that differ dramatically can be easily explored and compared with a computerized model with better decisions being the result.

Real estate and FM are about relationships and risk management, not only with the people involved, but also with the complex variables that impact daily operations. Just as economic models are driven by interest rates, consumer confidence, housing starts, currency valuations, etc., so, too, internal infrastructures and physical facilities are driven by many variables.

Managing infrastructure is all about managing RISK, which is defined as *the measure of probability and consequence of not achieving a defined goal.* Risk management is the process of identifying, assessing, and controlling risk. A manager can only avoid, reduce, transfer, and assume risk. By identifying and optimizing their key business processes, companies can get the maximum performance from their resources and achieve dramatic financial benefits. They can save money by avoiding unnecessary expenditures.

5.4.5.3 Case Study. Nippon Telephone (NTT DATA) based in Tokyo, Japan, was the first company to apply financial simulation modeling to its leasing management operation. By applying such

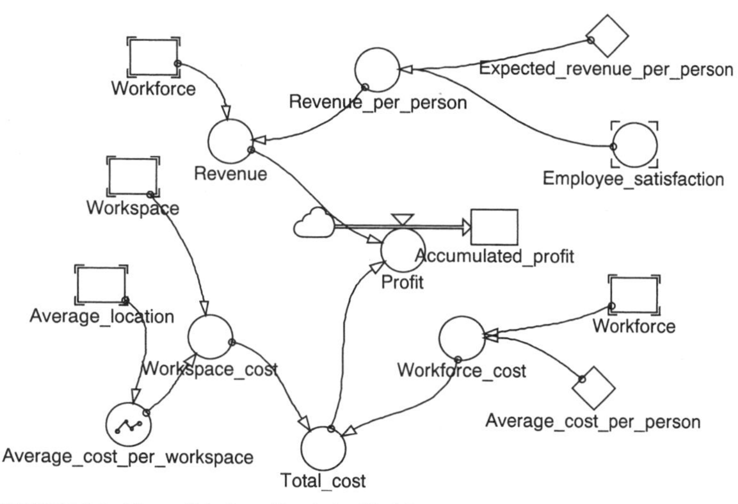

FIGURE 5.6 Nippon Telephone Simulation Model.

techniques, NTT was able to look at their space standards and decide what the optimal standard was in relation to profits, employee satisfaction, time to lease, etc. One of the NTT simulation models is shown in Figure 5.6.

Sensitivity analyses showed that the optimal figure was at $9m^2$, whereby the standard workspace maximizes accumulated profit. This observation resulted in the conclusion that when the space standard is increased to this level, the profits were optimized while other conditions stayed the same.

APPENDIX A WEB SITES FOR FINANCIAL MANAGEMENT

Other Service Industry Financial Analysis Sources

- http://www.healthedgroup.com/noframes/subjects/finance.htm
- http://www.epa.gov/opptintr/acctg/rev/tbl3-5.htm

Financial Calculators

- HP 12C, 17B, 19B, 38G, 48G: http://www.hp.com/calculators/purchase/financial.html
- Texas Instrument Business Analyst—Real Estate: http://www.ti.com/calc/docs/busi.htm

General Finance Analysis

- http://www.sandia.gov/pv/sysd/Economics.html
- http://www.nissd.com/sdes/papers/deslcc.htm
- http://www.iso.ch/9000e/9k14ke.htm
- http://www.tc207.org/home/index.html
- http://www.environment.gov.au/epcg/eeu/consumption/lessons.htm
- http://www.nap.edu/books/0309054419/html/29.html

APPENDIX B DEFINITIONS OF FINANCIAL TERMS

Assets. Real, intangible, and financial items the corporation owns or has a claim upon. Assets traditionally appear on the left side of the BALANCE SHEET.

Balance Sheet. One part of the FINANCIAL STATEMENT which lists the wealth of the corporation at a specific point in time. Traditionally, ASSETS appear on the left and LIABILITIES on the right and the two must be in balance. EQUITY is a form of liability representing what the corporation owes to the stockholders.

Breakeven. The breakeven volume is the number of units the company must sell to recover its fixed and variable costs of production, at a given price. The breakeven price is the price the company must charge to recover its fixed and variable costs, at a given sales volume.

$$\text{QBKEVEN} = \text{FC}/(\text{P} - \text{AVC}) \text{ or PBKEVEN} = (\text{FC}/\text{Q}) - \text{AVC}$$

Cash Flow, Cash Flow Statement. Cash flow measures the profit of the company over a given time period in cash terms (as opposed to accrual terms). A CASH FLOW STATEMENT is generated by beginning with an accrual-based INCOME STATEMENT, and making the following adjustments:

- add back in noncash expenses, such as depreciation
- subtract increases/add decreases in operating assets

- add increases/subtract decreases in operating liabilities
- subtract cash going to/add cash coming from capital investments
- add cash coming from/subtract cash going to financing activities

Debt. An obligation or liability arising from borrowing money or taking goods on credit. Forms of debt include bank loans, bonds sold to investors, and various types of "commercial paper." Debt holders are creditors of the company; stockholders are owners of the company.

Depreciation. A noncash expense entered on the income statement to recognize the reduction in value of FIXED ASSETS due to wear and tear.

Discount Rate. The rate at which future cash flows are discounted because of the time value of money.

Income Statement. One part of the FINANCIAL STATEMENT which lists the REVENUE, EXPENSE, and PROFIT (or loss) of the company over a period of time. Traditionally, revenue appears on the right and expenses on the left.

EDIT = Earnings Before Interest and Taxes. PROFIT before you subtract interest and tax expenses. It measures efficiency of operations independent of financial matters such as interest rates and tax rates.

EBITDA = Earnings Before Interest, Taxes, Depreciation, and Amortization. PROFIT before you subtract interest, tax, and noncash expenses. It measures efficiency of operations independent of interest rates, tax rates, and capital investments in plant, equipment, and intellectual property.

Equity. The "book value" of the company's stock. (Not the market value it is trading at on the stock exchange.) Equity holders are owners of the company.

Equivalent Annual Cost. The cost of obtaining and maintaining an asset annualized over its life (as if you were leasing it from someone else).

Expense. Payments made by the company to others for value received. DEPRECIATION is a "noncash" expense used to estimate the decrease in value of a FIXED ASSET over time.

Financial Statement. A comprehensive description of the company's finances including the BALANCE SHEET, INCOME STATEMENT, and CASH FLOW STATEMENT.

Fixed Asset. An asset such as property, plant, equipment, or intellectual property that has a long life and cannot be expensed in a single year.

Fixed Costs. Costs which do not vary significantly in the short term with the volume or production. Property, plant, and equipment are examples.

Income Statement. One part of the FINANCIAL STATEMENT which lists REVENUE, EXPENSE, and PROFIT (or loss) of the firm over a period of time.

Internal Rate of Return (IRR). Mathematically, IRR is the discount rate at which the investment has zero net present value (NPV). If that is a little too abstract, you can think of it as return on investment (ROI), although that is mathematically incorrect. IRR is used to compare and contract different projects that the corporation could invest in. Higher is better. There is no formula for IRR, but spread sheets can calculate it, iteratively.

Leverage. The ratio of total assets to total equity. A "highly leveraged" company will have high total assets in relation to low equity.

Liabilities. Any claims against the firm. Liabilities traditionally appear on the right side of the BALANCE SHEET.

Net Present Value. The initial cost of an investment (a negative number) plus the PRESENT VALUE of the future CASH FLOWS it will generate (a positive number).

Nominal Dollars. Dollars which are not adjusted for inflation or time value of money.

Opportunity Cost of Capital. The value of other opportunities you gave up to invest in the capital project in question.

Present Value. Future cash flows in terms of their value today, using a DISCOUNT RATE.

Real Dollars. Dollars at some other time adjusted to their value in current-day terms.

Revenue. Money (or other value) received for products sold to customers.

Time Value of Money. A dollar today is (usually) worth more than a dollar tomorrow, due to inflation, costs of borrowing, and risk. The DISCOUNT RATE measures the difference between current and future dollars.

Variable Costs. Costs which vary in the short term with the level of production. Raw materials are an example.

APPENDIX C ACRONYMS USED IN THIS CHAPTER

ABC	Activity-based costing
ABM	Activity-based management
AOR	Areas of responsibility
BPR	Business process reengineering
BSC	Balanced scorecard
CA	Condition assessment
CAFM	Computer-aided facility management
CFO	Chief financial officer
CIFM	Computer-integrated facility management
COLA	Cost of living adjustment
CSI	Customer satisfaction index
DfE	Design for the environment
EHS	Environmental, health, and safety
FAM	Facility asset management
FCA	Full-cost accounting
HVAC	Heating, ventilation, air conditioning
IAQ	Indoor air quality
IRR	Internal rate of return
IS	Information services
KPI	Key performance indicators
LCA	Life-cycle assessment
LCC	Life-cycle costing
MFL	Maximum foreseeable loss
MIS	Management information system
MTBF	Mean time between failure
NPV	Net present value
O&M	Operations & maintenance
PML	Probable maximum loss
RCM	Reliability-centered maintenance
ROI	Return on investment
RONA	Return on net assets
SPB	Simple payback
TCO	Total cost of ownership
TI	Tenant improvements
TVM	Time value of money

ENDNOTES

1. A word of caution: Different countries and different United States localities have unique financial regulations and laws that may be different from those offered in this chapter.
2. Campbell, J. 1995. *Uptime.* Productivity Press.
3. Schall L. D., and Haley, C. W. 1983. *Introduction to financial management.* McGraw Hill, p. 219.
4. Lev, B. *Financial statement analysis: A new approach.*
5. Means, R. S. 1998. *Means assemblies cost data,* 20th annual edition. Kingston, MA: R. S. Means Company.
6. An excellent reference on developing costs data and budgeting is available through the AACE (Association for Advancement of Cost Engineering through total cost management) and its web site at http://www.aacei.org
7. References for further reading on BSC:

   Norton, D. P. 1996. *The balanced scorecard: Translating strategy into action.* Cambridge, MA: Harvard Business School Press.

   Cooper, R. 1998. *Cost and effect: Using integrated cost systems to drive profitability and performance.* HBS Press.

   Harvard Business Review articles include: Cooper, R. 1998. *The promise—and peril of integrated cost systems,* (July–August) and Norton, D. P. 1996. *Using the balanced scorecard as a strategic management system.* (January–February).

   CD-ROM interactive simulation, *Balancing the corporate scorecard,* available from HBS Publishing.

   In 1994, HBS Management Productions produced a four-part videotape series, *Measuring corporate performance,* which presents concepts and companies' experiences with activity-based cost management and the Balanced Scorecard.
8. Sasser, W. E. et al. 1978. *Management of service operations.* Boston: Allyn & Bacon.
9. References for simulation reading:

   Harrel, C. et al. 1995. *Simulation made easy—a manager's guide.* Institute of Industrial Engineers.

   Profozich, D. 1998. *Managing change with business process simulation.* Prentice Hall.
10. Flanagain, R. et al. 1989. *Life cycle costing.* Oxford: BSP. pp. 44–45.
11. Fabrycky, W. et al. 1991. *Life cycle cost and economic analysis.* Prentice Hall.

CHAPTER 6
ULTIMATE CUSTOMER SERVICE

Stormy Friday
Founder and President
The Friday Group, a firm specializing in strategies and solutions for the facility management profession
E-mail: sfriday@erols.com

6.1 INTRODUCTION

Once they understand the critical role that ultimate customer service plays, it becomes a goal that most facility management (FM) organizations strive to achieve. Ultimate customer service not only determines and ensures the value that the organization adds to the corporate enterprise, but also adds value to the individual business units of that enterprise. The following text, with some minor modifications, comprises the substance of a module on customer service taught by the author as part of the course, *Achieving Facility Management Organizational Effectiveness*, at Michigan State University, Virtual University Masters Certificate Program in Facility Management. In order for FM organizations to achieve a combination of customer satisfaction and achievement of business objectives, a foundation of quality within FM organizations must be built and run on what the author defines as the "Five Pillars of Quality."[1] These are:

- The customer drives the process
- The FM manager must be committed to continual improvement
- Benchmarking and metrics are essential
- Front-line workers must be empowered and held responsible
- Marketing cannot be done by sitting in your office

As FM professionals inside a larger, corporate environment, we provide service to employees of the other organizational entities within that corporate environment. These employees—no matter where they reside in the corporate hierarchy, from secretary to senior executive—are, essentially, our *customers*.

Adopting the mindset of serving internal customers gives FM organizations the ability to see everything in a new light: that all organizational development efforts be viewed with an eye toward satisfying our customers. FM organizations need to focus on ways to have customers help determine the most flexible and dynamic organization structure that will facilitate the delivery of services to them as a customer. The role of the customer, therefore, is paramount to the planning and development of an FM organization.

6.2 IDENTIFYING CUSTOMER GROUPS

There are many different categories of customers, including:

- *Organizational Unit:* Each organizational entity can be considered as a discrete customer group.
- *Building Unit:* Customer category may be based on physical location. Whole organizational components or parts of organizations may be in different buildings. To further complicate this customer group, some buildings may be leased instead of owned, which would mean a very different brand of facility services, since the FM organization may not be the total provider of these services in all buildings.
- *Business Unit:* A company may be organized around business units that are responsible for certain product and service lines or organized as individual cost/profit centers.
- *FM Organization:* FM employees are themselves internal customers; they do not operate independently of the organization. Imagine that you, as facility manager, put each member of your staff out in the parking lot with a desk and a telephone and suggested they come up with facility services by themselves—it would be impossible!
- *Senior Management:* Top management is such an important and special category that it needs to be considered as a category unto itself.
- *External Clients of Numbers 1, 2, and 3 and Visitors:* Each of the categories named previously have *their own customers*.
- *Vendors:* External vendors are thought of more these days as "partners" in the service delivery arena. These people should have the same values and beliefs about customer service that you do.
- *Tenants:* If you provide tenant/landlord services, then tenants should be identified as customers.
- *Facilities:* This might be the most difficult concept to grasp: facilities themselves are also customers. Very often they are vocal customers that "speak" to tell us when something has gone wrong or needs repair. This is why we have preventive maintenance programs, conduct life-cycle costing analysis and develop strategic master plans, in order to proactively protect and satisfy the needs of facilities, *as a customer group*. FM may be the primary organization with the financial responsibility for taking care of the corporate real estate assets; therefore, thinking of real estate and facilities as customers really makes sense.

After determining exactly who your customers are, you will be in a position to provide ultimate customer service. The idea that customers are the only true judges of what constitutes quality and customer service is not a new concept, but it is one that needs to be frequently reinforced. Customer satisfaction is vital for several reasons, including:

- It breeds job security—within today's environment of outsourcing and "rightsizing" it is a way to show added value to a corporation.
- It dictates organizational trends—through knowing what the customer wants and needs.
- It provides a strong and vital relationship—customers can be our best allies. Satisfied customers will give you a distinct edge over any outside firm by being able to discuss the value your organization adds to furthering their business goals.

The next step is determining what your customers' exact requirements, needs, and wants are. Understanding and assessing customer perceptions and expectations for service is a critical factor in the customer service process. We have devised a workable tool for that assessment.

6.3 DETERMINING CUSTOMER PERCEPTIONS

Once you understand that *your customers' wants and needs* should be the basis on which your entire organization plan is founded, it is important to know how to go about determining these requirements and desires. How well we are able to assess those needs and develop our service plan accordingly will be the customers' measure of how your organization performs. The only true method for determining what your customers want is to *ask them.* Trying to guess at what they want or how they perceive you will waste valuable time and resources.

6.4 EXPECTATIONS AND PERCEPTIONS

The most proactive FM organizations demonstrate that we need to go beyond "traditional" needs assessment in order to get closer to our customers. In order to explore the value of learning about customers' expectations for the various services we perform in FM, we will want to expand the concept of needs assessment.

To develop a customer-focused service component to housekeeping services, for example, you need to ask your customers to describe their *perceptions* of how housekeeping services should be conducted within your company. After exploring perceptions within your specific company you should then follow up by asking customers to list their expectations for housekeeping services in *any* FM organization. Finally, you should have them list any housekeeping services your organization provides that surprised or pleased them beyond their expectations.

It is this combination of customers' predetermined expectations and their perception of your FM organization that governs their assessment of FM service quality. If there is a significant gap between customer expectations and customer perception, you have a service gap. If customer expectations are higher than their perceptions of the organization's abilities, you have a negative gap. However, if their perception matches their expectation you have customers that are satisfied. If their perception exceeds their expectation, you have the best of all FM service worlds in that your organization has delighted its customers. Whether your customers' perception is based in reality or not is totally irrelevant in this particular instance!

In addition to other customer groupings that you have identified, individual employees will bring their own set of expectations and perceptions about FM service with them. A "psychological contract" is developed between you and your customers[2] and, even if you do not use formal service agreements, all customers have a value system by which they judge how well your organization provides FM services. Listed below are some criteria that they will consider when evaluating their expectations for FM services:

- The purpose of the service being provided
- The degree of necessity for the service with respect to their jobs
- The degree of importance of the service
- Their view of the results of the service
- The relative costs of the service which may or may not be measured in monetary terms
- The risks involved

The following four factors, at a minimum, also contribute to the development of the customer evaluation criteria described above:

1. *Anticipation Factor:* A customer's expectation level is heightened by anticipation. Customers look forward to a certain outcome and develop a mindset based on anticipation. The FM organization sometimes unknowingly adds to their anticipation when the organization is vague about certain services and the way in which they contribute to customers' needs. If your customers are not given specific details about such things as service response, service levels, and service parameters, they will anticipate outcomes beyond your organization's control or ability to deliver!

2. *Previous Experience Factor:* How many times have you heard in FM that someone else did it differently and it was "better?" Customers will always have a yardstick of their own personal experiences either elsewhere in another company or with a former member of the FM organization by which they will measure the current service delivery. If their previous experiences, either on the outside or with your FM organization, were all positive, then they will expect the organization to completely satisfy their needs. If, on the other hand, some of their experiences were negative, the FM organization may be starting with a low level of expectation.

3. *Comparison Factor:* Customers all make comparisons of your FM service capability to some other FM service. Spouses, friends, and colleagues from other companies will all have their own stories and opinions about expectations for FM service. While not necessarily a fair comparison, it will influence their evaluation about how well your organization is performing. When your FM organization takes the time to understand what the organization is being compared with, there is a bigger pay-off for developing the most appropriate service delivery structure.

4. *Third-Party Information*[3]: This is the "bad news travels faster than good news" premise that we are all familiar with. Statistically, dissatisfied customers tell at least eleven people about their dissatisfaction, while only two or three people ever hear news of a satisfied customer. The FM business is particularly notorious for a third-party rumor mill about service. As an example, if you tell customers of an impending change in space configuration, they immediately come up with many horror stories of how difficult the experience was for other organizations within the company, and how other FM departments handled similar situations. This often-dangerous information has significant impact on how your customers form their opinions about your FM service organization.

Yet, even in the face of these customer challenges, customer satisfaction is possible and should still be a priority goal for a FM organization. As you can tell, we are building a customer database that will assist a facility manager in using the diagnostic approach to creating the most appropriate FM organization.

6.5 DEFINING QUALITY FM SERVICES AND STRUCTURING SERVICE DELIVERY ACCORDINGLY

Building a solid FM organizational structure that will meet customer requirements, perceptions, and expectations necessitates an institutional commitment to the concept of customer-defined quality. Without this commitment, an FM organization will spend an inordinate amount of time developing ideas about quality and service that may not necessarily be the same as customers' ideas about quality and service. *The customers' definition of quality is the only definition that should prevail in the minds of the FM staff.*

Too often, even in the current business environment, facility professionals talk as if they know what is best for the customer. In a true partnership, both parties must work together to determine need, define quality, and assess how an FM service adds value. It is critical for us to let our customers know that we are their partners in trying to achieve whatever goals they have established for themselves as a business entity. If an organization's senior management has told a business unit to increase sales, reduce overhead, and consolidate into smaller space, it is the job of the FM organization to partner with them to achieve that goal. In order to successfully achieve the articulated goals, it is incumbent upon the FM organization to ask the business unit what constitutes quality service in order to successfully achieve the articulated goals.

Not too many years ago, when facility organizations were in the formalization stages of their evolution, they had a fair amount of money and personnel, and thought they had a captive market. During that time, there was only one way to obtain facility services in most companies—through the facility organization. Most FM organizations had a monopoly since there were no alternatives available to the customer, which meant that customers had to accept the concept of quality as it was defined by the FM organization. At that time, the FM organization provided many services that did not meet customer needs or expectations, but the customer felt trapped. Because FM organizations had no competition, timeliness, quality, and appropriateness of service did not matter.

Today, however, the premise that the customer needs to define quality goes to the very core of quality facility services and must be a firm belief on the part of every facility professional. The concept that customers are the real judges of what constitutes quality in terms of the service processes, service outcome, and delivery structure is one that your FM staff needs to endorse wholeheartedly. It does not really matter if your organization thinks it is doing a good job, unless the customer can say the same thing.

6.6 INVOLVING CUSTOMERS IN THE SERVICE DELIVERY PROCESS

The provision of FM services truly is an interactive process in that from the moment a customer makes a request for service or asks for assistance in solving a work environment problem, the customer is involved in service delivery. Often the author equates FM service organizations with theatrical performers in that both are judged essentially by how well the live performance goes. Customers of FM organizations and stage actors evaluate performance "at the moment" it is taking place and there are numerous opportunities for mistakes to occur. Unlike a movie in which customers are not exposed to the takes and retakes of scenes that make the final production perfect, FM staff and stage actors are constantly "rehearsing" in front of the customer. They do not have an opportunity to stop the production and ask the customer to go away while they perfect their delivery by themselves.

Knowing that your customers are active partners and participants in the FM service delivery process provides the impetus for making them knowledgeable about your organization, its services, and the FM service staff. Marketing in general helps an FM organization direct customer involvement in a positive manner. There are four major components of a proactive campaign to actively involve FM customers in the delivery process:

1. *Anticipate and Stay Ahead of Customer Needs:* The FM organization must always be one step ahead of its customers by creating a customer database of requirements and needs that assists customers in making critical decisions about FM services. The creation of Customer Profiles is the vehicle that allows your organization to monitor customer trends and activities that guide FM decision making. If, for example, you know that one of your business unit customers has a product line that is seasonal, which requires the business unit to staff up with temporary help, you have valuable information to share with the customer. Since your historical trends data indicates that the business unit will need to reconfigure space on a temporary basis during the seasonal influx of business, the FM organization can anticipate this need and suggest alternatives to the customer in a proactive rather than reactive timeframe.

2. *Prepare Customers for What Lies Ahead:* All too often customers of FM organizations find themselves in a position of not knowing until the last minute that some FM organization event or activity will have an impact on their work environment. We need to involve our customers in our FM plans at a much earlier stage if we want them to react positively to our efforts and lend support throughout their organization. As an example, if we know two months in advance that we will be re-striping the parking garage for a two-week period and relocating all cars from a particular organization component to another level in the garage, we need to inform customers of our plans well in advance of the date. Advance preparation and information dissemination allows your FM organization to manage both customer reaction to an FM activity and the expectation about the outcome.

3. *Stimulate Interest in FM Projects/Activities:* When we undertake a significant project or activity, such as new facility construction, renovation of an existing building, or installation of a new infrastructure system that will enhance our work environment, we want our customers to share in our enthusiasm. The only way to accomplish this is to develop an active campaign to promote what we are doing. Our customers like to know what is happening and they like to receive reports on our progress. There is a reason why marketing experts recommend promotional activities in advance of a new product or service—to stimulate interest.

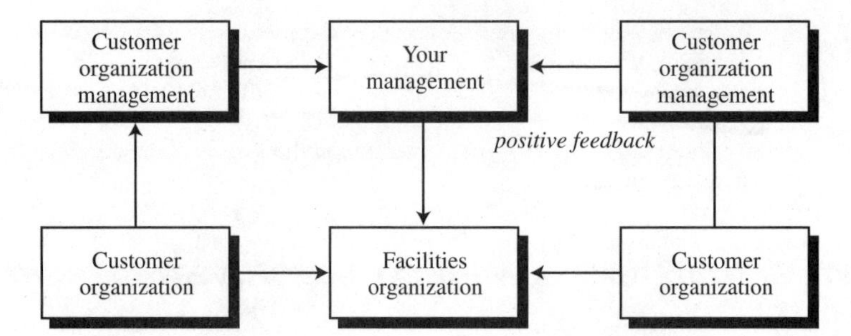

FIGURE 6.1 The Customer Role. *(**Source:** The Friday Group presentations on marketing and customer service.)*

4. *Prepare Customers for an Advocacy Role:* Our customers can be our biggest proponents or our worst enemies depending on how well they feel the FM organization is performing. We know that customers can be our allies, but they will not know that this is a role we want them to play unless we ask them. Your organization needs to remind customers that they should tell their peers and top management when the FM organization contributes something of value to their ability to achieve their business mission. Our customers often forget that we are in a business that is hard pressed to obtain compliments and kudos unless we direct them to do so. As advocates for FM, our customers play a critical role in what is described as the marketing chain. A graphic representation of the customer role appears in Figure 6.1.

6.7 ASSESSING CUSTOMER SATISFACTION WITH THE SERVICE DELIVERY STRUCTURE AND PERFORMANCE OF FM PERSONNEL

Assessing your customers' satisfaction level with your organization's services is a vital function for a healthy, effective, customer service relationship. When you are willing to have your organization measured against benchmarks and best practices, you are saying that you stand behind your efforts, recognize your shortcomings, are proud and aware of your strengths, and have a generally accurate sense of your customers. Subsequently, an unwillingness to measure your organization is a sign of insecurity. When not noted and judged against your standards, service is undefined, goals are intangible, and customer service is unknown and/or misunderstood. Thus, in today's competitive FM service environment, the healthiest of attitudes for an FM organization is to pursue being measured on a regular basis and use those measurements to establish your organization's service and performance standards.

The important things to consider when embarking on such an assessment are ensuring that your measurement system:

- Measures what is most important to your service
- Reflects the data accurately
- Can be done repeatedly without becoming so routine that its meaning is lost
- Can be administrated with no more than reasonable effort

Our approach to surveying customers is one developed with Dave Cotts and is used by the author with great success. It is based on the four basic principles of total quality management measurement: quality, delivery, cycle time, and waste. For FM organizations, those principles become:

- *Effectiveness:* How we get the job done from our customer's perspective
- *Efficiency:* The cost effectiveness factors which make us a "best-in-class" organization

- *Responsiveness:* How many times we complete a service in a timely fashion
- *Relevancy:* The extent to which the customers feel the services we provide are relevant and meet their needs

Customer satisfaction measurement instruments can vary widely within established frameworks, but they all can be used successfully to become the linchpin of an FM organization's structure, policies, procedures (personnel performance), goals, and trends data. An important decision is determining which services need to be measured daily, periodically, and annually.

You must also be sensitive to the dangers of over-surveying your customers. Using the same instrument too frequently, using an out-of-date instrument, or simply inundating customers with questions will backfire on you. Starting out with too extensive a questionnaire might discourage your customers on the very notion of surveys. It is better to begin by erring on the side of caution. It is also vital from the customer standpoint to have very relevant, concise, and unambiguous questions and concerns on your instruments. If customers ever suspect that you are wasting their time, this will affect your responses negatively, as well as allowing negative factors to influence this important relationship.

FM customer measurement systems need to incorporate the knowledge gained from the professional overall. Prior to developing survey formats for customers, you need to review "state of the industry" by analyzing the latest trends in customer service and other best practice instruments.

6.8 LINKING CUSTOMER FEEDBACK TO THE CONTINUOUS SERVICE DELIVERY IMPROVEMENT PROCESS

How facility managers link customer feedback on FM services and service delivery to efforts to strengthen the FM organization is the key to a continuous improvement strategy. In our quest to build facility managers' capability to implement a diagnostic approach to organizational development, our discussion about continuous improvement is a cornerstone for FM organizational success.

Customer feedback is designed to be used to strengthen an organization and provides no utility to facility managers unless the information is linked directly to the process of setting staff goals, as well as in the evaluation criteria for staff performance. Gathering customer feedback is only valuable to an FM organization if it is institutionalized as part of an annual planning process that ultimately is fed into the development of organization and staff objectives. In other words, collecting customer data on satisfaction with FM performance is not a one-time event, but part of an on-going process to improve service delivery.

To be fully utilized, customer satisfaction data should be translated into specific performance objectives for each and every staff member within the FM organization. Not only should FM staff have a generic performance goal of obtaining positive feedback on customer service attributes such as attitude, knowledge of FM services, policies/procedures, problem-solving capability, etc., but also specific performance objectives related to the timeliness, responsiveness, and quality of specific services performed by the individuals.

ENDNOTES

1. Friday, S., and Cotts, D. G. 1995. *Quality Facility Management: A Marketing and Customer Service Approach.* Wiley & Sons.
2. Sellerman, S. W. 1969. *Management by Motivation.* New York: American Management Association, Inc. p. 51.
3. Friday, S., and Cotts, D. G. 1995. *The Facility Management: A Marketing and Customer Service Approach.* Wiley & Sons.

CHAPTER 7
DISASTER RECOVERY PLANNING

L. David McDaniel
Chief Scientist
BMS Catastrophe Special Technologies Division
Fort Worth, Texas
E-mail: dmcdaniel@bmscat.com

7.1 INTRODUCTION

A disaster is defined as any event that creates an inability on an organization's part to provide critical business functions for some predetermined period of time. After a disaster strikes, the facility manager is central to the recovery process. He or she is expected to stop the damage—stabilize the loss. The facility manager will be expected to assess the damage, provide emergency resources and personnel, restore or rebuild the structure, recover the contents, and accomplish all of this within the time parameters defined by the business impact analysis, minimizing loss. In the case of a regional disaster, interaction with emergency personnel from the public sector will be needed. Access to your building can be controlled by public emergency response or law enforcement personnel. The contingency plan (also called business continuity plan, disaster recovery plan, or crisis management plan) is your road map to recovery. It defines the actions, resources, and personnel required for immediate response. Knowledge of your corporate disaster recovery plan is imperative. Many corporations have relegated the responsibility for generating and maintaining the plan to the facility manager. Often the disaster recovery plan is generated by risk management, safety, security, or human resources. It may exist at corporate headquarters. If you are not responsible for development, locate this plan so you can review it and can give your inputs.

7.2 BUSINESS IMPACT ANALYSIS

Unfortunately, corporations tend to write plans based on the probability of a specific disaster. What are the chances that a tornado, hurricane, flood, or explosion will hit them? Pretty infinitesimal at best. I would recommend that the mindset be planning not for the probability of a disaster, but planning for the eventuality! Something *will* occur to interrupt your business. The contingency plan priorities are set by a careful analysis of the cost per day of business interruption. At this juncture, no thought is given to the cause of interruption. This business impact analysis (BIA) is an intensive accounting exercise. Departmental priorities are set based on the cost analysis of the loss of departmental function. Does the loss of accounting cost the firm more than the loss of the call center or does a data processing loss impact more? Manufacturing organizations look at loss of facilities and the overall effect this loss has on the corporation. Single function loss is analyzed, as well as combinations and permutations. For seasonal businesses, time will be a factor. Accounting firms are very busy when federal income taxes are due, retailers have their biggest season around Christmas, etc. Disasters have an uncanny way of occurring at the worst possible time. With an understanding of the costs involved, the time required to restore the business before it is defunct is determined, as well as restoration priorities for different portions of the organization. This analysis focuses the proactive risk minimization efforts for the organization.

7.3 THE PLAN

At this stage, it is time to assess the risks faced by the facility. What natural disasters could occur? Depending on the geographical location there are tornadoes, hurricanes, flooding, lightning, tsunamis, forest fires, volcanoes, earthquakes (see Figure 7.1), and mudslides. What if a fire occurs? Who are your neighbors and what risk does your proximity to them raise? The fire department can provide you with any businesses with hazardous materials that might impact your facility. No business facility is immune from an explosion today (see Figures 7.2 and 7.3). In Oklahoma City, the attack was meant for the federal law enforcement agencies, not the childcare center. Proximity to major highways, railroads, and airports similarly bring risks. Utility failures, limited access for ingress and egress, communication cables, single points of failure, and loss of critical supplier or outside services should also be considered. Winter storms (see Figure 7.4) and civil disorder can close businesses. These risks can be handled proactively or reactively. Window film can reduce flying glass. Separating the UPS from the data center or separating the office battery supply (see Figure 7.5) from the switching equipment can prevent computer contamination if a battery shorts and overheats. Storing vital documents on the second floor can prevent flooding. A contingency plan for each occurrence should be drafted and tabulated.

Initially, DR planning began in the financial services sector. The plan was quite simple—identify your hot site vendor and sign up. If anything happens, grab the phone, declare, and move operations to the hot site. This plan is hardly sufficient for other business units that may incorporate manufacturing and distribution. In this case, the recovery must be done at the loss site. If a disaster happens in San Francisco, you cannot move San Francisco to San Diego for recovery. The plan

FIGURE 7.1 In an Earthquake, Unattached Files Will Be a Source of Great Distress.

FIGURE 7.2 An Explosion is a Fire that Occurs in Milliseconds. Here Soot Was Deposited on the 97th Floor by the Supersonic Forces Generated.

FIGURE 7.3 After the World Trade Center Explosion, Electronic Equipment on All Floors Required Restoration.

FIGURE 7.4 Heavy Snow Loads Allowed to Build on the Roof Can Lead to a Structural Collapse.

FIGURE 7.5 Batteries for UPSs and Office Power Should Be Isolated from the Rest of the Electronic Equipment to Prevent Corrosion from a Faulty Battery's Emissions.

should consider the essential functions within the organization and where they could be relocated, perhaps displacing other functions not so critical. In any case, you as the facility manager are charged with the restoration/recovery of the existing facilities. Vital outside suppliers and services should also be identified. Alternate sources must be identified. You should be assured that these organizations are also ready to recover if a business interruption occurs within their organization. For critical equipment, plans should be made to have the ability to replace this equipment within the time parameters defined in the BIA.

The plan must prioritize the functions (departments or divisions), equipment (see Figure 7.6), and documents which are critical to the operation of your business. Only you can define which subset of documents are vital and their location should be identified. The documents that are critical for

FIGURE 7.6 In a Production Environment, High Priority Critical Equipment Must Be Identified in the Disaster Recovery Plan for Immediate Attention.

the survival of the business will normally comprise 5–7% of all documents. They may be in the form of paper, microfilm, microfiche, or magnetic media. The most effective plan is corporate-wide with the support and backing of top administrators. To be useful, the plan must be simple and readable. Major decision tree pages should be marked with tabs, with the first on top. The response requirements should be no longer than one page with a simple bullet format. The plan must be made available to responsible employees and it must be reviewed, updated, and tested on a regular schedule. The top level of the plan deals with life, health, and safety issues, then stabilization and preservation of assets. The plan should provide for business continuity and an orderly return to normal operation.

The public response to a regional disaster is the incident command system. The local, regional, or state emergency operations center (EOC) is activated and the incident commander is in charge. Security perimeters are set up and maintained by local, county, state, and federal law enforcement personnel or National Guard troops. Access to the affected area is limited. Whether or not you are allowed entry is defined by your personal relationship with the public authorities and what arrangements you have made to interact with the EOC. To gain entry, you can act either proactively or reactively. Researching the public response disaster plan would expedite reentry. Who will be the incident commander? Where will the EOC be located? Which public officials will man the center? Check with local civil authorities and incorporate the public response with your plan. Local fire departments are a good place to start. You will need to be able to identify your response team, the location/locations where access is required, the functions to be performed, and the time required.

7.4 THE TEAM

Equally important is identifying the critical personnel who must be involved in a recovery. Just to assess the seriousness of any situation requires administrative, operations, facility, security, and expert personnel. Administrative personnel should include a person with the authority to declare an emergency and mobilize resources, a person with intimate knowledge of the insurance coverage, a person who knows the operations requirements, and a person who can write checks to provide emergency funds. Accounting should set up special case numbers to capture the cost of materials and services required in recovering the business. Internal labor should also be captured. All contractors must provide detailed invoices. The success in filing an insurance claim is dependent on this documentation. One individual should be named disaster recovery coordinator (quite often the facility manager) who manages the restoration/recovery project and another individual should be responsible for internal communications—activating the telephone tree. A single individual should be assigned to handle all questions from the local media and represent the company publicly. All other personnel should refer any queries to that individual. The facilities personnel know the building layout and location of utilities. They are frequently first on the scene, and must decide if escalation is required or if the problem can be handled internally. Security personnel will be needed to establish perimeters and protect assets. Often an outside security firm can be contracted to provide the increased levels of security needed. As soon as practicable, security needs to perform a sweep of the building to take custody of cash, jewelry, and other valuables. Expert personnel will include disaster restoration personnel who can determine what equipment, etc., is restorable, recommend steps to stabilize the loss until restoration can occur, and provide trained personnel dedicated to your recovery (see Figure 7.7). Other experts that might be required are structural engineers to determine the structural integrity of the building. Do not forget to define alternates. What would happen to your plan if one member of your team is on vacation? This composite group constitutes the disaster assessment team who will decide upon the appropriate level of response and will manage the restoration project. The actions will be implemented and the decisions communicated to the employees. Remember that you may have to do all of this notification/mobilization without access to your building. In case of a serious fire, the average length of time before you are allowed to enter the building is 48 h. If hazardous materials are involved or if the building is designated a crime scene, the time involved may be weeks. Notify your insurance broker immediately.

FIGURE 7.7 Precontracting with a National Restoration Company Can Bring Dedicated Personnel and Equipment in the Case of an Emergency.

In the case of a regional disaster, the normal communications infrastructure may be damaged or non-existent and key employees may be tied up at home making sure their family is cared for first. One solution to this problem is to contract with a national restoration firm so that their response is assured. This way, trained restoration technicians and equipment will be dedicated to stabilizing your building. The same is true of the other critical outside services you may require.

Since the priority of any plan is life, health, and safety (see Figure 7.8), this area should be placed foremost in the plan. Evacuation plans, safety shelters, meeting points, and medical aid should all be provided. Special attention is given to the means of communicating with employees through alarms and voice annunciation systems. These systems should be on emergency power systems. As gruesome as the details may be, the plan should include response to fatalities and serious injury. Do not forget to include plans for disabled employees.

FIGURE 7.8 If Lower Boxes in a Distribution Warehouse Are Exposed to Water, the Stacks Become Unstable.

FIGURE 7.9 If Untreated, Electronics Exposed to Heavy Concentrations of Acidic Soot Will Be Ruined within Hours.

7.5 WHAT TO DO?

Businesses today rely upon a complex combination of paper documents, microfilm, scanners, terabyte data storage systems, readers, printers, computer systems, and communication equipment. The critical computer hardware and the media containing vital documents must be stabilized to prevent loss. A fire often generates aggressive acidic corrosive gases which will be an integral part of the smoke. The smoke is analogous to fog and will condense on surfaces below dew point temperature in the building and contents. In this condition, bare metals will corrode. If electronic hardware is allowed to corrode, restoration may be impossible. Quick response is of the essence for stabilization (see Figure 7.9). Insurance companies pay more money each year for smoke damage than for thermal losses. Hurricanes often deposit salt water on equipment when the roof is damaged. The salts cause similar corrosion (see Figure 7.10). Floodwater may contain chemicals that generate the same damage.

FIGURE 7.10 Corrosion of Equipment Occurs Rapidly after Exposure to Sea Water Carried Inland by a Hurricane.

You, as the insured, are legally responsible for loss mitigation (stabilization) and you must take all reasonable and prudent precautions to minimize the cost of the loss. This must be done immediately. Service contracts and warranties are voided by the exposure of equipment to "a hostile environment." The process of recovery has three phases. The first is stabilization—stopping the damage. The second step is restoration—returning the equipment to pre-loss condition using appropriate cleaning protocols. The final phase is recertification. In the case of equipment this means tests, diagnostics, and service as required to reestablish the warranties or service contract. In the case of the building, the recertification is done by employing a certified industrial hygienist (CIH) to run the appropriate tests to recertify the building for re-occupancy. The building recertification is especially important in office buildings. The CIH should be made available to the office personnel to answer any particular questions that arise during re-occupancy.

The disaster recovery plan should define the steps for stabilization, and the equipment and materials required must be acquired and stored. The basic steps of mitigation are covered in the following subsections.

7.5.1 Electronic Hardware

The first warning signs will be the corrosion of mild steel hardware components. Objective chemical tests can be run to determine corrosion potential. *The most effective means of corrosion control is to lower the humidity to reduce the reaction rate.* Remove water. The steps are as follows:

WARNING: DO NOT ENERGIZE ANY WET EQUIPMENT—REMOVE POWER

- Open cabinet doors, remove side panels and covers, and pull out chassis drawers to allow water to run out of the equipment.
- Remove standing water with wet vacuum cleaners. Use low-pressure air (50 psi) to blow trapped water out of the equipment. Absorbent cotton pads (diapers?) can be used to blot up water. Use appropriate caution around header pins and back-plane wire wrap connectors to avoid bending.
- Vacuum and mop up water under any raised computer room floor.
- Equipment that contains open relays and transformers will require a special bake out before application of power.

7.5.2 Industrial Equipment

- Includes lathes, mills, other machine shop equipment, plastic injection molding machines, and other production equipment with precision ways and close tolerance metal surfaces.
- The corrosion process requires moisture, so the bare metal surfaces should be lightly coated with a petroleum-based lubricant. This process will have to be repeated as the lubricant evaporates.
- Tenting with desiccant dehumidification inside will also help protect the equipment from further damage.
- Complete restoration—removing the contaminants should begin immediately.

7.5.3 Magnetic Media

One important asset that must be preserved after a disaster is the critical data on magnetic media. Media that has been exposed to contaminants should be examined by a professional before any attempt is made to use them. If an attempt is made to use a floppy disk with hard particulate matter on the surface, damage to the oxide layer may destroy data as the floppy spins. Tapes must be dry and clean before any attempt is made to copy the data. Hard disk data can be partially saved—even after a head crash. This process is very labor intensive and requires special equipment in a clean room. Contaminated media is replaced with clean media. Restoration of data is a process involving the emergency cleaning of the media so that data may be copied onto other media. The original media will be discarded (or archived).

- If exposed to flood waters, keep tapes wet until they can be restored. Use *Ziploc* bags, pack in a plastic lined box, etc. Keep in a cool area (<65°F).
- Tapes must be cleaned within two weeks to avoid fungus growth.
- Do **not** attempt to dry with heat!!
- A 95–100% success ratio is possible. Predicated on 72–96 h response.

7.5.4 Microfilm, Unmounted Microfiche, X-ray Film

The most important thing to know about microfilm is that *once the film is wet do not let it dry!!* The film must be processed while still wet or the gelatin coating will stick to the next layer and the document information will be torn from the film. Here again, speed is of the essence.

- For short time storage, five gallon buckets can be used to store film with enough (preferably distilled) water to cover the film. *Ziploc* bags or *Saran Wrap* can also be used to package the film and prevent drying.
- Use gloves when handling wet materials and wash hands thoroughly to prevent infection from flood bio-contaminants.
- For longer storage than a few days, a conservator must add special gelatin hardening chemicals to the water.
- To minimize damage, store wet film in a refrigerated area with a temperature of 35–40°F.
- For long periods of time, film may be frozen to preserve it. It should never be freeze-dried, but thawed and wet processed.

SPECIAL NOTE (According to Kodak)

> Because the original quality level of film cannot be restored, the word "restored" is used to mean the microfilm is returned to a usable state so that it may fulfill the purpose for which it was primarily intended: the storage and retrieval of information. Its quality and appearance, however, will never equal that of the damaged film.
>
> In all restoration projects, it is important that you are aware that complete restoration cannot be guaranteed and that restored film *will not be of archival quality*. When archival quality is required, a silver duplicate film print must be made.

7.5.5 Optical Media, Magneto-Optical Media

- If optical media is wet, keep wet until it can be cleaned (similar to microfilm).
- Take care not to scratch the surface of the media.

7.5.6 Prints and Negatives

- Again, treated similar to microfilm.

FIGURE 7.11 Flooded Books Must Be Washed and Frozen to Prevent Further Damage.

7.5.7 Paper

If paper is wet (see Figure 7.11), the following steps are recommended:

- Keep air moving, lower humidity, lower temperature to avoid fungal growth. An industry expert, Dr. Peter Waters, states "To leave such materials more than 48 h in temperatures above 70°F and a relative humidity above 60% without good air circulation will almost certainly result in heavy mold growth and lead to high recovery and restoration costs."
- Freeze wet documents as quickly as practicable.
- Separate blocks of documents before freezing as required with plastic for possible later removal and segregation.
- If exposed to flood waters containing mud and silt, the document should be washed and sanitized before freezing.
- NOTE: The most effective means of restoration for paper records is freeze drying in a vacuum chamber. Sublimation dries the documents without having liquid water attacking the inks, surface finishes, and binding glues. This is the *only method* for glossy finish stock.

7.5.8 Microfiche (Aperture Card Mounted)

- Similar to paper. First freeze dried, and when dry, the film is cleaned and remounted if required.

7.5.9 Building Structure

- Wet buildings (see Figure 7.12) must be properly dried to avoid damage and microbial growth. Ever-present mold spores will bloom and grow in two days with the right conditions (see Figure 7.13). See Appendix A for more information. To discourage growth, decrease humidity, circulate air, lower the temperature, leave lights on.

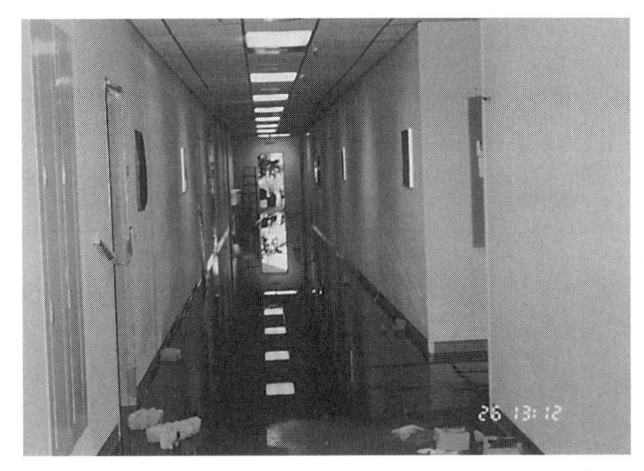

FIGURE 7.12 Two Inches of Water in a 20,000 Square Foot Facility is 25,000 Gallons.

- With the exceptions of musical instruments or museum-archived collections, the building cannot be dried too quickly. The faster the water removal, the less the damage. All water-soaked ruined contents should be removed and discarded before effective dehumidification is possible. In general, the most effective way to dehumidify is with large desiccant units sized to provide at least one air interchange per hour. Air movement is provided with turbo-fans.
- Proper instrumentation should be used to determine water content of walls, ceiling, and sub flooring. Progress of dehumidification is measured by monitoring humidity ratio, not relative humidity.
- Before return to the building, re-occupancy certification by a certified industrial hygienist will curtail many potential health concerns that tenants or employees may have.

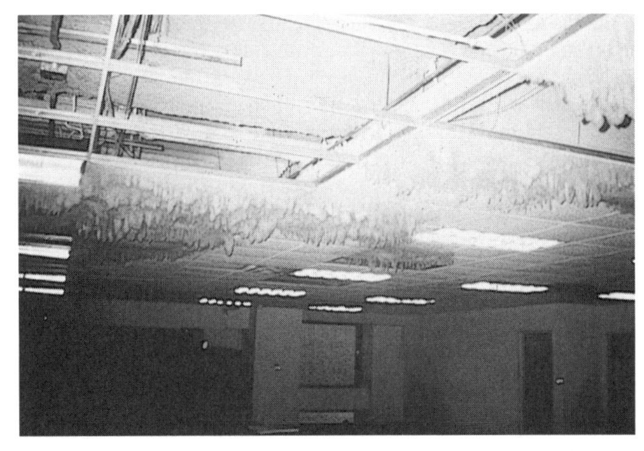

FIGURE 7.13 Fungus Growing on Water Saturated Ceiling Tile after Two Weeks Would Have Been Much Easier to Prevent Than to Clean.

7.6 CONCLUSION

At this point in the disaster recovery, the loss site is stabilized and the critical portions of the company are operating according to plan. As required, the employees are working at a hot site, alternate facility, or alternate building space. Communications have been rerouted, all critical information systems are operational, and we are ready for the next step: demolition, cleaning, restoration, replacement, and reconstruction as needed. This portion of the plan is too often only skeletal. The planning effort is focused on business continuity and the restoration and return to the original facility is not planned in detail. The business priorities defined should be followed and the recovery plan should provide for an orderly return to the original facility with no business interruption. This part of the project provides closure for the loss. The final result is that the company has returned to predisaster operation.

APPENDIX A A FRIGHTENING FUNGUS

Used by Permission of Fort Worth Star Telegram, May 23, 1997 by Karen Brooks

Health officials were puzzled by the illness of a suburban Dallas woman. For about two weeks after moving into a condominium, she had suffered flu-like symptoms. The "neighbors found her crawling around on her hands and knees complaining of earthquakes," said Ken McBride, a regional industrial hygienist with the Fort Worth-Tarrant County Public Health Departments.

Health officials investigating the case six years ago found the cause of her illness: fungus. Air-conditioning ducts at her condo were layered with mold.

"She was totaled. Her system was overwhelmed," McBride said. "She was in the hospital for 60 days before they got her cleaned out."

Public knowledge of fungi is replete with images of mushrooms, athlete's foot, and yeast infections. But researchers say that fungi carry a deadly potential, as well as the promise of health benefits. That's why the scientific community is reaching for information about one of the world's most ancient and mystifying life forms.

"In the last 10 years, mycology, the study of fungus, has become one of the most critical in all of medicine," said Michael Rinaldi, a mycologist at Audie Murphy Veterans Affairs Hospital in San Antonio.

In recent weeks, the fungus stachybotrys was found at some Birdville district schools after parents reported mold and mildew odors and allergic symptoms in their children. Health officials said the children were not in direct contact with the fungus and that there was no health threat.

The same type of fungus has been linked to illnesses among infants in Cleveland in 1992 and 1993. Twenty-one infants from extremely damp homes were hospitalized with bleeding lungs. Three died. The federal Centers for Disease Control and Prevention, which usually doesn't track fungal infections, is considering studying the fungus, officials said.

How much of a threat are fungi? The subjective nature of immune systems makes it difficult to nail down exposure limits, researchers say.

"Everybody has their own unique susceptibility in response to fungal spores—that's one of the reasons nobody has said, 'This is the dose of spores that will cause reaction,'" said Robert Garrison, a mycologist with IAQ Consultants, an industrial hygiene company in Southlake that helped investigate the Birdville incident.

Infants are more susceptible to infection because their immune systems haven't fully developed, immunologists said. The elderly tend to have weaker immune systems, too. People are more susceptible to a fungus when it is in a building. Asthma patients who are being treated with steroids are also susceptible, said Dr. Justin Radolf, associate professor of internal medicine and microbiology at University of Texas Southwestern Medical Center in Dallas.

Ironically, medical advances in treating diseases such as polio, cancer, and AIDS have created a population suddenly susceptible to unusual fungal infections.

"Anybody who is immune-suppressed, has cancer or is undergoing therapy for cancer, has had transplant surgery, HIV/AIDS, leukemia, lymphoma, Hodgkin's disease, hemotologic malignancies, people born with defects in their immune system, diabetics . . . those are the people that are at risk for fungal infections," Rinaldi said.

That's why the study of fungi has come to the forefront of medical science, he said.

"Fungi that were never connected with human medicine before—these are normal environmental creatures—we are now seeing that these fungi are killing these immune-suppressed people," he said.

Some fungi can cause asthma and bad breath. Some make fingernails fall off. Some can eat away a backbone or destroy brain capacity. Other times, fungi can save lives. Penicillin and other major antibiotics are produced by fungi, Rinaldi said.

Yet for the most part, fungi exist undetected after they make contact with the body. More than 200,000 species of fungus have been identified—only a tiny percentage of the variety—and the typical healthy individual is usually not at risk.

The constant exposure to small doses of fungi is beneficial to healthy people because, much like a vaccine, they build up the immune system.

"Every day of your life, you're walking around literally in a sea of microbes—you breathe them, they're on your hands, on your clothes, in the food that you eat," Rinaldi said. "And yet very few people die of infectious diseases, comparatively."

Most fungal infections can be treated with antibiotics, he said. But that doesn't mean they'll be eradicated soon.

"One reason they're such hardy survivors is that fungi rarely go hungry."

"There is very little on Earth that a fungus can't or won't eat," Rinaldi said.

"If you happen to wear eyeglasses and you set them on the ground and come back a year later, not only will the fungus have eaten the rims, it will have eaten the glass lenses, too."

"Compared to that, human tissue becomes . . . almost like dessert."

APPENDIX B PRE-LOSS RISK ASSESSMENT CHECK LIST FOR ELECTRONIC EQUIPMENT FACILITY

1.0 *Fire Suppression And Detection Systems*

- **1.1** Sprinkler Systems
  - **1.1.1** Water flow sensors—to detect flow and cut off power to EDP equipment to prevent electrical damage—precharged sprinklers. Controlled power-down sequence.
  - **1.1.2** Under computer room floor water detectors—to warn of impending water threat from flood, plumbing leaks, etc.
  - **1.1.3** Sprinkler heads in HVAC ducts—to prevent spread of fire.
  - **1.1.4** Sprinkler system reservoir charged with compressed air with flow value to limit accidental discharge of water.
  - **1.1.5** Routine use of EDP equipment and media storage covers during periods of non-use—to protect from accidental water exposure.
  - **1.1.6** Routine electrical shutdown of equipment during periods of non-use—electrically active equipment exposed to water may require extensive repair and retesting.
- **1.2** Halon Systems
  - **1.2.1** Back-up with sprinkler systems
  - **1.2.2** Needs to be checked routinely for facility containment duration. ASTM Door Testing.

1.2.3 Consider replacement with sprinkler system as Halons become more expensive and limited in use due to ozone depletion potential and Montreal Protocol.

1.3 Facility Modifications

1.3.1 Compare original facility layout with current layout—note all changes in wall locations, high shelf storage arrangements, types of material stored, new room areas, etc.

1.3.2 Re-evaluate fire suppression and detection system for suitability with facility modifications.

1.4 Fire Detection System

1.4.1 Location of detectors in high-risk areas—check all new areas for adequate coverage.

1.4.2 Check program for follow-up of manufacturer's instructions for periodic operational test and certification of detector system.

1.4.3 Upgrade of sensor devices—are current sensor devices properly matched with current risk? New technology—HCl and fuel leak sensors developed by BellCore Labs. Smart sensors.

1.4.4 Location of alarm system—external sites to alert area supervisor—use of phone lines to reach home locations.

2.0 *Facility Concerns*

2.1 Presence of dedicated HVAC systems for critical areas—containment and control of external fire by-products—prevention of cross contamination.

2.2 Use of HVAC duct located fire baffles and sprinkler heads to limit spread of fire.

2.3 Use of flame-retardant floor, wall, and ceiling materials in Computer Center construction.

2.4 Separation of Computer Center from high-risk manufacturing areas.

2.5 Location of Computer Center in high-risk flood areas and/or areas accessible to non-company individuals.

3.0 *Backup Capability*

3.1 Electrical power and services

3.1.1 Requirements for diesel generators and/or battery backed power supplies for backup to critical operations—EDP centers, communications, HVAC, services, maintain security systems, etc. Planned wiring/switching for external generators.

3.1.2 Location critical—separate UPS from data room, flood threat to generators?

3.1.3 Availability of auxiliary HVAC/dehumidification equipment to control post-loss environment.

3.1.4 Contacts with common carriers and power companies to provide for emergency backup services. Continuation of brokerage services, remote access capabilities, other financial system links.

3.2 Magnetic Media Backup

3.2.1 Routine program for daily media backup. Use of cartridges instead of reel to reel. Streaming tape media for PC hard disk back up recommended.

3.2.2 Remote off-site storage of backup media over 24 hours old. Virtual Private Network wide-band data vaulting to a secure vendor to avoid transporting tapes.

3.2.3 Availability of formatted blank copy and storage media for all EDP users at worksite.

3.3 Backup EDP equipment

3.3.1 In-house availability of similar EDP equipment and software. Specialized MICR equipment.

3.3.2 Contacts with external hot/cold sites and time share organizations for EDP services.

3.4 Backup Communications Equipment

3.4.1 Cellular phones, direct lines (convert FAX/MODEM lines), satellite link.

3.4.2 Radio

4.0 *Pre-Loss Assessment of EDP Assets*

4.1 Identification of EDP physical assets

4.1.1 Inventory listing by type, manufacturer, model, and serial numbers, date of purchase, replacement costs, unique configurations.

4.1.2 Inventory of software including purchased and custom programs—phone numbers, s/n.
4.1.3 Identify critical data/documents—source and archival documents inventory.
4.1.4 Copies of CSR sorted by DS3 and circuit number. Other outside network routing information. Equipment software configuration.
4.2 Location and preservation of vendor warranties, service maintenance policies, and third party service and repair policies, system configuration documentation.
4.3 Prioritization of hardware and software recovery processes. Specialized processing equipment: MICR, microfilm, microfiche, networks, security, etc. What replacements are available for this equipment should it be destroyed?

5.0 *Identification and Review of Outside Loss Recovery Services*
5.1 Restoration of media and/or data recovery from damaged magnetic media.
5.2 Hot/cold sites for EDP service backup.
5.3 Corrosion control and clean-up of water exposed equipment.
5.4 Environmental stabilization and moisture control.
5.5 Detection, analysis, and removal of hazardous materials.
5.6 Clean-up of EDP equipment and facilities.
5.7 Recovery of water-damaged source and archival documents.
5.8 Supplementary security personnel.
5.9 Experts—structural engineer to determine damage, archivist, CIH, others.

6.0 *Formula Disaster Recovery Plan*
6.1 Updated plan in place
6.2 Accessibility of plan to key employees. These employees should be identified and notified. Check list quarterly.
6.3 Testing of plan for feasibility and completeness. Fire scenario. Regional disaster scenario.

ENDNOTES

1. BMS Catastrophe. 1999. *Disaster restoration guide for disaster recovery planners*, 9th Revision. Fort Worth, TX: Special Technologies Division.
2. National Archives and Records Administration Instructional Guide Series. 1996. *Vital records and records disaster mitigation and recovery*. College Park, MD: Office of Records Administration.
3. Nyberg, S. 1987. *The invasion of the giant spore*. SOLINET Preservation Program. Leaflet Number 5. 1. (November).
4. Van Bogart, J. 1995. National Media Lab. *Recovery of damaged magnetic tape and optical disk media*. Presentation at Library of Congress Symposium. (September 21).
5. Waters, P. 1993. *Procedures for salvage of water damaged library materials*. Extracts from unpublished revised text. Washington DC: The Library of Congress. (July).
6. Reagor, B. T. 1991. *Smoke corrosivity: Generation, impact, detection and protection*. The Journal of Fire Science's Colloquium on Smoke Corrosivity, Baltimore, MD. (November).
7. An extensive library of Disaster Recovery Planning information resides at the Disaster Recovery Journal website http://www.drj.com and Contingency Planning & Management at http://www.ContingencyPlanning.com.

P • A • R • T • 3

ANALYSIS AND DESIGN

CHAPTER 8 ALTERNATIVE WORKPLACES

8.1 Introduction

To date, much of the alternative-workplace discussion has focused on space-planning standards and making them more flexible, responsive, and cost efficient. Alternative-workplace development is so much more. The objectives of this chapter are to represent the broader aspects of alternative work so that facilities managers can gain perspective, understand the components, and better integrate their service deliveries to their internal customer—business management and employees. Specific corporate and service provider examples are given so you can see these programs in practice.

8.2 Rethinking the Traditional Office

The evolution of alternative workplaces is examined. Work in the 1990s centers more around flexibility and innovation than production and sameness, so the traditional workplace model is outdated and too rigid. This section demonstrates why new models are better at facilitating work.

8.3 On-Site Alternatives Require More Than Changing Existing Space Standards

On-site alternatives require more than changing existing space standards. See what the industry experts have to say about the following:

- A new service delivery process for facilities planners
- New business drivers and needs
- Expanded corporate office space resources
- Expanded services beyond space planning
- How best to facilitate change in the transition

8.4 Planning for Virtual Services in Corporations

This section explains the "virtual" concept and the effects on space planning for facilities managers. Among the topics covered are software and application developments, strategy resorts, staff reduction, and corporate outsourcing.

8.5 Establishing Best Practices for Remote Work

To obtain the best alternative-workplace results, remote work practices should be deployed first. Remote practices include business management planning, supervisory administration, and employee operating practices. Specifically, employee operating practices refers to provisioning and support of work tools (remote access, last-mile connectivity, remote products, computers, software, furniture, etc.), corporate security practices, expense reimbursement conditions, home office planning, administration, etc. Find out more about all of these remote work practices in this section.

8.6 Business Centers—An Alternative Branch Office Strategy

The resources required to support and evolve a distributed workforce are primarily technology, administration, and training. Not only are business centers effective for locating office resource centers near customers and/or employees, but also for providing low-risk investment when corporations partner with domestic or global providers.

CHAPTER 9 FACILITIES CONDITION ASSESSMENT

9.1 The Facilities Audit

The facilities audit is an essential tool for the determination of conditions of an organization's capital assets, including buildings, infrastructure, and fixed and movable major equipment. This section describes the audit process purposes and goals and presents the case that the audit involves more than data collection; an audit program is an ongoing management tool for capital planning and management.

The audit process section provides an overview of the facilities audit process. An audit should be a comprehensive *quantitative* and *qualitative* assessment, collecting information on *adequacy* of facilities to support mission and program. The *quantitative* assessment evaluates the physical condition of building and infrastructure components; the *qualitative* assessment evaluates the capability of a facility to support a program through comparison to a common set of *quality* standards.

The audit phases section describes the four phases in the facilities audit process: 1) designing the audit; 2) collecting data; 3) summarizing results; and 4) presentation of findings. In this systematic approach, the scope of the audit is first determined, audit team selected, and inspection planned. In the second phase of the audit, data is collected through inspection of buildings and infrastructure and functional performance evaluations. In phase three, the data collected is evaluated, summarized, priorities set, and future audits planned. In the final phase, results are presented to various audiences.

In adapting the audit, we learn that although maintenance staff may collect and analyze some information to ensure the performance of buildings and infrastructure components, they do not typically collect planning data to assess long-term needs. Using the audit procedures in a systematic manner, either by in-house staff or by consultants, a database of information on facilities is produced that can be used as a base line for future condition inspections.

9.2 Technology for Capital Planning and Condition Assessment

This section introduces aspects of technology for capital planning and a comprehensive facilities condition assessment. Computer technology enhances the often intensive data management involved with a facilities condition assessment program. Technology can also automate the analysis of collected data, enhance an organization's ability to communicate findings, and also lower first-time costs of data collection involved.

The section on FCA technology overview: databases and the Web shows that a desirable goal for effective facilities management is an information system integrating space management with maintenance work planning, major maintenance, capital renewal, and deferred maintenance reduction projects. The combination of databases and Web technology has broad applications for facilities management and planning within an organization. This section describes the range of technologies available and imminent to automate the facilities condition assessment process and to assist in the subsequent financial decision-making.

CHAPTER 10 THINKING GLOBALLY—THE COMPETITIVE EDGE

10.1 Introduction

This chapter examines the potential challenges and opportunities presented to the facility manager by globalization. Globalization has so great an impact, for both transnational firms and local companies, that if you fail to realize the broad implications of globalization for your firm, you will not remain competitive for long.

10.2 Changing Political Boundaries and Interaction between Nations

The balance between nations is increasingly dominated by a single superpower—the United States. However, corporations exceed many governments in wealth and power. This section highlights the dynamics of the changing global power structure and the possible ramifications for the facility manager.

10.3 Changing Economic Boundaries and Interaction between Companies

Competition is no longer between nations, but between transnational corporations. Furthermore, global entities (e.g., The IMF, WTO, and the World Bank), rather than national entities, are now the dominant factor establishing global economic policy. This section describes some of the objectives and methodologies of these new global entities.

10.4 Changing Technology

Speed, not size, is now the discerning aspect of the successful company. This section identifies possible impacts of "fast" buildings for the facility manager.

10.5 Changing Social and Environmental Awareness

A growing and pervasive concern for social and environmental issues—a concern commonly shared by a company's employees and customers alike—has resulted in an increasing demand for transparency of operations, sustainable development, and green buildings. This section introduces possible ramifications of increased social and environmental awareness for the facility manager.

10.6 Conclusion

Successful facility managers are likely to be those who appreciate the big picture with all of its intricacies. Even if your company is not aiming to be a "global company," the impacts of globalization will affect you. Understanding the opportunities and challenges is imperative.

CHAPTER 11 SUSTAINABLE DESIGN

The consideration of environmental performance has been increasing steadily for decades. This movement has evolved from the early post oil embargo solar houses of the 1970s to highly integrated and efficient buildings of the 1990s. It is conceivable that buildings in the near future could even be able to produce more energy than they consume. Currently, it is feasible to design and build buildings that consume significantly less energy than allowed by the most stringent energy codes in the United States, and that greatly enhance the health and productivity of occupants. The environmental impacts of development may be reduced in terms of planning, site development, and the efficient use of materials and resources.

11.1 Introduction and Background

The concept of sustainability is based on the belief that we must meet our needs without compromising the needs of future generations. The construction and operation of buildings uses enormous quantities of energy and other resources, creating serious environmental impacts. Green buildings have many benefits, including reduced energy and operational costs, occupant health and productivity, and market differentiation.

11.2 Design and Construction

Start a project correctly by establishing environmental goals and through design integration. This chapter includes site and landscape design, passive strategies, climatic responsive design, and a discussion of whole building energy simulation and other design tools and methods. Daylighting issues, electrical lighting, and building systems are other aspects of green building design. Sustainable approaches to the use of materials and the construction process are included. Rating systems are now available for comparing the environmental performance of buildings.

11.3 Occupant Health and Comfort

People now spend a majority of their time inside buildings. Modern, conventional commercial buildings have led to several types of building related health problems, such as sick building syndrome. Indoor air quality can be improved through careful design and construction of HVAC systems. Alternative ventilation strategies, operable windows, and under-floor air systems are methods to improve indoor environmental quality.

11.4 The Future of Sustainable Design

As global environmental problems worsen, the need for sustainably designed communities will inevitably increase. The designers and architects who have been leading the charge toward sustainable development will be positioned to expand their leadership role. The integration required to create an ecologically balanced building is both "an aesthetic and spiritual challenge."

CHAPTER 12 SMART BUILDINGS, INTELLIGENT BUILDINGS

12.1 Defining Smart Buildings: Beyond Desktop Technology, Networks, and Central Management Systems

12.2 A Mandate for Buildings to Support Continuous Organizational and Technological Change

As corporate and federal executives plan future workplaces, there must be ongoing discussions of people, place, and technology. People are central to today's continuous organizational reengineering for productivity and innovation, shifting from hierarchical and linear work processes, to collaborative and dynamic. While modern organizations have become more organizationally and technologically dynamic, office buildings and their infrastructures remain stereotypically fixed. This section describes the organizational and technological evolution and the need for commensurate changes in workplace design.

12.3 Smart Buildings Require Flexible Infrastructures

To address issues of long-term productivity and organizational effectiveness, it is time to move beyond definitions of code-compliant buildings and even "high-tech" buildings to the creation of truly "motivational" buildings. Motivational buildings provide environmental performance at a level that consistently and reliably ensures health, comfort, security, and financial effectiveness, while supporting high levels of productivity with continuing organizational and technological change. This section describes the new design approaches to absorb change and avoid obsolescence. To avoid frequent environmental quality failures and long-term obsolescence, it is critical to invest in user-based infrastructures that are modular, reconfigurable, and expandable for all key services—ventilation air, thermal conditioning, lighting, data/voice, and power networks.

12.4 Smart Interior Systems

Interior systems in the intelligent or smart building should fully address the following three performance criteria: 1) organizational/spatial flexibility (continuous change without waste), 2) privacy and interaction/working quiet and teaming, and 3) ergonomics and environmentally appropriate finishes.

12.5 Smart HVAC Systems

Heating, ventilating, and air conditioning (HVAC) systems in the intelligent building should fully address the three performance criteria: fresh air for each individual, temperature control for each individual, and access to the natural environment for each individual. This section describes the five major HVAC systems innovations: 1) decoupled air and thermal systems, 2) individual control—task conditioning systems with broad-band ambient, 3) robust, micro HVAC systems and integrated grid-and-node conditioning, 4) load balancing—facade mechanical systems, and 5) multi-mode HVAC systems with natural ventilation.

12.6 Smart Enclosure Systems

Smart, sustainable buildings have an enclosure that ensures the highest thermal quality, air quality, visual quality, and long-term building integrity. A range of innovations in enclosure design worldwide offer significant opportunities to displace large energy loads in buildings, while simultaneously providing regionally innovative architectural aesthetics. This section describes innovative enclosure systems which effectively address seven major responsibilities: load balancing, solar control, heat loss control, daylighting, natural ventilation, passive and active solar heating, and enclosure integrity. At least three major enclosure innovations illustrate the next generation of "smart enclosure" systems: unplugged enclosures with minimum thermal loads, layered facades for daylighting, and enclosures as mechanical systems.

12.7 Smart Connectivity Systems—Data, Voice, and Power

Six major innovations illustrate the next generation of "smart connectivity" approaches, always addressing both electrical and electronic demands. This section describes the following innovations: distributed vertical risers and satellite closets to support distributed connectivity centers with central intelligence; accessible horizontal plenums—raised floors and open cable trays/baskets for horizontal distribution; upgradable wiring harnesses to provide a grid of service; reconfigurable nodes for data, power, voice, and environmental connectivity demanding modular outlet boxes homerun to distributed satellite closets; wireless data and voice for mobility; and technology pubs and multimedia conference hubs.

12.8 Smart Lighting Systems

Lighting systems in the intelligent building should fully address the demands of the dynamic organization for organizational/spatial flexibility (continuous organizational change without waste), functional flexibility (changing tasks and work tools without visual discomfort), technological flexibility (changing desktop and multimedia technologies without visual discomfort) and healthy, motivational work environments.

This section describes the following key innovative lighting design approaches: effective daylight utilization with controllable electric lighting interfaces, separate task lighting from ambient lighting, reconfigurable lighting with plug-and-play fixtures, and continuous change in lighting zone size and control.

12.9 Conclusions: User-Based Commissioning and User-Based FM

CHAPTER 13 LIGHTING

13.1 Introduction

Energy consumption, lamp and ballast maintenance, cooling load impact, glare complaints, and so on, can drive one to focus on minimizing first costs and/or maintenance costs—since this offers definitive control. However, employees' comfort, satisfaction, retainage, and productivity will be affected by lighting. Indeed, long-term corporate viability depends on the employees and their products. Lighting plays a significant role. This chapter outlines issues involved with lighting ergonomics, sustainability, energy efficiency, and sustained optimal lighting system performance. Exterior lighting (senses of safety and security as well as light pollution and light trespass) and interior lighting (daylighting integration, layers of light, lighting technologies, and retrofits) are discussed with convenient on-line resources referenced and checklists of key issues offered.

13.2 Lighting Ergonomics

For many decades, lighting was simply considered an intensity problem—and more light was considered better! Energy issues have forced reevaluation of lighting criteria. Employees as valuable resources need lighting that meets a host of criteria. These criteria are discussed here.

13.3 Lighting Codes

Foremost in current practice are energy codes and guidelines. These are introduced here.

13.4 Environmental Issues

Sustainability is hot—at the forefront of many corporate visions. Lighting design, procurement, maintenance, and disposal are all related to sustainability (or lack thereof). This section outlines these key areas and discusses the most prominent issues regarding sustainability.

13.5 Daylighting Strategies

Daylighting can help sustainability. Daylighting can ruin employee comfort and productivity. In this section, three key daylighting strategies are outlined. An in-depth discussion on the control of fenestration is included.

13.6 Interior Lighting

A general pattern of recessed lights alone no longer fulfills the goal of providing a quality work environment while minimizing energy expenditures. In this section, an approach using three layers of lighting is introduced—ambient, task, and fill (or accent) lighting are reviewed. Lamp limitations are discussed and light sources to avoid are cited. A checklist of key task lighting issues is presented.

13.7 Exterior Lighting

Safety and security of employees should be key considerations. Lighting criteria recently changed in the *Illuminating Engineering Society Ninth Edition Handbook*—but this new tack of "lowest common criteria" is not appropriate for most situations. This section outlines other more appropriate criteria resources and references. Read what recent research found regarding high-pressure and low-pressure sodium lamps for nightlighting situations and what lamps are more appropriate.

13.8 Lighting Technologies Today

Here is a succinct rundown of the latest technologies in lamps and ballasts.

13.9 Lighting Technologies Tomorrow

Here is a succinct review of emerging technologies which may offer commercial promise in the near future.

13.10 Retrofits and Simple Replacements

As lighting system complexity grows with efficient ballasts and lamps operating on electronics and with ever-improving luminaire optical performance, lamp and ballast replacement must carefully match performance characteristics of the hardware being replaced. This section outlines key points for consideration when undertaking mass retrofits or simply replacements.

13.11 Lighting Analyses

Don't be a slave to consultants! This section outlines what kind of marching orders should be issued to potential consultants and what you should expect to receive from any consultant, independent or vendor based, in order to assess various components' and systems' compliance. Some criteria for assessing a consultant are also offered.

CHAPTER 14 ERGONOMICS AND WORKPLACES

14.1 Introduction

Ergonomics is the science of fitting the task, the work environment, and worker into an integrated whole. As computers become part of almost every job, the importance of ergonomics in the workplace has risen. This section introduces ergonomics, discusses its goals and focus, and shows how ergonomics can be applied at both the individual and organizational level.

14.2 Ergonomics of Workspaces

The goal of ergonomics is to make the things people use and the ways and places in which they use them as safe, comfortable, and productive as possible. This section examines how ergonomics is applied to the design of workplaces by using body dimensions to define size and shape of workplaces and the placement of elements within workplaces. The use and placement of computer technology is critical to the comfort, safety, and performance of workers. Placement, configuration, and use of computer technology are also discussed.

14.3 Ergonomics of Seating

Most office work is performed in a seated position. Chairs are ubiquitous. Yet the challenges of providing a chair that adequately supports people as they perform their jobs are significant. This section provides research and recommendations for the characteristics of task seating.

14.4 Ergonomics, Performance, and Productivity

A large and growing body of research has examined the role of the workplace and workplace elements on performance and productivity. This section provides a summary of several key studies and the implications for the design and use of workplaces.

14.5 Ergonomics of Accessories

One of the more common ways in which workplaces are adapted to fit the worker is through the use of accessories. Devices such as keyboard trays, wrist rests, and copyholders are commonly used to provide an element of adaptability to "standard" workplaces. This section examines a variety of such devices and provides guidelines for their use.

14.6 Ergonomics and Macroergonomics

A comprehensive approach to ergonomics looks beyond the individual elements of the workplace and considers the content and context of the worker and workplace. A macroergonomic view includes such considerations as training, psycho-social issues, job design, and organization and stress. This section discusses these issues as they relate to the goal of ergonomics.

14.7 Conclusions

This section provides a list of considerations and strategies for achieving sound ergonomics in workplaces. The appendix "rules of thumb" shows how individuals can adjust and adapt their workplace to fit their specific circumstance.

CHAPTER 15 MANAGING THE NEW HEALTHCARE REAL ESTATE PORTFOLIO: A CASE STUDY

15.1 What Has Changed in Healthcare Facilities and Real Estate?

Since the failure of national healthcare reform in 1993, three major forces converged to change the typical healthcare real estate portfolio and the facility management needs of most acute care providers. This section describes how hospital mergers, acquisitions, and ambulatory care network development changed the typical healthcare real estate portfolio from single campuses focused on acute care to multi-site, multi-care level delivery networks that resulted in a host of new facility challenges. The synergy between rapidly advancing technologies in medicine, informatics, and telecommunications is explored because this convergence is creating "ubiquitous" medicine, able to be delivered in a much broader array of physical settings than ever before imagined. The impact of these technology trends on healthcare facilities is still underway. The same period from 1993–2000 saw a major squeeze on capital available for health facility investment, forcing hospital systems to explore new "off-balance sheet" capital financing mechanisms that demanded tighter investment analysis and accountability. These new demands were often beyond the expertise of existing in-house facility management skills. Given the pace and breadth of these changes, many healthcare facility managers were caught unprepared for their new challenges: getting an accurate "handle" on the expanded facility inventory, deploying capital equitably across multiple new system partners, and adding financial analysis and property management skills to the traditional hospital FM skill set.

15.2 How One Healthcare FM Team Responded

This section describes the case study focus, Partners HealthCare, which represents the merger of two academic medical behemoths—the Massachusetts General Hospital and the Brigham and Women's Hospital—along with several specialty and community hospitals and multiple physician practice sites. The FM response of Partners is described, including their rationale for developing a corporate-level office of real estate and facilities, the organizational chart, and staffing of that office, and the office's designated functions. This section identifies key FM issues that emerged and how they were addressed, including establishing balance between central and local facility authority, and developing capital priorities and distribution among entities. The four key components of the real estate office's FM data strategy are identified and the success (or failure) of their implementation is examined. Nine areas of concrete results from the Partners corporate real estate effort are explored, from development of a high-level strategic facilities plan and corporate office fit-out standards, to real estate consolidation savings and more.

CHAPTER 16 ORGANIZATIONAL READINESS CASE STUDY: IMPLEMENTING TECHNOLOGY AT ROCKETDYNE, A BOEING COMPANY

The earliest use of technology was just a plug and use approach where applications were developed to replace time-consuming calculations. As desktop systems evolved, more complex software applications emerged, requiring more understanding of the business processes, informational needs, and people's changing role in the workplace. This chapter examines the role and impact of technology within Rocketdyne's facilities organization and discovers that technology alone does not improve performance.

16.1 The Facilities Organization

In a manufacturing environment, the facilities organization is responsible for providing and maintaining the building infrastructure as well as production equipment. This section discusses the organizational changes that evolved as both technology and new management strategies were employed.

16.2 Process Change through the Use of Technology

This section describes one of the first steps in identifying the facilities role within a business. What services or products does a facilities organization provide that are key to the success of the business?

16.3 Technology

This section discusses the role of technology upon a process. A misconception is that technology will improve performance, but in reality technology is the enabler of a process.

16.4 Organizational Readiness

How do you implement an enterprise software package? This section discusses organizational readiness: processes, information, and people. Each facet must be addressed to successfully implement technology.

16.5 Driving Change

This section discusses the nature of change and what factors can drive change within the organization.

16.6 The Shopkeeper Model

At Rocketdyne, the facilities organization developed a model, a business within a business, by becoming the "shopkeeper." The model served as a new mindset for providing responsive, cost-effective service to its internal customers. This section discusses the model and how it affected the facilities organization.

16.7 Vision

To move the facilities organization toward a new model and processes, a vision is needed. The vision is both a statement and plan that drives people from diverse working groups to a common goal. This section discusses the need for a vision.

16.8 Developing a Framework for Organizational Growth

As Rocketdyne's facilities organization moved toward the shopkeeper models, it recognized that as any other business a long-term growth strategy must be developed. This section discusses how a facilities organization can position itself for growth and change.

CHAPTER 8
ALTERNATIVE WORKPLACES

Jeff Austin
Vice President Strategy
Corporate Real Estate
First Union Corporation
E-mail: Jeff.Austin@firstunion.com

Alan L. Bain
President
World-Wide Business Centers
E-mail: alb@wwbcn.com

Paul Heath
Principal
Space, LLC
E-mail: heathp@workplayce.com

Joel Ratekin
Director
Workplace Strategies
American Express Co.
E-mail: Joel.Ratekin@aexp.com

Ellen M. Reilly
President and CEO
Port Remote Services, Inc.
E-mail: reilly@PortStrategic.com

Eric Richert
Director
Workplace Operations and Research
Sun Microsystems, Inc.
E-mail: eric.richert@eng.sun.com

Christine Ross
Workplace Strategist
Cisco Systems, Inc.
E-mail: chross@cisco.com

8.1 INTRODUCTION

The variety of workplace arrangements is moving at the speed of technological change. And yet, not everyone is ready to commit to a new model. After all, the idea of a remote workforce is frightening. The image of a vacant glass office tower may come to mind. If you can't see your employees, how can you manage them? How can a corporation hold together if all the parts are scattering? How can you be sure the customer isn't falling through the cracks? These are all legitimate concerns, but ones that can be addressed through new, more-flexible work environments.

Alternative workplaces arose out of concern for employees, new global business opportunities, and remarkable advances in technology. Customers are using virtual interfaces at every turn. It only stands to reason that employees follow suit. If provided with enough support, flexibility, and resources, employees can work more productively and quickly—*from anywhere*.

Already there has been a shift of corporate overhead expenses away from real estate and toward information technology. Smaller branch network offices are sprouting up to accommodate employees and customers—both domestically and internationally. Corporate connectedness is happening through high bandwidth now, allowing employees to disperse globally and stay mobile. The result calls for sweeping changes from facilities planning and management.

8.1.1 Short-Term Objectives of Alternative Workplaces[1]

Adoption of alternative workplace designs is often motivated by two immediate needs: cost control and access to and/or retention of talented employees.

Cost Control. Adoption of alternative workplaces for the purpose of cost control generally takes one of two forms. The first is adoption of unassigned space schemes, often involving reserving office space on an as-needed basis. This "hoteling" concept was originally developed for large accounting and consulting firms that recognized the inefficient costs of office space (and related desktop technologies). This office space was generally unused because their "owners" were at customer sites. Subsequently, the concept has been adopted by sales organizations in a number of companies.

The second alternative-workplace scheme aimed primarily at cost control is the adoption of "universal office" designs. These schemes, at their extreme, are "alternative" because they dispose of all status expressions traditionally afforded through office size and amenities. They control facilities spending, because universal offices allow movement of people ("churn") without having to modify office size, structure or furniture, and the offices do not need reconstruction in response to occupants' promotions through management ranks.

Access to and Retention of Talented Employees. Increasingly, organizations seek to retain employees who need (or want) to work in locations prohibitively far from the central workplace. These employees may choose distant locations for lifestyle or family reasons. Or, a prospective employee may live far from a central work location and is not willing to move in order to be employed. In such cases, employers are increasingly allowing work from home or other chosen locations in order to hire or retain valued employees. Often, this version of alternative work arrangements is done on an ad hoc, as-needed basis. Practices that contemplate these arrangements as a long-term "norm" often emerge after a few business applications have been deployed.

8.1.2 Long-Term Objectives of Alternative Workplaces

Ultimately, the purpose of alternative workplaces is to support the rich variety of today's work and help organizations become more effective in conceiving, designing, producing, and delivering products and services to customers. The economic push for these kinds of workplace designs comes from both direct cost savings as well as harder-to-measure productivity gains. When work resources are aligned with the work that people do, those resources are also more likely to be efficiently and well used. And, when space and technology resources are designed in concert with each other, cost-efficient tradeoffs between resources are possible. Thus, alternative workplaces, in the long run, should cost less while providing workers with better support.

Alternative workplaces that fully support new ways of work will integrate the physical, technological, and organizational resources effectively. These results will require substantial change and change management. Not only will the way we work change, but also how we think about work. Managers, employees, and customers need to adjust not only to the new work environments and tools, but also in how they relate to one another. There will be resistance and adverse conditions. The changes proposed will take time to develop and need substantial, long-term commitment throughout the organization to effectively meet the new business paradigm. But, in the long run, as competitive advantages shift, they will be well worth the effort.

8.2 RETHINKING THE TRADITIONAL OFFICE[2]

8.2.1 A Historical Perspective

For the purposes of this chapter, "traditional offices," in their extreme, are those workplaces that are designed in accord with long-held (and generally unspoken) beliefs about the nature of work, employees, and the management of those employees. The strength and general acceptance of these beliefs preclude the need to investigate how accurately they reflect the work that is actually done in the traditional office. "Alternative workplaces," in their extreme, are designed in direct response to the work that is actually done, both by individuals and by teams of employees, with few preconceived notions about what constitutes correct workplace design.

Traditionally, work has been seen as a process of producing tangible goods. At one time, most tangible goods were produced by individuals who owned their own equipment and raw materials, and who worked for themselves. By the late 1700s, merchants began to act as "brokers," supplying individuals with raw materials then taking finished goods to market. This so-called "outwork" system "failed because of problems such as poor-quality, slack effort, waste, and theft." This failure led to "centralized workplaces—factories—established so that monitoring of the production process (and) use of materials" was possible.[3] In addition, centralized workplaces allowed the concentration of expensive capital assets and consequent economies of scale of production.

By the late 1800s, large centralized workplaces, housing large numbers of employees and enabling mass production of tangible goods, created unprecedented management problems. Employers were faced with "inefficiencies, careless safety, and arbitrary supervision." Employees "were thought lazy at best, uncooperative at worst, dumb in any case."[4] In the early 1900s, Frederick Taylor, "appalled at the wasted effort, exhausting work, long hours, petty dictators, arbitrary rules, inefficient methods, and goofing off" developed his principles of "scientific management" which sought to solve these problems by combining a high degree of work specialization, prescriptive work methods, and production incentives. Taylor's methods required that "all possible brain work should be removed from the shop and centered in the planning department."[5]

Mass production, narrow specialization, and centralized workplaces still exist, and managers are still seen as planners/decision makers, monitors, and controllers of work. The legacy of "scientific management" and industrialized work still resides in the walls of today's traditionally designed offices. Most people, however, would agree that much of today's work is knowledge-based, not capital-based. Work requires high degrees of individual adaptability and decision making at all levels of the workforce. Work happens in many locations, ranging from home to airports to customer locations to partner locations, in addition to workers' "headquarters" locations. And, today's work requires managers to be coaches, facilitators, and leaders more than controllers and monitors.

Against this background, then, "traditional offices" mean those workplaces that:

- Are centralized, requiring employees to convene in a single location
- Are configured to support production rather than innovation
- Emphasize individual work
- Assume employees to be in their single locations during fixed hours
- Reflect status, reinforcing positions of decision making and control
- Imply sameness of work, both across a workforce as well as over the course of an individual's work day or work week
- Constrain adaptability by individuals to reflect personal work styles or needs
- Utilize assignment of workplaces and work tools to individuals, under the assumption that what is assigned (e.g., the individual office) reflects what is needed (and all that is needed) to accomplish work

8.2.2 The Current Proposal

In contrast to traditional offices, "alternative workplaces":

- Include a variety of central and dispersed work locations for use by employees
- Seek a balanced focus between production and innovation work
- Seek to support both individual and group work
- Include flexibility regarding when and where work is done, and include necessary protocols and agreements that allow people to communicate and connect in the context of this flexibility
- Minimize design attributes that explicitly display status (though status attributes may emerge in other ways)
- Reflect the diversity of work across a workforce, as well as the diversity of an individual's work over the course of a work day or work week
- Enable adaptation of the workplace to individual and work group needs
- Utilize shared resources, allowing employees to use a variety of work resources when and where needed

In traditional workplaces, rarely are organizational, technological, and physical/spatial resources designed concurrently or in a coordinated way. In a sense, they do not need to be, since deep, unspoken beliefs about work and employees preclude the need to do so. That is, status issues, control over work hours, the need to convene in a single location, and so on, are constants in the traditional paradigm, so they do not need to be addressed. The function of technology in the traditional paradigm is not so much to help people work together in new ways to produce better products and services as it is to make traditional work practices more efficient, achieving more production with fewer resources.

Alternative workplaces break apart old organizational assumptions, so new ones need to be developed concurrent with the new workplace. Technology enables new work paradigms at the same time that the new paradigms demand more of technology. So, whereas the components of traditional offices (space, technology, and organization) can be considered separately, these components need to be "integrated" for a successful alternative workplace design.

These contrasts between traditional offices and alternative workplaces indicate several "dimensions" along which a workplace can be described. Rarely will a workplace be "purely" traditional or "purely" alternative. The forces of cost control, workforce coordination, and economies of scale, alongside the forces of knowledge work, global business, importance of customers, and rapid change are causing the creation of many hybrids of "traditional" and "alternative" offices and workplaces. The dimensions that can be used to describe the extent of a workplace's "traditional" or "alternative" characteristics are:

- Centralized—dispersed (work sites)
- Inflexible—flexible (work hours and work locations)
- Standardized—adaptable (space, environmental controls, furniture, and computing environment)
- Individual focus—team focus
- Production focus—innovation focus
- Assigned resources—unassigned/shared resources (space and technology resources)
- Segregated resource design—integrated resource design (space, technology, and organizational resources)

8.2.3 The Future of Work

The flexibility of new technologies is drastically changing the way work is and will be done in the future. Although the last 20–30 years have been significant in information automation, and this current period is valuable with regards to information networks, the future will be transformed beyond our imagination by work tools with cross-platformed software. The pace of change can be interpreted from the following statistics:

- In 1993, less than 1% of the U.S. households were on-line and in 1999, one-third were. In contrast, it took radio 38 years, telephone 36 years, television 13 years, and cable TV 10 years to achieve similar levels of penetration.[6]
- Home-office households on the Internet increased approximately 45% in one year (1997–1998).[7]
- E-mail infrastructure has been installed for 50% of corporate America—approximately 50 million employees—primarily in the last three years (1996–1999).
- 15–40% of the employees in large corporate organizations dial-in from remote locations to access corporate networks (1999). Remote access is expected to grow to 60–80% of the workforce.
- The majority of employees using remote access is no longer technology personnel, but general business employees.
- The number of IP messages in any given period (1999) now equals the number of phone and fax messages.[8]
- People with high-speed access search for information and make purchases on-line at approximately double the rate of those with lower-speed analog modems.[9]

In addition to cost control and employee retention, alternative workplaces are growing in popularity because new business conditions have placed a premium on *mobility* and *distance*. *Mobility* refers to workers' movement between workplaces, including different work settings within a building, as well as different work venues altogether. These different venues can include home (see Figure 8.1), customer locations, partner or supplier locations, airports and airplanes, satellite offices, and so on.

Distance refers to workers who need to work together, but who are distant from each other. Sometimes that distance is a product of mobility. Telecommuting, for instance, results in distance between individual employees and their work groups. Sometimes distance is a product of global markets and a global workforce. Vendors and customers continue to become more distant, as the customer interface becomes more and more virtual. An increase in mergers and acquisitions also contributes to this effect. Companies across the globe can operate under the same management without skipping a beat.

Mobility and distance are, of course, made possible by information and communications technologies, and as more people engage in mobility or distance work, they demand more functionality

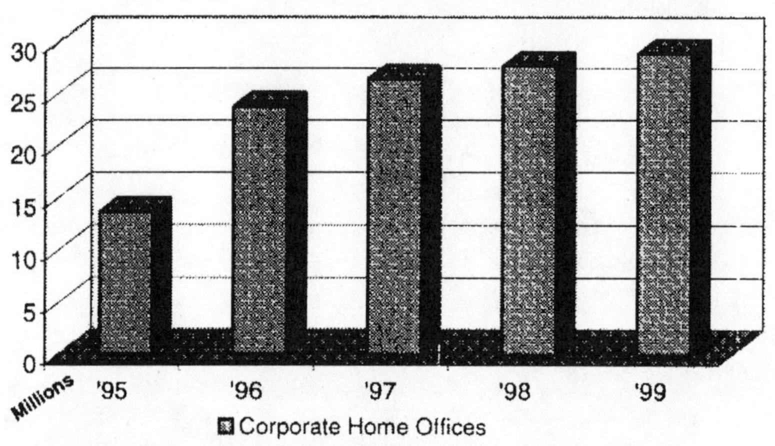

FIGURE 8.1 IDC Reports Corporate Home Offices.

from these technologies. In broad terms, the technologies that underlie the work tools of the present and future alternative workplaces are those that connect people and thus partially replace the need for workers to be physically proximate. These technologies are listed in Figure 8.2.

The use of these new work tools is dependent upon the cost of information networks and communication technologies, both of which are being reduced significantly as we write this chapter. Lower-cost communication, centralized storage systems, and cross-platform work tools (that in the next five years will become commodities) will encourage employees to relinquish ownership of

Data technologies	Voice technologies	Video technologies
• Networks that allow workers to post and have access to shared data, information, and messages both synchronously and asynchronously • Network appliances that allow workers to gain access to the network in a variety of ways • Software (groupware) that allows workers to work together synchronously, actively, and collaboratively creating and editing files, documents, drawings, and so on • Storage capabilities that allow large quantities of data to be stored and accessed from anywhere on a network	• Telephones that allow multiple parties to conference together from distant locations • Full duplex telephony with high quality microphones and speakers that allow conversations to proceed "naturally" • Other voice technologies including voice recognition, text-to-voice automation, voice with web-enabled support services, etc.	• Video conferencing equipment operating over high bandwidth channels that allow face-to-face meetings over distance, providing high-quality visual and voice connections • Document cameras in conjunction with video conferencing • Desktop-to-desktop video—stored, streaming, and interactive

FIGURE 8.2 Technologies that Enhance Distant and Mobile Communication.

their work tools and use centralized storage systems to access their data files. Although centralized storage systems have multiple operational and maintenance advantages, management productivity will be the key advantage. Less and less filing and reporting of work will be done in the future—assuming existing work processes are reengineered to automate the filing, reporting, and authorization requests. Managers will be able to review work as it is created. Where appropriate, customers will be able to review this work, too.

The need for mobility and development of sophisticated network systems will encourage less ownership of corporate on-site workstations. Employees may take temporary ownership of workstations to accommodate the need for workgroups; however, permanent ownership of workspaces may be reserved for a reduced portion of the corporate employee population. The flexibility in this future work condition is beneficial to both employees and business.

8.3 ON-SITE ALTERNATIVES REQUIRE MORE THAN CHANGING EXISTING SPACE STANDARDS

Dilbert is right! Scott Adams, Dilbert's creator, pokes fun at the cubicle in many of his comic strips. Terms like "prairie dogging" and "cube farms" have infiltrated our language because we all can relate to working in a standard open plan with lots of equal-sized boxes. The one-size-fits-all cubicle approach is easiest for facilities management (FM), but is it the most productive space for today's businesses? No.

The argument against fields of cubicles was well articulated in the early 1980s. Today mobile work (inside and outside the corporate office) makes high-density workstations a viable solution for portions of an employee organization. High-density cubicles, however, are not the answer for all employee job functions.

So what does this mean to the facilities manager? First, the business application will require an assessment of the business goals, operations, challenges, and real estate. Typically, the types of corporate space and services will be expanded to meet the business needs. Lastly, technology and communications should be upgraded to accommodate effective work processes. To this end, FM will need to consider the following:

- A new service delivery process
- New business drivers and needs
- Expanded corporate office-space resources
- Expanded services beyond space planning
- How best to facilitate change in the transition

8.3.1 Alternative Workplaces Require a New Process Delivery[10]

First Union, one of the nation's largest banks, has taken a participatory and integrative approach to alternative workplace strategies and their implementation. Projects had typically been done via a traditional methodology of applying a set of space standards and ratios to the business unit's programming requirements. After realizing that some strategic customers want more from the space planning process, First Union's corporate real estate group researched, developed, and deployed an integrated workplace strategy called Project Workout. Under this new strategy, an integrated work team that includes an organizational psychologist, strategic workplace planners, and internal consultants from corporate real estate, information technology, and human resources/training partner with business units who wanted the workplace to have a more progressive and responsive role in their business planning.

Although Project Workout began as a way to identify opportunities to deploy telecommuting, hotelling, and other off-site options for work, it has evolved into a broad-based process that seeks to align the workplace with business goals and needs. By understanding the leaders' business goals and the unit's work processes through a participatory process of consensus decision making and focus groups, the Workout team deploys integrated solutions that over time become the domain of the business unit. These projects shift the focus of projects from a purely occupancy cost issue to a broader vision of improving employee effectiveness and creating value for the business unit.

How Is a Participatory Process Deployed? First, the Project Workout team meets with the leaders of the business unit to gain understanding of the goals of the group, whether it be to shorten the decision-making time to approve loans, get the synergies of collocating dispersed departments, or foster more creativity and innovation in the organization. It should be said that none of the projects to date have increased the business units' occupancy costs and, in fact, many have saved costs, but these are the results for the process, not the goal. Once the business goals are understood, the Workout team and the business leaders agree on roles, time frames, and commitments from all parties and draft a formal agreement to proceed.

The next step is to engage a majority of the business unit's employees in focus groups to understand all the explicit and implicit drivers of their individual and group effectiveness. By the questions they ask, employees create a dialogue that expands the scope of "workplace" from only the physical attributes to include the practical, social, procedural, technological, and cultural aspects. Questions are also raised to better understand the details of their day-to-day processes—"a day in the life" analysis.

The Workout team analyzes all of this data and provides recommendations to the leadership, who signs an agreement to support changing the performance or culture of their unit via the workplace. Based on this contract, physical and social design teams are formed from within all levels of the employees in the business units, and the actual design work begins. In all cases, these solutions come with the required technology and training to maximize design performance. For instance, if more mobility is needed, portable voice and data technologies are integrated into the physical and social recommendations.

The physical and social design is accomplished within certain broad parameters known as "planning principles." These planning principles create the framework to ensure flexibility over time while allowing for maximum choice and involvement in the design process by the individual customer groups.

Physical planning principles include:

- Universally zoned floor plans with hard walls at core and open workspaces along exterior
- Integrated power/voice/data along fixed spines in the open workspaces
- Choices from a kit of parts that create the personal workspaces
- Activity and community based workplace planning
- Provision of alternate on- and off-site workspace

Some social-planning principles are:

- Encouraging employees to make decisions for themselves
- Redefining what work looks like and where, when, and how it is best accomplished
- Managing by results not by appearance
- Allocating space based on work process and function, not on hierarchy
- Achieving group qualities of collaborating, sharing, experimenting, and improving

Once the physical and social designs are completed, CRE's project managers and strategic alliance partners, with consulting from the Workout team, bring the recommendations to built form. While the construction is underway, the social design is implemented via widely deployed training programs and individually targeted learning sessions. Therefore, upon occupying the new workplace, the employees know how to best maximize their new environment.

The process does not end at move-in. Post-occupancy evaluations are done at 90-, 180-, and 270-day intervals to determine if the actual solutions met the intent of the physical and social design. Any changes to the environment are made as part of the management of the space. For instance, if design-intent criterion is maximum separation of circulation and workspace, but the heights of panels do not accomplish that, then they will be altered until the desired behavioral change is accomplished.

An integrated process requires open dialogue and forward thinking. To change the end user's expectation of FM, the facilities manager must broaden his or her skill base to include sales, customer service, and negotiation. In addition, the facilities manager must genuinely listen to the business's needs and requirements and be receptive to suggestions. Integrated solutions will also demand more seamless coordination between the facilities manager and other support groups, contractors, and other professional disciplines. Information technology, for example, must participate in facilities implementation and change management curriculum should accompany move management.

When instituting an alternative workplace solution, it may be hard to see long-term benefits, because short-term problems inevitably occur. The facilities manager, however, should always envision a structure in which employees will make facilities decisions for themselves. Doing so will promote new business opportunities but may result in less FM intervention over time.

Therefore, the success of alternative workplaces ultimately resides in the "culture" created. Both the internal business customer as well as the FM organization must be satisfied. How will the "basic tacit assumptions about how the world is and ought to be"[11] change for both customer and facilities managers when employees are part of this collaborative design and decision-making process? These alternative work strategy projects will never be fully successful until the corporate real estate/FM organizations shed their own traditional tasks of "watching, mandating and care-taking"[12] and become more strategic by being more responsive to the business and productivity goals of their internal business customers. Only then, when the "alternative" becomes the norm, will they be prepared to provide both strategic direction as well as quick, seamless, FM services.

8.3.2 Redefining Corporate Space Resources[13]

Think about the range of activities that comprise a "normal" workday. Every company, functional unit, and employee's day will vary based on their business, responsibilities, and work habits. At Cisco Systems, we have looked at how our employees spend their time when in the office. The chart in Figure 8.3 represents a worker's typical day.

Is a cubicle the most appropriate place to have a confidential discussion? Is there a great deal of informal learning when everyone sits in an office? Probably not.

At Cisco, we are not trying to change how people use workspaces. Rather, as technology and business processes change, we provide flexible work environments so that people can adapt how,

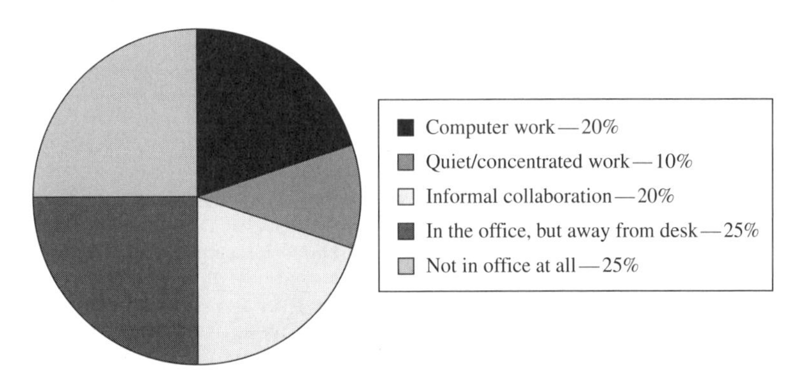

FIGURE 8.3 Employee Work Activities.

when, and where they work to best suit their needs. Furthermore, we understand how people change. The work environment should be flexible to adjust to those changes as they evolve.

In the past, the equipment needed to complete our work has primarily been tied to the desk. Power, data connections, and telephone lines were stationery, which resulted in most work being done in an individual's workspace. More and more employees now have access to mobile laptops, cellular phones, and wireless products, allowing individuals to work wherever they need to work—in and outside of the office. As globalization increases, virtual teams and remote work will also increase, necessitating more mobile tools and new business processes.

The combination of the change in business processes and new technology to support mobility means that a single workspace does not support all activities. Individual workspaces are still necessary, but more choices and flexibility are needed.

The evolution of technology allows for a closer look at how we provide workspace that better supports the variety of activities each worker participates in during the process of completing their work. At Cisco Systems, we offer a variety of work settings to provide a productive work environment for a wider set of tasks.

In addition to the activity spaces already included in most work environments—conference rooms, offices, and break rooms—Figure 8.4 describes other workspace settings we have added.

Workspace	
Just-in-time offices	These unassigned offices are available for small meetings, conference calls, concentrated work, and private conversations and meetings between managers and their staff.
Quiet rooms	A small, unassigned hard-wall office intended for quiet or concentrated work, conference calls, telephone work, and confidential conversation. Great space for a cubicle user who needs the privacy not afforded in a cubicle.
Touchdown carrels	Spaces ideal for short stays—good for checking E-mail and voice mail and a place to settle your stuff before going into meetings.
Team work areas	These vary from open cubicle environments in different sizes to hard-wall rooms dedicated to a team or projects. They provide for team-based and collaborative activities in an assigned or unassigned fashion.
Informal enclosed meeting areas	Hard-wall meeting rooms equipped with more informal furniture. Ideal for meetings, brainstorming, conference calls, and project work. These rooms are meant to break down barriers usually encountered in a more formal setting and are conducive to "creative thinking."
Informal open areas	Soft seating and/or small, open meeting spaces integrated throughout a space and used for informal discussions, socialization, reading, or just putting your feet up.
Resource areas	Hard-wall or open areas dedicated to support a specific resource, specialized equipment, library of manuals, or other reading material. These can be very informal or very structured like a library.

FIGURE 8.4 Workspace Settings.

When providing these additional options, many companies worry about potential decreases in space efficiency. Any or all of the above options can be coupled with changes to individual workspace (smaller or nondedicated workspaces) or telecommuting and satellite office strategies to provide flexibility without increases in space usage. The goal is to increase value to the organization and gain competitive advantage by leveraging space and maximizing productivity.

Offering a variety of options is not always intuitive to the user group. Training and participatory planning were key to Cisco's success. In many situations, during the discussions with the user community, the need for having these spaces wasn't clearly understood. Given the choice, in almost every project, the team would ask to give up some of the settings to provide for more individual workspaces. The implementation teams had to do a fair amount of cajoling to help the team see the value of the "nonstandard" spaces. After the project is completed and the employees have the opportunity to try out these spaces, we have found that the users find value in the "nonstandard" spaces. In many locations, we've actually had the user team approach us to add more of the varied settings because of the benefits the employees experienced. We continue to evaluate these spaces and evolve the settings as necessary.

Think about the office in the same way you think about a house. Do you cook in the bathroom or sleep in the dining room? Of course not. You have different rooms for different activities. Each room is uniquely set up to support the activities we all do in our homes. Why should the office be any different? Find a variety of options to support the activities appropriate to your business. Give people the options to decide where and how best to get their work done.

8.3.3 Utilizing Services to Affect Office Usage[14]

As alternative workplaces and technology networks expand, facilities and real estate management need to change their tack. They should no longer view themselves as the administrator of standards, but as a service provider to the internal business customer. In other words, they should not only provide offices and furniture, but services as well.

Although the following list is not meant to be all encompassing, FM should be prepared for the following eventualities:

1. *The Facilities Customer is Changing.* With a distributed workforce, FM's customer is no longer just business management, but the employees. As such, it will be important for FM to create different user interfaces that allow for direct communication between the employee and service provider, bypassing the central organization. This will accelerate processes like ordering of home-office furniture.

2. *Facilities Will Manage More and Different Types of Sites.* Alternative on and off-site workplaces will require the management of not only central sites, but also sites elsewhere (i.e., home offices, business centers owned by third parties, satellite offices, etc.). Furthermore, FM groups may need to develop minimum standards for both third-party-run business centers and home offices. For home offices, furniture orders (automated), furniture delivery, installation set up, and removal processes will need to be established. Many organizations have not addressed home office furniture standards. This needs to be done, since work from home is covered under most states' worker's compensation statutes. A short-term solution is a reimbursement allowance for furniture. Long-term, however, furniture standards that at minimum include desk chairs provided by the corporation, need to be established. Lastly, to avoid furnishing two primary offices, home office furniture offerings should only be supplied if the employee vacates office space on-site at the corporate office.

3. *Commitment, Accountability, and Improved Support Is Necessary for an Effective Remote Workforce.* If a corporation's goal is to decrease real-estate occupancy costs, the organization must show commitment to the distributed workforce in support, training, and resource services. If not, the organization risks repercussions beyond the traditional purvey of facilities, namely productivity and human resource issues. Although FM may not have direct responsibility for these areas, it is critical that they evolve a process that provides accountability within the organization. Effective and visible remote practices should be in place before extensive alternative workplaces are deployed. (See Section 8.5, Establishing Best Practices for Remote Work.)

4. *Build a Cross-Functional Implementation Team.* In many corporations, administration must now offer the internal business customer—business management and employees—integrated service offerings. Because corporate administration is subdivided functionally—real estate, human resources, and information technology—it's not unlikely that one of these groups will have more of an incentive to hurdle internal administrative barriers to accomplish a service goal. So, for example, with remote practices, corporate real estate groups may have the incentive or inclination to pay for implementation team hires (or outsourced services) that report functionally to both human resources and information technology. This integration, critical to the success of specific application, will require communication and buy-in among upper-level managers. Upper-level management buy-in can be time consuming in operational development; however, crossing traditional functional barriers is a means to excellent business or employee servicing.

5. *Utilize Unique Service Offerings to Educate Employees about New Services.* As creatures of comfort, employees will cling to familiar and often embedded work practices and patterns that exemplify the traditional office. If, however, they are enticed with attractive offers, they may wean themselves away from the known. Their acceptance is important for two reasons. One, they'll be more prepared to adapt to the new workplace environment. Two, they'll encourage others to join them. Here are a couple of offers you might use to entice employees: offer high-bandwidth technologies if employees telecommute two or three days per week and utilize shared office space; offer mobile workers one phone—a mobile phone—for both inside and outside the corporate office; or develop a contest—a free holiday to Bermuda, for instance.

6. *Service Offerings Might be Bundled.* In addition to FM services, consider offering human resources and information technology services as a bundled package. Consult with each of these groups to determine mutual objectives and develop bundled service offerings. For example, the company could offer compressed work weeks in conjunction with telecommuting. This would improve labor conditions, reduce real estate costs, and pave the way for future work arrangements.

7. *Club Membership Services Can be Developed at Hotelling or Business Center Sites.* Business center services might be offered on a club membership basis. Some possibilities include: pension advisory services, personal technology trainers, faster corporate network access, 24-h building access, drop-off child care, videoconferencing to minimize travel, corporate-paid monthly luncheons to meet other employees, catering services for conferences, athletic or social club activities, etc.

8. *Discounted Fees or Free Offerings Can be Used to Encourage Early Adopters.* It is always difficult to find businesses or employees that are willing to be early adopters of a new product or service. Facilities management may consider offering discounted services for a limited period of time to entice a unit or units to participate. For example, discounted construction fees, internal rent "rebates" or free, remote-work, classroom training.

9. *Corporate Communication is Changing for Facilities and Real Estate Groups.* Real estate/FM will need to develop a resource center, most likely on the corporate Intranet. There, marketing and sales campaigns can describe all of their service offerings. This will eliminate confusion, encourage consistency within service delivery, and reach out to employees directly. It is recommended that computer-based presentations (vs. brochures), electronic newsletters, database management systems, and resource libraries are made available to all employees.

10. *The Service Delivery Process Will Become More Automated.* Real estate and facilities groups are developing automated and real-time cost-accounting systems, product development, project management, and service support systems via the company Intranet and/or in conjunction with Extranet applications. Faster services via more direct lines of communication/processes will be expected between the end users and vendors/service providers. For example, offering a kit of parts for employee workstations requires direct access between the furniture dealer and employees. At minimum, web sites should allow business management or employees to view standards and pricing, make service requests, orders, and complain.

11. *Cost Accounting Measures are Being Reconfigured.* Collaborating, sharing, experimenting, and continuous improvement are desired business group qualities and may mean that shared spaces are not billable to single business units. Real estate and facility groups need to develop different cost-allocation plans to deploy equitable expense to business management. Common area charges may need to be expanded or a per user management fee may need to be created to cover this new way of working.

8.3.4 Cultivating Change Management[15]

In 1998, American Express began implementing a universal space standard for its offices worldwide. The company wanted to find new ways to enhance employee satisfaction, improve workspace utilization, and respond to changing requirements more quickly—while at the same time improving real estate economics.

In the past, American Express facilities reflected a variety of standards—geography and business cultures of different locales. Within this variety some common themes emerged—hard-wall structures, different office configurations, and, too often, archaic furniture styles and standards.

The new universal space approach uses modular and interchangeable workspace configurations that can be adapted to either open plan (30, 60, and 90 ft^2) or hard wall (120 and 240 ft^2) environments. The open plan uses cubicles positioned along the exterior walls and is designed to enhance acoustic and visual privacy. Hard-wall rooms of several sizes, located on the interior of the building, can serve as offices, conference rooms, storage rooms, and other public work areas. This configuration provides accessibility to windows and also minimizes distractions in the external cubicle environment.

Figure 8.5 is an example of a typical universal space standard layout.

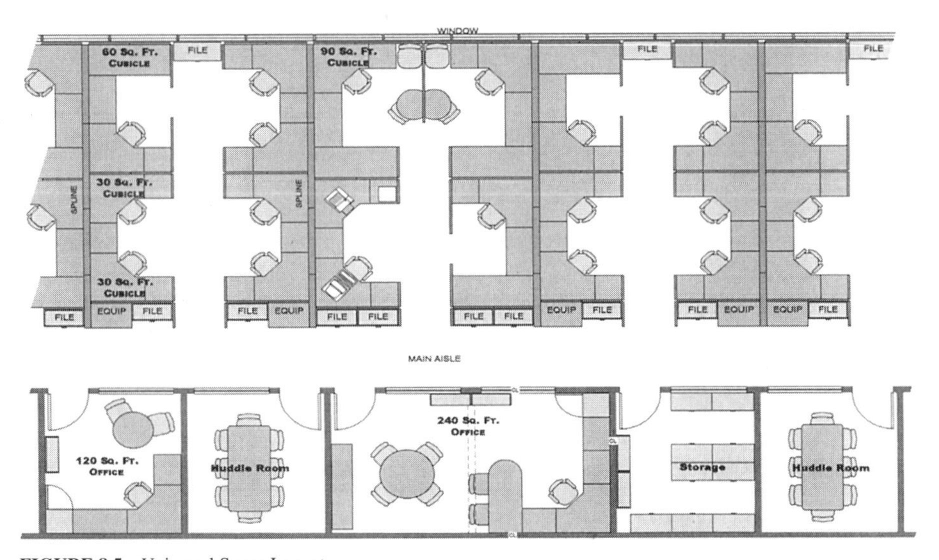

FIGURE 8.5 Universal Space Layout.

Implementing Change. Introducing the new standards presented American Express with two major challenges: first, to make sure employees understood the reasons behind the new environment and how it would affect them; and, second, to minimize disruptions in the workplace during implementation of the new standards.

Research suggests that emotions associated with relocation in the workplace are not dissimilar to those described in the well-known books *On Death and Dying,* by Elisabeth Kubler-Ross, and *Managing at the Speed of Change,* by Daryl Connor. When a manager keeps these various stages in mind, it helps employees maintain a focus on business priorities, thus minimizing a loss in productivity. The model in Figure 8.6 demonstrates the stages of emotional adjustment over time.

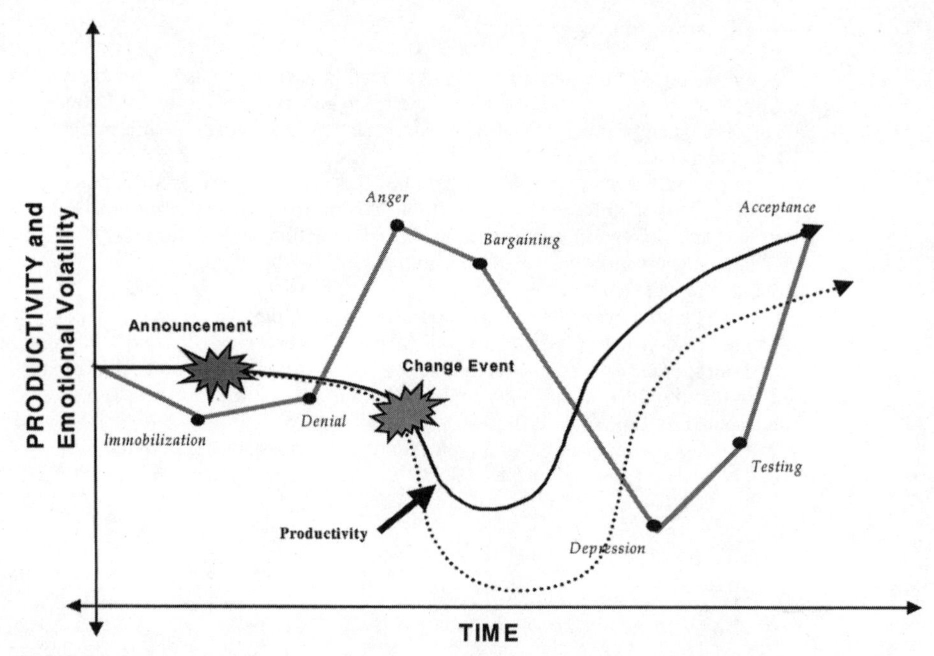

FIGURE 8.6 Emotions in Relation to Change. (*As adapted from "On Death and Dying," Elisabeth Kubler-Ross, M.D., Simon & Schuster, 1997; and Managing at the Speed of Change, Daryl Conner, Villard Books, 1993. Compiled by Nadasdi Consulting, 1998.*)

Employees respond more positively to change if they feel that the change is being done *with them* instead of *to them*. American Express developed tools to help employees contribute to the new workspace decision-making process. Types of participation included:

- Pre-move survey input
- Construction color-scheme selection
- Artwork theme selection
- Conference room naming
- Contests or raffles

In addition to employee participation, it is important to use effective communication methods during the transitional period. Two-way communication is used to exchange information when emotional volatility is high. One-way communication is more appropriate when volatility is low and people just want the facts. Figure 8.7 provides some examples of "active/passive" and "one-way/two-way" communication.

The universal space standard challenges employees to learn a new etiquette for working in open cubicles. The new work environment requires a wide variety of changes, such as:

- Moving from an office to a cubicle
- Loss of private window space
- Acoustical or visual privacy
- New "huddle room" conference room strategy

ACTIVE (INTERACTIVE) COMMUNICATION	PASSIVE (NONINTERACTIVE) COMMUNICATION
Examples of active communication include: tours, lunch and learn presentations, and town halls. Tools to help managers communicate personalized message are best used when emotions are high, especially during the anger and bargaining phases of the project.	Passive communication is used to deliver facts regarding the new standard or move instructions for the upcoming relocation. Passive communication includes: memos, e-mails, and voice mail broadcasts.
ONE-WAY (ACCESS TO) COMMUNICATION	**TWO-WAY (ACCESS TO) COMMUNICATION**
One-way communication is received directly by the person at his or her desk. Examples of one-way communication include: voice mails, e-mails, and newsletters. One-way communication is good for basic, repetitive instructions or guideline communication.	Two-way communication is actually sought out and obtained by a person. Examples of two-way communication include: discussion databases, toll-free call-in numbers, or help desks. Two-way communication is good for training and education purposes.

FIGURE 8.7 Communication Matrix.

- Ergonomic adjustment
- Cubicles with either shorter or taller panels

The Tool Kit. To help managers communicate to employees about the universal space standard, American Express created a tool kit and user's guide that allow real estate professionals to work with individual business leaders to build customized communications strategies. The tool kit is designed to help managers lead discussions about change with the use of various tools and contains the following items:

- A mockup of each tool, along with its description
- Bound copies of actual "town hall" and "lunch and learn" visual presentations
- Examples of past newsletters
- Examples of the etiquette guide
- A user guide comprising:

  Overview of change management

  Overview of focus groups

  Recommended application of each communication tool

  Project time lines with timing of each tool

  Examples of feedback and measurement (pre- and post-move surveys)

Business Strategy Sessions. Using the tool kit, real estate professionals meet with the business unit leader to select tools, determine the schedule, and discuss budget considerations. The timing within the communication strategy is critical to the success of the project. For example, if presented too soon, etiquette guidelines—a passive, one-way communication tool—could be viewed as negative "rules" instead of positive tools to facilitate work in the new environment. The time chart in

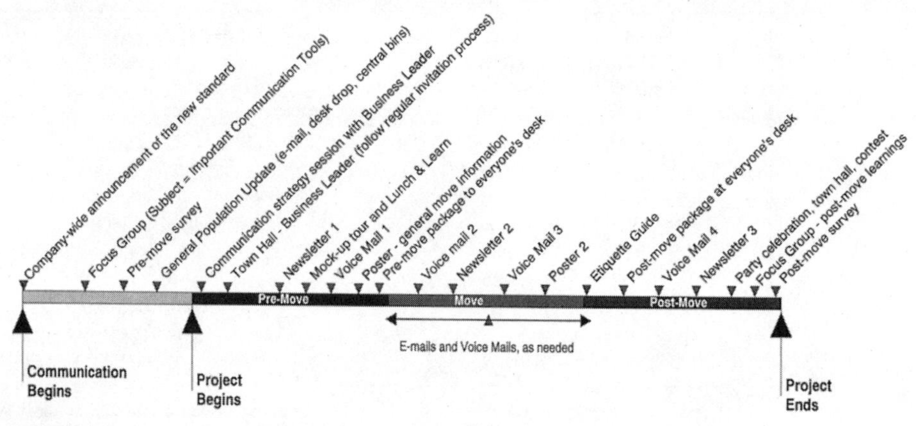

FIGURE 8.8 Communication Tasks and Schedule.

Figure 8.8 was created to help schedule the timing of the flow of communication into pre-move, move, and post-move phases. General company communication and research may occur prior to the business leader's actual project.

8.4 PLANNING FOR VIRTUAL SERVICES IN CORPORATIONS[16]

You've heard the terms "virtual companies," "virtual offices," and "virtual services," but your expectations of these may not match your experience. That's because the word "virtual" has an evolving, almost elastic definition. It's defined differently by different people, so don't take anything for granted. Make sure your definition matches that of your co-workers, customers, and colleagues. Generally speaking, and for our purposes, "virtual" refers to an entity, which may not physically exist, but does exist in essence or effect, often at another location (e.g., "virtual memory").

When discussing virtual entities, you should first establish where the employee sits in relation to the company and the company's servers. Typically, they will fall into one of four arrangements:

- Employees access information on servers located inside the corporate firewalls
- Employees access information on the corporation's virtual private network
- Employees access corporate networks from outside the corporate firewalls using the public Internet or via the corporation's private network (leased network) through a "last-mile" connection (remote employee to network via POTS, wireless, ISDN, DSL, cable, or satellite)

Whether the employee is operating "virtually" or whether the corporate network is, it's important that you anticipate and plan for changes to your work environment. Bear in mind that customers have been operating virtually for years now. ATMs, on-line ticketing, and stock transactions are just a few examples. Customers are leading the way, but, not surprisingly, employees are following suit. You need to be prepared for the "virtual" paradigm shift. Planning for virtual services requires that you take inventory in the following areas: software, facilities, staff reduction, and outsourcing.

8.4.1 Software and Application Developments

Innovations in software development, hardware storage, processing capabilities, and network alternatives are revolutionizing the service world. Software is now being developed by application service providers (ASPs) that provide software applications to companies from a server on an Extranet (outside the company). This means software is being written to run on servers that are accessed via a web-browser located on the employee's or customer's computing device, whether that device is a Windows-based PC, a dumb terminal, a palm pilot, or a TV. All the heavy-duty processing of the application occurs at the server site rather than on the computing device. This approach eliminates the need to distribute software to individual computers, makes software updates and upgrades easier to maintain, and, if designed well, creates a self-serve process between the employee/customer and the service provider.

Maybe even more importantly, new software applications are being rebuilt at a fraction of the cost it took five years ago and provide more flexibility in performance. With added flexibility and lower cost, corporations are faced with a difficult decision: developing and maintaining the software in-house or outsourcing the services entirely to an ASP. This latter alternative is appealing to companies when customization is less critical to deployment, services are not proprietary, and/or the service is cheaper than having to hire staff, maintain software, and invest in software updates.

Lastly, due to the modularity of software development, the ability for an off-site software application, an Extranet application, and on-site software application to interact together is more possible. Customizing your network is now, more than ever, cost-efficient and convenient.

8.4.2 Facilities

Software and communication advances may encourage you to rethink your strategic plan at both the portfolio and stacking-plan levels. As robust technology linkages become cheaper and more effective, a distributed workforce will likely increase production without incurring exorbitant overhead expenses. Therefore, when planning facilities, you must consider the following new-business conditions:

- Traditional corporate campuses may hinder labor recruitment practices
- Executive management, functions, and staff experts will disperse to run regional business operations around the globe
- Duplication of staff resources, like accounting, makes less sense with a distributed workforce
- Global operations that require 24-h servicing will increase facilities use and may require shared-office space
- An investment in high-bandwidth technology will make maintaining traditional adjacencies less important
- Common remote practices need to be developed and instituted to support an increasingly large distributed workforce
- The corporate office may become primarily resource centers for strategy development, virtual services, training, and cultural development
- Branch office strategy may shift from targeting customers to employees, since customers are doing more business virtually
- Distributed work can reduce real estate costs, but travel budgets may rise in order to establish and maintain strong relationships

With virtual employees connected electronically, the headquarters becomes a "strategy resort" of sorts. It houses a small "permanent" staff who lead and support the interactive corporate leadership activities from afar. In addition, headquarters may be redesigned to accommodate virtual employee or customer visits with "dormitory" or "hospitality" facilities. AT&T's School of Business has adopted this model to support strategy sessions for virtual employees. It allows easy transitions between work, dining, and informal relaxation—all within the same setting. Located adjacent to headquarters, it is a significant resource for international corporations in particular. Or, more commonly, corporations will install managed hotels on corporate sites, complete with hospitality services for the distributed workforce.

As the shift to virtual work occurs, the need for face-to-face interaction becomes paramount—only now the need is for periodic rather than continual face-to-face interaction. The workers need to spend enough time together at the start of a project to "get to know" each other and meet periodically to maintain the bonds that can only be generated through face-to-face interaction.

8.4.3 Staff Reduction

In the past, staff was needed to connect the dots between different "layers of information." Revolutions in software and networking technologies have made this system obsolete. For example, your expense reporting can now be electronically authorized by your supervisor and then routed automatically to the business cost center and the central accounting office.

Look to the "dot.com's" for the next steps in the employee/service-provider relationship. *Facilities.com, HR.com, audits.com*, etc. are in the works, and they will obviate the need for full staffing. We already have HR-by-800-number—they're linked to real time data and common policy interpretations, so the information is accurate! You can now administer your 401K plan on the Intra- or Internet, getting both immediate response and personal control. And, because it's networked, it's cheaper to administer. Administration can be done from anywhere and changes will reach a large distributed workforce or customer base with greater reliability and speed.

Training is another arena that will see a decrease in staffing. There is no reason to sit in a class and be fed information that can be more effectively distributed electronically and learned on a self-paced basis. Employees are more likely to stay on task if they can dictate the pace of their training. Face-to-face training promotes employee's feeling of being connected and can address questions outside the training scope, but for the most part, electronic training is a more than adequate substitute.

Sales forces, too, are becoming more streamlined. The largest Internet sales operation is not a "dot.com," but Cisco Systems, which sells 70% of its equipment through automated ordering on the Internet. Routine orders and re-orders are a direct connection between the customer and the delivery system. The new-model sales force can now focus on idea generation, problem solving, customer training and demonstrations, and customized orders.

When reporting, filing, and processing work are done electronically, it reduces the number of people who "touch" the work. Over the next 10 years, through the use of groupware, reporting and filing capabilities will require less paperwork and fewer people to monitor work progress. Authorization will be done in real time, electronically, and from multiple locations (if necessary).

8.4.4 Corporate Outsourcing is Expected to Grow Significantly

Given the new flexibility in software applications, corporations will need to consider outsourcing alternatives. Facilities planning must stay abreast of these opportunities, too. The capacity for collecting, analyzing, and distributing data through centralized applications is putting greater demands on system performance. The financial market sectors, for example, are beginning to demand that integrated systems be used for tracking standardized financial, real estate, human resource, and purchasing data to establish appropriate controls and identify savings opportunities.

Since systems can now differentiate routine bookkeeping tasks from decision making, they are in even greater demand. Functions that depend on centralized data or support features are most adaptable to outsourcing. Functions that rely upon decision making or interpersonal relationships will likely remain internal activities, but many other services have the potential for outsourcing. Each and every outsourcing decision will have to be examined for cost and risk potential, but here are some of the services you should consider outsourcing:

- Purchasing
- Invoicing
- Human resources

- Billing
- Postal
- Real estate
- Facilities management
- Information technology
- Administrative tasks

Each of the previous services has significant components of data storage and manipulation. Their management consists of routine activities that can be consolidated into centralized operations outside of the corporation. In the traditional workplace, the services were inseparable from the corporation. In the Virtual Age, it often makes fiscal sense to outsource these services to providers who specialize in them. For example, in late 1999 BP Amoco outsourced its U.S. accounting processes and related information systems to Pricewaterhouse Coopers (PWC) in a 10-year, $1.1 billion deal.[17] As compared to in-house staff and systems, long-term outsource contracts, such as the BP Amoco contract, are expected to save between 10–20% of corporate expenses.

8.4.5 Questions Facilities Planners Should Ask Business Management Regarding Virtual Services

1. What is the core business function (really)?
2. Is there currently a single- or multi-site emphasis? Any indication of change?
3. What competitive forces are driving location overall? Of discrete business units?
4. When do you think your support operation can function as a distributed workforce?
5. Are functional experts critical to your business and is physical adjacency necessary?
6. What demographic forces are impacting location both locally and remotely? Have you considered deploying alternative branch office strategies that are either (or both) customer or employee location dependent?
7. What internal services can be considered as having large components of "routine" activities? What new processes are currently being developed as virtual services; what processes in the future will be developed?
8. Do you expect to decrease staff due to virtual services or will business growth counter reductions?
9. Which segment of your business operation is most likely to be outsourced vs. retained in-house?
10. What groupware training development is planned to eliminate the need for reporting or filing records of work conducted by each employee or workgroup?
11. What are the elements of expense control being driven by corporate leadership? Are there other business operations that may effect our planning efforts, because this business operation could be merged, reduced, or eliminated?

Today, electronic and voice communication tools allow real-time interaction on shared data sets. The result is that the old workplace models that emphasize presence at a central location are no longer required. Now both individual and interactive work can be accomplished as part of remote or distributed workplace models. The tools are ahead of organizations' ability and willingness to utilize them fully. Shifts in demographics, and, more critically, required changes in business flexibility and strategy will drive that utilization. Remote interaction and remote work will continue to increase. There will be less and less emphasis on the central office, while the nature of the work done in the office changes. Layered with advances in the sophistication of centralized data functions and the ability to remotely interact with and communicate about this common data suggests that the distribution of work both inside and outside the corporation will continue to change. The balance will move toward greater concentration of focused activities in the corporation and greater distribution of customer and support functions outside of the corporation.

8.5 ESTABLISHING BEST PRACTICES FOR REMOTE WORK[18]

Remote work is here. In most Fortune 2000 companies, remote work is being performed by 15–60% of their workforce, yet very few companies have developed effective remote-work practices prior to rolling out extensive workplace options. Many companies recognize the need, but don't take appropriate action. American Express Co. and First Union Corporation are two of the exceptions that have benefited from their proactive approach. By deploying effective remote practices in advance of deploying new on-site alternatives, real estate is more successful and realistic about on-site alternatives.

Remote Work—What Is It? *Remote Work* is a term devised by Port Remote Services, Inc. and encompasses all types of work that occurs outside corporate headquarters; namely, mobile work and telecommuting. It also includes employees who travel, work at a customer site, or work from home, even if only occasionally at night or on weekends.

Establishing A Formal Remote-Work Program. More and more companies are providing employees with equipment and access to corporate information networks from remote locations. In order to maintain a credible, secure, cohesive, and quality organization, however, new operating practices are necessary to meet the needs of these remote workers.

How can you best support these remote workers? Provide them with the administrative support they need, give them the best technological and communication tools, and learn how to manage from afar. These practices are critical to the success of a remote-work program, especially during the transition. Fortunately, these practices also represent improvements in operating procedures, which better serve business management and the employee.

The Benefits of Mobile and Telecommuting Work. Remote work has the potential to increase the corporate shareholder's investment value beyond real estate occupancy savings, such as those reported by IBM ($100,000 million annually) in its North America operations and CISCO Systems ($7 million annually and initial build-out expenses of $30 million) also in its North America operations.

The real estate cost savings potential is dependent upon the asset real estate utilization ratio of gross square feet per employee. The asset utilization ratio is expected to decrease as the ratio of employees per on-site corporate workstations increases, as shown in Figure 8.9, assuming a larger telecommuting and/or more extensive mobile workforce. It is important to note that most corporations, no matter how large the proposed expense savings, are not deploying a distributed workforce because of real estate, but because they believe the competitive environment for their industry—both labor and operational—requires an effective distributed workforce.

The International Telework Association Council (ITAC) reports that a telecommuter can save 63% of the cost of absenteeism from that of an average on-site employee. Telecommuting also saves on job replacement and recruiting costs because of its versatility. In fact, many executives believe a future competitive advantage, from both a cost and operational framework, is an effective distributed workforce that demonstrates seamless integration with the customer, vendors, and corporate operations.

Many business units that operate remotely can provide some, or all, of the shareholder benefits listed:

- Increase employee effectiveness
- Improve customer services

FIGURE 8.9 Increased Real Estate Asset Utilization.

- Increase sales
- Attract and retain quality employees
- Reduce employee turnover expenses
- Reduce employee absenteeism
- Improve quality of life for employees
- Reduce space requirements for office personnel
- Reduce occupancy expenses—capital and operating expenses
- Quicken response time to business changes

Assuming it makes sense to build a quality, distributed workforce based on the previously listed benefits, the organization should begin to prepare and adjust by developing the following practices:

- Phase I: Remote Work Operating Practices—Developing Best Practices
- Phase II: Remote Management Tools—Integrating an Effective Remote Process
- Phase III: Support Services for Remote Work—Filling the Gaps

8.5.1 Phase I: Operating Practices—Developing Best Practices

Most organizations have been busy building new technology networks and have forgotten to communicate key processes to their internal customer—business management and employees. In many companies, you must know the right people in technology to get things done or life can be very frustrating!

Meeting Mutual Goals. The real estate or facilities groups are often the most motivated to undertake the development of corporate telecommuting or remote work practices, usually as a cost-reduction measure to justify alternative workplace developments. Human resources and information technology, however, are critical to developing effective best practices for remote work. Information technology's objective is to deliver a simple, cost-effective process. Human resources, on the other hand, would like to implement commonly known, visible, and equitable operating and management practices. Both groups typically want to limit corporate liability exposure associated with a dispersed workforce.

10 Questions Corporate Leaders Must Answer When Devising Remote Work Practices

1. What are our main objectives or motivations for adopting a remote work program?
2. Considering that the competitive operating trend is remote work, will we be at a disadvantage if we don't follow suit?
3. Will our internal customer, both business management and employees, be better served by a remote-work program?
4. What employee compliance requirements and operating practices should we install to make remote work run smoothly?
5. What executive and cross-functional leadership is necessary to accomplish our goals?
6. What corporate communication, processes, and systems need to be developed so that these new remote work operating and management practices become commonly known to all employees and are equitably implemented in the organization?
7. What support services are required to retain the amount of organizational cohesiveness needed to compete effectively in our industry?
8. What education or training is required to inform or support our employees through this transition and prepare them for the new work environment?
9. How are we going to assess the quality and effectiveness of the program over time?
10. How much does our company have to pay for program development, enterprise rollout and support, and what are the savings or benefits that will mitigate these expenses?

Best Practices. Remote work is an operating process that evolves with time and corporate commitment. It is the central organization's job to develop, commit, and make visible a series of "best practices." For example, corporate compliance, tools provisioning process, administration and specifications for equipment communications and software decisions, such as that shown in Figure 8.10, must be made.

The operating practices must speak to both telecommuters and mobile workers, whether they work occasionally or full-time out of the office. To this end, the corporation must provide how-to-guides that explain how to plan, administer, and operate in a remote-work environment. Specifically, employees need to know:

- How to get their work tool
- How to obtain equipment and software support
- How to be supported
- What security guidelines to follow
- Who pays for what expenses
- What responsibilities are the corporation's and what are their's

Traditionally, telecommuting has been structured as a corporate policy with a telecommuting agreement. Port approaches remote work as an operating practice that, if a policy is desirable, references the employee responsibilities typically found in a telecommuting agreement. A policy and formal telecommuting agreement, as traditionally deployed by human resources, is not necessary if the remote process is integrated into the organizational structure of operations for all employees.

Cross-Functional and Effective Leadership: Forming a Core Team. Senior management commitment to form a core team in advance of best-practice development will reduce the development to implementation time frame by approximately half. The core team should comprise three or four persons representing technology information, telecommunications, human resources employee relations/policy, and real estate. Although input from other corporate professionals will, of course, be requested during the development process, a large core team would not be an efficient use of pro-

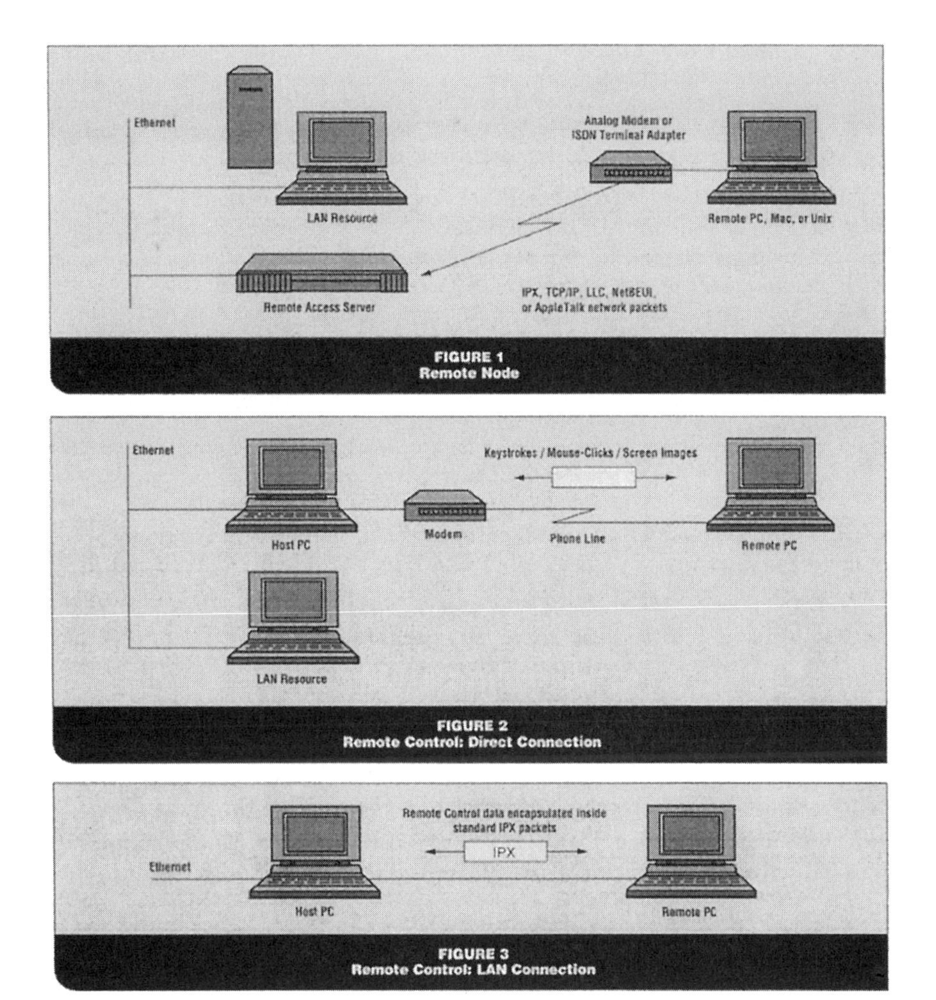

FIGURE 8.10 Equipment Provisioning and Support Practices.

fessional time. The core development team must have the respect of their functional senior management so that implementation is not delayed due to additional layers of review and changes. Typically, these managers are strategists with very effective project management skills who are willing to make recommendations on behalf of their divisions and are granted the authority to do so.

Comprehensive best practices for remote work can be developed in approximately two months. It may take another two months for business-unit implementation, however. The administration group that has the financial resources, which in many cases is corporate real estate (although occasionally information technology or human resources), typically pays for practices development.

8.5.2 Phase II: Remote Management Tools—Integrating a Remote Process

Senior management approval will not only be required to roll out a remote process, but also to obtain funding for remote-management tools necessary to ensure that telecommuting grows. Often, a successful remote-work program will only allocate funding for on-site, alternative-workplace strategies, but not remote-management tools. This can be a costly oversight.

Don't Oversell the Cost Savings. Real estate often oversells to senior management the ability to recoup occupancy cost expenses and then carries the burden to meet the objectives—without cross-functional support. Although real estate can target specific business groups to move employees remotely full-time, the majority of remote workers are occasional and hard to organize for a real estate cost reduction. It is estimated that the potential real estate savings, due to current information networks, will take three churn rates of corporate real estate, or approximately fifteen years, to realize.

Integrating an Effective Remote Process. Successful integration of remote work requires commitment and funding. In First Union's case, resources—staff and funds—were provided for implementation and support services. In our experience, this happens more often than not once organizations become familiar with our methodology—beware, however. Without sufficient funding for remote management tools, the practices developed in Phase I may be shelved and remote-work objectives will be difficult to meet. Funding should cover most of the tactics listed below:

- Create (or outsource) an implementation team separate from facilities managers to avoid cost-cutting resistance
- Reissue the corporate flex-place policy to reference the remote work process
- Conduct functional and business leadership meetings
- Sell and implement a process, NOT A PILOT PROGRAM
- Market a visible process by creating a web-based center
- Deploy an automated, integrated human-resource and information technology process for employees that want to move remotely
- Deploy tools that report and forecast your corporation's remote population
- Monitor remote-work conditions through surveys (preferably automated) and focus groups
- Deploy support services that fill organizational gaps (*see the next section, Support Services for Remote Work*)

Because real estate typically pilots facilities projects, they tend to do the same for remote work. We believe this is a mistake, because it is guaranteed to delay corporate implementation until after the pilot results are in,[19] which means you have to go back to executive management a third time. Ask executive management to approve the remote work process and get funds to market, monitor, support, and improve the program to get better results.

What About Productivity? We are often asked about productivity. Our response is this: "If there is no change in employee productivity, is remote work still a good direction for your organization?" Typically, the answer is yes. Productivity questions will need to be answered just prior to authorization of funds, so gather a list about other company success stories in preparation. This won't be hard, because far more succeed than fail. However, be prepared to defend the program even if there is no measurable increase in employee productivity.

Why? Because markers of increased productivity don't always come from employees. In fact, it is more likely that as a result of a distributed workforce, new automated processes, improved networks and software, not people, will bring increased production. Monitoring productivity is always advisable during transitional periods, but don't let the results carry undue weight. Remote work is a cultural evolution, similar to that of the 1980s when computers were introduced. What was true then still holds true today: *Operating practices will drive changes in management practices and proficiency comes only with experience, new ways of managing work and time.*

8.5.3 Phase III: Support Services for Remote Work—Filling the Gaps

Many organizations are not fully prepared to deploy all of the support services that are required for remote work because there is a certain amount of "wait and see" attitude. Support services are often slow in coming. Unfortunately, the need for these services is usually greatest at the outset. Following is a list of the support services that may not currently exist in your corporation, but should evolve if commitment to a distributed workforce is a serious goal:

- Process administration—Accountability and support is critical to quality service delivery
- Training—Continued management and technology training will facilitate cultural change
- Technology support services—To adjust and excel, remote workers need the appropriate tools and support services

Process Administration—Administering a Remote Work Process. The quality of service delivery from the corporation's central organization to its employees is dependent upon process visibility, ease of use, and maintenance of the process. To this end, an organization should assign, or outsource, a process administrator whose responsibilities will include process coordination in its entirety, cross-functional changes, program reporting and monitoring, program improvements and recommendations, etc. As keeper of the process, the process administrator can be held accountable for improving or changing the process if service quality is unsatisfactory.

Besides program coordination, another key responsibility for the process administrator is implementation services to internal business units. Although the practices designed should allow the business to implement without the central organization, the option to facilitate business implementation services should still be offered.

Training. Port's experience is that very little employee training happens at the point of transition into a remote work environment. With this in mind, corporations should provide strong, comprehensive reference materials regarding remote-work management practices. There are standard management practice materials available today so corporations may not have to create these in-house.

Although many employees will not participate in classroom training, we strongly recommend training if remote work is mandated and resistance is high. This training should occur after, not before, remote work begins. Since remote workers are more apt to feel alienated or isolated, they are more likely to benefit from the classroom environment in post-remote condition. Post-remote work training and focus groups that allow participants to participate in change management directions are very effective measures in evolving a remote workforce.

Technology Support Services. A distributed workforce requires integrated support services, whether they are located in-house or outsourced. The following are key technology groups critical to the deployment of a distributed workforce:

- Telecommunications—Administration of pre- and post-connectivity orders is required.
- Computer/software desktop services—Home installation for desktop computers is often arranged through this group. This group may also ensure that minimum computer and software standards are upheld.
- Security—Very often corporate security practices need to be upgraded for a distributed workforce. Maintaining corporate compliance objectives will require evolving security practices.
- Help desk services—A help desk can put out fires, reassure beginning telecommuters, and assist with technical and last-mile connectivity conditions.
- Network gateway groups—To assure adequate capacity planning, equipment at corporate gateways must continually be appraised for expansion.

Staying the Course. The author believes that remote work will evolve as information networks evolve. The first ten years of the twenty-first century will be a significant and historical evolutionary period for change in management practices; therefore, it is unreasonable to think that any one particular activity done today will solve *all* challenges. The goal is to make incremental cultural changes in ourselves, our work, and our management practices. Excelling at remote work must be the long-term objective, and living with less than perfect conditions is our current challenge.

8.6 BUSINESS CENTERS—AN ALTERNATIVE BRANCH OFFICE STRATEGY[20]

Traditionally, branch offices have been located to support customer sales or regional operations. Since more and more customers are using virtual interfaces, branch office strategies will also need to change. Furthermore, technology for the first time affords distributed back-office operations (i.e., telecommuting). Thus, smaller corporate resource centers are now being placed near employees, not customers. Rather than equip these smaller outposts with staff and equipment, corporations are turning to business centers. Naturally, this new practice will have a great impact on labor, branch, and back-office planning strategies. This section provides an overview of the business-centers industry.

Globally, there are approximately 5,500 executive-suites occupying 80,000,000 ft^2 and generating over $3 billion in revenues. Over two-thirds of these facilities are located in the United States. In the United States, the executive suite association (ESA) estimates that with the industry's efficient, cost-effective, flexible way of meeting office space needs, the number of square feet occupied by business centers will double over the next three years.

Executive suites, often known today as "business centers" because of the all-encompassing nature of their products and services, are offering solutions to the following issues:

- Soaring real estate rental costs and inflexible leasing practices
- Escalating lease-termination costs resulting from the proliferation of mergers, acquisitions, and consolidations
- Costly and diminishing pools of experienced and trained labor—domestic and globally
- Proliferation of, and increasing reliance on, sophisticated technologies and software applications that require on-site support and maintenance in the workplace
- Increased reliance on telecommuting and other alternative-office practices that require extensive physical and technical support to be effective
- Increased home- and virtual-office support-servicing needs
- Growth and competitiveness in international trade
- Increased cost of managing an administrative workforce
- Growth in corporate outsourcing
- Consolidation and vertical integration of real-estate businesses, including value-added business services for their tenants
- Growing dependence on the Internet and other technology networks

Companies no longer want to deal with the difficulty and expense of locating quality, small spaces in prime commercial buildings which offer little in the way of lease flexibility and demonstrate an unwillingness to accommodate uncertainty with respect to lease duration and/or space requirements. For example, Cisco Systems, Inc. locates new sales operations in business centers, if available, in new territories around the world. Their need to use business centers is short-term, until sales grow and new more permanent offices can be developed.

Nor are companies prepared to devote the time and expense necessary to negotiate the lease, plan the layout, design, decorate and furnish the space, select the equipment, and hire, train, and manage personnel. Wishing to avoid long-term commitments for space, people, and equipment, they are attracted to the business-center industry's products that eliminate these hassles and offer a number of side benefits as well. As a consequence, business centers, particularly the established international networks, are being aggressively sought out by large, well-capitalized companies in order to meet their branch-office needs. And, in response, these business-center networks are aggressively expanding and introducing new products and services to meet the demands of such sophisticated customers.

8.6.1 The Business Center Products and Benefits

Business centers are located in both urban and suburban locations. For a business center to be profitable and afford cost-competitive business services, it must be approximately 25,000–30,000 ft^2 or more in size, well-managed, and part of a larger corporate network. It takes approximately $2–3 million to build and furnish one business center. Thus, a well-capitalized, business-services company should be considered if a domestic or global partner is desired.

The following are the three primary products business centers offer:

Office or Workstation Rentals. For a fixed monthly fee, although in some cases per hour, clients have access to fully furnished, private offices or workstations, professional staff, cost-effective technology, kitchen, telephone, reception, and mail-handling services.

Virtual Office. For a fixed monthly fee, clients have access to telephone, mail handling, reception services, building-directory listing, and other business center services (i.e., conference space) as needed.

Business Center Services. Business center services include conference facilities, frequently workstations and support services. Support services include administration, concierge, mailroom, janitorial, photocopy, word and data processing, equipment rental, telecommunications (including high-speed Internet access), and videoconferencing.

Price depends primarily upon the size and location of the space to be leased, corporate credit, and the term of the contract. Lease terms have traditionally ranged from month-to-month to two years. Lease terms are extending, however, as corporate customers use the ability of the business center to manage the "hotelling" of its on-site staff. Business services typically have list prices that, depending on volume usage, may afford a discount price consideration.

The chart in Figure 8.11 summarizes the benefits and expense savings associated with a business center.

8.6.2 The Corporate Cost-Benefit Analysis

Although cost is not the only advantage favoring a business center, branch-office strategy, cost saving is an important factor. Following are some of the benefits to be considered.

Startup. In the start-up phase of a traditional corporate lease, up to ninety days can be allocated to finding and selecting suitable space in an appropriate location, negotiating a lease, designing, constructing, and finishing the space, and furnishing, staffing, and equipping the finished environment. Significant capital expenses are incurred and inflexible long-term commitments are made in the process. Start-up in a business center is instant. No capital costs are incurred and no long-term commitments of any kind are required.

Operations. Once in place, an independently maintained office incurs a significant ongoing financial as well as time commitment to its management. For example, lease occupancy costs (including utilities, cleaning, insurance, occupancy taxes, and furniture) are part of the monthly service fee of a business center.

Staff. Because the business center's staff also fulfills all back-office support functions, such as word and data processing, copying, facsimile and administrative, clerical, maintenance and technical support, clients have no need for staff to perform these functions. Additionally, because these staff members are available when needed and paid only for the time used, considerable savings are achieved since usage fluctuates according to the variable nature of the client's business activities. Consequently, clients of a business center operate at maximum efficiency and minimum cost to their company.

Business Center Advantages	
Corporate benefits	**Corporate expense savings**
• Offers access to office or conference facilities as needed. • Is convenient and immediate. • Creates a professional image. • Saves time in office selection and development—accommodates growth expansion easily. • Minimizes risks in new-venture locations. • Improves and accelerates business operations and enhances productivity with technologically advanced services. • Provides software and hardware support and maintenance. • Improves flexibility—responds quickly to business change opportunities. • Provides convenient locations—ideal for sales office drop-in support needs. • Includes access to competent office staff (well trained) without the hassles of hiring, training, and managing. • Offers specialty equipment not otherwise available (i.e., videoconferencing). • Permits establishment of local, business-community connections more quickly.	• Increases employee productivity through utilization of highly functional space, state-of-the-art equipment, and trained support personnel. • Avoids purchase or rental of back-office equipment. • Provides access to business services without the cost of the office. • Minimizes corporate, real-estate lease and operating costs. • Reduces capital and fixed operating costs. • Provides low-cost, instant global-expansion capability. • Eliminates tenant improvement costs. • Eliminates lease termination costs. • Eliminates need to pay for unoccupied expansion space. • Eliminates head-office support costs to hire, train, and manage back-office staff. • Eliminates equipment and furniture leasing expenses. • Saves money—no capital investment, no long-term obligations, and few fixed costs.

FIGURE 8.11 Business Center Advantages.

Equipment. Business centers' service-delivery capabilities are based on state-of-the-art hardware and software and include on-site staff to maintain and support them. Capabilities include advanced telecommunications, high-speed Internet access, video conferencing, word and data processing, high-speed and color copying, binding and fax services.

Business-center clients can totally eliminate their capital investment and reduce their ongoing operating costs by as much as 50% while enjoying increased efficiency as a result of access to a skilled on-site staff and the latest advances in back-office technologies.

The chart in Figure 8.12 compares the cost of an independently maintained office with essential amenities for two executives and appropriate back-office support with the cost of housing two executives at a business center. In each case, the rent is \$40 ft².

Cost–benefit analysis			
Item	Own office	Business center	Savings
Start-up	10,000	None	10,000
Capital investment	20,000	None	20,000
Annual rental	36,000	60,000	(24,000)
Annual operational cost	50,000	24,000	26,000
Annual personnel cost	60,000	None	60,000
Grand total	\$176,000	\$84,000	\$92,000

FIGURE 8.12 Business Center Cost–Benefit Analysis.

8.6.3 The Business Center Customer is Changing

Initially, customers for business centers were end-users, typically local entrepreneurs or self-employed professionals. Their needs were basic. The appeal was based on price, convenience, lack of cost-effective alternatives, and location. These attractions were enhanced as technology was introduced into the back-office support environment. Beginning in the 1980s, the client base exploded as major and middle-market corporations, as well as professional and service firms, came to recognize the value and benefits the industry offered.

With this change in end-user came a change in buying practices and expectations. Corporate buyers are typically not end-users, but representatives of an in-house professional department responsible for meeting their company's real estate needs. Their needs are different from those of the end-user, although they must be assured that end-user needs are satisfied. For example, for them, price is more important than location. They need access to information quickly to form a basis for evaluating competing options, a sense of security concerning the financial stability of the vendor, and assurance that the vendor can meet the corporation's technical, legal, and finance departments' expectations.

Initially, corporate real estate practitioners viewed the industry as addressing short-term occupancy needs only and thus the concerns previously described were tempered by virtue of the short-term nature of the relationship. Today, as they look to the business center industry to satisfy long-term needs as well, these issues are taking on increasing importance. For example, Fidelity utilizes business centers for smaller non-regional operations.

Accordingly, business center networks confer a variety of benefits, as described in Figure 8.13.

Network advantages	
Network partner benefits	Network partner attributes
Partnering has the potential to: • Reduce time and expense to find quality, well-located office space around the world. • Reduce roll-out cost for new products nationally or internationally. • Assure that contractual obligations are met. • Provide easy access to problem solving and cost management due to centralized back-office operations. • Reduce or eliminate multiple invoicing from separate landlords. • Reduce legal expense because a standard form of contract is accepted at all locations. • Achieve savings on multi-location contracts. • Ensure smooth, uninterrupted functioning of non-headquarters personnel and seamless integration with head-office systems and procedures. • Offer access to other locations on an as-needed and preferential basis.	Consider how the following network attributes make partnering preferred: • Immediate access to information regarding product availability, service capabilities, and pricing • National and international distribution. • Central reservations and sales. • Standard products and services. • Assurance of financial stability—business center profitability and growth to meet expanding corporate needs. • Products with a predetermined price range. • Central billing. • Uniform contractual terms. • Preferred vendor status. • Technical support capabilities that are compatible with their own internal systems. • Reciprocal agreements with other network offices.

FIGURE 8.13 Business Center Network Analysis.

8.6.4 Current Industry Players

Only recently has the business center industry begun to consolidate in any form. In the past three years, CarrAmerica Realty Corporation has bought OmniOffices and some of the HQ franchises. Alliance National, Inc. has bought Office Plus, certain HQ franchises, and more recently merged with InterOffice Holdings Corp. and Reckson Executive Centers to form a new company, Vantas. As this chapter is being written, CarrAmerica's executive suite affiliate, HQ Global Workplaces, Inc. has entered into a merger agreement with Vantas Incorporated.

Although all business centers can be valuable to corporations in the implementation of their branch-office strategies, the chart in Figure 8.14 outlines the four primary industry players in both domestic and global markets.

Business centers	Description
CarrAmerica Realty Corp. Washington, DC http://www.carramerica.com or http://www.omnioffices.com	A self-administered REIT, owning interests in over 250 office properties in over 100 locations. In addition, through an affiliated company known as HQ Global Workplaces, Carr owns/franchises approximately 175 centers acquired through acquisitions of HQ and Omni business centers beginning in 1997. Business Centers, located primarily in the United States are owned.
Regus *Staines, Middlesex* *United Kingdom and* Purchase, New York http://www.regus.com	Established in 1989, Regus is the leading European operator of business centers worldwide and has committed to building an extensive network of U.S. centers. All centers, approximately 200+, are owned (vs. franchised).
Vantas (member Alliance Business Centers Network) Costa Mesa, California http://www.vantasinc.com	Approximately 320 business centers primarily in the United States, but also worldwide. Approximately four-fifths of the centers are franchised centers; the remaining are independently owned.
World-Wide Business Centres New York, New York http://www.wwbc.com	World-Wide Business Centres, established in 1970, is the licensed operator of over 110 independently owned business centers located in 14 countries. It is the longest-running, business-center-network operator in the industry.

FIGURE 8.14 Key Industry Business Center Company Descriptions.

All of the business centers mentioned previously are offering occupancy arrangements for quality office space. Companies like Regus and Vantas are positioning themselves at the premium end of the market (occupying the newest building(s) in a given market), whereas World-Wide Business Centres is positioning itself at the value end of the market (emphasizing quality of service offered from Class A buildings). All five companies have concentrated their international activities in Europe, but all are actively seeking to expand into other geographic regions. Regus and World-Wide Business Centres are at this time considered truly international office networks. Although there are other business service companies, like Kinko's and MBE Business Express, they do not provide office space for rent. Going forward, corporations should look for new entrants, continued industry consolidation, and well-capitalized companies taking the lead as they compete for market share and corporate relationships through their evolving network structures.

ENDNOTES

1. Copyright © 1999. All rights reserved. Sun Microsystems, Inc.
2. Copyright © 1999. All rights reserved. Sun Microsystems, Inc. and Port Remote Services, Inc.
3. Pfeffer, J. 1994. *Competitive advantage through people,* Chap. 5. Boston: Harvard Business School Press.

4. Weisbond, M. R. 1987. *Scientific management revisited: A tale of two Taylors,* Chap. 1. San Francisco: Jossey Bass.
5. Taylor, F. W. 1911. *The principles of scientific management.* (reprinted 1993.) London: Routledge/Thoemmes.
6. Bane, P. W., and Bradley, S. P. 1999. *The light at the end of the pipe.* Mercer Management Consulting and Harvard Business School.
7. International Data Corporation. 1999. *IDC reveals home office Internet use reaches record high.* (September 15).
8. Bane, P. W., and Bradley, S. P. 1999. *The light at the end of the pipe.* Mercer Management Consulting and Harvard Business School.
9. Mercer Management Consulting. 1999. *Study of 1000 people in the Washington, DC area.* May.
10. Copyright © 1999. All rights reserved. First Union Corporation.
11. Schein, E. 1996. *Sloan management review.* (Fall). Cambridge: MIT Sloan School of Management.
12. Henning, J. 1998. *The future of staff groups.* San Francisco: Berrett-Koehler Publishers, Inc.
13. Copyright © 1999. All rights reserved. Cisco Systems, Inc.
14. Copyright © 1999. All rights reserved. Port Remote Services, Inc.
15. Copyright © 1999. All rights reserved. American Express Company.
16. Copyright © 1999. All rights reserved. Space, LLC and Port Remote Services, Inc.
17. Collett, S. BP Amoco outsources financial systems to Pricewaterhouse Coopers. *Computerworld,* Online News; 11/10/99 12:10 P.M.
18. Copyright © 1999. All rights reserved. Port Remote Services, Inc.
19. Assuming a facilities project, which may include piloting new on-site furniture, the delay in process approval could be an additional six to twelve months.
20. Copyright © 1999. All rights reserved. World-Wide Business Centres, Inc. and Port Remote Services, Inc.

CHAPTER 9
FACILITIES CONDITION ASSESSMENT

Dr. Harvey H. Kaiser
Founder and President Inc.
HHK Associates, Inc.
An architectural, facility management and urban planning consulting practice
E-mail: hhkaiser@worldnet.att.net

Thomas Davies
Assistant Director of Physical Plant
Design and Construction at Amherst College
E-mail: tdavies@amherst.edu

9.1 THE FACILITIES AUDIT

9.1.1 Purposes and Goals

The facilities audit is an essential tool for the determination of conditions of an organization's capital assets, including buildings, infrastructure, and fixed and movable major equipment. Organizations perform facilities audits for strategic planning purposes associated with financial planning of capital assets, for cost allocations and analyses of productivity of capital assets, and for strategic decisions on retaining or disposing of assets.

The audit has multiple purposes that include use by consultants and facilities management (FM) staff of an organization for the evaluation of buildings and infrastructure. Consultants retained to evaluate conditions for an owner can provide analyses that guide decision making for acquisition, renovations, or disposal. The audit provides a database that can determine the relative value of a capital asset by the effect of deterioration and other deficiencies that will require capital renewal for major systems and components, and corrections necessary for compliance with life safety, environmental, and accessibility regulations.

An organization's FM staff responsible for maintenance, capital renewal, and capital budgeting will find the audit useful to establish a baseline of conditions and identify specific deficiencies for corrective action. Circumstances may differ between organizations that undertake a comprehensive survey of all facilities for the first time, or those that have a limited set of goals for determining existing conditions. The basic principles and methodologies presented here can be used for all types and sizes of organizations, from a single structure to a facility consisting of multiple building complexes in dispersed locations. A continuous process of facilities audits, beyond a one-time program, provides up-to-date capital renewal priorities and can generate a significant portion of routine maintenance workloads. A related benefit is the development of a culture for maintenance staff personnel for routine inspections and reporting of facilities conditions. An effective audit program will extend the useful life of facilities, reduce disruptions in use of space or equipment downtime, and improve FM relations with facilities users.

The facilities audit is a process of:

1. Development of a database of existing conditions and functional performance of buildings and infrastructure by conducting a facilities audit
2. Assessment of plant conditions by analyzing the results of the audit
3. Reporting and presenting findings
4. Development of a plan for corrective actions as a component of a capital plan

The facility audit systematically and routinely identifies building and infrastructure deficiencies and functional performance of facilities by an inspection program and reporting of observations. The audit process assists maintenance management and the institution's decision makers by recommending actions for major maintenance, capital renewal, and deferred maintenance planning.

A well-designed audit has the overall goal of determining the *adequacy* of capital assets to support mission and programs and guide capital planning, including:

- Conduct *qualitative* and *quantitative* assessments of facilities to determine *adequacy* to support mission and programs
- Develop a baseline of current conditions of capital assets
- Determine priorities, costs, and phasing of correcting *adequacy* deficiency correction measures necessary for renovations and modernizations, and deterioration of systems and components and compliance with life safety, environmental, and accessibility requirements
- Restore functionally obsolete facilities to a usable condition
- Eliminate conditions that are either potentially damaging to property or present life safety hazards
- Identify capital renewal and replacement projects in order to improve *adequacy* of facilities to support mission and programs
- Forecast future capital renewal needs for capital budgeting and scheduling
- Provide for a routine inspection process for all facilities

9.1.2 Audit Process

A successful facilities audit program requires the support of senior financial management and FM. Planning for an audit program should incorporate their review of the process and form of results to ensure that requirements are met for capital asset planning and allocation of resources. Senior management's involvement in facilities audit planning can result in reliable sources of information on the plant conditions and in determining funding needs.

Conducting a facilities audit requires a clear set of objectives before committing staff and financial resources. Whether an organization has previously conducted facilities audits or is beginning one for the first time, there should be thorough preparation to assure understanding and support of all staff involved in the process. Management and inspectors must make a commitment to collect accurate data and identify deficiencies as objectively as possible. Cost estimates to correct deficiencies should be based on current local data and priorities set by established criteria. A key ingredient to success is the assignment of an audit manager for oversight of the process, whether conducted by in-house staff, consultants, or a combination of both.

An audit should be a comprehensive *quantitative* and *qualitative* assessment, collecting information on *adequacy* of facilities to support mission and program. The *quantitative* assessment evaluates the physical condition of building and infrastructure components; the *qualitative* assessment evaluates the capability of a facility to support a program through comparison to a common set of *quality* standards. Alternatively, the audit can be selective for specifically an evaluation of the physical condition of components of building types such as roofs, or a unique collection of information for safety or new regulatory requirements. The audit is encouraged to perform a comprehensive assessment as an integrated approach to gain a complete picture of funding requirements to correct all deficiencies for capital planning. Integration of findings from *quantitative* and *qualitative* assessments provides greater credibility for defined improvement projects.

Audit results are used to plan routine and major maintenance (capital renewal), urgent and long-term measures to correct facilities conditions through a deferred maintenance reduction program, and capital renewal and replacement budgeting and planning. For example, audit data collection forms can be used to provide a description of conditions for a special facility type. The audit methodology can also be applied for a survey of one or more systems and/or components that have a demand for which information is needed because of legal compliance requirements to meet new regulations or codes.

Quantitative Assessment: Condition Inspections. Quantitative assessments (condition inspections) are designed to provide a record of capital asset physical deficiencies and estimated costs to eliminate deficiencies. The *deficiency-cost* methodology requires training inspectors for a "self-audit" (or giving clear instructions to consultants) to produce an objective and consistent database for future reference. Incorporated into the training is the development of a process of continually observing and reporting deficiencies, and conveying results into maintenance work and capital planning. A successful audit program will introduce a culture of observing and reporting conditions not on a one-time basis, but as a regular part of supervisory and trades people work.

The condition inspection part of a facilities audit is a visual inspection of buildings and infrastructure systems and components. Table 9.1 outlines the systems and components for building and

TABLE 9.1 Building Component Descriptions

Primary systems	Secondary systems
1. Foundation and substructure Footings Grade beams Foundation walls Waterproofing and underdrain Insulation Slab on grade	5. Ceiling system Exposed structural systems Directly applied Suspended systems
2. Structural system Floor system Roof system Structural framing system Pre-engineered buildings Platforms and walkways Stairs	6. Floor covering system Floor finishes
3. Exterior wall system Exterior walls Exterior windows Exterior doors and frames Entrances Chimneys and exhaust stacks	7. Interior wall and partition systems Interior walls Interior windows Interior doors and frames Hardware Toilet partitions; Special openings: access panels, shutters, etc.

TABLE 9.1 Building Component Descriptions (Continued)

Primary systems	Secondary systems
4. Roof system	8. Specialties (examples)
Roofing	Bathroom accessories
Insulation	Kitchen equipment
Flashings, expansion joints, and roof hatches, smoke hatches, and skylights	Laboratory equipment
Gravel stops	Projection screens
Gutters and downspouts	Signage
	Telephone enclosures
	Waste handling
	Window coverings
Service systems	
9. Heating, ventilating, and cooling	12. Electrical lighting
Boilers	Lighting fixtures
Radiation	Wiring
Solar heating	Motor controls
Ductwork and piping	Motors
Fans	Safety switches
Heat pump	Telecommunications and data
Fan coil units	Emergency/standby power
Air handling units	Baseboard electric heat
Packaged rooftop A/C units	Lightning protection
Packaged water chillers	
Cooling tower	
Computer room cooling	
10. Plumbing system	13. Conveying systems
Piping, valves, and traps	Dumbwaiters
Controls	Elevator
Pumps	Escalators
Water storage	Material handling systems
Plumbing fixtures	Moving stairs and walks
Drinking fountains	Pneumatic tube systems
Sprinkler systems	Vertical conveyors
11. Electrical service	14. Other systems
Underground and overhead service	Clock systems
Duct bank	Communications networks
Conduits	Energy control systems*
Cable trays	Public address system
Underfloor raceways	Satellite dishes, antennae
Cables and bus ducts	Security systems
Switchgear	Sound systems
Switchboard	TV systems
Substations	
Panelboards	
Transformers	

TABLE 9.1 Building Component Descriptions (Continued)

Safety standards	
15. Safety standards	
Asbestos	Detection and alarm systems
Code Compliance	Disabled accessibility**
Egress: travel distance, exits, etc.	Emergency lighting
Fire ratings	Hazardous/toxic material storage
Extinguishing and suppression	

Infrastructure component description	
1. Site work	4. Utility delivery systems
Curbing and wheel stops	Central energy plants
Fencing	Chilled water distribution
Parking lots	Compressed air
Roads and drives	Distilled water
Walks, plazas, and malls	Domestic water
Water retention	Electrical distribution
2. Site improvements	Energy monitoring and control
Landscaping	Fire protection
Lighting	Heating hot water
Furniture: benches, bike racks, bus stops and shelters, waste receptacles, kiosks, signage	Irrigation water
	Natural gas
Emergency telephones	Steam and condensate
Flag poles and stanchions	Storm drainage
Sculpture, memorials, and fountains	Sanitary sewage
	Utility tunnel structures
3. Structures	Wastewater treatment and collection
Bridges	Water treatment and distribution
Culverts	
Retaining walls	

* Energy audit.
** Disabled accessibility audit.

infrastructure inspections. Primary systems include foundations, structural system, exterior wall system, and roofing system. Secondary systems include interior work that makes the facility usable: ceilings, floors, interior walls and partitions, and specialties. Service systems include all operating systems, such as HVAC, plumbing, and electrical systems. Safety standards, including life safety and code compliance, are grouped together. Decisions on where to record data for a unique system or alternatives to the component definitions should be flexible and decided in the initial organization of the audit and instructions to inspectors. Major infrastructure components are listed in groups that can be inspected by appropriate specialized staff or consultants. These lists should also be reviewed and flexibility retained until a final list is adopted.

Design of the condition inspection forms and methodology is based on how a building or infrastructure is constructed and how inspectors would logically proceed to make observations and collect data for deficiencies and costs of corrective measures. Building inspections begin with how a structure is placed in the ground, then travels upward to structural framing, exterior wall enclosures, and roof, and then moves to the interiors. Each service system—heating, ventilating, air conditioning, plumbing, and electrical—is inspected separately. A comprehensive condition inspection program provides a complete inspection of architectural, civil/structural, mechanical, electrical, and safety components of each facility. Infrastructure inspections are conducted in a similar, methodical manner.

Infrastructure is nonbuilding improvements that directly support the operating of a facility. Included are central utility plans, utility distribution systems, roads, walks, parking surfaces, culverts, drainage and watering systems, and other structural site improvements. Often representing a substantial investment, potential malfunction of infrastructure components can result in interruptions in normal operations, catastrophic failures that result in loss of life, and energy inefficiencies. Unlike buildings, infrastructure is generally not observable; that is, until a disruption occurs and immediate corrective action is necessary.

An infrastructure assessment methodology should provide an organization with a comprehensive overview of the current condition and code liabilities present, as well as a predictive model projecting the costs associated with the continuing degradation of these facilities. Desirable characteristics of a methodology are to provide information that avoids the cost of extensive forensic analyses or extensive direct observation of inaccessible utility components. The assessment should be integrated with a GIS for coordination of information in a database. In a campus setting, a survey based on photogrammetric mapping provided by aerial photography and confirmed by field observations can supplement information available from construction documents and maintenance staff knowledge and experience.

A proposed infrastructure assessment methodology is based on a series of actions for implementation that can progress simultaneously:

- Initial data gathering of existing conditions in a standardized format of mapping and infrastructure categories integrated with a GIS
- Calculations of the condition of unobservable infrastructure components remaining life cycle based on industry standards for life cycles of each type of component, and refined through collection and analysis of the organization's experience with regard to life expectancy of various infrastructure components
- Projections of the "expired" life cycle to represent capital renewal needs

The challenge of infrastructure condition assessment is that a significant portion of the infrastructure generally is not directly observable. Extensive forensic work is expensive and time-consuming and may not yield information more readily available from campus maintenance staff. Alternatively, a great deal of insight can be gained through the analysis of readily accessible existing utility information. A preferred approach for both observable and nonobservable infrastructure components is based on available documentation, interviews with maintenance staff most familiar with the components, and field inspections. Maintenance staff information and campus records can be supplemented by forensic techniques for unique conditions, as necessary.

Qualitative Assessment: Functional Performance Evaluations. A comprehensive facilities audit includes an assessment of *qualitative* aspects of functional performance of facilities. The *qualitative* assessment complements the assessment of physical conditions to complete the picture of *adequacy* of facilities to support mission and programs. Combined results of the two assessments support the preparation of a complete and credible capital plan.

The functional performance evaluation requires standards to guide the assessment. The standards represent a common set of characteristics that are a baseline quality level for the conduct of a program. The specific nature of a facility will determine the content of standards for space configuration, finishes, equipment, mechanical, electrical, lighting, communications, and other unique requirements. Application of the quality standards in an inspection process will highlight corrections of deficiencies necessary to meet criteria for the conduct of a program. Estimated costs include those improvements required to correct obsolescence and to bring facilities to contemporary standards to meet program needs.

Prior to summarizing the results of the condition inspection, functional performance of a facility to satisfy requirements for existing or planned activities should be integrated into a complete summary of facilities capital needs. For example, in the condition inspections, a building may by found to have significant deficiencies with costs of corrective measures exceeding replacement value. However, for historic, aesthetic, or other reasons, the building may be retained for remodeling and extended use. Major renovations resulting from functional performance evaluation can include all identified priorities in a single improvement program. On the other hand, a facility may be considered for remodeling, but demolition may be recommended because of conflicts with plans for future land use or sale as a source of revenues.

9.1.3 Audit Phases

There are four phases in the facilities audit process: 1) designing the audit, 2) collecting data, 3) summarizing the results, and 4) presenting of findings (Figure 9.1). In this systematic approach, the scope of the audit is first determined, audit team selected, and inspection planned. In the second phase of the audit, data is collected through inspection of buildings and infrastructure and functional performance evaluations. In phase three, the data collected is evaluated, summarized, priorities set, and future audits planned. In the final phase, results are presented to various audiences.

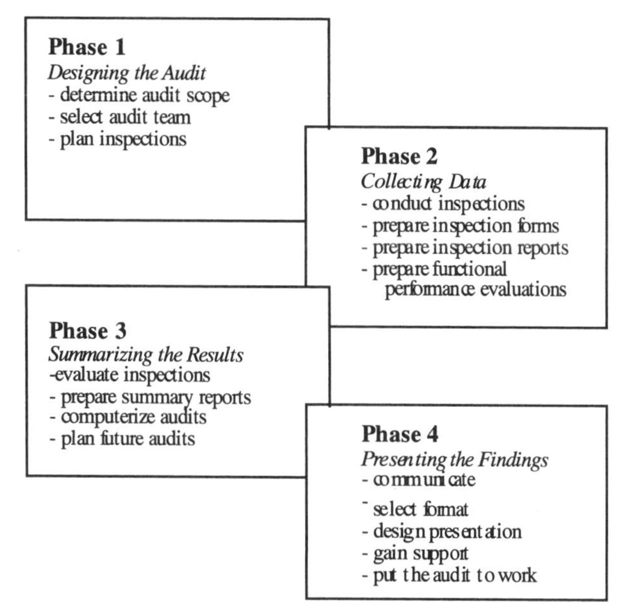

FIGURE 9.1 Facilities Audit Phases.

The audit is a method of collecting information on the current maintenance conditions and functional performance of a facility. Included in a comprehensive audit are:

- Inventory of facilities that provides descriptions of characteristics
- Inspections of existing buildings and infrastructure conditions
- Evaluations of functional performance
- Recommendations for correcting observed deficiencies

A parallel program of quantitative and qualitative assessments accomplishes the inspection process. The *quantitative* component—the condition inspection—is a systematic inspection of buildings, infrastructure, and major fixed and movable equipment by systems and components following the sequence of construction. The *qualitative* component, conducted by a comparison of facilities to a baseline set of quality criteria, is evaluated for renovations and modernizations necessary to meet contemporary standards. The methodology can be implemented in a comprehensive review of all facilities and infrastructure systems and components, or adapted to meet special conditions and needs of an organization.

Phase 1—Audit. The first phase of the audit process—designing the audit—includes:

- Determine audit scope
- Select the audit team
- Designate audit manager
- Plan the inspection

Building and infrastructure inspections and quality evaluations can represent a significant commitment of resources. The audit design should be prepared to adjust staff schedules for in-house inspectors and allow adequate time for managing consultants when they are used. Direct costs must be budgeted for reproducing plans, preparing building histories, laboratory testing, and producing other information. Cost-benefit analysis of in-house staff, consultants, or both should include schedule for completion, costs (including staff downtime or replacement personnel), and intangibles of retention of inspection knowledge and introduction of a culture of regular inspections by maintenance staff. Thoroughness of preparation, including staff training, will ensure the usefulness of facilities audit results.

An audit's scope is determined by goals and objectives for conducting condition inspections, methodology for inspections and assessment, use of quality evaluations, and intended use of audit results. The goals and objectives for conducting condition inspections and quality evaluations can include a broad overview of plant conditions and deficiencies, the determination of projects for major and minor maintenance, development of deferred maintenance reduction programs, and providing forecasts of future capital renewal and replacement.

The choice of specific facilities to be inspected depends on the audit's goals and is often influenced by deadlines, available resources (staff and budgets), building access, and seasonal conditions. In determining the scope of the audit, keep in mind that inspection methodology produces a database on building and infrastructure deficiencies and costs of correcting them. Information obtained from inspections should be tailored to staff ability and time to analyze results and implementing corrections for observed deficiencies.

An inspection program should be designed to assess all facilities on a minimum 3-year cycle. Facilities with special concerns (e.g., public assembly and unique structural design) should be inspected on a more frequent schedule and coordinated with safety inspections and other mandatory inspection requirements (e.g., sprinklers, elevators, load testing, etc.).

The condition inspection is based on an exterior and/or surface observations. Invasive or other destructive methods, including disassembly of materials for testing, is warranted only where conditions indicate the need for further examination.

The following checklist should be reviewed in planning an audit's scope:

- Define audit goals and objectives
- Determine inspection methodology (in-house staff and/or consultants)
- Prepare baseline quality standards
- Define the intended use of results and report formats
- Review available information on facilities to be inspected
- Prepare preliminary outline of facilities and components to be inspected
- Establish deadlines, availability of staff and/or consultants, and access to facilities

Methodology for inspection and assessment includes choosing forms used for data collection and deciding whether to use staff or consultants or both. Defining the use of information and planning how to computerize inspection results at an early stage of the process can avoid the need to redo work. The intended results should be defined and final report formats designed to assure that adequate inspection data is collected and analyses can be prepared.

Thorough planning of the field portion of building and infrastructure component inspections is essential to produce accurate, timely, and useful results. There are several critical factors to be considered in planning inspections, including:

- Scope of audit
- Scheduling inspections
- Responsibility for compilation and evaluation of results
- Information requirements for inspections assignments
- Training
- Tools and equipment
- Notification to building occupants
- Emergency work

Phase 2—Data. The second phase of the audit is collecting condition data by building and infrastructure system and components, quality data by inspecting facilities in comparison to a set of baseline criteria, and preparing the inspection reports. Collecting data for the facilities audit in the deficiency-cost methodology should use standard inspection forms that are completed for each of the suggested fifteen building and four infrastructure components (Table 9.1) and quality characteristics. The design of standard forms provides a consistency for the inspection process, enabling modifications to the number or terminology of components.

Building and infrastructure inspections provide observations of conditions to be summarized on inspection report forms for individual components. A sample condition inspection form (Figure 9.2) contains three sections:

1. Facility inspection data
2. Component deficiency description
3. Component deficiency evaluation

Use of the form provides a reference source for capital renewal planning and project management. Section #1 of the form identifies the background on the name of the facility and component, inspector's name, and date of inspection. Section #2 describes the deficiency, location, and suggests a corrective action and the need for specialized testing. Section #3 provides information, including priority rating, suggested year of implementation, and cost estimates that can be incorporated into the summary of projects for a capital improvement program.

The condition inspection thoroughly examines the component and documents all deficiencies requiring corrective measures. In a deficiency-cost condition audit, the inspector has the responsibility to record deficiencies, assign a priority, and estimate the cost for corrective measures. Inspectors use the component form as a guide in making field notes of observations describing the deficiency and defining corrective measures. Photographs or video recordings can supplement the inspector's field notes when preparing the report. Drawings of building floor plans or infrastructure installations should be noted to locate the deficiency. Each deficiency is assigned a sequential number corresponding to the written description, designating it on the drawing, and providing a reference for job planning and future audits. Any inaccurate information on the component description should be noted and corrected in the information base. Conditions observed by inspectors requiring testing or specialized inspection skills, such as removal of material or inspecting hidden conditions, should be identified for follow-up work.

1. FACILITY INSPECTION DATA			
Facility #:		Facility Name:	
Component #:		Component Name:	
Inspector:		Date:	

2. COMPONENT DEFICIENCY DESCRIPTION
#1.
#2
#3
#4
#5
#6
#7
#8

3. COMPONENT EVALUATION

Project Title	Proj. #	Priority	Year	Project Estimate		
				Const $	"Soft" $	Total $

FIGURE 9.2 Sample Condition Inspection Form.

The condition inspection report form is prepared after fieldwork is completed. The inspection form is the basic information-collecting instrument, requiring a simple process of organization before issuance to inspectors. The form is designed so that it can generate data for a variety of summaries and is easily adapted to simple word processing or spread sheet formats. Report preparation following fieldwork enables the inspector to compare notes with other members of the audit, request any specialized testing, and consult resources available for estimating costs of corrective measures.

A comprehensive facilities audit includes an evaluation of building functional performance that can be conducted simultaneously with a condition inspection. A functional performance audit of a facility is a survey of the programmatic capability and adequacy of building interior spaces based on *quality criteria*. The goal of this audit component is to provide information that enables an evaluation of the use and usefulness of existing space to meet contemporary standards and possible alternative uses. In combination with the facilities audit, the functional performance audit can guide

1. FACILITY			
Building #:	________	Building Name:	________
Gross Square Feet	________	Building Use:	________
Net Assignable Square Feet	________	Date:	________
Year Occupied			

2. QUALITY CRITERIA (Sample Classroom)		
Characteristic	Quality Criteria	**Score (1-5)**
a. Functional Relationships	Space configuration and size satisfies teaching/learning requirements for primary activities.	
b. Architectural	Room finishes are appropriate for space function, easily cleaned and maintained.	
c. Mechanical	Heating and cooling systems with adequate controls provide a comfortable year round teaching and learning environment.	
d. Electrical / Data / Telecom	Adequate electrical capacity, data and telecomm connections.	
e. Lighting	Adequate quality and level of lighting provided with controls to adjust levels as required for specific needs.	
f. Plumbing	Adequate services as required for space activities.	
g. Acoustics / Noise / Vibration Control	Architectural finishes with appropriate sound and vibration absorption and reflective qualities sufficient to provide a teaching and learning environment free of distracting noise and vibration levels.	
h. Instructional Support	Spaces equipped appropriately with chalkboards, whiteboards, projection screens, and a full range of audio-visual equipment.	
i. Furnishings / Major Equipment	Seating, demonstration tables and other major equipment appropriate to the space configuration and activities.	
j. Accessibility	Compliance with ADA standards to meet program accessibility standards.	
k. Other		

***4. COMMENTS** (Identify major deficiencies and recommended corrective action(s)*

____

____

____

____

____

____

____

____

5. OVERALL BUILDING /PROGRAM RATING

____	(A)	Optimum Space	____	(D)	Poor Space
____	(B)	Adequate Space	____	(E)	Unsatisfactory Space
____	(C)	Fair Space			

Prepared by: ________________ Date: ________________

FIGURE 9.3 Sample Quality Criteria Evaluation Form.

decisions on the appropriateness of an investment in a facility for renovations and modernizations to eliminate obsolescence, or alternatively, to consider a replacement.

The quality criteria evaluation form (Figure 9.3) provides a record of the characteristics of a facility with a choice of five ratings for programmatic adequacy (optimum, adequate, fair, poor, or unsatisfactory). Individuals in an organization knowledgeable about functional requirements should provide information to assist the inspectors in the evaluation. The completed form is used in the project prioritization process and can help decide whether a deferred maintenance project should go forward as a high priority or be delayed until program decisions are finalized. This sequence ensures that major maintenance and deferred maintenance backlog work will be appropriate for the projected use of a facility.

A project prioritization rating system is necessary for standardized and consistent capital planning and budget management. The reality of limited funding for capital projects requires that

TABLE 9.2 Suggested Priority Rating System

Priority 1: Immediate concerns
Urgent needs to be completed within one year, such as correcting a safety problem, eliminating damaging deterioration, or complying with life safety and building electrical and environmental codes.
Priority 2: Short-term concerns (1–2 yr)
Potentially urgent deficiencies that should be corrected in the near future to maintain the integrity of the building, including systems which are functioning improperly or not at all and problems which, if not addressed, will cause additional deterioration.
Priority 3: Long-term concerns (3–4 yr)
Deficiencies that are not potentially urgent but which, if deferred longer than 3–4 years, will affect the use of the facility or cause significant damage. Included are building and infrastructure systems and components that have exceeded their expected useful life, but are still functioning.
Priority 4: Delayed projects
Required or desirable to bring the facility to perform as it should, including systems upgrades and aesthetic issues. Included are projects that are not time sensitive and can be delayed indeterminately, e.g., work to conform with codes instituted since the construction of the building and therefore grandfathered in their existing condition and that can be addressed in any major renovation effort; unresolved program use of a facility; a facility under consideration for removal from the building inventory, etc.

priority rating systems sort out the relative importance of each project; that is, when a project should occur. Projects given a high priority are those that entail a recognizable risk or significant benefits, for example, enhancing life safety; preventing loss of a component, a system, or an entire facility; meeting legal or code requirements; conserving energy; or saving other costs. Low-priority projects are those that are not time-sensitive.

A preferable approach is a rating system that has multiple uses for field inspections of facility conditions for both buildings and infrastructure, and capital improvement program budget preparation. A suggested priority rating system is shown in Table 9.2.

Phase 3—Results. The third phase of the audit assesses results of the condition inspection and functional performance evaluation, reviews the effectiveness of the inspections and evaluations, and develops summary reports of the collected data and observations. The audit manager reviews the completed forms and plans or maps with indicated deficiencies and discusses the process with the inspectors. Any testing or laboratory work results are evaluated to confirm the recommended corrective measures and accuracy of cost estimates.

Thoroughness and consistency of the inspections are important considerations for the audit manager in reviewing results. A random selection of inspections and visits by the audit manager can confirm results or suggest reinspections. Subjectivity of the inspector/evaluator is a factor that should be considered in evaluating recommended corrective measures. The manager should suggest any needed improvements in inspector's training programs.

The database provided by the inspection/evaluation reports can be the source for a variety of summary reports. Using data collected in an audit of buildings and infrastructure conditions and functional performance, various report formats offer information on the overall *adequacy* of facilities, facilities types, individual facilities, systems, or components. Further sorting of data can identify projects by priority, cost ranges, use of in-house crafts or trades, or contractors. Reports can be designed for presentations for different purposes and are limited only by requirements determined by senior facilities and financial administrators.

Summary reports should be more than facts and figures. An executive summary should provide an overview and highlights of findings. A narrative should describe the audit process, objectives, and priority selection criteria. Conclusions should include a summary of overall facility conditions, a needs assessment for capital renewal/deferred maintenance projects, and an assessment of the adequacy of operating budgets for maintenance. Audit summaries can be organized in many ways: by all facilities audited or individual facilities; by all building systems or individual building systems; by all building components or individual building components; by project costs; and by project schedules.

Comments from the audit are also helpful in producing feasibility studies of changes in building use that can result in renovations or replacements. The inspections and reports of building and infrastructure conditions and functional performance evaluation allow critical questions to be addressed: Is the facility suitable for its current uses, or will it require remodeling? What is the actual cost compared to a new building and is relocation to another building feasible and desirable? Is disposal (sale, lease, or demolition) a preferred alternative?

A comprehensive facilities audit provides data on deficiencies that can be used in several types of analysis. A method of measuring the relative condition of a single facility or group of facilities is useful in setting annual funding targets and the duration of deferred maintenance reduction. The facilities quality/condition index (FQCI) serves this purpose. The FQCI is the ratio of cost of remedying facilities deficiencies to current replacement value (CRV).

$$\text{facility quality condition index (FQCI)} = \text{deficiencies/current replacement value}$$

Deficiencies are the total dollar amount of corrective measures to achieve desired *adequacy* of facilities identified by a comprehensive facilities audit. *Current replacement value* is the estimated cost of constructing a new facility containing an equal amount of space which is designed and equipped for the same use as the original building, meets the current commonly-accepted standards of construction, and also complies with environmental and regulatory requirements.

The FQCI provides a readily available and valid indication of the relative condition of a single facility or group of facilities. It also enables the comparison of conditions with other facilities or groups of facilities. The higher the FQCI, the worse the conditions. Suggested ratings for comparative purposes are assigned FQCI ranges as follows:

FQCI range	*Condition rating*
Under 0.05 (5%)	Good
Between 0.05–0.10 (5–10%)	Fair
Over 0.10 (10%)	Poor

In addition to data on current conditions and functional performance, the audit provides data to enable projections of capital renewal for ongoing facilities funding. Often absent from FM annual capital planning practices, capital renewal for facility components that deteriorate on varying life cycles require a pool of funds available on an annual basis to supplement annual maintenance budgets. A guideline for an annual capital renewal allowance, *in addition to operations and maintenance funding*, is 1.5–3.0% of a facility's current replacement value. Data available from the facilities audit and the calculated FQCI enables scenarios of various alternatives to address reduction of capital renewal backlogs—deferred maintenance—by testing alternative funding levels and goals for an acceptable FQCI over differing time periods to reduce backlogs.

Phase 4—Findings. The final step in the facilities audit process is presenting the findings. The facilities audit is one of the most valuable tools available for FM in performing its responsibilities when well-developed and presented. But even a flawlessly performed evaluation of audit findings is useless unless the information can be communicated to audiences in a readily understandable format. Careful consideration should be given to the audience, its interests, knowledge of the subject, and the issues it faces as a decision-making group. Conclusions and recommendations should be able to stand on their own merits.

The facilities audit will provide documentation that will be of more than passing interest because of its effect on an organization's financial planning. Thorough preparation is necessary in the design of presentations to establish and maintain credibility for FM and the reliability of its information. Before beginning the audit process, consider the presentation format and potential audiences. The wide array of summaries available from the inspections should focus on priorities and costs, with supporting material keyed to concise presentations. If the report of facilities conditions is to be submitted in print form only, without oral presentation, consider what graphic material would be helpful. The report itself may be presented as a brief statement of facts with graphics or as an extensive narrative that includes background, description of methodology, findings, and conclusions.

In all cases, the chief facilities officer should provide material that is concise, easily understandable, and attractively presented. It should be free of jargon, confusing terms, or acronyms that are not self-explanatory. It should not be over-simplified for readability, but designed with the documents readily cross-referenced.

Material should be developed in anticipation of the sharpest minds in the organization receiving the information. The documentation must be meticulous in detail and accuracy. Simple arithmetic errors and broad generalizations should be avoided through thorough checking of financial data, priority selections, and cost-benefit analysis. Expect the unexpected. Be prepared to answer the question: "What will happen if we postpone or don't do the work at all?" Organize the presentation so that the train of thought can be followed. Above all, keep the presentation simple and to the point.

9.1.4 Adapting the Audit

Although maintenance staff may collect and analyze some information to ensure the performance of buildings and infrastructure components, they do not typically collect planning data to assess long-term needs. The audit goes beyond a single snapshot of current conditions for planning of capital renewal and elimination of deferred maintenance. Using the audit procedures in a systematic manner, either by in-house staff or by consultants, a database of information on facilities is produced that can be used as a baseline for future condition inspections. Facilities managers and/or consultants should incorporate the special characteristics of an organization and its facilities into an individualized facilities audit. Design of forms for recording inspections is adaptable to unique building and infrastructure systems and components not included in the illustrative forms (Figures 9.2 and 9.3).

Careful design and preparation of a facilities audit enables the facilities manager to predetermine the level of information to be obtained and ensure that information gathered is appropriate for the projected application of the findings. However, the selection of approach must be driven by the nature of an organization's facilities, budgeting methods, and organizational structure. Formats for reporting the audit findings should be tailored to match the input requirements for the maintenance work order system and capital planning process.

As the facilities manager and staff gain experience with the audit, they will recognize its potential for familiarization with building and infrastructure condition, functional performance of facilities, and creation of a database for maintenance and capital planning. The audit contributes to overall effectiveness of the FM organization by developing *quantitative* and *qualitative* assessments as a routine part of operating activities.

9.1.5 Summary

Although some information may be collected and analyzed by maintenance staff to ensure the working order of a facility, planning data needed to assess long-term needs is not typically included. The facilities audit process goes beyond maintenance planning. By following the audit procedures and using the suggested forms, an information database is created on facilities and infrastructure deficiencies that is a baseline for future condition surveys and capital renewal planning and budgeting. Audit methodology tasks are summarized for input, analyses, and findings in Table 9.3.

TABLE 9.3 Audit Methodology Tasks

Input	Analysis	Findings/databases
Space inventory	Facility condition assessment	Condition deficiencies
Building plans	Quality assessment	Facility condition index
Program description/activity	Cost estimates	Quality deficiencies
Quality standards	Project impacts	Facility quality index
Design standards	Program implications	Facility condition/quality index
Finishes, HVAC, electrical, lighting, etc.	Financial planning	Renovation/modernization/ replacement needs
Laboratory standards		Prioritized projects
IT standards		Cost estimate
Current replacement value		Phasing schedule
Cost estimate model		

As discussed in the following section, a relational database makes it easier to conduct special audits from the database. For example, surveys of hazardous materials, conditions of components (roofs, roads, etc.), or building or space types (housing, classrooms, etc.) are possible with less start-up effort and will contribute to expanding a database. Feasibility studies for acquisition, remodeling, or consideration of disposal of a facility can be based on the information provided by a facilities audit. The audit process provides a powerful tool for capital planning and allocation of resources and offers a method for regularly incorporating information obtained with a high level of confidence into FM decision making.

9.2 TECHNOLOGY FOR CAPITAL PLANNING AND CONDITION ASSESSMENT

As described in the preceding section, the facility condition assessment (FCA) process involves more than data collection; an FCA program is an ongoing management tool for capital planning and management. Technology significantly enhances most FCA programs by lowering the cost of the program and by enabling better use of the FCA data collected. Computer technology enhances the often intensive data management involved with an FCA program, automates the analysis of collected data, enhances an organization's ability to communicate findings, analyses and directives, and can also lower first-time costs of data collection involved in an FCA.

Historically, facilities condition assessments have under-utilized technology. Resulting utility has been limited, cost has been high, and analytical potential for the accumulated data has been low. Early attempts to automate the FCA process, primarily by the federal government, involved expensive software development efforts which, while useful for their direct purpose, were not scalable or adaptable for application on a more generic basis. Today, there are a number of commercially available specialized software packages for facilities condition assessments offering a range of capabilities from basic functionality to sophisticated financial modeling and integration with other FM software systems. Increasingly, use of web technology and the World Wide Web itself offers new capabilities for FM software and for capital needs assessments and the subsequent use of the information created through an assessment.

The case to utilize technology in an assessment is clear when the cost of the technology is considered in the context of the total program cost, both first cost and ongoing cost, as well as the time savings that use of technology affords. In some cases, a low-tech approach of pencil and paper remains an appropriate method for assessment; for other situations, a mid-tech approach of a simple database makes sense. However, the time to recognize the limitations to any approach to the technology used for an FCA is in the planning stage of the process. Unfortunately, many organizations plan and conduct an FCA only to find that the technology used was not adequate to fully realize the benefit of the work done. It is inherently difficult to foresee the future uses of the FCA program within an organization if that organization has no experience with FCA programs. What appears adequate at the outset can ultimately prove inadequate as the planners do not initially perceive the full range of uses of the data that they amass. Often FCA planners are unaware of the benefit of integration with other FM databases, or simply are focused on the efficient (and cost-effective) collection of data, not the future analysis of that data or the ongoing costs of the FCA data management. Fortunately, a modest amount of planning will reduce the potential for waste and lost potential. Even if a high-tech approach is not initially pursued, a transition plan from the low or mid-tech approach to a high-tech one will enable a potential upgrade at a later date, protecting the initial investment in data collection.

The following section describes the range of technologies available and imminent to automate the FCA process and assist in the decision making that follows the FCA process. Appropriate technology for a given situation, however, is more a function of the particulars of that situation than it is a function of the state of available technology. As such, the various FCA program drivers are mapped and their implication for appropriate technology selection is described to assist FCA planners in choosing appropriate technology for their particular program. Like most technology questions today, the issue is not "if" but rather "how much" and "how soon" to apply software automation.

9.2.1 FCA Technology Overview: Databases and the Web

A desirable goal for effective FM is an information system integrating space management with maintenance work planning, major maintenance, capital renewal, and deferred maintenance reduction projects. A relational database has broad applications for facilities and financial management and other applications within an organization (Figure 9.4). At the minimum, commercially available services and software solutions can semi-automate the facilities audit process of inspecting and reporting building and infrastructure conditions. At the cutting edge of this technology is found sophisticated financial modeling capabilities, web-enabled automated data capture across geographically dispersed facilities, automated analysis and decision support for capital planning, and web-enabled statistically significant benchmarking throughout the FM world.

The facilities audit process and methodology are readily adaptable to automated applications through software applications known as capital planning and management solutions (CPMS). Use of relational database programs allows the integration of the audit into an FM information system. Organizations have developed automated facilities audit programs with internal resources with varying degrees of success. Consulting firms provide various options, including digitizing drawings, performing facilities audits, preparing condition assessments, and training staff in software and assessment techniques. Services and software products offer automated data collection and estimates of cost deficiencies, cross-referencing data to drawings, and graphic presentation of results. Products range from the single user desktop PC program running on an analogously simple database, to automated questionnaire-based programs running over the Web, utilizing powerful back-end database engines. To date, very few firms specialize in FCAs and even fewer in FCA technology, though almost all design and construction firms are involved in conducting FCAs in some manner. Evaluation of appropriate methodologies, techniques, and applicability of products is recommended in selecting technology assistance from consultants.

Automated applications of the facilities audit introduces a powerful tool for the facilities manager. Important benefits are readily retrievable information and capability for updating. Computerizing makes it easier to change the perspective of FM from collecting data for a one-time or occasional evaluation of conditions to continually updating facilities conditions on a continuing basis. Thus, the assessment can become an operational tool, useful for predicting needs and a valuable component in decision making for maintenance management and prioritizing capital improvement projects.

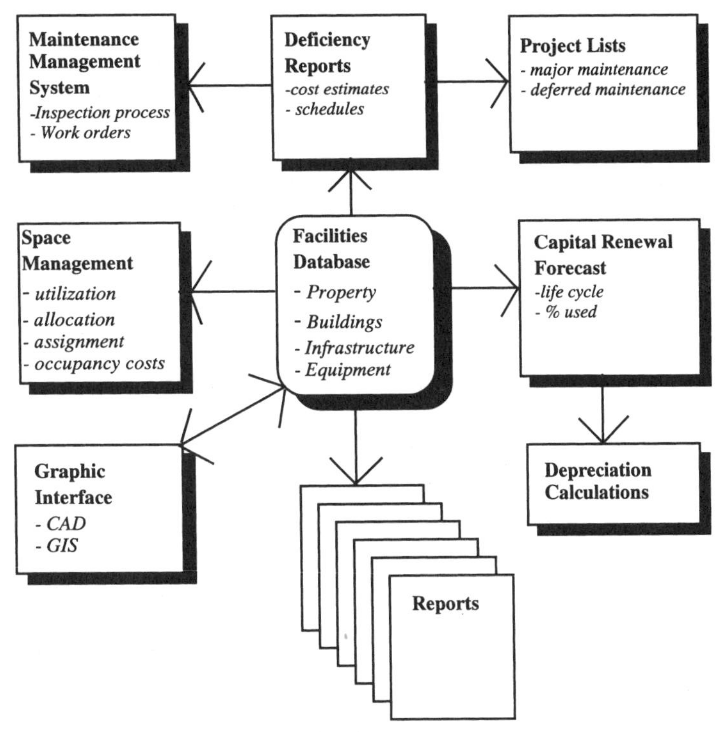

FIGURE 9.4 Relational Facilities Database.

The growing application of web-based technology throughout the software world dramatically increases the benefits of using technology for assessment programs. Web technology and "the Web" are actually two different aspects of the same development. Together or separately, they benefit the assessment program through facilitating distributed data collection efforts over the Web and through better access to data and analytical tools via web-based solutions.

In Figure 9.5, consistent base data inform decision making at a variety of levels of the organization simply through appropriate levels of data aggregation. Minutia of detail is required for a deferred maintenance manager to plan and manage individual correction projects or for an environmental health and safety officer to input newly identified hazardous conditions. A high level of data aggregation is appropriate for multi-year capital budgeting by the higher levels of the organization. The critical aspect to this upward data feed is that the data remain consistent. There is no "telephone game" effect as information is communicated in a multi-step process, neither is the information based on impressions or other less reliable sources. The downward arrow indicates the communication of organizational goals and priorities from the top down. By communicating organizational goals (i.e., Priorities 1 and 2 life safety issues shall be corrected within eighteen months) the individual decision makers at each level of the organization optimize their efforts consistent with the organization's priorities instead of assuming what those priorities are. The use of the Web makes all of this technology available anywhere, across a geographically dispersed organization or from 35,000 ft traveling between those locations. As this is written, wireless Internet connectivity is becoming commercially viable for an increasing number of applications. Downloading of assessment information from hand-held devices via the Web over a phone line is currently available; real-time wireless downloading and uploading between portable computing devices and a Web server will be commercially viable in the near future.

Web technology is increasingly lowering the cost of distributed software application functionality, such as that useful for conducting assessments and utilizing the resulting data. As indicated in

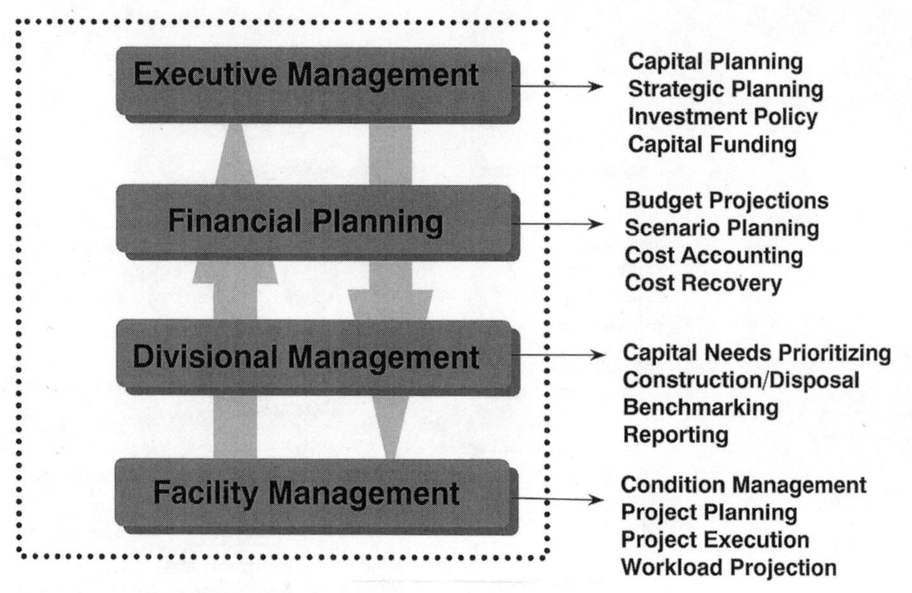

FIGURE 9.5 FCA Data Communication throughout an Organization.

Figure 9.6, Web technology enables access to data collection tools from any computer browser without that computer necessarily having any assessment-specific software loaded onto it. Similarly, data analysis can be performed "remotely" on any connected machine. This reduces the cost to maintain the technology, as software is maintained on one machine only. Moreover, this lowers the cost to use the technology as the interface is within a standard browser, reducing training required. Importantly, Web technology is very flexible, enabling affordable customizing of the inter-

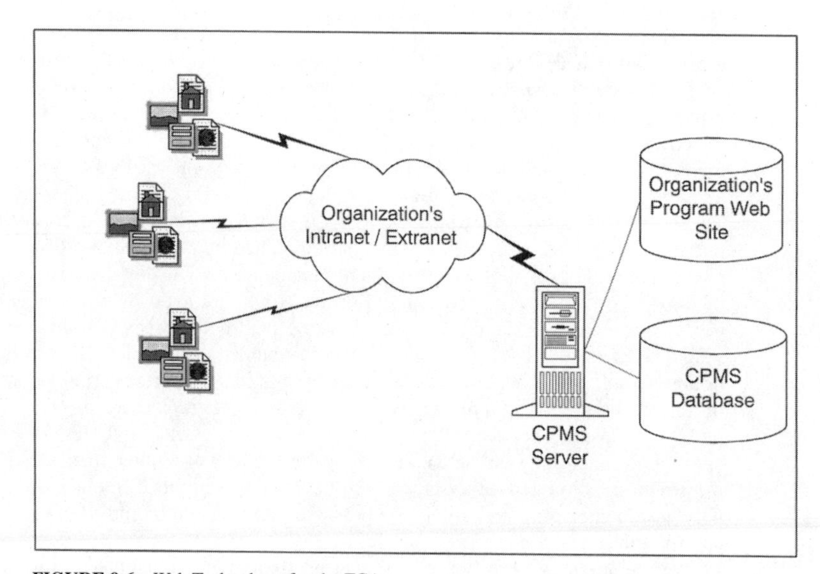

FIGURE 9.6 Web Technology for the FCA.

face for specific user types or tasks. An assessor's interface is tailored specifically to their task of data input while a program manager's interface is tailored to their task of analyzing construction methodologies. This adaptability of interface is commonly seen in executive information system (EIS) interfaces, which are simply data views and tools organized to support executive-level needs.

9.2.2 FCA Technology Planning—Program Drivers

With such a broad range of technology solutions available for the FCA, a rational analysis of each unique situation enables planning for the correct technology. As with any planning analysis, one must step back from the decision, in this case what technology to employ for a FCA program, and analyze the program's drivers.

Driver 1—Scope of the Assessment. *"How many facilities are we including in the FCA?"* is the first question that will help determine appropriate technology to apply to the program. This chapter generally assumes a multi-building or large building assessment scope, although FCAs are often conducted on small sets of facilities or single facilities for particular needs. Even single building analyses will benefit from a "mid-tech" solution of a simple database system in some situations, though word processing and spreadsheets may suffice and eliminate the need to develop a simple database system and associated reporting. Larger portfolios translate into more data collection and analysis, more data and therefore more data management.

Generally the larger the portfolio to be analyzed, the greater the benefit will be to employ more sophisticated technology to the FCA program. As indicated in Figure 9.7, the benefit of FCA technology grows exponentially as the size of the portfolio increases. Size can be thought of as the total gross square feet or the number of facilities. Beyond one million square feet, the benefits are tangible; beyond ten million the benefits are very significant. Benefit of the use of technology is considered to be the additional effort that would be required to maintain the same FCA program without the benefit of technology.

Driver 2—Perceived Need for the FCA. *"Why are we going to collect FCA information?"* may not be such a trivial question to answer. However, organizations generally are initially motivated by one of the following fundamental drivers for FCA programs:

1. Funding justification
2. Risk mitigation

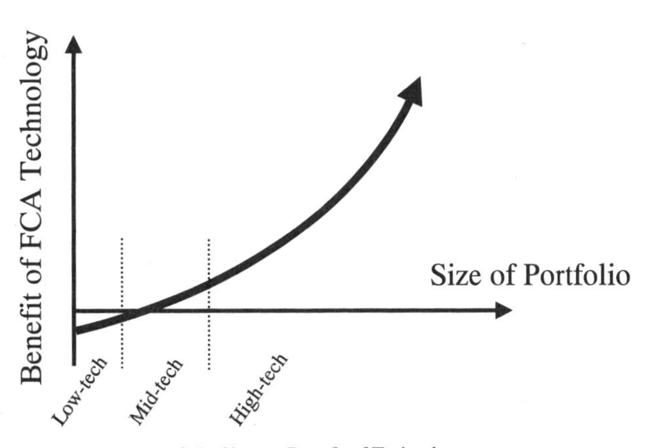

FIGURE 9.7 Portfolio Size vs. Benefit of Technology.

3. Better informed capital planning
4. More efficient project expenditure

These drivers should answer the questions that follow, though often it is only through the answers to these questions that an organization can determine the real driver for their FCA program:

- Do we need gross information only or detailed?
- How defensible does the costing need to be?
- What needs are to be included?
- Should we go beyond condition issues to program needs?
- Should we go beyond budgeting information to provide information that will help us manage the issues identified?
- Should we use the FCA program to enhance the efficiency of our internal communication and processes?

Each of these fundamental assessment drivers has associated with it a logical level of detail for the program (depth of analysis) as well as scope of the program (breadth of analysis). Funding justification alone does not require a deep analysis, but a broad analysis in order to capture all funding requirements, not just those related to deferred maintenance and renewal. Risk mitigation, such as risk of shutdown of critical operations, may require deeper analysis in those aspects of the facility that could fail and cause shutdown. Identification of only the risks may be appropriate, without detailed corrective methods or cost estimates. This may be appropriate if the risk avoided is dramatically more costly than the corrective action required. Better informed capital planning requires a wide breadth of analysis and somewhat deep analysis, as planning depends on the interrelations between competing needs and their costs. Finally, if more efficient project expenditure is the goal, both depth and breadth of analysis are required in order to fully inform the process, thus enabling more fully informed decision making related to capital expenditures.

Of course, different technology solutions are appropriate for each of these combinations of depth and breadth of the assessment and subsequent data analysis. Very often, an organization begins with one of these reasons as the overriding driver, and gradually integrates the other drivers into the program as the potential for the FCA program becomes understood. As such, erring to a more sophisticated technology solution is advantageous. At a minimum, a flexible modular technology solution is recommended so that additional depth and/or breadth of capabilities can be added as they become recognized as valuable.

With funding justification as a primary driver, the approach taken for the FCA and the technology used must provide adequate credibility of the findings for the organization's governing authority and their decision-making process. Credibility is typically achieved through demonstration of a rigorous program with appropriate detail (particularly for issues that involve significant capital resources), as well as programs that are impartially conducted and based on defensible costing information. Increased detail of analysis again translates into more data and thus increases the benefits of using a high-tech solution for the FCA. Defensible costing information can be gained through detailed construction cost estimating utilizing industry standard cost data accessed directly through a FCA database software system.

Risk mitigation as a primary driver typically is best accommodated by a more sophisticated technology solution primarily because risk mitigation programs are generally long-term solutions. Long-term versus one-time assessments are discussed in the following sections. Risk mitigation may not involve any cost estimating requirements for the FCA; however, comparison of risks (risk to operations and/or life safety risk) throughout a portfolio is required to enable a prioritized plan for corrective actions that mitigates risk most effectively.

If better capital planning decisions are the driver, a primary requirement is the association of various elements of capital need both within any given building and across a portfolio. As such, the FCA program's breadth would include all aspects to capital need including deferred maintenance and renewal, as well as code issues and programmatic needs. Integration with program needs tracked separately in a program, such as a space management information system, could be highly beneficial. When planning capital projects, the planner would have access to all aspects of capital need within a given facility. Moreover, analysis across the portfolio's needs can identify opportunities for aggregation of projects which reduce overall cost through lowered percentage general condition both external and internal to the organization, as well as lower unit costs through bulk purchasing and economy of scale. Technology can enable the analyses across the portfolio that are required to capitalize on the FCA program, effecting more efficient capital expenditures over time.

Driver 3—Availability of Existing Data. *"What existing data can feed the assessment?"* is another critical question when planning an analysis program and its technology. Typically, an extensive amount and variety of existing data are available throughout a facilities organization, but the data are inconsistent or of unknown completeness and accuracy. Until validated through systematic review, all existing data is suspect and should be tracked as such. However, if good, this information can dramatically reduce the time, effort, and cost of the first round of the assessment program. A validation process, part of the assessment planning process, is recommended. If existing data exists in only paper format, transfer to electronic format may not be appropriate, at least not until that data is determined to be accurate and useful for the FCA program. More typically, the mix of electronic data available in various formats would suggest that the use of a database application for the collection and normalizing of existing data is appropriate, as described in the following sections and shown in Figure 9.8.

Data Validation Phase 1—Data Conversion. A wide range of existing data are typically available in electronic form, however, in a wide variety of formats. The first step is to create consistency of data in order to assess validity. Existing databases and spreadsheets can be ported to a consistent data structure, preferably a database. Paper-based information sources must be reviewed to include important and useful data. Unfortunately, even with character recognition software, converting any

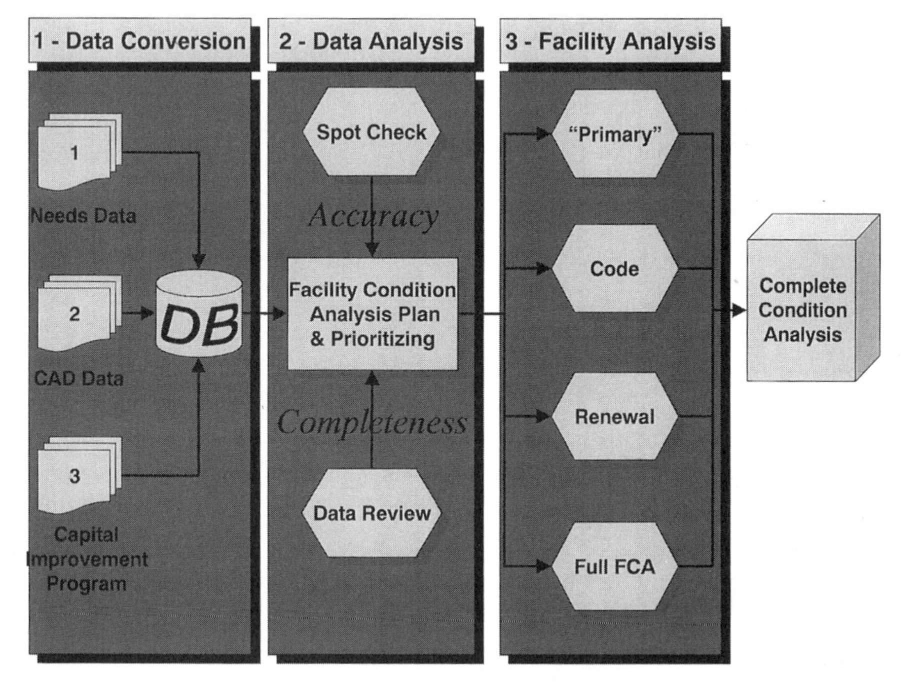

FIGURE 9.8 Validation of Existing FCA Data.

significant amount of paper information into a useful database format is fairly labor-intensive. Scanning images of paper data and creating links to those images is of limited utility, as the data remains abstract and undefined by data field and therefore is of limited value in analysis. CAD files, digital photography, scanned images of reports can all be added to this consistent database in this preliminary phase of data conversion. This process is beneficial even if no assessment follows, as data from various sources can be integrated and analyzed together, and becomes available to all in the organization, not just the keepers of certain "silos" of information. Moreover, it is typically not very costly to do this conversion.

Data Validation Phase 2—Data Analysis. Once the existing data is collected in a consistent and accessible format, the compiled information can be validated. This is a two step analysis for accuracy and also completeness. Accuracy answers how good the existing data are; completeness answers what percentage of current deficiencies or capital needs are represented by the existing data, even if inaccurately. These are two aspects of data usefulness and must be addressed separately.

The data review for completeness is sequentially first and involves a comparison of the existing identified needs against benchmark norms for similar facilities at peer institutions. The data are reviewed for breadth of inclusion. For example, the data may indicate that a particular facility has a number of building envelope problems that have been well-documented, but there is an absence of code noncompliance deficiencies documented. This indicates either that there are in fact no such deficiencies in the building, or that the analysis previously conducted was performed by someone with inadequate building code knowledge. By comparing the existing data for each facility to expected problems based on benchmark information, the gaps in the existing data can be estimated. Systematic analysis should be conducted for completeness for each system or each building type and for each existing data source. Typically analyzing 15–20% of the total existing data is appropriate to estimate the overall completeness of existing data prior to the second step below.

These facility condition information gaps are then investigated as part of the second step, estimation of the accuracy of the existing data. A cost-effective method for estimating the accuracy of the existing data is a "spot check" approach of each building system in various types of buildings across the portfolio. It is recommended to distribute the "spot-check" efforts over as many different facilities as possible, getting a representational subset of the buildings. Distribute spot checks by building type, age, systems, and type of issue. Further distribution is possible for low overall cost and effort by analyzing different building systems in each representative building, but not analyzing any building in its entirety. For a large portfolio, a detailed analysis of 5% of the total portfolio will provide enough comparative data to estimate the validity of the existing data sources. Combining the estimation of accuracy of the existing data with estimation of its completeness, it becomes clear where existing data can be used to reduce the FCA effort and where existing data must be supplemented or replaced with valid analyses.

Regardless of the technology employed in the FCA, it is imperative that the source of each datum is tracked. Relational database structuring of data provides the ability to sort and query by data source (i.e., 1993 life safety code review by Smith Associates). This facilitates the removal of data from a source found to be too inaccurate or, more typically, wholesale adjustment to data from a source that is found to consistently cost estimate too high or too low.

If significant existing FCA data exist, particularly if in electronic format, a high-tech tool is increasingly beneficial. Figure 9.9 approximates the exponential advantage gained as more existing information is available. The primary benefits are derived from the ability to integrate disparate data sources into a single location. This facilitates analysis across all data sources, giving a more complete picture of all the issues relative to any given facility or facility system. Moreover, aggregate comparisons to averages and norms can graphically highlight inconsistencies in data completeness. It is important to note that the effort required to translate electronic data from an existing database or spreadsheet into an integrated data source is dependent on the number of sources, not on the size of any given source. The effort required to "port" a data source is essentially the same regardless of the number of records in that source. As such, the benefit of combining existing data sources into a common data source is also higher if the data sources are fewer and larger.

Drivers 4 and 5—Schedule and Longevity. The other fundamental FCA program drivers are temporal: schedule and longevity. When does funding need to be justified, risk mitigated, capital plan-

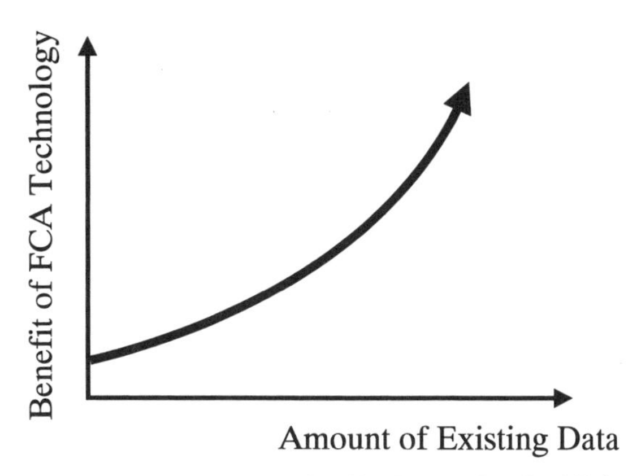

FIGURE 9.9 Extent of Existing FCA Data vs. Benefit of Technology.

ning informed, or project expenditure more efficient and for how long? Often organizations initially perceive an FCA as a one-time solution to a pressing problem and only subsequently commit to the value of maintaining FCA data in an ongoing program. Automation of the data collection process and subsequent data analysis process is increasingly beneficial as the schedule pressure for the completion of the FCA increases. Figure 9.10 describes the benefit of employing technology as a function of FCA program longevity. While some benefit is seen in a one-time program due to time savings through automation, significantly greater benefit is derived from a perpetual program. Once the first cycle is completed, the bulk of the data are in place, as well as analytical reports and the in-house familiarity with the system being used. This initial investment is leveraged in an ongoing program as benefits of the program continue with relatively little ongoing investment.

Other Drivers. Secondary factors will enter the technology decision analysis as well, such as availability of in-house staff, their access to and proficiency with computers and/or the Web, funding available for the FCA technology, whether other FM technologies are in use or planned, and the geographic distribution of the facilities in question. These secondary factors are all critical elements to consider in successful planning for the technology of the FCA, but should not be considered primary drivers. Unfortunately, the issue of funding can become a primary driver if not addressed proactively. As with any under-funded program, under-funded FCA programs, particularly under-

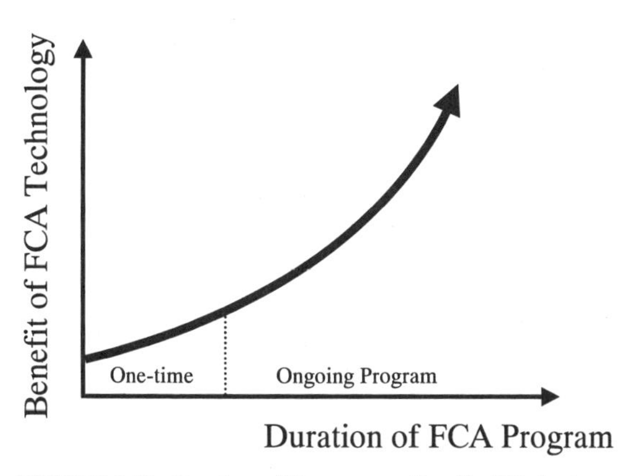

FIGURE 9.10 Duration of Program vs. Benefit of Technology.

funded technology solutions, are rarely successful. Considering the low percentage of technology cost to the total cost of an FCA program, and the savings possible via automation and better analysis capabilities, justification of FCA technology funding is clear.

9.2.3 FCA Technology Features

This section offers a brief review of the various features generally available for FCA technology solutions. As with any computer-based technology, the pace of development does not allow for a specific analysis of the particular features of competing technologies. The span between the writing of this chapter and the publishing of this book is long enough to antiquate any such discussion of "cutting edge" FCA software or hardware technologies. As such, the following briefly describes the generic benefits to utilizing the available technologies' features, again attempting to map these features into the specific needs and drivers of any given organization.

Data Capture Technologies. Data capture, the manner that FCA data is collected and organized, ranges from simple to highly automated. The vast majority of data capture in the field is low-tech pencil and paper, jotting notes down on "book plans" attached to a clipboard. Even in most cases when a sophisticated application is employed for data management and analysis, the data generally start as notes on paper. This low-tech approach is easily complemented with digital photography, which can be downloaded into a database structure and associated with building records or capital needs records. Digital voice capture is sometimes used and converted to text via speech recognition software. (As of this writing, this approach has very limited practical application due to the limitations of speech recognition software and background noise present during an FCA field inspection, such as that in a mechanical room.)

Direct to digital data capture in the field is available, but appropriate applications are limited. The low price and ease of use of the ubiquitous personal digital assistant (PDA) make their application to the field inspections of an FCA a natural use. These units are inexpensive, easily transported, and can be shoved into a pocket in order to climb a ladder up onto a roof. However, the use of a PDA is limiting in the detail and specificity of data collected. The PDA is very appropriate for a "pick list" approach to the FCA where the inspector is identifying which issues are present in a building from a list of preestablished capital needs. Two limitations are presented: difficulty in establishing a need that is not included in the preestablished list and difficulty adjusting a preestablished item to the specifics of an individual issue. Still, the use of PDAs is very appropriate in an overview analysis of an FCA where the facilities are very repetitive and a nearly comprehensive preestablished "pick list" can include almost any issue that may be encountered in the field. Tablet computer hardware provides freedom to input specific information describing the particulars of a given field situation, but the units are large, expensive, cumbersome, and fragile. The aggressive environment in the field process of an FCA makes this an inappropriate application for such a device.

Cost Modeling. Identified capital needs resulting from field analysis of an FCA are associated with cost estimates ranging from order of magnitude cost to fairly detailed estimates. Technologies are available to assist throughout this range. Order of magnitude estimating is assisted via systems that provide building system costs by square foot for various building types. A number of applications provide generic costing of typical needs by square foot of some other unit of measure. More detailed estimating is assisted by systems which create estimates through built-up line items that can be specific to a particular situation. In some cases, the underlying cost data is adjusted for local labor rates and costs and can be based on published industry standard cost data. Importantly, technologies are available that adjust cost estimates over time, enabling the estimates to remain accurate as labor and material costs change over time. This is critical to a long-term FCA program, as many of the identified and cost estimated items will not be addressed for a number of years. As these programs continue, the underlying cost basis can be updated, resulting in continued accuracy of the initial cost estimating. In addition, an organization's actual costs can be input in order to refine cost estimates to reflect actual experience with similar projects.

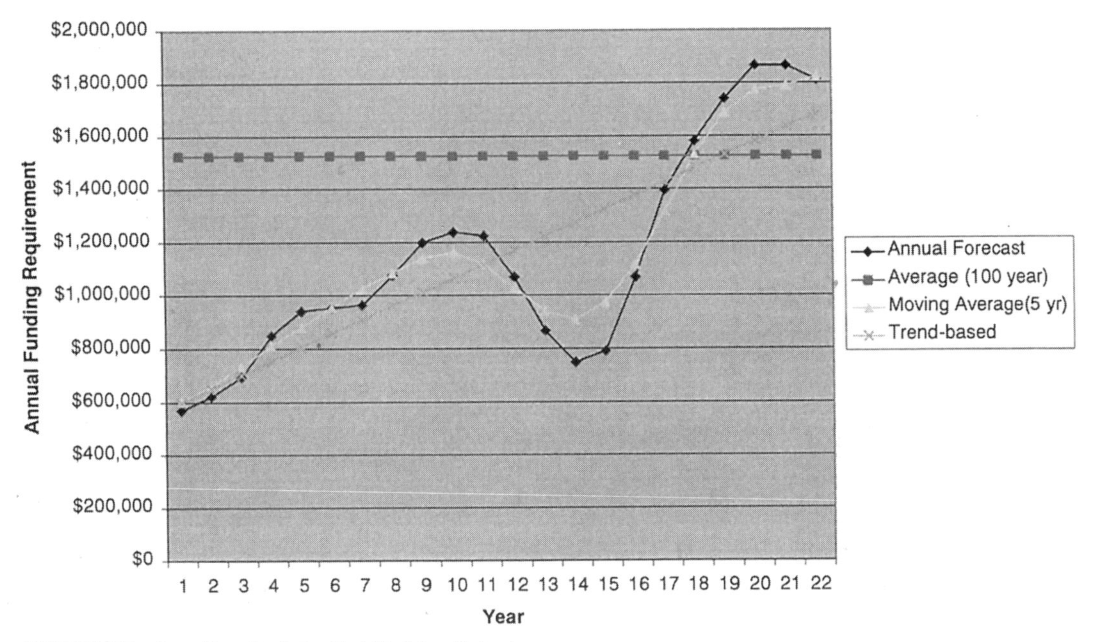

FIGURE 9.11 Long-Term Predictive Cost Modeling Technology.

Long-term predictive cost estimating is assisted through applications which use algorithms to project capital costs for renewal of facility components as they reach the end of their useful lives and need to be replaced or refurbished. Applications for this purpose range from very low-cost stand-alone tools through to part of a more comprehensive FCA solution. As seen in Figure 9.11, long-term predictive cost modeling provides order of magnitude costs, but provides an indication of the magnitude of expenditure required by an organization to maintain operability of its facilities. As such, it is useful when funding justification is a primary driver for a FCA program.

Capital Planning Scenario Analysis. Combining the long-term predictive cost modeling with specific needs known to exist at present time, algorithms can model the implications of various funding scenarios across a large portfolio and over time. The National Association of College and University Business Officers (NACUBO) was first to introduce this idea, preceding the technologies available now to fully implement such analyses. Modeling the feedback loops associated with the escalation of costs from deferring renewal enables analysis of various funding streams and their effect on facility condition, functionality, and ability to accomplish programmatic mission. Figure 9.12 models the implications of various funding streams in terms of the preservation of the value of the capital assets. The aggregate investment over multiple years (in present value) is associated with change in total plant value. This is one measure of how well a capital investment strategy performs.

Capital Project Building. FCA technologies can also provide automation and decision support assistance in the implementation of the projects identified through the FCA program. FCA data can provide the initial information for scoping, budgeting, and tracking of the resultant projects. A number of applications are available that provide such extension of the value of the FCA data. Some

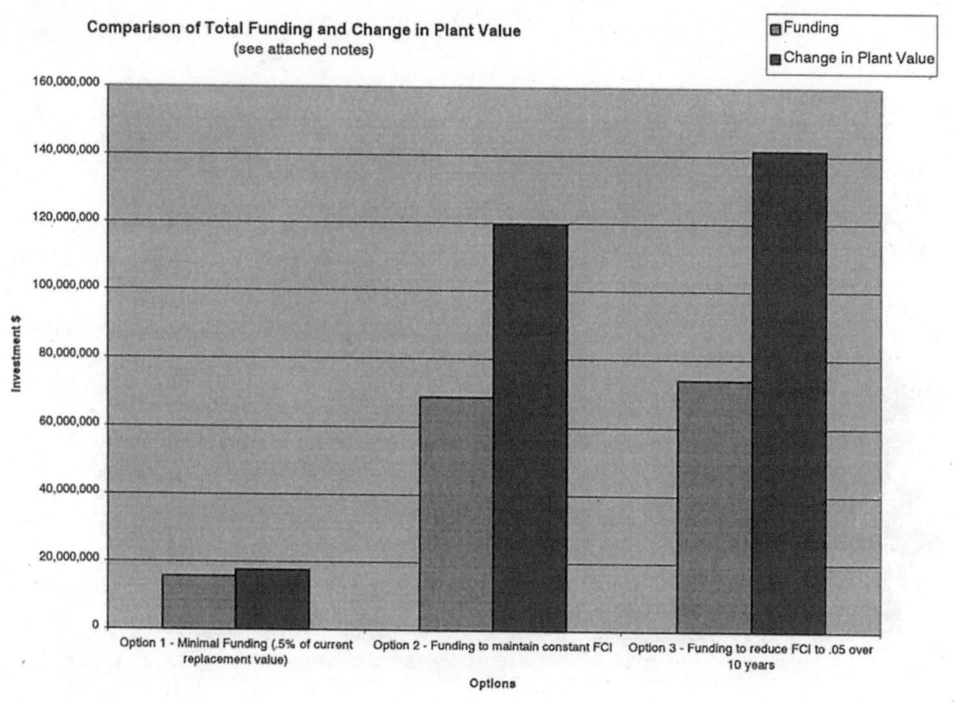

FIGURE 9.12 Asset Value Preservation Analysis.

rely on the transfer of FCA data to project management software applications; others provide functionality that is specific to the use of FCA data across a large portfolio. Analysis in such applications enables logical grouping of capital projects, applying constraints such as annual budget allowances of various types (accessibility improvement, code compliance upgrades, etc.) or by cost account limits (departments or regions.) By logically grouping projects, a large organization can more effectively utilize its purchasing power through the economy of scale enabled.

9.2.4 Conclusion

As with the application of technology to any need, the primary question to be answered is what is the motivation for the program. A wide range of technology solutions with a variety of optional features is commonly available in the marketplace. Of course, any given vendor will indicate that their solution is the most appropriate for essentially all situations. The successful selection of FCA technology is overwhelmingly dependent upon an organization's ability to understand the drivers for the FCA program and map those drivers onto the various options available to them. Technological determinism, the notion that if a technology exists it will be used, should not sway a deliberate thoughtful analysis of the particular requirements of a given organization. It is wise, though, to plan for more technology in the future of your organization's FCA program.

CHAPTER 10
THINKING GLOBALLY—THE COMPETITIVE EDGE

Dru Meadows, AIA, CCS, CSI
Principal
theGreenTeam, Inc.
E-mail: dmeadows@busprod.com

10.1 INTRODUCTION

Think globally, act locally was a rallying cry of the early environmental movement. It is ironic that such a prophetic statement has been reduced to a well-worn, tattered bumper sticker. That decades-old environmental wisdom offers an exquisitely concise working principle for the next millennium. It provides a direct response to the reality of what the corporate world has labeled *globalization*.

There is no aspect of life today that is not affected by globalization. Facility management (FM) is no exception. Whether your company is a transnational corporation or a local firm, the fact of globalization has an impact on you. Globalization has so great an impact, that if you fail to realize the broad implications of globalization for your firm, you will not remain competitive for long. Increasingly, international dynamics are dominating local practices. The utilization of international standards such as ISO 9000 and ISO 14000, for example, are changing expectations and can affect how a company operates, impacting everything from the widget that is produced to the facility in which it is made.

There are some obvious, tangible impacts of globalization on facility managers. Communication is the example that most immediately comes to mind. The exchange of information via improved (and continuously improving) technology such as the Internet, cell phones, and DVD systems is a necessity for every company today, so facility managers must constantly monitor available technologies for integration in the structure of the building. These are tangible impacts of globalization. Facility managers must monitor the communication requirements of the company. Who requires how much hardware? What kind of software? Facility managers must match demand with affordable and available technologies. What level of flexibility in the building design is available to install fiber optic cables? How soon will the facility be wireless? Facility managers must prevent interruptions to the greatest extent possible and anticipate the appropriate emergency response just in case. How can consistent delivery of service be secured? What happens if a server goes down? What is the back up?

The tangible daily requirements of the company must be satisfied; however, a successful facility manager is not merely a technician responding to demand, but a visionary anticipating possibilities. In the near future, it is entirely possible for ownerships to change rapidly and often. Who will own the facility next year? Who will occupy the building next week? A proactive facility manager will not only be aware of the needs of the company in terms of the daily routine of the employees, but also in terms of the potential long-range opportunities for the company as a whole. The tangible becomes inseparable from the intangible.

The shift from technician to visionary means combining a solid working knowledge of the practical with an understanding of the market forces that are driving the practical. The facility manager who simply responded to a memorandum requiring a video system and invested in Betamax VCR technology failed to appreciate the bigger picture. The facility manager failed to *think globally*.

Facility managers are in a unique position within a company. To perform their role well, they must understand a little about what everyone is doing. They should also have an appreciation for the overall goals of the firm. By combining that information, the local information, with the global information, facility managers can help advance their firms ahead of the competition.

A building is a tool. It is a piece of equipment that allows a company to deliver efficiently and effectively the company's product or service to its customers. A building is a large piece of equipment and a hefty investment. One company may own several such pieces of equipment in various locations. It may have assembled the equipment on a large campus headquarters. Regardless of how many or how big, a building is still simply one more tool. The facility manager is responsible for the smooth, uninterrupted operation of this tool, and for making sure that the particular model is appropriate to the company's needs by predicting and implementing "upgrades" as appropriate. By considering the performance requirements of the tool abstractly, a facility manager can help develop creative and competitive approaches to the company's operations. Office "hotelling" is an excellent example of such facility creativity. Hotelling may not be appropriate for every firm, but, for some, it is an ideal physical response to the need for speed, flexibility, and service. It is also indicative of the dynamic and competitive playing field—a field that is changing due to globalization.

Everything has changed. Anyone operating under the old system might as well be using a rotary phone and shipping out product by pony express. The rules of the game have changed. The term *globalization* is often used quite casually to refer to how small the world seems now or to the new international markets suddenly open for trade, but, the term is more powerful than that. It encompasses some significant changes to the way in which we all interact with each other. This chapter will review some of the implications of the following aspects of globalization, each of which can provide insightful information for the facility manager:

- Changing political boundaries and interaction between nations
- Changing economic boundaries and interaction between companies
- Changing technology
- Changing social and environmental awareness

Certainly these aspects are not discrete facets of globalization. The definitions are not that simple. Globalization itself is complex. The children's stories about the snowballing effects of a small event are pretty accurate, but only part of the picture. Such children's stories demonstrate direct, lineal cause-and-effect relationships. A butterfly lands on a flower, knocking a small seed off of the plant, that drops on the head of another small creature that is startled and falls against a tree, knocking down an apple. The connections may continue with increasing magnitude until an entire town is in disarray. That is part of the picture. In the realm of globalization, however, the relationships are not lineal, they are more akin to a multi-dimensional spider's web. They resemble the ripples spreading out in all directions from a small pebble tossed casually into a pond. The ripples get larger and the wave action more pronounced the farther from the initial, small "plop" of the pebble. One small, inconsequential event can start a ripple effect, impacting many different people in many, many different ways. These aspects are intricately connected to each other and are broken apart here only for purposes of discussion.

10.2 CHANGING POLITICAL BOUNDARIES AND INTERACTION BETWEEN NATIONS

Where once communism and democracy established political boundaries, with changes and interactions measured as adjustments to the balance of the two superpowers, now there is only one national superpower, the United States. The ramifications of having a single superpower are not yet entirely clear, nor is the reaction from consumers, employees, media, governments, or business. This is uncharted territory. This is the new business of peacetime, and it is global (with a few localized exceptions). With the collapse of the cold war inhibitions, the void is being filled with a new economic exhuberance and the world is suddenly permitted an opportunity to confront larger social and environmental issues. Individually, many are able to consider quality of life issues beyond the basic need for food, clothing, and shelter. As the awareness of our new condition becomes more prevalent, how will we respond? What expectations will we have in recognition of the new opportunities and new challenges?

In particular, how is the only superpower going to respond? Is the United States to be the global police? In some respects, it has certainly behaved that way, but always cautiously, conscious of the threat of inciting a potentially devastating global confrontation. After all, even relatively poor countries—not to mention individuals—have access to weapons of mass destruction. Perhaps the United States should be a global benefactor. It is rich enough, but it is commonly recognized that the wealth of the United States is often obtained at the expense of other, poorer countries. Consequently, the imperial might of the United States is more resented than welcomed. "Historically, the United States either has been isolated and aloof from world affairs or has been compelled to enter deeply into the world as part of a moral crusade. . . . [Today] the United States is the biggest beneficiary and sole superpower. . . this is the world we have . . ."[1]

Facility managers of American firms should find this shift from a cold war philosophy to a new economic era a particularly pivotal time. American identity is founded in a strong belief in the individual and contains a profound sense of leadership relative to international affairs. Often, upon entering other countries, American firms assume that the American way is best and proceed to "educate" the natives. Obviously, this is not the most thoughtful or considerate approach. Increasingly, it is not the correct approach. In the shifting global arena, there are many examples of political/economic models fashioned on a more controlled, state-regulated, less-individualistic philosophy. Witness the successes of Japan and Germany over the last several decades. Not withstanding the "Asian Crisis," Japan is economically strong, as is a now-united Germany.

Informally, the political boundaries have changed perhaps even more dramatically. "Corporations have emerged as the dominant governance institutions on the planet, with the largest among them reaching into virtually every country of the world and exceeding most governments in size and power. Increasingly, it is the corporate interest . . . that defines the policy agendas of states and international bodies . . ."[2] Developing countries are anxious to court the industrial base that they believe will help propel them into the ranks of developed countries. They want the technology, the investment, and the jobs. Often, they get more.

Transnational corporations are often accused of exploiting resource-rich and technologically poor developing countries. Certainly, there are many, many examples of low wages, long hours, and poor working conditions. The social fabric of traditional cultures is rent irrevocably by the invasion of industrial processes and expectations. Cultural craft items are replaced by sterile, plastic fixtures. Giant swaths are cut through forests to open an area for "development," devastating the local (and global) environment and altering forever the relationship between local cultures and the earth; however, in many circumstances, the ripples of the pebble expand in different directions. The social fabric is altered, yes, but not always for the worse. Consider the young women working in a new factory. They represent the first generation of women in their culture to earn their own income, to live and work outside the home. For women in cultures that regularly imprison them under the guise of domesticity and religion, the change is welcome. The sword cuts both ways.

Corporations, for their part, are as interested in entering developing countries as are the developing countries in having them. Corporations have come to recognize that the continuous improvement of productivity includes a two-pronged approach of: 1) identifying new markets and 2) lowering operational costs. New markets will provide an outlet for the surplus of goods/services that are produced at lower and lower cost. Conventional economic theory asserts that profits will rise accordingly. The identification of new markets often includes expansion into developing countries. Such expansion is intended to provide the added bonus of improving stability and flexibility. It also increases the complexity of the job description for the facility manager, as well as the number of buildings that the facility manager must coordinate.

The simple premise is that you do not put all of your eggs in one basket. Your grandmother taught you this. If currencies are suddenly devalued in Eastern Europe, or if a hurricane suddenly devastates an Asian base, or if there is a sudden political reversal in the South African arena, the corporation can readily shift operations to another facility. By taking advantage of the increasingly open political boundaries, corporations can position themselves to operate in areas of lower cost, establish convenient production facilities adjacent to growing market bases, and hedge their bets against isolated disasters. Always listen to your grandmother.

The role of the facility manager in such circumstances is to be proactive in coordinating the operations of the facilities individually and collectively. At the top, the facility manager would coordinate the shifting of operations between far-flung facilities—not just after a currency devaluation, climatic event, or political upheaval, but before. The visionary facility manager will have a finger on the pulse of global current events and be able to anticipate the impact of various, diverse potentialities. The fire drill herding employees down the appropriate stairwells in an orderly fashion is not going to be an adequate response to the possible "fires" in this new age of globalization.

10.3 CHANGING ECONOMIC BOUNDARIES AND INTERACTION BETWEEN COMPANIES

During the cold war and shortly thereafter, economic competition and trade were perceived as being between nations. National protectionist measures often resulted in tariffs or trade restrictions. Such national policies are becoming more and more difficult to impose. Corporations are no longer multinational, operating as separate entities in different countries and reporting back to the national headquarters. Corporations are now transnational, and competition is not between nations, but between other transnational corporations. Their loyalties are not to any particular nation, but to their own brand name. Corporations, especially transnational corporations "are a strong force behind global economic integration. In 1970, there were some 7,000 transnational corporations. Today, there are nearly 54,000."[3] These corporations' operating budgets can exceed the GNP of many nations.

Governments, on the other hand, are increasingly viewed as ill-equipped to address the complex demands of globalization. According to a renowned international corporate advisor, "governments have become the major obstacle for people to have the best and the cheapest from anywhere in the world."[4] The role of governments, according to such strategists, is to provide the infrastructure for corporations to succeed in their mission of trading goods and services. This argument favors corporate nationalism as a social and economic organizational approach to replace outdated governmental institutions.

Reflecting their economic might, transnational corporations, not governmental bodies, are increasingly able to direct international economic policy. Informally, corporations impact international economies via basic contractual negotiations with governments eager to promote the relationship. Governments will offer a buffet of tax breaks, incentives, and benefits in their bid to secure industry facilities—and the jobs that go with them.

Once the exclusive stratagem of the governments of developing countries, the approach is now common in the wealthy, developed countries (including the United States) as states and municipal governments seek to obtain the residency of transnationals. The economic picture for many in the United States does not correspond to the prosperity calculated under the GNP and the Dow Jones industrial average. Working-class Americans are more likely to have lost a job in the wake of NAFTA than to have realized profits from a mutual fund. Further, most Americans have witnessed a decline in real wages for the better part of the last three decades. As globalization continues, the migration of jobs has begun to include high tech and white-collar positions. Consequently, states and municipalities in the United States are vying for the privilege of subsidizing a new facility. Alabama, for example, awarded Daimler-Benz nearly $300 million in benefits for locating a new Mercedes factory in Tuscaloosa.[5]

Corporations also have access to more formal channels for impacting global economies—primarily through the International Bank for Reconstruction and Development (World Bank), the International Monetary Fund (IMF), and the World Trade Organization (WTO)—the successor to the General Agreement on Tariffs and Trade (GATT). These organizations not only facilitate the flow of money internationally, they also help to establish a common language. They promote reliable business practices and accounting practices.

One of the primary mechanisms with which organizations accomplish this is by rewarding participants in the global economy that comply with certain standards. The "push for standards is part and parcel of the emergence of economic webs. Standard processes, platforms, components and systems allow companies to connect and separate. Strong standards had to be in place before the most rudimentary steps could be taken in outsourcing, shared services, and remote teamwork."[6] Previously, standards addressed a product, delineating performance requirements. Because of the incredible speed with which new products, whether hardware, software, or other goods are developed, standards for products are often obsolete before they are issued. The trend now is toward standardizing process. As the most prominent international standards for process, ISO 9000 and ISO 14000 delineate expectations for quality management and for environmental quality management, respectively.

Standards are extremely powerful. For example, in 1997 the WTO ruled that member nations could not impose health standards restricting farmers' use of growth hormones in beef cattle more stringent than those recommended by CODEX Alimentarius, an international food standards agency. This ruling came about as a result of a U.S. government challenge of European Union (EU) health standards that blocked the import into Europe of beef that had been injected with hormones. Despite EU regulations prohibiting the sale of beef containing growth hormones, international standards had precedence and the EU is now obliged under the rules of the WTO to allow the import of beef containing growth hormones.

Standards form the basis of governmental regulations, building codes, and private contracts. They form the basis for marketing strategies and labeling claims. Originally, the federal government established most of the standards used in the United States. Over the years, this role has been assumed by private industry. The single greatest force in the development of standards in the United States is the American Society for Testing and Materials (ASTM). And, because the United States is arguably the most powerful and richest nation in the world, housing the most powerful and richest transnational corporations, ASTM greatly affects international standards as well. Many countries adopt ASTM standards because they cannot afford to generate their own standards and because they want to manufacture and export products and provide services to the United States. Furthermore, ASTM establishes the United States' position for the majority of standards developed by the International Organization for Standardization (ISO).

While there are countless groups developing standards globally and thousands of individual standards, ISO and ASTM are, undeniably, the major standards for developing groups—for nearly every type of tangible good and service. Consequently, it is vital to have your interests represented within ASTM and ISO. While participation in these standards development organizations tends to be open, it is often expensive and, historically, without any obvious payback for participation.

The rules have changed. Standards are a significant tool of global commerce and impact your company's products, services, and operations. For example, someone from your company—probably the facility manager—ought to be a member of ASTM E6, the committee on the performance of buildings. You certainly want to monitor, if not commit to participation, in ISO TC (Technical Committee) 59, building construction.

Relationships between corporations and governments have changed as corporations have grown in wealth and corresponding power; however, the relationships between corporations themselves are also changing. Costs for improved efficiencies, narrower profit margins, and new technologies are constantly escalating. Given the extraordinary expenses associated with research and development, many corporations have developed new relationships—strategic alliances—which allow them to share the costs and the risks. In varying degrees, every industry has experienced this evolution from competitor to partner. Toshiba, Siemens, and IBM joined forces in 1992 in development of the 256 M-Byte chip.[7] Alliances and mergers in the telecommunications, pharmaceuticals, and automobile industry are rampant. On January 10, 2000, the $172 billion merger of AOL, the world's biggest Internet company, with the Time Warner media empire was announced. In times past, such a deal would have warranted, at a minimum, a courteous "heads up" to Washington officials. This time, the White House heard the news on CNN, just like the rest of the world. Corporations rule.

A corresponding building response might involve strategic alliances for construction and operation of facilities. If the firms involved were small, perhaps the relationship marries similar interests to achieve economies of scale. Small producers joined together to purchase the expensive, but more efficient, manufacturing equipment and distribution facilities. While such an approach may not fit with the individualistic ethic embraced in the American culture, it is utilized successfully in many European countries. In limited applications, it is even evident in the United States. For example, independent consultants in nearly every American city can lease office space that includes access to "shared" services, such as receptionist/secretarial and security services.

If the firms were large, perhaps the relationship unites tangential operations that are able to identify some symbiotic advantages. In environmental terms, such a relationship is classified as "industrial ecology" and mirrors the mutually beneficial relationships between different organisms in an ecosystem. Simply put, the waste from one entity is food for another. Everything cycles. In the summer, the carbon dioxide that we exhale may be utilized by the leaf of the tree, which will fall to the ground in autumn, decompose under the winter snow, and nourish our vegetable garden next spring. This practice has been used, with limited success, in ecoindustrial parks where waste heat from the processing facility might power the adjacent office building whose waste paper might provide feedstock for the printing press whose waste chemicals . . . etc.

Marrying strengths and weaknesses offers the promise of synergistic market positioning. But, while strategic alliances open new opportunities, they are not without a certain degree of risk. How do competitors suddenly agree to share R&D knowledge? How can confidentiality be maintained if competitors share facilities? It is impossible to address these questions within current paradigms. Ultimately, successful alliances will depend upon a complete restructuring of accepted corporate practices to create true partnerships, not just the shared liabilities of joint ventures. The facility manager can have a significant role in structuring such alliances.

10.4 CHANGING TECHNOLOGY

New technology has resulted in increasing the speed of communication, production, delivery, . . . everything. If the gossip grapevine in your facility was highly developed prior to e-mail, it is exponentially greater now. Furthermore, it is connected to the rest of the world. Information is shared swiftly and informally. Sometimes the information is good and sound, sometimes not. Nevertheless, opinions and expectations are altered by it. Individual employees, customers, and competing *smaller* firms (a particularly exciting change if you happen to be a smaller, local firm) are empowered by new technology. Consulting firms can operate from hotel rooms, airplanes, and (if desired) poolside. "Virtual corporations" link hundreds of individuals together from their homes across the continent. Retail facilities can exist—profitably—entirely on the Web.

Speed, not size, is now the discerning aspect of the successful company. The challenge for facility managers under the new measurements of globalization, then, is how to make their building "fast." How can the building, as a tool, support the company's need for increased speed and flexibility?

Speed is also an issue with the revisions to existing technology. New software is constantly required, as older versions become incompatible. New hardware, like hand-held PCs with Internet capability, eclipse cell phones, which have displaced pagers. How is the facility manager to balance the need for competitive technology with a realistic budget? Some companies are beginning to explore the possibilities of a "green lease," so named because it is ostensibly an environmental concept. "Green leasing is a new, but dramatic shift in the traditional perspective of leased equipment. Under a green lease . . . when the customer no longer requires the use of the particular product . . . the manufacturer would be obligated to reclaim it and refurbish it or disassemble it for recycling . . ."[8] Green leasing has been beta-tested by some major equipment companies, including Apple and Xerox. Carpet manufacturers and manufacturers of RTUs have also entered the dialogue, exploring the potential for green leases. The visionary facility manager has an incredible opportunity to take advantage of the preliminary groundwork and position their firm strongly for speed and flexibility.

All sorts of technology that improves speed of operations may have unintended consequences for the facility manager. Consider the potential impacts of voice recognition technology. Voice recognition will provide additional speed to certain tasks. It may become as ubiquitous as the cell phone and personal computer. This technology is likely to mean replacement purchases or upgrades to existing equipment for the facility manager. However, it also has building impacts. How can such technology work properly if the acoustics in the space are inappropriate? Will too much background noise impair the proper functioning of the technology? Will inadequate acoustical insulation or poorly placed cubicle walls mean that the equipment for one employee may be triggered by the telephone conversation of an adjacent employee?

In addition to the tangible aspects of changing technology, there are also some very real intangible aspects. Changing technology and the need for customization has altered expectations. The ability to manipulate one's environment has been evolving exponentially since the beginning of human history. Within the lifetime of many of us, the most practical method for cooling a building was to open the windows for cross ventilation. Climate control of interior space is rapidly being replaced by "individual" controls. The office worker now expects that he or she will be able to adjust the temperature, humidity, lighting, and ergonomics of his or her 100 ft^2 cubicle. It is impossible to envision what the next generation will expect. Certainly, the average six-year-old is fearless in front of a computer monitor. Physical limitations are not acknowledged. The question is no longer "if . . ." but "how . . . ?" When the six-year-old of today hits the workplace of tomorrow, facility managers must be prepared to address the technological expectations for equipment and comfort.

It is important to keep up with technology. The adage, *you either lead, move out of the way, or go slow and get run over*, applies more than ever in today's businesses. Conventional wisdom maintains that current technology is essential for maintaining an edge, for improving productivity and quality. Certainly, once your competition is faster and better, you too must seriously consider making an equivalent investment. However, changing technology has implications for staffing as well. Corporations, and their facilities, have a greater ability to migrate than ever before. The entire world is available to them. This is one of the glories of globalization—there are no boundaries. However, your employees can migrate just as easily. If you want to keep them, you must keep them happy.

10.5 CHANGING SOCIAL AND ENVIRONMENTAL AWARENESS

Globalization is not just economic anymore, there is a growing understanding of social and environmental issues that has resulted in an increasing demand for transparency of operations, sustainable development, and green buildings. This translates into socio-environmental corporate responsibility in all aspects of a firm's operations. This is a function of globalization on two different levels: the *think globally, act locally* environmental concern for the quality and quantity of earth's limited resources, and the ability of individuals to get information from all over the world, process it, and respond to it.

Individuals certainly are empowered via technology to demand that facilities respect issues of personal concern. What a single person can accomplish with a laptop is impressive. As an example, in 1997, Jody Williams won the Nobel Peace Prize for her contribution to the international ban on landmines. Ms. Williams orchestrated 1,000 different human rights groups on six continents . . . from her home, via the Internet. The labor unions of the previous era are nothing compared to a single person with a networked laptop.

Younger individuals are not only empowered by technology; they are molded by it. Youngsters raised on the Internet have decidedly different perspectives than their parents. "Those reared in the passivity of the television age are not used to talking back. . . . But the X Generation, trained by interactive games, wants to connect with television, not just watch it. Their children, the Millennium Generation, are being reared on the computer and will demand interactivity even more."[9] Facility managers, prepare thyselves—the workforce is changing along with the technology.

Social and environmental power has also come of age organizationally and is quickly becoming a contender with the new political and economic organizations. The non-governmental organization (NGO) is the new defender of environmental and social concerns. NGOs respond to ". . . gaps left by government in meeting social needs, including adequate and equitable provision of health care, education, food, shelter, and environmental protection. . . . Growing from under 1,000 in 1956, the number of international NGOs now tops 20,000."[10]

NGOs generally have credibility due to their long history of altruistic pursuits. In contrast to industry organizations whose primary goals are to facilitate commerce and promote the products of their membership, many NGOs have the altruistic goals of protecting society and the environment. In contrast to standards organizations that are demarcated into isolated specialties, NGOs regularly address complex, holistic concerns adequately. In this arena, the NGOs are, without question, the experts. It is entirely possible that the NGOs, not ASTM, ISO, or other established standards groups, will lead the practical shift to new ways of thinking, new standards, and new measurements. For example, the Forest Stewardship Council (FSC), an international non-profit organization that trains, accredits, and monitors independent, third-party certifiers for well-managed forests around the world, is gaining an acceptance in segments of the construction industry. This is unheard of for NGOs in previous decades.

10.6 CONCLUSION

Thinking globally and acting locally is imperative. The rules have changed dramatically and are continuing to change as the world settles into the new era of "globalization." Speed and flexibility are prerequisite for the successful firm. Speed and flexibility can be either hindered or promoted by the actions of the facility manager. Successful facility managers are likely to be those who appreciate the big picture with all of its intricacies and can apply that understanding locally, within the company. The real value that FM can contribute to a company or client exists in anticipation, interactivity, and the ability to customize. Even if your company is not aiming to be a "global company," the impact of globalization will affect you. You can participate in how they change—or not, but you ought to become informed about the direction that they are taking.

Welcome to the new millennium.

ENDNOTES

1. Friedman, T. L. 1999. *The Lexus and the olive tree: Understanding globalization*. New York: Farrar, Straus and Giroux, p. 372.
2. Korten, D. C. 1995. *When corporations rule the world*. West Hartford, CT: Kumarian Press, p. 54.
3. Brown, L., et al. 1999. *Vital signs 1999: The environmental trends that are shaping our future*. New York: WorldWatch Institute, p. 25.
4. Ohmae, K. 1999. *The borderless world: Power and strategy in the interlinked economy*. New York: HarperBusiness, p. 11.
5. Greider, W. 1997. *One world, ready or not: The manic logic of global capitalism*. New York: Simon and Schuster, p. 93.
6. Davis, S., and Meyer, C. 1998. *Blur: The speed of change in the connected economy*. New York: Ernst & Young LLP, Perseus Books, p. 95.
7. Greider, W. 1997. *One world, ready or not: The manic logic of global capitalism*. New York: Simon and Schuster, p. 177.
8. Spiegel, R., and Meadows, D. 1999. *Green building materials: A guide to product selection and specification*. New York: John Wiley and Sons, p. 32.
9. Morris, D. 1999. *Vote.com: How big-money lobbyists and the media are losing their influence, and the Internet is giving power back to the people*. Los Angeles: Renaissance Books, p. 108.
10. Brown, L., et al. 1999. *Vital signs 1999: The environmental trends that are shaping our future*. New York: WorldWatch Institute, pp. 25–26.

CHAPTER 11
SUSTAINABLE DESIGN

David Lehrer, AIA
Senior Associate
Gensler
E-mail: davidlehrer@earthlink.net

11.1 INTRODUCTION AND BACKGROUND

Imagine a building that produces more energy than it consumes, costs less to build and operate than a conventional building, and in which occupants feel energized, healthy, and productive. Although this may appear to be a utopian vision, this is the future of architecture and construction seen by the leading practitioners and researchers in the realm of green building and sustainable design.

The green building movement in Europe and the United States is currently going through a period of expanding and maturing. Progressive design leadership, new building science research, a growing number of successful well-documented case studies, and environmental business strategies are bringing ecological considerations into the mainstream of architectural practice. In addition, the benefits of good environmental stewardship are leading an increasing number of companies and institutions to support initiatives in many areas of their planning and operations. In the design, building, and operation of commercial facilities, there are a multitude of ways by which building owners can reduce their environmental impacts while at the same time enhancing the workplace environment.

A number of descriptions are commonly used to describe design for environmental responsibility. The terms "green" and "sustainable" are used interchangeably, though there are shades of meaning implied by each. Today's conceptions about sustainable design have roots in the ecological movement of the 1960s and the post oil-embargo era of the 1970s. The concept of sustainable development was formally established in 1987 by a special commission of the United Nations commonly referred to as the Bruntland Commission. The commission's report, *Our Common Future*, describes a pro-growth development approach that takes the requirements of future generations into account. Sustainability is a goal that allows for the continuing improvement of standards of living without irreversible damage to resources we need to survive as a species.

Sustainable development is development that meets the needs of the present without compromising the ability of future generations to meet their own needs. According to some environmental advocates, the Bruntland report's concept of sustainability allows us to avoid the true dangers implicit in the world's population growth and increasing standards of living. Other terms used to describe ecological approaches to architectural design include "high-performance design" and "resource efficient design," descriptions that may be used to diffuse potential opposition by those that are skeptical about the dangers of environmental degradation and the need for reform. Finally, we have "ecological design," a description that has been employed for this design approach, coined and described eloquently as "any form of design that minimizes environmentally destructive impacts by integrating itself with living processes."[1]

The oil crisis of the early 1970s resulted in numerous efforts to make buildings more energy efficient. Although this led to innovation, these "first generation" energy efficient buildings were experimental and had many shortcomings. For example, in the attempt to reduce energy losses due to infiltration (i.e., air leakage), designers made buildings that were very airtight. This strategy led to unexpected problems related to poor indoor air quality. Today's green buildings respond to a variety of issues in addition to energy efficiency. Major considerations include occupants' health and comfort, impacts on urban infrastructure and building sites, and environmental impacts of building materials, construction, and demolition. Establishing priorities among these considerations has become one of the challenges of sustainable design practitioners as there is no singular authority in these matters and various design goals may be in conflict. In general, however, building occupants' health and comfort is viewed as one of the highest priorities, since increased productivity carries the highest potential benefit for building owners and managers, and problems related to occupant health carry the highest potential liability.

11.1.1 Environmental Impacts of Buildings

To work toward the creation of sustainable development, one must understand the environmental impacts of buildings and their relative importance. The major and best understood impact is caused by energy use from building operations, accounting for over 30% of the United States' annual energy consumption and 60% of the electricity utilized. If we include the energy used in the extraction, processing, and transportation of the materials used in construction, buildings account for 40% of the world's annual energy expenditure.[2] The primary energy loads in buildings are created by lighting, space heating and cooling, equipment, and domestic hot water. The production and consumption of this energy contribute to air pollution, acid rain, global climate change, and the impacts of fossil fuel extraction and refinement, such as oil spills.

While it is relatively straightforward to quantify the environmental effects of building construction and operation, community and regional planning issues carry an equal or greater significance than the buildings themselves. Even the most innovative suburban green building is extremely energy intensive if one considers the energy required for people to commute to the site in single occupant vehicles (SOVs). Transit and pedestrian-oriented developments, as well as urban sites, are inherently more energy efficient.

Outside of the building footprint there are environmental concerns related to the design of the site. For better or worse, the majority of new development and construction in the United States is suburban, creating substantial site-related impacts. In undeveloped areas, the majority of rainfall filters into the soil and into the natural aquifer which feeds streams. On developed sites, however, the majority of rainfall leaves the site as surface runoff. The large quantity of runoff from paved areas and roofs introduces contaminants into water resources, creates erosion along natural watercourses, and causes water tables to drop as the natural aquifers' sources are diverted. This runoff may create major problems for municipalities that combine sewage treatment with storm drainage. Excessive runoff during times of peak rainfall can exceed water treatment capabilities. When these systems flood, the result is the release of untreated sewage into rivers, lakes, and beaches, creating serious health and environmental problems. This increased development also damages or destroys natural habitats and ecosystems. It has been estimated that the current rate of development and deforestation worldwide is causing a mass extinction of species at the rate of 27,000 species each year, or the equivalent of 74 per day.[3]

The urbanization of suburbs and cities also increases the temperature of those areas, a consequence described as the "urban heat island effect." The effects of shade and evapotranspiration of plants is reduced, and heat is absorbed and reradiated by building and paving materials, especially dark materials, such as asphalt. As a result, the temperature in urban areas is typically 4–5°C higher than surrounding rural areas, and may be as much as 6–8°C higher.[4] This heating of developed areas adds to air conditioning loads and energy consumption and creates uncomfortable microclimates in regions which might otherwise be comfortable.

Construction materials also have environmental impacts, from the extraction of raw materials, processing into building products, transportation and installation of materials, and disposal at the end of their useful lives. The sum of the energy required to extract, process, transport, and install these materials is described as "embodied energy," and there is a growing body of research attempting to quantify the relative environmental impacts of various materials. Construction and demolition debris makes up 15–40% of municipal waste, depending on the region.[5] The cost of landfill disposal is leading some major contractors to find creative ways to deal with waste by separating, recycling, and reusing materials whenever possible. Logging of forest products for structural uses and finishes is leading to deforestation on a massive scale worldwide. Environmental impacts from logging include destruction of habitat, loss of topsoil, loss of biodiversity, and impacts on water resources.

These negative environmental impacts are not generally taken into account during the building design and development process. Most architects and development managers are aware of minimum environmental requirements such as state energy codes and environmental impact report (EIR) requirements, but few projects are designed to exceed these minimum standards. As these effects are external to the operation of the project itself, they may be described as *externalities*. Although there are methods to account for the total environmental effect of a development, the complexity that this accounting would create have led proponents of green development to focus instead on the many benefits of green design.

11.1.2 Benefits of Sustainable Design

The most effective way of overcoming the obstacles to sustainable development is to stress the benefits of the approach to building owners, operators, and users. Researchers at the Rocky Mountain Institute (RMI) have been collecting and documenting case studies for several years, persuasively illustrating these benefits. One of the most compelling benefits is the potential enhancement of employee productivity when the workplace environment is more healthful, comfortable, and controllable by the users. Studies have shown productivity gains of 6–16% and reduced absenteeism that resulted from improved interior environments. Correlating productivity with the physical environment and normalizing for all the associated human factors is a complex exercise. However, because employee salaries comprise the vast majority of a companies' operating budget, even a small increase in productivity can easily pay for potential added costs of environmental enhancements. By evaluating the total cost of constructing and operating a commercial building over a 30-year period, the initial construction cost accounts for just 2% of this total; operations and maintenance account for a mere 6%. Personnel costs, however, account for 92% of the cost of operating the facility.[6] For a typical business, a mere 1% increase in productivity could result in savings that exceed the company's entire energy budget.[7]

Although the productivity gains that result from a high-performance workplace may be hard for the average facility director to measure, energy conservation is not. For this reason, cost savings from energy efficiency remains one of the most easily documented benefits of green design. One approach for evaluating energy cost savings is the "simple payback" method, derived by dividing the added cost of the energy efficient system (if there is an added cost) by the projected annual savings. The result is the time required for the energy savings to "pay" for the increased first cost; however, this simple payback method is not a truly effective measure of these savings, as it may not take into account the effects of tax incentives, utility rebates, and other programs which are available in many localities. An alternate method for evaluating the economics of energy efficiency is to consider the return on investment for the additional cost of the energy conservation measures, as this employs terminology which is familiar to financial decision makers.

Sustainable developments also benefit from their unique niche in terms of public relations, marketing, and approvals. Unusual approaches and design features typically generate more interest in media, resulting in free publicity, and differentiate green projects from competing developments. In some cases, a developer's willingness to go beyond minimum environmental compliance may cause building and planning officials to be sympathetic and expedite the approval process, saving developers' time and expense.

In today's business environment, a growing number of companies understand the benefits of environmental stewardship, realizing that these values are transmitted to employees and customers. There is also a growing body of research that correlates environmental performance to stock portfolio performance. Ratings of companies' environmental performance are now being documented by mutual fund managers for funds that invest only in socially and environmentally responsible companies. A 1996 study indicated that corporate environmental leadership could result in investors' perceptions of lower risk and a stock valuation enhancement of 5%.[8]

11.2 DESIGN AND CONSTRUCTION

Design professionals have access to new tools and approaches for creating innovative projects incorporating ecological principles. Some strategies, such as daylighting, natural ventilation, or designing for solar orientation and climate, represent a return to simple but effective approaches that have been largely ignored in recent decades. Other design strategies utilize new tools, such as computer simulation and state-of-the-art building systems. The design of any project will benefit from design integration and having all stakeholders represented, informed, and supportive of the environmental goals of the project.

11.2.1 Team Building and Goal Setting

The creation of a "green team" is the first phase of creating a successful project. Designers, architects, and consultants who are willing to explore alternative strategies and think "outside the box" are preferred to those who would rather use the same approaches that they have successfully used in the past. Design consultants who have invested in their knowledge base and have green building experience will add to the success of the project. Although a growing number of design professionals are becoming familiar with the issues related to sustainable design, in many cases having an independent sustainability consultant on the team will contribute information on new or alternative materials and methods. This specialist can also work to establish performance objectives and ensure that the objectives are pursued by the owner and design team through the design and construction process.

While project teams are accustomed to working with program requirements and constraints for budget, schedule, and codes, establishing environmental and performance codes may be unfamiliar territory to some consultants. To define realistic, achievable goals that can be maintained throughout the design and building process, early input from the entire team is required. Formal workshops including all consultants may be set up at a project's initiation to explore and define project goals. Such workshops or "charettes" typically are lead by facilitators or green building consultants familiar with the process. Consultants may be broken down into issue-specific facilitated teams for brainstorming on specific topics (e.g., mechanical systems, site issues, or the building envelope). The results of these workshops should be summarized into a document to guide the design team and be referenced at predetermined intervals to verify that the project's objectives are being actively pursued.

11.2.2 Design Integration

In conventional practice, the architectural design is frequently allowed to advance ahead of the engineering of building systems. The unfortunate result of this practice is that opportunities to optimize and integrate the building envelope with systems are lost. One of the ways to enhance building performance and constrain costs is by integrating the design process. Researchers at RMI promote "whole systems thinking" and "front loaded design," advocating that through informed early integration of the entire design team a building might be designed to perform at significantly higher levels while minimizing or eliminating additional costs. It has been estimated that when only 1% of a project's upfront costs are spent, 70% of the project's life-cycle costs have been established. When 7% of the costs are spent, 85% of the life-cycle costs have been established.[9] For example, if a highly efficient glazing system is selected for a building, it may be possible to reduce the size of the mechanical system, helping to offset the cost of the more expensive glazing. If this decision is made late in the design process, the mechanical system design may have already been too far developed to allow for appropriate down-sizing or "right-sizing" to the reduced load. Attempting to add on design enhancements late in the design process generally adds cost and is less effective than if those decisions had been made early in the design process.

11.2.3 Site and Landscape Design

In conventional design practice, potential building sites are analyzed in terms of geometry, views, proximity to streets and infrastructure, and other issues that impact the costs and practicalities of the site's development. It is still common to see a new site completely bulldozed of any vegetation and natural contours erased to create large flat building pads, even in hilly terrain. The design issues related to site design are then delegated to civil engineers and landscape architects. In the creation of a sustainable project, this process must be greatly improved upon. The principle goals of sustainable site design are to minimize the impact upon the site and enhance the natural benefits the site provides. This requires a thorough documentation of a site's natural resources be made prior to the initiation of design. This site analysis should include documentation of trees and plant resources, water features, runoff patterns, wetlands, solar access, microclimates, and habitats. Project design should proceed with the preservation of these site resources in mind. Manmade features, such as existing buildings, roads, and other improvements, should also be included in the site analysis and considered as resources that may potentially be used and/or adapted in the new design.

Although a great deal of consideration may be given to the objective of making green *buildings*, the means of transportation to and from a site may have an environmental impact greater that that of the buildings themselves. This is especially true with facilities in suburban locations. Appropriate site selection, which reduces the distances that employees, customers, and suppliers must travel, is a first means to reducing transportation impacts. This is often most easily achieved by utilizing available sites in developed areas and may reduce infrastructure costs as well. Reducing the reliance on automobiles (especially single occupant vehicles or SOVs) is a primary objective of all sustainable developments. This may be achieved through a variety of strategies, such as employer subsidized vans or carpools with privileged parking, vouchers for public transportation, shuttle services to and from public transportation, and providing bicycle parking and shower facilities. Companies should avoid policies that may encourage driving, such as the practice of providing individual or paid parking as an employee benefit. Automotive dependence can also be reduced (and employee convenience enhanced) if amenities such as day care, fitness, and dining facilities are provided on-site. In some cities if a commitment to providing alternatives to SOVs is made, it may be possible to provide fewer parking spaces than required by local planning codes, saving costs and potentially reducing the area within the site to be dedicated for parking.

As noted previously, the hydrology of an area may be adversely affected by development. A number of design strategies can mitigate the effect of development on water resources by allowing water to filter back into the ground and recharge aquifers. This may be achieved by minimizing the amount of built and paved areas and including infiltration swales that hold rainwater to allow it to slowly percolate into the ground.

Alternatives to conventional paving for areas that are not subjected to heavy parking loads include "reinforced turf" systems—plastic or concrete grids filled with layers of gravel and soil and then planted with grasses. There are now pervious mixes of asphalt or concrete that allow water to filter through and are still suitable for parking areas. Subsurface groundwater recharge basins may be built under parking lots and similar structures that allow collected storm water runoff to be stored and percolate slowly into the soil. Other types of permeable paving such as decomposed granite may be used for pedestrian areas.

Another effect of development, which may be mitigated through green design, is the urban heat island effect described previously. This effect may be mitigated through several measures, including minimizing paved areas, using light colored paving and roofing materials, providing shade trees in parking and other paved areas, and maximizing the amount of vegetation on unbuilt areas of the site. Green roofs—roofs that are planted with natural plant materials—represent another strategy to reduce heat islands. The Gap's 901 Cherry office building in San Bruno designed by William McDonough and Partners and Gensler utilized a grass roof planted with native grasses to mitigate the effects of the building's development. The grass roof insulates the building thermally and acoustically, protects the roof membrane, slows runoff to storm drains, and acts much like an undeveloped site in terms of the heat island effect.

The selection of appropriate plant materials is another goal of sustainable design. The use of native species is preferred, as these plants will have the best chance for survival with the minimum amount of water and maintenance and will support the native wildlife indigenous to the site. Xeriscaping—the use of drought-resistant plants that require little or no irrigation—reduces water use and can reduce both maintenance costs and initial costs, as permanent irrigation systems may not be required. For the landscaping of two AT&T office parks in the Chicago area, turfgrass was replaced by indigenous tall-grass prairie grassland. The new landscape was installed for less than $2,000-per-acre, the amount that AT&T had been paying annually for maintenance of the existing turfgrass. The cost to maintain the new prairie grassland is less than $500-per-acre, less than one-quarter of the previous cost.[10]

11.2.4 Design for Climate

Historically, architects and builders required a thorough understanding of climate to create comfortable shelter for human habitation. Diverse building traditions evolved throughout the world by which materials and architectural forms were adapted to solar orientation, temperature, and wind patterns to create comfortable spaces. Buildings in arid climates with wide diurnal temperature swings, for example, tended to have massive walls that could absorb and mitigate heat during the day and retain it at night when the exterior temperature falls. Buildings in tropical or humid climates were traditionally built with lightweight materials and architectural forms that shade walls and openings, allowing for maximum ventilation to keep the interior cool.

In the years following World War II, several innovations in building technology led to dramatic changes in the design of commercial buildings. When whole building HVAC systems and fluorescent lighting became economically feasible for large commercial buildings, architects were freed from the necessity of designing with the local climate in mind. They could now provide large areas of glass with nonadvantageous orientations, delegating the need to provide comfort to the mechanical engineers who now could design and size mechanical systems for practically any configuration of building form, perhaps even relishing the challenge of making an impractical design habitable. The new buildings evolved to have large floor plates that were completely dependent on electrical lighting and HVAC systems, with building occupants far from windows and natural light sources. While these advances in technology have given architects much greater freedom in design, the technology has not generally been applied in a manner that either enhances building performance, energy efficiency, or occupants' comfort.

11.2.5 Passive Design Strategies

One of the first considerations in designing a climatically responsive building is consideration of appropriate *passive* design strategies. Passive design refers to the creation of building elements and configurations that take advantage of the physical environment of a specific site to either heat, cool, or ventilate a building. With most commercial buildings, it is not possible to entirely eliminate the need for electrical lighting and mechanical systems; however, passive design is very effective in reducing the reliance or loads on these systems. Case studies illustrate that it is possible to reduce energy consumption by 30% by merely optimizing for the appropriate orientation and massing.[11] To apply passive design strategies, an analysis of the local climate and specific conditions of the building site must be made. Such an analysis would include local climate data (monthly averages and extremes for temperature, humidity, and wind), site-specific microclimate information, and solar access. Climate data is available for this use from several sources, though much of the available data is collected at weather collection locations such as airports, that differ from the conditions found within the built environment. Consequently, an analysis of the site's microclimate is necessary to compare the characteristics of the site to the collected weather data. A review of the site's surroundings and the use of latitude-specific sun path diagrams, also available from a number of sources, can be used to determine the effects the sun will have on the site. Investigation of traditional buildings in the area may also provide ideas that can be adapted for modern use.

With the understanding of the site that a climate analysis yields, the design team may proceed to establish appropriate passive design strategies. These strategies include the massing and orientations of the buildings, window sizes and locations, overhangs, shading devices, use of insulation, and in some climates, natural ventilation. In some climates, it may also be advantageous to utilize and optimize the building's *thermal mass* (i.e., the heat absorbing capacities of materials such as concrete and masonry) to take advantage of, or mitigate, diurnal temperature swings.

11.2.6 Building Simulation

Moving beyond the passive strategies that may be used to reduce loads upon the building, integrating the envelope design with the building systems will allow for the optimization of the entire project. To achieve a high level of integration and performance the design team must utilize appropriate design tools and techniques. Computer simulations are fundamental to prediction of energy performance and may be utilized for analysis of daylighting strategies, electrical lighting, and other performance variables. Simulations can be used to test and evaluate various design alternatives easily in an *iterative* process, allowing the design team to refine the design and determine the cost effectiveness of the alternatives. The process of modeling, analyzing, and selecting alternatives should be repeated at successive intervals during the design process as the design is developed to document the performance and integration of the design solution.

The software available for analyzing building performance varies in cost and complexity. One of the most commonly used software tools is DOE-2, a powerful tool which is typically used by energy analysts or in some cases by mechanical engineers. Other specialized software applications may be utilized for analyzing daylight or electrical lighting designs. Although not specifically designed for this purpose, three-dimensional architectural modeling programs may be used for sun and shadow studies, and to evaluate "direct beam" solar penetration into buildings. Regardless of which software is used, it is critical that the process of design and simulation be initiated early in the design process before the major design decisions have been finalized.

11.2.7 Building Envelope Design

In addition to the passive strategies described previously, the major elements that determine the performance of the building envelope are the window, wall, and roof assemblies. The insulation values and thermal mass of roof and wall assemblies will have major effects on the thermal performance of a building and may be tested in the whole building simulation already described. The exterior roof color will also contribute to the energy performance of the building, with white or light colors preferred in temperate and warm climates.

Windows and glazing systems contribute to the energy performance of a building, to the safety and comfort of the occupants within, as well as to the character of building facades. Consequently, the window design and glazing selection are critical to the performance and appearance of the building envelope. The newest advances in glazing technology are high performing *spectrally selective* products. These glazing products allow for higher transmittance of visible light than that of the heat-contributing infrared portion of the spectrum. This spectrally selective quality may be achieved through use of certain tints, coatings, or both. *Low-e* coatings are metallic coatings that reflect certain types of radiation and transmit others. (Ironically, it is the automotive industry that drives glazing research and development, and the building industry benefits from new developments.) Performance can be further enhanced by "tuning" the glazing for different facades, as the environmental considerations in terms of light quality and solar intensity vary greatly between orientations.

To assist in gaining an understanding of glazing properties it may be useful to provide some definitions of commonly used terms and properties:

Shading Coefficient (SC). This represents the amount of heat a particular glazing allows to penetrate into a space, expressed as a ratio. A high number, such as 0.8, allows a lot of heat to enter a space, a low number such as 0.3 or 0.4 lets less heat into a space. In commercial buildings that are generally in a cooling mode it is highly preferable to use glazing with a low shading coefficient.

Transmittance—Visible (VL). This is the amount of visible light that passes through the glazing, expressed as a percentage of the light striking the exterior. For daylighting objectives this number should be relatively high, perhaps as high as 50–70%, though it must be balanced against the potential for heat gain, which is measured by the shading coefficient. The potential for glare is another important consideration, explained in greater detail below.

Transmittance—Solar (IR). This is the percentage of the infrared light (which is invisible and contributes heat to a space) which passes through the glazing. For energy performance and comfort, this number should be a low value.

Tints. This is the integral color of the glass itself (not to be confused with reflective coatings). Some tint colors, such as azurite, are inherently spectrally selective.

Reflective Coatings. This is an applied solar reflective coating which reduces the heat gain of glazing. While these coatings may reflect high amounts of visible and infrared light, the tradeoff will be lower light transmission through the glass.

Ceramic Frit. This is an applied silk-screened pattern that will reduce light transmittance and heat gain.

11.2.8 Daylighting

In commercial buildings, electric lighting is typically the largest energy load and can account for 40–50% of the total energy load. Heat generated by lighting adds 3–5% to the total energy consumed. Properly designed daylighting strategies can reduce the demand for electrical lighting by 50–80%.[12] By reducing the lighting load in a space, the associated heat is reduced and the mechanical systems may be reduced in size and cooling loads reduced. Ironically, the greatest demand for lighting is during the business hours when ample daylight is available. Successful incorporation of daylighting linked with efficient lighting and lighting controls can reduce the demand for electrical lighting and cooling, and provide benefits to the building occupants as well. Recent studies have correlated daylight in buildings with increased retail sales and higher student test scores. Retail stores with skylights boosted sales by an average of 40% over nondaylit stores, and students' test scores increased by 15–26% when classrooms had larger windows, skylights, or higher levels of daylight than control cases used in the study.[13]

However, using the sun as a light source for illuminating a building interior is like trying to get a drink of water from a fire hydrant; it is easy to get too much. If not designed for correctly, daylight will add excessive heat to a space, negatively impacting both energy consumption and occupant comfort. Architects' affinity for large expanses of glass (including skylights) has led to the design of spaces that overheat, in spite of massive air conditioning systems. Occupant's visual comfort is achieved by providing the appropriate illumination level, individual control, and the reduction of glare. Although we usually perceive glare simply as a light source that is too bright, glare is actually caused by excessive contrast within a space. The human eye has the ability to adapt to widely varying levels of brightness over time; however, our eyes cannot adjust to great contrast within its view. The gradient of daylight within a space generally varies from the highest level adjacent to windows to the lowest level at the location most distant from windows. Successful daylight reduces, or flattens, this gradient to reduce the potential for glare by controlling the brightness at the windows, increasing the light levels in the darkest part of the space, and reducing the contrast adjacent to windows.

Daylight approaches fall into two categories: sidelights and toplights. Building configuration is a major factor in developing the best daylight solution, with inherent opportunities and challenges with each approach. Toplighting with skylights may be most appropriate for low rise, industrial, and atrium buildings. While skylights allow light to enter the central portion of a building, the amount of sunlight that is incident on the skylight is greatest in the summer when the potential for overheating is greatest. The light incident on the skylight is lowest in the winter when the demand for daylight is at its maximum and heat gain may even be preferable. Clear glazed skylights which allow direct solar radiation to enter a space typically result in excessive glare and heat gain, and are only feasible if fitted with louvers or other means of control. Clearstory windows on roof "monitors," saw-tooth configurations with north facing windows (assuming a north latitude), or skylights with translucent glazing allow for the best toplight solutions with control of direct solar radiation.

Sidelights may be tuned for their respective orientations. Because east and west orientations are most difficult to control, designers often minimize these exposures. When this option is unavailable, vertical fixed shading devices or operable shades are effective solutions. North elevations allow for more daylight without potential for heat gain, and south elevations may be shaded with overhangs which, if sized properly, will provide maximum shading in the summer months due to the sun's higher position in the sky. Light shelves are effective means for shading glazing and bouncing light onto ceilings and deeper into the interior of a building.

A number of tools are available for the analysis of daylighting design strategies. Software programs such as Radience and Lightscape may produce realistic representations of a space and create illumination "maps" of surfaces. Because light "scales" perfectly (i.e., the light qualities in a scale model will accurately represent that in the final built space), physical models may also be used to predict daylight illumination levels and qualities. For such models to yield accurate results, care must be taken to accurately represent interior finishes, building geometries, and window details. Due to the need for such accuracy, such models are generally built at a rather large scale, such as $\frac{1}{2}'' = 1'0''$. The models may be mounted on a heliodon to simulate any hour or day of the year and photographed or videotaped to verify the qualitative characteristics of the daylight within the space. Light levels may also be measured within the modeled space using miniature light sensors.

Finally, daylighting schemes must be incorporated both into the base building architecture and into the interior or workplace design. Open plan layouts, partitions with glazed transoms or sidelights, and light-colored finishes will enhance daylighting effectiveness.

11.2.9 Building Systems Design

After a building's design has been optimized in terms of envelope and daylight design, the building's performance will be determined by the efficiency of the building's heating, ventilating, and air-conditioning (HVAC) systems. The energy consumed by these systems is typically between 40–60% of a building's total energy use. Before the energy crisis in the United States in the 1970s, mechanical systems were typically designed for heating and cooling simultaneously, resulting in poor energy efficiency.[14] Current "best practice" for HVAC design balances the need for temperature and humidity control, ventilation for indoor air quality, and energy efficiency.

Several HVAC design tools and system innovations allow for enhanced system performance. Mechanical engineers may utilize computer simulation to test and evaluate projected energy use for alternative system designs. Energy management systems are available to coordinate the operation of system components and monitor building performance. A number of efficiency enhancements to HVAC components may be considered, including premium-efficiency motors, variable-speed drives, direct digital control (DDC) systems, and advanced control strategies. Through well-integrated design, equipment may be appropriately sized (not oversized based on rules of thumb) for part-load efficiency and reduced initial cost.[15]

11.2.10 Building Commissioning

From experience with numerous case studies, design practitioners have learned that building systems are not always installed and operated as designed. This is especially true with buildings that employ sophisticated systems for energy efficiency and occupant comfort. Common problems and complaints include indoor air quality, lack of thermal comfort, loss of efficiency, and maintenance problems. To ensure that systems are operating at their maximum performance, a sequence of commissioning activities should be established. Mechanical systems, plumbing, electrical systems, controls and fire safety systems are a few of the concerns of the commissioning process.

A commissioning program should begin during the early design phases to be most effective. The first steps include determining the scope of the commissioning process and responsibilities among the design and construction team. The process is typically most intensive toward the end of the construction process before the building is turned over to the owner for occupancy. At this point, final adjustments of systems ensure proper operation and efficiency. Training of the building operators is another important aspect of this process. A final commissioning report is then prepared for the owner to document the completed checklists, approvals, and system manuals. The commissioning process should then be continued on a regular interval for approximately 12 months after occupancy to ensure continued operation and performance.[16]

11.2.11 Design for Change and Flexibility

As the British theorist Charles Handy has said, "change isn't what it used to be."[17] The rate of change in today's economy means that businesses can no longer assume that they will be doing the same thing in 5 or even 2 years that they are doing today. Companies must be nimble to implement new strategies and adapt to changes, consequently their facilities must have built-in flexibility to readily accommodate growth and reconfiguration. The manner in which change in the workplace is accomplished may have significant implications in terms of both cost and environmental impact. Typical renovations for tenant improvements (TIs) include changing "permanent" partitions (typically gypsum board and metal studs), ceilings, electrical wiring and fixtures, mechanical ductwork, and wall and floor finishes. These types of changes are costly, create large amounts of construction waste, and the materials involved generally end up as landfill material.

Several design approaches allow for TI reconfiguration, which reduce or eliminate the construction waste associated with conventional approaches, and result in considerable savings. By minimizing built partitions and using modular office furniture systems, changes are quickly and cheaply accomplished using only in-house facilities resources and not relying on outside contractors. When private offices and enclosed work areas are required, the use of demountable partitions allows for the reuse of partitioning materials. Several manufacturers offer demountable systems, and although these systems have a higher initial cost, they may be cost effective when churn costs over time are considered. Fixed spline systems offer another means to create a flexible workplace. Permanent partitions or "splines" with data and power infrastructure are installed at established modules to create separate bays, typically with a module of 18–30′ (verify with WP group). Furniture systems then infill between the splines to create the required workplace configuration.

Raised floor systems provide many benefits both in terms of flexibility, occupant comfort, and indoor air quality. These systems consist of 24″ × 24″ concrete-filled panels supported on steel pedestals that are adhered to the structural slab. Mechanical diffusers and data/power floor boxes are built into the floor panels or "tiles," which are easily relocated for reconfigurations. With raised floor systems, churn costs may be significantly reduced. When Owens Corning completed their

275,000 ft^2 to headquarters in Toledo, Ohio, they had a churn rate of 130% in their first year of occupancy. The access flooring system allowed for these moves to be made inexpensively, saving a half-million dollars in relocation costs in the first year, a savings of $1.80 per ft^2.[18] The benefits of raised floor systems in terms of indoor air quality, comfort, and energy conservation are described in the section on ventilation approaches that follows.

11.2.12 Energy Efficiency and Alternative Energy Sources

Design teams typically use existing energy codes such as ASHRAE/IES Standard 90.1-1989 or California's Title 24 as benchmarks against which to measure energy performance of buildings. As noted previously, by simply optimizing a building's orientation and massing, it is possible to reduce energy consumption by as much as 30%. When effective passive strategies are coupled with energy efficient building systems, it is possible to create buildings which are as much as 50–90% more efficient than a conventionally designed building. This fact is supported by many well-documented case studies.

Green building proponents envision going beyond the mere reduction in energy loads to a future in which buildings are "net energy exporters"; that is, buildings will actually produce more energy than they consume. One strategy for achieving this is through the use of photovoltaics, solar cells that convert sunlight into electrical current. Photovoltaics (PVs) may produce twenty times more energy over their lifetime than is used in their creation and do not create greenhouse gases or other pollutants.[19] While the cost of PV systems is often prohibitive, the costs have been decreasing steadily and are expected to continue to do so. Electricity generated by PVs was recently reported to have a cost of around $0.30 per kW, compared to the average cost of $0.08–0.10 per kW for electricity derived from fossil fuels.[20] As with many new technologies, as demand and production of PVs increase, the costs are expected to become competitive with other forms of electrical production. According to some estimates, this may occur within the next 5 years. Government and utility incentive programs may further reduce costs. In twenty-three states "net metering" programs are in place which mandate that electrical utilities must purchase excess power from consumers which have PV arrays or other forms of electrical generation.[21] The most progressive of these laws require that the utility purchase the power at the same rate as the utility charges. The fact that PVs produce their peak when the sun is at its most intense—which coincides for the peak demand for electricity—adds to their feasibility.

A current trend in PV design is to integrate the panels with other building materials such as roofs and curtainwalls, systems that are described as building integrated photovoltaics (BIPVs). These systems help offset the cost of PVs, as they replace other materials which would have been required, and in the case of roofs, can help to protect the roof membrane from thermally induced movement and ultraviolet degradation.

Although PV systems represent an emerging solar technology, thermal solar systems have been used in the United States and throughout the world for heating domestic hot water, pool heating, and space heating. However, the majority of these applications is of a scale more suited for domestic use as opposed to commercial applications.

Fuel cells represent another emerging technology that, in spite of its current high cost, holds promise for on-site electrical generation. Truly a "space age" technology, fuel cells were first developed and tested in the U.S. space program in the 1960s and continue to be used for power production in space.[22] Fuel cells create electricity as the result of a chemical reaction, with the only emissions from this reaction being pure water and heat, which conveniently may be used in buildings that require domestic hot water. The fuel cell requires inputs of hydrogen (H_2) and oxygen (O_2), which are combined by the action of a catalyst, often made of platinum. The hydrogen may be produced from a variety of sources, including natural gas, methane, and methanol; in fact, there are a variety of hydrogen-producing methods of great interest and potential for creating efficient clean energy. However, at this time, most building installed fuel cells use natural gas as a source due to its availability. This type of application utilizes a fuel reformer to convert the natural gas into a hydrogen rich fuel stream that supplies the fuel cell.

While still in many ways a demonstration technology, several manufacturers now produce fuel cells for building applications. Four Times Square, one of New York City's most prominent new green buildings utilizes two 200-kW fuel cells, testimony to the owner's commitment to sustainable design. For the electrical upgrade of a New York City police precinct in Central Park, the design team determined that a fuel cell installation would be less disruptive and costly than trenching for new electrical service from existing supply lines. A 200-kW fuel cell will be used to supply the upgraded police building and is expected to be used to charge the police department's electric vehicles.[23] One of the most viable applications for fuel cells may be in cases where electrical power quality and reliability is a concern. In such cases, the cost of the fuel cell may be partially offset by eliminating the need for an uninterrupted power supply (UPS).

11.2.13 Lighting Controls and Efficiency

As noted previously, electrical lighting constitutes a major component of building energy use. Its impact can be reduced through the careful implementation of daylighting strategies. However, no daylighting strategy will actually save energy if all the lights are still turned on. To take advantage of daylighting as an energy conserving method, lighting controls must be incorporated which limit electrical light usage. (Unfortunately, building occupants are not a reliable means for turning off lights, especially in commercial buildings.) Occupancy sensors are the most failsafe method of control and provide maximum savings by reducing the lighting consumption to zero when rooms are unoccupied. Timers may also be incorporated which automatically turn off lights at the end of the business day. Daylight sensors combined with dimmable or stepped ballasts and lamps allow lights to be dimmed when there is sufficient daylight within a space. Although these systems are available from a number of manufacturers, they must be carefully engineered, installed, and calibrated to work correctly. Sensors must be correctly located and calibrated to accurately measure daylight levels. Consequently, the lighting control systems must be part of the building commissioning process.

Lighting controls should also be zoned appropriately so that individual users may control and turn off lights on demand. With the large number of visual display terminals (VDTs) in use today, individuals may prefer to have less light than in the past. Having the ability to dim lights, or step down by 50% (this feature is required by some state energy codes), allows individuals to fine-tune light levels to their preference. In businesses that are highly computer-intensive, many people keep lights off and window shades closed the majority of the time.

Several design strategies are available to reduce the energy use of electrical lighting systems. Although current building codes allow light loads to be anywhere from 1.5–2.5 watts per ft^2, a lighting power density as low as 0.65–1.2 watts per ft^2 can provide a fully functional lighting design.[24] Designing to appropriate lighting levels for specific tasks and locating lighting where it is required is more efficient (and may yield more interesting spaces) than a uniform spacing of light fixtures. The use of "task/ambient" systems—providing low levels of light for general illumination and individually controlled lighting at desks and workstations—is another commonly used strategy. Also, the number of high quality energy efficient lamps has grown in recent years, with highly efficient T8 and T5 lamps, compact fluorescent lamps, and electronic ballasts typically preferred for energy efficient installations. Although many designers still believe that incandescent lights provide a better light quality, the color characteristics of fluorescent lights have been greatly improved.

Indirect lighting systems which illuminate by throwing light onto the ceiling and reflecting it to the space below are often preferred for the soft quality they create. Although these systems were originally considered to be a more expensive option, competition between manufacturers has caused these systems to be cost competitive with conventional down-lights.

11.2.14 Water Management

Water consumption in the United States amounts to over 340 billions of gallons each day. Enormous opportunities for water conservation have been illustrated in the industrial applications in the past. Due to efficiencies in industrial systems, U.S. industries use 36% less water than they did in 1950.[25] A number of design strategies may be implemented on commercial buildings to conserve water and reduce the amount of waste that is released into rivers, streams, and other bodies of water. Toilets consume the majority of water in residential and commercial buildings. Low-flow toilets for commercial use which consume less than 1.6 gallons per flush are available and required in many states. Waterless urinals are also available for commercial use and have been installed in numerous locations.

Although still relatively rare, a number of alternative water conserving systems have been installed in buildings in the United States and Europe. Gray-water systems are dual pipe systems that store "gray" water from sinks, showers, and other nonsewage waste for use in toilet flushing and landscape irrigation. Cistern systems collect and store rainwater runoff from rooftops for these same uses. On-site biological waste treatment systems utilize constructed wetlands which process wastewater (gray-water or sewage, also known as black-water) through the action of microorganisms and plants. California and Texas have implemented standards to encourage gray-water use, and some progressive cities are beginning to require that new buildings be dual plumbed for use with planned municipal gray-water supply systems.

A case study in water conservation and wastewater management, the C. K. Choi Institute for Asian Research at the University of British Columbia does not require a connection to the university's sewer system. The 30,000 ft^2 building, designed by Matsuzaki Wright Architects, includes waterless composting toilets to save nearly 1,500 gallons of water a day. City water is only required for lavatory faucets and kitchen sinks. A cistern collects rainwater to further reduce water consumption. Gray-water is treated on-site through a simple biological wastewater treatment system, and sanitary waste is treated through the use of composting toilets which produce a nutrient-rich humus suitable for landscaping.[26]

11.2.15 Materials

In the selection of building materials, environmental impacts should be considered in addition to the normal questions about cost, aesthetics, and durability. The *Environmental Building News* writes independent reviews of materials and products and has published a list of the most important considerations for material selection.[27] One of the primary considerations, according to *EBN*, is how the materials affect the building's energy performance. For example, insulation, glazing, and mechanical systems should be selected to enhance the building's energy performance. Another concern is how the material selection will affect the health of the occupants. Many architectural finishes and furnishings may potentially emit odors or toxic emissions. Although the most dangerous known materials, such as asbestos and lead-containing paints, are now widely prohibited, many paints and adhesives contain volatile organic compounds (VOCs) which are still generally unregulated and may adversely affect occupants' health. VOCs also react in sunlight to contribute to the creation of smog and ground-level ozone, a pollutant that is not to be confused with the ozone in the stratosphere which protects us from ultraviolet radiation.[28] California long ago enforced regulations limiting the maximum VOC content in paints; the Environmental Protection Agency began phasing in similar regulations for the entire country in 1999, and many manufacturers now offer low-VOC and zero-VOC paints.

Other important environmental considerations for material selection include durability and maintenance, as early replacement and disposal of poor quality products with short life spans is a wasteful use of resources. The amount of energy required to extract, process, transport, and install is termed the "embodied energy" of the material. Although there is much interest in conserving energy resources by reducing the embodied energy of a building by selecting low energy materials, it remains a complex issue with no clear answers. For example, in comparing the embodied energy of a steel structure with a concrete one, it is clear that the production of steel is more energy intensive than concrete production. But steel often contains high amounts of recycled steel content and the steel may be recycled indefinitely after the building is demolished. Although there are no clear answers for material selection, green building advocates typically have several key considerations in mind to reduce the environmental impacts of materials. Materials are preferred which are produced locally to the building site to reduce energy for transportation. Natural materials produced from renewable sources that do not create hazardous by-products in their manufacture are also preferred, and there are a number of alternate materials now available with this goal in mind. For example, natural linoleum produced primarily from linseed oil and cork is preferred over other types of flooring which are petrochemical based and create toxic by-products. Some products utilize recycled materials as a feedstock to reduce the need for virgin materials, and many products are now available that are largely produced from what otherwise would be landfill waste. The most advanced manufacturers are now able to use reclaimed building products that may be used as feedstock to create the same product. This is preferable to the practice of "down-cycling" in which reclaimed materials are used for lower quality products.

Due to frequent publicity related to deforestation and endangered wood species, designers have become aware of the environmental issues regarding wood specification and use, perhaps more so than other materials. Recent progress with third party certification programs has made environmentally responsible specification of wood products more straightforward and reliable than in the past. Wood products that come from forests managed in accordance with the standards of the Forest Stewardship Council (FSC) may be certified and specified by architects and designers. A growing number of wood products are available from these certified sources. Home Depot, one of the largest wholesale buyers of wood products, has announced that it will begin to phase in certified wood products, creating a new awareness and demand for sustainably harvested wood.

The increased interest in green buildings has caused many manufacturers to reevaluate their industrial processes and environmental impacts. A number of alternative green products have been growing to meet the increasing demand. Although there is the danger of *greenwashing*, the practice of making false claims about environmental responsibility, there is a growing body of information available to assist architects and designers in making material choices.

11.2.16 Construction Phase Activities

The construction process has environmental impacts in terms of noise, air and water quality, disturbance of habitats, and waste sent to landfills. To control and reduce these impacts, project specifications should include environmental performance requirements for construction. Such specifications should include noise and erosion control, indoor air quality, and waste management. Many municipalities now have programs to encourage recycling of construction and demolition debris (C&DD), which is a major contributor to landfill sites. Materials that are typically recycled include metals (which are of the highest value to recyclers), wood, earth, gypsum board, cardboard, and rubble from brick, asphalt, and concrete.

In many locations, contractors are faced with high costs for disposal of this waste (referred to as tipping fees), making the recycling of this material cost effective. Portland, Oregon has been a leader with respect to controlling construction activity waste, and has implemented a policy for city-sponsored projects requiring aggressive recycling of C&DD. During the construction of the arena for the Portland Trailblazers, Turner Construction worked with local recyclers to recycle 95% of all debris. With the project 60% complete, the contractor had saved over $190,000 through this effort.[29]

11.2.17 Green Building Rating Systems

Sustainable design is a complex field of interconnected and sometimes conflicting demands. For example, the need for daylighting and ventilation for occupant health and comfort may be in conflict with the ideal design for energy efficiency. Analyzing the various sustainability priorities and design strategies is also not a process that most building design teams and owners are familiar with. To help designers and their clients incorporate green building strategies, the Leadership in Energy and Environmental Design (LEED) green building rating system has been introduced. The system was created by the U.S. Green Building Council (USGBC), a membership organization of building industry groups and companies. LEED identifies six areas of concentration—site design, energy efficiency, materials and resources, IAQ, water, and the design process—and establishes a system of assigning points, or credits, for meeting prescribed requirements or levels of performance. Buildings must first comply with a number of ten mandatory prerequisites, and based on the credits earned (of forty-four possible) can qualify for a bronze, silver, gold, or platinum rating. The system is applicable to new and existing commercial, institutional, and high-rise residential buildings. The USGBC is currently developing versions of LEED for use with interior (or tenant improvement) projects and residential development.

In the United States the LEED system is becoming the standard for comparing green building performance and as a design checklist. At the time of this writing, a pilot version of LEED is being applied on several dozen buildings and the results should be available for the design community in the near future.

11.3 OCCUPANT HEALTH AND COMFORT

The design of buildings has effects on the environment at large and the health and well-being of building occupants. From the many publicized examples of sick buildings in the recent past, people are much more aware of the potential health effects of hermetically sealed buildings. Indoor air quality, ergonomics, visual and thermal comfort, and occupant control of the environment can enhance or detract from occupants' health, productivity, and well-being.

11.3.1 Indoor Air Quality (IAQ)

Some estimate that Americans spend over 90% of their time indoors. Although it may seem surprising, studies have shown that air quality inside of buildings is generally worse than outside air, even in urban areas with high levels of pollution. In addition, the U.S. Environmental Protection Agency cites problems from poor indoor air quality as one of the leading health risks.[30] Adverse health effects related to building occupancy fall into several categories. Sick building syndrome (SBS) is the term used to describe a collection of symptoms that affect building occupants as the result of exposure to any number of airborne pollutants.[31] These effects are often short-term and may disappear after individuals leave the building. Symptoms may vary greatly, but may include eye and respiratory irritation, headaches, dizziness, fatigue, and allergy-like symptoms. Building related illnesses (BRI) include more serious health effects that do not subside after leaving the building and pollutant source. Cancer and Legionnaires' disease are examples of BRIs, verifiable diseases that are caused by specific pollutant sources. Less well understood is the illness described as multiple chemical sensitivities (MCS), which may produce symptoms similar to SBS. However, individuals with MCS symptoms may be affected by exposure to low levels of pollutants, making detection of the source difficult or impossible. Some people with MCS develop heightened sensitivity to a number of contaminants and require specially designed and built environments to relieve these symptoms.

The legal and economic consequences that can potentially result from IAQ problems create an enormous liability for architects, designers, and building owners. These consequences include occupant health problems, loss of building use, and in extreme cases, even death of building occupants. Surprisingly, many design professionals are relatively unprepared to address IAQ issues beyond the most basic concerns. The strategies to improve IAQ include control of the sources of pollutants, ventilation control, mechanical system design, commissioning, and maintenance. The first of these strategies, source control, requires that architects or designers evaluate materials and furnishings in terms of their potential for off-gassing of volatile organic compounds (VOCs) or other harmful chemicals. While this may seem to be an enormous challenge, it is advisable to initially focus on the major potential sources: carpet and flooring, paints, ceilings, and furniture. Material safety data sheets (MSDS) may be requested from manufacturers to compare potential products, and in some cases, test chamber data may be available to show the off-gassing rates from products or assemblies over time. Microbial contaminant sources such as molds and mildew result from excessive moisture levels in materials, finishes, and mechanical systems, and can also cause a number of health problems. The growth of these contaminants may be reduced by specifying materials that are resistant to microbial growth, especially in areas prone to excessive moisture. Encapsulation of insulation or other materials, which might support microbial growth and correct placement of vapor barriers, are additional measures. (Although this varies by climate and should be carefully confirmed with mechanical engineers, the vapor barriers are typically located on the "warm-humid" side of the assembly.)

Mechanical system design and construction also have significant roles to play with respect to providing good IAQ. Because there will always be some level of indoor air pollutant (often created by the occupants themselves) mechanical systems must be designed to meet or exceed minimum ASRHAE guidelines for ventilation rates in terms of the volume of outside air supplied per person. The location of outdoor air intakes must be carefully considered to avoid possible contaminant from outside, including loading docks and trash areas, traffic areas, food service, building exhaust, birds, etc.

During construction, ducts should be protected to prevent dust and other contaminants from collecting within. Duct insulation should be completely encapsulated to prevent fiberglass or other fibers from entering the airstream, and if the encapsulation is damaged in the construction process it should be repaired. The installation of materials must be carefully phased so that those that emit VOCs and other pollutants are installed well before materials that may absorb them and then reemit these pollutants. (This has been described as informally as "wet" before "fuzzy.") The practice of providing a "bake-out" phase prior to occupancy (raising a completed space to a high temperature) is no longer advised, as the abnormally high temperature causes off-gassing from materials that would otherwise not occur, and the pollutants are absorbed by ceiling tiles, carpets, and furnishings, and then continue to emit pollutants for long periods of time. A "flush-out" with tempered 100% outside air may be beneficial prior to occupancy, provided that humidity is controlled, and should be considered in some instances.

11.3.2 Alternative Ventilation Strategies

The ability to control ventilation is critical for providing occupant comfort. While there are several ways to give people control of ventilation, operable windows remain the most intuitive and simple way to give people control over their environment. It has been pointed out ironically that the only people who see any advantage to sealed windows are mechanical engineers. Although there is currently a renewed interest in operable windows, there are several challenges to incorporating them into commercial buildings. Many mechanical engineers need new methods to design for operable windows and still maintain energy efficiency. Also, the latest UBC-based building codes require that high rise buildings include smoke management as part of the life safety system. The smoke control system must utilize differential pressure between floors to prevent smoke from migrating between floors, and the system must be approved by local building officials. At the time of this writing, there is no common interpretation to the new code requirements and in some cases local code officials have created unreasonable obstacles to using operable windows. Other obstacles to the use of operable windows include locations with noisy exterior conditions, pollutants, or particulates, and extreme climates in which the open windows create a large penalty for energy efficiency.

Other controllable ventilation strategies include window systems with operable mullions or vents, and task/ambient systems that provide occupant control at the desktop. Operable diffusers that are part of underfloor air supply systems, described in the next section, are another type of task/ambient system.

11.3.3 Underfloor Air Systems

Access floor assemblies have been used for many years for cabling in wire intensive applications such as radio and television studios and computer rooms. Such pedestal supported floor systems have been improved in recent years and provide a solid surface underfoot without the hollow acoustical quality traditionally associated with them. Underfloor air systems integrate the air supply into this type of raised pedestal floor system. The application of this system has been slowly growing for years in the United States and is sparking much interest as the benefits of this type of system are documented. Some of the benefits of this system are:[32]

- Flexibility is increased as mechanical diffusers and electrical/data floor boxes may be quickly changed by moving floor tiles.
- Individuals control the volume of airflow in their workspaces by adjusting floor-mounted air diffusers (including open office areas).
- Energy efficiency is improved by conditioning the lowest occupied portion of the space, allowing heat to rise and stratify. By introducing air in contact with the slab it is possible to utilize the thermal mass of the building's concrete slabs to increase energy efficiency.
- Indoor air quality (IAQ) is improved because pollutants tend to rise to the top of a space where they are exhausted. Conventional overhead systems continuously mix pollutants with supply air.
- Conditioned air is supplied at a higher temperature, allowing for chillers to be downsized and increasing the use of economizer cycles (i.e., cooling the building with outside air that has not been cooled).
- In some cases, underfloor systems allow for the elimination of the dropped ceiling, reducing costs and creating higher ceilings.

Although the floor and pedestal systems typically cost about \$5 per ft^2 installed, it is possible to offset this cost through savings in chiller sizes, ductwork, wire management, and eliminating the need for electrified furniture. The greatest potential cost savings will be realized by the tenant who may benefit from reduced cost and time needed for reconfiguration.

11.4 THE FUTURE OF SUSTAINABLE DESIGN

With the world population now over six billion people and still growing, the environmental impacts of human activity will continue to grow evermore critical. Building design professionals, owners, and operators may find that environmental performance will soon be a large part of the measure and definition of a successful project. Design professionals that take the lead in learning about and driving new design and technological advances in green development will be well positioned to take advantage of this growing area. With ever better design tools for modeling and comparing building performance, governmental programs and controls, and leadership from the private sector, the attention paid to sustainable development will certainly continue its evolution and advancement.

> Green buildings do not poison the air with fumes nor the soul with artificiality. Instead, they create delight when entered, serenity and health when occupied, and regret when departed. They grow organically in and from their place, integrating people within the rest of the natural world; do no harm to occupants or the earth; foster more diverse and abundant life than they borrow; take less than they give back. Achieving all this hand in hand with functionality and profit requires a level of design integration that is not merely a technical task but an aesthetic and spiritual challenge.[33]

ENDNOTES

1. Van der Ryn, S., and Cowan, S. 1996. *Ecological design*. Washington DC: Island Press, p. 18.
2. Roodman, D. M., Lenssen, N., and Peterson, J., Eds. 1995. *A building revolution: How ecology and health concerns are transforming construction*. Washington, DC: Worldwatch Institute, p. 25.

3. Hawken, P. 1993. *The ecology of commerce*. New York: Harper Business, p. 29.
4. Langston, C., Ed. 1997. *Sustainable practices: ESD and the construction industry*. Sydney: Envirobook Publishing, from essay by Rick Best, *Environmental impact of buildings*, p. 120.
5. Rocky Mountain Institute, et al. 1998. *Green development: Integrating ecology and real estate*. New York: John Wiley & Sons, Inc., p. 299.
6. Gottfried, D. 1996. The economics of buildings. *The Sustainable Building Technical Manual* I–13.
7. Rocky Mountain Institute, et al. 1998. *Green development: Integrating ecology and real estate*. New York: John Wiley & Sons, Inc., p. 17.
8. Durante, G. 1999. *Current approaches to measuring added value* (presentation at 1999 Silicon Valley Environmental Health and Safety Conference, Redwood City, CA) from original study by IFC Kaiser.
9. Rocky Mountain Institute, et al. 1998. *Green development: Integrating ecology and real estate*. New York: John Wiley & Sons, Inc., p. 43, from original source by Joseph Romm, *Lean and clean management*, 1994.
10. Rocky Mountain Institute, et al. 1998. *Green development: Integrating ecology and real estate*. New York: John Wiley & Sons, Inc., p. 140.
11. Rocky Mountain Institute, et al. 1995. *A primer on sustainable building*, p. 39.
12. Abraham, L. E. 1996. *Daylighting. The Sustainable Building Technical Manual* IV.7. (This publication is available at: http://www.sustainable.doe.gov/pdf/sbt.pdf)
13. Wilson, A. 1999. Daylighting: Energy and productivity benefits. *Environmental Building News* 8(9): 12–13.
14. Bisel, C. 1996. HVAC, electrical, and plumbing systems. *The Sustainable Building Technical Manual* IV.43.
15. For a detailed list of HVAC recommendations, see Bisel, C. 1996. HVAC, electrical, and plumbing systems. *The Sustainable Building Technical Manual* IV.43–50.
16. Bernheim, A. 1996. Building commissioning. *The Sustainable Building Technical Manual* IV.81–87.
17. Duffy, F. 1997. *The new office*. London: Conran Octopus Limited, p. 51.
18. Wilson, A. 1998. Access floors: A step up for commercial buildings. *Environmental Building News* 7(1): 8.
19. McQuillen, D. 1998. Harnessing the sun. *Environmental Design & Construction* July/August, 20.
20. Ibid., p. 21.
21. Ibid., p. 20.
22. Wilson, A., and Malin, N. 1999. Fuel cells: A primer on the coming hydrogen economy. *Environmental Building News* 8(7): 1.
23. Lee, J. 1999. AIA, Fuel cells generate green power. *Environmental Design & Construction* May/June, 52–53.
24. Bisel, C. 1996. HVAC, electrical, and plumbing systems. *The Sustainable Building Technical Manual* IV.50.
25. LEED Reference Guide, U.S. Green Building Council, 1999, p. 123.
26. Lee, J. 1998. AIA, Campus's first green building serves as a model. *Environmental Design & Construction* July/August, p. 40.
27. Malin, N., and Wilson, A. 1997. Material selection: Tools, resources, and techniques for choosing green. *Environmental Building News* 6(1): 12–14.
28. Malin, N. 1999. Paint the room green. *Environmental Building News* 8(2): 12.
29. Rocky Mountain Institute, et al. 1998. *Green development: Integrating ecology and real estate*. New York: John Wiley & Sons, Inc., p. 305.
30. Ibid., p. 16.
31. Bernheim, A. 1996. Indoor air quality. *The Sustainable Building Technical Manual* IV.63.
32. Wilson, A. 1998. Access floors: A step up for commercial buildings. *Environmental Building News* 7(1): 8–14.
33. Hawken, P., Lovins, A., and Lovins, L. H. 1999. *Natural capitalism: Creating the next industrial revolution*. Boston: Little, Brown, and Company, p. 110.

The author wishes to thank Huston Eubank of Rocky Mountain Institute Green Development Services for his review of the manuscript.

CHAPTER 12
SMART BUILDINGS, INTELLIGENT BUILDINGS

Vivian Loftness, AIA
Professor of Architecture
Head
School of Architecture
Carnegie Mellon University
E-mail: loftness@andrew.cmu.edu

Dr. Volker Hartkopf
Professor of Architecture
Director
Center for Building Performance
Carnegie Mellon University
E-mail: hartkopf@cmu.edu

Stephen Lee, AIA
Professor of Architecture
School of Architecture
Carnegie Mellon University

Dr. Jayakrishna Shankavaram
Research Associate
School of Architecture
Carnegie Mellon University

Azizan Aziz
Research Associate
School of Architecture
Carnegie Mellon University
E-mail: azizan@cmu.edu

Center for Building Performance & Diagnostics
School of Architecture
Carnegie Mellon University &
the Advanced Building Systems Integration Consortium[1, 2]

12.1 DEFINING SMART BUILDINGS: BEYOND DESKTOP TECHNOLOGY, NETWORKS, AND CENTRAL MANAGEMENT SYSTEMS

Traditionally, the "smart" or "intelligent" building has been defined by the introduction of a long list of the latest products in telecommunications, electronics, security, automation, and building control systems. Although the United States has been a leader in the development and packaging of high-tech products and images, the buildings to house these technologies have not advanced significantly. Existing building infrastructures, in the form of mechanical, electrical, telecommunications, and interior systems, are so inflexible that they are incapable of accommodating the rapidly changing organizational and technological requirements for buildings today. What is needed, instead, are buildings with dynamic, user-based infrastructures where individuals can themselves reconfigure the spatial, environmental, and technological services on a continuing basis. Only through team design processes, involving the facilities managers and occupants of buildings, can we arrive at the effective "flexible-grid, flexible-density, flexible-closure systems" for *user* commissioning of environmental conditions and services—with expert central system response.

A better definition for smart or intelligent buildings would stipulate: Intelligent buildings will support on-going organizational and technological dynamics in appropriate physical, environmental, and organizational settings to enhance individual and collective performance and human health, comfort, and motivation.

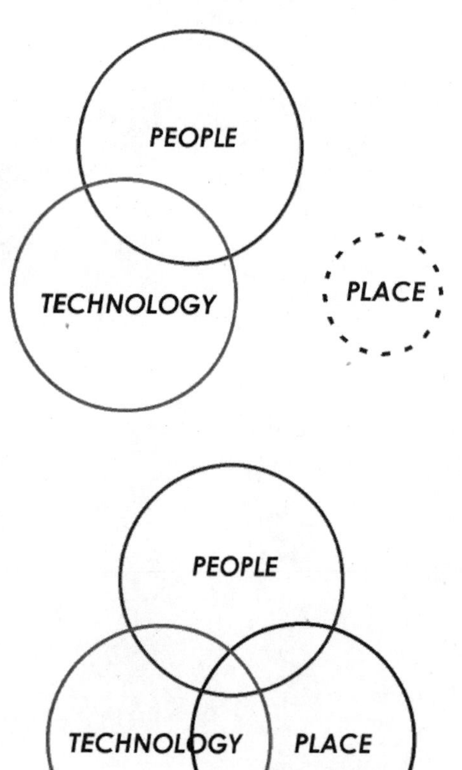

FIGURE 12.1 The Integration of All the Three Critical Assets: People, Place and Technology is Key to Organizational Effectiveness and Productivity.

12.2 A MANDATE FOR BUILDINGS TO SUPPORT CONTINUOUS ORGANIZATIONAL AND TECHNOLOGICAL CHANGE

As corporate and federal executives plan future workplaces, there must be on-going discussions of people, place, and technology. People are central to today's continuous organizational reengineering for productivity and innovation, shifting from hierarchical and linear work processes, to collaborative and dynamic. Technology is also central to organizational reengineering for productivity and innovation. Executive commitment to technology is clearly evident in our annual investment in the latest tools, networks, software, and training—now between $8,000–10,000 per worker per year (Forrester, 1995). Place, however, is an uncertain partner (Figure 12.1). While a few executives believe that place matters in planning the future of work, most regard real estate as a pure cost-center, unrelated to productivity and innovation. For most, the "strategic asset management" goal for real estate is to reduce, reduce, and reduce. It seems that the case has not been adequately made as to the full interdependencies between all three—people, *place,* and technology.

12.2.1 Organizational Reengineering Changes the Workplace

The commitment to organizational reengineering is growing worldwide and the reevaluation of buildings that support the "dynamic" organization (Figure 12.2) has already resulted in a number of major spatial changes, creating new demands on the workplace—the buildings, infrastructures, and interior systems (Tu and Loftness, 1998):

1. Workstations are getting smaller (64 ft^2 is now generous)
2. Occupant densities are going up (3% a year at least)
3. Partitions are coming down (from 80–64″ to 52–42″)
4. The walls are also coming down (from 30% closed to none?)
5. Teaming spaces are proliferating (growing from 3–15% of gross ft^2)
6. Floor plates are getting bigger (growing from 10,000 ft^2 to over 60,000 ft^2)
7. Workspace ownership is under negotiation (free address, hoteling increases)
8. Abandoned shopping malls and warehouses will do?
9. And, if less than 20% could have a window, why not take it from all?

While some of these changes result from commitments to increasing collaboration and communication, the majority are driven by reducing the cost of real estate. While some of these changes include significant investments to ensure the quality of workplace environments, an equal number assume reduced investments.

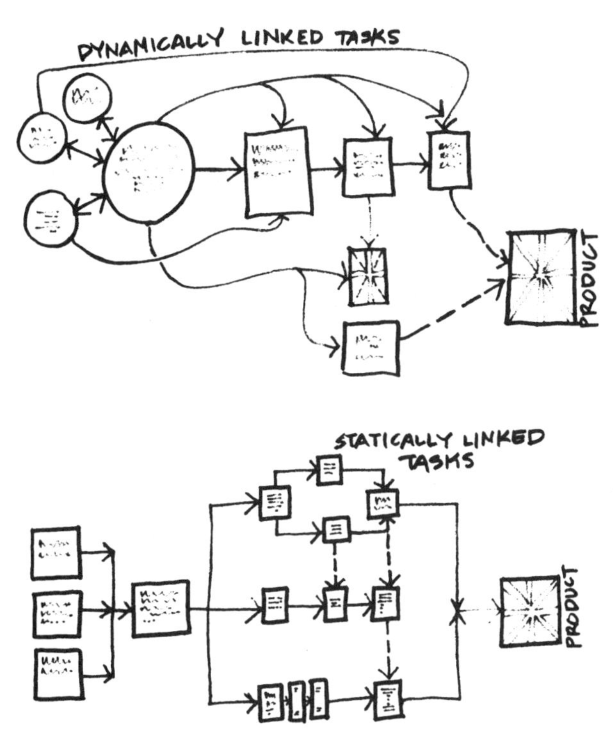

FIGURE 12.2 Continuous Organizational Reengineering Reflects Changing Concepts of Workflow, Time-to-Delivery, and Collaboration—Changes which Necessitate Flexibility in Buildings, Infrastructures, and Interior Systems.

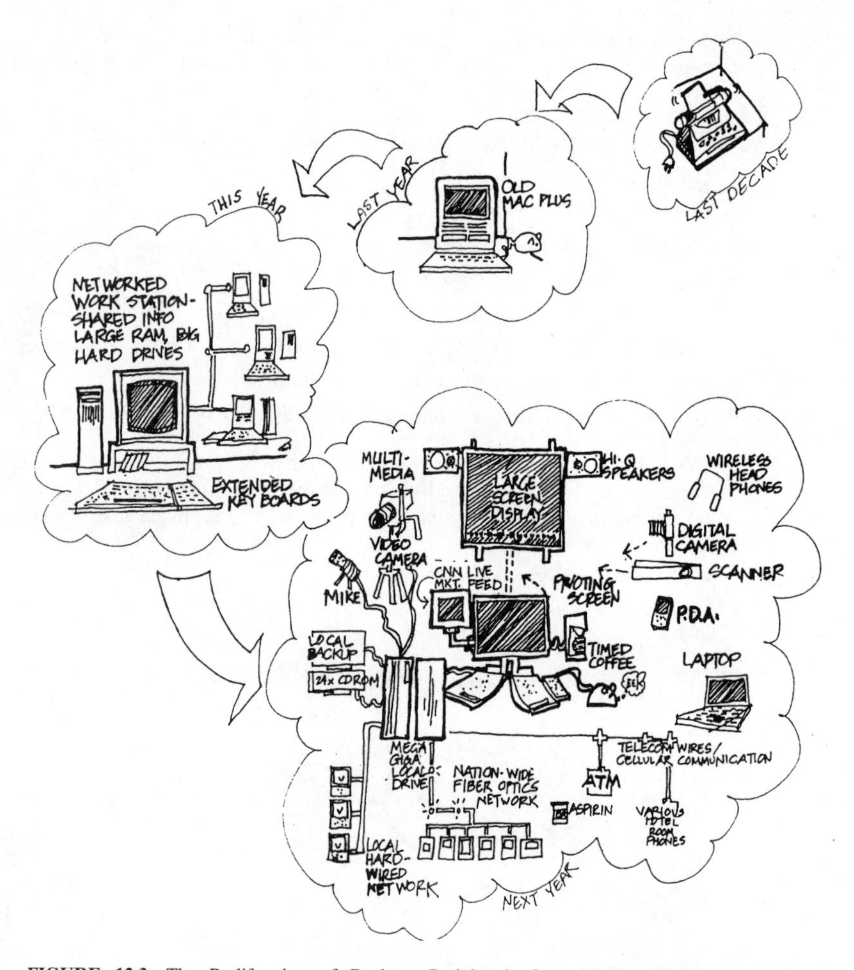

FIGURE 12.3 The Proliferation of Desktop Peripherals from Multiple Drives, to Printers, Fax/Modems, and Next Generation Video Input/Output Devices, Demands New Approaches to Worksurfaces and Storage, Local Shared Equipment Areas, Acoustics, Lighting, Ventilation, and Thermal Management.

What must be clear in this list, however, is that each of these changes has implications for the design/engineering of physical space. A doubling of occupant density combined with a doubling of desktop technology increases power, cooling, and ventilation loads significantly—well beyond the capacitance of most buildings. Moreover, these approaches must be seen as clearly temporal, with the obvious swing of the pendulum still to come. The question is not which organizational reengineering is right, but for how long, and whether the "place"—the building and its infrastructure—can support the change.

12.2.2 Technological Reengineering Changes the Workplace

The commitment of organizations to continuous investments in technological advances is even more pronounced than their commitment to organizational reengineering (Figure 12.3). These investments in technology also create new demands on the workplace—the buildings, infrastructures, and interior systems.

1. Technology densities are increasing (with space implications)
2. Connectivity demands are increasing (data, power, and voice)
3. Visual rules are changing (light levels and worksurface position)
4. Spatial rules are changing (new individual and collaborative ergonomics)
5. Acoustic rules are changing (new noise generators)
6. Air quality/thermal rules are changing (heat generators and new air patterns)

While the number of desktop technologies has gone from as few as one peripheral (a phone) a decade ago, to an average of six today (a phone, drive, screen, fax, printer, and modem, to name a few), the consequences of this explosion on physical place is still undervalued. Ensuring continuously adequate connections (data, power, and voice), and appropriate lighting, ergonomics, thermal and air quality, as well as acoustic conditions, is still in need of comparable investments in the workplace.

12.2.3 Buildings and Their Infrastructures Remain Rigid and Idiosyncratic

While modern organizations have become more organizationally and technologically dynamic, office buildings and their infrastructures remain stereotypically fixed, designed to meet early space planning requirements. Buildings and infrastructures continue to be designed with the assumption that the facility needs of organizations do not differ significantly over time.

Class A designations are given irrespective of the density, quality, or reconfigurability of major HVAC, lighting, or electrical systems (Figure 12.4). The resulting signs of a building's inability to respond to changing organizational and technologically needs are clearly measurable, including inadequacies in lighting, thermal, and air quality, as well as connectivity. The HVAC systems in buildings, for example, often have inappropriate cooling capacitance, supply air volume, and diffuser density to support the technological and organizational changes in offices today. Moreover, zone size (or the number of people sharing a thermostat) is unacceptably large for the dynamic office, with interior HVAC zones even larger than perimeter zones. The lighting is often inappropriate to support the technological and organizational changes underway. Field measurements in numerous Class A and B buildings reveal high connected lighting loads ($>$1.5 watts/ft^2 or 16 watts/m^2), high light fixture density, poor brightness and glare control, and poor locational control (lighting the aisles and cabinet tops rather than worksurfaces). Access to electric and telecommunications networks is also inadequate for the changing workplace, in density of service, location, and type of outlet. The number of outlets provided averages at seven (four power, two data, one voice), less than the demand for connectivity. Fixed outlets (poke-throughs and tombstones) are still prevalent, followed by furniture-based outlets, while raised floors and modular, relocatable outlets are still a rarity.

The interrelationships of all three building infrastructures (HVAC, lighting, and connectivity) reveals a serious level of mismatch with evolving workplace layouts. In both plan and section, there are signs of a lack of systemic approach to coordinate and integrate various building services into a coherent service entity to meet workplace needs. The layered, "idiosyncratic" and fixed solutions to introducing major building systems results in inefficient system operation; difficulty in maintaining and modifying system components; poor individual and organizational satisfaction with service; and greater costs in the plenum "real estate" due to inefficient plenum utilization. An in-depth analysis by Dr. Kung-Jen Tu of ten offices across North America, 2,319 workstations, and 557 surveys have revealed statistically significant relationships between organizational workplace dynamics and building infrastructure flexibility for delivering the appropriate environmental and technical quality in offices (Figures 12.5a and 12.5b; Tu, 1997).

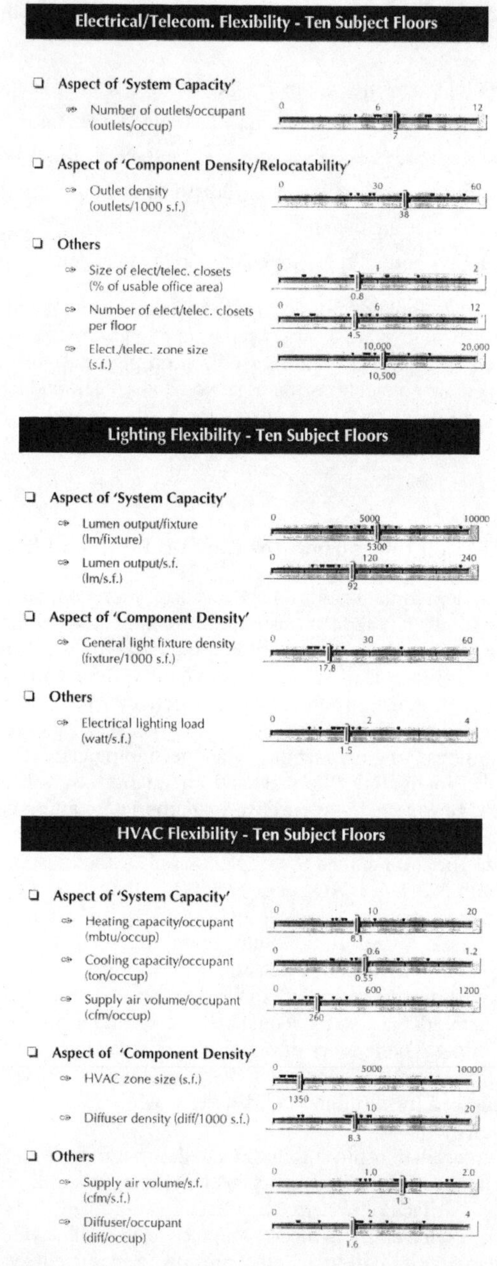

FIGURE 12.4 In Class A Designations, Building Infrastructures Remain Unrecorded in Capacitance, Zone Size, and Flexibility (Tu, 1997).

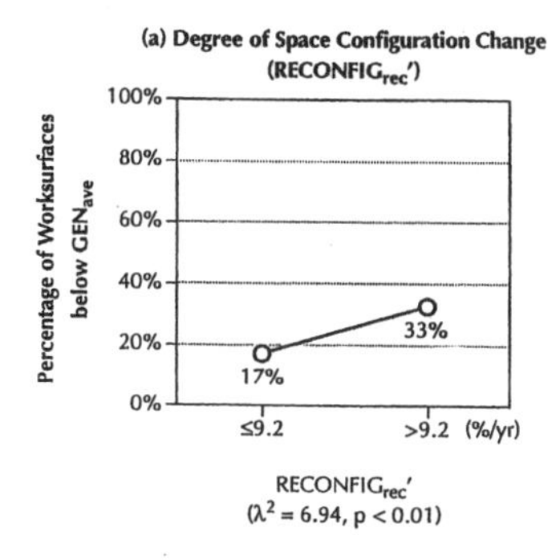

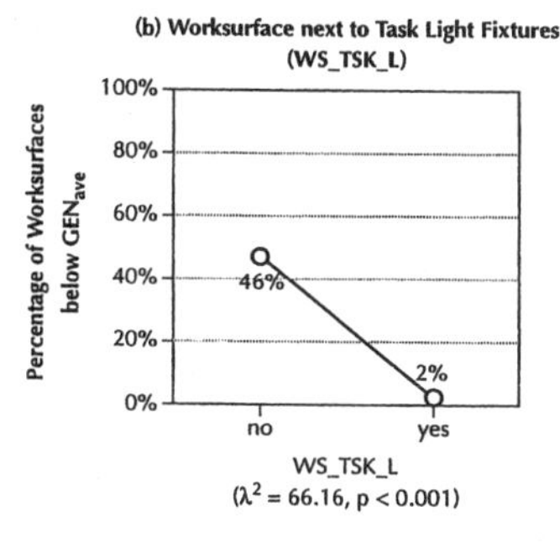

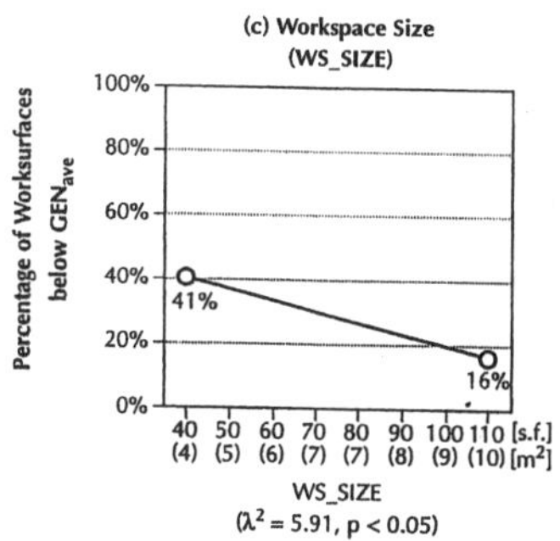

FIGURE 12.5a The Percentage of Worksurfaces Whose Light Levels are *Below* the Average Acceptable Light Level (WS < GEN_{ave}) Correlates with (a) the Degree of Space Configuration Change, (b) the Provision of Task Light Fixtures, and (c) Workspace Size (Tu, 1997).

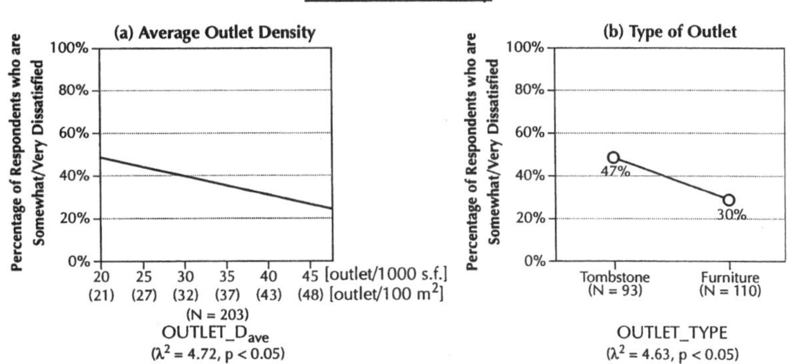

FIGURE 12.5b The Percentage of Respondents Who Felt that Outlet Locations are "Somewhat or Very Inappropriate" is Related to (a) Average Outlet Density, and (b) Type of Outlet (Tu, 1997).

It is important at this point to emphasize that the inflexible infrastructures leading to performance failures in the dynamic workplace are not confined to older buildings. Although new buildings often have greater central system capacitance (cooling, power, and telecommunications), the vertical risers, horizontal plenums, zone sizes, and user interfaces in today's least-cost new buildings are just as embedded, inflexible, and idiosyncratic as their predecessors. Indeed, reinvestment inside of beautifully crafted, historic buildings could ensure greater infrastructure flexibility to absorb today's rapid organizational and technological change than moving to newly built space in today's market (Loftness et al., 1996a).

12.3 SMART BUILDINGS REQUIRE FLEXIBLE INFRASTRUCTURES

To address issues of long-term productivity and organizational effectiveness, it is time to move beyond definitions of code-compliant buildings and even "high-tech" buildings to the creation of truly "motivational" buildings. Motivational buildings provide environmental performance at a level that consistently and reliably ensures health, comfort, security, and financial effectiveness, while supporting high levels of productivity with continuing organizational and technological change. In contrast to present practice, motivational buildings rely on guarantees that every building occupant, at their individual workstation, will be supplied with critical services. The following list enumerates seven basic infrastructures every occupant/workstation needs individually:

1. Fresh air
2. Temperature control
3. Lighting control
4. Daylight and view, reduced isolation from outdoors
5. Privacy and working quiet
6. Network access to multiple data, power, voice connections
7. Ergonomic furniture and environmentally appropriate finishes

By ensuring these seven mandates over time, "smart" or intelligent buildings can provide productive environments that attract the best workforce, offer personalized infrastructure and control, and support continuous change in organizational and technological configurations through infrastructure flexibility.

What is actually supplied at the workstation in old and new buildings, however, does not mirror this obvious list of environmental and technical needs for today's workers. Since the 1950s, we have been investing the minimum amount possible in our building's hidden infrastructures—from least-cost thermal zoning to minimum lighting control to jerry-rigged network connections. The following list enumerates what every occupant/workstation actually gets collectively:

1. Variable air supply, dependent on thermal demand
2. Blanket supply of cooling, large zones for fifteen people (average)
3. Uniform, high-level lighting
4. Rare daylight and view, isolation from outdoors
5. Rare working quiet and privacy control
6. One data connection, nonrelocatable; two power connections, nonrelocatable; one voice connection, nonrelocatable
7. Precomputer furniture, non-ergonomic; unmeasured indoor pollutant sources

The least-cost "blanket" conditioning and networking offered in present day buildings emphatically cannot accommodate organizational and technological changes. The rapid increase in desktop technology requires multiple connections to data, power, and voice networks and increased cooling. The rapid exploration of new space planning concepts to reflect new organizational structures and "teaming" work approaches (which radically redistributes the density of workstations, equipment, and space enclosures) requires new approaches to networking, HVAC, lighting, and control systems.

Yet, both new technologies and new space planning concepts are often introduced into buildings without any modification of the buildings base systems—cooling, ventilation, lighting, networking, or ceiling/acoustics—with disastrous results. In corporate eagerness to try new organizational concepts, there is little corresponding discussion of the need for each workstation to sustain key independent services, with serious concerns and failures occurring with each spatial renovation. Following is a list of new technologies and new space planning concepts with potential stresses in existing subsystem and service:

1. Cooling and thermal quality: capacitance, diffuser grid density and location, control
2. Ventilation and air quality: zoning, diffuser grid density and location, control
3. Acoustic quality for individual and collective work
4. Lighting and visual quality: grid density and location, control
5. Access to window, building enclosure control
6. Rank, territoriality, and personalization
7. Voice connectivity
8. Data connectivity
9. Power connectivity
10. Wall systems, spatial modification, and material reuse
11. Ceiling systems and closure, acoustics and light
12. Work storage and access

If fixed, open plan concepts such as "universal" or box/cubicle workstations are adequately serviced for each occupant at the outset (not a given), the potential stresses in relation to existing systems and services will be low. However, if dynamic workplace concepts are being considered, if an organization is intended to evolve in size, mission, and structure, or if workspaces are being planned for changing tenant users and equipment densities, then major shifts in the selec-tion of HVAC, lighting, enclosure, and networking subsystems and service must be pursued (Figure 12.6).

12.3.1 New Design Approaches to Absorb Change and Avoid Obsolescence: Flexible Grid—Flexible Density—Flexible Closure Systems

To avoid frequent environmental quality failures and long-term obsolescence, it is critical to invest in user-based infrastructures that are modular, reconfigurable, and expandable for all key services—ventilation air, thermal conditioning, lighting, data/voice, and power networks. The dynamic reconfigurations of space and technology typical in buildings today cannot be accommodated through the existing service infrastructures—neither the "blanket systems" for uniform open-plan configurations or the idiosyncratic systems for unique configurations. Instead, what is needed are flexible infrastructures capable of changing both location and density of services. Flexible grid—flexible density—flexible closure systems are a constellation of building subsystems that permit each individual to set the location and density of HVAC, lighting, telecommunications, and furniture, and the level of workspace enclosure (ABSIC/CERL, 1995). These services can be separate ambient and task systems where users set task requirement and the central system responds with the appropriate ambient conditions, or they can be fully relocatable, expandable task/ambient systems.

HVAC: LEVELS OF ZONING & CONTROL

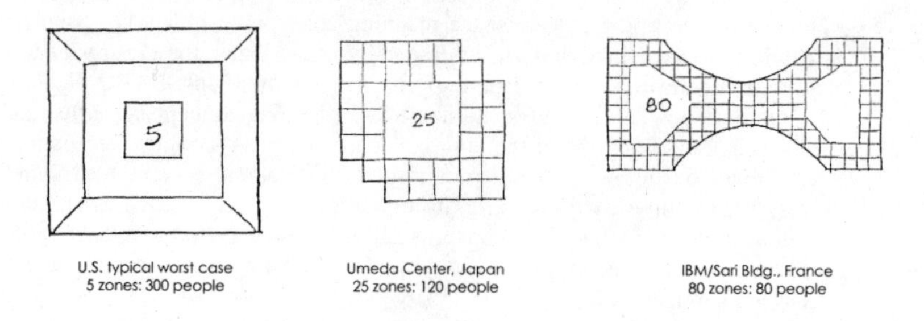

LIGHTING: LEVELS OF ZONING & CONTROL

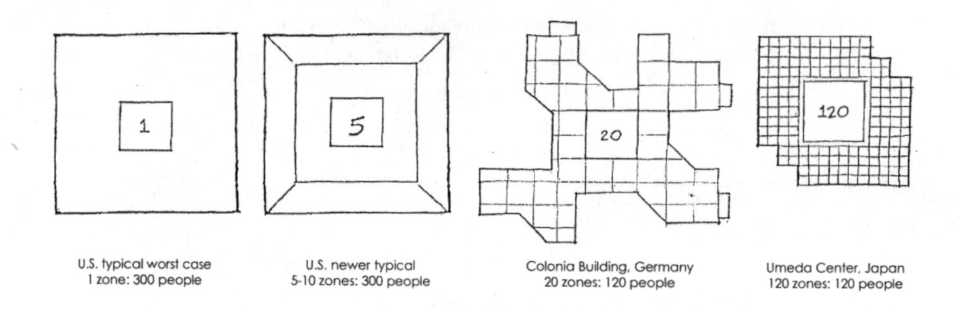

POWER/DATA/VOICE: LEVELS OF CONNECTIVITY
HORIZONTAL DISTRIBUTION OF NETWORKS

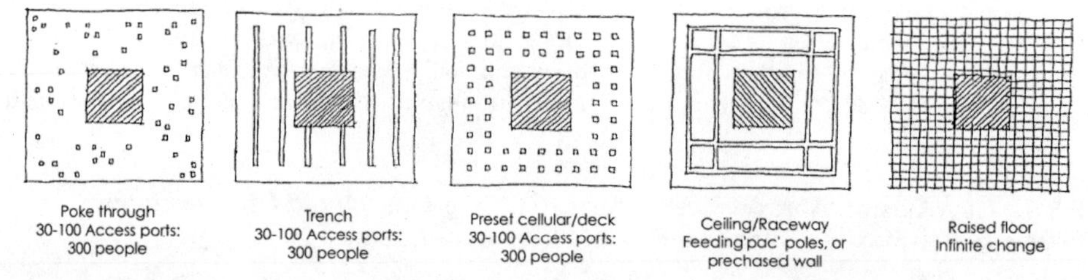

FIGURE 12.6 Unlike International "Smart" Buildings, Most U.S. Examples Suffer from Excessively Large Zones for "Blanket" Heating, Ventilation, and Cooling, "Blanket" Lighting, and "Blanket" Networking, with Serious System and Performance Failures Occurring with Each Spatial Renovation [ABSIC International Studies (United States, Germany, France, United Kingdom, and Japan) 1988–1993].

Services Needed at Workgroup	Services Needed at Workstation
HORIZONTAL	
GRIDS	NODES
Data	Data Outlets (1-4 per person)
Voice	Voice Outlets (1-2)
Video	Video Outlets (1)
Power	Power Outlets (1-10)
Structural Columns	Structural Beams
Furniture	Worksurface: Horizontal + Vertical
Ceiling Grid	Ceiling Tiles
Ambient Lighting	Lighting Fixtures (1-6)
Floor Grid	Floor Tiles
Thermal Service	Diffusers/Radiators (1-6)
Ventilation	Diffusers (1-6)
Plumbing	Kitchenettes
Security	Doors
Fire	Sprinklers
Environmental Control/Zones	Sensor/Controllers (1-4)
Acoustic/Sound System	Speakers, Acoustic Materials
Windows	Viewing Cone
Core to Shell Distance	Wayfinding, Access to Vert. Service
VERTICAL	
Floor to Floor Height	Horizontal Plenum Size and Access
Horiz. Plenum: Ceiling, Floor, Furniture	Size and Access to Services
Floor to Ceiling Height	Light & Air Distrib., Service Access
Panel / Wall Height	Light & Air Distrib., Service Access

FIGURE 12.7 While Grids of Service Would Ensure the Overall Level of Capacitance Available (or Expendable) in Reliable Locations, the Nodes Will Support the User Control of the Location, Density, and Type of Service (Loftness et al., 1996b).

12.3.2 The Concept of Grids and Nodes: Ensuring Seven Basic Needs for Each Individual

Access to all of the basic needs for a healthy, productive workplace—air, temperature control, daylight and view, electric light control, privacy and working quiet, network access, and ergonomic furniture—can only be provided by a shift away from blanket and centrally controlled infrastructures to the concept of grids and nodes (Loftness et al., 1996b). The "grids" establish the overall level of capacitance available to support the working group or neighborhood (fresh air, cooling, power, and network capacitance). Then, the "nodes" or user interfaces must be flexible in terms of location, density, and type of service offered (Figure 12.7).

These grids and nodes should not be dealt with in isolation, but as compatible assemblies and, in some cases, integrated systems. "Plug-and-play" technologies developed in other markets (unlike the building industry) assume distributed capability, user differences in customization, and the ability for end users to help themselves if systems are not meeting requirements. This also reflects the introduction of home and car functionality in the office, something clearly affordable today, though not often seen in the workplace.

12.3.3 Flexible Infrastructures Begin with Accessible and Expandable Vertical Service

First, the central capacities of power, data, phone, cooling, and ventilation must be reexamined. In many buildings, the central capacity and backup is inadequate in central power capacity and reliability; data/voice capacity and ease of reliability; central chiller capacity and the ability to increase zone densities; central ventilation capacity independent control from thermal conditioning; and riser space and easy access for modification.

One way to ensure that these central capacities and vertical distribution capacities will be adequate over time is to clearly label allowable floor densities. Just as an elevator or firestair has a load or occupant limit, each building floor should have posted occupant and equipment load limits based on central capacities, vertical distribution capacities, zone sizes, and diffuser/node densities on the floor.

Alternatively, central systems will need to be reconceived as modular and expandable to support the increased loads that are inevitable in a smart building. Moreover, there should be a significant shift toward distributed systems to support local control by organizational units with differing equipment and occupant densities, or with different work schedules, ensuring appropriate technical and environmental service without excessive costs.

12.3.4 Flexible Infrastructures Require Collaborative Horizontal Plenum Design and Relocatable "Nodes" of Service

The worst example of noncollaborative design/engineering may be above the ceilings in American buildings. From 3–5 ft (1–1.5 m) of plenum space is dedicated to linear, nonintegrated decision making, guaranteeing poor access, poor flexibility, and in most cases, poor performance. Systems are totally idiosyncratic, tangled, and designed to serve a specific floor plan at a specific point in time—quickly obsolete. As a result, air quality and thermal complaints abound. The density of service nodes (air diffusers, controllers, outlet boxes, and lights) is entirely inadequate for the occupant and equipment densities common in today's workplace. Moreover, abandoned wiring in plenums and slabs has made many U.S. buildings a greater potential source of copper than copper mines.

A number of advanced buildings today demonstrate that floor-based servicing may more effectively support the dynamic workplace. Since networking, ventilation, and thermal conditioning needs to be delivered to each workstation, services at floor level or at desktop offer a greater ease of reconfiguration (York, 1993). In addition, electrical and telecommunication cabling and outlet density can be continuously updated to meet changing needs. Today, a number of industry partnerships are forming to offer collaborative solutions to flexible infrastructures—floors, data/voice, power, thermal conditioning, and ventilation. With these modular, floor-based services, the ceiling can become more playful and elegant—as a light and acoustic diffuser—defining working groups, neighborhoods, and landmarks.

12.3.5 Flexible Infrastructures will Finally Support Reconfigurable Workstations and Workgroups

Once the appropriate modular, relocatable infrastructures are provided, the continuous recreation of workstations and workgroups can finally be achieved with assurance of thermal comfort, lighting quality, air quality, and connectivity. It is critical, however, to design the furniture/wall system to support rapid changes between open and closed planning, between individual and teaming spaces, as well as rapid changes in occupant density, equipment density, and infrastructure/service to match these configurations. While the grid of service and the number of nodes will determine the maximum densities on a floor, the ability to continuously rethink the organization should be fully supported by modular, reconfigurable desks, storage, walls, doors, and ceiling components.

In conclusion, we need to move beyond embedded technologies in buildings to end-user technologies. Both the next generation of new buildings and the revaluing of existing buildings must explore the attributes of microzoning and user modifiable systems—through neighborhood service grids and individual, user-responsive nodes—to support the dramatic changes in technologies and organizations. The manufacturers of building components and subsystems will have to develop products that are compatible in open architectural systems; user-modifiable; expandable and relocatable through modularity; and support multiple vendor plug-in capability. Justifications for user-based systems are growing, from measurable productivity to measurable reductions in operating and renovation costs, to significant increases in building longevity with life cycle investment values. The development of life cycle costing techniques should fully recognize buildings not as nonperforming assets, but as enabling environments.

12.4 SMART INTERIOR SYSTEMS

Interior systems in the intelligent or smart building should fully address the following three performance criteria:

- Organizational/spatial flexibility (continuous change without waste)
- Privacy and interaction/working quiet and teaming
- Ergonomics and environmentally appropriate finishes

A number of innovative design approaches and interior components and systems illustrate the next generation of "smart interiors."

12.4.1 Organizational and Spatial Flexibility through Kit-of-Parts Design: Ensure that Workprocess and Furniture Decision Making Is Both First and Last

In a break from conventional design documents, designing the office of the future must begin with *multiple* floor plans, envisioning years of organizational reengineering and technological change (Figure 12.8). These floor plans will provide the basis for determining the flexible infrastructures needed, the grid of service, and the number of nodes needed for air, cooling, light, data, voice, and power. These floor plans will also provide the basis for selecting the furniture "kit-of-parts" that will support the dynamic organization without waste. Only the right workplace infrastructure guarantees flexibility with higher user satisfaction, technological and organizational change, and environmental sustainability.

Today's constant organizational reengineering will demand the regeneration of workgroup and workstation solutions during the months of infrastructure design, engineering, bidding, and construction. With the design of infrastructures that are capable of supporting multiple layouts, and the selection of grids of service with adaptable and relocatable nodes in an accessible plenum design, it is possible to address workplace issues both first and last in design.

As a result of this design process, the final selection of the full range of furniture components, worksurfaces, partitions, chairs, task lights, storage, and teaming spaces could be selected at the latest possible moment to meet the needs of the latest individuals and workprocesses. These furniture components should ensure maximum environmental performance (visual, acoustic, thermal, air quality, and integrity) while supporting future organizational change through components that are relocatable, scalable (can be added and subtracted), user customizable, and compatible with the other building systems. Moreover, these furniture components should represent the best industrialized products affordable through comparable performance specifications and competition.

It is critical that the design team be well-versed on innovation in the furniture industry, international space planning approaches, and the latest research on work process and its relationship to workplace planning. There are at least ten major interior design decisions that must be made in a collaborative process:

1. Neighborhood clarity and shared services
2. Layers of ownership, multiple work environments

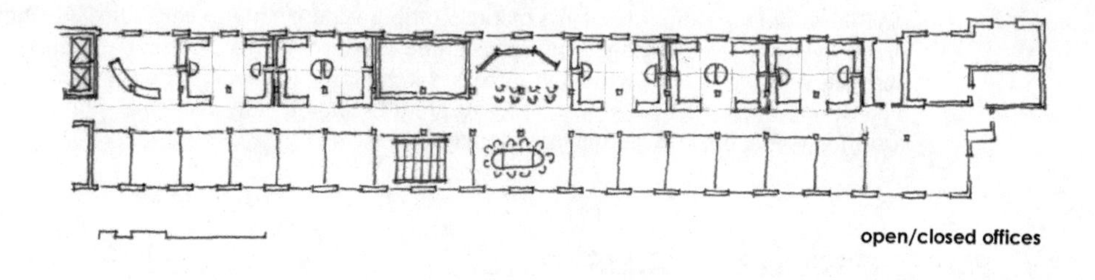

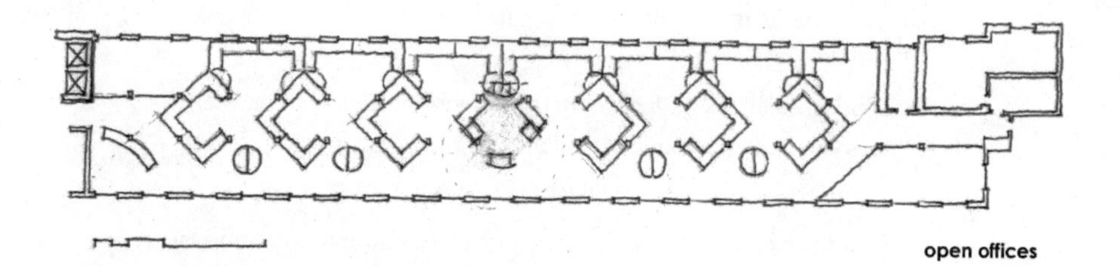

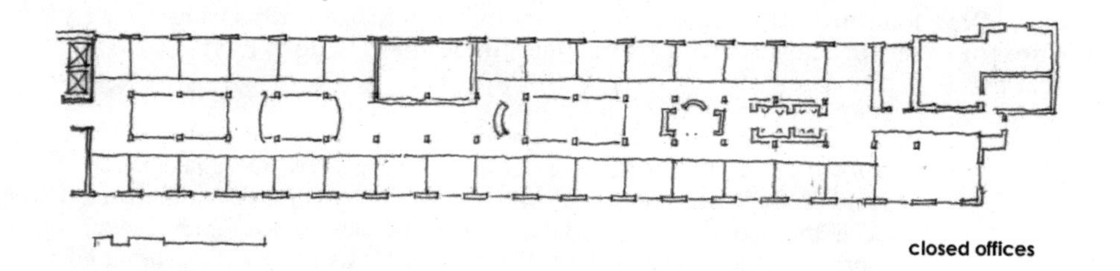

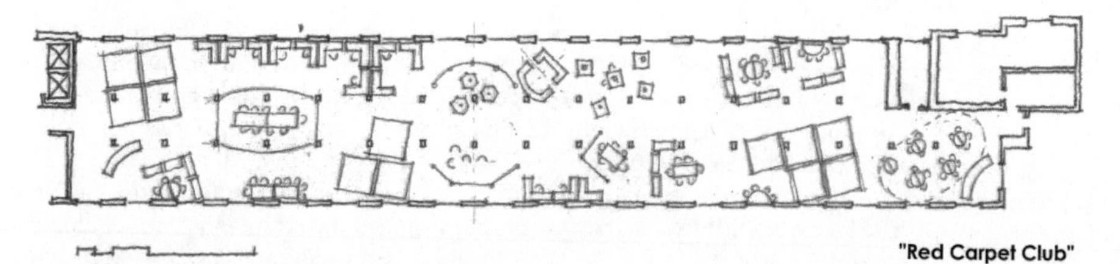

FIGURE 12.8 The CBPD Project Designer's for General Services Administration's Adaptable Workplace Laboratory Generated Multiple Workgroup Layouts, for Designing Infrastructures Capable of Supporting Continuous Organizational Dynamics (Loftness et al., 1998; Hartkopf et al., 1999).

3. Functional support for shared work processes
4. Functional support for individual work processes
5. Layers of closure, open/closed variations
6. Layers of mobility
7. Levels of personalization
8. Infrastructures to support environmental control

9. Infrastructures to support technical control
10. Healthy, detailed, aesthetic environment

To make these decisions effectively, with the entire design team and client group, a series of 3-D CAD alternatives are required, based on a range of modular furniture components. The drawings need to be three-dimensional (instead of 2-D line drawings) because neighborhood clarity, layers of closure and mobility, and infrastructures are only apparent in three dimensions. The path out of "Dilbertsville" is not the elimination of square corners, which serve a major purpose, but the introduction of dynamics in the third dimension (heights of walls and elements) and the introduction of playful, mobile worktools and personalization components that the industry is rapidly developing today. The development of on-line alternatives can also allow for the rapid substitution of manufactured components to ensure that multiple, competitive products can deliver the solution. Finally, the on-line alternatives—made up of a kit-of-parts of interior components—can visualize the capability of the space to evolve after move-in to support changing work processes and individual or group needs.

12.4.2 Collaboration and Teaming as Well as Privacy and Working Quiet

Corporations around the world are discovering that organizations need both stronger collective work processes and more productive individual/concentrated conditions. To achieve both of these goals, interior space plans are shifting to dynamic combinations of closed and open spaces, micro workstations, mobile workstations, and team rooms. Numerous furniture manufacturers have been developing components that support the configuration of small, partially closable individual offices; mobile furniture pieces that can be taken to alternate work locations; and a growing array of shared work area furniture for conferencing, relaxing, concentrating, teaming, sorting, or presenting work.

To ensure effective individual as well as collaborative work, it is important to avoid the pitfalls of least-cost decision making. Individual productivity can be seriously compromised by least-cost strategies such as downsizing the workstation, lowering partitions, eliminating workstation ownership, and open teaming spaces directly adjacent to "concentration" spaces. There is little question that valuable focus will be lost for individual effectiveness if workers have to spend time searching for materials or are frequently interrupted by adjacent workers. Smaller workstations no longer support the one-on-one meetings that are most common. The loss of individual or group ownership of workspaces makes assembling teams difficult for spontaneous developments. Teaming spaces are often eliminated in value-engineering, or left open-plan, compromising both collaborative and individual productivity. We should select furniture systems that can support evolution and individualization for a range of work processes as well as absorb major redistributions between individual and collaborative work.

"Smart" workplace and workstation furniture systems should reflect the highest quality affordable, given life-cycle criteria, including:

- Stackable panel systems with acoustically absorbing panels and appropriate light reflection characteristics; the capability to stack up to a ceiling level, potentially support doors and a level of closure that will enable the organization to evolve from open to closed planning
- Relocatable floor-based worksurfaces with ergonomic keyboard supports, ergonomic chairs, and articulated arm task lights, of elegant and sustainable materials
- Innovative solutions to a number of shared office amenities—the business center, the coffee pub, the multimedia conference room, the hoteling center, and the reception area
- The creation of outdoor work areas that recognize the need for access to the natural environment

12.4.3 Ergonomic Furniture and Environmentally Appropriate Materials

In the past 5 years, significant attention has been addressed to the improvement of spatial quality in the individual workplace. Much of this work has come under the title of ergonomic design (see

Chapter 14) with significant advances that include ergonomic chairs; adjustable supports for keyboards, screens, and paper copy; variable height work surfaces; task lights; and integrated cable management. To ensure a uniformly high quality of furniture purchase and layout planning, performance guidelines should include:

1. Anthropometric workstations
- Adequate worksurface, reconfigurable
- Adequate storage, storage walls
- Adjustable height/position keyboard, screen, and document support
- Floor-based worksurfaces, modular, L or U configurations
- Significantly increased storage in workstation and workgroup

2. Ergonomic chairs
- Adjustable seat height, with locking mechanism
- Swivel on five- to six-caster base
- Adjustable back height/lumbar support
- Adjustable height/width armrests

3. Relocatable infrastructures
- Levels of acoustic and visual control
- Modular, stackable wall systems, variable enclosure for privacy and interaction
- Dedicated lights, adjustable levels, relocatable
- Dedicated air
- Dedicated thermal control
- Dedicated networks—data, voice, power, video, and environmental control
- Environmentally benign materials and finishes

Given the redistribution of work responsibilities and related telecommunications equipment, there is a critical need to shift away from the traditional allocation of space size and furniture by rank to allocations by task or function. Although secretaries traditionally receive only 20–30% of the workspace of their boss (who in turn spend significantly less time in their office), the secretary's workspace must often accommodate more technology and materials to support that boss's activity. Although a salesperson may be able to effectively function with a phone and a "touchdown" worksurface, with files maintained in a central file bank or even their car, a researcher relies on extensive files and books and computer networking to complete their tasks. Consequently, the "smart" office design must consider the range of tasks and effective workstyles in the determination of workplace size and furniture options (Figure 12.9). The continuous urge to reduce workstation size in the United States is counterproductive and unnecessary given one of our greatest commodities—land.

At the same time, it is extremely important to select environmentally benign materials and finishes. These include materials that do not deplete nonrenewable resources or fuse primary materials in such a fashion that they cannot be recycled. This also includes selecting materials that are not a primary pollutant source for the occupied space (emitting particulate, volatile organic compounds (VOCs), or mass pollutants), or a secondary pollutant source by absorbing and reemitting dust, mold, and other pollutants. The interior architect also carries the responsibility to ensure that individual workers are given environmentally appropriate work environments, which means avoiding the pitfalls of inadequate mechanical and electrical infrastructures, as well as avoiding the least-cost pressures to use substandard buildings (oversized, windowless, uninsulated, with inadequate HVAC) such as abandoned shopping malls or warehouses as the "office of the future."

12.5 SMART HVAC SYSTEMS

Heating, ventilating, and air conditioning (HVAC) systems in the intelligent building should fully address three performance criteria for each individual:

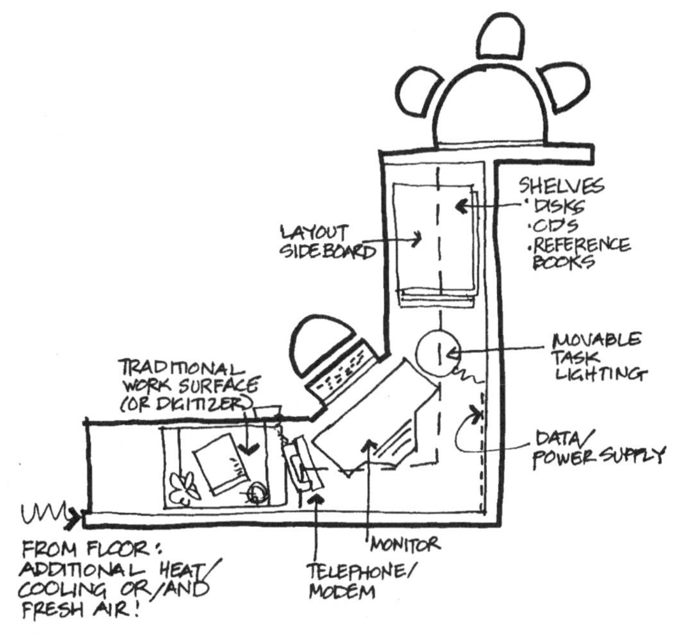

FIGURE 12.9 The All-Capable Knowledge Worker Who is Expected To Be Creator, Producer, Manager, and Promoter at Once Needs Multiple, Diverse Work-surfaces and Storage, As Well As Adequate Space to Multitask and Collaborate Effectively.

- Fresh air
- Temperature control
- Access to the natural environment

At least five major HVAC innovations illustrate the next generation of "smart HVAC" systems.

12.5.1 Decoupled Air and Thermal Systems

There are significant advantages to separating thermal conditioning (heating and cooling) from ventilation (breathing air supply). Combining air conditioning with ventilation (air quantities of 10:1 proportionally) has consistently resulted in an albatross of ducts and large thermal zones. These solutions have created both air quality concerns and poor thermal comfort in commercial buildings, which will be further aggravated by the dynamic changes in occupant and equipment densities, as well as changes in space closure. Combined thermal and ventilation systems have also led to pressurized buildings which has eliminated the opportunity for operable windows, not just in high-rises, but in low-rise offices, schools, community centers and more. The advantages of separating or decoupling the delivery of ventilation air and thermal conditioning are multi-fold.

Energy Efficiency and Space Flexibility. With a constant volume of 100% outside air for ventilation only (conditioned as needed to appropriate temperatures and humidities), much smaller ventilation ducts can be utilized in new and retrofit construction, with a guarantee of adequate outside air—regardless of season and internal thermal loads. Since air is a very inefficient thermal transport media, the constant volume ventilation system can be coupled with a wider range of energy efficient thermal conditioning options to meet localized, dynamic cooling demands with greater energy efficiency and comfort: variable air volume (VAV) thermal conditioning, water-based heat pumps, fan coils, radiant heating and cooling systems, as well as load balancing facades (which use waste heat to eliminate perimeter heating loads).

The Hines Partnership has introduced double duct systems in their tenant office buildings. Combining large cooling ducts for VAV supply with small ventilation ducts with constant volume (not hot and cold ducts) has ensured unmatched 97% occupancy rates in their "healthy air" buildings. The Europeans have introduced *displacement ventilation* systems with water-based heating and cooling. The displacement air systems introduce conditioned 100% outside air in large volume—low velocity ducts to provide silent delivery of conditioned breathing air. Heating and cooling is then provided through an independent system—typically water-based fan coils, heat pumps, or radiant elements.

Local Control of Ventilation Rates and Purge Cycles. Offering local control of fresh air quantities independent of thermal loads will also allow user response to high pollutant loads, high occupancies, smokers, and individual user needs, without in any way hurting the effectiveness of the overall system. In the SARI buildings in Paris, occupants can independently request more outside air, including a "purge" button calling for 100% outside air for a 5-min period, in order to clear the air after intense working (or cleaning) sessions. The 1,800 fan-coil units in the SARI building, bundled into four mechanical rooms per floor, then ensure individual thermal control to complement the independent control of ventilation rates.

Operable Windows. Finally, the separation of ventilation air from thermal conditioning will enable operable windows to be reintroduced in commercial buildings. While a dedicated, constant-volume ventilation system guarantees the levels of outside air at the desk (regardless of window aperture), the decoupled thermal conditioning system can be shut off to avoid simultaneous heating and cooling. In the Ministry of Finance and Budget in Paris, a constant volume supply of 100% outside air ensures ventilation requirements (despite changing wind conditions). When any of the 12,000 occupants opens a window to cope with local overheating and pollution build-up, or to enjoy a perfect day, the perimeter fan-coil units will shut off to avoid energy waste, while the constant volume ventilation air supply continues (Figure 12.10).

Decoupled thermal and ventilation systems offer significant gains for improving thermal comfort and air quality, increasing energy efficiency through zoning and load matching, as well as reopening the opportunities for operable windows in the workplace.

12.5.2 Individual Control—Task Conditioning Systems with Broad-Band Ambient

A second approach to improving indoor air quality (IAQ) and thermal comfort has been a shift away from blanket or uniform ceiling-based systems to workstation or task-based systems. The level of control offered by these individual conditioning systems varies. The HVAC control for each individual includes:

- Decreased zone size to one zone per person
- User control of air direction, air volume, air speed
- User control of air temperature, radiant temperature
- Split ambient and task temperature control
- User control of outside air quantities

Microzoning. The first and most valuable step toward providing increased comfort in buildings is a higher level of thermal zoning. Zone sizes in U.S. buildings range from averages of 1,000 ft^2 to as high as 10,000 ft^2—with as many as 200 people sharing a thermostat. Variations in occupant and

FIGURE 12.10 In the Ministry of Finance in Paris, Operable Windows Provide Occupants with Outside Air As Desired, but Avoid Energy Waste by Shutting Off the Perimeter Heating/Cooling Unit (ABSIC; France, 1991).

equipment densities in the modern office make this "blanket" conditioning strategy untenable, with a critical need to be able to add and relocate thermal zones to support organizational change. To guarantee environmental conditions in today's dynamic workplace, thermal zones should be no larger than 4–6 workstations, with capability for some level of control by each individual. If HVAC systems are designed to be plug-and-play, thermal zone boxes and controls could be purchased on a just-in-time basis (in combination with variable speed drive fans and pressure sensors). As zones are added over time, the building will eventually achieve a level of service that could be called micro-zoning—a zone per person—dramatically increasing the environmental quality and long-term value of the building.

User Control of Diffuser Location and Air Direction—Task Air Systems. Control over the direction of the air supply may be the least costly strategy for improving thermal comfort. The ability to relocate diffusers can compensate for the frequent mismatch between diffuser and furniture layouts, and for the comfort differences between cold air and warm air distribution patterns from the same diffuser. Since air distribution patterns from ceiling diffusers will be affected by changes in furniture layout, as well as occupancy and equipment densities, control of air flow direction would be a real benefit to maintaining effective air distribution without drafts. The introduction of relocatable diffusers and/or operable vanes in overhead diffusers would allow users to change the direction of the airflow (without shutdown) to match the season and the needs of the dynamic workplace.

To improve thermal comfort and air quality in commercial buildings, a significant number of innovative, underfloor task-air systems have been introduced in the past 10 years. These floor-based systems offer continuous control over the location and density of supply air diffusers, as well as in some cases, the volume and direction of airflow. The various floor-based air distribution systems have been designed with the assumption of a minimum of one diffuser per person, and an optimum of six diffusers per person. In the most affordable mode, these user-controlled VAV diffusers are bundled with a pressurized plenum (not more than 50 ft from the ducted source). In the United

FIGURE 12.11a Hiross Flexible Space System™—The Flexible Space System™ Provides a High Freedom of Access to Its Components, to the Floor Void, and All Facilities for Easy Cleaning and Maintenance. Distributed Relocatable Fan Boxes Support Volume Control along with the Location and Directional Controls (Courtesy: Hiross, Inc.).

States, companies such as Tate, Maxcess, and Interface, offer a plenum floor system with relocatable air diffusers in the floor, in collaboration with York, Titus, and Krantz. These systems enable the density and location of air diffusers to match the needs of changing workstation layouts and occupant and equipment densities. Even higher air quality and thermal performance may be achieved through partially ducted and ducted systems, or through the use of distributed fan-driven diffusers.

User Control of Air Speed and Volume. Unlike ceiling distribution systems, many of the floor and furniture-based air supply systems also provide user control of air speed and volume (Figures 12.11a and 12.11b, Tate's Task Air Module™, Hiross's Flexible Space System™). The Personal Environmental Modules™ from Johnson Controls, the Argon Personal Air Control System™ from Argon, and Climadesk™ by Advanced Ergonomic Technologies, Ltd., have introduced fan-driven desktop diffusers that can be coupled with underfloor or ceiling air supply, to provide the maximum

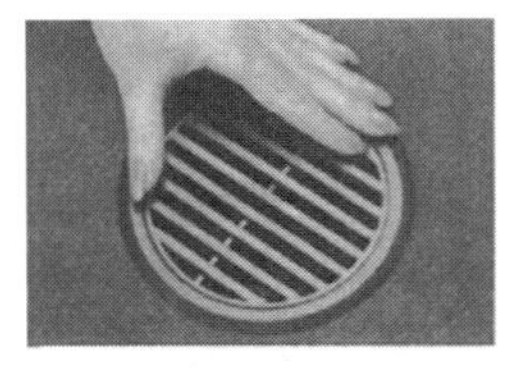

FIGURE 12.11b Tate Task Air Module™—Four Independent Grills Rotate a Full 360° for Complete Control of Air Direction. An Easy-to-Use Thumbwheel Fan Speed Control Provides Adjustment of Air Volume from Off, to Low, Medium, and High Speeds. Quick Connect/Disconnect Electrical Connectors Make Installation of a Task Air Module™ Easy and Simple (Courtesy: Tate, Inc.).

level of individual control at the worksurface. A legislated increase in ceiling air diffusers (to a minimum of one diffuser per workstation) could support the introduction of control over air volume and direction in the ceiling, with occupancy sensors and variable speed fans in the central system to ensure maximum energy efficiency.

User Control of Air Temperature. The introduction of individual temperature control, within a band of ±2°C, has consistently eliminated calls to facilities management about overheating and drafts. The most prevalent example of individual control over supply-air temperature is found in a new generation of high quality fan-coil units, heat pumps, and induction units. In each of these systems, the user can set supply air temperature, either through water flow control or air mixing control. Whenever large numbers of individuals are calling for maximum heating or cooling, the central system can respond with modified supply air or supply water temperatures. Johnson Control's Personal Environmental Module (PEM™) provides individual control of air temperature by putting a mixing box at every desk. This module allows conditioned outdoor air to be delivered directly to the desk without energy penalty, with the end-user deciding the quantity of filtered room air that will be mixed in for thermal control (Figure 12.12). Not only is the percent of outside air more reliably delivered, the combined outside and room air is locally filtered with two staged filters to more reliably eliminate particulates and VOCs. Given the desktop location of the diffusers and horizontal and vertical directional control, the age of air is also significantly younger than ceiling and floor-based systems (Faulkner et al., 1993; Mahdavi et al., 1999).

These task-air approaches have demonstrated measurable improvements in thermal comfort and indoor air quality, as well as in user satisfaction studies (Bauman and Arens, 1996; Bauman et al., 1997; Drake et al., 1991; Kim and Homma, 1990; Kroner et al., 1992; Hedge et al., 1993; Shute, 1992), by reducing age of air, supporting changing space layouts with continued levels of HVAC

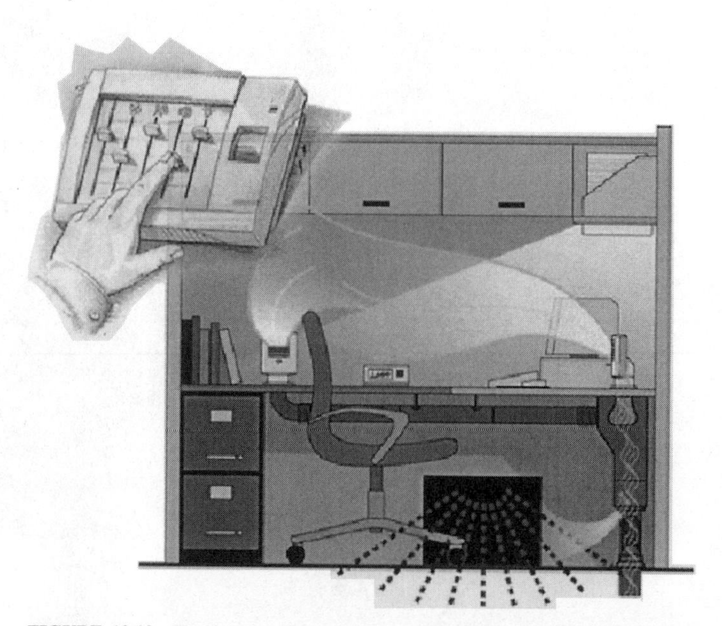

FIGURE 12.12 The Johnson Controls Personal Environmental Module (PEM™) provides a Mixing-Box at Every Desk, Allowing Each User to Individually Control Ventilation, Air Speed, Direction, and Temperature, As Well As Outside Air Quantities (Courtesy: Johnson Controls, Inc.).

delivery, and providing an unprecedented level of user control, albeit still far short of the level of control we expect in our cars. By combining individual or workstation-based thermal conditioning systems (driven by occupancy sensors) with ambient thermal conditioning to much broader standards of comfort, maximum energy savings will accompany the gains in individual comfort.

12.5.3 Robust, Micro HVAC Systems and Integrated Grid-and-Node Conditioning with Plug-and-Play HVAC Systems are Accessible, Maintainable, and Modifiable

A third major innovation to improve air quality, thermal comfort, and energy performance in buildings has been the development of smaller, packaged HVAC systems or components with plug-and-play characteristics to continuously maintain or improve performance delivery. While yesterday's packaged heat pump, fan-coil, and window air conditioning units have received very poor scores from facilities managers due to maintenance problems, noise, and inefficiency, the next generation of expandable, plug-and-play technologies may be the key to improving indoor environmental quality in dynamic buildings.

The leased IBM-headquarters building by Sari in La Defense, Paris, has a fan coil for every occupant in the modern high rise. Over 1,800 fan coils are distributed in two mechanical rooms on every floor, with short flexi-duct runs to adjustable ceiling air diffusers (Figure 12.13). Not only can each occupant set personal temperature conditions within a ±2°C band, they can also call for a 5-min 100% outside air purge cycle to eliminate local pollutants (such as cigarette smoke) during a meeting. The maintenance of this newly engineered Carrier system is far easier than conventional systems, with plug-in electronic diagnostic equipment used on a twice-a-year cycle. Whenever a fan-coil unit is not performing acceptably, it is pulled from its quick-connect mount and replaced by one of several spares. This allows the facility manager to ship malfunctioning equipment to the factory for repair rather than attempting on-site maintenance. This also allows the key HVAC equipment to be replaced on an appropriate cycle (<50 years), much as components of a car are replaced.

Induction systems, heat pumps, and split heat pumps have also progressed since their packaged beginnings. The Dutch Embassy in Washington, DC demonstrates the use of heat pumps under each window throughout the facility, with specifications and quick-connects for ease of maintenance.

FIGURE 12.13 The Introduction of a Fan Coil Per Person in IBM/Sari—France Demonstrates a Highly Modular, Distributed HVAC System That Successfully Supports Individual Comfort, "Smart" FM, and Energy Efficiency (ABSIC; France, 1991).

These distributed heat pumps and a constant volume ventilation system allow windows to be opened by the occupants, and give individual temperature control with major energy efficiencies. While the Sari fan-coil system merges ventilation air in distributed mechanical rooms before the ceiling-based delivery of conditioned air to the workspace, the Dutch heat pump system maintains a separate ceiling-based ventilation system with greater conditioning potential. Both maintain the constant delivery of air from dedicated outside air ducts, which may be key to the long-term promise of indoor air quality.

It is becoming increasingly clear that the delivery of thermal comfort and air quality depends on the robustness of the "integrated system." The present delivery of over five independent system components to the site—generation (central heating, cooling, and ventilation units), distribution ducting or piping, terminal units, sensor/controllers, and building automation systems—has led to consistent idiosyncrasies and failures. Tomorrow's smart building owner will mandate the delivery of thermal and air quality *service*—looking for robust, integrated solutions from the lead manufacturers. The development of prototyped and tested HVAC conditioning service (systems not components) for each 10,000 ft^2 workgroup—tailored to your climate and your level of space dynamics—will offer unparalleled improvements in cost and performance quality.

FIGURE 12.14 Lloyd's of London by Piano & Rogers/OveArup Demonstrates the Use of a Triple-Glazed Enclosure as a Return Air Duct, Comfortably Dissipating Excess Internal Heat Gains and Eliminating Perimeter Loads (ABSIC; U.K., 1991).

12.5.4 Load Balancing—Facade Mechanical Systems

Given the consistent increases in internal loads in commercial buildings, there is a real justification for increasing the periphery of buildings with the façade designed as an integral part of the mechanical system. Both airflow windows and water flow mullion systems enable excess heat from the core—heat from occupancy, lights, and equipment—to be effectively dissipated through the facade. By taking return air through the "glass duct" of a triple glazed airflow window, core cooling loads are dramatically reduced and perimeter heating is almost eliminated (as seen in the Comstock building in Pittsburgh, PA, GEW in Cologne, and Lloyd's of London; Figure 12.14). Indeed, the perimeter heat pumps installed at Lloyd's have never been used because return air through the façade effectively eliminated the perimeter loads and ensured thermal comfort.

Water mullions (thermally broken from the outdoors) can also use "waste" heat from the core to minimize loads. In this system, waste heat from building cooling or power generating systems can eliminate perimeter heating requirements and radiant imbalance, while allowing an increase in building periphery for views and light at every workstation. Indeed, the building facade could be seen as the natural dissipater of energy, a "circulatory system" resembling that of a healthy human—with appropriate surface to volume ratios.

One other facade mechanical system of note is the use of double envelopes to provide north-south-east-west load balancing in climates where solar loads are significant and beneficial. The Occidental building in Niagara Falls is one of the earliest examples of a double envelope construction (Figure 12.15). When solar energy is received on the east, a natural convective loop of solar heated air wraps the entire building (Rush, 1986). This continues throughout the day to eliminate simultaneous heating and cooling, maximize passive solar contributions to the heating load, and ensure excellent mean radiant conditions. In summer, the double envelope is vented to the outside, and precautions are taken for fire protection. The Occidental building uses less than 30% of the heating and cooling energy of a conventional office building in upstate New York (Bazjanac, 1980).

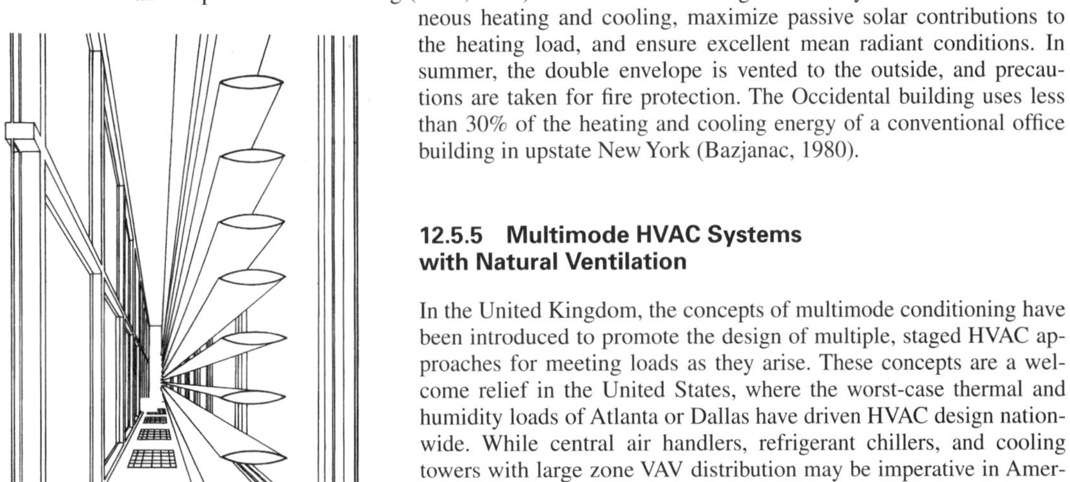

FIGURE 12.15 This Section Perspective Shows the Air Space, Louvers, and Double-Glazed Skin That Act as a Nine-Story Passive Solar Collector and Daylight Source for the Occidental Headquarters. The Louvers Serve as Both Thermal and Visual Shutters, and the Double Envelope Effectively Insulates the Building with Natural Energies (Occidental Chemical Company Corporate Headquarters; Rush, 1986).

12.5.5 Multimode HVAC Systems with Natural Ventilation

In the United Kingdom, the concepts of multimode conditioning have been introduced to promote the design of multiple, staged HVAC approaches for meeting loads as they arise. These concepts are a welcome relief in the United States, where the worst-case thermal and humidity loads of Atlanta or Dallas have driven HVAC design nationwide. While central air handlers, refrigerant chillers, and cooling towers with large zone VAV distribution may be imperative in America's hottest, humid climates, they shut out consideration of alternative HVAC and architectural approaches to thermal conditioning and air quality in other regions. The singular, sealed building approach also makes architects sloppy environmental designers and eliminates early design partnerships with engineers.

Rather than a dual-HVAC system that may be implied in a multimode concept, we would like to introduce the concept of "ascending environmental conditioning strategies." In an "ascending strategy" to mechanical system design for cooling, for example, the design team would begin with the fundamental principles of "architecture unplugged" (see Section 12.6: Smart Enclosure Systems). Massing, orientation, shading, and building mass would be optimized to minimize the cooling loads and maximize the number of months for which no cooling would be needed (could be 9–12 months in Pittsburgh, for example). Then, passive strategies such as cross-ventilation, stack ventilation, fan-assisted ventilation, and night ventilation would be introduced, which requires the full involvement of mechanical engineers in the early design process. Passive cooling would be followed by desiccant cooling when humidity levels exceed the effective comfort zone. Desiccant cooling would require that the windows be closed and would lead to desiccant-evaporative cooling cycles. As outdoor temperatures or indoor heat loads exceed the capability of these systems, then absorption and finally refrigerant cooling would be introduced, first at a task comfort level. Only the last stage of this ascending conditioning system would be the task-ambient central-system refrigerant cooling that is pervasive today throughout the United States despite energy, indoor air quality (IAQ), and thermal limitations.

A complement to "ascending" energy strategies is the introduction of "cascading" energy strategies designed to make maximum use of limited natural resources (Hartkopf and Loftness, 1996). In a cascading system, small fuel cells and photovoltaics might be bundled for the building's power generation; waste steam could then be used to drive the desiccant, then absorption and then refrigerant systems; and finally the resulting waste heat could be used for space heating and hot water. The bringing together of innovations in ascending conditioning strategies (load reduction, passive, then active) with innovations in cascading conditioning strategies (the waste from each energy process can support the next) may offer the most environmentally sustainable approach to designing and retrofitting buildings in the future (Hartkopf, 1999).

12.6 SMART ENCLOSURE SYSTEMS

Smart, sustainable buildings have an enclosure that ensures the highest thermal quality, air quality, visual quality, and long-term building integrity. The enclosure design clearly demonstrates advances in enclosure components and assembly design, as well as their effective integration with interior systems, structural systems, mechanical conditioning systems, lighting systems, telecommunications, and power systems. Sustainable enclosures effectively address seven major responsibilities, indeed areas of major innovative opportunity, in the pursuit of aesthetics, comfort, and resource conservation:

1. Load balancing
2. Solar control
3. Heat loss control
4. Daylighting
5. Natural ventilation
6. Passive and active solar heating
7. Enclosure integrity

A range of innovations in enclosure design worldwide offer significant opportunities to displace large energy loads in buildings, while simultaneously providing regionally innovative architectural aesthetics. At least three major enclosures innovations illustrate the next generation of "smart enclosure" systems—"unplugged" enclosures with minimum thermal loads, layered facades for daylighting, and enclosures as mechanical systems.

12.6.1 Regional Enclosures—Architecture Unplugged

Smart commercial buildings must continue to support the productive organization through the brownouts and rolling black-outs expected in our energy future. The sustainable architecture required to move through power unreliability must be entirely regional in character, carefully reducing the liabilities of the specific climate (HUD, 1978) while maximizing the natural conditioning opportunities such as day–night load balancing, daylighting, passive solar heating, and natural ventilation.

The most critical early design decisions relate to building massing—the height, depth, and orientation of buildings—which has a major impact on indoor environmental quality and energy loads. Neither the tallest building in the world nor the largest building under one roof offer sustainable work environments in a power failure. Indeed, these buildings guarantee significantly higher energy loads in almost every climate since they eliminate any use of daylight, natural ventilation, or natural dissipation of internal heat gains (Mahdavi et al., 1996). They guarantee that the building must be abandoned in a power outage. If air quality, comfort, and energy are a driver for building form, then the next generation of buildings will strive toward controlling building depth, height, and orientation to achieve environmental comfort for a maximum percentage of the year—unplugged.

Regionalism would also require a shift away from the pervasive sameness of building enclosures, favoring neither the international styles with unshaded glass and unthermally-broken concrete, steel, and aluminum, nor the post-modernism of today. "Architecture unplugged" would require enclosures that demonstrate serious attention to the management of solar gain, heat transfer, moisture migration, and day–night load balancing. These mass, color, venting, and thermal insulation characteristics are key to energy conservation in buildings, requiring entirely regional solutions. Heavy capacitance masonry facades will be seen in desert climates with large diurnal swings; heavily shaded and highly operable facades will be seen in warm, humid climates; and solar exposed, yet well-insulated, facades will be seen in cold climates. In each climate, the massing, orientation, and selection of façade materials will be regional in character, providing the maximum natural comfort before mechanical systems need to be introduced.

12.6.2 Layered Facades for Daylighting Windows for Workers (and *not* Windows 2000™)

Regional design also requires far more serious attention to opening size, location, materials, and controls. Windows are both the weakest elements in an enclosure for heat loss, solar gain, and infiltration, and the most critical to heat dissipation, natural ventilation, and daylighting—for both energy conservation and health.

In the United States, commercial building energy use is evenly distributed between lighting, cooling, and heating (EIA, 1995). Effective window design can offset over 80% of the electric lighting loads with daylighting, 50% of the ventilation and cooling loads with shading and natural ventilation, and utilize passive solar energy for 20–80% of the heating loads. The Robert L. Preger Intelligent Workplace (IW) laboratory at Carnegie Mellon University demonstrates these savings with significant improvement in indoor environmental quality for workers (Figure 12.16). Moreover, the windows and building envelope of the IW take advantage of load balancing through water mullions, technologies that could minimize or eliminate the simultaneous heating and cooling loads in most commercial buildings today.

FIGURE 12.16 In the Robert L. Preger Intelligent Workplace, 75% of the Facade Area is Glazed with High-Visibility, Low Shading Coefficient Glass, Maximizing Effective Daylighting throughout the Year. The Windows Are Operable To Allow Natural Ventilation Cooling for 9 Months and Water "Mullions" Dissipate Waste Heat for Energy Efficient Perimeter Heating. (Architects Bohlin Cywinski & Jackson, Pierre Zoelly, & the CMU Center for Building Performance.)

Access to Daylight and View for Each Individual. The first step to maximizing the benefits of "smart enclosures" for natural conditioning is to increase building periphery so that each workstation is guaranteed a view and daylight access. This view of the outdoors should be with content—views of pedestrians and trees and community life—to critically maintain our sense of time and season. The view should not be obscured by highly reflective glazing, which reduces visibility below 35%, demands electric lighting during the day by default, and reflects those lights as glare, further obscuring the occupants' view. One of the great competitive advantages of a land-rich nation is the possibility of providing its workforce with adequate workspace for individual and group needs, all with seated access to the natural environment.

Numerous buildings have been completed in the last decade that demonstrate even, effective distribution of daylight through bilateral lighting, atria, and skylights, in combination with light shelves. These buildings have reduced electric lighting loads to well under 0.5 watts/ft^2 (and this is typically linked to night work), and introduce the full range of daylight attributes into commercial spaces, as previously described. Justifications for increasing access to windows for building occupants go well beyond energy savings, however.

Hospital studies have shown that patient recovery time is quicker in rooms that have windows with views (Ulrich, 1984). School studies reveal statistics that link windows and daylight to higher test results, better student health, and faster physical growth (Nicklas and Bailey, 1996; Heschong/Mahone, 1999a). The importance of windows, daylight, natural ventilation, and views for mental and physical health has led the European community to mandate that all occupied spaces should have operable windows, and that no worker should sit more than seven meters from a window (NKB, 1991). In the United States, the best cost-justification for reintroducing windows and contact with the natural environment into commercial buildings could be gains in organizational productivity, such as sales and customer satisfaction. Heschong/Mahone found that sales were 40% higher in daylit retail stores in a cross-sectional study of 80 stores in northern California (Heschong Mahone, 1999b). In tight employment markets, workplaces with windows—better yet workplaces with operable windows and views—may be the critical condition to improve both the attraction and retention of knowledge workers.

Sunshading and High Visible Transmission Glass. New developments in glazing enable architects to specify high visual transmittance with controlled solar transmittance. For effective daylighting and maximum views, visible transmittance should always be above 50%, rather than the 15–20% common in office buildings today. It is possible to achieve light and view transmittances over 50% while maintaining shading coefficients under 45%, in order to keep solar cooling loads as low as possible for the internally load-dominated office buildings. In addition to high-transmission/low-shading coefficient glass, external shading devices and light redirection devices are the most effective measures to minimizing overall energy loads with maximum environmental quality in the workplace. Already 20 years ago, a study from Lawrence Berkeley Laboratories demonstrated that exterior operable shading devices, and even interior shading devices, were more cost-effective and energy/resource-effective than highly tinted glass in all regions of the United States, from Los Angeles to Chicago to Miami (Figure 12.17; Winkelmann and Lokmanhekim, 1981).

Daylighting and Light Redirection. Since electric lighting loads and additional cooling to eliminate the heat from lights are two of the largest components of peak demand in commercial buildings, the use of daylight without solar heat gain is a key strategy to long-term environmental quality and energy effectiveness. The most cost-effective demand-side management solutions are those that directly eliminate lighting loads through daylighting (Sullivan et al., 1992). In the Lockheed building in Sunnyvale, California, the innovative daylighting design uses deep light shelves, sloped ceilings, and a top-lit central atrium to provide 70% of the required ambient illumination throughout the year in a deep open plan office space (Figure 12.18). Unexpectedly, this shift to a daylight work environment resulted in a 15% reduction in absenteeism (Romm and Browning, 1994), more than paying for the daylight features in the first year.

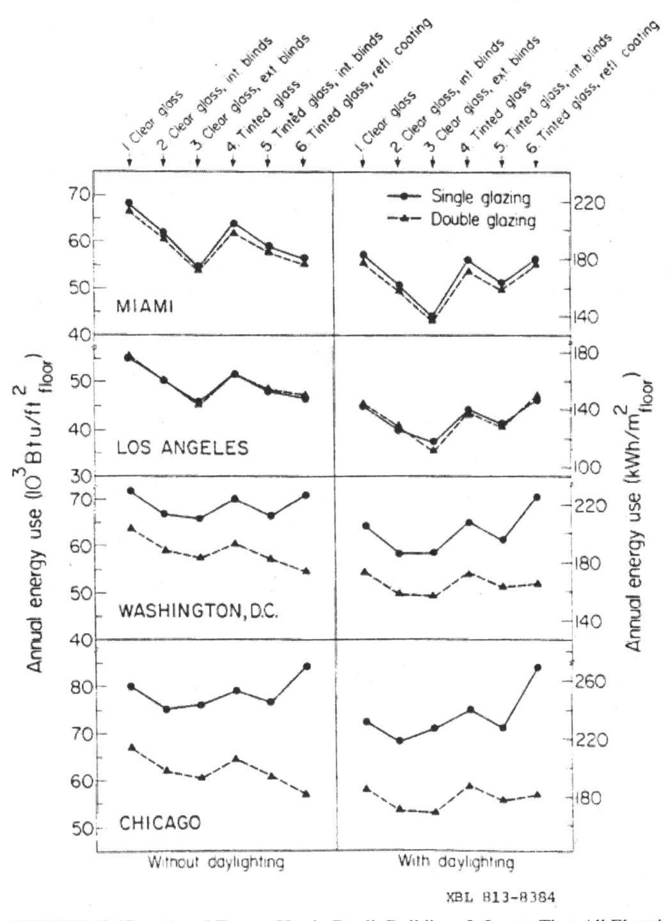

FIGURE 12.17 Annual Energy Use in Daylit Buildings Is Lower Than All Electrically-Lit Buildings in Four Climatically Different Cities with Six Fenestration Options (Winkelmann and Lokmanhekim, 1981).

Although the rules for effective daylighting in offices have long been established, relating to percent of aperture, ceiling height, room depth and color, and sunlight redirection devices at the window, a key design change required is the commitment to a layered facade. These facades have dynamic light redirection and shading devices that can be external to the façade, integral to the glass assembly, or interior devices. The intelligent workplace introduces a layered facade (Figure 12.19) that includes external louvers and interior inverted venetian blinds with automated and manual controls to allow for glare-free light redirection or full shading as the environmental conditions demand. The design of layered facades enable seasonally and daily dynamic control of the light, heat, and ventilation energies in the natural environment. By displacing mechanical and electrical loads, these facades provide near-term savings, sustainability, and long-term environmental satisfaction.

Smart Glazing Technologies for Daylight and Shading. There are also a range of exciting new technologies that will support effective daylighting, glare control, and shading of commercial

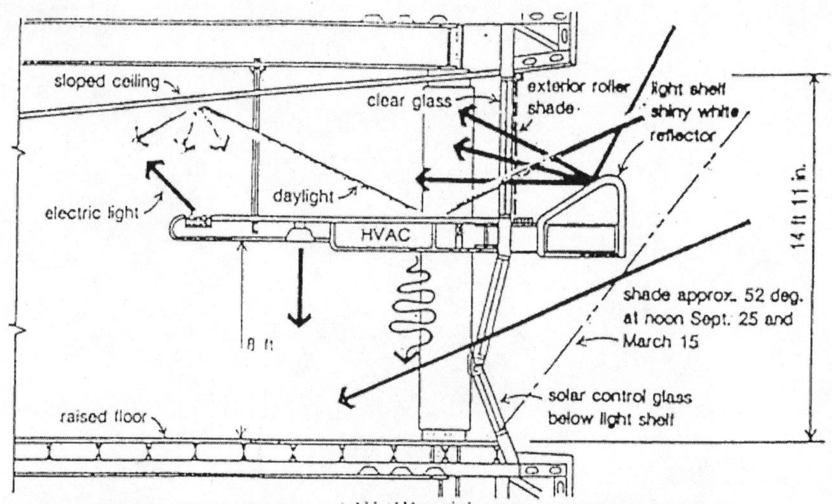

FIGURE 12.18 The Lockheed Building in California by Leo Daly Architects Demonstrates Innovative Use of Daylighting with Light Shelves and Sloped Ceilings to Provide Effective Lighting Deep into the Open Plan Space (ABSIC; North America, 1990–1991).

buildings. Electrochromic, photochromic, holographic, and prismatic glazing materials are all under development, which would enable facades of different building types in different climates to be "tuned" to the energy flows that are desirable on a daily and seasonal basis (Figure 12.20; Selkowitz and Lampert, 1992). These technologies may make the use of daylight the primary daytime light source for commercial buildings again, with significant environmental benefit. Light pipes that take daylight from the building enclosure to "fixtures" throughout the building may also become economically viable in the next few years, improving on both enclosure and interior design details.

12.6.3 Enclosures as Mechanical Systems

Four major innovations in enclosure design offer significant opportunity for the facade to become an integral part of the mechanical system: dynamic enclosures for natural ventilation, enclosures for day–night load balancing, enclosures for passive and active solar heating, and enclosures for power generation. Since day–night load balancing enclosures have been discussed in the Smart HVAC text, and passive and active solar heating has generated extensive articles and text books, this section will discuss the remaining two major innovations in smart enclosures.

Dynamic Enclosures for Natural Ventilation. A number of architects have recently taken on the challenge of designing commercial office buildings without any mechanical system or with a minimum of mechanical systems. These efforts have required collaborative design processes and a major rethinking of the depth, height, cross-section, and orientation of the buildings. The mechanical system that is typically eliminated first is cooling, and, secondly, central ventilation. The Leicester Engineering school in the United Kingdom, by Short and Ford, is a university building without refrigerant cooling. The building is designed for cross-ventilation from prevailing winds and stack ventilation from vertical chimneys—creating both a dramatic architectural form and an effective conditioning system for the cooler English climate (Swenarton, 1993).

FIGURE 12.19 The Robert L. Preger Intelligent Workplace Demonstrates A Layered Enclosure, with Dynamic Light-Redirection Louvers, Operable Windows, High-Visibility Glass, and Internal Shading Devices, To Maximize Daylighting and Solar Control As Well As Support Natural Ventilation (Façade engineered, manufactured, and donated by Josef Gartner & Company).

At the same time, other projects represent a balance of natural conditioning and mechanical conditioning approaches. The Commerzbank by Norman Foster/Ove Arup is the tallest building in Frankfurt and yet has operable windows and daylit office spaces. Through the ingenious design of stacking open-air courts, this high-rise breaks down both wind speeds and stack effects to allow office windows to be opened as desired, with refrigerant cooling used only a minority of the time (Herzog, 1996). The IBM tower by Ken Yeang in Malaysia represents another approach to introducing natural conditioning in large commercial buildings. By removing the large volume of vertical

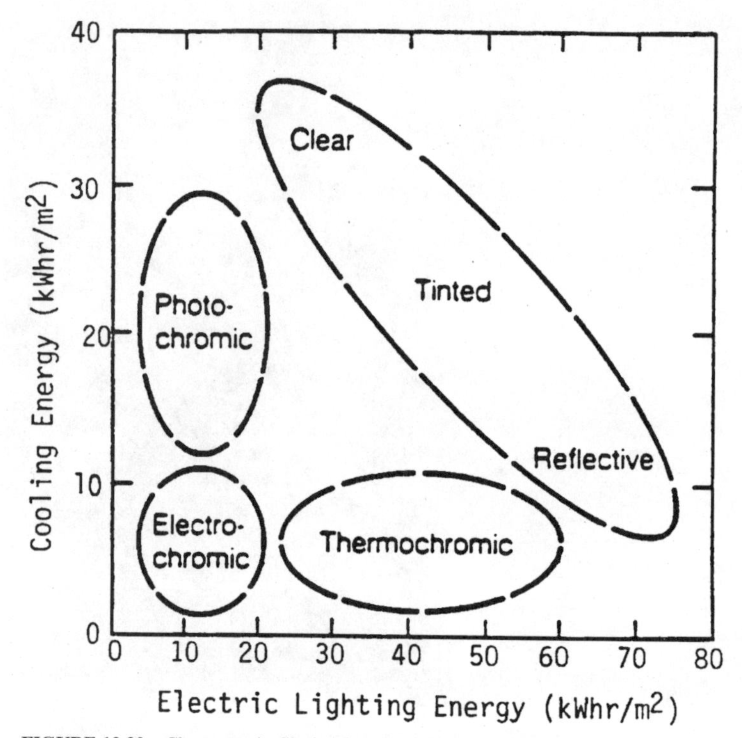

FIGURE 12.20 Chromogenic Glazing Strategies for Dealing with the Relative Requirements for Cooling and Lighting Will Offer Innovative Regional Solutions to the Smart Enclosure (Selkowitz and Lampert, 1992).

circulation from the conditioned core to perimeter location, the elevator and stair towers are naturally conditioned and strategically buffer the east and west facades.

Night ventilation through the facade or the building's mass is another smart enclosure strategy being pursued by architectural engineering teams. From the State Office projects in California under architect Sim Van der Ryn in the 1970s to the Farsons Brewery by Short and Ford in Malta (Figure 12.21; Rickaby, 1991), night ventilation strategies have displaced a large percentage of refrigerant coolings loads. Lloyd's of London is possibly the best-known building that utilizes night ventilation of the structure to effectively eliminate cooling for most of the following day—even with the massive influx of 4,000 trading agents each morning (ABSIC, U.K., 1989).

Smart Glazing Technologies for Power Generation and Conditioning. The vast areas of roof and facade that absorb or reflect long hours of sunshine can be turned into an asset even in hot climates. The use of innovative, photovoltaic (PV) materials for facade and roof enclosures enable solar energy to be turned into power for meeting plug loads and heating and cooling loads. For windows, new PV glazing assemblies can provide effective shading, daylight distribution, and a competitive power source (10–15 kWh/m^2). For roofs and walls, the early PV assemblies that were rigidly sandwiched behind glass, have been replaced by newer flexible PV materials that can replace common building materials with 3- to 10-year life-cycle justifications. A United States Department of Energy report by Steve Strong identifies the range of materials where photovoltaics will play a major role in generating distributed power and providing energy effective solutions for lighting, heating, and cooling (DOE, 1997).

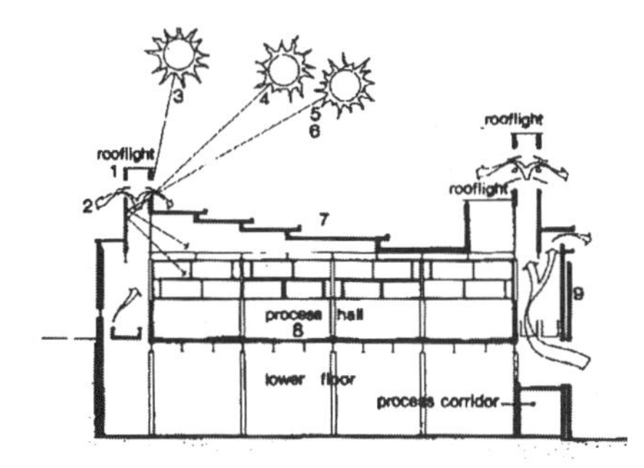

FIGURE 12.21 The Farsons Brewery in Malta by Short & Ford Eliminates the Need for a Refrigerant Cooling System by Using Wind Towers for Natural Ventilation. *Top:* Sectional Sketch; *Bottom:* Exterior View (Rickaby, 1991).

12.7 SMART CONNECTIVITY SYSTEMS—DATA, VOICE, AND POWER

There is very little desktop technology that does not require access to both power and data/voice connections, with continuously changing demands as to location, density, and capacity of service. At least six major innovations illustrate the next generation of "smart connectivity" approaches, always addressing both electrical and electronic demands.

12.7.1 Distributed Vertical Risers and Satellite Closets to Support Distributed Connectivity Centers with Central Intelligence

Distributed, accessible vertical risers are necessary to support the dynamic connectivity demands of a "smart building." These risers support the backbone of services for data, voice, power, environmental control, security, and future media, and support user-customization in multiple satellite closets. The risers and satellite closets tie together distributed services for power and uninterrupted power supply (UPS), for data and voice, wide and local area networks (WANs and LANs) as well as file servers; for environmental controllers; and for wireless communication.

The distributed satellite closets should be modular and relocatable, customized for working neighborhoods of 35–50 people, and networked with a backbone that ensures reliability and external connectivity. Modular racks, typically 19-in., can support fiber step-down, data and voice patch panels, routers, servers, environmental control modules, and security modules (Figure 12.22). Instead of permanent mounts and hardwiring, this plug-and-play approach to connectivity allows for continuous change in hardware density and functionality, without delay or obsolescence, for today's most dynamic building infrastructure. The modularity and expandability of these racks can also support just-in-time purchasing of user nodes and functionality, in lieu of redundancy and obsolescence.

12.7.2 Accessible Horizontal Plenums—Raised Floors and Open Cable Trays/Baskets for Horizontal Distribution

The demands of dynamic work settings and dynamic technology can only be served by accessible, spacious, plenum designs. Yesterday's least-cost poke-throughs and even cellular decks and trenches have proven inadequate to support the dynamic changes in service capacity, wiring innovations, and user interfaces (including outlet requirements). The most flexible "smart buildings" have consistently incorporated raised floors with open cable trays or baskets to feed reconfigurable services to reconfigurable outlet boxes.

Given the immense quantity of existing office area, ceiling distribution systems of HVAC and telecommunications will always maintain a significant market. The effective use of ceilings as horizontal distribution plenums is dependent, however, on the development of integrated ceiling/cable trays and prechased modular wall and furniture systems to bring the network capability down to the

FIGURE 12.22 The Intelligent Workplace Demonstrates Distributed Satellite Closets Which Are Modular and Relocatable, Customized for Working Neighborhoods of 35–50 People, with Modular Racks Supporting Fiber, Data, and Voice Patch-Panels, Routers, Servers, Environmental Control and Security Modules. While Most 10,000 ft^2 Neighborhoods Might Need Two Modular Closets, This Laboratory Demonstrates Extensive Power Submetering and Competitive Control Systems Requiring Six Closets.

desk. Meanwhile, internationally, there is a growing emphasis on raised floor technologies for the horizontal distribution of cables and for conditioned air. The raised floor distribution provides ease of access for service, expansion, and reconfiguration.

12.7.3 Upgradable Wiring Harnesses to Provide a Grid of Service

Upgradability, expandability, and reconfigurability are key to successful connectivity. In addition to distributed vertical risers and open horizontal plenum distribution spaces, the selection of the wiring harness and its flexibility is important. Ideally, all connectivity wiring is "home-run" from the individual workstation to the satellite closet for easy reconfiguration and maintenance. For data, environmental control, security, and voice, the cable type should provide the maximum speed and reliability available (Category 5+ or higher) and the flexibility to be used interchangeably for data, voice, or video demands with simple modifications in the satellite closet. For power, no less than six outlets per employee should be supported, using a power distribution module to maintain "homerun" configurations for all workstations, while shortening the length of the tether for relocating the outlet boxes (Figure 12.23).

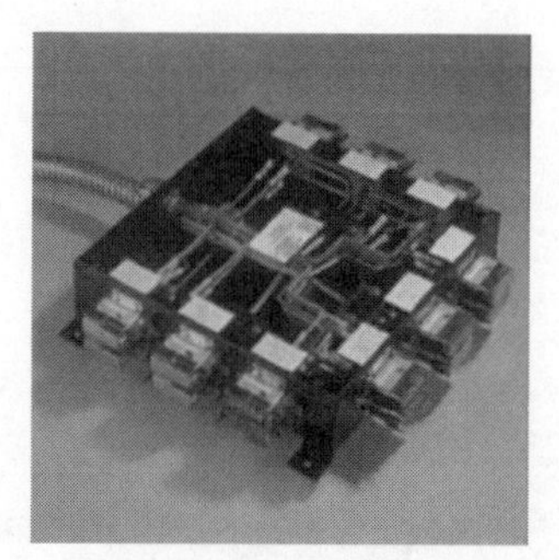

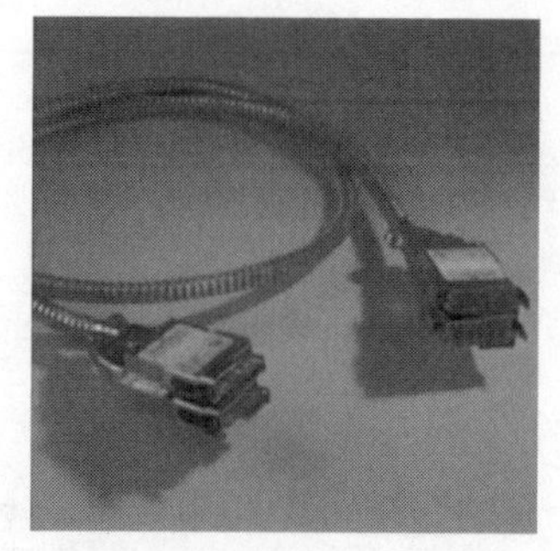

FIGURE 12.23 TateFlex® Modular Wiring—Tate Has Introduced a Modular Wiring Package for Raised Floors Which Allows Users To Change the Density and Location of Power (Courtesy: Tate, Inc.).

These connectivity infrastructures must be developed collaboratively by the data, voice, power, environmental control, security, and fire professionals. The team must select the wiring and cabling type, design harnesses that can be prefabricated for roll-out installation, establish the grid of service required for move-in densities and the mode of expanding, replacing, and relocating service to accommodate workplace dynamics. Again, almost all desktop technology today requires access to both power and data/voice connections, with continuously changing demands as to location, density, and capacity of service.

12.7.4 Reconfigurable Nodes for Data, Power, Voice, and Environmental Connectivity Promotes Modular Outlet Boxes Homerun to Distributed Satellite Closets

Each individual requires multiple data, voice, and power outlets, with significant variations in density and functionality to fully support today's constant layout, activity, and technology changes. As a result, modular floor or desktop outlet boxes are needed, with interchangeable outlets for variations in data, phone, power, video, security, and environmental controls to provide reconfigurable infrastructures without waste. Data, voice, and power wiring should no longer be independent activities by independent unions. Outlet boxes should no longer be in fixed locations with fixed numbers of power or data or voice ports. The boxes should harness all connectivity services, be relocatable by the end-user, and support modifications in density and type of outlets over time (Figure 12.24). These relocatable connectivity boxes (nodes that can plug into the grid of service) also offer the advantage of just-in-time purchasing. This allows facility managers to purchase fewer outlet boxes of higher quality (greater modifiability), and move them around for maximum utilization, only purchasing additional nodes of service when the existing infrastructure is fully loaded—just-in-time.

12.7.5 Wireless Data and Voice for Mobility

Today, wireless capabilities are improving so rapidly that many argue that wiring in buildings will be unnecessary. However, capacity and speed still remains one generation behind wired connectivity, and power continues to rely on wired infrastructures. Moreover, wireless infrastructures require as much hardware and design response as wired infrastructures.

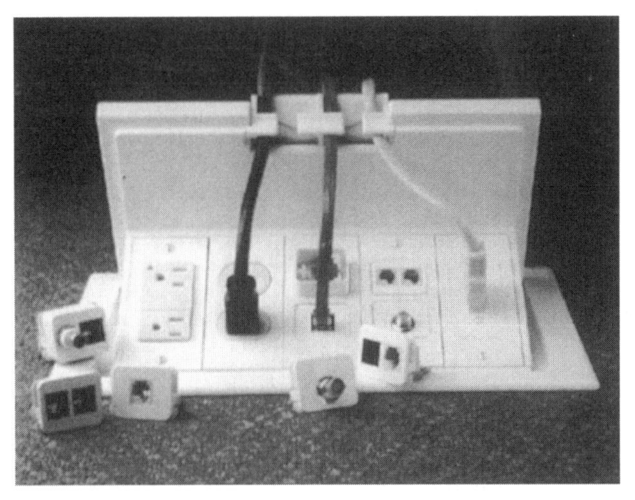

FIGURE 12.24 The AMP Access Floor Workstation Module® Provides Access to Power, Data, and Voice Networks, with an Easily Customizable Number of Modular Outlets (Courtesy: AMP, Inc.).

FIGURE 12.25 Conference Hubs and Technology Pubs Will Support the Dynamic Advances in Multimedia Equipment for Collaborative Multimedia and Face-to-Face Exchange.

Nonetheless, wireless technologies offer a level of mobility for knowledge workers that is critical to the next generation of collaborative work. "Smart buildings" must support both wired and wireless infrastructures fully. Wireless antennas will rely on the same satellite closets that serve 35–50 people, and investments will have to be made in receivers for each desktop technology. In the near term, independent wireless infrastructures will be needed for voice and for data/web access, though these should eventually merge. The continued improvements in power batteries for laptops and phones may lead to greater utilization of the wireless infrastructure for communication than the wired infrastructure, though both will be central to "smart connectivity."

12.7.6 Technology Pubs and Multimedia Conference Hubs

Two major technology-driven additions to office planning today are distributed multimedia meeting spaces and additional shared technology spaces. In defining the flexible infrastructures needed in buildings, it is important to anticipate the major increase in teaming and multimedia spaces—distributed on-demand *throughout* the workplace—and the corresponding needs for ventilation, thermal conditioning, and connectivity to support collaborative multimedia and face-to-face exchange. At the same time, equipment centers or "technology pubs" must be supported throughout the workplace to provide shared access to the latest copiers, printers, scanners, and other new hardware, as well as to provide a "natural" social center for information exchange in the workplace (Figure 12.25).

12.8 SMART LIGHTING SYSTEMS

Lighting systems (see Chapter 13) in the intelligent building should fully address the demands of the dynamic organization for:

- Organizational/spatial flexibility (continuous organizational change without waste)
- Functional flexibility (changing tasks and work tools without visual discomfort)

- Technological flexibility (changing desktop/multimedia technologies without visual discomfort)
- Healthy, motivational work environments

A number of innovative design approaches and lighting components and systems illustrate the next generation of "smart lighting systems."

12.8.1 Effective Daylight Utilization with Controllable Electric Lighting Interfaces

Daylit work environments are one of the most important directives for the smart or intelligent building. Daylight and natural ventilation are two attributes of quality work environments, critical to attracting and retaining the high-tech worker. As a result, the design of window control systems and electric lighting control systems are critical design challenges. Manual or automated window controls must ensure effective shading and glare control, as well as contributing to light redirection for effective daylight distribution. Manual or automatic lighting control systems should support the use of daylight as the dominant light source by enabling dimming or off-switching for perimeter lights.

Effective daylight utilization with controllable electric lighting interfaces offer significant savings in lighting power demands and cooling costs. These savings could quickly offset the first costs of lighting control systems, presently from \$0.60–1.20 per ft^2, and even offset the costs of more sophisticated shading and glazing systems. In New Haven, Connecticut, the corporate campus of Blue Cross/Blue Shield receives 30% of their interior illumination from the sun, saving the company more than \$100,000 annually in lighting and cooling costs, and contributing to higher employee morale (Dubbs, 1991). As mentioned in the Smart Enclosure section, the introduction of daylight in commercial buildings has contributed to reduced absenteeism, reduced operational costs, higher user satisfaction and health, as well as improved sales—major cost-benefits for smart, daylit buildings (Heschong Mahone, 1999a; Heschong Mahone, 1999b; Mahdavi et al., 1995; Romm, 1999; Romm and Browning, 1994; Rubenstein et al., 1998; Shavit and Wruck, 1993).

12.8.2 Separate Task Lighting from Ambient Lighting

The uniform grid of high-level lighting that is common in North American buildings is an unacceptable strategy for the smart building. The high light levels required for reading fine print are too high for computer work, and the shadowing caused by partitions and storage elements make "unplanned" task lights a pervasive reality. An excellent alternative for achieving proper light levels that can support changing tasks, tools, and layouts is the shift to low-level ambient lighting with planned task lights at every workstation (Steelcase, 2000).

These split task and ambient systems should have daylight response for the ambient lighting and user control of task light location, density, and on/off switching. Performance selection of the task lights will be critical to visual quality and energy conservation. Each workstation should have a minimum of one or two task lights, they should be relocatable by the user to match worksurface configuration and use, they should have adjustable arm/directional control of the light distribution, and there should be occupancy sensors for automatic shutdown when the workstation is unoccupied.

The introduction of an ambient lighting system (often indirect for shadowless spatial reconfiguration) in conjunction with task lights, can significantly reduce office lighting energy demands, reduce the air conditioning load, and provide light levels that are better matched to today's diverse work activities.

12.8.3 Reconfigurable Lighting with Plug-and-Play Fixtures

In conventional office design, changing the density or location of ceiling-based ambient or task-ambient fixtures is often a costly and destructive procedure. Multiple unions must disrupt worksta-

FIGURE 12.26 The Colonia Headquarters Has a Modular Reconfigurable Ceiling with Plug-and-Play Capability for Relocating Light Fixtures with Each Workstation Layout Change (ABSIC; Germany, 1989).

tions, remove ceilings, walls, and light troffers, rewiring fixtures and switches into new settings. As a result, neither the density nor the location of fixtures is typically changed unless a total renovation is underway. In compensation, the fixture grid is often overdesigned initially to ensure adequate working light levels in a wide range of reconfigurations. To eliminate this energy and material waste, one smart lighting approach is reconfigurable ceiling lights where density and location can be changed by the occupant or in-house staff.

Austrian lighting designer, Dr. Bartenbach, has designed such a system for the Colonia Insurance Headquarters in Cologne, Germany (Figure 12.26). In the Colonia building, octagonal acoustic ceiling tiles are interchangeable with lightweight light fixtures to enable the simple relocation of fixtures along with each desk or room reconfiguration (ABSIC, Germany, 1989). The density of fixtures can also be simply modified, since the pigtail connections (male–female plugs) allow fixtures to be added or subtracted from any circuit and its corresponding light switch. Track lights also offer this level of plug-and-play flexibility. Lightolier and Zumtobel have developed office lighting tracks that support high-voltage and low-voltage fixtures for plug-and-play uplighting, downlighting, and accent lighting. With this grid of service, the nodes—or light fixtures—can be purchased on a just-in-time basis and continuously reconfigured to support organizational, functional, or technological changes.

12.8.4 Continuous Change in Lighting Zone Size and Control

As electronic or "smart" ballasts increase in sophistication and decrease in price, the payback of individual fixture controls over traditional "blanket" on–off controls is less than 2 years, given energy savings alone. Supplying advanced electronic ballasts for each fixture will support a variety of control options, from individual on/off switches to timers, occupancy sensors, daylight/photocell readers, lamp depreciation controllers, and/or peak load shedding strategies. Indeed, individual tenants and facility managers can vary or assemble combinations of these six control strategies, sending "intelligent commands" to low-voltage data network controllers for maximum energy performance and user satisfaction.

The most critical reason for individually ballasted fixtures (with smart ballasts) is the ability to continuously change control strategies to match spatial and functional changes. The facility manager at Lloyd's of London, for example, can redefine lighting control on-line in minutes. By "lassoing" the desired number of fixtures on a reflected ceiling plan and assigning a data address for the controllers to support changing tenants and tenant layouts, the time and financial costs for lighting changes are nominal (Figure 12.27). With individually electronically ballasted light fixtures, facility managers are able to quickly redefine the size and shape of control zones and their digital/infrared switches to accommodate continuous spatial and organizational change.

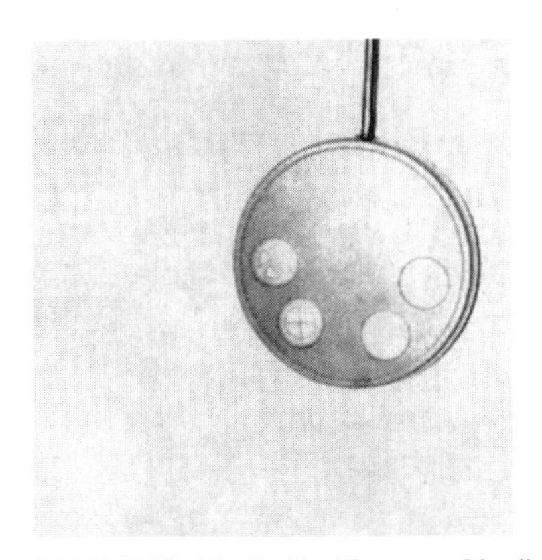

FIGURE 12.27 The Facility Manager at Lloyd's of London Can Reassign Lighting Controls for Tenant Reconfigurations in Minutes On-Line. This Building Combines Micro-Zoning with Individual Hand-Held Controllers to Enable the Occupants to Control a Redefinable Number of Lights (On/Off), Air Diffusers (On/Off), and Heat Pumps (On/Off), to Provide Individual Comfort and Energy Efficiency (ABSIC; U.K., 1989).

12.9 CONCLUSIONS: USER-BASED COMMISSIONING AND USER-BASED FM

The advantage of distributed systems, micro-zoning, and user controlled services is that individuals can configure and reconfigure their own environmental and technical conditions. Calling a facilities manager each time there is a need to turn on or off the heat or lights, to plug in a new piece of hardware, or to relocate zones—dampers, diffusers, thermostats, and switches—is an antiquated concept defying the potential of smart or intelligent buildings. At the same time, further "automating" inadequate blanket and idiosyncratic services already in place also defies the concept of intelligence. The critical direction for smart buildings is toward distributed, user-configurable services and central "intelligent" response.

A look at the density and reconfigurability of the nodes proposed in this definition of an intelligent building and the system/user control opportunities they suggest, typically raise corporate questions of anarchy in space planning and system interface. Should each individual be allowed to set up furniture, lights, data/voice connections, and air supply on a continuously changing basis? George Weller of Steelcase gave the best answer to this concern: "When Alexander Graham Bell first announced the telephone, he projected the vision that there would be one in every home (how about one on every belt?). The shocked response was that this impossible vision would require every person to be a switchboard operator!"

Facilities management (FM) and commissioning would by no means disappear with these smart or intelligent building systems. We are not even near the vision of flexible-grid, flexible-density, and flexible-closure servicing of buildings that would allow for dynamic location and density of service to support continuous organizational reconfiguration. It will take another 10 years of building subsystem improvements and assessments supported by facilities managers to develop effective reengineering approaches and to iteratively test and improve flexible solutions.

Once relieved of daily complaints about individual control of services, there are numerous roles for FM in the user-controlled intelligent building: in preventative maintenance; in feedback to owners, designers, and manufacturers; and in the development of expert systems for building management. A number of advanced buildings now demonstrate the design and operational strategies that can evolve from interactive learning between managers, designers, and occupants of buildings (Hartkopf, 1992).

It is clear, however, that the dynamics of space and technology in today's work environments cannot be accommodated through the existing building infrastructures—neither the "blanket systems" planned for uniform open-plan configurations, nor the idiosyncratic systems planned for unique office configurations. What is needed are flexible-grid, flexible-density, and flexible-closure systems—a constellation of building systems that permit each individual to set the location and density of ventilation and thermal conditioning, lighting, telecommunications, and furniture, including the level of workspace enclosure. Only the move to relocatable, user-based infrastructures will support individual comfort and productivity, organizational flexibility, technological adaptability, and environmental sustainability.

ENDNOTES

1. The Advanced Building Systems Integration Consortium is an NSF IVCRC international university–industry–government consortium dedicated to improving the quality of the built environment.
2. We would like to gratefully acknowledge the continued support of the following Advanced Building Systems Integration Consortium members: AMP Incorporated; Armstrong World Industries; Dutch Building Agency; Interface, Inc.; Johnson Controls, Inc.; Honeywell, Korea; LTG Aktiengesellschaft; Siemens Energy and Automation, Inc.; Steelcase, Inc.; UTC/Carrier/Otis; Zumtobel Staff Lighting, Inc.; U.S. National Science Foundation; U.S. Department of Defense; U.S. Department of Energy; U.S. Department of State; U.S. Environmental Protection Agency; U.S. General Services Administration; and VISCOMM. Past members include: American Bridge Co./Continental Engineering Corp.; Bechtel Corp.; Bell of Pennsylvania; Duquesne Light Co.; PPG Industries; Westinghouse Electric Corp.; and The Knoll Group. Corresponding members include: Bosse Design; Grahl Industries, Inc.; Hüppe Form; Josef Gartner & Company; Mahle GmbH; Mori Biru La Foret Engineering; Nucor; Bayer USA; and Osram/Sylvania.

BIBLIOGRAPHY

ABSIC Japan. 1988. Hartkopf, V., Loftness, L., Drake, P., Dubin, F., Mill, P, and Ziga, G. 1991. *Designing the office of the future: The Japanese approach to tomorrow's workplace.* John Wiley & Sons, Inc.

ABSIC Germany. 1989. ABSIC U.K. 1989. ABSIC North America. 1990–1991. ABSIC France. 1991. *Field studies of advanced technology and intelligent buildings: Research report series.* Center for Building Performance and Diagnostics, Carnegie Mellon University, Pittsburgh, PA. Advanced Building Systems Integration Consortium Research Team: Hartkopf, Loftness, Mahdavi, Mill, Drake, Rainer, Dubin, Posner, Rosenheck, Ziga.

ABSIC/CERL. 1995. Flexible grid—flexible density—flexible closure—The intelligent workplace. USACERL Contract DACA88-93-D-0004, Phase I, January. Center for Building Performance and Diagnostics, Carnegie Mellon University, Pittsburgh, PA.

Bauman, F. S., and Arens, E. A. 1996. *Task/ambient conditioning systems: Engineering and application guidelines.* Center for Environmental Research. Berkeley, CA: University of California. Publication Number: CEDR-13-96.

Bauman, F., Baughman, A., Carter, G., and Arens, E. 1997. *A field study of personal environmental module performance in bank of America's San Francisco buildings.* Center for Environmental Research. Berkeley, CA: University of California. Publication Number: CEDR-01-97.

Bazjanac, V. 1980. Architectural energy analysis. *Progressive Architecture,* April.

DOE. 1997. *Photovoltaics in the built environment: A design guide for architects and engineers.* Washington, D.C., U.S. Dept. of Energy, 254 pp. prepared by Solar Design Associates, Harvard, MA, and the National Renewable Energy Laboratory, Golden, CO, for the U.S. Department of Energy, September.

Drake, P., Mill, P., and Demeter, M. 1991. Implications of user-based environmental control systems: Three case studies. *Healthy Buildings,* IAQ '91, pp 394–400.

Dubbs, D. 1991. Blue Cross cuts cost with daylighting. *Facilities Design & Management* 25, October.

Faulkner, D., Fisk, W. J., and Sullivan, D. P. 1993. Indoor airflow and pollutant removal in a room with desktop ventilation. *Ashrae Transactions* 99(2).

Forrester. 1995. Computing strategy. *The Forrester Report* 12(3).

Hartkopf, V. 1992. Whole building performance in the international arena. IFMA Conference, Philadelphia, PA. March.

Hartkopf, V. 1999. Building as power plant. Research, Development and Demonstration Initiative. Research report. Pittsburgh, PA. Center for Building Performance and Diagnostics, Carnegie Mellon University.

Hartkopf, V., Loftness, V., Aziz, A., Shankavaram, J., and Lee, S. 1998. The GSA adaptable workplace laboratory. In N. A. Streitz, J. Siegel, V. Hartkopf, and S. Konomi, Eds. *Cooperative buildings: Integrating information, organization and architecture, lecture notes in computer science 1670,* Pittsburgh, PA. Springer: Heidelberg, 12–28.

Hartkopf, V., and Loftness, V. 1996. Global relevance of total building performance. CIB-ASTM-ISO-RILEM 3rd International Symposium: Applications of the Performance Concept in Building, organized by the National Building Research Institute. Tel Aviv, Israel. (December).

Hedge, A., Michael, A. T., and Parmelee, S. L. 1993. Reactions of office workers and facilities managers to underfloor task ventilation in offices. *Journal of Architectural and Planning Research* 10(3), 203–218.

Herzog, T. 1996. *Solar energy in architecture and urban planning.* Munich: Prestel.

The Heschong Mahone Group. 1999. *An investigation into the relationship between daylighting and human performance.* Condensed report. Fair Oaks, CA: Heschong Mahone Group.

The Heschong Mahone Group. 1999. *Skylighting and retail sales: An investigation into the relationship between daylighting and human performance.* Condensed report. Fair Oaks, CA: Heschong Mahone Group.

HUD. 1978. *Regional guidelines for building passive energy conserving homes.* Prepared by the American Institute of Architects for the Office of Policy Development and Research, U.S. Department of Housing and Urban Development, in cooperation with U.S. Department of Energy.

Kim, I. G., and Homma, H. 1990. Possibility for increasing ventilation efficiency with upward ventilation. *Ashrae Transactions* 98(I), 723–729.

Kroner, W., Stark-Martin, J. A., and Willemain, T. 1992. The west bend mutual study: Using advanced office technology to increase productivity. Center for Architectural Research, Rensselaer Polytechnic Institute. Troy, NY.

Loftness, V., Beckering, J. J., Miller, W. L., and Rubin, A. 1996. *Re-valuing buildings: Invest inside buildings to support organizational and technological change through appropriate spatial, environmental, and technical infrastructures.* Grand Rapids, MI: Steelcase Inc.

Loftness, V., Hartkopf, V., Mahdavi, A., and Shankavaram, J. 1996. Flexible infrastructures for environmental quality, productivity and energy effectiveness in the office of the future. Presented at the International Facility Management Association (IFMA) Intellibuild '96 Anaheim, CA (June 17–20).

Mahdavi, A., Mathew, P., Kumar, S., Hartkopf, V., and Loftness, V. 1995. Effects of lighting, zoning, and control strategies on energy use in commercial buildings. *Journal of the Illuminating Engineering Society* 24(1), 25–35.

Mahdavi, A., Brahme, R., and Mathew, P. 1996. The 'LEK'-concept and its applicability for the energy analysis of commercial buildings. *Building and Environment* 13(5) 409–415.

Mahdavi, A., Cho, D., Ries, R., Chang, S., Pal, V., and Ilal, E. 1999. *A building performance signature for the intelligent workplace.* Submitted to the Conference on Intelligent and Responsive Buildings, Brugge, Belgium. March 29–30.

Nicklas, M., and Bailey, J. 1996. Daylit students shine brighter. *SunWorld* 20(3).

NKB. 1991. Indoor climate—Air quality. Nordic Committee on Building Regulations. NKB Publication, No. 61E, June.

Rickaby, P. 1991. The art of energy: Peake, Short & partners in Malta. *Architecture Today* 14.

Romm, J. J., and Browning, W. D. 1994. Greening the building and the bottomline: Increasing productivity through energy-efficient design. Rocky Mountain Institute. December.

Romm, J. 1999. *Cool companies—How the best businesses boost profits and productivity by cutting greenhouse gas emissions.* Washington, DC. Island Press.

Rubenstein, F., Jennings, J., Avery, D., and Blanc, S. 1998. Preliminary results from an advanced lighting controls testbed. Proceedings, 1998 Illuminating Engineering Society of North America Annual Conference. San Antonio, TX. August 10–12.

Rush, R. 1986. Occidental chemical company corporate office building—Case study. *The Building Systems Integration Handbook.* John Wiley & Sons, Inc.

Selkowitz, S., and Lampert, C. M. 1992. Application of large-area chromogenics to architectural glazings. Lawrence Berkeley Laboratory, Energy and Environment Division. Berkeley, CA. Report nos: LBL-28012/OM-270.

Shavit, G., and Wruck, R. 1993. Energy conservation and control strategies for integrated lighting and HVAC systems. *Ashrae Transactions.*

Shute, R. L. 1992. Integrating access floor plenums for HVAC air distribution. *ASHRAE Journal* October: 46–51.

Steelcase. 2000. *Life cycle comparisons of direct and indirect lighting for offices.* Draft research report, Collaborative Research Project between Steelcase North America, Center for Building Performance and Diagnostics, Carnegie Mellon University, and Gary Steffy Lighting Design, March.

Sullivan, R., Lee, E. S., and Selkowitz, S. 1992. *A method of optimizing solar control and daylighting performance in commercial office buildings.* Lawrence Berkeley Laboratory, Energy and Environment Division. Berkeley, CA. Report no. LBL-12298.

Swenarton, M. 1993. Low energy gothic: Alan Short and Brian Ford at Leicester. *Architecture Today* 41.

Tu, K. J. 1997. The effects of organizational workplace dynamics and building infrastructure flexibility on environmental and technical quality in offices. Ph.D. dissertation. School of Architecture, Carnegie Mellon University. Pittsburgh, PA.

Tu, K. J., and Loftness, V. 1998. The effects of organizational workplace dynamics and building infrastructure flexibility on environmental and technical quality in offices. *Journal of Corporate Real Estate* 1(1): 46–63.

Ulrich, R. S. 1984. View through a window may influence recovery from surgery. *Science* 224, 420–421.

Winkelmann, F., and Lokmanhekim, M. 1992. *Life cycle cost and energy-use analysis of sun-control and daylighting options in a high-rise office building.* Lawrence Berkeley Laboratory. Berkeley, CA. Report no. LBL-12298.

York, T. R. 1993. Can you afford an intelligent building? *FM Journal* September/October, pp 22–27.

CHAPTER 13
LIGHTING

Gary Steffy, LC, FIALD, IES
President
Principal Designer
GarySteffyLightingDesign Inc.
E-mail: grs@gsld.net

13.1 INTRODUCTION

Lighting can be a missed opportunity in both human and natural resources' expenditures, effectiveness, and productivity. This chapter will offer insights into leveraging the lighting system, both exterior and interior. Exterior lighting affects worker safety and security. Interior lighting—that is both daylighting and electric lighting—influences HVAC operations; annual energy costs; annual maintenance replacement costs; annual disposal costs; and sustainability efforts. Most importantly, lighting affects workers' moods, comfort, task visibility, and, ultimately, performance. Lighting ergonomics embodies comfort and task visibility simultaneously. Sustainability embodies design, procurement, energy, maintenance, and disposal without negatively impacting ergonomics. Ergonomics and sustainability will underpin discussions on lighting technologies and practice. Regulatory requirements are referenced. Current lighting practices are discussed and proposed lighting practices are introduced. Room surface colors, glazing selection, and furniture layout as they impact energy and operating costs as well as worker comfort are also discussed. Technologies, current and emerging, are identified and their positive and negative attributes outlined. Checklists and tabular summaries are used to quickly identify lighting issues, practices, and technologies.

13.2 LIGHTING ERGONOMICS

Vision is very forgiving, or so it seems. People adapt. Problems may not be identified as vision or vision related. Sore necks, headaches, and poor posture can be symptomatic of poor vision situations. For sighted people, vision is used nearly to the exclusion of all other senses in performing work. Cutting corners on lighting can be very detrimental to worker comfort, productivity, and satisfaction, and yet never found culpable. What's worse, lighting budget cuts can only account for perhaps a half a percentage point of a facility's entire construction and/or operating budget.

Yet, resulting bad vision conditions may lead to increased break time, increased sick time, or just slightly less overall productivity—any one of which yields costs significantly greater than the small amount saved with lighting system cost cuts. Workers' vision needs should be addressed by lighting systems (including both daylighting and electric lighting). This seems obvious, but is seldom fulfilled. People respond to light intensity, the ratios of intensities across the field of view, to the color characteristics of light, and to reflections—all layered with people's expectations and previous experiences.

Lighting criteria associated with worker satisfaction and productivity can be categorized into quantity and quality issues. Quantity alone will not ensure appropriate lighting for work or comfort, let alone both. Quality, while sold as an elusive design aspect that is the exclusive realm of designers, is nothing more than assessing/addressing *all* lighting quantity criteria while simultaneously addressing equipment (lighting hardware) aesthetics. Here is a significant distinction. Many lighting design analyses and system audits limit reviews solely to illuminance (light intensity) on worksurfaces or floors. Lighting quantity not only encompasses this horizontal illuminance, but also must include horizontal illuminance uniformity; vertical illuminance; vertical illuminance uniformity; luminaire luminances; and ceiling, wall, and worksurface luminances and respective luminance uniformities. Finessing luminances and luminance uniformities (sometimes called luminance or brightness balancing) can indeed influence the style of a space, and as such, is a feat which successful designers have mastered.

Until recently, color of light and color rendering were relegated to design taste—and considered expendable criteria. Research over the past 10 years, along with growing bodies of experience, show that color of light and color rendering have a demonstrable impact on workers' levels of visibility and comfort. Table 13.1 outlines salient lighting criteria that should be addressed in any situation.

The facility manager's charge to the design and engineering team should require addressing these criteria—with a request for specific documentation citing criteria parameters and respective references (e.g., Illuminating Engineering Society of North America) and how the criteria are being met on the project of interest. Lighting technologies are capable of meeting a lot of criteria simultaneously. Later in this chapter, a review of current lighting technologies outlines their capabilities and limitations. After the detailed technical analyses necessary to meet the criteria outlined in Table 13.1, value–engineering and simple cost cutting can render the lighting system inappropriate. This is never cited by the proposers of such cost-cutting—for several reasons.

First, lighting does not constitute a significant portion of most projects' budgets, except lighting retrofits—so how could saving a quarter percent of the project budget seriously affect the entire project's success? Most people assume it can't and won't. Second, because of the forgiving nature of vision, problems espoused by workers tend to be more physical—sore necks, aching backs. Third, the link between productivity and vision has yet to be well-documented in research in a quantifiable, measurable way. It can only be estimated that lousy lighting reduces productivity a percent or two, if that.

Unfortunately, the reduction in productivity sounds like a small value—until the math reveals that over the 20+ year life cycle of the lighting system, and accounting for salaries, benefits, cost of money for facility financing, energy and the like, this is a big, big item. Fourth, and finally, those involved in value–engineering and cost-cutting rarely have a depth of understanding of the lighting issues pertinent to the project of concern.

Hence, criteria issues carefully established and prioritized by the design team to match the vision requirements of workers, as well as the corporate "vision," are weakened or lost entirely. So, if lighting was developed according to the previously cited criteria and comes under scrutiny during value–engineering, the value–engineering team should be held to offering solutions meeting the same lighting criteria and providing sufficient backup documentation proving the same.

13.3 LIGHTING CODES

Lighting and lighting-related codes are established by the federal, state, and local governmental authorities and can be classified as life safety, energy use, and waste disposal. Reviews of project-specific municipality, state/region, and/or country codes are necessary to establish life safety lighting criteria. Web searches can be helpful in this regard.

TABLE 13.1 Lighting Criteria

Some criteria	Meaning	Rationale
Horizontal illuminances	Lighting intensities on horizontal surfaces: • Worksurfaces • Laps • Floors	• Task visibility • Leverage wages/benefits
Vertical illuminances	Lighting intensities on vertical surfaces: • Merchandise • Walls • Objects on walls • Faces • Computer screens	• Task visibility (note: in some cases less light is better, such as computer screens) • Psychological impressions (improve senses of overall brightness and spaciousness) • Leverage wages/benefits
Luminaire luminances	Light fixture brightnesses	• Direct glare • Reflected glare • Leverage wages/benefits
Surface luminances	Surface brightnesses	• Adaptation • Transient adaptation effects • Leverage wages/benefits
Color characteristics	Color temperature	• Whiteness of light • Sense of normalcy • Leverage wages/benefits
	Color rendering	• Accuracy of colors • Vividness of colors • Sense of normalcy • Leverage wages/benefits
Power budget	Also known as connected load—quantity of watts per unit area necessary to operate the lighting system	• Limit resources expended on lighting
Cost budget	Design fees and initial hardware and installation costs	• Limit capital expended on lighting
Productivity/Satisfaction	User pleasure, comfort and/or productivity over the life of the installation	• Leverage wages/benefits (was it worth doing in the long run)
Life cycle	Issues and costs associated with manufacture, procurement, operation, and disposal	• Limit resources expended on lighting
Maintenance	• Relamping frequency • Reballasting frequency • Difficulty of either	• Limit capital and resources expended on lighting

Adapted from Time-Saver Standards for Architectural Lighting, McGraw-Hill, 2000, with permission.

TABLE 13.2 Lighting Resources

Lighting issues	References	Website address
Quality/quantity	• Illuminating Engineering Society of North America (IESNA)	http://www.iesna.org/
Product safety	• Underwriters Laboratories (UL)	http://www.ul.com/
Energy	• American Society of Heating, Refrigerating and Air-Conditioning Engineers (ASHRAE)	http://www.ashrae.org/
	• California Energy Commission (CEC)	http://www.energy.ca.gov/title24/index.html
Life safety	• Building Officials and Code Administrators (BOCA)	http://www.bocai.org/boca_codes.htm
Installation	• National Electrical Code (NEC)	http://www.nfpa.org/Codes/index.html
	• National Electrical Contractors Association (NECA)	http://www.necanet.org/
Disposal	• US EPA	http://www.epa.gov/epahome/rules.html#codified
	• State agency	[references available on EPA website]
Document searches	• National Standards Systems Network (NSSN)	http:www.nssn.org/
	• Global Engineering Documents	http://global.ihs.com/

Adapted from Time-Saver Standards for Architectural Lighting, McGraw-Hill, 2000, with permission.

The 1992 United States EPAct (Energy Policy Act) dictates minimum energy code requirements. Many states have adopted the ASHRAE/IESNA Energy Standard for Buildings Except Low-Rise Residential Buildings (the latest of which is ASHRAE/IESNA 90.1/1999) to comply with EPAct. Meeting these requirements necessitates careful design planning (e.g., over-lighting is no longer an option); the use of controls and daylight integration; and the use of low wattage/high efficacy (high efficiency) light sources. California's Energy Efficiency Standards for Residential and Nonresidential Buildings (commonly referenced as Title 24/Section 6; revised in 1999) is a model lighting energy code. The ASHRAE/IESNA 90.1/1999 standard closely parallels the California Title 24 lighting document. Table 13.2 cites Internet addresses for respective lighting references.

13.4 ENVIRONMENTAL ISSUES

Twenty-five years ago, energy codes were mandates responding to oil embargos. Now, they are intended to limit both oil dependence and environmental destruction, including the greenhouse effect. A more recent environmental issue, and one which is not yet codified, is sustainability; that is, care in the selection of materials (reducing or eliminating the need for highly toxic materials); how resources are managed to create products (reducing or eliminating the need for highly toxic processes to develop the materials into products); how design/planning strategies use products (reducing or eliminating the need for products which use highly toxic materials and/or production processes); and how/if such products are disposed/recycled.

So, connected lighting loads are now legislated at quite low levels. Generally, 1.1–2.0 w/ft^2 depending on the space function—commercial facilities at the lower end and retail facilities at the higher end. Low wattage, high-efficacy lamps, electronic ballasts and well-designed luminaire optics must be the norm. The use of fewer components is desirable for sustainability (this minimizes natural resource use, initial cost, and solid waste). Lighting needs to be layered—accent/architectural, ambient, and task—in order to meet workers' vision needs while simultaneously addressing energy and sustainability issues. Surface reflectances and finishes need to augment lighting, since they can reduce lighting energy by as much as 20%. Daylight must supplant (at least for some of the day) or supplement (for much of the day) ambient lighting in work areas near windows and/or at skylights. Lighting controls that interface daylight with electric light are necessary to optimize energy effectiveness without sacrificing worker comfort and productivity. Halogen infrared (HIR) mains voltage (120v) and low-voltage (12v) lamps or ceramic metal halide PAR lamps (for higher ceilings) have to be the norm where accent lighting is desired. Triphosphor Triple Tube Compact and Tubular Linear T8 or T5 fluorescent lamps operating on electronic ballasts are necessary for most ambient and task lighting.

General lighting intensities are designed to moderate-to-low levels and local task lighting is used to provide sufficient intensities on work areas or areas of interest. Task lighting is based on low-wattage triphosphor fluorescent lamps with multilevel user controls. Value is placed on lighting as a leverage of salaries and benefits—maximizing user satisfaction and productivity over the long term.

13.5 DAYLIGHTING STRATEGIES

Daylighting and its management can, if not properly understood and planned, wreak havoc. Exterior view alone is a big benefit to workers—a connection with nature, a comprehension of the weather, an ability to judge time, and an eye relaxer (offering distant-view breaks from the close-vision work associated with reading hardcopy and computer screens). Unfettered daylight/view, however, creates significant brightness balancing (harsh contrasts) and glare issues. The challenge to a design team, then, is to:

1. *Control Fenestration (Daylight Openings).* This is done by analyzing all building elevations over various seasonal and time-of-day situations. A combination of architectural treatments interior or exterior, both fixed or adjustable, and glazing which has a transmittance less than 30% is typically required in most areas of the United States. Controlling fenestration will limit glare issues, but won't eliminate them.

2. *Integrate the Daylighting with the Electric Lighting.* Develop a strategy and implement an approach that takes advantage of daylight by either dimming up/down or selectively switching off/on electric lights in the zone of daylighting. Integrating daylighting and electric lighting will yield a more sustainably appropriate approach to lighting.

3. *Balance Luminances.* Perimeter and/or skylit zones that are excessively bright compared to the rest of a space will cause serious user adaptation issues and glare. This requires careful attenuation of daylight while simultaneously addressing luminances (surface brightnesses) in nondaylit zones. Luminance balancing will mitigate harsh contrasts.

13.5.1 Controlling Fenestration

Fenestration (any daylight opening, such as windows, clerestories, skylights, roof monitors, and the like), while allowing exterior view, is first and foremost a lighting liability—no different than deciding to use 1,000-watt lamps! Unless well-shielded and/or significantly dimmed or filtered, users will come to hate it. Taped-up or tacked-on cardboard shields, dark gray acetate sheets, or blinds are typical user reactions. Headaches and watery eyes are typical reactions. Although building orientation can reduce the severity of the problem, fenestration issues occur on most any building face.

North exposures are considered least problematic; however, partly cloudy and lightly overcast skies reflect sufficient light back to the users along these exposures that harsh contrasts and glare will exist without proper attention to glazing and glazing treatment(s). Here are considerations for fenestration development:

1. *Limit Fenestration Orientation.* In new construction (in the United States), limit fenestration to north exposures, northeast exposures, and northwest exposures. This may include unique window configurations on west and east faces which help orient/limit views to the northwest and northeast, respectively.

2. *Limit Sky View.* Use architectural techniques/treatments (overhangs, horizontal baffling, awnings, and the like) to limit the view of the upper sky. A view of just the horizon means less chance of viewing the sun or solar disc and less chance of viewing large expanses of bright clouds, both of which contribute to glare and harsh contrasts.

Less permanent solutions include interior blinds and shades. Except in private offices, blinds should be horizontal, since vertical blinds always have a "bad" orientation for some users unless fully closed. Shades need to be low transmittance and roll-down from above to limit sky view and brightness. Image-preserving shades with transmittances as low as 3% are available. This allows users some exterior view, but to be successful, these must be of dark value (dark gray, dark brown, or black).

3. *Limit Glazing Transmission.* Double- and triple-glazed systems offer many opportunities for light transmission reduction. Tints alone can reduce transmission to as low as 25%. Green and green–blue tints are popular, as these seem to offer the users a more "real" or "true" viewing experience of landscapes. Recognize, however, that a greenish hue is cast onto interior surface finishes. Bronze tints, while adding a pink hue, minimize the sense of gloom more common with gray tints.

Coatings, when combined with tints, can reduce transmission to as low as 6%. Coatings will make the glazing mirror-like as viewed from the exterior during the day and as viewed from the interior during dark-day conditions or at dusk and night.

Frit coatings (ceramic coatings baked onto the glass) can be used alone or in combination with tints and coatings. Frit is available in several colors and shades of gray (white to black). Grayer/darker frits are preferable as these limit light transmission. White or light frits are so translucent that they can create glare situations on bright days or if they are exposed to direct sunlight.

After-market films are available for existing situations, which are problematic. Installation is crucial to success—bubbled applications and/or incorrectly trimmed applications are very disturbing visually.

One pitfall to avoid: sandblasted glass or translucent films on glass or in the lamination. This situation not only increases the glare potential from glazing systems, but does not allow view—which is the only real reason workers want windows. Another pitfall to avoid: using high transmittance glazing systems without any shade control on the premise that this is healthy for the worker and sustainable for the earth. Neither can be substantiated. While seasonal affective disorder (SAD) is somewhat common in the northern latitudes, research indicates this can be mitigated with several hundred to 1,000 footcandles of light exposure over a relatively short time (perhaps half an hour to an hour or so each day). An all-day dose of these intensities will yield other health problems—mainly that folks won't be able to stand the glare and/or see the work which they are to be performing! Sustainability has more to do with carefully balancing the needs of the workplace with the materials and resources used in the workplace to accommodate work. So, high-transmittance glazing which may result in overheating of the space (and thus require greater cooling capacity) isn't saving energy resources. Similarly, high-transmittance glazing which yields glare and is ultimately shaded with heavy shading media for much of the day, and which precludes views, results in a waste of resources needed to make such glazing in the first place (if workers can't appreciate the view in such situations anyway, why put in windows?).

Daylighting studies should be part of any project scope. Understanding the extent of available daylight in any given region and then knowing how the particular site and evolving architectural design will influence daylight in the building is paramount to designing fenestration openings, setting their orientations, and selecting glazing and glazing treatments—this is not a decorative art, but is a science requiring analyses and/or mockups. On sunny days in summer, it is quite possible to have in excess of 7,000 footcandles (70,000 lux) on a building's roof in suburban settings. East, south, and west vertical exposures may experience several thousand footcandles. North vertical exposures may experience hundreds of footcandles. On cloudy days throughout the year, several thousand footcandles are likely on the roof, with many hundreds of footcandles on all vertical exposures during some of the day. For interior work settings where an overall ambient light intensity of 30 footcandles (300 lux) is considered appropriate for workers viewing most electronic and paper tasks, these daylight intensities will be serious detriments to comfortable, productive work.

13.6 INTERIOR LIGHTING

Interior electric lighting design must account for any daylighting and needs to consist of several techniques or layers. Layering lighting is paramount to achieving quality lighting and energy efficiency. There are three layers of lighting in a successful installation: ambient (or general) lighting, task lighting, and fill lighting. No two layers can meet all lighting needs in a work setting. Deleting or short-changing one lighting layer will diminish the success of the installation. As a general rule on any lighting layer, using fewer, but higher-wattage sources is courting trouble. These layers of lighting cannot be successful without appropriate room surface reflectances—89% ceiling reflectance; 30–60% architectural wall reflectances; 30–50% furniture partition reflectances; and 10% or greater floor reflectances.

13.6.1 Ambient Lighting

Ambient lighting, whether direct, indirect, or indirect/direct is intended to provide a low-to-moderate intensity of light uniformly throughout a space. This allows for great flexibility in rearranging furnishings and for general circulation throughout open plan work areas, while keeping overall connected load at a reasonable level. Even the best of ambient lighting systems, however, may not provide for a sea-change in workstation approach. Low or no partitions today, but moderate or tall partitions tomorrow (and vice versa) can wreak havoc on the ambient lighting. So, an understanding of workplace culture and/or the likelihood and number of changes over the anticipated life of the facility or lighting system are necessary to establish the need or degree of flexibility of an ambient lighting system. This should not influence what kind of ambient lighting technique is used (direct, indirect, or indirect/direct), but should influence the kind of hardware and hardware connections considered (modular, hardwired, plug-and-play, electronically addressable, and so on).

Ambient lighting techniques abound. The more traditional approach of recessed, direct lighting has been long considered the "value" approach to lighting. It is—if initial cost is the extent of the definition of value. Indirect and indirect/direct (mostly up light with some downlight) lighting has, because of increasing popularity, become more affordable. The popularity of these systems is also driven by their ability to provide open, spacious, bright-looking environments while simultaneously providing good to excellent viewing conditions for computer screens—no matter where the screen is located and/or how it is oriented.

High-lumen package light sources should be avoided in most lower-ceiling office situations. Hence, where ceiling heights are less than 10 ft, avoid the 40- and 50-watt biaxial (sometimes called large compact) fluorescent lamps and avoid the newer high output T5 fluorescent lamps. In general, HID (high-intensity discharge) lamps (including high-pressure sodium and metal halide) of wattages greater than 50 watts should be avoided for ambient lighting regardless of ceiling height.

13.6.2 Task Lighting

Local task lighting, when combined with an appropriate ambient system, will provide the user with sufficient light to read hardcopy, but without glare and heat. At least three factors must be considered: extent of intensive task work area; extent of overhead storage or shelving, if any; and extent and height of partitions, if any. The worksurface—the desktop surface(s) on which reading hardcopy is likely—should be fitted with task lighting. Where overhead storage and/or shelves are used above the worksurface, task lighting should be integrated into the bottom of the storage or shelving to fill in shadow areas and to offer a softly bright background against which to view negative contrast computer screens (the screen has a bright background with darker text and/or graphics). If no overhead storage or shelving is anticipated, then freestanding or rail-mount task lighting is appropriate (some furniture systems exhibit a rail that is mounted a foot or so above the worksurface onto which a variety of accessories can be affixed). Task lights need to be ubiquitous. If some of the worksurface is intended for paper storage and only occasional reference and/or if partitions are nonexistent, however, then ambient lighting without the supplement of task lighting in those areas is likely sufficient. Task lights should be locally switched so that the user has control over their use. Task lights should also be motion-sensor controlled to limit energy use and extend lamp life (although there is a point of diminishing returns if users are coming/going quite frequently—switching fluorescent lamps on/off every 15–30 min will result in excessive wear and premature lamp failure).

Halogen task lights should be avoided—these use short-lived lamps and operate at very high temperatures, requiring safety precautions because they can result in skin burns and/or a hot working environment. Halogen task lights are also typically glare (either on direct view or when reflecting from glossy paper and/or worksurface materials).

Attempting to provide all lighting from the ambient lighting system will, more than likely, lead to user complaints. This is also very likely to lead to an increase in power consumption both to operate the users' own task lights (which will be brought to work to fill in the heavy shadows on worksurfaces) and to account for an increased cooling load from these worker-supplied task lights which are typically of high wattage incandescent sources—not to mention the visual clutter and potential hazards (e.g., are they all UL-listed and labeled for such an application?) of these varied-style, varied-age, varied-vendor user-supplied task lights.

Task lighting is one of the few products that are offered by light fixture (luminaire) vendors as well as furniture systems vendors. Typically, the luminaire vendors have a better product for visual ergonomics—the optics on the better models are well-designed to limit glare and distribute light uniformly across the worksurface and onto the tack-up surface (where partitions exist). Luminaire manufacturers also have a broader line of task light sizes and ballast and lamp combinations to meet many situations. Furniture systems vendors offer lights which are mostly afterthoughts, or so-called value-adds with little to offer in lighting ergonomics.

To get the furniture order, cheap task lights are offered with little or no optical control, old-technology lamping and ballasting with or without any options on lamping and ballasting. Task lights with current-technology lamps of wattages and colors to match ambient lighting lamps and with low output ballasts or multilevel switching ballasts are considered specials or deluxe models by many furniture vendors and result in exorbitant lead-times and/or unrealistic pricing. Pressure from facilities managers responsible for large furniture systems' installations can help change these practices. Review a variety of options when procuring task lighting—from furniture vendors' OEM offerings; to furniture vendors' installing luminaire vendors' task lights purchased by facility management (FM) or an electrical contractor; to facility electricians or electrical contractors purchasing/installing task lights from luminaire vendors. Here's a checklist of key items or issues regarding task lights:

- The task light should provide 25–35 fc (average, maintained) over the worksurface
- Maximum light at any given point from the task light should not exceed 50 fc
- Task light intensity gradient should increase from front to back
- A local switch should be integral to each task light
- If task lights cannot be procured meeting above items, then consider task lights with low–high switching or low–medium–high switching
- Use motion sensors to control task lights at individual workstations
- Total connected load for all task lighting should be 0.25 w/ft^2 or less
- Use electronic ballasts with high power factors and low harmonic distortions
- For small, freestanding task lights, use lower wattage compact fluorescent lamps (e.g., 13 watts)
- For linear task lights use rapid start or programmed-start, triphosphor or triphosphor deluxe (which are most efficient) fluorescent lamps of T5 or T8 size
- Avoid striplights or unlensed white-box task lights as these are quite glary and cause serious hot streaks on the worksurface
- Avoid linear task lights with an aperture less than 2.5 in. in width and/or which use a standard stippled lens, as these are also quite glary and cause serious hot streaks on the worksurface

13.6.3 Fill Lighting

Fill lighting should be used to fill the darker voids in a lighting situation and/or as accenting to provide visual attention or attraction to artwork, special wall finishes/features, or colored surfaces/objects. This visual attraction is beneficial physiologically by offering eye rests—establishing visually distant focal points onto which workers viewing hardcopy and computer screens can gaze from time to time. Further, this accenting can help soften or reduce any harsh contrasts caused by darker zones. Accent lighting works best when relatively soft.

13.7 EXTERIOR LIGHTING

Safety and security are the primary drivers for exterior lighting. Aesthetics and/or marketing (by virtue of a lighted structure or signage) are perhaps secondary drivers for exterior lighting. Safety relates to the users' abilities to maneuver on site to/from the entry while avoiding obstacles (undulating sidewalks, objects on the sidewalk, and the like) and incursions with vehicular traffic. Security relates to the users' sense of threat (or lack thereof) from crime.

Safety issues can be addressed with horizontal and vertical illuminance and uniformity criteria. With regard to parking lot lighting, however, the latest Illuminating Engineering Society of North America (IESNA) guidelines are woefully lacking in guidance. Consider reviewing and adopting criteria from the IESNA RP20–1985 or the eighth edition IESNA Handbook and Reference Volume (1993). Further, safety issues can be enhanced with whiter light. Studies over the past 7 years indicate that whiter light sources (incandescent, mercury vapor, and metal halide) offer better visibility under very low light levels than their yellower counterparts (high-pressure and low-pressure sodium—HPS and LPS, respectively). Indeed, the research suggests that HPS and LPS are perhaps only half as efficient as, say, metal halide in very low-light (exterior) situations.

Security issues are also addressed with horizontal and vertical illuminance and uniformity criteria. Vertical illuminance is crucial to establishing lighting that evokes a sense of security. With vertical illuminance and uniformity, figures and facial features can be recognized from some distance. Color rendering is a key criterion with security lighting. Perpetrators operating in the monochromatic environments of HPS and LPS lighting have less chance of being identified—clothing colors and skin coloration are altered significantly by HPS and LPS sources. Higher color rendering ceramic metal halide lamps or compact fluorescent lamps offer significant improvement in color detection and determination. Finally, silhouetting techniques help limit the "cover of darkness" situation perpetrators prefer. Highlighting selective building features and/or landscaping, even on a large campus setting, will help provide a softly lighted backdrop against which figures will be silhouetted.

Light pollution and light trespass are common difficulties with exterior lighting. First and foremost, however, is to recall the primary reason for exterior lighting—the safety and security of employees. Nevertheless, prudent use of exterior lighting should limit the impact on light pollution and minimize light trespass. Light pollution relates to spill light up into the atmosphere which, when the environment is relatively dirty and/or moisture-laden, results in an overall sky glow. This sky glow arguably limits night sky viewing. Light trespass relates to light falling onto neighboring properties and/or buildings which may interfere with the adjoining properties'/buildings' activities. Here are some tips for implementing lighting more sensitive to light pollution and light trespass issues:

- Don't exceed IESNA guidelines if using fluorescent or metal halide lamps
- Use low wattage sources (100 watts or lower) for uplighting
- Use shoebox/cutoff optical packages for primary postlights
- Use relatively short poles (35 ft or less) with relatively low wattage sources (250 watts or lower) for primary postlights
- Use decorative pedestrian-scale postlights which direct nearly all light outward/downward (about 80% down and perhaps 20% up)

- Use short poles (12 ft or less) with low wattage sources (100 watts or lower) for decorative post-lights
- Use lower reflectance ground materials (asphalt or colored concrete versus typical white concrete)
- Use well-controlled and shielded accents for façade and/or landscape highlighting
- Light façade and/or landscape judiciously (don't floodlight the whole façade or an entire landscape scene, but rather accent key façade features and/or landscaping (such as ornamental plantings)
- Avoid lighting glass facades since nearly all light is simply reflected skyward (while the glass remains dark in appearance)
- Keep all adjustable lights appropriately aimed, lamped, and shielded
- Light only manicured landscaped or property areas (leaving natural areas as darkened preserves)
- Don't aim lights outward from property toward other properties

13.8 LIGHTING TECHNOLOGIES TODAY

For best efficiencies, the following general categories of technologies should be considered for new, retrofit, or replacement projects

Electronic Ballasts. For fluorescent and HID (high intensity discharge) lighting, electronic ballasts offer best efficiencies with no audible hum and no flicker effects. Many ballast manufacturers make tray assemblies when retrofitting the smaller electronic ballasts into platforms on existing luminaires that were originally intended for the larger electromagnetic ballasts. An advisory: when retrofitting, match ballast output performance with lamp input requirements.

Triphosphor Fluorescent Lamps. The deluxe triphosphor lamps are the most efficacious white-light sources available. These lamps are available in T2, T5, T8, and compact sizes. The triple-tube compact lamp is available in five wattages and offers a wide range of applications. An amalgam in these lamps results in optimal performance over a wide range of temperatures, including exterior applications, such as bollards and steplights. An advisory: the amalgam in the triple-tube compact lamps will result in a warm-up period. Hence, while providing instant light when switched on, full output may take 10–15 min even in interior applications. Further, for exterior applications, the ballast must also be rated for low-temperature starting conditions (e.g., 0°F or –20°F). An additional advisory: fluorescent dimming is least problematic with linear lamps, somewhat problematic with the larger biaxial lamps, and quite problematic with small compact fluorescent lamps. All fluorescent lamps on dimmers are to be "seasoned" or burned-in at 100% full output for 100 h prior to initiating any dimming. Not following such practice yields reduced lamp life and may result in short ballast life. This is most serious an issue with small compact fluorescent lamps and is not advised in large applications until lamp and ballast manufacturers perfect the dimming process.

Ceramic Metal Halide. HID lamps are notorious for poor color, too much light, and lots of hum. Ceramic metal halide lamps eliminate poor color and hum as obstacles. Even the 39-watt lamps, however, are quite intense for most interior, low-ceiling commercial applications, except retail. In atria or other multistory interior spaces, however, these lamps offer excellent light intensities, with incandescent-like color and rated lamp life of between 10,000 and 15,000 h.

In exterior applications, these lamps offer excellent high-color rendering to landscapes and architectural features and skin tones are near true.

Halogen Infrared. Halogen infrared or HIR technology has breathed new life into incandescent lighting. These lamps, offered in PAR envelopes and in MR16 low voltage configurations, are more efficacious and longer-lived than their tungsten and halogen counterparts. For nearly half the wattage, light output is as good as or better than standard incandescent lamps. Depending on wattage and voltage, lamp life is 50–100% greater than standard incandescent counterparts. An advisory: these lamps are hot-shock sensitive; that is, if aimed, adjusted, or lamped while electrified, the vibration may short the filament.

13.9 LIGHTING TECHNOLOGIES TOMORROW

Some emerging lighting technologies are worthy of note. These offer great promise in energy efficiency and/or lamp life without sacrificing ergonomic issues. Some of these are in early-release and are therefore expensive and/or difficult to obtain and/or maintain. Others have been around for some time, but have failed to enter main markets. Within the next 5 years, however, some of these may become mainstream technologies.

Electrochromic Glazing. Electrochromic glazing is glazing in which an inner layer of material is formulated and electrified so that it can be manually or photoelectrically switched from transparent to translucent. Primarily a privacy feature, however, with opaque-to-translucent switching, this could be more practical for presentation rooms or other areas requiring blackout conditions from time to time.

Photochromic Glazing. Photochromic glazing automatically responds to light intensity and dims transparency to a predetermined level. This could be useful on north, east, and west elevations where periodic daylight intensities are too high.

Addressable Lights. Lights which are addressable from a computer terminal or by a handheld device are called addressable lights. Addressing includes on/off functions and dimming—primarily for fluorescent lighting. This offers great potential in flexible office areas where groups and/or functional requirements change over time without rewiring. Further, this allows for local daylight integration without rewiring.

LEDs (Light Emitting Diodes). LEDS are small light sources with no mechanical/physical filament, resulting in extraordinary lamp life (up to 100,000 h). Light output is quite low, as is power input. LEDs show promise for marker lights (path lighting, steplights), and with higher output versions (now being studied), could influence task and fill lighting.

Dimmable Metal Halide. Dimmable metal halides are now being studied and some early-releases are available. Very poor color shift issues and limits on dimming range are serious problems.

13.10 RETROFITS AND SIMPLE REPLACEMENTS

A great deal was learned in the 1980s and 1990s when retrofitting fluorescent and incandescent lighting systems. Here are some key points to consider when performing retrofits:

- For general lighting applications using fluorescent lamps, replace lamps with identical light output versions (equivalent lumens)[1] of same envelope size and of equal or better color characteristics
- For general lighting applications using R or PAR lamps, replace lamps with identical candlepower versions of same envelope size and of equal or better color characteristics (center beam candlepower and beam spread should be within 10% of original lamping for a similar look)[1]

- Use equal- or lower-wattage lamps
- Screw-in fluorescent lamps are only a short-term alternative to screw-in incandescent lamps and, to date, have not been shown capable of providing equivalent intensity and/or color regardless of light output
- Electronic ballasts should be the norm

For lamp replacement, either spot or group relamping, consider that the original design was based on a host of criteria. Lamps and luminaires in combination were used to best approach or meet those criteria. Many times, luminaire reflector designs and/or socket positions are optimized for specific lamp envelope configurations. To maintain optimal performance, lamps should be replaced with identical lamps; that is, lamps of same wattage, envelope shape, candlepower, lumen output, color temperature, color rendering, and voltage (preferably from the same vendor). Off-brands or "superior-life" or "full-spectrum" lamps or lamps of other shapes and styles will not best use resources to maintain lighting criteria and should be avoided.

13.11 LIGHTING ANALYSES

Lighting solutions can only be as good as the analyses surrounding their development and selection. A priority should be given to establishing appropriate lighting criteria and agreeing to these criteria. Based on these criteria, then, simulations can be done to assess various lighting hardware for compliance. These simulations should be performed using commercially available software. Simulation techniques should account for objects such as workstation partitions for interior environments and should include a host of lighting system and environmental parameters such as:

- LLD (lamp lumen depreciation)
- LDD (luminaire dirt depreciation)
- BF (ballast factor)
- RSDD (room surface dirt depreciation for interiors)

All assumptions made regarding these items should be cited in studies, reports, or presentations made to facility management. For example, an LLD of 0.95 for triphosphor fluorescent lamps is based on an assumption that group relamping will occur at about 70% of rated life. If group relamping never takes place, and lamps are only replaced as they burn out, then an LLD of 0.9 is appropriate (this means that the quantity of proposed lighting hardware will increase by about 5% to accommodate for the lower light output anticipated due to lamps remaining in place until burnout).

Qualified personnel for lighting analyses should have, at the very least, lighting certification credentials from NCQLP (National Council on Qualifications for the Lighting Professions—see http://www.ncqlp.org). This is represented by "LC," and signifies that the individual has a measure of educational background and experience and has passed the half-day certification examination. Additionally, look for persons with membership in the Illuminating Engineering Society of North America (IES) and in the International Association of Lighting Designers (IALD—see http://www.iald.org).

ENDNOTE

1. If more light is desired and/or is part of the rationale for the replacement, then lumen output and/or candlepower should be greater than the original lamp—confirm with original luminaire manufacturer that replacement/retrofit won't void any UL listings.

CHAPTER 14
ERGONOMICS AND WORKPLACES

Timothy Springer
President
Hero, Inc.
E-mail: hero_inc@ameritech.net

14.1 INTRODUCTION

The goal of ergonomics is to support people in the way they work so that they are safe, comfortable, and productive. The primary focus of this chapter is on the tools and technology people use (computer workplaces). Ergonomics requires consideration of the things people use and the ways and places they use them to accomplish their job duties. Effective ergonomics focuses on both the individual and organizational level. At the individual level, ergonomics gives attention to details of equipment and workplace design, such as worksurface height and seat dimensions (so-called *micro*ergonomic considerations, see Figure 14.1). At the organizational level, attention focuses on global issues such as training, job organization, locus of control, and stress, as well as interactions of various elements including physical, psycho-social, and organizational considerations (*macro*-ergonomic considerations).

What follows is a discussion of the ergonomic considerations for workplaces in which people use computers. The discussion proceeds from the details of individual workplaces to the more global, macroergonomic issues. Where relevant, research support is provided for recommendations. In certain instances, these recommendations mirror design guidelines, but are couched in ergonomic terms.

14.2 ERGONOMICS OF WORKSPACES

Most office workers use a variety of tools in a variety of places. Chief among their tools is the computer. Consequently, the main thing their workplace should do is support computer interaction. But office workers also use a lot of paper. So, workstations must also support these tasks and materials. Certain elements of workspace design are driven by the characteristics of the technology—computer size, for example. The primary function of workspace design is to support the needs of the people who use the technology.

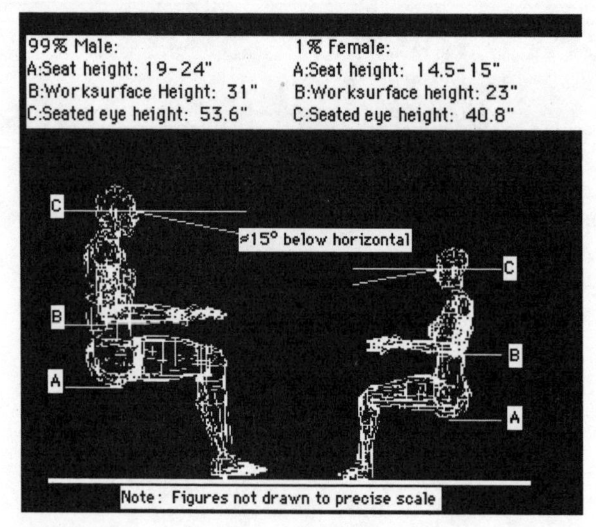

FIGURE 14.1 Diagrammatic Representation of Key Ergonomic Dimensions.

14.2.1 General Workstation Ergonomics

Much of ergonomic concern in workstation design is shaped by the characteristics of people. Human dimension data (anthropometrics) such as size and reach help determine dimensions and placement of elements in a workplace. Generally, people work in a roughly semi-circular space directly in front of them. This primary work area is defined by what people can easily reach (see Figure 14.2).

This type of information is important in determining where to place critical materials, tools, and technology in the workplace (see Figure 14.3). Figure 14.3 illustrates the shape of the worksurface and how the placement of the tools and equipment can interact to provide efficient use of space and ease of use for the worker.

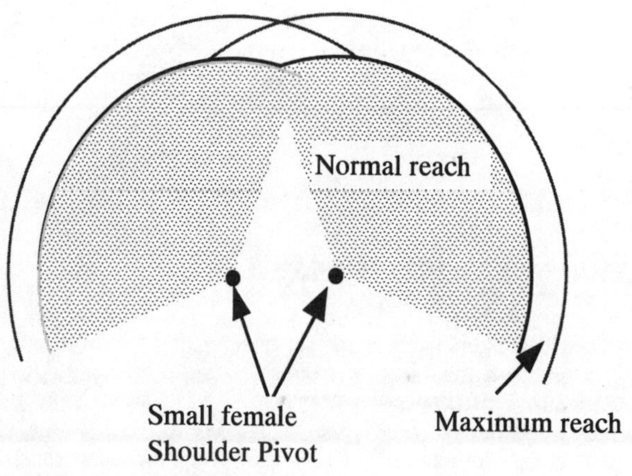

FIGURE 14.2 Overhead View of Horizontal Reach Envelope.

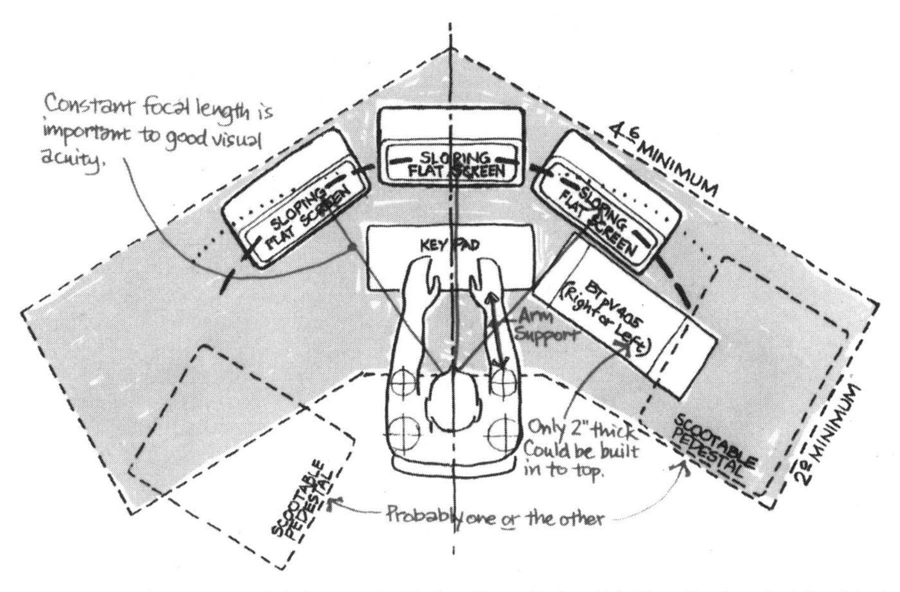

FIGURE 14.3 Fitting the Workplace to the Worker Places Tools within Easy Reach and at Consistent Viewing Distance. (M. D. Tatum, 1997.)

Altering the shape and configuration of the desk to better fit the worker and their work behavior allows exploration of alternative configurations of groups of desks and workplaces (see Figure 14.4).

It is important to remember that space exists three dimensionally, so comparable information for vertical reach envelopes also should be considered (see Figure 14.5).

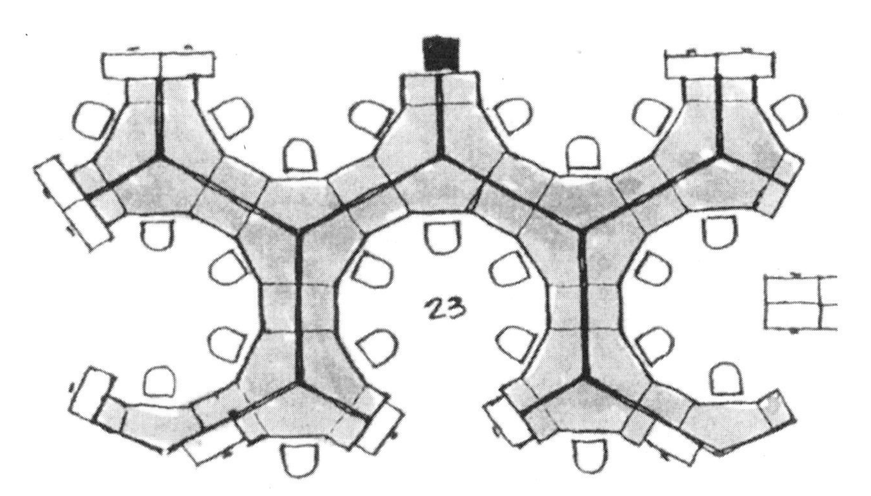

FIGURE 14.4 Shaping the Workplace to Fit the Worker Allows Changes in Configurations of Groups of Workplaces. (M. D. Tatum, 1997.)

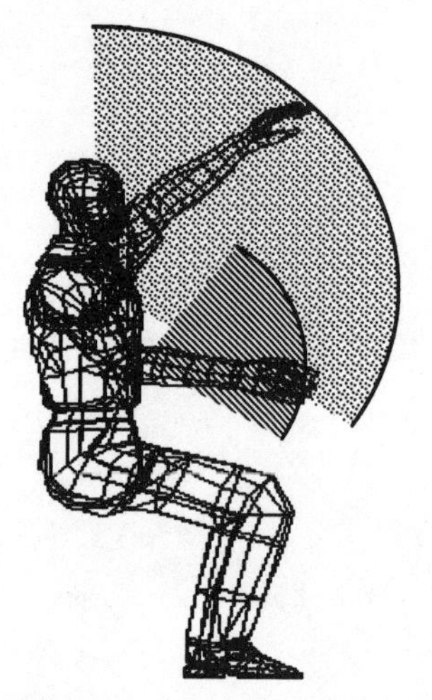

FIGURE 14.5 Vertical Reach Envelopes.

14.2.2 Technology

Ergonomics tends to deal with technology in terms of the nature of and need for control and display. In the case of computer technology, these two elements translate into control requirements with keyboard and cursor control devices (touchpad/mouse/trackball) and display monitor interaction.

Keyboards, Mice, Touchpads, and Trackballs. The keyboard, mouse, touchpad, or trackball are the control devices provided or used with most computers to manage interaction with the computer. These devices are the components of the computer that require physical interaction. Consequently, the location and support of these devices is the primary ergonomic consideration in workstation design.

Keyboard Use. How the keyboard is used is an important consideration in determining the configuration of the workstation and accommodating the support requirements of keyboard use. Unlike using a typewriter, where the majority of workers were "touch typists" engaged in sustained, rapid, two-handed "typing" on the keyboard, the majority of computer users are not touch typists. Many office workers engage in intermittent keying, use special function keys more frequently and compose, read, edit, and think in front of the computer. Consequently, for these people, there is a tendency to rest the hands and wrists during pauses in keyboard use. This "resting" behavior must be properly accommodated since pressure against the underside of the wrists is one of the contributing factors in development of cumulative trauma disorders. Ways in which this requirement can be accommodated will be discussed in the following sections.

Cursor Control Device. The human computer interface, in most instances, involves controlling the cursor on the display screen. A variety of techniques have evolved to serve this purpose. Most common is the mouse; however, trackballs, touchpads, and pointing devices are also used. A trackball is essentially a mouse turned upside down where the ball is operated by the fingers. Touchpads, as the name implies, go one step further than the trackball and provide a sensitive area that is activated by the user's finger. In the case of the mouse, it must be close at hand and sufficient room must be provided to allow the user to maneuver. The mouse requires desktop space to move about, whereas the trackball can be placed in one position and requires less hand and arm movement and less desktop "real estate." Touchpads are most often found as part of mobile (laptop) computers, but they are also available in certain configurations as part of the desktop computer keyboard or as an optional peripheral. Task activity requires the operator to move frequently and easily from keyboard use to cursor control use and occasionally simultaneous use of both.

There are several components to the proper support of a person using a keyboard and mouse:

- Support surface height
- Support surface and keyboard angle
- Horizontal field

Support Surface Height. The height of the support surface is critical in determining healthy working postures at computer workstations. Support surface height directly influences hand, wrist, and forearm positions, angles, and strain. The height of the keyboard is related directly to the seated task posture. For example, an erect sitting posture with a flat seat pan (0° angle) results in a requirement of lower keyboard height than a more reclined posture (Grandjean et al., 1984, 1983; Weber et al., 1984). A chair with a forward seat angle resulting in a more erect posture requires a higher seat height and higher keyboard support height (Mandel, 1984, 1985).

FIGURE 14.6 Wrist "In-Line" Position Provides Greatest Strength and Least Stress.

Worksurface support height should allow the computer user to keep an "in-line" wrist position (see Figure 14.6).

This posture is a function of the seat height, chair armrests (if provided), and the keyboard thickness. The forearms should be approximately parallel to the floor; however, the angle at the elbow will generally be greater than 90° (Chaffin, 1991). Support of the forearms in the area between the elbow point and 2–3 cm behind the wrist may allow the forearms to be raised slightly. In any case, one should avoid placing the support surface at a height requiring significant deviation of the wrists and hands from the in-line position (see Appendix A, "Rules of Thumb").

Display Height. The human eye evolved to function in a very specific way related to survival. We look up to see far off (4′ or more—"the lions are coming!") and we look down to see things close at hand (8–36″—"the fruit is on the ground," "don't step on that snake"). "When sitting at rest, the eye focuses 15–20° below horizontal (0°), as shown in Figure 14.7. This is one of the reasons people often hold reading material in that general area of the visual field. Our visual survival characteristic presents certain challenges in placing a visual display in the middle distance (16–42″) where most computer displays usually are located.

FIGURE 14.7 Sight Angles.

Another factor increasing as a consideration in determining display placement is the visual acuity of the worker. As we age, our ability to change focal distance quickly begins to decline. Many people past the age of forty require glasses to aid in reading, or increasingly common is the use of bifocal and trifocal eyeglasses. When Benjamin Franklin invented the half-moon reading glasses that became the basis for bifocals, he was intent on reading paper—primarily newspapers held comfortably at arm's length and in the 15–20° "resting zone" (see Figure 14.8).

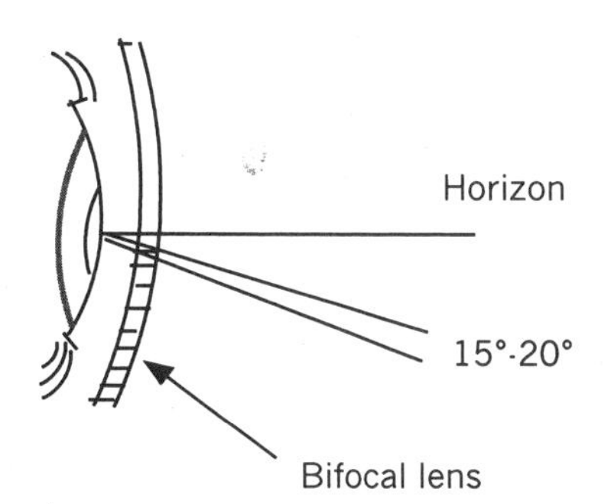

FIGURE 14.8 Bifocals.

Using these types of eyeglasses with computer displays that are at or above the typical worksurface height requires the worker to tilt their head backward and bring their trunk forward in order to bring the image into focus in the lower third of their eyeglass lens. They then reverse this action to read paper on the desktop, to see the keyboard, or to change their view of other items or materials. This change in posture and position can happen many hundreds of times a day depending on the nature of the task. The "bifocal bob" is the source of much physical discomfort and strain among the population of bifocal wearers (see Figure 14.9).

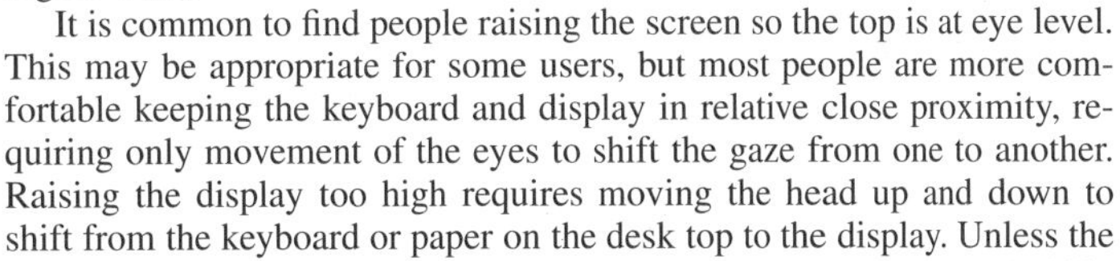

It is common to find people raising the screen so the top is at eye level. This may be appropriate for some users, but most people are more comfortable keeping the keyboard and display in relative close proximity, requiring only movement of the eyes to shift the gaze from one to another. Raising the display too high requires moving the head up and down to shift from the keyboard or paper on the desk top to the display. Unless the person is a touch typist, eye shifts between the keyboard and display screen can average 200 shifts per hour for normal "programming" use (Price, 1992). Most office workers spend from 2 to 4 h a day focused on the computer; this could approach 800 head movements in a day. This can lead to extreme neck and shoulder pain, headaches, and eventually to disabling conditions (Collins et al., 1990).

If one optimizes the worksurface height for the principal tasks (i.e., keyboard use), changes to display height can be accommodated in a variety of ways (see accessories).

Support Surface and Keyboard Angle. Many keyboards are designed with a forward slope; that is, the rear edge is higher than the front edge. Many computer keyboards provide some device for raising the back edge of the unit to tilt the keyboard toward the user. Evidence is emerging to suggest that a lower slope is preferred for certain task postures. Some researchers are suggesting a support surface with a negative slope (i.e., angled away from the user) (Hedge and Powers, 1991). In practice, a flat keyboard angle is generally recommended. Personal preference coupled with informed decision making will enable people to make adjustments to this.

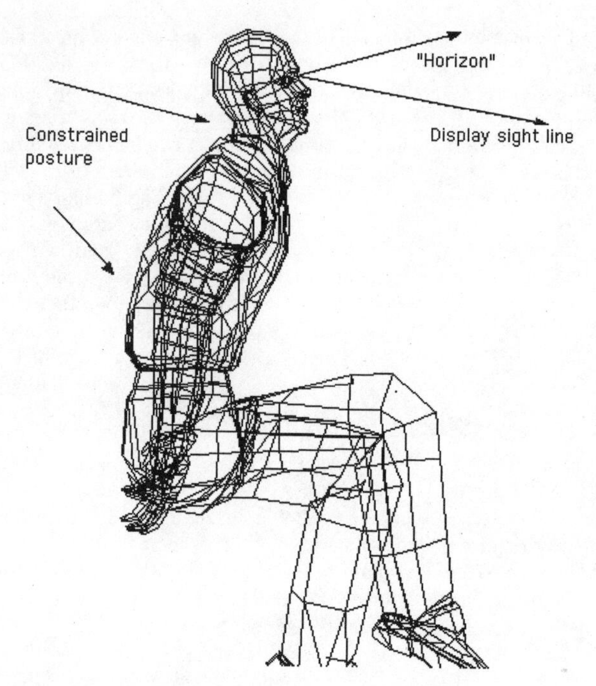

FIGURE 14.9 The Bifocal "Bob."

Horizontal Field.

Keyboard and Mouse Space. Increased power, speed, and sophistication of computer systems, Graphic User Interfaces (GUIs), and multitasking have increased the need for sufficiently large worksurfaces to accommodate a larger keyboard, a mouse and its area of use, and paper and reference materials all in the same horizontal plane. This requirement is based on several observed behaviors:

- Use of keyboard and mouse in rapid succession or simultaneously
- Reference to paper materials while using the keyboard
- Larger displays and increasing screen "desktop real estate," which translates into increased space needed to maneuver the mouse.

Consequently, a broad worksurface equivalent to user's reach envelopes (48–66″) is warranted when using large monitors with a mouse or trackball.

Display Viewing Distance. Research suggests a viewing distance of 24–36″ (Mean = 30″) for visual display terminals (Grandjean, 1983). The size of the display, resolution, and character size will influence appropriate viewing distance within this range.

Paper Handling. Observation of work behavior uncovers many varieties of work habits; however, one constant emerges. Knowledge workers use paper in conjunction with the computer. To accommodate this, the worksurface needs to be both broad enough and deep enough to accommodate the monitor, keyboard, and mouse, and at least one open reference binder. This paper "work in process" must be readily accessible and able to be "stored" on the worksurface.

Research by Lansdale (1991) supports the notion of handling and storing paper work in process in the horizontal plane. This study supports the need for large worksurfaces as well as providing additional support for "pile filing." Pile filing is the natural tendency for people to arrange information in stacks and piles. Research and experience has shown that people are remarkably adept at "mental mapping" using "visio-spatial" cues (e.g., "I know it is in the second stack from the left, and I put it on top of a bright orange piece of paper"). Pile filing behavior is not well accommodated by traditional file folders and file drawers. It can be accommodated in a variety of ways. One approach is a series of shelves raised above and arrayed about the primary worksurface that allows the user to see and reach "hot files" while keeping the main part of the worksurface clear. If this approach is used, it is important to remember the vertical reach envelopes (Figure 14.5) and place the most important materials within easy reach. Another approach is a cabinet the size of a two drawer lateral file with shelves instead of drawers. The shelves pull out in a cascade effect where the lowest shelf is fully extended and subsequent shelves extend only part of the way. This arrangement consolidates stacks of papers, yet makes them visually accessible.

14.3 ERGONOMICS OF SEATING

Computer use, by its nature, requires people to sit to use the technology most effectively. The chair has become the foundation for most jobs and nearly all office jobs. Consequently, the chair is a critical ergonomic element of the workplace. A good chair should do three things well: provide a stable base, support the person anatomically, and support task postures.

14.3.1 Stability/Mobility

The challenge of providing a mobile seating platform that is very stable has been addressed by nearly every office chair on the market. The most common way in which this goal is accomplished is by providing five-legged bases with casters. This feature allows the chair to be rolled over a cord or pencil and still have at least three legs touching the floor, thus making it more difficult to tip these chairs than those with fewer legs. It also provides a base that is resistant to tipping when the operator shifts their weight around. The stability of such five-legged, "star-based" chairs is dependent on the length of the leg, size of the caster, and the height of the seat above the base. These factors affect the center of gravity and thus determine the stability of the platform.

14.3.2 Anatomical Support

This requirement is individually specific. Since people vary widely in size and shape, chairs must vary widely as well. Most chairs are shaped to support the greater "middle-size" portion of the general population. They feature curved front edges to avoid cutting off circulation in the legs. Backrests are also curved and padded to provide support for the curvature of the spine. However, many chairs do not adjust sufficiently to support the full range of users. Short people are often forced to sit with their legs not touching the floor. Large people do not fit comfortably on many chairs because the seat is too narrow or shallow to support them. Very tall people often find chairs require them to sit with their hips below their knees resulting in blood pooling and the tingling sensation referred to as a leg or foot "going to sleep."

Because of the wide variety of sizes of people and the physical limits of the technology used to make adjustments, it is almost impossible for one chair to adequately fit the entire adult working population. Some furniture manufacturers have recognized this need and now provide different sizes of certain models of chairs. Organizations should be prepared to purchase several different sizes or models of chairs to meet the needs of all of their workers. From an aesthetic point of view, use of the same fabric or same color palette will provide a consistent appearance.

14.3.3 Task Postures

To support people at work, it is not sufficient to provide a chair in which they can merely sit. Chairs must support people in the particular postures from which they work. Postures and positions may vary considerably for the range of tasks people perform throughout a day or week. Operating a computer keyboard imposes different requirements than handling paper (reading and writing), collaborating with others, or speaking on the telephone. Since computer operation is the primary task of most knowledge workers, the focus will be on supporting these postures. Characteristics of chairs that support task postures include the seat pan, the backrest, and arm rests.

Seat Pan. The seat pan should be placed at a height that does not cause pressure on the thighs of the seated worker. The wide range of body dimensions among people who work in offices dictates that any chair must be adjustable in height. The determinant of seat height is a measure of the length of a person's lower leg measured from the floor to the underside of the thigh (popliteal length). The range of popliteal length in the population is 13.1–20.0″. Adjusting for shoe thickness and considering research data, the functional range for appropriately adaptive seat height is 14.5–24″ (Diffrient et al., 1981; Springer, 1982).

The relationship between the seat pan and the backrest has been the subject of many guidelines and several research studies. Generally speaking, much of what is recommended in regard to "ergonomic" seating is wrong. In particular, the "90° rule" is incorrect. This rule states that people who use computers should sit in an erect posture with their feet flat on the floor. The angle between the upper arm and forearm, trunk and thigh, thigh and lower leg should all be 90°. *This is wrong.* The so-called "90° rule" stems from traditional measurement techniques in which people were measured in this position. It also represents the center of the range of motion for most limbs and thus represents the position of greatest strength; however, it does not represent the position of greatest comfort or least strain. Research (Keegan, 1953; Bendix, 1984; Mandel, 1984, 1985; Schoberth, 1978) has shown that the stress and strain on the vertebrae and muscles of the lumbar spine are greater in the 90° posture than when the angle between trunk and thighs is opened up (120–135°). A field study of VDT operators by Grandjean et al. (1983) supports this notion and goes further in recommending a high back chair that allows and encourages operators to assume a more reclined position.

In regard to seat pan angle, the best conclusion to draw from the research is to provide users the opportunity to determine what works best for them (Bendix, 1984). Doing so allows users to open the angle between trunk and thigh. Providing a range of seat pan angles allows them to sit more erect or more reclined and to change positions depending on the need. A recommended range of seat pan angles is −8° (backward) to +15° (forward).

Backrest. A medium- to high-level backrest (one which supports the back up to the level of the shoulder blades) is recommended for most intensive computer work. The shape of the backrest should be forwardly convex in the lumbar region and gently merge into a plane or slightly concave surface in the upper region. An excessively curved backrest is worse than one that is flat. The population exhibits considerable variation from the so-called normal spine. Scoliosis, lordosis, kyphosis, and other deviations, in all three axes, from the "normal" or "healthy" spinal curve requires accommodation in the shape and adjustability of the chair backrest. A variety of approaches are available, including movable cushions, pneumatic bladders, and adjustable springs. The appropriate solution depends largely on the range of deviation from the norm exhibited in the population.

Armrests. Personal preference is very important in regard to the use of chair armrests. Armrests can provide appropriate support for people using keyboards, but they can also hinder movement and performance. It all depends on individual tasks and work style. Just as one size shoe does not fit every one, armrests that are fixed at a particular position are not adequate for a range of users. Armrests can pose particular problems for people who are at the upper and lower ends of the population distribution in regard to height and breadth. Adjustments in distance between armrests (width) are generally necessary for larger and smaller people to comfortably use most office chairs. If armrests are preferred and warranted, it is strongly suggested that they be adjustable, at least in height. Many manufacturers are recognizing this need and have provided this feature in their newer model chairs. Armrests, if appropriate, and positioned properly can provide sufficient support to relieve strain on the hands and wrists.

14.4 ERGONOMICS, PERFORMANCE, AND PRODUCTIVITY

14.4.1 Support Surface Height

Research investigating the effect of appropriate keyboard height on operator performance has tended to use sustained, high output jobs (data entry and word processing). Cushman (1984) found that operators made fewer errors (approximately 36% fewer) when the keyboard was "at about the right height." Contrary to many recommendations, Cushman's results suggest a keyboard height of 5–10 cm above elbow height. The accompanying Appendix, "Rules of Thumb," provides simple-to-follow suggestions for determining a starting point for adjustment of support surface height.

14.4.2 Seating

Several studies by Grandjean and colleagues (1983, 1984) measured key postural elements such as wrist and arm position in addition to trunk and thigh relationships. This research supports the conclusion that attention to proper chair position allows and encourages better overall working posture. Of particular import are data showing a proper relationship between the person, the chair, and computer workstation yields appropriate (nonstraining) positions of the wrists and hands. None of these results support the guidelines for 90° angles at any of the joints in the hands, arms, hips, or legs. This research is particularly important in light of the rising costs associated with cumulative trauma disorders and concerns over workers' compensation injuries attributable to keyboard use.

A different approach to providing support of task postures is offered by Mandel (1984, 1985). His work gave rise to several chairs that feature forward seat tilt and eliminate the backrest. A study by Rogers and Thomas (1990) showed an average performance improvement of 7% using a chair with forward seat inclination and a low-height backrest.

Worker control over seating position and adjustment was responsible for a 4–6% improvement in performance in a study involving an interactive computer task (Springer, 1982). A noteworthy artifact of this study is the accuracy people demonstrated when making chair adjustments in repeated trials ($\pm < 1''$).

Personal preference coupled with appropriate training in how to use the adjustments, plus orientation on the benefits of good support when seated, is strongly recommended.

14.5 ERGONOMICS OF ACCESSORIES

Considerable attention has been focused on the use of accessories to modify traditional or nonadaptive workplaces in an attempt to accommodate computing technology. Devices such as adjustable keyboard trays, display "cradles," footrests, wrist pads, document/copy holders, and back rest cushions are purchased with the hope that they will provide the individual with the ability to adapt their workplace to meet their needs. None of these devices is inherently good or bad. Some are very appropriate and are used correctly in certain situations. Others exacerbate problems or may solve a particular problem, but create one or more new ones if used inappropriately.

Accessories can be classified into two categories:

1. Personalization accessories are devices that do not significantly alter the functional characteristics of the workplace, but support a particular desire of the individual. A copy holder is a good example. These devices are appropriate for certain types of documents and certain types of tasks. Most are easily positioned by the operator and should be provided on the basis of personal preference.
2. Prosthetic accessories are devices intended to substitute or replace functional characteristics of the workplace. Generally, these devices are designed to solve one particular shortcoming of workplaces. Often, they are used in a way that solves the original problem, but creates one or more new problems. An example of this type of device is the adjustable keyboard tray.

Several of the prosthetic accessories are discussed and evaluated in the following sections.

14.5.1 Adjustable Keyboard Trays

The arguments surrounding the placement of computer terminals and keyboards has led to the development of the adjustable keyboard tray as an accessory. The manufacturers claim a number of advantages:

- It can be added to any worksurface
- It offers easy height and angle adjustment
- Depending on its location it can provide some lateral movement
- It stores out of sight when not in use
- It doesn't take up valuable desktop space

These claims may hold true under certain conditions; however, careful consideration of the device and the approach it embodies illuminates the following:

- Many are not big enough to accommodate a full size, full function keyboard
- Few accommodate a mouse, trackball, or digitizing puck
- Very few allow the height of the keyboard to exceed the worksurface to which the device is attached
- Most are not solid and vibrate when used
- Many present knee obstructions when stored under the worksurface

Alone, these factors give cause for concern when considering the use of adjustable keyboard pads, but the most critical argument against their use is the simple fact that these devices force the user to move away from the worksurface in order to operate the keyboard. By so doing, the operator cannot easily reach a mouse, any paper documents, reference materials, or writing surfaces without considerable effort and contortion. In addition, when placed in a corner arrangement (an increasingly common practice driven in part by monitor size), this device substantially increases the floor space used by computer terminal operators without providing adequate tradeoffs of supporting the task.

A better approach is to provide a solid worksurface at the appropriate height on which the keyboard is placed. If the monitor needs to be adjusted in height, this can be better accommodated through accessories. By following this strategy, the accompanying worksurfaces are available for complementary task support. This need for worksurfaces adjacent to the computer workplace was a key requirement to support productive work that emerged from the State Farm studies conducted by Springer (1982).

14.5.2 Monitor Stands

Raising the display monitor can be effective and inexpensively accomplished. The issue of display height has been discussed elsewhere; however, it is important to reiterate the recommendation that displays be placed 10–20° below eye height. If people wish to raise their monitors, it is easy to do so by using accessories such as the monitor stand placed on a solid worksurface. Should frequent changes in position of the display be warranted, more sophisticated devices that provide easy height, distance, and angle adjustments are available. However, appropriate design and layout of the individual workstation should alleviate this need in the vast majority of cases.

14.5.3 Wrist Rests

The use of wrist rests should be a personal preference. These devices can be used quite effectively and should exhibit the following characteristics:

- A minimum depth of 3.5″—support the wrist and not just the palm
- A surface that is nonabsorbent
- A surface that is firm but not hard
- Does not present any sharp or hard edges
- Is $\frac{1}{4}$–$\frac{3}{4}$″ thicker than the front edge of the keyboard

Generally, wrist rests are not required if the configuration of the workstation provides comfortable support of task postures. They can be effective when tasks require intermittent keyboard use and the hands are rested in the "ready" position. Wrist rests can pose problems when they encourage deviation of the hand and wrist from the "in-line" position and when they encourage people to rest their hands and wrists heavily on the device while operating the keyboard. Both of these situations have been demonstrated as contributing to the development of cumulative trauma disorder, of which carpal tunnel syndrome is most common.

The best approach is to design workplaces that support the arms and allow the hands and wrists to assume a "neutral" in-line position without requiring direct support or resulting in fatigue. The key to this approach is appropriate position and relationship between the chair, the support surface, and the keyboard and mouse.

14.5.4 Conclusion

Accessories should be used to personalize a workplace, not accommodate shortcomings in the fundamental, functional, design of the workstation. These devices can be cost effective when considered independently, but the interrelated nature of the workplace elements means it is impossible to change just one thing without affecting other parts of the system. Consequently, a low cost accessory solution to one problem often results in creation or aggravation of multiple high-cost problems. Accessories should be used judiciously.

14.6 ERGONOMICS AND MACROERGONOMICS

As mentioned in the introduction to this chapter, *macro*ergonomic considerations involve more global issues, such as training, job organization, locus of control, and stress.

14.6.1 Training

Meeting the need for appropriately adaptive workplaces must be supported by providing adequate training of employees in the proper use of the support equipment. Industries spend billions to train workers how to do their jobs or operate sophisticated equipment. Similarly, the increasingly sophisticated workplace equipment (chairs, for example) require an understanding of their operation, use, and benefits to maximize their impact. Unfortunately, these devices and products suffer from the "familiarity breeds contempt" phenomenon. Everyone knows how to sit in a chair; however, very few people know how to use a highly adjustable chair to maximum effect. To do so requires training.

Research supports this notion. Verbeek (1991) reported closer approximations to recommended adjustments for employees who received training in the use of their chair and desk when compared with employees who had not received the training. In a finding of greater potential economic impact, proper training when combined with ergonomically sound equipment was found to significantly reduce experienced discomfort and increased the use of the adjustment. Those people who suffered from cumulative trauma disorder and did not receive instruction in the use of the equipment were far less satisfied with their workplace, used the adjustments less often and less appropriately and reported it was more difficult to change equipment features (Green and Briggs, 1989).

14.6.2 Psycho-Social Factors

A whole class of factors that affect people's perceptions of the work environment are classified as psycho-social variables. These include stress, perceptions of control over their work and workplace (locus of control), and job organization.

Research shows that characteristics of the job and supervision contribute to perceptions of stress and job satisfaction. A longitudinal study showed correlation between work-related stressors and worker strain changed over time (Carayon, 1992). In three separate annual surveys, the profile of job-related stressors changed from heavy workload, work pressure, and lack of supervisory support, to clarity of task, and concern over job security.

Locus of control, or perceived control over the work and nonphysical elements in the workplace are important to worker health and well-being. Among workers with equally well-designed "ergonomic" workstations, those whose jobs were poorly organized were three times more likely to report neck and shoulder pain than those who had more control over their job and nonphysical working conditions (Linton & Kamwendo, 1989).

One source of stress is concern over safety at the workplace and risk of workplace injury. Research shows that in addition to ergonomic equipment, an appropriate exercise program is effective in alleviating the symptoms of cumulative trauma disorder. In a study by Dyrssen et al. (1989), workers who lifted 3 kg weights for $\frac{1}{2}$–1 hour three times a week experienced less pain that those who did not. Weight training alleviated discomfort of repetitive stress injury and tendonitis sufferers. General exercise increased strength, but *did little to alleviate discomfort.*

14.7 CONCLUSIONS

The point to all of the research cited is to illustrate the importance of a total systems approach to ergonomics and workplace design. Design recommendations that embrace this approach may appear to be odd or unusual. They are unique because every situation is unique. They are also sound and grounded in a careful analysis of the ergonomic, technological, environmental, and spatial requirements of client organizations and their employees.

14.7.1 Strategies for Ergonomic Workplaces

To summarize, a comprehensive approach to workplace ergonomics embraces the following strategies:

1. *Adapt the Workplace to the Worker.* Appropriately adaptive equipment relieves strain on the worker to adapt to shortcomings in the workspace. Provide adjustable furniture and equipment to support the wide range of sizes and shapes of people in the workforce.
2. *Optimize Support for the Primary Task.* Worksurfaces that are solid and large enough to support the primary task of computer workstation interaction are required.

3. *Provide User Control.* A sense of control over the workplace is important to workers' sense of satisfaction and performance. The individual worker must control workplace adjustments. These adjustments must be easy to perform.
4. *Provide for Personalization of Workspace.* Accessories that complement, as opposed to supplant, the function of the workplace allow the user to "fine tune" their workspace to meet their individual preferences.
5. *Emphasize Ease of Use.* Adjustments, control motions, connects, and disconnects should be easy to use. For example, access to power, network, and telecommunication ports should be at desk height or belt-line.
6. *Support Work in the Way It Is Done.* Appropriate support of work styles and practices should be provided. For example, if folks prefer storing paper information in "piles," use of horizontal storage is recommended. Similarly, providing easily moveable and adjustable support surfaces should accommodate portability and mobility of nomad technology.
7. *Train People in the Proper Use of Equipment.* Training that demonstrates the technique and benefits of appropriate adjustments are required. The best workplace is only effective if people know how and why to use it.

It must be noted that without appropriate full-scale mock-up and testing of specific design recommendations in an appropriately controlled and measured setting, it is very difficult to postulate regarding the potential impact on performance and productivity. However, one thing is clear—research supports the approach offered. Research also suggests that when a total systems approach is followed, the impact, in terms of performance improvement and return on investment, is considerable.

APPENDIX A RULES OF THUMB FOR ERGONOMICS OF WORKSPACES

In ancient days, the unit of measure was determined by the size of the king's foot and his thumb to the first knuckle. (This is the origin of the English standard unit the inch.) These human referenced measures gave rise to the expression "rules of thumb." What follows are steps one can take to establish their own personal referent "rules of thumb" by which an individual's workspace can be appropriately adjusted to fit.

Ergonomics, simply put, is the practice of making the tools, equipment, and places of work safe, easy to use, and comfortable for the people who use them. By so doing, people can be more comfortable and more productive.

Preparation

- *Warm up.* It is important that people be willing to take charge of their posture and comfort when at work. Remember, "they don't call it work for nothing." So, take a few moments before beginning work to "warm up" with some stretching, loosening motions. This can help your body as well as your mind get ready to work. Also, if adjustments need to be made to your workspace, a few moments to adapt the furniture and equipment will avoid later discomfort.
- *Check up.* When was the last time you had your eyes checked? Using a computer terminal is a visually demanding task. So, it is important that your vision be the best it can be. If you wear eyeglasses, make sure your prescription is up-to-date. When you see your eye care professional, tell them you are or will be working on a computer terminal.

The "Rules." The following are simple, general "rules of thumb" for adapting workspaces to fit individual workers.

1. *Posture.* Generally good posture and good working positions are very similar, but it is important to note that only you can determine what is comfortable for you. For most people, sitting erect in a chair will cause the least fatigue and discomfort over the long term. It is important to pay attention to your posture when you work. We all develop bad habits that, if not corrected, can lead to problems. Try to avoid slumping too much or leaning heavily on your wrists or forearms. Your body is wonderfully designed to support you in a variety of positions. Just as your arm or leg "goes to sleep" if you hold a position for too long, working in the same position without change can lead to similar feelings in hands, arms, feet, and legs. It may be necessary to change positions to avoid discomfort. Shifting your weight or changing seated positions or getting up and moving around within your workspace can do this. Remember alternate activity—one where you are not using the same muscles or not using them in the same way—is often as good as a rest. While it's not always necessary to "rest" in order to recover, taking breaks from any extended activity is recommended.

2. *Chair Height.* Since most office work is done from a seated position, a good chair is the foundation of a good workspace. A good chair should do three things:

- Provide a stable base
- Fit your body
- Support the positions you assume to perform your work

To adjust your chair:

- Stand facing your chair
- The front edge of the chair should bump your leg 1.5–2″ below your kneecap (see Figure 14.10)
- Sit down in your chair. You should be able to sit with your feet comfortably on the floor.
- If you cannot do this, you may need to use a footrest to support your legs and feet

FIGURE 14.10 The Front Edge of Your Chair Should Bump Your Leg below Your Knee.

Many chairs offer other adjustments, such as backrest angle, backrest height, seat angle, arm rest position, etc. Remember the three characteristics of a good chair and use these adjustments to ensure the chair supports your body and your task postures. It may be necessary to change these adjustments periodically as you change positions or tasks.

3. *Worksurface Height.* Once your chair is adjusted to a comfortable position:

- Sit facing your work surface
- Turn 90° to one side. Let your arms hang comfortably at your side.
- Your worksurface should be at or below the crease in the inside of your elbow (see Figure 14.11). If it is not, adjust the worksurface up or down. (*Note:* Try not to adjust your chair height.)

This applies to keyboard heights as well (i.e., home row should be at elbow height.) A check on this is to rest your fingertips lightly on the keyboard (Don't rest your wrists!). Your forearm should be about parallel with the floor and your hand, wrist, and forearm should be in a straight line, not bent. Think of typing at a keyboard in the same position as a concert pianist plays the piano—their wrists are straight, they arch their fingers and strike downward—not "splay-fingered."

4. *Positioning Things On Your Desktop.* Generally, you want to place the things you handle most often, like paper, your keyboard, a computer mouse, or calculator, within easy reach. Easy reach means you don't have to stretch your arm from a comfortable, bent elbow position. Chair armrests can help support your arms in this position.

In the case of a computer mouse, it is not uncommon for people to reach and stretch to use them. Some accessories are available that move the mouse off the desktop and allow it to be used in a position that does not involve stretching and reaching. Another alternative is to use a trackball—essentially a mouse turned upside down, or, more recently, touchpads. These units do not require movement of the device, demand less movement of the wrist and arm to operate, and consequently, can be positioned more readily in an easy to reach and operate position.

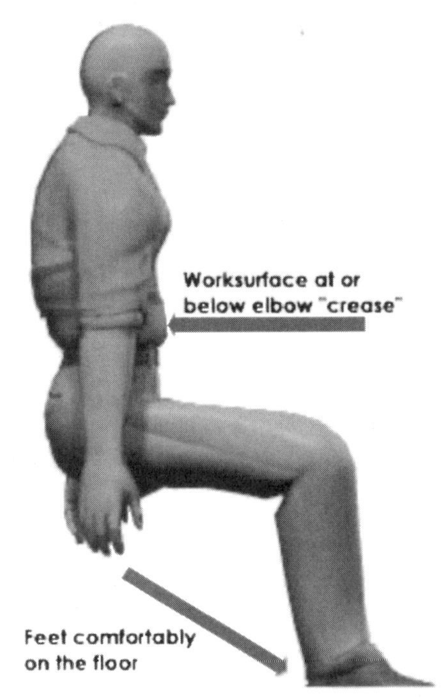

FIGURE 14.11 Body Reference Point for Worksurface Height.

5. *Display Position.* Both paper and computer monitors (visual display terminals = VDT) are displays. It is important that the thing you will be looking at most often is placed directly in front of you. Doing so prevents you from repeatedly turning your head to the side or holding your head in an awkward position for long periods of time. If you must look at two things frequently, try to position them side by side so you can see them easily by moving your eyes rather than your head or neck.

6. *Display Height.* As humans evolved, our eyes became one of our primary sense organs. We learned to look down to see things close at hand ("Oh, the fruit is on the ground.") and to look up to see things far away ("Yikes! The lions are coming!"). As we moved into buildings and began reading, those characteristics continued. We look down to read and see things close-up, we look up to see things across the room. This inherent trait of our visual system should translate to where we place things we look at while at work—like displays.

The most comfortable position for the eyes to view material in the near to mid range (16–30″) is 15–20° below horizontal. This means that the top of the display, whether paper or computer monitor, should be about equal to the height of your chin when sitting erect (see Figure 14.12).

7. *Viewing Distance.* As noted previously, displays are most easily read in the near to middle distance of 16–30″. Thus, your monitor should be about arms length away. To check BOTH monitor height and distance:

- Sit comfortably erect facing your monitor
- Raise your arm until it is straight out from your shoulder and about parallel with the floor
- You should be able to easily touch the top of the display screen with your fingertip

8. *Glare.* Computer displays are reflective surfaces. If you can see light sources, either lamps or windows, in your screen you may want to change the angle of the display (and thus change the angle of reflection) to help you see what is on the VDT. Some monitors are equipped with tilt and swivel stands. Others will fit on add-on accessories that provide the same features. If neither is an alternative, you can prop documents or books under the front or rear of the monitor to change the angle. **Caution:** Do not try to change the angle too greatly as this might cause the monitor to be unstable and may result in it falling.

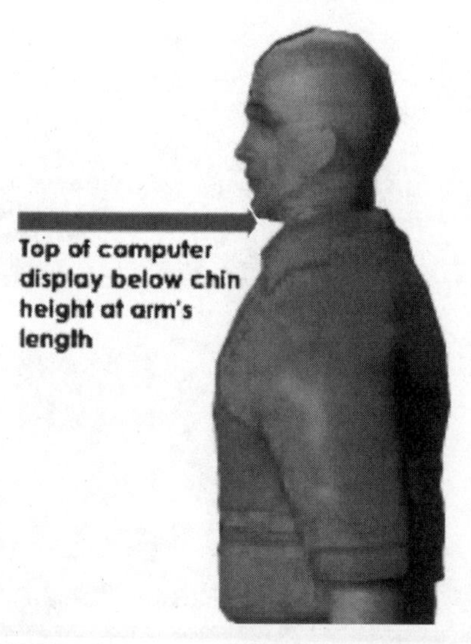

FIGURE 14.12 Top of Display Should Be at About Chin Height.

It is especially important to keep your monitor clean so you are not trying to see images through a "dirty window." Use an approved method of cleaning your monitor to keep it free of dust, dirt, and fingerprints that can obscure images and make information look fuzzy. Also, it is important that the brightness and contrast of your VDT be adjusted to meet the lighting levels in your workspace. Consult an operating manual or technical support if you are not familiar with how to make these adjustments.

9. *Accessories.* It may be necessary to make adjustments to your workspace to adapt to your needs. If the range of adjustment does not bring the workspace elements into a position where you can work in comfort, you might need to use accessories to supplement the adaptability of the workspace. These accessories can include, but are not limited to, footrests; padded palm, wrist, or forearm rests; back cushions; document stands and copy holders; monitor stands; telephone receiver cradles; headsets; trackball or touchpad; and a variety of other "add-on" equipment to make your workspace fit you.

BIBLIOGRAPHY

Grandjean, E., Hünting, W., and Piderman, M. 1983. VDT workstation design: Preferred settings and their effects. *Human Factors* 25(2), 161–176.

Grandjean, E. 1984. Postural problems at office machine workstations. In E. Grandjean, Ed. *Ergonomics and Health in Modern Offices.* London: Taylor & Francis.

Weber, A., Sancin, E., and Grandjean, E. 1984. The effects of various heights on EMG and physical discomfort. In E. Grandjean, Ed. *Ergonomics and Health in Modern Offices.* London: Taylor & Francis.

Mandel, A. C. 1984. What is the correct height of furniture? In E. Grandjean, Ed. *Ergonomics and Health in Modern Offices.* London: Taylor & Francis.

Mandel, A. C. 1985. *The seated man.* Third edition. Klampenborg, Denmark: Dafnia Publications.

Chaffin, D. B. 1991. Biomechanical strength models in industry. In *Ergonomic Interventions to Prevent Musculoskeletal Injuries in Industry.* Chealsea, MI: Lewis Publishers.

Price, S. 1992. Ergonomics—who really cares? *Canadian Facility Management,* June.

Hedge, A., and Powers, J. 1991. An experimental test of a design prototype of the Protex™ system. Cornell University, December 1.

Lansdale, M. W. 1991. Remembering about documents: Memory for appearance, format, and location. *Ergonomics* 34(8), 1161–1178.

Diffrient, N., Tilley, A., and Bardagiy, J. 1974. *Humanscale 1,2,3,4,5,6,7,8,9.* Cambridge, MA: MIT Press.

Springer, T. J. 1982. *VDT workstations: A comparative evaluation of alternatives.* Bloomington, IL: State Farm Mutual Automobile Insurance Company.

Keegan, J. J. 1953. Alterations in the lumbar curve related to posture and seating. *Journal of Bone & Joint Surgery* 35A(3), 589–603.

Bendix, T. 1984. Seated trunk posture at various seat inclinations, seat heights, and table heights. *Human Factors* 26(6), 695–704.

Schoberth, H. 1978. *Correct sitting at the work place.* Frankfurt, Germany: University of Frankfurt, Ostee Clinic.

Cushman, W. 1984. Data entry performance and operator preferences for various keyboard heights. In E. Grandjean, Ed. *Ergonomics and Health in Modern Offices.* London: Taylor & Francis.

Rogers, C., and Thomas, R. 1990. The effects of chair-type and workstation configuration on work performance. *Proceedings of the Human Factors Society 34th Annual Meeting,* pp 674–678.

Verbeek, J. 1991. The use of adjustable furniture: Evaluation of an instruction programme for office workers. *Applied Ergonomics* 22(3), 179–184.

Green, R., and Briggs, C. 1989. Effect of overuse injury and the importance of training on the use of adjustable workstations by keyboard operators. *Journal of Occupational Medicine* 31(6), 557–562.

Carayon, P. 1992. A longitudinal study of job design and worker strain: Preliminary results. In *Stress and Well Being at Work.* Washington: American Psychological Association.

Linton, S., and Kamwendo, K. 1989. Risk factors in the psychosocial work environment for neck and shoulder pain in secretaries. *Journal of Occupational Medicine* 31(7), 609–613.

Dyrssen, T., Paasikivi, J., and Svedenkrans, M. 1989. Beneficial exercise programme for office workers with shoulder and neck complaints. *Third International Conference on Musculoskeletal Accidents and Injuries in the Workplace,* November 7–12. Bangkok, Thailand.

Collins, M., Brown, B., Bowman, K., and Carkeet, A. 1990. Workstation variables and visual discomfort associated with VDTs. *Applied Ergonomics* 21(2), 157–161.

CHAPTER 15
MANAGING THE NEW HEALTHCARE REAL ESTATE PORTFOLIO: A CASE STUDY

Tom DeChant
Space Diagnostics, Inc.
E-mail: Tadechant@spacedx.com

Meredith Spear
Space Diagnostics, Inc.
E-mail: Mespear@spacedx.com

John Messervy
Partners HealthCare Systems
E-mail: jmasservy@partners.org

15.1 WHAT HAS CHANGED IN HEALTHCARE FACILITIES AND REAL ESTATE?

Ignited by the failure of federal healthcare reform efforts in 1993, the pace of change in the U.S. healthcare system rapidly accelerated through the remainder of the last decade, resulting in profoundly new and different facility management (FM) challenges for this industry segment. In this chapter, we will explore through a case-study approach how one major multi-hospital academic medical system has coped with the "new game" of healthcare facility and real estate management. First, however, it may be helpful, both for those within and outside this industry, to review the forces that changed the rules for healthcare facility managers.

15.1.1 Restructuring

With the failure of public healthcare reform efforts, the mutual cost containment concerns of private employers and health insurers ran headlong into growing consumer demands for choice and convenience, ushering in the era of disgruntled managed care in which we still find ourselves. Responding first to the tighter reimbursements of this new economic environment, healthcare providers engaged in a flurry of organizational restructurings and new financial arrangements to better shield themselves from the vagaries of this changed medical marketplace.

One common provider response was to merge with or acquire other providers in order to shore up and expand market share. These organizational restructurings resulted in rapid and continuing transformation of many traditional hospitals from a single campus entity with a singular building type into multi-site, regional delivery networks with a significant variety of real estate holdings. Since market share capture was and remains the primary driver of these acquisitions, the quality of the facility assets being acquired has often been an afterthought. As a result, healthcare facility managers are inheriting many properties of dubious value. Buildings may have serious deferred maintenance, code deficiencies, or very limited flexibility for potential reuse. Sites may lack strategic usefulness due to their poor location, weak image, or limited size. Many of the newly acquired properties are leased rather than owned, forcing facility managers into landlord, and property management roles for which they may have little preparation or available staff. And many new and unfamiliar facility types are being added to the traditional hospital portfolio—residential facilities like assisted living, free-standing surgery centers and medical offices, fitness centers and other alternative healthcare settings, and even retail spaces.

The same economic forces that encouraged massive provider restructuring also helped accelerate the shift of many diagnostic and treatment procedures from their traditional hospital inpatient settings to ambulatory or outpatient facilities. These lower cost settings were encouraged by payers and readily embraced by consumers as more convenient and user-friendly. The rapidity of this shift was dramatic. Whereas in the early 1980s, a progressive community hospital may have done 20–25% of its surgeries on an outpatient basis, by the late 1990s, a 65–75% outpatient surgery rate was not uncommon. The shift to outpatient surgery, paralleled by similar changes in the settings of other diagnostic and interventional services, had a profound impact on the type and location of health facilities needed. Beds, once the most common measure of a hospital's size and importance, rapidly became excess real estate as payers mandated shorter lengths of stay. And the smaller, "lighter," more user-friendly ambulatory spaces that both physicians and consumers came to prefer were hard to carve out of the old inpatient-oriented hospital campuses, resulting in many misfits between available spaces and the new ambulatory functions. Entering the twenty-first century, hospitals found themselves saddled with "main frame" facilities in a portable PC world—with the wireless revolution just around the corner.

15.1.2 Technology

While the rapid escalation of technology had impact on virtually all building types through the last decade, health facilities were particularly affected because they are the physical locus where multiple information systems and complex medical technologies converge. Focusing first on the massive revolution in information technologies, health facilities began to respond to the advent of telemedicine, digital radiology, and consumer and provider Web access, while continuing to expand the capacity and connectivity of their existing health information systems both within hospitals and external to them. These new information capabilities not only made physical demands on existing facilities, but more importantly, expedited the "unbundling" of the traditional hospital complex. The standard rationale for fixed physical adjacencies within hospitals has largely evaporated, and the convergence of ubiquitous electronic information coupled with the demand for greater patient convenience has driven many health services to more accessible, distributed settings—everything from individual patient homes to shopping center store fronts to free-standing centers (for surgery, wellness/fitness, imaging, and birthing) to comprehensive "hospital without beds" ambulatory complexes. As wireless web access and web-based information system protocols continue their exponential growth, the trend toward delivery of healthcare in virtually any physical setting or location will expand on the same trajectory.

The trend toward "medicine anywhere" has been complemented and further fueled by the parallel revolutions in medical technology and science that have progressed from a smolder to full combustion over the last decade. Development of smaller, lighter, faster, less-invasive and more portable machines cut across virtually every area of medical diagnosis and treatment. Mobile and ultra-fast imaging, endoscopic surgery, point-of-testing lab results, remotely read microsensors tracking multiple anatomical functions—these are but a few of the medical "hardware" innovations that changed the rules of what could be done where. These new technologies enabled a fundamental shift in models of care—letting providers bring services directly to the patient rather than moving patients to services. Such a shift profoundly changes the way we might now address the basic circulation and infrastructure components of health facilities—everything from elevator locations and capacities to patient room sizes to equipment storage configurations, and more.

On the horizon are still more formidable medical science breakthroughs that will not only change the physical where and how of healthcare delivery, but will totally reformulate the very paradigm which has guided western medicine throughout the last century. "Diagnose and repair" will rapidly transition to "predict and prevent"—a shift that will be enabled by the likely convergence of three rapidly emerging technologies: genetic engineering, nano-technologies, and artificially intelligent robots. Together, these mutually supporting technologies could usher in a brave new world of self-replicating, self-learning, gene-machines that could place the "hospital" of tomorrow right within the android-like human body itself. If this sounds like far-distant science fiction, just read the essay by SunMicrosystem's Bill Joy in the April 2000 issue of *Wired* magazine to be convinced that it's not.

15.1.3 Finance

Interwoven with healthcare's changing technologies and organizational structures was yet another force that impacted facilities—the methods and incentives for capital financing. Starting early in the decade, the federal Medicare program began diminishing its direct capital "pass-through" funding for health facility costs, gradually shifting to a more prospective funding mechanism based on average costs rather than facility specific costs. This paralleled Medicare's shift in reimbursement for medical care, from a fee-for-service model to prospective payment based on fixed diagnostic groupings (DRGs). The Balanced Budget Act of 1997 put a further across-the-board squeeze on Medicare reimbursement that led many facilities, especially larger academic hospitals, to lower profitability or actual losses. This frequently resulted in lowered bond ratings and diminished debt capacity, further curtailing access to affordable capital while increasing reliance on philanthropy, funded depreciation, and alternative sources of capital for financing major projects. Financing vehicles that were previously little known or used in healthcare circles became increasingly common toward the end of the 1990s. Among these were real estate investment trusts (REITs), master and synthetic leases, shared equity positions in projects with other owners, often physicians, and similar off-balance sheet financing tools. With capital being tighter and more creatively attained, capital projects also began to undergo tighter financial scrutiny and analysis, with explicit return on investment (ROI), return on asset (ROA), or contribution to margin studies being required. Several systems even established explicit internal hurdle rates that every major capital investment had to meet before financing was approved. This type of analysis and accountability for capital investment returns entailed a new level of financial sophistication on the part of healthcare facility managers that was often not part of their existing skills repertoire.

15.1.4 Conclusion

Given the rapid pace and broad scope of the changes experienced by the U.S. healthcare delivery system in just a brief 10 years, few FM teams were fully prepared to meet their new challenges. Strategic opportunities were seized quickly in the mergers and acquisitions arena, meaning new facilities were added to inventory with little preparation for their impact on facilities staffing, capital budgets, or maintenance. These newly expanded portfolios encompassed many new facility types, a wider geographic distribution of sites, a mix of leased and owned properties, and an abundance of maintenance and repair needs—all confronting the healthcare facility manager with a host of new management demands. First among them is the need to get a "handle" on exactly what they've got; that is, to thoroughly inventory their new portfolio—its physical condition, strategic usefulness, and capital requirements, both for maintenance and repair and long term. Other issues that quickly emerge are the need to develop consistent and equitable methods of investing capital across different system entities; the need to find alternative capital resources and creative financing arrangements to cover the expanded demand for capital; and the need to deploy facility resources much more rapidly than in the past in order to take advantage of more competitive windows of opportunity. The need for electronic tools to manage and document projects also surface rapidly as the numbers of construction and renovation projects multiply and project cycles accelerate. While other sectors of the economy have long understood the need for agile and scaleable facility responses to meet rapidly changing business needs, this is a radically new concept for a healthcare universe that has traditionally brought buildings on line (or decommissioned them) with glacial speed. Finally, new real estate, financing, and property management skill sets are required immediately to meet higher-level project analysis and broader portfolio management demands.

Despite this often-dramatic escalation in healthcare FM responsibilities, few systems allocated time or adequate resources for anticipating these demands or for thoughtfully restructuring the FM organization and operations to better address them. And few FM models existed within the healthcare industry to guide others through these changes—healthcare facility managers were largely on their own in figuring out how to cope. For these reasons, the review of one large system's response to this new facility environment may provide both enlightenment and solace to those still struggling with these challenges.

15.2 HOW ONE HEALTHCARE FM TEAM RESPONDED

15.2.1 A New Boston Provider—Partners HealthCare

Late in 1993, the Massachusetts General Hospital (MGH), a 1,045-bed academic medical center affiliated with the Harvard Medical School; the Brigham and Women's Hospital (BWH), another Harvard affiliate; and North Shore Medical Center, based in Salem, formed the Partners HealthCare System. This was an aggressive strategy to create a dominant integrated delivery network within the Boston healthcare market. With the goal of a million and a half "covered lives" or insured participants, Partners continued its growth strategy through the balance of the decade. First steps included fully integrating the hospitals already affiliated with MGH—McLean Hospital, a neuropsychiatric facility located in the Boston suburb of Belmont, and the Spaulding Rehabilitation Hospital located in downtown Boston in close proximity to the MGH campus. Spaulding also included a second, smaller inpatient campus on Cape Cod. Newton-Wellesley Hospital has since joined Partners and a 1,000-physician primary care network (PCHI) has been created in eastern Massachusetts. These affiliations, particularly of primary care providers throughout the region, resulted in the addition of a significant number of small, leased medical office spaces to the system's facility inventory. By 1998, Partners had assembled a portfolio of over 250 leased or ancillary properties, in addition to its 11 million ft^2 of owned facilities on seven campuses throughout the greater Boston region.

15.2.2 Partners' FM Organizational Response

Obviously, this rapidly expanding real estate portfolio demanded an equally rapid and appropriate management response. To its major credit, one of Partners' first steps was to create a real estate function at the corporate level. A Vice President of Real Estate and Facilities was hired from outside the healthcare industry, and had joint reporting authority to the system's CEO and Vice President of Operations. The leadership in the Partners' alliance had the foresight to develop this corporate-level real estate function right from the system's outset, in part because several key stakeholders had backgrounds and great interest in this field. In particular, the board president of the MGH was a national corporate real estate developer. This fortuitous circumstance meant that the highest level of system leadership fully grasped the magnitude of what was at stake in assembling such a vast portfolio of holdings in one of the countries most expensive real estate markets. While many health system facility managers are still trying to get their voices heard within the executive management team, Partners started out with strong facilities advocacy and understanding at the highest corporate level.

This corporate vision was further focused by several specific situations occurring in the Boston real estate market in the mid-1990s, including commercial office rents in the $35 per square foot range, a dearth of readily developable land near any of the system's large in-town campuses, and the rumored availability for acquisition of at least one major competitor. Partners' leadership also anticipated that it might need to engage in a number of real estate transactions sooner rather than later, because it had identified that many of the system's facility assets were past their useful lives and would likely be candidates for disposition in the near future. This identification of "to-be-excessed" properties was made through application of a high-level facility condition evaluation tool that was applied consistently to all buildings on the five major campuses. Application of this common tool resulted in a valuable database that was used to identify the "highest and best use" of each facility asset, as well as to guide the appropriate level of future investment in each building based on its condition and potential for reuse. While Partners' early development of a corporate-level real estate function demonstrated unusual foresight, and its use of a consistent facility evaluation approach jump-started development of a corporate facility database, it would still be fair to characterize the system's initial facility function as more reactive than proactive, since it was fundamentally driven by response to market opportunities that were perceived at the time.

In developing the initial corporate real estate office, only selected functions were incorporated in order to preserve a balance between the prerogatives of the individual system entities and the new corporate structure. The functions "peeled off" to the corporate level initially included:

- Planning (including setting broad project priorities across the system)
- Capital allocation and budgeting at the per-entity level
- Construction (including in-house CM function)
- Working with external authorities (finance, Boston Redevelopment Authority)
- Leasing off-site property for the entities as well as Partners' corporate
- Utilities contracting and energy conservation

To support these broad functions, the corporate real estate developed a staff model that selectively integrated new leadership with existing facility and real estate expertise culled from the two principal Partners' entities—the MGH and the BWH. The formal staffing organization for the corporate office is represented in Figure 15.1.

Within this new leadership structure, many staff groupings from the existing entities were redeployed. For example, existing facility units from both the MGH and the BWH were reestablished under the Vice President of Real Estate and Facilities. To further enhance continuity between the old structure and the new, the newly created position of Director of Capital and Facilities Planning was hired from within and was the former MGH Director of Planning and Construction.

At the individual entity level, each hospital within the system retained its own facility staff and made its own facility staffing decisions. Each entity also retained its own FM organizational model, including local project management and outsourcing decisions, and those models varied widely between the different organizations in the system. For instance, the MGH had a 50-year tradition of institutional facility planning—including the retention of an in-house architect—and so it had subsequently developed a much greater internal capability in front-end project planning, programming, and preliminary design. Likewise, the MGH had traditionally completed a substantial amount of its internal renovation projects with its own construction forces. In contrast, the BWH's approach to project planning relied much more on the use of external consultants and architects, and renovation projects were generally implemented by external construction labor with the BWH facility staff providing project management. These different planning and construction models remained in place within the new Partners' structure, but coordination and some integration was provided across all entities by the new Director of Capital and Facility Planning.

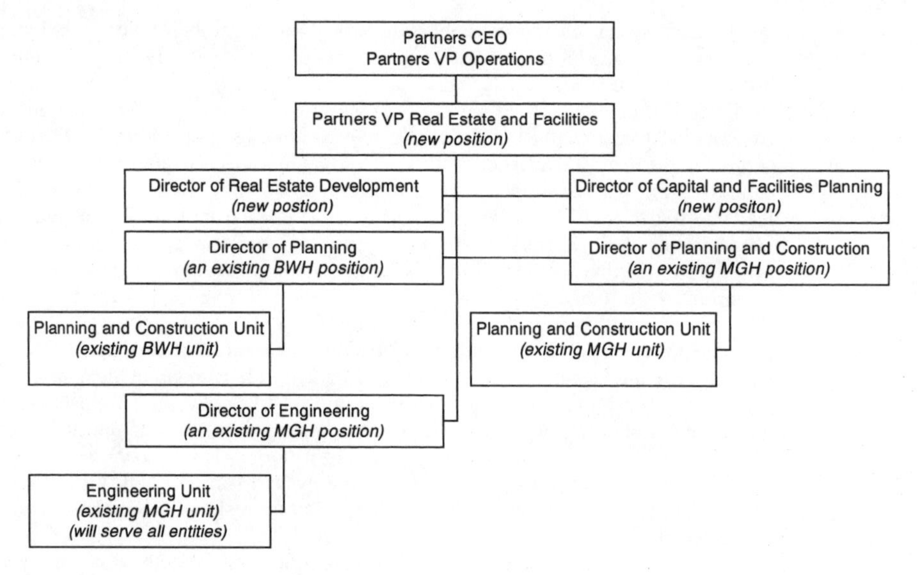

FIGURE 15.1 Partners Healthcare Organizational Chart.

15.2.3 Issues Within the New Partners Corporate Real Estate Structure

In any corporate merger or acquisition, one of the most demanding challenges is meshing and integrating the distinct cultures of the organizations being brought together. This was certainly true at the broadest institutional level of the Partners' system, and filtered down to the specific functions of facilities and real estate as well. The MGH and the BWH both had long and distinguished reputations as two of the premier academic medical centers in the nation. They were rivals in the Boston area and had evolved very different management and decision-making cultures. The MGH traditionally processed most major decisions at the institutional level, whereas the BWH had a history of strong academic departmental autonomy countered by a more centralized and top-down approach to administration of the hospital.

A large and still on-going cultural struggle within the Partners' facility arena is the balance of power and autonomy between the local entities and the corporate office. Since the MGH–BWH affiliation within Partners is viewed as "an association of equals," and since it does represent a true asset merger between the two institutions, from the outset the facility structure was designed to be "loose." Issue by issue, the individual entities could generally opt in or out of immediate participation in the new corporate real estate model, and day-to-day facility operations remained completely local at the start. Even under this loose arrangement, some entities chafed. The BWH initially wanted to retain local control over real estate decisions and on-site project management, while agreeing to let property management and long-range master planning functions move to the corporate level. All entities wanted to retain local autonomy over priority setting among capital projects, which the "loose" model did and still does support, albeit with increasing reluctance.

Capital allocation and priority setting among projects remains perhaps the most persistent and contentious balance-of-power concern among the entities, in large part because of the nature and source of capital funding within the system. While capital is pooled, and Partners is the bonding authority for all major financed projects, the revenue stream used to support a bond issue comes from the specific entity undertaking the project. In addition, philanthropy levels and specific donors vary widely among the entities, so this source of capital is not easily pooled since that would dilute donor appeal that is based on loyalty to a specific institution. So, while capital is broadly allocated among the entities by the corporate level, the competition for "a share of the pie" remains an aggressively fought annual contest among the individual institutions. The prerogative of priority setting among projects remains fiercely held at the local level. While at present there is no *integrated* priority setting *across* entities at the corporate level, a new Partners' Vice President of Finance is forcing that issue by insisting on the development of a rolling 5-year capital and financing plan identifying current and future major capital commitments. On a similar front, and emerging from a recommendation in the 1998 Partners' Facility Master Plan, a Partners' Facility Council has been established that includes CEO representation from each hospital to review entity master plans and all proposed projects over $2 million. In addition to moving forward with its charge, the council is emerging as a vehicle for tackling some very difficult issues brought on by the continuing growth of the downtown hospitals in particular, such as the geographic distribution of services to other Partners' hospitals and potentially also to new sites.

The current "free market" system of competition for capital and local setting of priorities may be ready to evolve, with the anticipated outcome being a more "federalist" model that provides mutual checks and balances between local autonomy and system-level authority. This would permit the corporate level to give more specific strategic direction to its capital investments, while perhaps offering to the individual entities a broader pooling of finance risk. Allocation of capital for system-level IS/IT investments, new acquisitions, and other strategic initiatives will continue to be made by the corporate office. As hospital margins shrink, Partners is taking a harder look at capital priorities within the individual entities. Major capital projects are now being reviewed in terms of return on investment (ROI) or return on asset (ROA), and at a minimum each project has to demonstrate its necessity and its contribution to margin through a rigorous and consistent presentation of facts to the corporate parent.

15.2.4 Partners' FM Data and Documentation Strategy

Partners' first corporate Vice President of Real Estate and Facilities came from a development background, as opposed to one in healthcare. From his "outsider" viewpoint, he brought to the new organization a philosophy that real estate was a significant, but often poorly managed corporate asset, and that decisions regarding it should be supported by hard facts and data, not just dreams and desires. This perspective was further reinforced by the fact that most Partners' executives and decision makers were physicians—a group of professionals that traditionally relies on data-driven research to reach conclusions, clinical or otherwise. Based on this convergence of philosophy and needs, the corporate real estate office decided early on to develop a deliberate and consistent approach to assembling its facility documentation and databases.

The first step in this approach was to consult with other real estate experts to pinpoint which data could provide the most beneficial "yield" and to identify the types of software applications that would best organize that data into useful information. One of corporate's first priorities was simply to inventory its real estate holdings in order to understand the total value of its portfolio, since the system had grown in short order to encompass six major hospital campuses and over 250 other leased or owned properties. Responding to that immediate need, as well as to longer-range system information needs, consultants recommended that the corporate real estate office invest in four specific facility document and database applications:

1. A CAD/CAFM system to accurately document buildings and their space uses
2. A lease tracking system to inventory and manage the large portfolio of rented spaces
3. A capital asset tracking system to better manage the cost of capital projects
4. An occupancy cost system to understand the individual occupancy costs of each asset

This collection of software applications was deemed to be the baseline needed to adequately inventory and manage the Partners' portfolio.

To select the specific applications, a form of triage was employed. First, the corporate real estate office reviewed software already in use for each purpose at the two major entities (MGH and BWH), regardless of whether the applications were developed in-house or purchased from external vendors. If existing applications did not meet the need or none existed, then external vendors and products were reviewed. The point of this process was to leverage existing software and data investments wherever possible while building political support for the new corporate real estate function by acknowledging existing institutional expertise. This systematic applications review resulted in the following decisions regarding facility information systems:

1. For CAD, it was determined that the MGH's investment in field-measured AutoCAD™ drawings for the 4 million plus ft^2 of its physical plant set an appropriately high, though costly, standard for accurate as-built drawings that should be emulated by the other Partners' institutions. Regarding CAFM systems, the MGH also had invested in a highly customized version of the ARCHIBUS/FM space management module that produced specific outputs for Medicare cost reporting and other purposes. In both cases, the software involved was viewed as industry-standard and well-supported, and the investment incurred was already yielding value for the parent institution. So, these applications were suggested as appropriate for the entire Partners' system, although the cost of developing or converting data and drawings for these applications had to be borne by each of the individual entities. This budgetary barrier has slowed enterprise-wide implementation of a common CAD/CAFM system, but the two largest institutions are now on board, and they represent the vast majority of the system's physical assets.
2. For lease tracking, no Partners' entity had an application in place that met the broader needs of the entire system, so a fresh search of vendors and software was required. After careful review, a California-based vendor and product was selected and implemented to manage the 200 plus leases in Partners' rapidly assembled portfolio. While the software product performed as anticipated, providing timely intelligence to the internal property management function, its vendor ultimately withdrew from the marketplace, leaving the application unsupported. The database of lease information has now been transferred into a modified version of the real property and lease management module of ARCHIBUS/FM. As part of the implementation protocol, it was established that Partners' Office of General Counsel would input all new lease information to ensure legal accuracy, and this has heightened the application's credibility and usefulness.
3. While significant resources were committed by Partners to integrating financial systems across the network, it would be at least 5 years before the capital asset and project management modules of the new PeopleSoft system would be available. As an interim solution, corporate real estate determined to utilize the systems developed by the planning offices at the BWH and the MGH. The BWH, with the help of an external consultant, had developed a fairly robust project-tracking system and the MGH had developed a project accounting system for capital projects. By integrating these two systems, most of the information elements needed for corporate level project management would be available. The modified software now provides in-house project managers with "real time" views of their individual capital project schedules, budgets, and expenses, while providing the corporate office with a system-wide overview of all capital project activity on a monthly basis. The system has the capability to generate purchase orders and to accept "downloads" from the work order systems of other hospital departments supporting capital projects, such as telecommunications, buildings and grounds, and environmental services. Project manager timesheet information is also picked off and posted against projects.
4. None of the Partners' entities had developed an application to compute and track occupancy costs for their owned assets, and no "out-of-the-box" solution was readily available, so a consultant firm was hired to develop this application. Significant investment was made in constructing this application and particularly in collecting and entering the initial data set. Identifying, standardizing, and assembling occupancy cost components across the different entities was a cumbersome task, yet in the end the yield from this exercise was questionable. Initially projected uses of the data, such as for internal space charge backs and to better calculate indirect costs for funded research, either were not implemented or proved to have little incremental value. Since the cost of maintaining and updating the data for this application is particularly high and the information yield low, this fourth component of the initial data plan is no longer supported by the corporate real estate office.

Selection of these four FM data applications was completed relatively quickly, but the pace and coverage of implementation varied. One of the first implementation hurdles was simply staffing. Once the scope of effort required was clear to all, corporate real estate chose to hire a full-time Manager of Facility Information Systems. The individual selected for this position had both an IS and a real estate background, although no previous experience in the healthcare industry. Next, priorities needed to be set. For all four applications, the initial goal was to get the two largest campuses, MGH and BWH, up and running since they represented by far the greatest asset value as well as square footage within Partners' real estate portfolio. Roll-out to the other hospital campuses would be slower, in part because funding for implementation had to be budgeted by the individual entities, and also because the applications were presented as guidelines and not mandates. So, the pace of implementation was governed by the local entity, not the corporate office.

To date, the two main campuses are both up and running with AutoCAD drawings that employ similar (though not identical) layering and other conventions. Both institutions' drawing files are also linked to a customized ARCHIBUS space management module that provides common room-level information and space reporting for both campuses. The other hospitals in the system are in various states of transition to the corporate CAD/CAFM guidelines, with several of the older institutions, such as McLean, still lacking any CAD database whatsoever. The property management function resides at the corporate office level, so once the system's 250 leases were entered, implementation of this component for all entities was essentially completed. Lease information is updated through the corporate counsel's office, and the system is now being enhanced with the addition of digital photos and basic site/facility condition and capacity information for each leased property. Likewise, the project management software was implemented corporately, so once the "bloody" software modification process was concluded—standardizing financial data and project manager requirements across the board was no mean feat—all capital projects across the system can now be tracked in common fashion. The occupancy cost effort was essentially completed for just one campus, the MGH, and then dropped, based on its perceived limited value.

About one year ago, in an effort to broaden access to the extensive facility data now assembled, corporate real estate began migrating its three facility databases to the Partners' intranet, which linked all hospital campuses. A major criteria for the facility information Web-site design was that it be "self-maintaining"; that is, that it require no intervention from facilities staff to handle the queries, since no "Web-master" was available. This resulted in a design that gave users direct, view-only access to the live databases. Since two of the applications reside within the ARCHIBUS software family, that vendor's "Web Central" product was used to provide a rapid web interface. It took a while for folks to realize that this trove of facility data was now readily available, but its constituency is steadily growing as people discover it. Groups and departments who have begun to use the data include project managers, security, telecommunications, research affairs, housekeeping, and a few intrepid administrators. The next step in the evolution of facility data management is currently under exploration: the adoption of a vendor-supported Web-based project management collaboration tool. Several have already been tried in conjunction with specific design projects, and a standard may be forthcoming.

15.2.5 Tangible Outcomes

A system-wide Partners' Master Facility Plan was prepared in 1998 modeled on several market scenarios and subsequent growth and revenue consequences. Preparation of the plan required broad, high-level support from all elements of the Partners' system, and though a set of recommendations was fashioned and adopted, their implementation has not been consistent. This is less a consequence of any short-comings of the plan itself, and more a reflection on the natural maturation of the Partners' system—that it behave more like a system and less like an association of hospitals.

As support services moved to the corporate level, such as human resources, information systems, finance, legal, real estate, and others, there was a desire to consolidate like services from the different hospitals in a single location. In all, Partners' corporate has more than 1,500 employees and occupies over 400,000 ft^2 of space, all leased. Since employees were coming from many different officing situations, it was determined that standards should be established for the types of space that would be available. A working group with senior managers from each support service met over a 6-month period to establish the standards. While space allocation by job function was the driving force, there were many beneficial by-products extending to outline specifications for architectural finishes and a build-out cost per FTE. A consistent application of the space standards has removed both the inequities and dissatisfaction that often accompanies forced relocation; it has also resulted in a reduction in the square feet per employee compared to preconsolidation space usage. These standards are also being adopted by the individual hospitals for use within the hospitals.

The buying power of seven hospitals was not lost on Partners' real estate. There have been several initiatives to purchase services and supplies on behalf of the member hospitals, the most successful being furniture and equipment. Working groups developed corporate standards for office systems' furniture and equipment that now apply across all entities. Proposals were solicited and interviews held with national vendors and their local representatives and a 3-year agreement offering significant discounts, dedicated customer representatives, and a "quick-ship" commitment on selected pieces was negotiated. Annual savings of $2–3 million a year are estimated by this one action alone.

Significant variation in the cost to fit-out off-site ambulatory clinics and physicians offices caught the attention of one of the hospital CEOs and Partners' real estate was requested to analyze and report back. Analysis surfaced two problems—first, sites were being leased solely for their location and often without any review of the existing building conditions or ability to accommodate the proposed use, and second, project scope was sometimes increased to include the replacement of major building systems and the upgrading of public areas in buildings where the hospital may occupy only 20% of the space. These revelations led to a number of changes, including centralizing leasing activities within Partners' real estate, undertaking a thorough evaluation of building systems and suitability for its proposed use prior to committing to a lease, and negotiating responsibility for repairs with the owner. Working with the MGH and the BWH ambulatory clinic managers, Partners' real estate developed build-out standards for off-site space, and also established a number of scope and budget thresholds that should only be exceeded with the knowledge and concurrence of the COO/CFO. While savings are difficult to quantify, by a number of measures the range of unit costs has narrowed considerably since the ambulatory standards were adopted.

Given the annual volume of leases up for renewal and new leases initiated, Partners' real estate solicited proposals from real estate consultants to assist with site identification, lease negotiations, and permitting. Partners selected a firm that specializes in assisting owners with their real estate transactions and that does not broker real estate on behalf of others. While the consultant was asked to identify and negotiate leases on Partners behalf, traditionally broker's services, they agreed to be compensated on an annual retainer basis rather than on a transaction basis. This served Partners' purposes by removing any incentive for the consultant to promote a more expensive transaction and because it also freed Partners to request different services from transaction to transaction. Through the first 2 years there have been significant savings to Partners as a consequence of this arrangement.

While the centralization of real estate services by Partners has brought about a consistent approach to projects and identifiable savings, it has also brought a more complicated decision-making structure for capital projects that may at times slow the delivery process. A consultant's master plan recommendation was that Partners needs to be "nimble" in making facility and real estate decisions, especially in a market with a low-vacancy rate and rising rents. One approach has been to look to alternative project delivery systems to harness the efficiencies available in the commercial marketplace and at the same time use those delivery systems to stretch the resources available within Partners. Over the past 5 years, Partners has committed several major projects to alternative delivery systems with excellent results. A $46 million proton therapy center was procured through design/build and a $42 million research lab is presently being developed under the "turnkey" method. Several smaller design/build clinical fit-out projects have also proceeded. A key to their success is that they are discrete, stand-alone projects. Harnessing private market experience and innovation to implement well-packaged projects can certainly help Partners be more nimble.

The rationale for a common CAD/CAFM system was that Partners would want/require "apples to apples" comparison of space assets for various reasons, like Medicare cost reporting, space allocation, asset valuation, and day-to-day manager knowledge of "what's going on" related to space use. There was strong belief that since Medicare reimburses hospitals in part for their capital costs, they would demand equal comparison of spaces across Partners' entities. That hasn't been the case to date. In fact, Medicare, NIH, and others who require that space cost and use be substantiated and that changes in use be closely tracked one year to the next approach each hospital in the Partners' system separately. Nevertheless, the availability of space and use data across the BWH and MGH and, to varying degrees, the other Partners' hospitals, does make for significantly easier internal analysis as well as consultant support. While the initial cost is high, it is Partners' contention that the payback is short given the multiple uses to which the information is put and the alternative, which is to pay a consultant to prepare an existing conditions CAD plan each time a section of the hospital is being analyzed or redesigned.

On the issue of nimbleness, the lease management system (now in ARCHIBUS) seems to have proved its value—there is genuine enthusiasm for this system at the corporate level. One example: for a recent property acquisition study, Partners' real estate was given one hour to assess the status of all Partners' leases within the proposed acquisition. This application did the job, letting management see "the whole picture" of current lease liabilities quickly and completely.

The project management system is also proving very valuable. Every project manager can get a "real time" picture of each of their projects' schedules and costs-against-budget on a daily basis. Corporate can compare vendor performance historically (there are now 3 years of corporate-wide project data in the system to analyze). Project manager workloads are easier to predict and balance. Smarter purchasing could result if construction item purchases were pooled, which is now possible. No specific dollar savings from this system have been documented to date, but the general feeling is that it is a good investment, it does save construction dollars, and it enables Partners and the hospital-based project managers to maintain excellent relations with their vendors.

CHAPTER 16
ORGANIZATIONAL READINESS CASE STUDY: IMPLEMENTING TECHNOLOGY AT ROCKETDYNE, A BOEING COMPANY

Carolyn Castillo
Program Manager
Metrics, Systems and Process for the Space and Communications Group business segment of the Boeing Company
E-mail: Carolyn.Castillo@West.Boeing.com

This chapter deals with the implementation of technology within a facilities organization supporting manufacturing operations in the aerospace industry. In this case study, the author discusses the approach undertaken by Rocketdyne's facilities organization in implementing technology. In the early 1990s, Rocketdyne's facilities management (FM) was striving to understand the expectations of a process-driven organization, and transformed the organization by establishing visions, new process models, and driving change through the use of a Computer-Aided Facilities Management (CAFM) system. As the organization started its journey toward automating its processes, it realized that a new computer system alone does not improve performance.

Rocketdyne, part of the Boeing Company, is a manufacturer of propulsion systems, including rocket engines, space station electrical systems, and high-energy laser systems. The company, headquartered in Los Angeles County, has approximately 4,500 employees. Its three facilities consist of 1.9 million ft^2 of office, factory, laboratory, testing, and warehousing areas. The plant produces rocket engines for both reusable and expendable launch vehicles. In addition, it manages a rocket test facility located in the Santa Susanna Hills. The most widely recognized rocket produced by Rocketdyne is the Space Shuttle main engine.

16.1 THE FACILITIES ORGANIZATION

The facilities organization is responsible for space planning, facilities design, coordinating and managing construction projects, as well as maintaining equipment and building systems. Rocketdyne, a defense contractor, manufactures and tests rocket engines and propulsion systems. Since the 1970s, the business was *program* oriented; that is, a government entity, such as NASA or the Air Force, awards contracts to fund various programs. The programs have a specific objective and life cycle. For example, Rocketdyne's main product during the 1980s and 1990s was the Space Shuttle's main engine. The program included the development, manufacturing, and testing of the Shuttle main engine. The facilities organization supported the manufacturing operations by planning and designing the factory as well as test stands.

The manufacturing of rocket engines requires specialized equipment, which ranges from very close tolerance numerically controlled machine tools, such as lathes and grinders, to sophisticated x-ray cells, and vacuum and furnace chambers. The facilities organization was responsible for performing the material flow analysis, the installation of production and test equipment, as well as the maintenance of company property. Industrial, mechanical, structural, and electrical engineers worked either in facilities engineering or maintenance. Typically, there was a division of work between the engineering and maintenance organization with little integration. The facilities engineering group managed various capital projects, such as specifying, selecting, and installing production and building equipment. In addition, facilities engineering managed building construction projects ranging from the design and construction of a building to minor facilities modifications. The maintenance organization was responsible for maintaining the building infrastructure, as well as production equipment. Maintenance work was primarily reactive—"Don't fix until broken" philosophy; however, in prior years maintenance began instituting equipment preventative maintenance programs for various types of equipment that included both machine tools and building systems.

By the 1990s, Rocketdyne was facing two primary challenges: lower sales as programs were completing their life cycle and noncompetitive labor rates reflecting a culture where the cost of the product was based on a cost plus contract. Rocketdyne's customers, experiencing spending cuts and other fundamental changes in the marketplace, compelled Rocketdyne to shift their pricing strategy to attract fixed price commercial contracting. To capture this new business, Rocketdyne became focused on reducing costs, increasing efficiency, and improving its quality through process management. Eventually, a new management philosophy touting process flows, variation reduction, and continuous process improvement (CPI) emerged, including a number of initiatives; for example, concurrent engineering, activity-based management, and design to cost. Rocketdyne was on the path to rightsizing, cost cutting, and improving performance. Functional organizations began to better understand their work flows, process flow charts began replacing procedures, and process performance was beginning to be analyzed in terms of cycle time and variation.

The facilities organization, in concert with the business objectives, struggled to improve its own productivity. One of the earliest attempts to improve productivity through automation was a pilot project, the installation of a Computer-Aided Facilities Management (CAFM) system. The CAFM system would provide two functions: first, it would account for space by producing area-reporting statistics as well as occupancy utilization data, and second, it would automate the move management process. Theoretically, once occupancy information was collected, the information would be reused instead of being discarded. This alone would improve productivity in terms of the industrial engineer's time used to collect information such as department numbers, occupant names, phone numbers, and associated assets for personnel moves. Also, the updates on asset location and ownership could be fed into property accounting systems. The pilot project was only a partial success. The facilities engineering organization was able to produce area-reporting statistics by location and space type to the department level.

The move management capability, though demonstrated, was never implemented as originally envisioned. It was expected that once the industrial engineers were trained on the system, data would be collected and entered into the system. The industrial engineers would use the system and the data would be maintained as part of the move process. It should be noted that the organization had no understanding of how to implement a facility system. In the past, CAD systems were just installed and personnel trained on the use of the equipment and software. This type of implementation can be coined a "plug–and–use" approach. In the past, it worked because the systems were stand-alone. The question now becomes: why didn't the move management work? This first facilities automation pilot, only a partial success, forced the author to reconsider how to successfully implement technology. There was a lesson learned—the need for an overall plan for implementing integrated computer systems.

The facilities organization, slow to understand the nature of process management, experienced a significant emotional event, an organizational change. Its superb "rocket test stand" to "industry

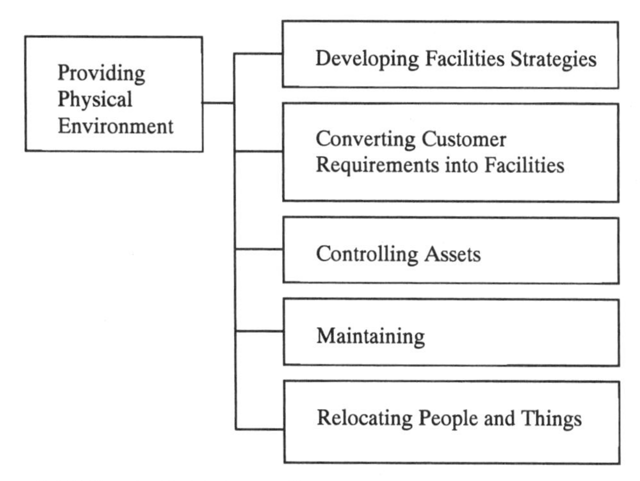

FIGURE 16.1 Facility Process Trees.

standard" capable design functionality, was eliminated. Design services were to be subsequently outsourced. It was a question as to whether more of the organization would be outsourced. This was the beginning of a new era for the facilities organization. The management staff, working together, forged a framework to explore the characteristics of organizational change and its tools.

16.2 PROCESS CHANGE THROUGH THE USE OF TECHNOLOGY

"We have always done it this way." "Why change?" "What if we fail?" Rocketdyne's FM viewed themselves as performing optimally and meeting the business goals, but did they really understand the challenges within their own organization? The business need was to become competitive in the marketplace by reducing operating costs, improving performance, and focusing on its core products and services.

The first question the FM team attempted to answer was the role of facilities within Rocketdyne. After several meetings, process trees were drawn where a primary core process and five subprocesses were defined (Figure 16.1). The facilities organization defined their primary function as providing a physical environment to conduct business and manufacture products. For the facilities organization to provide that physical environment, they performed a number of processes:

- Development of facilities strategies dealt with understanding the business's long-term goals in term of production capabilities and capacities, as well as space requirements over the life of a specific program or product
- Converting customer requirements into facilities and equipment involved evaluating customer requests, such as space needs and proposed capital equipment, and converting those requests into a facilities project
- Controlling assets required performing equipment utilization studies, tracking asset locations and ownership
- Maintaining equipment involved performing service and repairs on production equipment, building equipment, and systems
- Relocating people and things involved the planning and implementation of moves in support of a business initiative or program life cycle

The FM team believed that once processes were analyzed, a framework could be developed to improve performance. In addition to examining existing processes, the FM team decided to evaluate industry trends, specifically, technology.

16.3 TECHNOLOGY

"Why use technology?" When Rocketdyne FM decided to use technology, the application was identified: a computerized maintenance management system (CMMS) that would replace an obsolete maintenance work order system. As mentioned earlier, initial attempts to implement technology were only a partial success. What could be done differently? It soon became apparent that there was a lack of understanding of the role of technology and its relationship to processes and information.

Thomas Davenport's[1] book, *Process Innovation,* provides insight into the role of technology in process base management. Davenport addresses three aspects not easily understood: first, information technology enables a process, it is a tool; second, technology is a radical change to the process; and third, the process will allow the organization the capability to achieve its business goals. There is a relationship between the tool, process, and goal (Figure 16.2). The tool enables the process to achieve goals.

As depicted in Figure 16.2, the tool enables the process. The tool for the maintenance group was the CMMS. The facilities team was assigned to recommend a new computer system and struggled to identify a process. Unknowingly, the facilities engineers started to flow chart the work order process and lost sight of the goal, recommending a computer system. It is important to emphasize that the selection of a process is critical—start with the process not the tool. The process defines the tool.

The CMMS must be financially justified because it constitutes a major investment. The cost of a computer system commonly includes software, hardware, training, and implementation. The business case for Rocketdyne's proposed maintenance management system hinged on demonstrating a cost savings, which can only be achieved by reducing material and labor. The facilities organization had a number of initiatives, such as *just in time* delivery of maintenance stock, improvement of equipment effectiveness-downtime, and process cycle time. Based on those initiatives, the facilities engineers focused on specific processes: inventory management, failure analysis, and performance metrics, to identify potential cost savings. The primary areas where the team felt savings could be achieved was to reduce the count of spare parts and inventory. Also, as we later discovered, it was difficult to identify specific labor savings, but gains in work efficiencies could easily be identified. The work efficiencies would appear over time in the reduction of work order backlogs and improved cycle time.

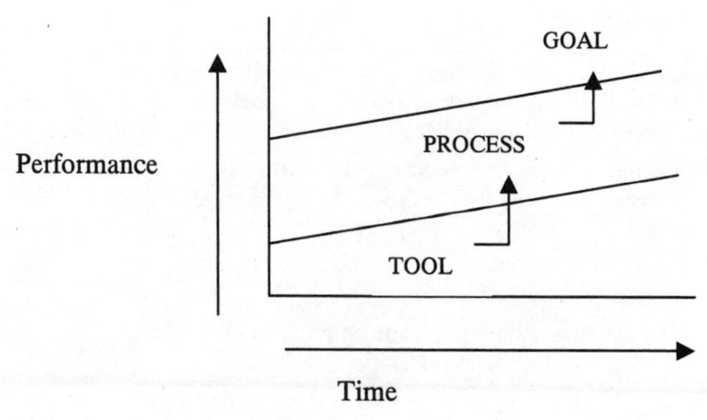

FIGURE 16.2 Relationship Chart.

16.4 ORGANIZATIONAL READINESS

Most organizations blindly purchase software packages expecting that, once installed, people will use the system. The product often does perform according to specifications, but very few understand how to implement an *enterprise* software system. Hence, the so-called shelfware effect—software bought but never used or not used optimally. The problem is not necessarily in the software, but rather in organizational readiness. Organizational readiness involves three factors: processes, information, and people.

16.4.1 The Process

A process is the delivery system: the sequence of steps, tasks, and activities that must be performed to satisfy the customer request. The starting point, or input, is the customer request. The output is the customer's request being satisfied. Process performance is how the product or services met the customer's expectation in terms of responsiveness, quality, and cost.

As a starting point, the facilities engineers had developed a process flow chart that included subprocesses. To illustrate how the new process model evolved, we will examine a simplified process. The work order process flow is as follows:

1. The customer request is received.
2. The request is logged into the system. Requester's name, department number, phone number, charge number, and problem is recorded. If the problem is associated with a piece of equipment, the property number is also recorded.
3. The request is forwarded to the manager for review and, if accepted, converted into a work order.
4. The work order is assigned to the maintenance worker.
5. The maintenance worker performs the work and completes and closes out the work order.

After the process was flow-charted (Figure 16.3), performance was investigated from a customer perspective: how long did it take to complete a work order, and was the customer satisfied? Because there was little or no information on performance, the facilities engineers listed the causes for any potential delays. What prevented the work from being completed sooner? The reasons follow:

- Incomplete or incorrect information: department number, charge number
- Asset information not available: property identifier, location, repair manuals, and spare parts listing
- Lack of materials or tools not available to perform work: spare parts or material not on-hand or unable to locate.
- Lack of skill to perform repair or maintenance: employee does not possess the information, appropriate knowledge, and in some cases, the training to repair the equipment

The list correctly identified that critical information was missing and the material to perform the work was not always available. This was key to the process analysis. In the ideal environment, when work is assigned to the maintenance worker, the worker can locate the equipment, then identify and fix the problem by replacing broken parts. Since we did not have the ideal environment, we needed to create processes that allowed the maintenance worker to perform the work the first time through. As further process analysis was completed, it became evident that a critical process, the planning and scheduling of work, was lacking. The new process model now included a new subprocess: planning and scheduling work. The new work order process is as follows:

1. The customer request is received, logged into the system, and routed to the team leader
2. The work order is received and evaluated

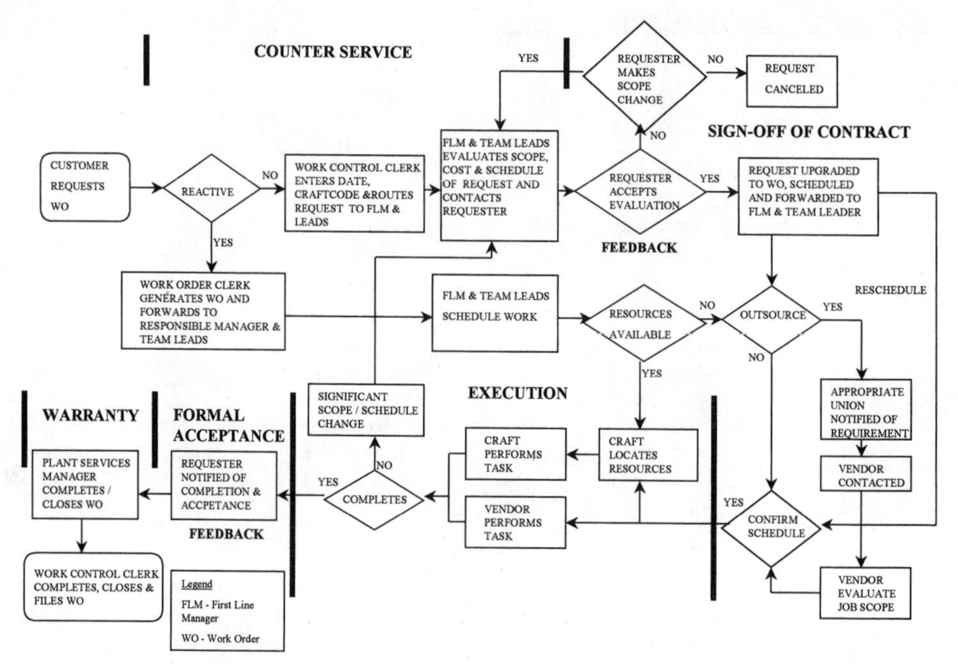

FIGURE 16.3 Work Order Process Flow.

3. The customer request is accepted and converted into a work order
4. The work order is scheduled by checking the availability of equipment, spare parts, and the appropriate maintenance worker
5. The maintenance worker receives the work order, analyzes the problem, and obtains all the necessary parts, which have been kitted in the stock room
6. The maintenance worker performs the work and completes and closes out the work order

How did the new sub-process, work scheduling, change work? As shown in Figure 16.4, planning and scheduling involved availability—having the right person, the right part, and access to the equipment all at the same time.

Figure 16.4 demonstrates the importance of having the right spare part and material on hand for the maintenance worker. An examination of the stock room operation found the stock listing was poorly maintained and there were no records to link the spare part to the equipment. Equipment repair manuals were missing and preventative maintenance instructions were not detailed enough to provide equipment-specific information. It was surmised that most maintenance workers faced the following scenario:

- The worker receives the work request and attempts to research the equipment history
- After searching, the equipment is located
- Without the aid of a manual, the worker must perform tests to determine the cause of the equipment failure
- Once the failure is determined, a number of parts are needed for repairs that hopefully are in the stockroom
- Unable to find the part, a purchase order request is initiated to order a new part
- After management approval, a purchase request is assigned to a buyer

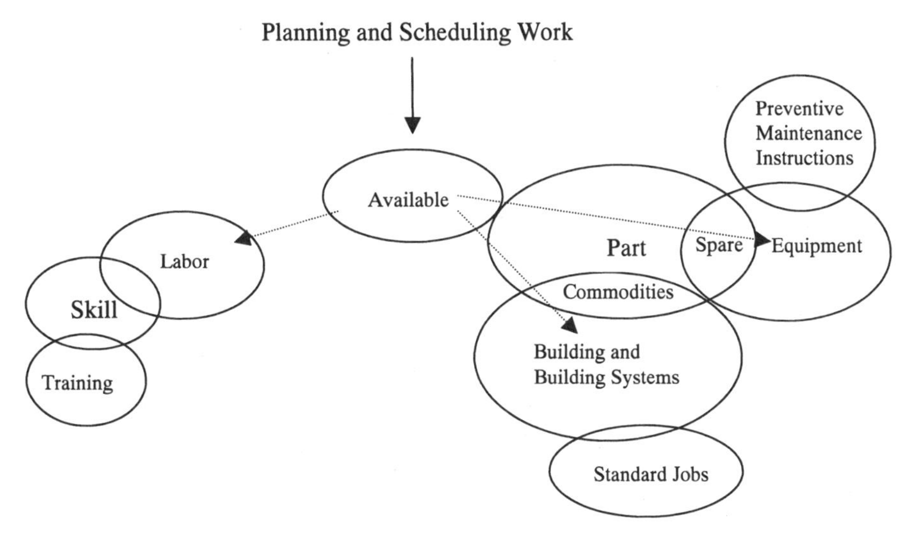

FIGURE 16.4 Planning and Scheduling.

- The buyer must then locate the part without knowing the correct part number or supplier
- The part is ordered and received days/weeks later
- The maintenance worker is notified that the part has been received
- The maintenance worker then checks on the availability of the equipment—if down, work is performed. If not, work is rescheduled when the equipment is available
- After work is completed, the maintenance worker closes out the work order

The scenario confirmed that time was wasted. To reduce wasted time, equipment information must be available and accurate, spare parts must be stored at their assigned locations and linked to the equipment and repair manuals.

For *commodities*, different sets of problems existed. For example, it was discovered that when construction specifications were developed, the designer would specify different types of materials or use instructions such as "match existing." As a result, the stock commodity ordered would vary. In the case of poorly defined material for construction, the craftsman would have to determine what materials to order to "match existing." Whatever the case, the outcome was the same—an increased stock inventory count consisting of similar parts, different manufacturers. A good example of this would be floor tiles or paint where different colors were used in different areas within the building. This problem could only be solved by standardizing the material and providing the design engineers with a list of standard materials that would be referenced on construction drawings.

Additionally, a new element emerged—standard job instructions. Historically, job instructions had been used to develop equipment preventative maintenance (PM) documents. The PMs developed internally used the manufacturer's information by providing check-off instructions to the craftsman to maintain equipment based on a predetermined schedule or equipment running time. However, the use of standard jobs to provide general instructions and materials had to be created. For both the PM and standard job, a labor standard had to be established to provide an estimate for the amount of time expended. The estimated time is used to schedule the labor resources.

In summary, the new process model defined informational requirements and relationships essential for maintenance through the work order system. Through a number of discussions regarding the work order system, questions arose about the differences between the engineering and maintenance work order system. Eventually, it was concluded that the new process model would be expanded to serve the entire facilities organization. The opportunities for cost savings were much greater with the integration of specific work processes and the sharing of information, such as asset location. The engineers were no longer focused on a maintenance management system, but rather a Computer-Aided Facilities Management (CAFM) system.

16.4.2 Information

To operate, a system requires *information.* Information can consist of a drawing, a table, diagrams, or documents. The formats can vary between graphics, databases, or word documents. Systems' implementers must define the information requirements, as well as how the information will be moved into the new system. The informational requirements (Figure 16.5) provide a visual diagram of the elements that must be considered. The facilities organization uses database information and graphical information, as well as accesses the company's business systems.

For the facilities organization, the information requirements (Figure 16.5) can be categorized into three main elements: space planning and location information, asset information, and business structure.

Space planning and location information provides real property information: property location, size, and ownership. This information, a property portfolio, lists the business properties and gross square footage. If the property is leased, lease information, such as termination dates, can be collected. Typically, the data format is a combination of graphical information coupled to a database.

The building layouts provide a graphical depiction of each building floor, walls and partitions, equipment and furnishings located at each property address. Most facilities organizations break down the type of space used into categories such as office, laboratory, factory, warehousing, and miscellaneous. This information, used in strategic facilities planning, involves the identification of appropriate space requirements to support the business's objectives. In the building layout, specific equipment locations are shown, as well as organizational boundaries. Used for decades, building layouts are a facilities tool used to illustrate operational flows and indicate equipment locations. The missing element has been occupant information. The CAFM system allows the linkage between an office space to an occupant. One critical aspect that must be discussed is the relationship between graphical information and the database. It has only been recently that technology has evolved to allow the updating of drawing information via links to a database. This is a break-

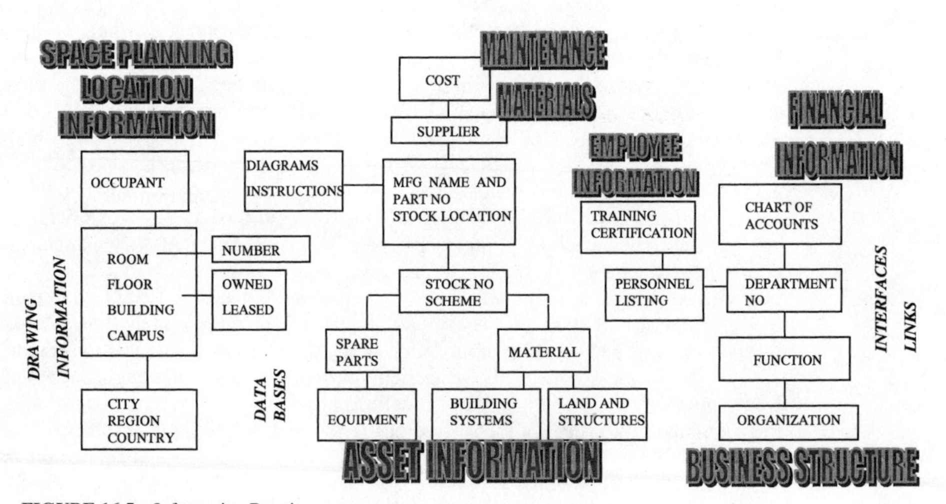

FIGURE 16.5 Information Requirements.

through. Facilities information can be maintained though the database more efficiently than updating a drawing. Today, CAFM software provides bi-directional updating of drawing and database information.

Asset information is a listing of building structures, building systems, and equipment. Businesses depend on property accounting systems to maintain asset information, such as a description, manufacturer's name, model and serial number, property identifier, value, age, location, and ownership. For the facilities organization, the asset information must be expanded to include spare part information, and repair and maintenance instructions.

Normally, building materials and commodities are not found in property listings, but rather developed by the FM organization. This information is collected and maintained in a database. Equipment information can be migrated from the property systems. All other information, such as stock number, spare part stock number, used for, stock location, and cost, must be migrated from existing systems or entered manually.

Business structure, such as administration, marketing, operations, human resources, finance, and engineering, refers to an organization's internal structure. The facilities organization requires information, such as personnel listing by departments, property listings, and ledger account information. This information is usually "owned" by other functions that have their own systems. Rather than duplicate the information, an interface to the financial and human resource systems is recommended. One aspect of the business structure information that must be included in the facilities system is the listing of facilities personnel along with their skills, knowledge, credentials, and certifications. This information is used to assign and schedule facilities personnel to the appropriate work.

Summarizing the informational requirements, it is important that the system be populated with complete information. The data and interfaces must be operational and tested before people can actually use the system.

16.4.3 People

"People" refers to those employees who will be involved or impacted by the implementation of new processes or technology within the organization. People perform work. Each person brings specific knowledge and experience to perform the appropriate task or activity. As jobs were created, specific tasks and responsibilities were defined. A new process model, resulting in a dramatic productivity improvement, is referred to as reengineering. Originally, reengineering addressed the process, but not necessarily the people. In *The Reengineering Handbook,*[3] people issues were addressed by considering the social design of work; that is, what are the jobs that will be changed and how will that change affect employees. Jobs will be redefined, skills and staffing will change, as well as the management structure. Management must recognize early that they must be involved with the implementation of technology to plan any job transition. There is a perception that jobs will be eliminated by technology, but, in most cases, new jobs are created. Employee's roles and responsibilities are expanded.

The concept of *organization readiness* evolved over the last few years and, if followed, will certainly reduce implementation costs and shorten the implementation schedule. An assessment of readiness can be developed and should be tailored to the organization in terms of processes, information, and people. For Rocketdyne, a questionnaire was developed. Typical questions are listed below by category:

- *Organization.* Are the processes identified, flow-charted, and documented? Does the management support a learning environment? Are process owners assigned to oversee and manage the process?
- *Space.* Does the organization have a layout department? How is the CAD system used to identify type and use of space? What types of area reports are produced? Who uses the report?
- *Work Management.* Is there a single work management system for the organization? How is work assigned and prioritized? What performance metrics are reported?

- *Parts.* Is an inventory management system used? What is the inventory count of items in the stock room? Is there a process in place to receive, issue, and eliminate stock? Is an annual physical inventory conducted?
- *Finance and People.* What systems are used? What reports are generated and how often? What type of costs are tracked and monitored?

When completed, the assessment will identify the performance gaps and indicate the level of effort needed to implement the software. Simple metrics such as cycle time, backlogs, and inventory part counts can be used.

16.5 DRIVING CHANGE

"Why change? We have always done it this way." The problem is that new technology and new process models drive change. Few businesses understand the nature and drivers of change. Davenport[1] best characterizes change as linear or radical (Figure 16.6). A linear change is equated to continuous improvement. Continuous improvement focuses on reducing variance and eliminating valueless activities. Improvement is considered incremental since there is little change to the process other than changing the sequence of tasks or eliminating some activities. Cost savings achieved are minimal.

A radical change is a one-time event, usually driven by management or an extraordinary event, cross-functional, and results in a significant cost savings and/or satisfaction of other strategic business objective. A radical change occurs when there is a delivery system redesign or new mind set. A *delivery system redesign* takes place when a process model is expanded and integrated cross-functionally to change the business framework. Another approach undertaken by many businesses is to use technology as the basis for designing a new process. Without a new process model, a business can drive change by imposing a new technology and forcing the attendant establishment of a new mindset. This approach takes strong leadership to instill a vision and develop a change strategy. A new mindset can include restructuring the business to "stay" in business. The reason some organizations take this approach is due to a significant trauma, such as a performance gap leading to lower sales, a loss in market share, or a threat of a takeover. This change strategy is immediate/discontinuous, a step function, and is meant to bring about required change and/or improve performance immediately.

16.6 THE SHOPKEEPER MODEL

At Rocketdyne, the facilities leaders experienced a significant trauma with the elimination of their design department. They also sensed that the organization needed to adapt to other realities of the marketplace to significantly improve its performance and develop a different type of relationship with the rest of the company (customers). The new mindset, the shopkeeper model (Figure 16.7), established the facilities organization as a "business within a business." The premise of the shopkeeper is that facilities must operate as an independent business. As a business, we had customers, sold products, or provided services. As owners, we had to know our capabilities, operational cost, market our services, get customers, perform, and satisfy our customers' needs. To stay in business, we needed to be cost effective—cheaper than our competitors; provide quality service—warrant our work; and be responsive—deliver our services within a reasonable time. Lastly, we needed to measure our performance: cost, responsiveness, and quality.

Once the shopkeeper model was established, the facilities leadership had to convey the new mindset throughout the organization. As indicated earlier in the chapter, the FM had identified its "core processes." As processes were analyzed, a process owner and a team were assigned to these specific core processes. The process owner and his/her team became responsible for carrying out

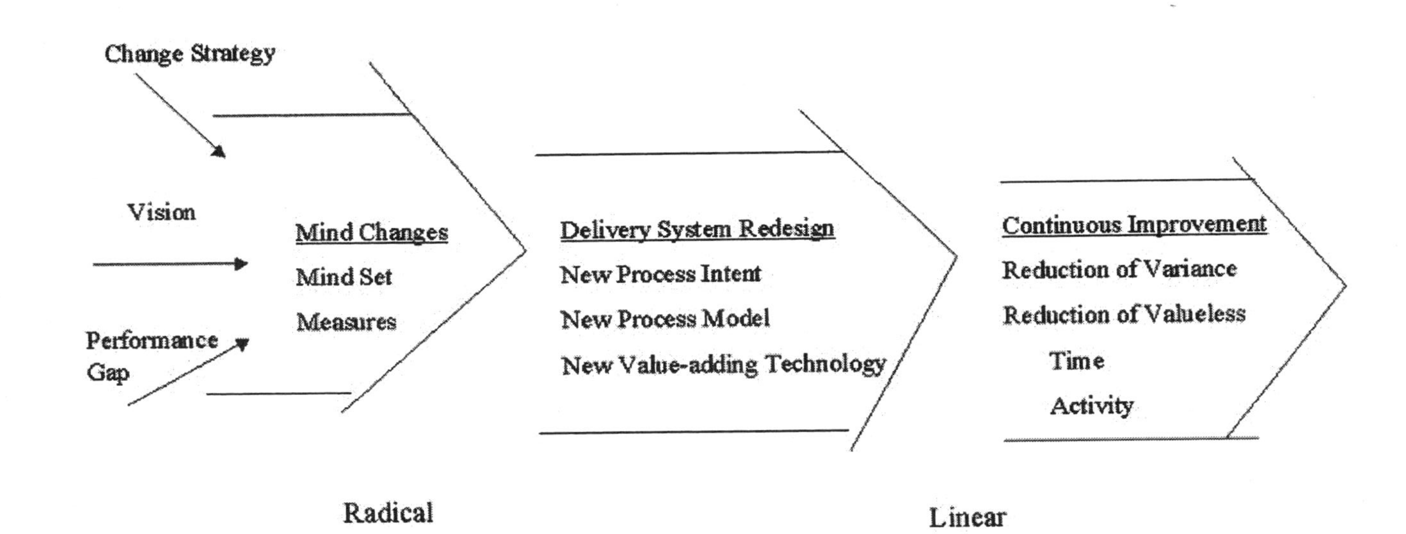

FIGURE 16.6 Process Change.

1.0 STRATEGIC BUSINESS PLAN	2.0 COUNTERSERVICE: PRODUCT DRIVEN ORGANIZATION PLAN	3.0 COUNTERSERVICE: PROVIDING PRODUCT/SERVICE	4.0 MEASUREMENT PERFORMANCE
1.1 Overall Shopkeeper Metrics	2.1 Maintain Req'd Long Term Plans	3.1 Select and Establish Definitized	4.1 Measure Key Indicators and
	Land and Building	Agreement with Customer	Ratios
1.2 Select and Establish Products	Seismic	Verbal	Actuals vs Budget
Core	Factory Machine	Written	Cycle/Process Time
Non-Core		Terms and Conditions (Includes	Transaction Rates
Existing	2.2 Select and Establish Product Cost/Billing	dealing with disputes, changes	Performance of
Proposed Future	Defining and Establishing Reqmts/Cost	and cost adjustments)	Employees
	Schedule		Equipment
1.3 Cost of Doing Business	Price	3.2 Providing Product/Service	Subcontractors
Marketing	Transport Delivery	Warranty	Customer Surveys
Goodwill	Backorders/Returns	During	Employee Surveys
Legal Reqmts	Billing/Final billing	At Delivery	
Employees/Shareholders		Post Delivery (Comebacks)	
	2.3 Select and Establish Processes		
1.4 Strategic Actions/Plans	supporting the Product/Service		
Benchmarking	Staff Mix		
Overall Training Strategy	Project/Service Organization Plan		
Special Projects	Legal Requirement (Includes Health/Safety)		
Organizational Redesign	Integrate Initiatives		
Facilities/Resources			

FIGURE 16.7 Shopkeeper Model.

continuous process improvements by developing the process flow chart and deciding how and what performance to measure. Employees were briefed on the shopkeeper model's five elements:

- Counter service prior to sign-off of the contract
- Sign-off of contract
- Execution
- Formal acceptance
- Warranty

The shopkeeper model elements were superimposed on the process model, as shown in Figure 16.3. *Counter service* can best be described as being the person behind the store counter—the recipient of the request. The process owner/team member/agent at the counter has a responsibility to evaluate the request, obtain customer clarification, provide a cost for the product or service, and commit to a completion date. *Sign-off of contract* is the "meeting of the minds" where the owner and customer reach agreement on the product specifications or the service provided, cost, and scheduled completion date. This step allows the customer to obtain other competitive services or decline the offer. *Execution* is the performance of the work. *Formal acceptance* is a review of the work performed by the customer. The customer knows that the work was performed and has an opportunity to accept or reject the product and/or service. A rejection should be viewed as a quality problem. After the customer agrees to the work being completed, the work order is closed and is equated to a sale being complete. *Warranty* is characterized as a guarantee of the work, and if a breakdown occurs within a specific time period, legitimate warranty work will be performed at no cost. Occasionally, audits are conducted to ensure continued smooth operation. When Rocketdyne's facilities organization employed this model, the focus shifted to being customer-oriented.

The shopkeeper model is divided into two segments: one focused on the day-to-day tactical operations and the other focused on future business and strategic issues. Each of the tactical operations managers would have a number of separate process teams within their organization reporting to them. It was understood that the boundaries surrounding each of the operations managers' units would be flexible to ensure the customer's request could be met. The processes existed under a specific operations manager umbrella because the task responsibility fell within that manager's charter—the processes were placed where it made sense. Additionally, the process team would rely on support from key persons within the organization. It is also required that process owners understand the methodology and possess knowledge regarding their subprocesses and system performance; that is, they must understand the entire delivery system model and the relationships between the subprocesses. The performance of each subprocess will affect the entire delivery system.

The other aspect of the shopkeeper model is that of addressing organizational strategic components, including the strategic business plan. The strategic issues and business plan deals primarily with the organization's effectiveness, both present and future. Did we know our core processes and did we perform those processes well? Noncore processes were candidates for outsourcing. What was the cost of doing business? Was our performance comparable to our competitors? As the facilities leaders addressed these questions, it became evident that we needed to predict the cost of maintaining our buildings and infrastructure. This is where a facility assessment turns into a critical tool for a facilities organization. A facilities assessment is an audit on the condition of your building based on the life cycle of the asset, both structure and infrastructure, and cost to maintain that asset to a specific condition level (see Chapter 9). The concept is very similar to maintaining your house. Over a period of time, the walls need painting, the roof needs to be replaced, the piping needs to be replaced, and wiring needs to be upgraded. The facilities organization started to collect this information and referred to the document as a reserve report. The reserve report was intended to list, by building, the expected life of major equipment and systems and replacement cost. This report would allow the capability to predict the timing when major renovations were required. Capital funding requests could then be planned around this information.

As the shopkeeper framework developed, new leadership roles were defined. The facilities director role was to become the interface to the "outside world," provide strategic direction, and align the processes to maximize its delivery system. This is a major paradigm shift that must take place to move the organization from a reactive to a planned environment. Also, as processes mature, process owners will replace managers and the role of the manager will change. The manager's new role will focus on "patching" the organization by reassembling teams and being a system's integrator to meet the customer's need. We know that to be competitive in the marketplace of the future these processes must be dynamic and managers/integrators and their process teams must be able to work as an agile organization.

According to the shopkeeper model, the facilities organization must anticipate its customer's needs by examining their contributions to the organization, and understanding how external factors, such as deregulation of electrical utilities, new technologies, and growth, affect facilities and operational costs. Areas of investigation and concentration include:

Sense of Urgency

- Emphasis on helping employees be more productive and retaining them
- Emphasis on new product development (e.g., Northern Telecom has 80,000 employees and $\frac{1}{4}$ are R&D)
- Speed to market
- Responsiveness

Team Work

- Major concern is how to get people of various disciplines to come together and work as a team—engineers and scientists of different specialties are considered especially difficult
- Team centered, colocation by mission, process driven

Focus on Workspace as a Tool (Not Just Places)

- Less allocation of individually assigned space to common shared space
- Technology addressed (current and projected)
- Local adjacencies of high cost shared equipment
- Office/lab—must be addressed on the component level and the system level together
- Space allocation by task, not by title
- Consolidation, dual use of space; for example, cafe immediately outside of elevator doors
- Cost effective approach to growth, evolving processes, changed circumstances, flexibility, and adaptability; for example, rugged modules and other facility/tool assets

- Fixed/standards
- Flexible infrastructure
- Customizable components
- Utilization of maximized, reuse of existing, budget sensitive

Appropriate Support of Related Processes Including

- Acquisitions/divestiture
- Alliances
- Outsourcing

"Cool" Place to Work

- Quality workplace
- Alternative work environments
- Pride in work and workplace

16.7 VISION

In the beginning of the chapter, Rocketdyne had identified a process that could be automated—the work management system, for the maintenance organization. Earlier, we discussed that the new process model was changed to include planning and scheduling process. The work management system was also expanded to include the facilities engineering organization so that the customer has one point of contact and work would be routed internally to the appropriate recipient. The organization was then trained on process management and process analysis, such as cycle time. Everything seemed to be in place; however, the most critical change driver was missing—a vision. A process will not undergo change unless the people understand and support this vision.

A vision is both a statement and a plan. A vision provides direction, is long term, and aligns people to a common goal.[4] When leaders talk about a vision, it radiates passion—to be the best, to be world class, to be cutting edge. The vision statement developed by the facilities organization describes facilities as "the best service providers as judged by its customers." The key word is service providers. As the service provider, facilities provides architects, designers, and construction managers to build and modify structures. In addition, facilities provides services to maintain equipment, systems, and structures. Process owners responsible for core processes can then launch their own process visions. As an example, the relationship chart was adopted as the format for the maintenance vision plan (Figure 16.8). In the illustration, a linkage was established to relate the tool, a CMMS, to a process, inventory management, to achieve the goal of a reduction of the inventory count. This approach serves as a communication mechanism for the process owner to align diverse operations toward achieving an organizational goal. As the vision plan is developed, the process owners must take into account the time required to change the operation.

In summary, the implementation of a new process model coupled with a new mindset is the foundation for driving change. The change mechanism is a vision followed by the strategy to initiate process change.

16.8 DEVELOPING A FRAMEWORK FOR ORGANIZATIONAL GROWTH

Technology can be the driver for organizational change since internal processes are changed. However, as the process changes occur and cost savings are realized, the organization must develop a

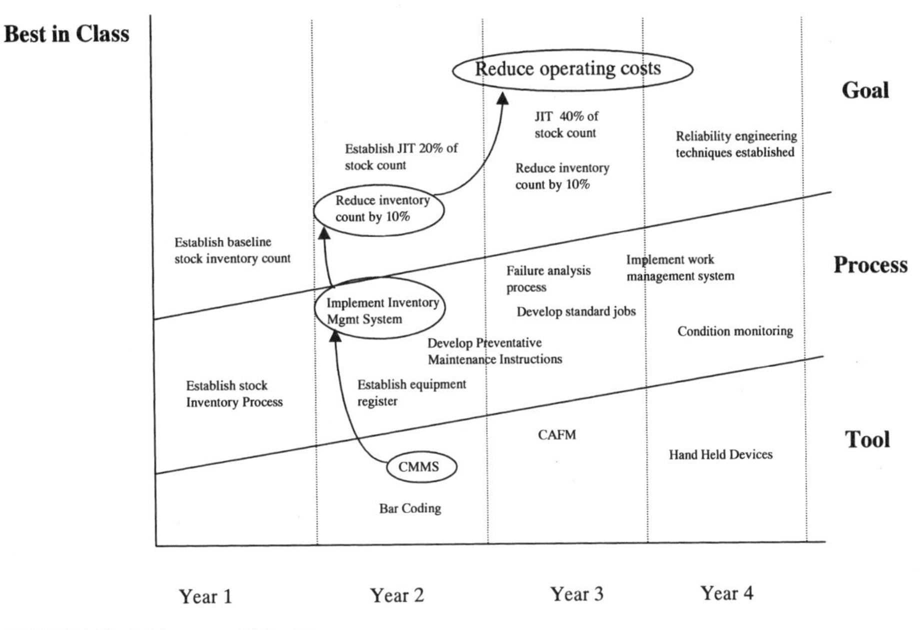

FIGURE 16.8 Maintenance Vision Plan.

framework for continuous growth and change. Typically, a 5-year plan is considered a normal business cycle.

The business model illustrated in Figure 16.9 is designed to provide constant business growth. A business must reinvent itself by analyzing market trends and developing new products that will increase their market share and ensure continued profitability through increased sales. The business develops specific strategies by understanding their customers' needs. Plans are developed, communicated, and implemented throughout the organization. After a period of time, actions are measured by the business leaders and evaluated. Based on the evaluation and market analysis, new strategies are developed and the cycle continues. Similar to the shopkeeper's model, in which Rocketdyne's FM positioned their organization to operate as a business, the facilities organization must employ the business model to reinvent itself. New business opportunities and technologies will change the facilities organization. So how does the facilities organization establish that framework?

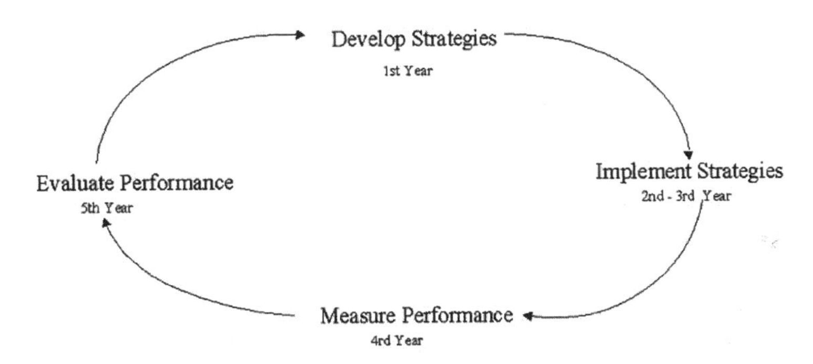

FIGURE 16.9 Business Cycle.

One approach recommended is adapting Thomas Jackson's model,[5] as indicated in Table 16.1. What is a model facilities organization? There are three cornerstones for organization growth which encompass strategy, structure, and strengths. For each cornerstone, specific elements were developed (which are shown in the second column), such as *Customer focus, Organization,* etc. The third column, *Model organizations should* provides guidelines on what the organization should consider in "running their business." *Investigate,* the fourth column, provides an insight on what to investigate and aligns the questions to the elements. The fifth column, *Best practices,* is a combination of defining some organizations' visions, as well as what best practices were found by benchmarking other facilities organizations.

As the framework is developed, stategies are identified. The leadership team should decide what elements are critical for their own growth and focus on developing their own change strategies. As the strategies are implemented, performance is measured and the FM can monitor progress. Similar to the business model, the facilities leaders can evaluate organizational performance and then forge a new framework. As a review, refer to Table 16.1 to examine the elements of what the model facilities organization should do.

The key activity, under *Strategy,* is *Customer focus.* Since facilities is an internal organization to the business structure, its growth strategy is customer-oriented. Instead of researching market share,

TABLE 16.1 Model Facilities Organization

What is a model facilities organization? What are the performance measures?

		Model organizations should	Investigate	Best practices
Strategy	Customer focus	Understand customer expectations and requirements. Focus on meeting customer demands.	What is the average time to respond to a customer request? How many requests/month? How is it characterized?	Minor request cycle time 5 days. Major request cycle time 60 days.
Structure	Organization	Work performed to maximize value delivery system.	How is work processed? Use value added analysis.	Work steps minimized to perform activity/tasks.
	Supplier partnership	Develop a partnership on those processes/services that are best served by a supplier.	What types of supplier relationships are in place today? Outsourced/out-tasked.	Design–build supplier % of work send out. 85% JIT—stock provider count # JIT/total item count.
	Use of technology	Use technology as an enabler to improve processes.	How is information collected and shared within organization? Any work analysis performed?	A data infrastructure for sharing and analyzing information.
Strengths	Productivity	Target on a specific process improvement goal.	Are there any target goals identified and communicated?	30% process improvement; cycle time on 1–2 processes annually.
	Cost reductions	Strive for an annual operating cost reduction goal.	Are there any goals identified and communicated?	$2K/employee/year.
	Quality improvement initiatives	Target on first time through for all processes; eliminate inspections.	Any quality measure monitored?	First time through: no defects.
	Employee empowerment	Provide employees information and resources to meet company's objectives.	Is work structured to allow employee high degree to empowerment?	Self-directed teams. Know cost of services provided.
	Inventory management	Inventory levels maintained to support work in process.	Inventory levels and how managed.	Inventory levels valued at 10% of operating budgets.
	Maintenance	Maintain land and buildings to a "good" facility condition on index. Maintain production equipment to minimize downtime by employing advance technologies.	% of work unplanned vs. planned. Type of work related to maintaining building vs. customer requests.	75% of work performed (labor) is involved with planned work.
	Safety programs	Provide safe workplace and employ safe work practices.	Employee's attitudes and work practices toward safety.	0 Lost work days

the facilities organization should understand its customer's needs by meeting with the customer and understanding their expectations. Customer surveys provide an indication of customer satisfaction in relationship to facilities performance.

Structure looks at the organization in terms of how it delivers its services. Are people organized to optimize its delivery system? For those processes that are not core, are supplier partnerships established? How is technology employed? Is there a mechanism for sharing information throughout the organization? Normally, strategies regarding structure are limited to one aspect at a time since it is difficult to implement too many initiatives concurrently. The facilities leadership would identify the most critical need and set targets. For example, in today's environment, most facilities organizations do not have a system (data infrastructure) to manage its facilities. Without this capability, the facilities organization will not be responsive to the business needs.

Strengths are based on what is important to the organization in terms of performance. Cost, quality, employee empowerment, and safety are critical elements that should be addressed annually. Facilities leadership, setting annual targets, should inspire continuous learning. Productivity and cost reductions translate into doing work differently. One should not be afraid of taking a risk and approaching an old problem with a new solution. Quality and safety are attitudes. We should have pride in the way we approach our work. We should do it well with safety in mind. Employee empowerment is the future. We must provide employees with the opportunity to take the lead and demonstrate their capabilities to meet organizational goals. Finally, core processes must be visible. The business organization must understand that its facilities organization is critical to its continued success by maintaining the buildings and infrastructure cost effectively.

In summary, the facilities organization must develop a framework of continued growth. One aspect that must be understood is the development of an organizational model that includes the use of technology. As you build the model, it is critical that the organization understand that technology is the enabler for organizational change by establishing a value delivery system that aligns its processes to meet its customer needs. In addition, as your delivery system engages critical processes, new informational needs are defined. Facilities must adapt a new mindset that utilizes a basic business model to focus on growth strategies based upon performance.

ENDNOTES

1. Davenport, T. H. 1993. *Process innovation reengineering work through information technology.* Boston, MA: Harvard Business School Press, p 337.
2. Hammer, M. 1996. *Beyond reengineering.* New York: Harper Business, p 285.
3. Manganelli, R. L., and Klein, M. M. 1994. *The reengineering handbook, a step-by-step guide to business transformation.* New York: American Management Association, p 318.
4. Kotter, J. P. 1996. *Leading change.* Boston, MA: Harvard Business School Press, p 187.
5. Jackson, T. L. 1996. *Implementing a lean management.* Portland, OR: Productivity Press, p 162.

P • A • R • T • 4

IMPLEMENTATION AND MANAGEMENT

CHAPTER 17 PROJECT MANAGEMENT AND INTEGRATION

17.1 Introduction

Project management has a well-developed body of knowledge, best practices, case studies, and recommended approaches. Recent failures in large capital projects, as well as information technology implementations, suggest that even well-managed projects often fail when they are not fully integrated into the business mission, organizational structure, and individual perceptions, characteristics, and proficiencies. Project integration as a concept takes project management to a new and deeper dimension. As the new economy transforms organizations and individuals from an industrial age to an information rich and networked age, information technology in combination with project integration has the potential to dramatically improve the quality and delivery of projects.

17.2 Project Failure

Findings of a Business Roundtable study suggest that major projects often fail to meet their objectives.

17.3 Information Technology as a Driver in Project Integration

Case studies of a new chemistry building and a project integration center suggest that Web-based collaboration platforms as well as physical spaces can be drivers for project integration.

17.4 Organizational Modeling as a Means to Project Integration

A new realm of research, arising from the field of anthropology, is making significant progress in understanding the nature of hierarchical organizations as compared to the informal and implicit "networks" that form the fabric of all human interactions. Case studies of companies and enterprises analyzed with new research and information technology tools suggest that networks are far stronger and more effective, and a definite indicator of organizational health. Projects that recognize these network structures and integrate an understanding of them into the project plan and execution are far more likely to succeed.

17.5 Project Integration Trends on the Horizon

Software simulation tools, both for spatial characteristics of buildings and functional relationships, and project management relationships in terms of critical path method scheduling, are now being used to more fully integrate an understanding of the project team and its effectiveness. Integration of team dynamics is seen as critical to successful projects.

17.6 Conclusion

The synergy of project integration and information technology has the potential to enhance the current body of knowledge and best practices of project management.

CHAPTER 18 REAL ESTATE PORTFOLIO MANAGEMENT

Most corporations approach management of their real estate portfolio on a reactionary, single transaction basis. This chapter discusses the importance of developing a portfolio-wide approach to real estate management, introduces the importance of supply–demand forecasting, and outlines the factors that should be considered when developing a comprehensive, portfolio-wide occupancy plan.

18.1 Occupancy Planning—The Art and Science of Projecting Real Estate Needs

Effective occupancy planning requires the right balance of space, time, business groups, cost, and workplace tools. This section also discusses the pros and cons of developing a comprehensive occupancy plan. The options of developing an internal process for planning or using an external consultant are compared. Often, a combination of the two are the most practical, cost-effective approach. Three key factors in determining the right space are identified (size, location, and type) and criteria that should be considered in developing specific requirements are outlined. The importance of timing in the delivery of space is reviewed. Real estate markets have historically run in cycles, but timing the market in real estate is as risky as trying to time the stock market. A better strategy is to tie your real estate commitment to the timing requirements of your business. With today's critical shortages in labor supply, identifying the right business groups to match both current and projected labor supply within a defined geographic area is essential to successful occupancy planning. Components required in a comprehensive demographic analysis are discussed. Although typically not viewed as the number one criteria in the occupancy decision process, cost is still a critical factor. This section explains why context is important when considering costs. It is not enough to provide space for our businesses, but more than ever they have to be provided with the right workplace tools. The "tools" are defined in this section and the reader is referred to Chapter 8 for a more comprehensive discussion of the topic.

18.2 Portfolio Management—The Supply Side

While the previous sections focused on the space demand side of the demand/supply equation, this section discusses the importance of the supply side. Lease–own decisions, impact of REITs on future corporate space decisions, and commitment of space vs. control of space are all discussed in this section. Which type of companies are good candidates for leasing and owning and the impact real estate investment trusts (REITs) are having on the leasing issue are topics discussed in this section. Without realizing it, most companies are attempting to achieve the right balance between "committed" space and "controlled" space. The terms are defined, the various types of commitments to real estate are categorized, and the issue of balance in the portfolio is explored.

CHAPTER 19 SUPPORTING THE MISSION OF THE ORGANIZATION: AN APPROACH TO PORTFOLIO MANAGEMENT

19.1 Introduction

Facilities are part of the resources used by an organization to support its mission. When setting the overall business strategy of the enterprise, facilities are considered by forward thinking managers as one of the "factors of production." It is therefore essential to link the real estate strategy to the business strategy of the business units and of the enterprise. In this case study, portfolio management is placed in the context of the decision making and information flows about facilities. Projects are used to illustrate this focus and how the ASTM standards for Whole Building Functionality and Serviceability are used to meet these objectives.

19.2 Performance and Serviceability—Background

This section traces the development of a performance-based approach to managing facilities and the development of the ASTM standards.

19.3 Tensions between Quality and Cost

This section addresses the tension between cost and quality, and the way real estate groups are organized in response to trends and pressures.

19.4 Portfolio Management: Essential Link from Facilities to the Core of the Enterprise

Traditionally, the mission of organizational units has seldom been linked to the infrastructure resources required. Facilities, in particular, were rarely considered. This is changing. This section presents the overall process of planning, delivering, and managing real estate assets.

19.5 Functionality, Serviceability, and Suitability Tools for Measurement: ASTM Standards on Whole Building Functionality and Serviceability

Measures of satisfaction and measures of quality are different from each other, but complementary. Both are useful. They should not be mistakenly used in lieu of each other. This section focuses on measures of quality. A set of standard processes and tools, approved by ASTM (American Society for Testing and Material) and recognized by the American National Standards Institute (ANSI), gives managers and facility users alike the means to determine functional requirements, assess the capability of facilities provided, and verify that those facilities match the stated requirements.

19.6 Applications of the ASTM Standards on Whole Building Functionality and Serviceability

The U.S. State Department (A/FBO) is upgrading the way user requirements for its embassies are defined, so that they can be more responsive to the needs of the staff at the posts. This section highlights benefits and hurdles to be overcome during implementation. Steps in the process are illustrated, as well as results to-date. Other projects are used to illustrate multiple uses of the ASTM standards.

20.1 Introduction

The task of design and construction uses a set of skills that may be new to the facilities manager. There are many phases to the effort, many sets of players, and many considerations that are unique to each project. Skilled facilities managers need to know how to navigate the changing conditions in each project and in each phase. The manager also has to know how to adapt his or her personal style to the situation at hand; and this varies greatly by phase. This chapter discusses the major phases, project components, and team building considerations needed for success in this challenging role.

20.2 The Task: From Need to Occupancy

The facility-planning task on a new project calls on a set of skills vastly different from day-to-day facility management. Expertise is assumed in building technology, finance, aesthetics, and politics. The circumstances of each project—both the physical characteristics and the business landscape—are always different.

20.3 Assembling Resources

Resources include land, financing, professional services, and management attention. They all need to be identified, acquired, and managed.

20.4 Phases of Work

There are four principal phases of work on any project. All need to be considered, none are simple, and decisions in each lead to results in the next. The phases are formulation, design, construction, and operations. The facility manager and his or her team have different considerations, risks, and management needs in each phase.

20.5 The Players: The Facility Manager in the Middle

The facility manager is the gatekeeper in the middle of a universe that includes not only vendors, like architects and contractors, but also the peers of the manager within the organization, and also the internal constituencies of the organization. These internal "clients" can be numerous, powerful, demanding, and not sympathetic to facility issues. An exhibit lists the key players, motivations, and suggestions for picking the team.

20.6 Elements of the Project and What to Consider

Each project has the same basic elements: site, foundation, structure envelope, mechanical systems, finishes, and furniture and equipment. The elements contain different levels of risk and uncertainty, and the management of each can vary with those levels. The elements and what to look for are described.

20.7 Building the Team: Value Added, Risks Accepted

The facility manager will be a key person in building the team. The team members have a range of value-added skills depending on the project and depending on their roles. They also will accept (or not accept) risk in a variety of ways. This section discusses the major players, what they bring, and what to look for in their respective business arrangements.

20.8 Conclusion

The successful facility manager will enter a design and construction project with a good understanding of the phases of work, the skills and risk profiles of the players, the physical characteristics of the building, and the different skill sets needed to survive and thrive as the orchestrator of complex sets of tasks.

CHAPTER 21 SPACE AND ASSET MANAGEMENT

21.1 Financial vs. Facility Asset Management

One hurdle to understanding facility asset management is defining what it is. The term "asset management" has been in use for a long time by financial managers and is a term that they clearly understand. However, when facility managers usually think about managing facility assets, it might be more proper to refer to them as "facility resources" since they are often not "assets" as defined in a purely accounting sense. The most significant example of this is leased space. There are six features that distinguish financial assets from facility "assets." These distinguishing features must be understood in order to make sense of the role of a space and asset management system as implemented by a facility manager.

21.2 Goals of Facility Asset Management

A facilities space and asset management system provides information about facility resources that can be used by management to increase the productivity of staff and the resources they use, and to lower the long-term cost of occupancy of those facilities. Just as importantly, a facilities space and asset management system provides an information foundation for facilities strategic planning, which allows the strategic planner to support and facilitate the business plan of the organization.

21.3 System Principles

Much of the current written material on facilities space and asset management focuses upon concerns like outsourcing of implementation or use of off-the-shelf vs. custom written software. These concerns are really secondary to understanding how a space and asset management system works in the context of an organization. Experience has shown that there are several considerations that cannot be overlooked if a facilities space and asset management system is to produce maximum benefits at the least cost. This includes factors in selection and application of an appropriate system; the approach to starting and growing a system; why, when, and by whom database information must be updated; how to keep costs down by using "data exhaust" from other processes; identification of the critical stakeholders in a system; and keeping the system healthy over the long term.

CHAPTER 22 OPERATIONS AND MAINTENANCE

22.1 Purpose and Goals

The traditional focus of a CMMS on equipment maintenance is changing. The software vendors are expanding the scope of their product to focus on a facilities enterprise asset management system. This section looks at the growth of the products and what they can offer to a facilities management (FM) organization to assist in work management. It defines the scope of CMMS and the value that it brings to an organization for work management. In gathering information through a CMMS you build an institutional memory. How can you use performance metrics to enhance your organization?

22.2 The Evolution and Application of Work Management

The Digital Information Age is upon us. How will it affect operations and maintenance? This section looks at the potential for the "paperless work order." All the topics discussed are available today. Software vendors are working to bring them together into a plausible workflow environment.

22.3 CMMS Implementation for Operations and Maintenance

Even the best software can fail when poorly implemented. This section provides some basic project management guidelines to prevent a failed implementation.

22.4 The Anatomy of the CMMS

At the core of operations and maintenance (O&M) work management is the work order. This section examines in detail the issues surrounding two types of work orders, those created based on demand or service calls and those based on scheduled or preventive maintenance. The flow of information and the interrelationship with the business processes defines the value of the information collected. Recording cost information and the outcome are critical to developing a data resource to assist in enhancing productivity through the efficient use of resources.

22.5 Conclusions

Throughout this chapter, examples of current systems illustrate how the facility manager can benefit from CMMS. The FM is expected to be accountable for the work performed from both the life safety aspect, but also the fiscal tracking of resources.

CHAPTER 23 ENERGY MANAGEMENT

23.1 Energy Management Defined

Energy cost is usually an area that can be significantly reduced in a very cost-effective manner, but it is complex. Energy management has come to the forefront of facility manager's cost control efforts, primarily through the necessity to manage budgets effectively. This chapter gives a different look at the subject by taking a "nuts and bolts" approach to a complex process and showing facility managers how to cut through the confusion and get real understandable cost savings through energy management.

23.2 Utilities Invoice

The utilities invoice is the starting place to get significant savings. You cannot make an intelligent decision about any phase of energy management without knowing what your current usage is from the invoice. The invoice defines your load profile, cost per energy unit, peak demand, and gives you the most important piece of information to make any energy decision by—kilowatt-hours used per month. We analyze a typical invoice to enable us to understand all of the important data, from where the kilowatt-hour information is located, to the competition transmission charges that are starting to appear in deregulated states.

23.3 Deregulation

Electrical deregulation is a fast moving target with major changes occurring. Facility managers need to be aware of these changes. Utilities are reinventing their businesses, as they must convert from monopolies of power generation to the competitive marketplace where power generation is now a commodity. As the utilities restructure, there is a major effort to sell other services besides power and to become their customer's major energy service providers. This change in the marketplace has created many opportunities and just as many pitfalls for the energy consumer. We have approached this section from the standpoint of knowing what questions need to be asked from deregulated power suppliers and how a facility manager separates fact from fiction. There is a lot of fiction and significant consequences in making the wrong decisions about the deregulation marketplace.

23.4 Benchmarking Usage

This segment shows facility managers how to benchmark their energy usage against known standards. We explain how to develop the most important thing to know about energy consumption, the existing usage in kilowatt hours per square foot per year. From this starting point, the existing usage can be compared to standards such as the EPA, energy star building benchmarks to determine if your facility is efficient or wasteful, and, if inefficient, by how much. This is one of the most valuable pieces of information for making any energy decision.

23.5 Energy Audit Benefits and Costs

After benchmarking your energy costs and finding out how much potential energy cost reduction is available, the next most important step is to perform an energy audit. The energy audit quantifies and qualifies energy savings. The audit becomes the plan of where you're going to get energy savings and what the cost will be. An audit is a set of facts that are derived from measurements of existing conditions and calculations that verify real savings. This segment shows what is involved in performing an audit, how the calculations are developed, and cost vs. benefit analyses.

23.6 Performance Contract

After all of the benchmarking, audit analysis, cost budgeting, and implementation strategies have been formulated, it eventually comes down to implementing the project and ultimately realizing the energy savings projected. A performance contract becomes extremely important at this point because, if you don't get the energy savings promised, there must be a consequence to the contractor making the commitment. This segment outlines the necessity for these types of contracts and how to write them. Performance contracts are an absolute necessity in the energy management business. If you don't know about them or don't use them, you have no assurance that you will get any energy cost reductions.

23.7 Conclusion

CHAPTER 24 SECURITY

24.1 Contemporary Security Threats

Security technology to counter conventional threats to persons and property, such as ballistics attacks, breaking and entering, theft and physical assault, is addressed in many security texts. This chapter focuses on new security threats, such as bomb attacks and bomb threats, as well as chemical and biological attacks.

24.2 Weapons of Mass Destruction

The first atomic bomb dropped on Hiroshima in the waning days of World War II introduced us to the concept of a weapon of mass destruction and to the damage that such weapons could cause.

Today, a comparable capability to destroy exists when large explosive weapons are used in dense urban neighborhoods. This chapter describes the properties of explosive weapons and their use in urban environments.

24.3 Historical Defense Measures against Explosive Weapons

Confronted with large explosive weapons, security administrators quickly gravitated to rudimentary, low-tech defense approaches. More security personnel were hired with only the expectation that this expensive defense measure would be effective. Simple, static vehicle barriers, many disguised as planters, were deployed to keep vehicles, potentially bearing large bombs, at as great a distance as a crowded urban setting would allow. Very little high technology was available to quickly apply to bomb defense.

24.4 Explosive Weapons

Explosive weapons can be easily assembled from many manufactured or improvised materials. The destructive potential of an explosive weapon can be estimated from its weight and from the nature of the explosive used. The physical size of a bomb, an important element in bomb defense, is also addressed.

24.5 Quantitative Characteristics of Explosive Weapons

The analytics of explosive weapons and the blast loads which they produce on structures and people are discussed. Specific blast effect computation techniques are illustrated along with various by-products of explosive detonations. This section will permit the reader to make virtually all important blast wave calculations.

24.6 Bomb Attacks—A Historical Retrospective

The use of large explosive weapons in urban settings is illustrated with four actual bomb incidents: the World Trade Center and Murrah Federal Building in the United States, the Bishops Gate bombing in the United Kingdom, and the San Isidro bombing in Peru. Details of the weapons used and the resulting damage and injuries are presented.

24.7 Defense against Bomb Attack

There are three basic components of effective bomb defense: prevention, mitigation, and recovery. Technology can play an important role in implementing each of these defense components; however, mitigation is the area where technology deployed by the facility manager can have its greatest positive impact. Technical approaches involving structural screening and glazing design are discussed along with integrated security development and personnel training. The section ends with a brief discussion of recovery issues.

24.8 Chemical and Bacteriological Weapons—A Look to the Future

A nascent threat of growing concern is the use of chemical and bacteriological (C/B) weapons in attack scenarios where they will work as weapons of mass destruction. There are a great many C/B agents making detection very difficult. The most effective defense approach today involves limiting dispersion of the agent. Unfortunately, it appears that future C/B attacks are likely to be much more severe than past attacks.

CHAPTER 17
PROJECT MANAGEMENT AND INTEGRATION

Stephen R. Hagan, AIA
Project Knowledge Center
GSA's National Capital Region from 1998–2000
E-mail: stephen.hagan@gsa.gov

17.1 INTRODUCTION

New models are emerging for project management (PM) that fully integrate the project within the context of an organization's mission and its bottom-line results. Integrating the project across the enterprise life cycle, from strategic facility planning through occupancy and facility management (FM), is fast becoming a critical skill. In addition, it is becoming an increasingly customer-focused world, and understanding the mission of an enterprise, and therefore, all of the projects that the enterprise executes, in the context of the customers' needs and wants, is critical to an organization's success.

Projects fail more often than anyone expects or desires. Recent studies by the Business Roundtable have begun to develop an insight into the reasons for project failure, and suggest new strategies and best practices to increase the chances for success. Information technologies hold the promise of providing the critical connection and collaboration platforms for this integration to occur. Three parts of this chapter will illustrate the concepts of project integration and information technology:

- The National Institute of Standard and Technology's (NIST's) new $75 million Advanced Chemical Sciences Laboratory (ACSL) provides an excellent case study of a design–build approach to project delivery. It is also an example of integrating the project team by means of information technology. Both tools were success factors enabling NIST to deliver a state-of-the-art chemistry building on time and within budget.
- On the scale of a large capital development program, GSA's National Capital Region is utilizing the concept of a project integration center to provide a focus for its project initiation and ongoing project review of a $3 billion capital program.
- At the organizational level, new project integration tools, incubated in research centers such as Stanford's Center for Integrated Facility Engineering and UCLA, will facilitate this integration for project teams, and are now becoming commercially available.

To understand project integration, it is necessary to understand the context of project management within the overall framework of organizations, processes, individuals, and how information technology can facilitate the integration of projects more successfully into the fabric of the enterprise.

17.2 PROJECT FAILURE

In 1996, the construction committee of the Business Roundtable (BR) began to see the declining state of project execution as a critical issue that needed to be studied. *The Business Stake in Effective Project Systems* was the outcome of a study by the Business Roundtable in cooperation with the Independent Project Analysis (IPA) Corporation of Reston, Virginia. From a database of over 2,000 projects, the IPA and the BR looked at sixty companies in terms of their performance in delivering projects. A surprising statistic emerged " . . . more than two-thirds of major projects built by process industries in the past five years have failed to meet one or more of the key objectives anticipated at authorization."

More importantly, as a metric to the bottom line, Ronald Howard, director of construction for the Business Roundtable, in a speech to a government/industry forum on capital facilities and core competencies (Federal Facilities Council, March 25, 1998), noted that:

> . . . The best-performing company in the study can take an industry average of 15 percent return on an investment project and turn it into a 22.5 percent return. The poorest performer in the study started with a 15 percent average return on investment and drove it down to 9 percent.
>
> The best-performing company in the study spends 72 cents of the industry-average dollar for the same functional scope. The company with the shortest delivery time takes 70 percent as long as the industry average to bring a project from a business idea to an operating facility. Finally, the company with the best track record in start-up and operation achieves 6 percent more production from their assets.
>
> *"If these three companies could be taken together at their best, the return on investment for a capital project would increase from 15 to 25 percent."*

The conclusion was that dramatic changes in project processes were necessary if enterprises were to achieve the full potential of the projects they undertake. The Project Management Institute provides a graphic that illustrates the various processes involved in project execution (see Figure 17.1).

What becomes clear from the Business Roundtable study is that all of the project processes, from initiation through execution, must be integrated into the fabric of an organization in order for the project to be successful for the enterprise. This becomes especially important in an era of downsizing, rapidly changing business requirements, and an increasingly competitive global environment. Howard notes the necessity for integrated functions:

> Companies that have succeeded in this new environment have fundamentally changed the way they view the business world and the capital project system. They now see capital proj-

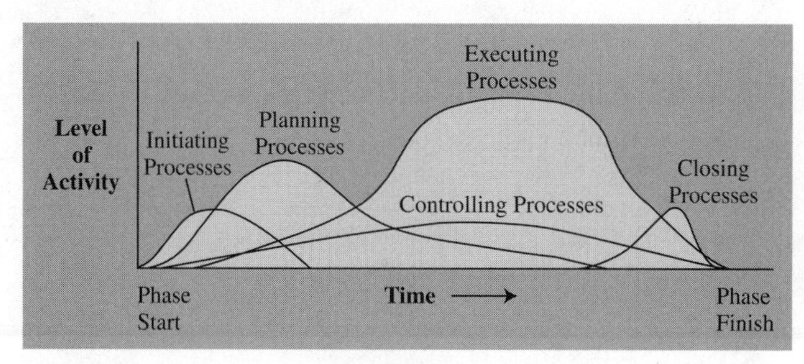

FIGURE 17.1 Project Processes (from PM Body of Knowledge).

> ects as a principal way that the company's capital-assets base are created. They also see technology and engineering as key elements in the supply chain resulting in competitive projects, not merely as nonintegrated functions.

One of the most important findings of the BR study is the importance that must be placed on the planning stage of a project. This is where projects are defined, budgets are established, and schedules are developed such that minimal changes are required during the execution stage. The increasingly customer-focused economy is driving this planning stage focus, as Howard notes:

> The supply chain begins when customer need is identified and translated into a business opportunity. Following this is the critical planning phase of the project. Business opportunity is explored in the first stage of the planning phase and alternate methods of meeting the defined needs are investigated, including some noncapital assets. The most successful companies are using their technical resources in this business development process.
>
> As the business plan focuses on a capital project as a solution, the project management professionals are added to the team to work with the business leaders in the facility planning stage. This is the stage at which the broad project objectives are honed into a particular project at a particular site with a particular technology configuration and schedule.

Howard eloquently articulates the process that links a business opportunity and its strategic planning objectives with a particular project that has been formed. What becomes evident by this is a new diagram that places the project and the project processes clearly at the crossroads of an enterprise (see Figure 17.2).

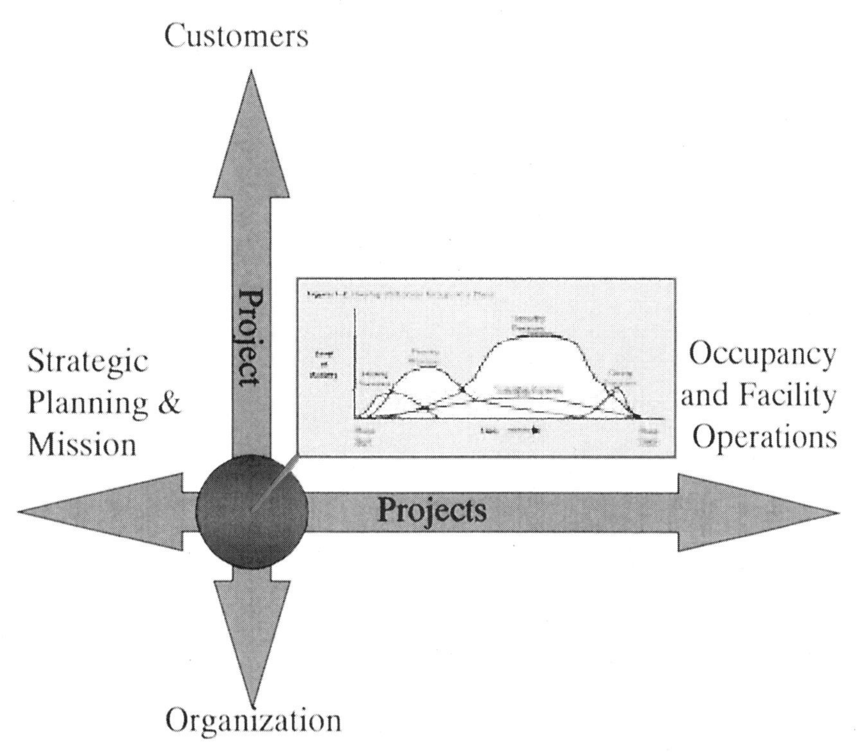

FIGURE 17.2 Project Integration of Project Processes.

Howard summarized the Business Roundtable findings with a renewed emphasis on integration as essential to success:

> The importance of the advance planning process cannot be overstated. Analysis of the IPA database showed that 49 out of 50 projects that achieved best practices in advance planning met all of their objectives. The project supply chain required an integration of the business, technical, and manufacturing functions into teams that can create a project uniquely fills the business need. . . . What are the characteristics of the best capital project systems? In addition to using fully integrated cross-functional teams, they actively foster a business understanding of the capital project process.

Project integration, then, incorporating project processes within the context of an organization's business needs and customer requirements, can be instrumental in achieving project successes and minimizing project failure.

17.3 INFORMATION TECHNOLOGY AS A DRIVER IN PROJECT INTEGRATION

Information technology as a means of implementing successful project integration is illustrated in two case studies: a new chemistry building constructed by means of a design–build process and a project integration center being utilized to manage an enterprise-wide capital construction program. Two elements illustrating how project integration can be enabled by information technology are included in this section:

- Web-based collaboration platforms for executing a project
- Physical spaces, such as a project integration center, that is integrated with a Web-based program reporting tool and an organizational process that supports the physical center and the virtual center.

Construction of the National Institute of Standards and Technologies (NIST) Advanced Chemical Sciences Laboratory in Gaithersburg, Maryland, was completed and occupancy commenced in March 1999, 4 years after the project's inception in March 1995. It is a state-of-the-art chemistry building including 165 laboratories for 150 researchers in NIST's Chemical Science and Technology Laboratory (see Figure 17.3).

A design–build approach to project delivery was formulated, based on the preliminary assumption that up to 8 months of project time and $2.3 million of project costs (notably, escalation costs)

FIGURE 17.3 NIST Advanced Chemical Sciences Laboratory, and Web-Based Site Cam during Construction.

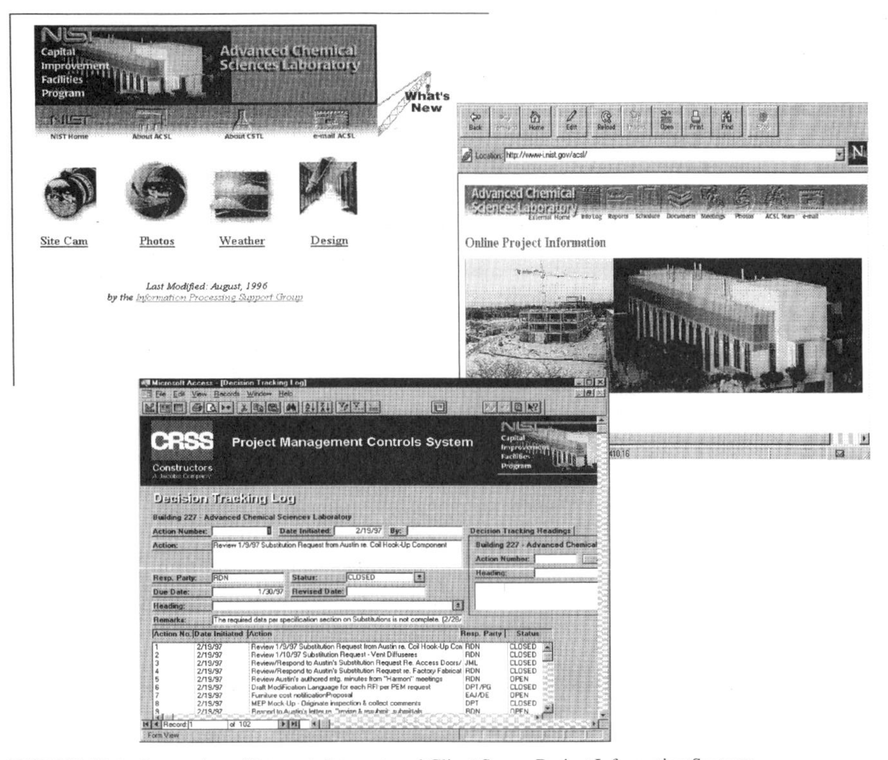

FIGURE 17.4 Integration of Internet, Intranet, and Client Server Project Information Systems.

could be saved. The successful dedication and occupancy in March 1999 proved these assumptions correct. Contributing to the successful execution of this project were integration strategies that brought together the CSTL researchers and laboratory management, NIST management, project designers, laboratory planners, the construction manager, and the design builder into an effective management structure. As mentioned in the Business Roundtable findings, the ACSL project structure included management on the team from the very beginning, so that the business goals were aligned with the project delivery goals and objectives.

A Web-based project information system, established early in the project, provided a communication and collaboration platform for the project team. An external Web-site was established that provided a live site cam of the construction, weekly updates of progress photos, and other project information. Internet users from all over the world visited the site, as well as members of the project team. The CSTL management actually downloaded photos from the Web-site for use in management presentations and brochures. The geographically diverse project team, which included Earl Walls and Associates in San Diego, Jacobs Engineering in Philadelphia, CRSS Constructors (now Jacobs Engineering) on the project site in Gaithersburg, Maryland, and The Austin Company in Cleveland and on-site, all benefited from the use of this Web-based technology (see Figure 17.4). It facilitated the creation of a project community and permitted communication across all members of the project team, including NIST management.

The information technology platform the ACSL team utilized included an external internet site, internal (NIST only) intranet, and client/server application that was developed by CRSS Constructors (now Jacobs Engineering).

Project community, at the broader scale of a multi-million dollar capital program, is being nurtured and implemented by GSA at its project integration center in Washington, DC. An intriguing

FIGURE 17.5 Evolution of Project Integration Center and Web-Based System at GSA's National Capital Region.

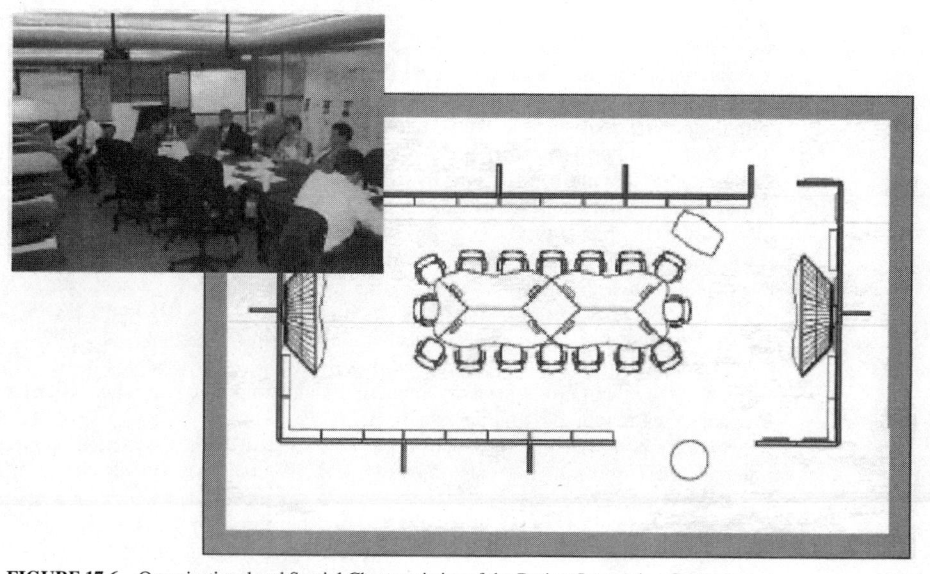

FIGURE 17.6 Organizational and Spatial Characteristics of the Project Integration Center.

FIGURE 17.7 New Models for the Project Integration Environment.

aspect of this approach is the integration of an evolving information system, a physical space, an organizational structure, and individual project teams and approaches.

The information system evolved from large 4′ × 9′ poster boards that contained paste-ups of fact sheets, photos, budget documents, and project schedules. These boards lined the room and were "rolled out" once a week for a project review meeting where project managers and their service delivery team directors discussed project issues and actions. The posters became printed reports, then evolved to an on-line project information system. Now the project information is updated directly to an on-line Intranet system (see Figure 17.5).

The project integration center itself is a study in the integration of furniture, write-able wallboards, supply storage, computer projectors, and smart boards into a state-of-the-art environment for communication and collaboration. The success of both the project reporting system and the project integration center itself has lead to discussions about taking this approach to other venues and levels of interest throughout the organization (see Figure 17.6).

New models of the project integration center include a reorientation of the project teams away from a "table-based" environment, where participants are focused on papers and notebooks in front of them. The new concept is a "pedestal" environment, where all chairs are oriented toward the "smart boards" and the visual display is the focus of the meeting (see Figure 17.7).

17.4 ORGANIZATIONAL MODELING AS A MEANS TO PROJECT INTEGRATION

The project integration center mentioned previously illustrates the concept that projects are executed in the context of an organization's physical environment. Spatial characteristics of "war rooms" and periodic project review meetings can contribute to knitting the organization together. But at a deeper level, the underlying structure of an organization and personalities of individuals within the organization can directly affect project success. Modeling organizational and individual behavior can lead to insights into lessons learned in project successes and opportunities for improvement where projects fail or fail to meet desired objectives. Two elements illustrating how organizational modeling enables project integration are included in this section:

- *Analytical tools* developed at Stanford University's Center for Integrated Facility Engineering (CIFE) link an organization's structure to project execution processes, illustrating project constraints and opportunities not previously apparent.

- *Software tools* arising from organizational studies from an anthropological standpoint developed at UCLA examine the relationship of an enterprise's hierarchical organization structure as compared to its network structure, where work and relationships often have more power and influence on a project's success.

VITÉ is a software and services company that grew out of academic research by Stanford Professors John Kunz, Raymond Levitt, and Yan Jin, who, in a paper written in 1998, developed a concept of the virtual design team (VDT). The concept included a computational simulation model of project organizations. The genesis of the VDT concept grew out of the limitations perceived by the authors of organizational theory to address increasing complexity of enterprises in a global and highly competitive environment:

> Faced with increasingly competitive global markets and tight-fisted taxpayers, many private and public organizations now "reengineer" themselves to improve their products or services and to reduce time between receipt of a new order and delivery of a requested product or service to a satisfied customer. When managers change existing work processes to reduce schedules dramatically, interdependent activities that were previously performed sequentially must then be performed concurrently. Organization theory predicts that the coordination of concurrent interdependent activities is significantly more difficult and costly than coordination of the same activities performed sequentially. Yet traditional organization theory can neither predict the magnitude nor the specific actors and activities that require incremental coordination, even though coordination load and rework can grow exponentially as there is great concurrency of complex, interdependent activities performed in parallel.

What grew out of this analysis was the development of *The VDT Micro Theory of Project Organizations.* The researchers further developed this theory as follows:

> The basic premise of the VDT model is that organizations are fundamentally information-processing structures. . . . In this view, an organization is an information-processing and communication system, structured to achieve a specific set of tasks, and composed of limited teams (called "actors") that process information. Actors send and receive messages along specific lines of communications (e.g., formal lines of authority) via communications tools with limited capacity (e.g., memos, voice mail, meetings, etc.). Thus, for example, each modeled manager has specific and limited (boundedly rational) information-processing abilities. Managers send and receive messages to and from other actors along pre-specified communication channels, choosing from a limited set of communication tools. In the organizational literature, coordination load is the complex set of requirements for coordination among the various actors in an organization.

The VDT model develops activities, communications, actors and information processing, exceptions and decision making, and incorporates these various components into an object-oriented, discrete event simulation. Figure 17.8 illustrates how a typical team member receives information coming into his "in-box," processes the information, transmits it out to team members and others in the organization, and executes activities in a project schedule.

Case studies utilizing the VITÉ software have illustrated the significant value that this analysis can bring to analysis of project execution. In the case of construction of an overseas assembly test facility, a U.S. electronic manufacturing company engaged VITÉ to analyze retrospectively the first of a series of manufacturing plants that would be constructed with shorter and shorter time frames. Figure 17.9 illustrates the sort of organizational to project timeline analysis that was accomplished.

VITÉ developed a retrospective model of the project processes utilized in constructing the first manufacturing plant, and then compared the VITÉ model to actual project data. According to VITÉ:

> VITÉ's simulation tool correctly predicted the activities and groups that had experienced the largest backlogs and the resulting schedule and quality risks. After validating the model,

Vité Simulates Project Participants Working and Communicating

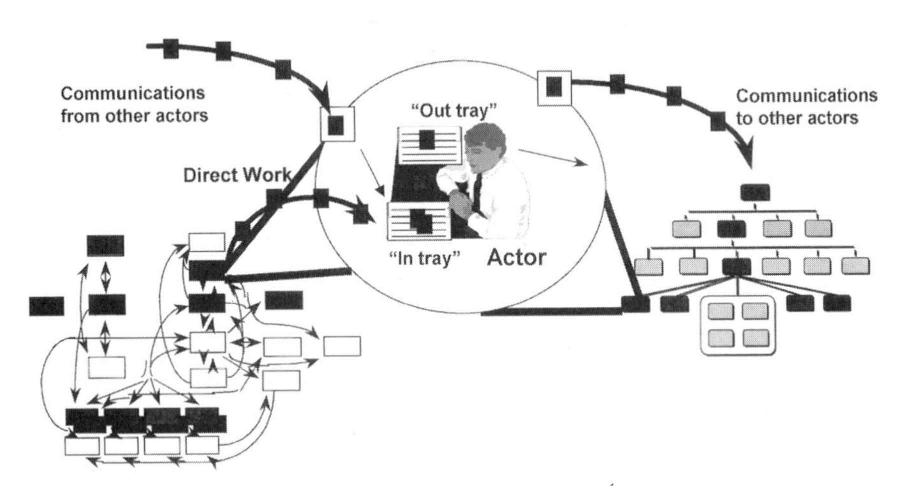

FIGURE 17.8 Modeling of Project Scheduling and Organization in VITÉ Software Simulation.

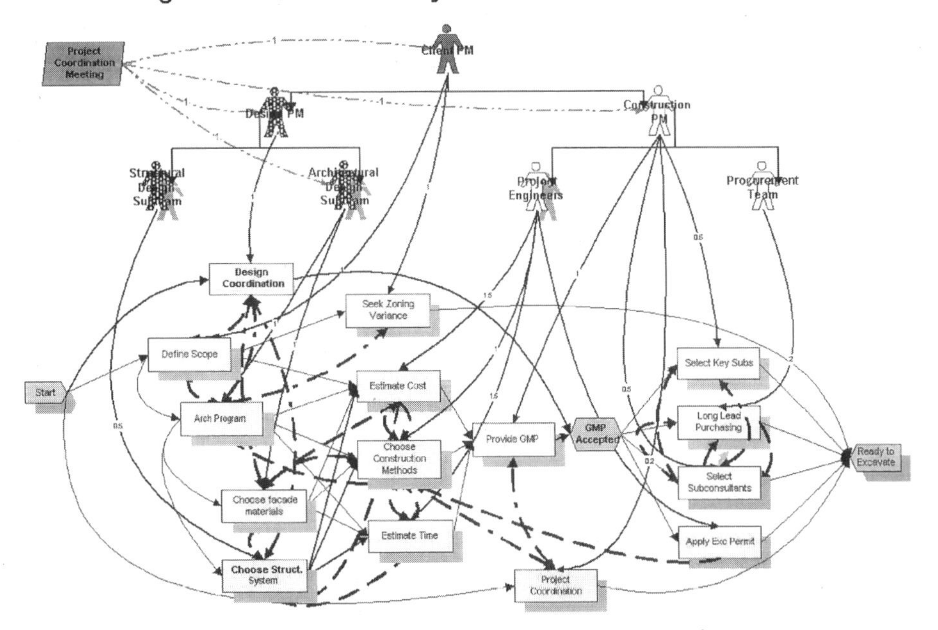

FIGURE 17.9 Integration of Project Scheduling and Organizational Structure in VITÉ Simulation.

> VITÉ ran multiple simulations to find ways to shorten duration and enhance quality with minimal additional staff. Specifically, VITÉ's analysis provided the following insights and recommendations through the model's quantitative output.

Company internal groups, which manage the subcontracting relationships, were identified as being among those with the largest backlogs and greatest risk of schedule slippage and quality problems. A company representative commented that contractors complained regularly about the company delaying their work, which lends anecdotal support to the simulation tool's quantitative output.

"What if . . . ?" analyses suggested that during periods that required high support for contractors, schedule and quality risks could be dramatically reduced by decentralizing decision-making responsibility and adding a small number of staff with appropriate skills to a few of the company's internal groups.

Additional analyses examined the effects of enhancing the skill sets of key design consultants and reducing the degree of formalization of communications among organizational participants. This analysis produced a series of reports that, for each scenario studied, illustrated the relative levels of different kinds of risks for all project activities and groups. VITÉ provided the company's construction project managers with:

- A systematic way to design their projects so as to plan for and minimize the most significant risks from the project's inception
- A strategy for initiating additional interventions to manage risks as the project evolved

The VITÉ model actually resulted in a template that could be utilized by the company on future projects to simulate organizational constraints that might affect project success. An equally innovative approach is the work of Karen Stephenson and her company, Netform. Stephenson has studied both the rigid organizational structures of enterprises, as well as the networks that exist beneath the surface. She describes networks:

> What is a network? In today's popular literature and business press, there is a lot of talk about social and organizational networks and the role they play in fomenting change. Typically, this literature focuses on the notion of "networking" as action-oriented; that is, network as a verb. In this light, networking behavior produces relationships and those relationships provide greater access to instrumental resources; for example, in politics.
>
> There is a second meaning to network, however, and it is far more profound than the first. If one imagines a network as a noun; that is, as an organizational structure, then the next logical question to ask is: Are there any recurring or predictable structures and what role do these structures play in forming or maintaining a company or country culture? We are all well aware of the existence of communications channels or networks that honeycomb organizations. Messages and judgments course silently and unseen in networks which connect people and organizations. We have all been a part of one and surprised by a few. What if these networks have predictable patterns, and what if they could be accurately identified and diagnosed so that tasks could be accomplished efficiently and effectively?

The diagram that Stephenson utilizes to illustrate the mechanism of networks is like a kite (see Figure 17.10).

In every network, there are three central positions: hubs, gatekeepers, and pulsetakers. These three positions are the "cultural carriers." Every culture comprises networks and every network will have these three nodes. The individuals who occupy these positions do so by virtue of their deep-standing (not necessarily long-standing) relationship of trust with most members of the organization.

Stephenson's Netform software makes the underlying networks visible. Figure 17.11 illustrates how organizational structure is transformed by the addition of network lines of communication and connection.

TREAT THE CAUSE,
NOT THE SYMPTOMS OF CULTURE

•Hubs are people who are socially connected to the nth degree. With the highest number of direct ties they are the opinion leaders.

•Gatekeepers function as a human way station on critical pathways between parts of the organization.

•Pulsetakers: "unseen but all seeing". They are subtle influencers.

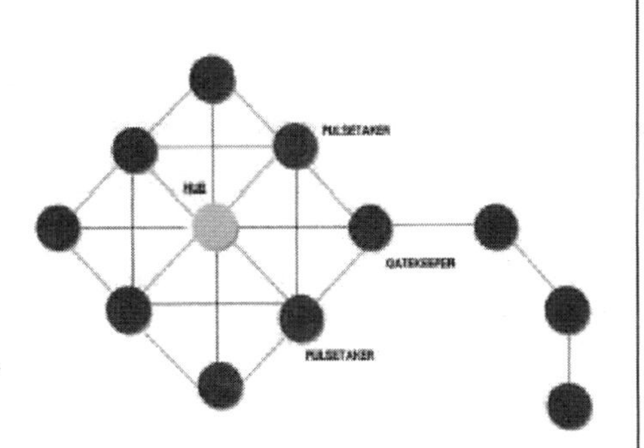

Together these network positions account for the stability and flexibility of organizational culture.

NetForm

FIGURE 17.10 The Topology of Networks.

NetForm

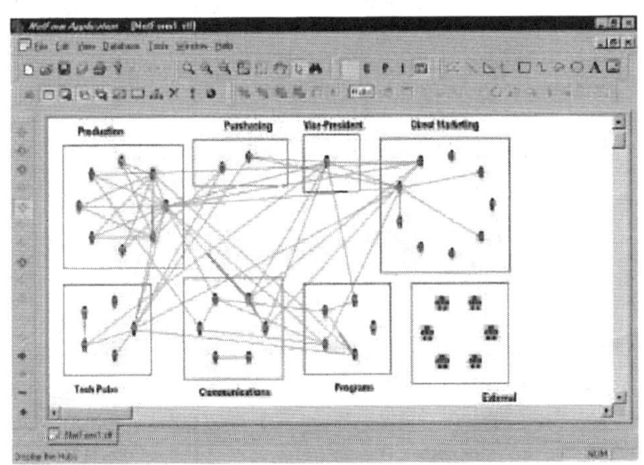

FIGURE 17.11 Netform Software Modeling an Organization in Terms of Its Networks.

What becomes important in Stephenson's analysis is the trust that develops between individuals. The network factor is profound in power and pervasiveness, and essential to understanding how a project is integrated in an organization. The factor that becomes most important, perhaps not so surprisingly, is trust.

Networked organizations are a millennia-old response to coping with unpredictable change. These organizations need a different form of governance. We have proposed a governance model which emphasizes coordination, context, and commitment, and which utilizes the following prescriptions:

> To be effective as a virtual team or organization, one must plan the time it takes to develop trust because even after three thousand millennia, the face of culture is still a human face. Use experiential learning to catalyze trust and shorten the time it takes to get trust established.

Consensus is catalyzed by trust, but contexts change and, when they do, you need a management system that is adaptive: sensing and responding to unpredictable changes. Often problems arise in introducing consensus management where hierarchical command-and-control was a preexisting condition. A common example is when commitment is negotiated horizontally, but accountabilities are assigned and counted vertically. This interaction effect between networks and hierarchies divides responsibility from authority and can lead to bureaucratic bungling.

Globalizing or virtualizing operations means recognizing that other people do things differently than the "home" office, and that differences are authentic. This means that accountabilities between people and cultures must be negotiated, not assigned. And it means that commitments can be made and then renegotiated later. As some uncertainties are resolved, others might arise, giving birth to a context that is different from the one in which the original commitment was forged. When this happens, it is incumbent upon the person who first recognizes that events have invalidated the context of a commitment to inform the other partners. This will not happen without trust.

Both VITÉ and Netform illustrate new information technology tools and services that are illuminating the organizational dimensions essential to integrating projects into the enterprise.

17.5 PROJECT INTEGRATION TRENDS ON THE HORIZON

The increasingly competitive ".com" world of venture-funded information technology firms has added hundreds of new companies addressing the need to integrate projects within an enterprise. VisionPlanner is a software tool being developed in mid-2000. Its concept, of taking the knowledge of an organization and integrating past experience into an overall project infrastructure, is profound (see Figure 17.12). This knowledge base within an organization is then implemented at a project level by a "New Project Wizard," which asks a series of questions about cost, schedule, and quality, and results in a detailed project management plan on-line (see Figure 17.13).

Microsoft Project, a scheduling application achieving its first decade of use in 2000, has evolved from a stand-alone productivity application to an enterprise-wide solution to knowledge management (see Figure 17.14).

With the addition of collaboration environments developed by Microsoft entitled "Project Central," and third-party software enhancements that focus on integrating the project into the overall enterprise, the opportunity to fully integrate projects within the framework of enterprise-wide deployment is becoming a reality (see Figure 17.15).

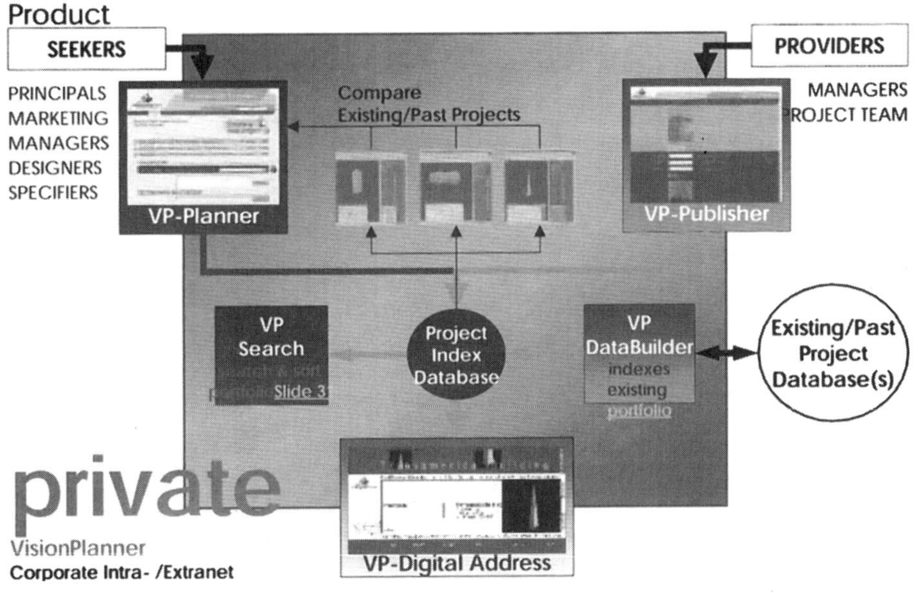

FIGURE 17.12 Vision Planner Development of a Project Infrastructure.

FIGURE 17.13 Screen Shot of Corporate Intranet "New Project Wizard."

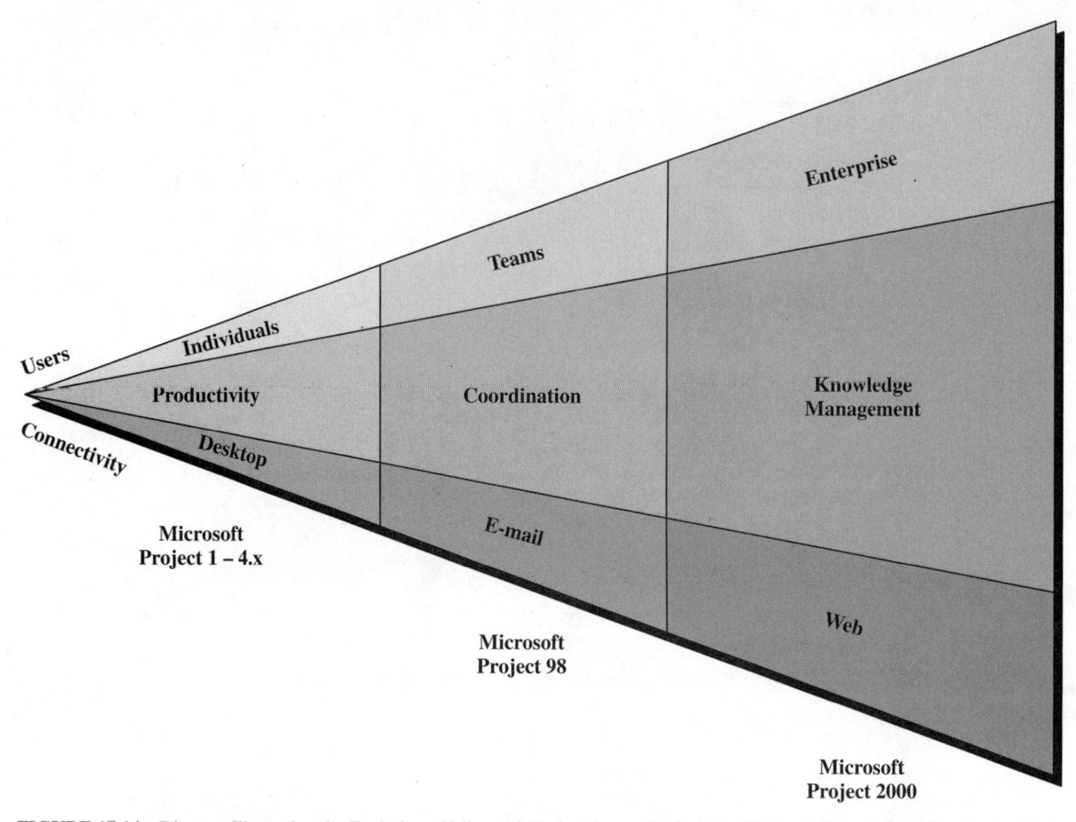

FIGURE 17.14 Diagram Illustrating the Evolution of Microsoft Project from a Productivity Tool to a Knowledge Management Tool.

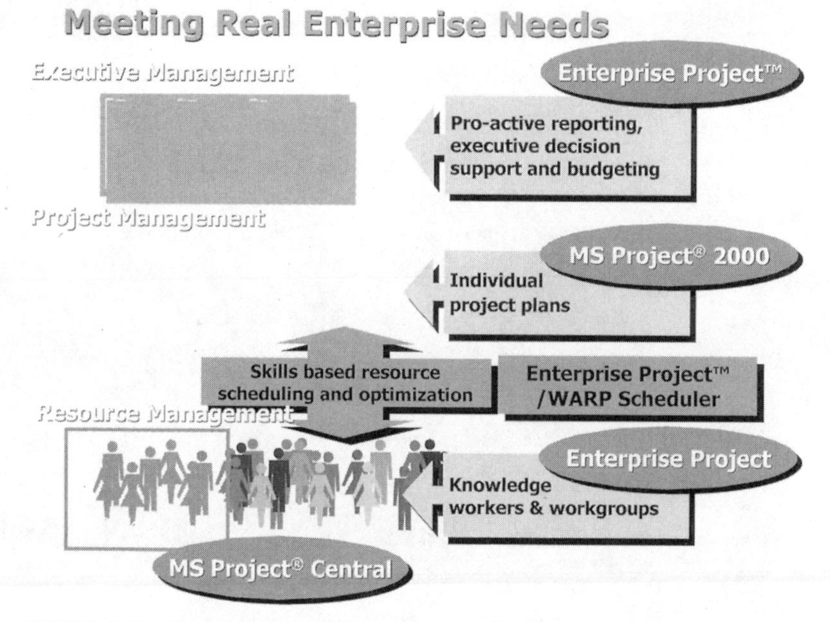

FIGURE 17.15 Enterprise View of Microsoft Project with Enterprise Project and MS Project Central.

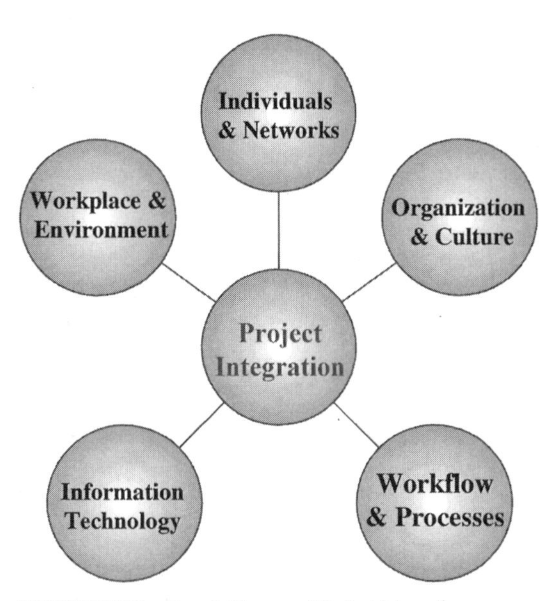

FIGURE 17.16 Overall Diagram of Project Integration.

17.6 CONCLUSION

In spite of the great strides made in development of new software tools and organizational management perspectives, what remains is the hard work of integrating all of these to bring a project into focus within an organization. The fundamental factors remain that all must be integrated: individuals and networks of individuals, organization and culture, workflow and processes, information technology, and workplace and the work environment. The successful enterprise is one that is able to integrate all of those factors within a project-centered world (see Figure 17.16).

ENDNOTES

1. *The Business Stake in Effective Project Systems,* Construction Committee of the Business Roundtable, 1996.
2. Information and white papers on VITÉ are available from www.vite.com. Further information on the entire range of activities at Stanford's Center for Innovative Facilities Engineering (CIFE) and, in particular, its e-Commerce summit which took place on April 8, 2000, is available at http://www.stanford.edu/group/CIFE/ecommerce.summit.html.
3. The Netform software and white papers on the methodology underlying Karen Stephenson's work are available at: http://www.netform.com.
4. New technologies for project integration such as Vision Planner and Enterprise Project are available at http://www.visionplanner.com and www.elabor.com, respectively.

CHAPTER 18
REAL ESTATE PORTFOLIO MANAGEMENT

Stephen Bell
President
Fidelity Corporate Real Estate L.L.C.
Fidelity Investments
E-mail: stephen.bell@fmr.com

Developing an effective, flexible real estate strategy and implementing that strategy on a portfolio-wide basis is the focus of this chapter. In narrowing the focus for this chapter, we chose to concentrate on those issues which have received significant press of late and are considered to be areas of high value added for today's corporate real estate manager: occupancy planning and portfolio management. Portfolio management, for purposes of this chapter, is defined as "managing real estate as a group of properties in order to achieve greater corporate benefits"[1] and occupancy planning (see Figure 18.1) is broadly defined as strategically matching real estate demand to supply.

18.1 OCCUPANCY PLANNING—THE ART AND SCIENCE OF PROJECTING REAL ESTATE NEEDS

Academically speaking, occupancy planning in the corporate environment is easy. There are not any difficult formulas or tough scientific concepts that need to be applied to the process. It is all pretty basic and involves just five variables: space, time, business groups (or employees), cost, and workplace tools. Simple. Wrong! The challenge facing the corporate real estate (CRE) professional responsible for strategic occupancy planning is finding the right combination for these five variables. The fully integrated real estate strategy should have the right amount of space at the right time for the right business groups at the right cost with the right workplace tools. Anyone who has tried formulating a strategic occupancy plan knows that not only do you have to have these five Rs, but as soon as the plan is complete (and in some cases before), it's obsolete.

To plan or not to plan, or more accurately, how much to plan, is the question constantly facing the real estate professional. The level of real estate planning in corporations runs the gamut from no plans (which is itself a plan by default) to detailed 5-year plans that are constantly updated. Several fast growth hi-tech companies have opted for the development of less detailed plans that are flexible and allow them to react quickly to changes in business direction. Rather than spend a lot of time and effort on developing all five of the "R's" they focus on developing an understanding of the business and providing physical environments that can change rapidly as the needs of the business change. A shortcoming of this approach is the fact that it takes time to bring real estate on-line and by the time a business requires a new space, often there's not enough time to either build it or even

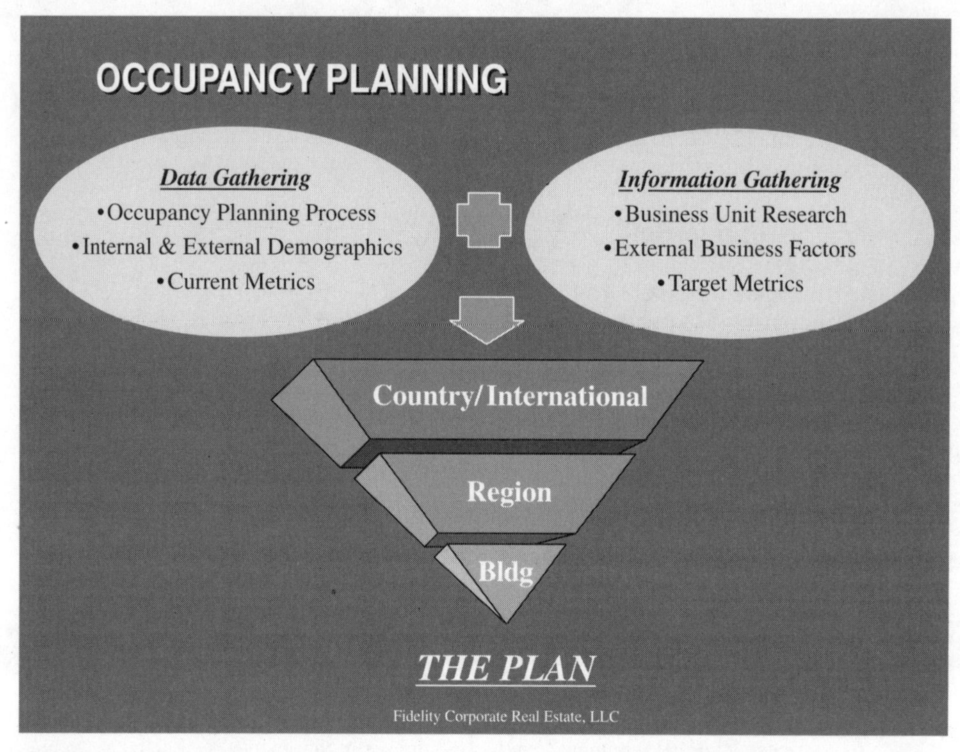

FIGURE 18.1 The Occupancy Planning Process Combines Internal and External Data with Knowledge of the Business.

lease it. In the hot Silicon Valley real estate market, fast growth firms cannot afford to be too particular about where they locate or even the type of building they will occupy since there is very little real estate product available. In the current economic environment, occupancy planning can be a strategic advantage since it can put the real estate group in a position to make a decision quickly when a business needs additional space, thus saving both time and money and potentially avoiding a bad location decision.

Occupancy planning is dynamic. The amount of change within a business varies greatly and is primarily dependent upon the level of maturity of the business (e.g., start up, growth, mature, and declining). Even within these levels, rate of change may vary. To further complicate the planning task, many companies have multiple businesses that fall within several of these levels. For example, one real estate strategy for all the business within many of today's mega-merged companies is not realistic. Imagine arguing that your strategy for the Internet company within TIME/AOL should be the same as for the print business or the entertainment business. Even within those groups, strategies will vary. Level of maturity of the business, profitability, rate of growth, competition, technological advances, and availability of labor all have an impact on the real estate requirements of the business groups.

18.1.1 The Planning Process

An occupancy plan not only needs to be dynamic (i.e., able to change quickly as business needs and market conditions change), but the process of developing a plan needs to be dynamic. Each one of the five variables of the occupancy plan puzzle must be considered in context with the four other variables during the formulation and revision of the plan. In the 1980s, corporations began to hire consultants to assist them in developing a 3–5-year strategic plan. The consultants, recognizing the importance of understanding the business and the vision of the corporate leadership, would interview all critical parties, document existing business practices and organizational structures, as well as the real estate portfolio, and then develop several real estate scenarios based on the information collected. In a business environment of slow change, this strategy can work, but in today's fast-paced environment, the plan is often out of date before it is complete. If consultants are used in the development of a real estate strategy, there are two alternative methodologies that can be followed, either of which will work. The first strategy is to hire a consultant to assist in the data gathering and process development stage while the in-house staff continues to refine the methodology, gather new data and information, and develop scenarios after the consultant has finished his or her work. The key under this approach is the development of appropriate tools that will later be used by the CRE staff in the on-going planning effort. Fidelity Investments used this approach in 1995 when it retained IA (Interior Architects) to develop the planning process, gather important business information, and develop the initial corporate real estate strategy. Although the consultants are no longer involved, the CRE staff has continued to refine the methodology, has automated the process and linked it to the CAFM and human resource tracking systems, and has developed a high level of credibility with senior management, thanks in large part to those initial efforts of the Interior Architects.

The second approach is to hire consultants for the front-end data gathering and development work as well as the on-going occupancy planning and scenario development work. Under this alternative, the consultant group replaces the in-house planning staff. Key to the success of this approach is the integration of the external consultants in the daily operations of the corporation and the development of a trusting relationship between those running the business and the consulting group. The Rise Group, a strategic planning and program management group, was retained by General Motors to sit on a team of consultants and help coordinate the real estate strategy for their central Michigan operations. The consultant team not only developed the process to be used in the planning, but also collected the data, developed the plan, and then was retained to implement the plan. The consultants have integrated themselves into the fabric of the corporation and are trusted with sensitive business knowledge that they use in modifying and carrying out the strategy.

18.1.2 The Right Space

There are three primary criteria in determining the right space: size, location, and type. Of the three, size is usually the most difficult, since the commitment for space is usually for a multi-year period, which requires businesses to project their space needs over that planning time horizon. In today's business environment, even with the most sophisticated of planning tools, those projections are usually far from perfect and are subject to constantly changing business conditions. Even in mature, slow growth industries where long-term commitments to space used to be the norm, many companies are opting for flexibility because of business uncertainty. Using real estate ownership as a measure of required flexibility (assuming an inverse correlation between ownership and flexibility), large corporations once thought to be slow moving and more likely to own than lease, are turning to alternative commitment options once reserved for start ups. Sprint Corporation, AT&T, Applied Materials Corporation, and AMEX have all shifted their ratio of leased space to at least 60% of their portfolio, while IBM and First Union also continue to move in the same direction.[2]

Once the commitment to space is made, the real estate professional must still continue to monitor the business occupancy issues and real estate market conditions for any significant change and be prepared to take advantage of new opportunities. A financial services company committed to a 10-year lease in Toronto then negotiated two expansion options for additional space that coincided with the projected growth of the business. The first option was available in the third year and there were at least 2 years of growth built into the initial space. Plans changed, and by the first year the company was out of space and continuing to grow rapidly. The business had exceeded all expectations and the original projections had proven too conservative. In addition, the company had made a decision within the first year to enter a new business that would add 25% more headcount to the already exceeded projections. The real estate group had continued to monitor the real estate market, company growth, and changing business conditions and had built in what initially appeared to be enough flexibility in the lease. Before the end of the first year of occupancy, the CRE group realized they needed to rethink their strategy because of the changing business conditions. Instead of starting from scratch, they modified the existing plan to include additional space in a building close to the original site that could easily be linked technologically. They also made sure that the lease for the additional space coincided with the lease term of the initial space (including termination and expansion options) and began developing a longer-term consolidation strategy to be implemented when the leases expire or are terminated. The lesson here is to develop a planning process and occupancy plan that allows for changes as business conditions change and to realize that plan flexibility is key to a successful real estate strategy.

18.1.2.1 Determining Size Requirements. Determining the size projections for a business is a very in-exact science. For businesses with more than a few thousand employees, up until very recently, projections were most often done on a macro basis by using square footage calculations. For example, at Fidelity, we determine the "ideal" rentable square foot metric for each business. This figure includes not only personal space, but all shared spaces, such as team space, conference rooms, training facilities (if specific to the business), circulation, and so forth. This metric in turn is multiplied by the growth in personnel to arrive at the size of the required space (see Appendix A, The Metric Model).

This method assumes there are space guidelines for individual and team workspaces and that building layout efficiency is assessed at the time you obtain the space, since that efficiency factor is not included in the metric calculation. Unless the business is relocating or their space is going to be reconfigured, the base square footage is assumed to be what they currently occupy. If the business is relocating, then the metric is applied to all existing personnel as well as the projected growth.

With the advent of more sophisticated space identification and tracking systems, more companies are tracking and projecting space by seats instead of square footage. Today, it's possible to assign a number of attributes to a space including its use, the business using it, and whether or not it is occupied. Using this method, it is easy to identify all space that is not occupied or is underutilized, which greatly increases the accuracy of occupancy planning over the square foot method since there's no longer an "approximation" of vacant space, but an accurate accounting. As units expand or contract, actual counts of vacant seats allow the CRE planner to take the guess work out of determining available growth space within the business unit's existing space envelope. Matching available vacant seats to seat demand, the planner can easily identify the additional seats required. This method of matching demand and supply requires robust information systems that are constantly updated (at least as often as occupancy scenarios are required).

New workplace strategies that are blurring the line between personal space and shared space will also make projecting space and seats more challenging since the basis for today's projection methodology demands detailed information about how space is currently used. As individuals use more shared spaces and individual spaces are reduced, planning metrics will have to be adjusted and real-time, automated employee tracking systems will have to be implemented if vacant seat count information is to be accurately maintained.

18.1.2.2 Location Decisions. Location decisions are usually somewhat easier to make than decisions about size commitments. More than 40 statistical indicators (see Figure 18.2) which can be classified into seven categories are often used in determining location decisions. Labor, taxes, occupancy costs, transportation, quality of life, incentives, and communications are usually the primary concerns of most businesses, with the importance weighting each one receives directly related to the type of business looking for space. In today's competitive labor market, finding workers is often the most critical of all the criteria. Whether the new location is to house a new business group (small but growing) or a relocated existing group, finding workers will impact the decision. Concern over losing existing critical workforce often means the location options are narrowed initially to within commuting distance of the existing workforce. But once made, location is considered a

New regional site search

Criteria ranking

- Labor
  - Costs
  - Education
  - Growth (general labor and FIRE) industry
  - Recruiting potential—competition
- Occupancy costs
  - Real estate taxes
  - Property taxes
  - Utilities
- Taxes
  - Corporate
  - Personal
  - Sales
  - Communication
  - Unemployment
- Transportation
  - Air
  - Public
  - Distribution hub
- Quality of life
  - Cost of living
  - Median income
  - Weather
  - Crime
- Incentives
  - Job tax credits
  - Training
  - Tax incentives cover under tax category
  - Recruitment assistance
- Communications
  - Fiber, ISDN & Sonet ring availability

FIGURE 18.2 Demographic Information Listing.

long-term commitment until a new business is formed or the existing business outgrows the local labor market, which takes years if the initial location decision was made correctly.

The right space also deals with the type of building the business needs. Some functions within every company are considered critical to the successful operation of the core business and must be housed in facilities with a high level of infrastructure support and often require power and communication redundancies. Company image, customer access, building and interior systems agility (easily moved at low cost), public transportation, parking access, and space efficiency are all-important considerations in determining the right type of space for a business. For example, very few law firms who want to project a conservative, established image are found in large open warehouse space, but that type of space may be perfect for a young start-up company.

18.1.3 The Right Time

Most real estate commitments are for a multi-year period. Leasing is usually considered more flexible than owning, at least in the United States, where typical lease terms run from 3 to 10 years. Determining the lease or ownership term has as much to do with the anticipated real estate market as it does with the growth projections of the business. Since the primary objective of the CRE professional is to provide the right space at the lowest possible cost, and history has shown that the real estate market runs in 4–7 year cycles (Figures 18.3a and 18.3b), timing can be critical in meeting the lowest cost objective. Companies that committed to space in Dallas or Boston in the late 1980s or early 1990s for at least a 10-year period enjoyed very low rental rates throughout their leasehold. Now, however, they are finding themselves on the opposite side of the table and having to pay rates that in many cases have doubled during their lease term. On the other hand, companies that built their own facilities during the late 1980s and early 1990s, although few did, are still enjoying low real estate costs with the ability in many cases to sell their real estate at high-return rates because of today's demand.

The length of time that the business will require the space correlates directly with the issue of the right space. Size, location, and type of space all impact the timing decision since the space ideally will accommodate the business throughout the target period while minimizing excess or vacant space. Periods of high growth or rapid downsizing may dictate shorter holding periods,

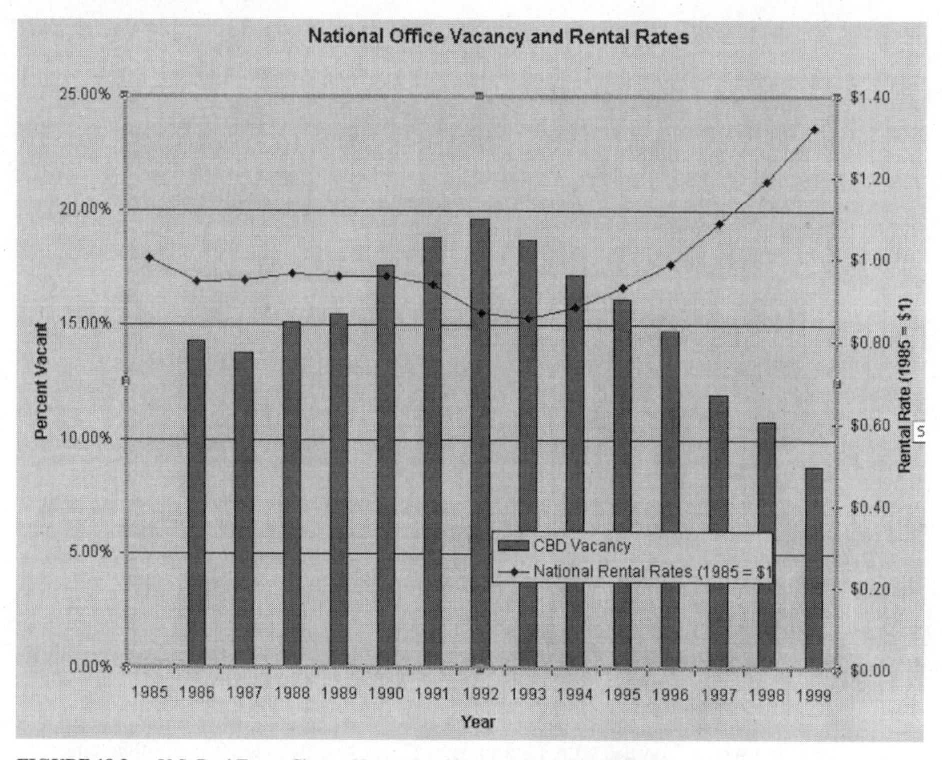

FIGURE 18.3a U.S. Real Estate Chart—Vacancy and Rent over the Past 30 Years.

while periods of slow growth or insignificant downsizing would indicate a longer holding period. Again, the key is to match the length of term to the size requirement as closely as possible. The more uncertain the business plans, the more flexibility needed in the timing and size commitments.

18.1.4 The Right Business Groups

In a small business of a few hundred employees, selecting the right groups for a space is not an issue. However, in larger companies with multiple business groups, this is often the most controversial of all space decisions. For example, a business is based in New York City, has outgrown its space, but has made the strategic decision to keep a core group of employees in the current location. Now it needs to select the groups that must relocate. In considering this question, the criteria of "right space" must be brought into the mix, since type of building and location will be dependent on the type of business tapped to move. If the location is predetermined, then it is a matter of matching the business demographics to the location demographics. If the business is determined first, then the reverse needs to occur.

The key to matching business to location is having an understanding of the business. Business unit profiles should be developed and maintained for each of the key businesses within a company if CRE is to plan effectively (see Figure 18.4). There are five areas that should be covered in a profile: 1) service or product, 2) organizational structure, 3) group functions, 4) key tools and technology, and 5) labor force.

Torto Wheaton National Office rent growth

Year	National rent growth (%)	National rental rates (1985 = $1)	Non CBD vacancy	CBD vacancy
1981	12.5			
1982	7.0			
1983	−0.9			
1984	5.5			
1985	2.1	$1.00		
1986	−5.5	$0.95	23.0%	14.3%
1987	0.5	$0.95	22.2%	13.8%
1988	1.5	$0.96	21.3%	15.1%
1989	−0.5	$0.96	21.0%	15.5%
1990	−0.1	$0.96	21.1%	17.6%
1991	−2.0	$0.94	21.4%	18.8%
1992	−7.2	$0.87	20.4%	19.6%
1993	−1.6	$0.86	18.9%	18.7%
1994	3.0	$0.88	17.0%	17.2%
1995	5.5	$0.93	15.2%	16.1%
1996	6.0	$0.99	12.9%	14.8%
1997	10.2	$1.09	11.3%	12.0%
1998	9.3	$1.19	11.2%	10.8%
1999*	11.0	$1.32	11.3%	8.8%

* 1999 represents forecasted amount

† Rental growth source, Torto Wheaton; Vacancy source, Trammell Crow.

FIGURE 18.3b U.S. Real Estate Table—Vacancy and Rent over the Past 30 Years.†

The first point addresses the business of the business. Questions that should be answered include:

1. Who are the customers?
2. What are the business drivers and key business indicators used to measure the group's success?
3. What changes are on the horizon, such as new products/services, new customers, or new ventures with other companies?
4. Are there key dependencies (both within and external to the business) with other groups?
5. What is the business strategy (e.g., low cost provider, high quality provider, first to market, etc.)?

Organizational structure addresses specific responsibilities in delivery of products and services, the current size and reporting lines for the group, and the history and evolution of the business. Often, a discussion of the evolution issue will provide insight into where the business is headed over the next several years.

Interactions with other business groups in the company; type of work; work flows; compliance, regulatory, and tax issues; can all fall in the group functions category. Much of this information will be important not only in determining geographic location but also the type of space required by the group. Whether the business group can be isolated as a stand-alone unit or needs to be near other groups in the company or near suppliers or customers depends on their function within the company.

Information on key tools and technology needs to be gathered at least at two levels. The more strategic level requires knowledge of the computer and communication systems the business needs.

I. *What service/product is provided?*
- Product or service line?
- Internal customers?
- External customers?
- What are the business indicators or drivers?
- Units of measurement and direction of business?
- Measurements of success?
- Key dependencies?
- Time sensitive?
- New services/venture/products?
- Key competitors?
- Overall business strategy?

II. *Organizational structure*
- Organizational chart
- Business leaders' responsibilities
- Business history

III. *Group function (functional units)*

Departments (aka functional units) work together

Functional Unit A:

Work group design
Deliverables and lead times
Group dependencies
Group interaction
Other group relationships
If "team design" please define team requirements
Work flow
Compliance issues
Confidentiality requirements
Contingency ratings
Security requirements
Physical barrier requirements

IV. *Key tool sets and technology*
- Technology
- Special infrastructure requirements
- Software applications

V. *Labor force*
- General categories of labor
- Most highly skilled labor group
- Other skill sets
- Hiring statistics
- "Feeder" organization
- Hire college level entry labor or more seasoned/experienced labor
- Recruiting or retaining
- Training
- External additional training
- Hourly shifts
- Participate in alternative work practices
- Current labor market
- Any additional information not covered above

FIGURE 18.4 Business Unit Profile Chart: Outlines the Basic Information That Should Be Captured and Maintained If Occupancy Planning Is To Be Effective.

Must the systems be on-site or will remote access work for the majority of workers? Practically all businesses today require high-speed data communication links, but not all have their own mainframe systems and more and more companies are moving to outsourced server networks. Moving these critical support systems out of the business to outsource providers lessens the requirement for $24 \times 7 \times 365$ support (24 h per day, 7 days a week, 365 days a year) which in turn broadens the type of space available to the business group. A few companies are now experimenting with wireless communications (Sun Microsystems, Eli Lilly, IBM, Fidelity Investments, and Cisco Systems[3]) for both voice and wireless communications, which can have some impact on the type of facility required because of areas of coverage with these systems, but will have more impact on the layout of the space itself. Once wireless becomes a standard and is developed enough that it can handle all communication applications. Building infrastructure issues, such as type of cabling (category 2 vs. category 5, for example) and underfloor delivery of wiring systems, will no longer be an issue. If the business requires on-site critical rooms (data and voice communication systems), then power and mechanical systems' redundancy must also be addressed.

In today's labor market environment and for the next 7–10 years, finding enough qualified personnel will be one (if not the most) difficult criteria to satisfy when locating a business. Information on key job functions and required skill sets, coupled with the hiring history, turnover, career paths, and promotional opportunities within the business, are necessary to formulate a sound labor strategy. This information can then be used to develop potential targeted locations by matching required labor to available labor supply. Hiring and turnover statistics are important in determining the size of the market. Call centers which have notoriously high turnover, often running at 35–50% annually, need to be located in a labor market with enough available labor to satisfy the continued demand for new labor. Initial hiring requirements for these centers is only the tip of the iceberg. A 300-person call center with a relatively low-turnover rate of 35% and a hiring ratio of six applicants for each hire (also low) would require an annual labor pool of at least 630 people.

18.1.5 The Right Cost

The right cost is usually the lowest cost on a present value basis for the type of space required. Once again, however, cost cannot be considered the only criteria and, most of the time, it's usually not the first of the five criteria to be considered. Only after the group to occupy the space and their requirements has been determined can the type of space be identified. Once the type of space is identified, then costs become an important consideration. Using cost as the first criteria can negatively impact a company's business, and what started out as a valiant effort can lead the business down a path from which it may be difficult to recover.

For example, let's take the case of a software development company that is located in the high technology corridor in Boston. The company is growing by 40% per year and is already out of space at its current location. Attracting and retaining qualified workers is critical to the company's success. Determining location demographics of the current workforce, labor profiles of alternative locations, and commuting distances or transportation and communication infrastructure if the possible new locations are in different areas of the country, are among the first issues that should be addressed if future locations are to have a positive impact on the company. Choosing locations based upon lowest cost before considering these demographic issues could lead to location commitments that could not support either the existing workforce or projected workforce. Once geographic location and type of facility have been determined, then cost can become the focus.

It is important to recognize all the costs associated with space when comparing cost information. Too often, companies consider only leasing or acquisition costs. They do not include the cost of designing and fitting up the space, the cost of relocating the employees, furniture and equipment costs, financing costs, or in the case of establishing a new geographic location, hiring, training, and incentive relocation packages. In my 17-year corporate real estate career working with some of the best brokerage and corporate service consultants in the industry, I have yet to find one who always includes all transaction and annual costs in their financial analysis without a reminder from the CRE executive.

The important questions of whether to purchase or lease include:

- Type of purchase
- Type of financing
- Balancing the portfolio

These issues are addressed in the portfolio management section of this chapter.

18.1.6 The Right Workplace Tools

Recently much has been written about the importance of workplace tools in today's work environment. Michael Brill and his research team of BOSTI[4] recognized as early as the late 1970s that the work environment of the knowledge worker has a significant influence on worker productivity and, thus, business performance. However, hard statistical data on the actual impact of tools in the workplace has been primarily limited to manufacturing environments until very recently. With the movement from the Industrial Age to the Information Age then subsequently to the knowledge environment, and with most new jobs and companies entering the service sector today, the subject of workplace impact on the worker has become a hot topic. Within the past several years, a new group of consultants has sprung up in an attempt to guide businesses into this new arena. These "workplace consultants" focus on current and future business processes, the need for agile work environments (easily changed at low cost), measurement of business drivers as impacted by the workplace, change in recruitment and retention of workers, and shifting workplace configurations as a response to shifting organizational structures.

Although not as integral to the occupancy planning process as the other four Rs, the right workplace tools are still an important component of workplace strategy. The CRE organization that continues to provide the same kind of work environment to its customers regardless of work processes is not meeting the needs of its customers and is in fact "inhibiting" rather than "enabling" the work of the business.

Because of the importance of workplace tools to today's knowledge workers, this topic is addressed in Chapter 8, "Alternative Work Strategies."

18.2 PORTFOLIO MANAGEMENT—THE SUPPLY SIDE

Demand forecasting is only half the picture. Now that we have looked at the issue of trying to predict the amount of space a corporation will need, which groups will need it, where they will need it, and most importantly, when they will need it, it is time to look at how the space will be supplied. In this next section, we will explore in more detail the concept of matching supply to demand and how to hedge against uncertainty by balancing committed and controlled space. We will also review the lease vs. buy question, percentage of real estate ownership, and its relationship to shareholder value. Finally, this section will also touch on those metrics of most importance in determining the effectiveness of portfolio management.

Total value of all commercial property in the United States has been estimated at between $3.5 and $4.5 trillion, with ownership by corporate users at roughly half of that amount—some $2 trillion.[5] The focus of this section will be on corporate real estate in the nonretail sector. It has been well-documented by a number of researchers that in most companies, real estate accounts for approximately 25% of real property assets,[6,7] an amount equivalent to 40–50% of net operating income,[8] 5% of total operating expenses, and 4–8% of gross revenues.[9,10] Because of the size of the asset on the balance sheet alone, strategic management of real estate assets deserves a high level of attention.

The debate for the past 20 years revolves around how much of a corporation's capital should be tied up in real estate assets. Investment analysts suggest that capital tied up in corporate real estate assets is not earning the return it could be if it were deployed in core business activities. This argument suggests smaller real estate holdings and a movement toward leasing and off-balance sheet financing vehicles as more appropriate strategies for today's corporation. What these arguments often fail to recognize is the corporate real estate provider is not in the business of maximizing investment return in real estate, but, rather, is in the business of providing a raw material (plant and equipment) to facilitate the operation of the business. Without space, the business cannot survive. (True, there are today a few "virtual" companies that do not have corporate real estate holdings, but we are talking about the vast majority of companies in our discussion.) As important to the corporation as the real estate itself is how the space is developed and managed. Previous chapters in this book have discussed the importance of space configuration, effective infrastructure systems, matching property services to corporate requirements, and optimizing the utilization of real estate as a resource.

As Kevin Deeble of TriNet Corporate Realty Trust has said, "the fundamental mission of a raw materials procurement process is to ensure availability of high quality materials at an acceptable cost to the firm, while preserving flexibility to reduce or terminate procurement commitments as business conditions change."[11] These four objectives accurately describe the mission of the corporate real estate portfolio manager. The CRE professional has the task of ensuring a high reliability of real estate, or making sure that it is there when the corporation needs it.

Earlier in this chapter we discussed the importance of timing, and this objective reinforces the importance of timing in the space equation. Maximizing flexibility is the second objective and is usually sited as the most important criteria in managing a real estate portfolio.[12,13] Deeble defines flexibility as the degree to which real estate assets can be reduced at no cost as business conditions change. I would suggest a broader definition to include both reducing and adding real estate assets at an acceptable cost to the corporation as business conditions change.

The quality objective continues to take on more importance as new and better building and interior systems lead to better work environments and improved productivity in the workplace. There is a lot of discussion, and some research, underway today by academics, researchers, and consultants (including Mike Brill of BOSTI, Michael Jaroff of MIT, and Barry Varcoe of Johnson Controls Corp.) on the development of more agile work environments and the means to measure their impact on corporate profits.

The final objective of minimizing costs should include not only real estate acquisition cost, but the longer term operating costs as well. Portfolio management plays a critical part in the cost objective in two ways, first through the development of strategic real estate plans that accurately project demands, and second through the effective management of the asset to minimize on-going operational costs.

18.2.1 Lease vs. Own

As long as corporations require real estate, the debate over leasing vs. owning will continue to rage. There is obviously no "right" answer that fits all circumstances and all corporations. Factors that should be considered in the own–lease analysis include:

- The type of industry
- Life-cycle stage of the business
- Speed of change in the industry
- Speed of change within the company
- Level of criticality of the functions housed and their relationship to the success of the business
- Availability and cost of capital

Start-up companies with a fast growth curve and high business uncertainties are not strong candidates for ownership since rapid expansion–contraction and short-term commitments are usually high priorities for new businesses. Even well-financed "dot coms" would be well-advised to focus on leasing instead of buying and putting capital to work in the core business where returns are typically much higher than those found in real estate. More mature businesses with more predictable growth are typically considered more viable ownership candidates. Even companies providing software or hardware to the Internet market, such as HP, Oracle, Cisco Systems, Sun Microsystems, and Lucent Technologies may choose to own rather than lease for some very sound reasons.

Even though most of these companies are enjoying double-digit growth, there is still a degree of predictability to their growth, and given the nature of their business, they would perceive themselves as long-term players in their respective markets. Critical back-office operations, production centers, and computer facilities may be considered essential to their business success and therefore targeted for ownership. In his research titled, *The Coming Disposal of Corporate Real Estate,*[14] Peter Linneman argues that even firms in early stage and mature growth industries should consider leasing instead of owning. This frees up capital which can be invested in core business activities which, according to Linneman, is where their shareholders expect them to invest. I would suggest that this theory holds true when the pool of available capital along with the appetite for debt is not large enough to cover both core business investment and real ownership. On the other hand, if financial analysis indicates purchase is the most prudent decision, the commitment duration is long term (more than 7 years), and there is enough capital for both core operations and real estate ownership, then there should not be a negative impact on shareholder value since investors are still realizing their expected return from the company's core business.

Impact of REITs on the Lease–Own Decision. As the real estate investment trust (REIT) industry continues to grow, an argument can be made that leasing opportunities with more flexible terms and more competitive long-term rates will make it more cost effective as well as more flexible (expansion and contraction) to lease. From 1992 through 1998, the total capitalization of REITs grew from $15 billion to an estimated $128 billion with their share of the commercial real estate equity market now estimated at approximately 37%.[15,16] In 1996 and 1997, 50% of all individual office property sales and 75% of office portfolio sales went to REITs, and the industry is expected to grow at an annual rate between 20–25% for the next decade.[17] At that growth rate, REITs' share of the commercial real estate equity market will top 50% by 2001. However, even at the 50% level, REITs will still only be approximately one-third the size of the total commercial property portfolios of America's corporations.

Wall Street analysts, real estate academics,[18,19] and some corporate real estate practitioners argue that REITs, given their dominance in the leasing arena, will offer corporations more flexibility for slightly higher costs than owning (estimated at 100–150 basis points—Pederson, 1998). However, most corporations will continue to have a capital borrowing advantage over most REITs since credit-based debt has traditionally been lower (100–300 basis points) than mortgage debt. As long as that spread exists and as long as a corporation has enough capital or borrowing capacity to fund core initiates, ownership of commercial real estate will continue to be dominated by corporations.

18.2.2 Commitment vs. Control

Kevin Deeble has offered the theory that effective portfolio management is about maintaining the right balance between committed and controlled space. Committed can be defined as the real estate for which the corporation is obligated to pay. Most obvious are leases that specify quantity of space, rental obligation, and duration. Not as obvious are structures owned by the corporation whose initial costs may be fully depreciated, but still represent an asset on the company's books due to continued upgrading/maintenance and on-going depreciation write-off of these improvements. Leases containing future obligations ("must take" space options, for example) also fall into the commitment category.

Control is defined as space the company has the contractual right to occupy, but not the obligation. In its simplest form, future options in leases are considered controlled space. Both additional space options (as long as they are subject to the sole discretion of the lessor) and options to terminate space are important to include in the control category. Space that can be built on existing owned property is also controlled space. For example, a company may own its own office building that has expansion capability. The company is committed to the existing facility but controls the right to expand.

Achieving the Right Balance. So why is it important to develop and achieve the correct balance of control vs. committed space? In an ideal situation, all of a company's space would fall into the controlled category with time commitments determined unilaterally by the occupying company. If it needed more space, it could have it when it needed it; if it needed less space, it could exit with no notice and no remaining obligation. Although real estate market analysts argue this model is one the market is moving toward (as REITs grow and become more important in providing corporations with space), market forces will always require some level of commitment from occupants.

Because corporations need a level of assurance that space will be available when they need it and many corporations want the flexibility to either take more or vacate some of their space at a reasonable cost in a reasonable period of time, corporations will always (at least in this author's work life) be required to maintain a balance between committed and controlled. Figure 18.5 compares various types of commitments an organization can make to the variables of quality, flexibility, and cost. Even though the suggestions of the chart are not accurate 100% of the time, it forms a reasonable guide.

For example, office suites offer a short term (1–12 months) which offers flexibility, but generally at a higher cost and less quality than may be acceptable to the corporation. They are also usually smaller spaces. Several providers of this type of space are exploring ways to offer more customized, higher quality, large space in these short-term commitment arrangements, but to-date costs are significantly higher than those under longer term commitments. Cost of construction,

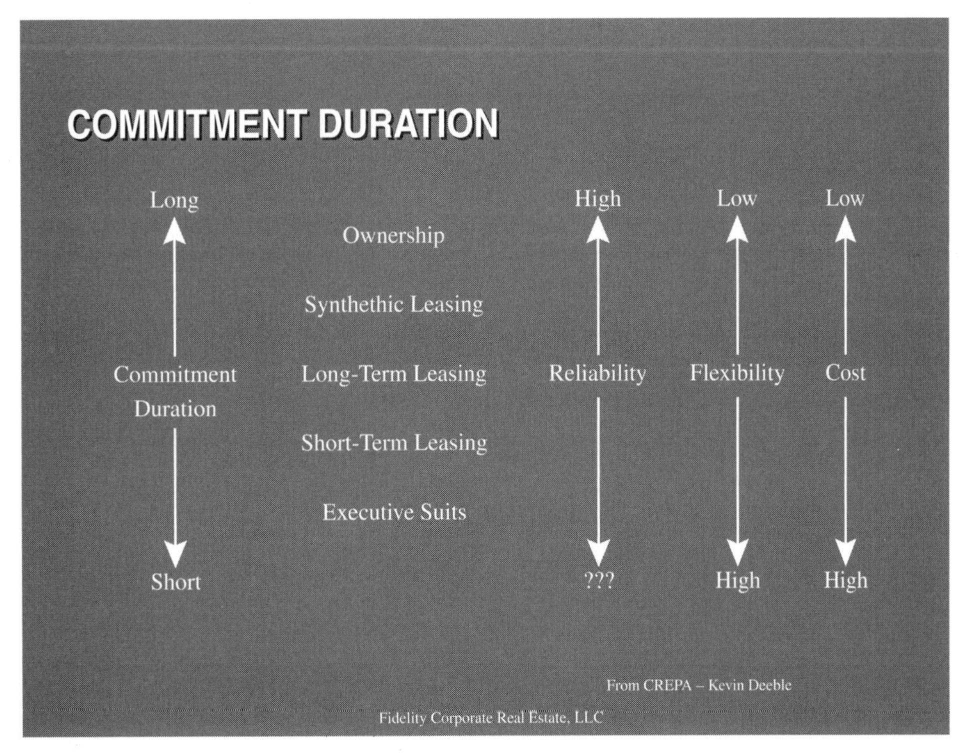

FIGURE 18.5 Space Commitment Duration Compared to Type of Commitment.

accounting and tax depreciation right-off rules, and landlord uncertainty about vacant periods all drive up the cost. Until these variables change or can be mitigated, corporations will have to commit to longer terms (at least 3 years and typically 5–7 years) if cost and quality are a higher priority than flexibility.

Earlier, we discussed the importance and the definition of reliability, flexibility, and cost. Now, we will relate these factors to the control/commitment balance. Reliability, or the assurance the space will be available when needed by the corporation, and cost reduction, are highly correlated to longer commitment duration (see Figure 18.5).

The caveat is the potential to overcommit on costs if business conditions change and space has to be vacated earlier than planned. Flexibility is usually maximized with shorter commitments, while reliability and costs are generally higher the longer the commitment.

Reliability can be maximized without increasing costs by enhancing control. When lease options for additional space or longer duration can be obtained at no additional cost and when expansion options on owned facilities are available in existing facilities, the corporation benefits from increased reliability.

When balancing control and commitment, the CRE professional must manage commitment to balance flexibility and cost and to match the level of commitment to the duration of demand. Each real estate transaction must be evaluated based on its contribution to the overall portfolio commitment level to insure the appropriate balance. The control factor should be used by the corporate real estate executive to maximize reliability of space and should always attempt to increase control at every opportunity. When control is not free, the probability of being able to acquire space when needed and the consequences if it cannot be acquired must be weighed against the additional cost of control.

APPENDIX A THE METRIC MODEL

INSTRUCTIONS

The metrics model is a tool to help analyze space utilization and anticipate the real estate impacts of growth and change in Fidelity functional units. The model utilizes normalized ratios and dimensions of typical program spaces to calculate a target amount of rentable sf per person. However, the metrics are only a guide; each location, each building, and each workgroup has its own distinct characteristics.

As the model calculates the various workspaces required for each functional unit, notice many of the numbers in the '# required' column are not round numbers (e.g. '6.7 small conference rooms'). Small functional units may only show fractions of certain kinds of spaces (e.g. '0.3 videoconference rooms'). This does not mean you should build a room at 30% size! What it does suggest is you should either 1) share a videoconference room with one or more other units, so you can use it about 30% of the time, or 2) design one of your other conference rooms to double as a videoconference room.

Line

1.0 The typical Fidelity functional unit has one type of basic (nonmanager) cube: phone centers have 5 x 5s, operations units have 5 x 7s; and most other units have 7 x 7s. When you enter the total headcounts on line 1.0, please sort them by the size of basic cube each one utilizes. Include the managers and directors in each group.

2.1 & 3.1 Enter the percentages of the total headcounts in each group you expect to be housed in private offices (2.1) and in 7 x 10 managers' cubes (3.1). It may be helpful to think in terms of ratios of offices and managers' cubes to basic cubes; for example, if a unit has a typical team or work group size. You can use the automated formula below to calculate percentages from these ratios. Enter values in the two boxes and the formula calculates the percentages of total headcount. (You must enter values in <u>both</u> boxes!)

Ratio of offices to basic cubes: 1 :	10	Percentage of offices:	8 %
Ratio of managers to basic cubes: 1 :	8	Percentage of managers:	10 %

7.1 Most 5 x 5 and 5 x 7 groups do not require visitor offices. However, if yours do, enter the number required in the '# reqd' column.

8.1 The standard Fidelity training room (8.1) is 30 x 40 , which accommodates up to 20 students at 36" x 36" PC work stations. The model provides two breakout rooms for each training room. All training facilities should be allocated to the functional unit which operates them, in order not to skew the metrics of other units.

However, you should also run a 'check' on training facilities for the B/U <u>as a whole</u> to make sure they are reasonable. A reasonable 'benchmark' for Fidelity B/Us <u>as a whole</u> is one standard training room per 150 persons (one per 100 for phone centers).

8.3 List any other training-related spaces (e.g. library, self-directed lab) on schedule A, and enter the total net sf in the box on this line.

9.1 Enter size and number of labs. If you have more than one size lab, list them on schedule A and enter total net sf on line 9.2.

9.2 List any other tech lab-type spaces (e.g. hardware test, human interface test) on schedule A, and enter total net sf in the box on this line.

10.4 List any other telecom spaces (e.g. call log room, wire service room) on schedule A, and enter the total net sf in the box on this line.

13.1 List any other types of specialized spaces (e.g. library, records imaging) on schedule A, and enter the total net sf in the box on this line.

15.0 The standard core & circulation multiplier was derived from prototype layouts of a representative sample of Fidelity B/Us. It includes the space required for open files, open copy/fax stations, and closets, as well as circulation, toilets, and core mechanical and electrical.

18.0 The functional unit metrics include floor common spaces, but not building common spaces. The latter vary significantly by location, building and function, and require case-by-case judgements by the planner.

METRICS SUMMARY

Company / BU:		RSF Reqd		Headcount		Target Metric
Functional Unit #1:	0	0	+	0	=	#DIV/0!
Functional Unit #2:	#REF!	#REF!	+	#REF!	=	#REF!
Functional Unit #3:	#REF!	#REF!	+	#REF!	=	#REF!
Functional Unit #4:	#REF!	#REF!	+	#REF!	=	#REF!
Functional Unit #5:	#REF!	#REF!	+	#REF!	=	#REF!
Functional Unit #6:	#REF!	#REF!	+	#REF!	=	#REF!
Functional Unit #7:	#REF!	#REF!	+	#REF!	=	#REF!
Functional Unit #8:	#REF!	#REF!	+	#REF!	=	#REF!
Functional Unit #9:	#REF!	#REF!	+	#REF!	=	#REF!
Functional Unit #10:	#REF!	#REF!	+	#REF!	=	#REF!
Company / BU - Wide Metric:		#REF!	+	#REF!	=	#REF!

METRICS WORKSHEET

Company / Business Unit: Functional Unit:

	5 x 5 GROUPS				5 x 7 GROUPS				7 x 7 GROUPS			
WORK STATIONS	**SF* Each**		**# Reqd**	**Net SF Reqd**	**SF* Each**		**# Reqd**	**Net SF Reqd**	**SF* Each**		**# Reqd**	**Net SF Reqd**
1.0 **Total headcount**			**0**				**0**				**0**	
2.0 Private offices	155				155				155			
2.1 % private offices		0 %				0 %				0 %		
2.2 # private offices			0	0			0	0			0	0
3.0 Manager cubicles	74				74				74			
3.1 % manager cubicles		0 %				0 %				0 %		
3.2 # manager cubicles			0	0			0	0			0	0
4.0 Basic cubicles	28				38				52			
4.1 Headcount			0	0			0	0			0	0
OTHER SPACES	**SF* Each**	**Ratio per HC**	**# Reqd**	**Net SF Reqd**	**SF* Each**	**Ratio per HC**	**# Reqd**	**Net SF Reqd**	**SF* Each**	**Ratio per HC**	**# Reqd**	**Net SF Reqd**
5.0 Reception	200	1 : 300	0.0	0	200	1 : 300	0.0	0	200	1 : 200	0.0	0
6.0 Conference												
6.1 Open conference	155	1 : 35	0.0	0	155	1 : 35	0.0	0	155	1 : 25	0.0	0
6.2 4-6 person room	155	1 : 100	0.0	0	155	1 : 100	0.0	0	155	1 : 50	0.0	0
6.3 20-30 person room	500	1 : 200	0.0	0	500	1 : 200	0.0	0	500	1 : 150	0.0	0
6.4 Videoconference	500	1 : 300	0.0	0	500	1 : 300	0.0	0	500	1 : 300	0.0	0
7.0 Customer/visitor												
7.1 Private office	155			0	155			0	155	1 : 50	0.0	0
7.2 Conf/presentation	500	1 : 500	0.0	0	500	1 : 500	0.0	0	500	1 : 500	0.0	0
7.3 Client resource center				0				0	1200			0

No.	Item	SF	Ratio	Qty	NSF	SF	Ratio	Qty	NSF	SF	Ratio	Qty	NSF
8.0	Training												
8.1	20 person room				0				0	1200			0
8.2	Breakout room			0.0	0			0.0	0	230	2 : TR	0.0	0
8.3	Other *(show TOTAL SF of other spaces in box; list each space on Schedule A)*												
9.0	Laboratory												
9.1	Software lab				0				0				0
9.2	Other *(show TOTAL SF of other spaces in box; list each space on Schedule A)*												
10.0	Telecom												
10.1	Console	800			0	800			0	800			0
10.2	FTR/LAN room	300	1 : 150	0.0	0	300	1 : 150	0.0	0	300	1 : 150	0.0	0
10.3	PTR	1200	1 : 1000	0.0	0	1200	1 : 1000	0.0	0	1200	1 : 1000	0.0	0
10.4	Other *(show TOTAL SF of other spaces in box; list each space on Schedule A)*												
11.0	Shared services												
11.1	Kitchen	300	1 : 100	0.0	0	300	1 : 100	0.0	0	300	1 : 100	0.0	0
11.2	Cafe seating	300	1 : 150	0.0	0	300	1 : 150	0.0	0	300	1 : 200	0.0	0
11.3	Copy room	100	1 : 150	0.0	0	100	1 : 150	0.0	0	100	1 : 100	0.0	0
11.4	Open copy/fax station	**				**				**			
12.0	Shared storage												
12.1	Storage room	155	1 : 150	0.0	0	155	1 : 150	0.0	0	155	1 : 100	0.0	0
12.2	Files room	155	1 : 300	0.0	0	155	1 : 300	0.0	0	155	1 : 200	0.0	0
12.3	Open files	**				**				**			
12.4	Coat closets	**				**				**			
13.0	Other specialties												
13.1	Other *(show TOTAL SF of other spaces in box; list each space on Schedule A)*												
14.0	Subtotals program spaces				0				0				0
15.0	x core & circulation factor				1.75				1.75				1.75
16.0	Subtotals RSF				0				0				0

17.0 Total HC: 0 **Total RSF: 0**

18.0 Functional Unit Metric #DIV/0!

* *Measured to centerlines of panels & walls* ** *Included in core & circulation factor* ☐ *Variables input by planner*

SCHEDULE A: OTHER SPACES

Company / Business Unit: ______ Functional Unit: ______

The Metrics Worksheet has a few lines which allow you to enter special 'other' types of space which are unique to a few - or perhaps only one - functional unit. Use this form to list those spaces, and enter the total net sf in each category on the Worksheet

Other Training Spaces:	Net SF	Other Tech Lab Spaces:	Net SF
Total: Enter on Worksheet Line 8.3	0	Total: Enter on Worksheet Line 9.2	0

Other Telecom Spaces:	Net SF	Other Specialized Spaces:	Net SF
Total: Enter on Worksheet Line 10.4	0	Total: Enter on Worksheet Line 13.1	0

ENDNOTES

1. As defined by the Corporate Real Estate Portfolio Alliance, a consortium of 25 companies, led by the McMahon Group, in the exploration of effective Portfolio Management practices.
2. Workplace Consortium Benchmarking (WPC) Survey, 1999. The WPC is a group of 25 diverse companies and government agencies with the common interest of developing workplace strategies that have a positive impact on workplace productivity.
3. Ibid.
4. BOSTI is the Buffalo Organization for Social and Technological Innovation located in Buffalo, NY.
5. Linneman, P. The coming disposal of corporate real estate. Wharton Real Estate, *Review*. p 1. A research paper prepared and supported by the Research Sponsors Program of the Zell/Laurie Real Estate Center at Wharton.
6. Nourse, H. O., and Roulac, S. E. 1993. Linking real estate decisions to corporate strategy. *The Journal of Real Estate Research,* p 475.
7. Zeckhauser, S., and Silverman, R. 1983. Rediscover your company's real estate. *Harvard Business Review.* January–February.
8. Nourse, H. O., and Roulac, S. E. 1993. Linking real estate decisions to corporate strategy. *The Journal of Real Estate Research,* p 475.
9. Checkijian, C. 1998. Comment: Corporate properties and capital markets. *Journal of Corporate Real Estate* 1(1), 5.
10. Duckworth, S. 1993. Realizing the strategic dimension of corporate real property through improved planning and control systems. *The Journal of Real Estate Research* 8(1).
11. Deeble, K. 1999. Financing corporate real estate: A raw materials procurement approach. A research paper prepared for the Corporate Real Estate Portfolio Alliance, p 2.
12. Varcoe, B. 1999. The five dimensions of real estate and facility management. A research paper presented to the Corporate Real Estate Portfolio Alliance, p 2.
13. Wilson, R. 1991. *Strategic positioning and facilities planning: Reviewing business plans and facilities strategies.* Document #50490. Prepared for the Internal Development and Research Council.
14. Linneman, P. The coming disposal of corporate real estate. Wharton Real Estate, *Review*. p 1. A research paper prepared and supported by the Research Sponsors Program of the Zell/Laurie Real Estate Center at Wharton.
15. McIntosh, W., and Whitaker, W. 1998. What REITs Mean. *Corporate Real Estate Executive* January–February, 24–27.
16. Pederson, R. 1998. Corporate real estate and capital markets. *Journal of Corporate Real Estate* 1(1).
17. McIntosh, W., and Whitaker, W. 1998. What REITs Mean. *Corporate Real Estate Executive* January–February, 24–27.
18. Pederson, R. 1998. Corporate real estate and capital markets. *Journal of Corporate Real Estate* 1(1).
19. Linneman, P. The coming disposal of corporate real estate. Wharton Real Estate, *Review*. p 1. A research paper prepared and supported by the Research Sponsors Program of the Zell/Laurie Real Estate Center at Wharton.

CHAPTER 19
SUPPORTING THE MISSION OF THE ORGANIZATION: AN APPROACH TO PORTFOLIO MANAGEMENT

Francoise Szigeti
Vice President
International Centre for Facilities
Vice President
TEAG—The Environmental Analysis Group
President
Serviceability Tools & Methods, Inc.
E-mail: icf@icf-cebe.com

Gerald Davis, AIA, F-ASTM, F-IFMA, CFM
President/CEO
International Centre for Facilities
President
TEAG—The Environmental Analysis Group
E-mail: icf@icf-cebe.com

19.1 INTRODUCTION

19.1.1 Portfolio Management in Context

Facilities are used for a purpose, in support of the mission of the organization that owns or leases them. It is therefore essential to define and forecast the demand for facilities, assess the capabilities of the facilities to meet that demand, and then be able to assess how well they match.

For owner–occupiers of large portfolios, real estate is a major long-term infrastructure investment. Costs need to be managed and benefits need to be assessed in terms of support to the core business. Senior management increasingly expects workplace facilities and corporate real estate (CRE)–facilities management (FM) service providers, whether in-house or outsourced, to deliver comprehensive, rigorously defined, consistent levels of functional support to users. Tools are required to allow management to judge how their real estate portfolio meets the needs of the business units.

This chapter focuses on the priorities of building occupants, owners, managers, and other stakeholders. It places portfolio management within the value chain, links it to the business strategy of the enterprise, and follows the decision-making flows. It does not deal with the financial aspects of portfolio management *per se,* but with the capability of facilities to meet the requirements of the occupant groups, and therefore to affect the bottom line.

This chapter presents and discusses ways to define the requirements of the occupants and to assess how well facilities serve their users and provide support for the mission of the enterprise. To balance their portfolios, managers need to answer questions such as:

- How well do facilities meet the needs of occupants and visitors?
- Do facilities match organizational objectives?
- Can we measure and compare the functional capabilities of different facilities?
- How do we know that we are getting good value from facilities? What are the benchmarks?
- Can we compare our facilities with those of others like us? What is typical?
- How well do facilities satisfy and serve their users?
- How can we demonstrate and illustrate those results with simple graphics?
- When is further investment justified?
- Can we assess the "functional" condition of a portfolio, as well as its physical condition?
- Can we use such assessments when there is a need to set priorities and decide which properties to keep or dispose of, to renovate or adapt now or later, or to keep as is?
- Can we measure and correlate the satisfaction of the users with the quality, functionality, and serviceability of the facility they occupy?

Measures of satisfaction and measures of capability are different from each other, but complementary. Both are useful. They should not be mistakenly used in lieu of each other. This chapter will focus on measures of capability.

19.1.2 Link Business Value of Facilities to Corporate Strategy

The relative *business value* of an individual facility, or of a portfolio of real property, depends on how well it supports, enhances, or impedes the strategy of the business and the functioning of the occupants. Business value is not real estate market value. It is quite separate from how the real estate market prices a particular piece of land and the buildings on it. Standard procedures for appraising the real estate market value of a building are accepted in the real property industry. Until recently, there was no standard way for evaluating the relative business value of a property.

Most of the time, managers use their best judgment when comparing buildings as a special type of "tools of production." These informal estimates are not adequate for benchmarking the business value of a portfolio, or for comparing portfolio strategies based on business value. At the same time, building occupiers increasingly expect workplace facilities and service providers to meet rigorous, consistent, and measurable demands for quality or functionality.

19.1.3 Managing Assets as a Factor of Production

Facilities have traditionally been planned, procured, designed, and managed on a "one-off" or one-at-a-time basis. For owners or managers of large portfolios, it is becoming evident that this traditional approach does not work any longer. The discipline of portfolio and asset management is now recognized as a must. Large organizations, private and public, are developing methodologies for creating formal "asset management plans," which include building condition reports (BCRs), serviceability ratings, financial reports, site plans and drawings, photo surveys, space recaps, etc. Key information from these plans are then rolled-up to the portfolio level, as a basis for overall assessment and strategic analysis.

In the future, assets will be recognized for what they are—a support for the mission of the enterprise across the whole range, from small businesses to multinational corporations and all levels of government. This perspective is starting to be recognized both in the private and public sectors, as well as by those people who look at real estate and infrastructure assets from a financial and investment perspective. As a consequence, asset management needs to be placed in the context of the decision and information processes that affect it.[1]

19.1.4 Using Standard Tools

The ASTM (American Society for Testing and Materials) standards on whole building functionality and serviceability, recognized as American National Standards Institute (ANSI), is a new set of standard processes and tools which responds to the needs of senior management and provides measures of quality, functionality, and capability. These standard processes and tools gives managers and facility users alike the means to determine functional requirements, measure the quality and serviceability of facilities provided, and verify that those facilities match the stated requirements. Because these are publicly available, open, and consensus standards, they are well suited for use with total quality management (TQM), ISO 9000, benchmarking, and other quality management systems.

This methodology, diagrammed in Figure 19.1, uses two matched, multiple-choice questionnaires. One is a set of scales for setting functional requirements (demand) using nontechnical words. The other is a set of scales for rating the serviceability of buildings and building-related facilities (supply) using technical and performance terms to describe combinations of building features.[2]

These scales cover more than 100 topics of serviceability and assess more than 340 building features. Each set of scales can be used independently of the other. The scales address the primary concerns of occupants, the concerns of property, and the concerns of its management. The scales use a 9 to 1 gradation, 9 representing *more* and 1 representing *less*, rather than good to bad. For example, a building may provide more or less security, be more or less easily identified by the public, or be more or less flexible and able to cope with change. Level 5 of the rating scales is calibrated to reflect what one would expect to find in a commercial building in a town of 50,000 people that the Building Owners and Managers Association International (BOMA) would call a Class B building. The processes are participatory, consensus-building, and include all stakeholders.[3]

These scales are part of a kit of tools diagrammed in Figure 19.2. As diagrammed in the previous figure (Figure 19.1), the serviceability levels for demand and supply can each be graphed as bar charts and compared by overlaying bar charts to show differences, whether supply is more or less than demand. Other tools include a consistent method (practice) for using interviews and focus groups to develop an organizational profile, a practice for determining quantities of space required, and links to decision making and cost and value engineering. Through the link to Uniformat II, translation is facilitated into outline specifications (Davis et al., 1993).

The ASTM standards have many uses in support of portfolio and asset management. When managers provide a better fit between the facility and the needs of its users, substantial savings and/or increases in productivity are possible, because the cost of the workers whose effectiveness is enhanced far overshadows the cost of the facility.

Recently, the U.S. State Department (Administration/Foreign Building Office (A/FBO)), has implemented a process for upgrading the way user requirements for its embassies are defined, so that they can be more responsive to the needs of the staff at the posts. For this, A/FBO uses the ASTM standards at several stages of the planning, procurement, design, and evaluation process, with the methodology known as "serviceability tools and methods." This chapter summarizes the steps followed and the results to date. It highlights the benefits and hurdles to be overcome during imple-

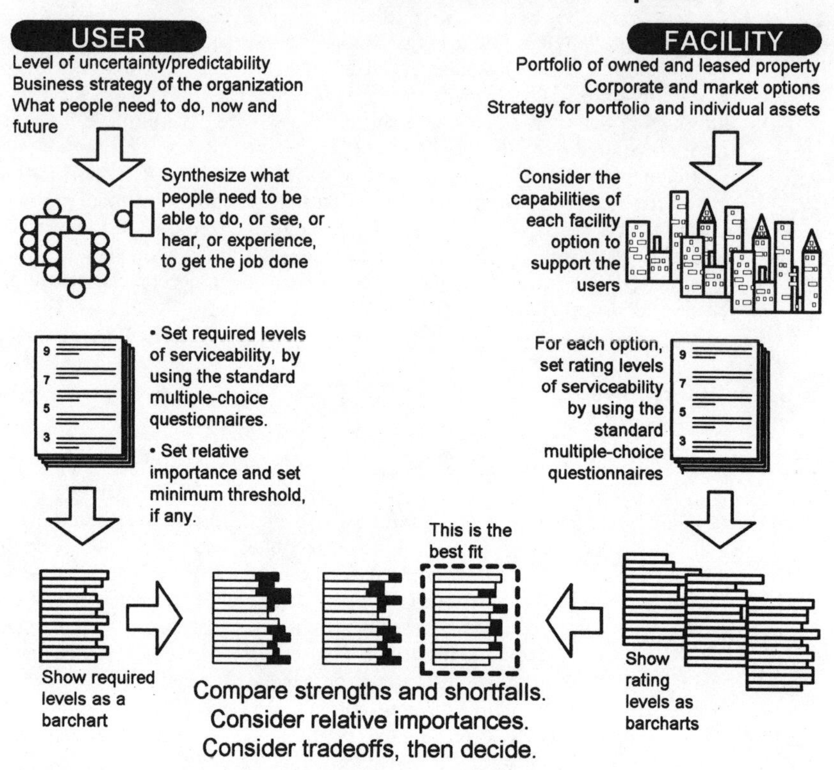

FIGURE 19.1 Using Standard Tools for Portfolio Management.

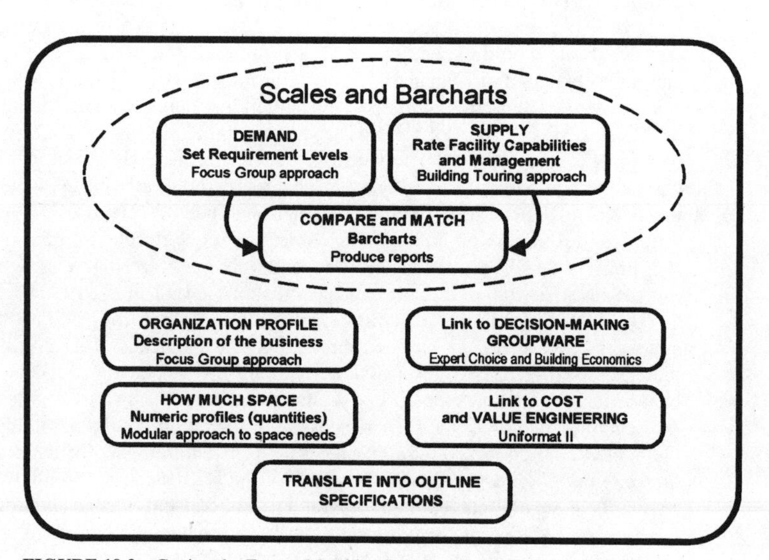

FIGURE 19.2 Getting the Facts—Matching People and the Facilities They Need.

mentation of a new functional programming approach that is more objective, structured, and systematic than the current, traditional, "one-off" approach. It complements and builds on other well-known tools, such as Post Occupancy Evaluations (POEs) and customer satisfaction surveys.

19.2 PERFORMANCE AND SERVICEABILITY—BACKGROUND

Traditionally, the mission of the organizational units has seldom been linked to the infrastructure resources required, in particular as it pertains to facilities. To introduce the main content of this chapter, this section contains a brief history of functional programming and evaluation, and the evolution of the performance concept in building. Thereafter, the process of portfolio, asset, and project management will be diagrammed within the context of their overall organization or enterprise. The role of defining functional needs will be placed in that context. Projects will be used to illustrate this approach.

19.2.1 Purpose of Facilities and Other Infrastructure Assets

Facilities are used to support the mission of the organization that owns or leases them. It is therefore essential that the demand for facilities be defined and predicted, that the capabilities of the facilities to meet that demand be assessed, and that the two be compared and matched. The same applies to other infrastructure assets.

Senior management increasingly expects workplace facilities and CRE–FM service providers to deliver consistent, agreed, specified levels of functional support to users. Management needs tools to validate that their real estate portfolio meets the functional needs of the business units. On the other hand, facilities are still mostly considered as overhead and managed as a cost center, rather than as an essential support to the mission of the business units. Facilities' impact on the productivity and effectiveness of the core business is starting to be recognized, but is not yet widely accepted, and is still difficult to prove. However, the notion that the occupants are to be treated as valued customers by the providers of the facility they occupy is slowly but surely taking hold, whether the provider is inside the parent organization, or an outsourced contractor. Recent uses of a customer satisfaction survey illustrate this point.

19.2.2 Measurements of Capability Complement Customer Satisfaction Survey

Since 1995, the U.S. Public Buildings Service (PBS) of the General Services Administration has commissioned the Gallup Organization to conduct a customer satisfaction survey of the occupants of its buildings, using as a starting point the survey instrument prepared by the International Facility Management Association (IFMA). A large portion of the 7,550 buildings in the PBS portfolio were covered by the customer satisfaction survey. The self-administered questionnaire was left on the desk of almost two-thirds of all the occupants of the PBS buildings, and over 45% of the occupants responded, a total of approximately 250,000 respondents. PBS is using the results of the survey to identify the aspects of their services and of the facilities that the occupants care about.[4]

In parallel, PBS has also been testing the effectiveness of the ASTM standards, described in paragraph 19.2.3, and the links between levels of serviceability for their buildings and the scores from the customer satisfaction survey for a given building (Davis et al., 1993). The two instruments were created independently of each other. They are complementary, but, at the present time, they are not fully matched. Nonetheless, where they do match, each confirmed the findings from the other. The serviceability tools provided the added information about the building that allowed the local staff to prepare an action plan backed by objective, comprehensive, defensible data. The results were a clearer set of priorities, the discovery of some lack of communication between PBS and the occupants, that could easily be remedied, and decisions about what to do next.

The target of these initiatives is to improve the quality of the facilities where it counts and to understand how to deliver services and facilities in the most effective manner. In one building, the solution to improving the relationship between PBS and its tenant groups has been to locate an FM officer in the building. That person is described as "the face of PBS" in that building. In another building where personal security was noticed as a problem, a single uniformed police officer now patrols the corridors and building lobby. Reports of petty theft in that building dropped by 75%. More or better hardware is not always the solution!

A second interesting example comes from the new Ciba-Geigy laboratory in Tarrytown, NY. In this case, the design–build team for the project agreed to incentives for schedule, cost, and satisfaction. Payment was subject to winning at least 75% of the vote from all the employees on a 15-question ballot regarding satisfaction, administered 90 days after move-in. The design–build team won 84% of the vote. The client is so pleased with the results that this team is being asked to do the next project for Ciba-Geigy.

19.2.3 Looking Back

The ASTM standards offer a macro, broad-brush set of tools appropriate for strategic decision making in facility procurement and portfolio management. They were developed as a way of dealing with the relationship between users and their facilities and to bridge the language gap between the occupants and the professionals involved. Architecture and design typically focus on each building project as a unique event and rarely go back to capture the lessons learned from one project in order to feed forward to the next. In contrast, clients needed a systematic, consistent, and comprehensive approach to deal with repetitive facility projects (Davis et al., 1993).

19.2.4 The Performance Concept in Building

The serviceability tools have strong foundations. This work does not stand in isolation—it builds on and links with the work of many others.[5] The serviceability tools have been founded in part on the performance concept in building, which has roots before World War II in the United States, Canada, and overseas. In the United States, in the 1950s and 1960s, the U.S. Public Building Service (a component of the U.S. General Services Administration) funded the U.S. National Institute of Standards and Technology (then known as the National Bureau of Standards), to develop a performance approach for the procurement of government offices, resulting in the "Peach Book" series of publications. An early marketplace application of the concept was the project of several California school districts (School Systems Development Corporation), which purchased component building systems for schools by specifying and testing performance. Similar applications to school construction occurred in Canada, in the provinces of Ontario and Quebec.

Founded in 1946, ASTM Committee E06 on performance of buildings has developed standard performance test methods for building components. Starting in the early 1980s, the performance concept was applied to facilities for office work and other functions by ASTM Subcommittee E06.25 on whole buildings and facilities.

In England, leadership for development of the performance concept has been provided by the government's Building Research Establishment (BRE). In Canada, the effort to develop norms for the physical setting of a "productive workplace" were led by: the former building use section of the National Research Council; the Department of Public Works of British Columbia (succeeded by the British Columbia Buildings Corporation); the government of Alberta; and Public Works Canada (now known as Public Works and Government Services Canada).

19.2.5 Performance and Serviceability

By 1985, the importance of distinguishing between *performance* and *serviceability* had been recognized and standard definitions for facility and facility serviceability were developed. *Facility performance* is defined by ASTM as the "behavior in service of a facility for a specified use," while *facility serviceability* is the "capability of a facility to perform the function(s) for which it is designed, used, or required to be used."[6]

Serviceability is more appropriate than performance for specifying the functional requirements for a facility, because the focus of performance is on a single specified use or condition at a given time. Indeed, the *ranges of performance* specified constitute the capability of the building to respond and, thus, its *serviceability.*

19.2.6 Programming and Briefing

The term *program*, meaning a statement of requirements for what should be built, was in common usage in the mid-nineteenth century by architects and students at the Ecole des Beaux Arts in Paris, and came into use in American universities as they adopted the French system for teaching architecture. In North America, the architect's basic services include architectural programming; that is, "confirming the requirements of the project to the owner," but exclude setting functional requirements, which is the owner's responsibility. In Britain and parts of Canada, the term *briefing* includes programming, but the distinction between functional, architectural, and technical programming is not often made.

19.2.7 Programming and Evaluation for Organizations with Large Portfolios of Similar Facilities—Getting the Facts

Governments at all levels, certain types of organizations and institutions, and many large corporations have hundreds or thousands of street addresses, yet only a few functional categories of facility. They sign or renew many leases each month, and launch many similar construction projects. Fast, economical programming and evaluation should be part of their continuing, repetitive, asset management process. The serviceability tools were developed to meet their typical need for capturing and using the institutional memory of what works best, and what does not work well, and for ensuring that their portfolio of property, leased and owned, best matches the functional needs of their core business groups—matching people and the facilities they need.

To be able to manage a portfolio of properties, it is important to have accurate and comprehensive information. This need was recognized in the 1970s by a few key individuals. One executive of Public Works Canada (now PWGSC), J. M. (Derm) Dunphy, had a key role in the development of the serviceability tools. In the early 1970s, he co-authored with Professor Doug Shadbolt, a report for the design and construction branch of PWC, which included recommendations for improved asset management, building programming and evaluation, and an enhanced response to occupant requirements. The purpose was to improve the products, reduce cost overruns, and deliver on time.

In 1976, Doug Shadbolt, then head of the School of Architecture at Carleton University in Ottawa, and Guy Desbarats, then assistant deputy minister of Public Works Canada (PWC), teamed to launch a graduate degree program in the management of facilities for public and institutional organizations, to be funded by the tuition fees from mid-career civil servants in need of training and an added degree. The Carleton/PWC project was aborted because of budget cuts, but PWC continued its leadership.

In the mid-1980s, as the first assistant deputy minister of the accommodation branch of PWC, Dunphy initiated a nationwide competition to select a group to assist the branch in becoming a "knowledgeable client" of its architectural, engineering, and realty colleagues. Providing "productive workplaces" to the tenant departments was one of the goals of that effort.

In 1987, a contract was awarded to develop standards, methods and tools for programming and evaluation that would become part of portfolio and asset management. The ASTM standards for whole building functionality and serviceability are based in a large part on work performed under that contract.[7] After rating hundreds of owned buildings in the early 1990s, a contract for comprehensive "asset management plans," including a serviceability rating, was recently awarded by the Canadian government.

It is remarkable that over the last 30 years, Canada, the Netherlands, New Zealand, and the United Kingdom have played a role in innovative developments quite disproportionate to their size. In each case, a combination of individuals with imagination and foresight commission or lead the work, in an organizational context able to provide the necessary resources to do the work. A comparative study of portfolio management of public properties has been undertaken by a student at the University of Delft, with the support of the governments of Canada, the Netherlands, the United Kingdom, and the United States, starting in 1998.

19.2.8 Considering Buildings and Facilities as a Whole

In the early 1980s, a task group was first established within ASTM Committee E06 to consider standards that would deal with buildings and facilities as a whole and attempt to define their functionality and quality. By 1983, as stated previously, this task group had become the ASTM Subcommittee E06.25 on whole buildings and facilities. The concept of functional requirements of occupants and serviceability features for *buildings taken as a whole* was a novel concept in the early 1980s. Now, at the turn of the millennium, the world of standards has recognized the need for those kinds of standards. The worldwide acceptance and success of the ISO 9000 and ISO 14000 families of standards are examples.

19.3 TENSIONS BETWEEN QUALITY AND COST

Some CRE executives define end users as property professionals who are involved in managing property assets for corporations whose primary business is not real estate. These executives, in fact, are end users to the *building industry*, and only surrogates for the real end users who, as noted previously, are the building occupants.

19.3.1 Input-Oriented Cost Containment

In the *input-oriented* perspective of end users, CRE organizations are suppliers of workplaces, providing complex products (buildings, furnishings, and technology) and services. As companies and governments strive to reduce expenditures, cut staff, economize, and minimize, this focus on the inputs to a business process leads naturally to cost control and expense cuts. Many participants in a survey said their organizational culture and priorities drive them to this focus on cost containment and real estate values.[8]

19.3.2 Output-Oriented Management

In contrast, *output-oriented* facilities organizations focus on helping the core business to be more effective and competitive. To them, the building with its furnishings and support services are means, not ends. Economy, efficiency, and skillful management of physical assets are still necessary, but will no longer suffice.

Output-oriented workplace service providers are committed to outcomes, rather than inputs. Top priority goes to meeting the needs of the occupants, by adding value to core business processes, and responding to changes in the core business workforce, all to best contribute to organizational success.

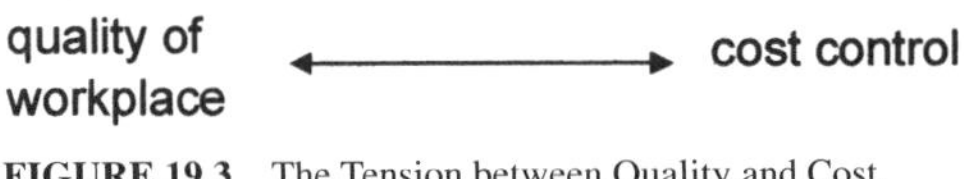

FIGURE 19.3 The Tension between Quality and Cost.

quality of workplace ←→ yield on real estate

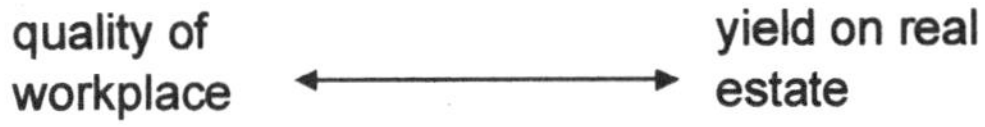

FIGURE 19.4 The Tension between Quality and Yield.

19.3.3 Managing the Tension Between Quality and Cost

In most organizations, the essence of FM seems to be to manage those tensions between quality of the workplace and cost control, as shown in Figure 19.3.

Then there is the issue of management of the portfolio of real estate. The ownership of buildings should yield income! New tensions arise, as can be seen in Figure 19.4.

These tensions create new problems for FM organizations. These tensions can be diagrammed as shown in Figure 19.5.

This idea can be developed into a profile of the real estate/facilities organization of a corporation or a government.

19.3.4 Top Executives Pay More Attention to Adding Value to the Organization as a Whole

Some top executives focus on how to increase the added value to the primary process, or core business, of their real estate organization. These executives realize that this triangle of tensions needs to be managed not only by developing and implementing a strategy, but also by showing leadership in balancing those three very important issues (see Figure 19.6).

In the private sector, some leading-edge facilities organizations are focusing on adding value to the core business processes of their parent organization, while most are still in cost-cutting mode. In the public sector, the political pressure has been to focus on the bottom side of that triangle. Indeed, some

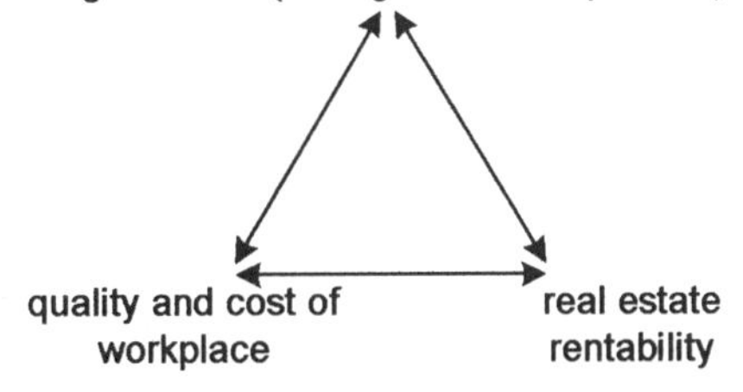

FIGURE 19.5 The Tensions to be Managed.

FIGURE 19.6 Strategy for Balancing Tensions.

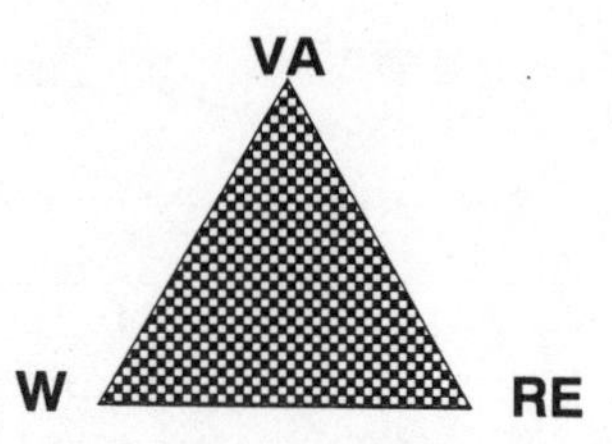

FIGURE 19.7 Strategic, Strong CRE/FM, Seen As An Asset to the Corporation, Present on the Executive Level.

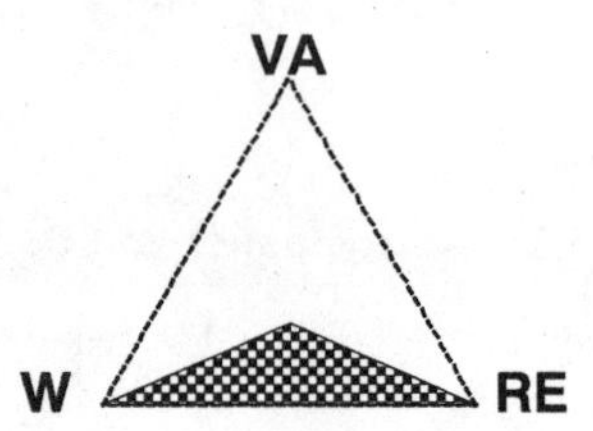

FIGURE 19.8 Typical Real Estate Organization in Government, Not Present in Major Decision Making.

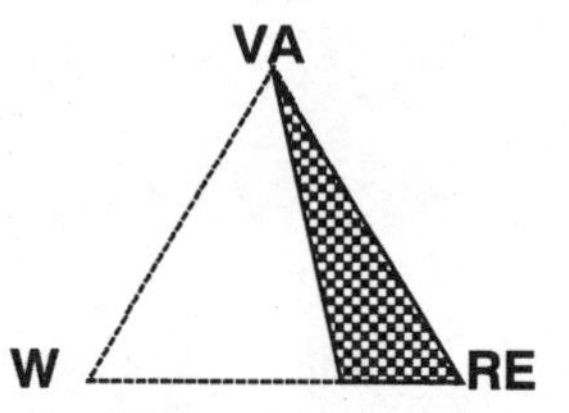

FIGURE 19.9 Real Estate Organization of a Bank Where Strong Pulls Are for Buildings to Serve Core-Business Tenants, As Well As to Make Money on Investment.

organizations excluded from their mandate any responsibility for value added to the primary processes or policies of government. Decisions that affect added value are explicitly left to tenant organizations.

The three triangles (shown in Figure 19.7) each represent the profile of a real estate/facility management organization with a different balance to those three tensions.

In the three diagrams, Figures 19.7, 19.8, and 19.9:

- **VA** means the pull to add value to the primary process of the organization
- **W** means the pull to manage quality and cost of the workplace
- **RE** means the pull to enhance real estate rentability or asset value

Thus, the tension from these competing demands is shifting.[9]

19.3.5 Comprehensive Outsourced Portfolio Management: The Pendulum is Swinging Back and Forth

Within the group of organizations surveyed, comprehensive outsourcing of the provision of workplace facilities, support services, and building operations to a private sector company is not typical. One of the organizations surveyed has done it. This organization has retained in-house a core team with expert capability both in the professional side of managing workplace services and facilities, and high-level management competencies. For quality management, their outsource contract includes performance indicators quoted directly from the standardized serviceability scales (ASTM standards).

In other organizations, however, we see some of the potentially serious problems when the process of outsourcing is not well managed. One company with which we are familiar (*not* an orga-

	Centralized at HQ	Distributed by function	Distributed by geography	Devolved to core BUs
Workplace services by own in-house staff				
Mixed: in-house plus some out-task, as needed				
Out-source managed by in-house expertise				
Total out-source: no in-house expertise				

FIGURE 19.10 Which Box Best Fits Your Organization?

nization surveyed) failed to retain adequate in-house expertise, corporate memory about workplace management, or mastery of its computerized facilities database. Furthermore, their new contract outsource firm was unable to attract even half the cadre of former company employees they had planned on needing to complement an expected but nonexistent robust in-house capability. With neither side having adequate human or knowledge resources, quick impacts loom for the bottom line and for core operations.

As well, many organizations surveyed know horror stories about organizations that have outsourced and realized afterward that the expertise needed to manage the external providers was no longer available to them. We see frustration that the newly appointed outside providers do not know and understand the core businesses. When several business units of a single organization each have separate outsource providers, and these providers are not centrally coordinated, we hear of the units or their providers bidding the price up in a single geographical area.

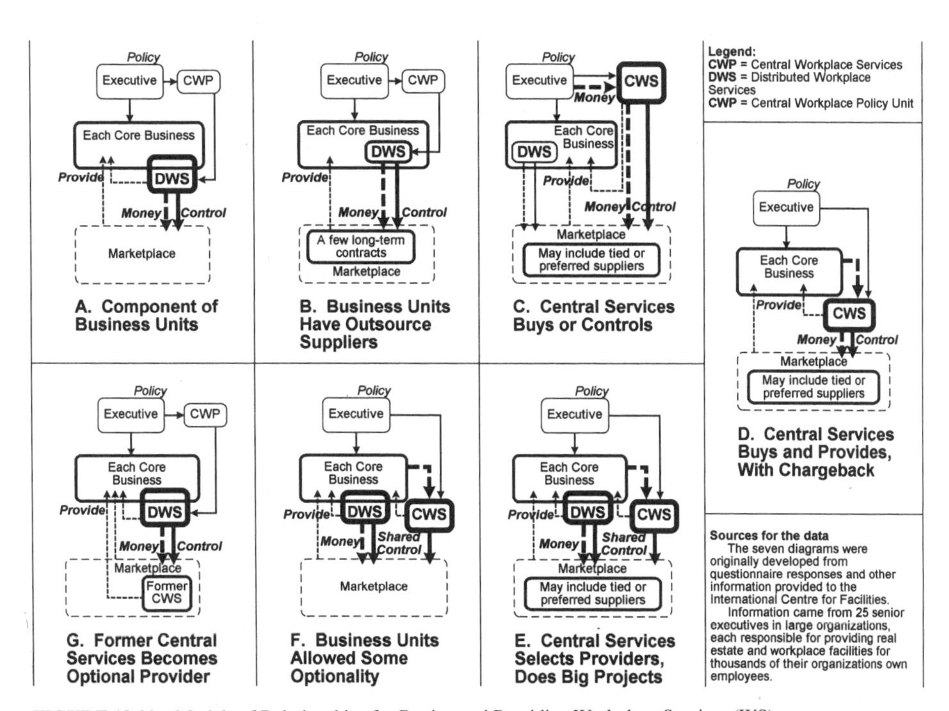

FIGURE 19.11 Models of Relationships for Buying and Providing Workplace Services (WS).

Business units have less buying power when they go to the market separately for professional services, furniture systems, and supplies. Building occupants, the internal customers, are coming to appreciate and value, that their facilities support group had acted as their skilled surrogates when buying from the marketplace and had kept the institutional memory about what works and what works less well, in the organization's facilities. Indeed, in some places, the drive to retain expertise and experience is being driven by the core business units.

To see how these changes affect the organizations surveyed, we asked, "In this table (Figure 19.10), which box best fits your organization?"

Based on the responses to this and other questions, a set of seven diagrams were developed to represent how large organizations are setting themselves up to provide and manage workplace services, facilities, and other support functions. Trends driven by the tensions described previously include: centralizing control while outsourcing; decentralizing or devolving FM; devolving FM functions into core business units; devolving control to core business units while outsourcing the functions; choosing preferred providers; drastically reducing the number of providers; and so on.

The underlying issue in the matrix shown in Figure 19.11 and in the diagrams on the following pages is whether the people who provide workplace facilities, services, and support should be in a separate, central unit or integrated into the core business processes of a company or government. If the latter, how is professional and technical expertise to be available to support and enhance the value of those core processes?[10]

19.4 PORTFOLIO MANAGEMENT: ESSENTIAL LINK FROM FACILITIES TO THE CORE OF THE ENTERPRISE

19.4.1 Portfolio Management: The Overall Process (Diagrams and Table)

A definition of portfolio management was developed in the Corporate Real Estate Portfolio Alliance research project, in 1999, as follows:

> Managing real properties as a group in order to achieve greater corporate benefits from them as productive working environment assets, financial assets, and strategic assets, above the benefits derived from managing them individually.

This project included firms with some of the largest market values today, including computer software and hardware giants, high-tech manufacturing, investment management, communications companies, insurance, and financial institutions.[11]

As we have said before, real property assets are occupied by an enterprise to conduct its business. Whether it is private sector or public, an enterprise uses facilities to provide workplaces for its staff, to provide places to meet its customers, and for a range of other business purposes. Therefore, the needs of the enterprise are the starting point for portfolio management. Workplaces and other facilities are factors of production, analogous to machine tools, delivery trucks, and computers. They are all part of the infrastructure of the enterprise. The principle is the same, whether the property is owned, rented, or occupied under some other form of tenure.

For an enterprise, the primary value of a workplace lies in its functionality. Buildings and other workplaces are usually far more important for their effect on enterprise achievement of mission, and consequent profitability, than for their market value. Managing a portfolio of facilities is clearly different from managing a portfolio of financial instruments, such as bonds, equities, or mortgages. Financial instruments can be bought and sold in moments for their market value, taking into account their yield, their risk, and the prospects for change in their market price.[12]

Portfolio management is the function that links real estate and facility operations to the core functions of an enterprise. Portfolio management decides what workplaces should be in the portfolio, and what projects are needed to get the right mix in the right places. Projects may be to dispose of property or to acquire it; to build, buy, or rent; or to finance or rehabilitate it.

A corporate view of portfolio management, asset management, and subordinate operating functions is diagrammed in Figure 19.12. This is reprinted from the executive report of the CRE Alliance Research Project.

FIGURE 19.12 Corporate View of Real Estate Functions. *(**Source:** Executive Report on Research Findings, Special Report, Fall 1999, ©The McMahan Group for the Corporate Real Estate Portfolio Alliance.)*

In this diagram (Figure 19.12), portfolio managers provide infrastructure to support organizational strategy. They give financial direction to asset managers who are in charge of major individual facilities and clusters of facilities. FM is seen as one of the technical operating functions that support portfolio and asset managers. This diagram presents the perspective of some unusually thoughtful, forward-thinking, high-ranked real estate executives. During the course of the alliance project, the consensus shifted from treating a portfolio of corporate real estate from a financial perspective to valuing their real property assets as tools of production.

This enterprise-based view of real estate is diagrammed in Figure 19.13. The cross-hatched area at the left represents the broad range of enterprise-wide functions, whether the enterprise be a manufacturer of widgets, an accounting firm, a software developer, or a government department. This area also represents the core businesses of the enterprise, the internal infrastructure that supports it, and the staff who conduct the business. At the top left, the box for external environment represents their external competitive market environment, with business cycles and legislative changes, fast-changing information technologies, and emerging global pressures. At the bottom, the box for other stakeholders and investors reflects the roles of shareholders, customers, and the general public interests.

Information about core functions of the enterprise flows to those who manage the real estate strategy for providing physical workplaces for the enterprise, and to those who manage the finances. Together, these real estate managers and financial managers develop the strategy for the real estate portfolio, and related financing. Within that framework, the team of portfolio managers takes direction from the business units that will need workplaces, identifies projects, and budgets for them. Portfolio managers also ensure that the portfolio as a whole meets the current and future needs of the enterprise. This is a crucial part of the value chain in support of the enterprise.

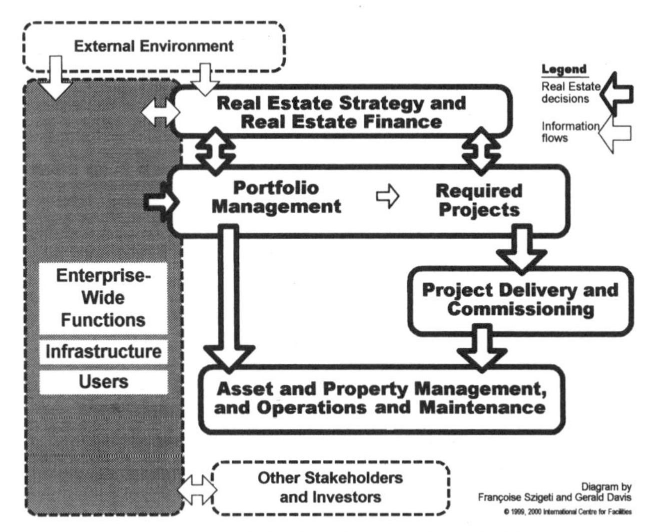

FIGURE 19.13 Framework for Real Estate and Portfolio Management—A.

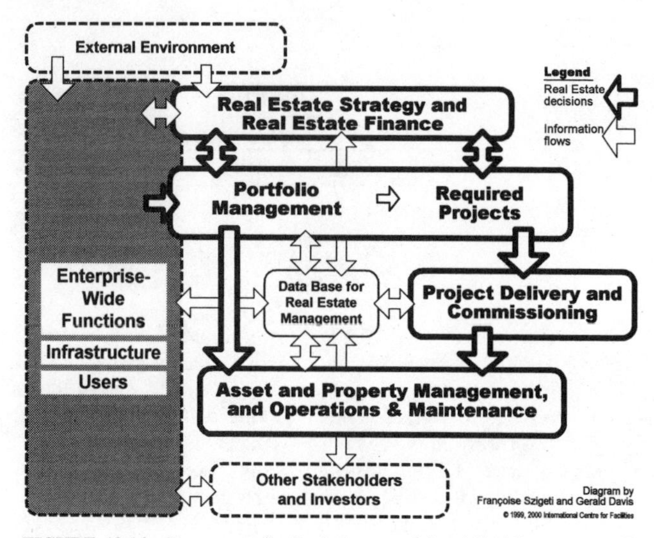

FIGURE 19.14 Framework for Real Estate and Portfolio Management—B.

The box for project delivery and commissioning encompasses all activities after a project has been defined and funded until a facility is occupied and commissioned, or terminated and disposed of, as the case may be. The box, asset and property management, and operations and maintenance, receives direction from two main sources. First, it responds to what was provided in the project—this is its "given," unless funds are provided for remodel or refit. Then, throughout the service life of a facility, it is operated and maintained under the direction of managers who receive direction and funding through the portfolio management function.

Figure 19.14 fills the hole in the middle with a crucial link: the database for real estate management. Arrows with a thin border show that all the boxes need shared access to a comprehensive database. Unfortunately, in many large enterprises, this database does not exist. Too frequently, data is incomplete, and what should be components of a single comprehensive data system are instead kept in separate databases, often in formats not compatible with each other.

The database should include more than the specific data about each real property asset, whether leased or owned, and data about each project under way. It should also include the summary data, or roll-ups, that are created from the raw data and serve as intermediate data tools for portfolio management.

Not only do most large organizations lack a comprehensive database; they also fail to develop an institutional memory of lessons learned. They are too often dependent on what best practices have been recognized and remembered by individual real estate and facility staff members, and passed on informally to their subordinates and successors. For that reason, wise firms keep the members of their successful project teams working together on successive projects, so that when the balance and chemistry in a team work well, lessons learned are carried forward from project to project, and teaming skills are enhanced. This is diagrammed in Figure 19.15. As each facility project is commissioned, whether it is new construction, remodel, or refit, both the facility and the process for executing it should be evaluated. Each phase of each project should be considered as a potential source of lessons, including programming, design, construction, and commissioning.

Many firms review the project file after completion and note whether the project was completed within budget and on schedule. Some firms assess how well each new or remodeled facility meets the need of the business users who occupy it. This essential knowledge can be captured as a formal institutional memory of what works well, what works best, and what should not be repeated.

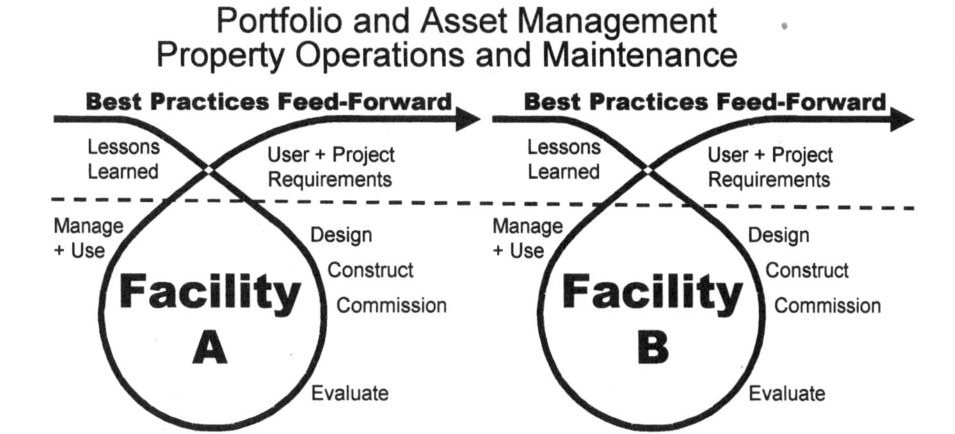

Diagram by Francoise Szigeti and Gerald Davis

Based on a diagram by John Zeisel

FIGURE 19.15 Feed-Forward of Lessons Learned.

This can be done with the new generation of multiple-choice questions introduced previously. The questions use indicators of functional capability to assess how well a proposed design or an occupied facility meets the functional requirements specified by the business units and facility occupants. Now that these tools have been accepted as national standards, their use is gaining momentum among large organizations. (See Section 19.5 of this chapter for a more detailed description of these tools, and Section 19.6 for a description of their application.)

Figure 19.16 points out that all enterprises need a comprehensive knowledge base of shared data and supporting detailed data. Even a small business, with only a few dozen staff, needs to capture and conveniently access the key facts about its workplaces, how they are used, and lessons to apply next time.

A gap analysis of the portfolio identifies the need for facilities, and projects are launched. The main phases of project delivery are diagrammed, with commissioning as the bridge into operation and maintenance. The evaluation phase is diagrammed, both soon after completion, and later in the life cycle. After evaluation, at some point a decision may be taken to dispose of an asset, to retrofit it, or to re-use it for some different purpose.

Figure 19.17 starts a series of diagrams about where the authority to make strategic decisions about a portfolio of corporate real property is usually best situated, and about the follow-on decision flows to implement the strategy. This diagram shows the main elements from the life cycle. In this diagram, decision flows are shown with heavy arrows, and information flows with thinner, dashed lines.

The facility operations group needs an explicit and verifiable process to set priorities for which projects will be required in each budget cycle, and to set priorities for funding, including for major capital projects, subprojects, and smaller line items of repair. Projects are required and launched because of decisions flowing, explicitly or implicitly, from portfolio strategies and initiatives. In a sense, the work of project delivery and commissioning includes making decisions about property operations and maintenance, because once a facility is created or remodeled, it must then be operated and maintained until it is changed through another project.

Portfolio strategies and initiatives also provide the framework, policies, and specific directives for asset management. Asset managers create and maintain an asset management plan for all large

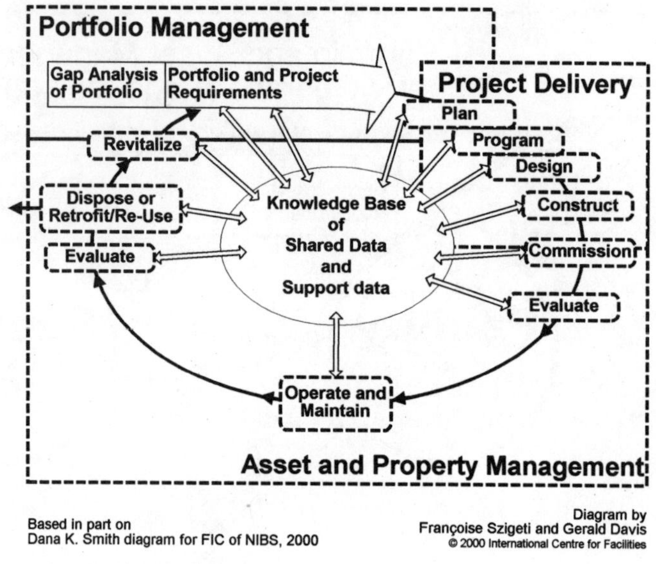

FIGURE 19.16 Life Cycle of Facilities.

or significant properties. Decisions about funding and investment in a property implement that asset management plan.

Figure 19.17 also shows, as dotted lines, the information flows needed to manage and operate a facility throughout its life cycle. At the center of the diagram, as in the life cycle of facilities diagram (Figure 19.16), is the asset database about each individual property that is maintained throughout its life cycle. The database incorporates information such as the levels of service it pro-

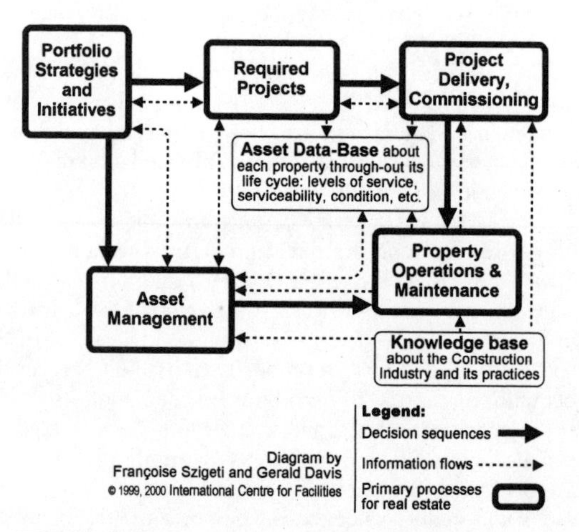

FIGURE 19.17 Decision and Information Flows for Project Delivery and Property Management.

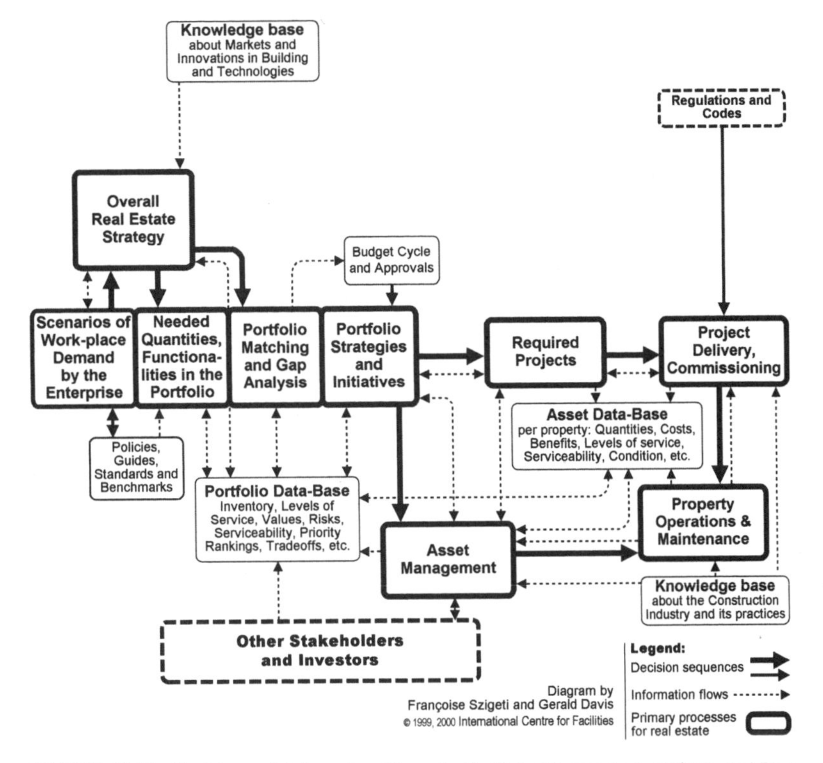

FIGURE 19.18 Decision and Information Flows for Portfolio Management, Projects, and Property Management.

vides, its serviceability, its physical condition, as well as costs to build or acquire, operate and maintain, the condition and remaining service life of building systems and materials, and so on.

The next diagram (Figure 19.18) adds the decision and information flows within portfolio management. It starts at the left with scenarios of workplace demand by the enterprise. What is the range of likely futures for which workplaces will be required, to a horizon of 3 or 5 years, and for the longer term? In some organizations, a set of several scenarios is updated for each annual budget cycle, and also when abrupt major changes occur in the environment external to the enterprise, such as a market crash or a political upheaval in a key market area. Estimates of probability or risk are attributed to each scenario.

As Martha O'Mara points out in her book, *Strategy and Place*, the wise workplace provider will respond with an appropriate portfolio strategy, such as incremental change to the portfolio under conditions of high uncertainty, or standardization of facilities in an enterprise that is relatively stable, or value-based when change and uncertainty are moderate.[13]

Unfortunately for the workplace providers, most organizations, large and small, do not have such scenarios available for the workplace provider. More typical are projections of sales or other volumes for the next year or several years, often with related overall staff sizes, but lacking the information needed to estimate what workplaces, of what general capabilities, will be needed in what locations or regions. Often this information is only available for individual departments or business units, with no overall or coordinated estimates for the enterprise as a whole: managers may describe the enterprise as a set of silos. This is often true, even in organizations with relatively stable external environments, and typical in companies undergoing rapid change, such as high-growth technology companies.

When a reliable basis for demand projections is not available to the multi-year horizons of real estate construction or leasing, portfolio managers rely on a range of forecasting methods to manage the risks associated with workplace provision. The portfolio manager may even create straw-case scenarios for the enterprise and test them for plausibility with occupant managers, as a basis for demand projections. In organizations where departments operate in "silo-mode," without coordination and sharing among departments, the facilities group may be better placed to project the overall corporate future than the individual vice presidents do. In many cases, the best understanding of what is likely to happen across the whole organization can be built from a roll-up of the "guerilla intelligence" available in the real estate and facilities department. In our own experience, a multinational corporation used staffing projections developed by the facilities team as its hiring plan for the next 3 years, and another multinational company got projections of enhanced sales volumes from its facilities programmers as the rationale for proposed investments in more supportive workplaces.

An overall real estate strategy takes direction from scenarios of workplace demand and relies on a portfolio-level database that contains roll-up totals and summaries such as: inventory of facilities, considered by size, geographic region, functional category, age, tenure (own, lease, etc.); level of service each provides compared with requirement; values, including book, replacement, etc.; risks associated with each facility or group of facilities; serviceability (capabilities whether needed or not at this time), including building condition; priority rankings for projects to maintain serviceability; considerations for tradeoffs on investment; and so on. Senior real estate managers bring to the corporation a current understanding of real estate markets, and innovations in the building industry and

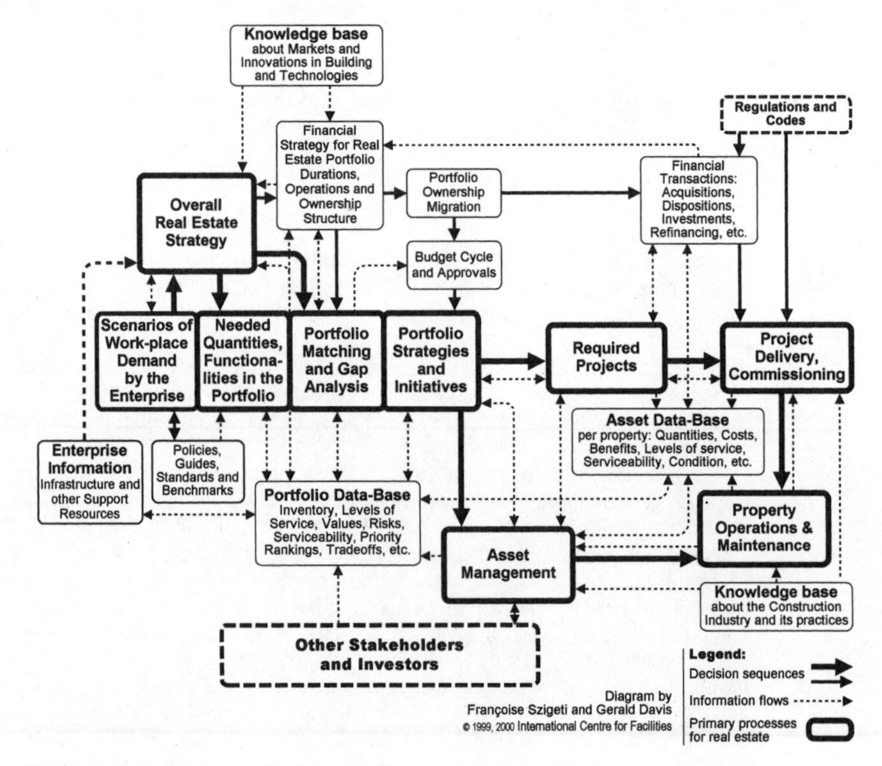

FIGURE 19.19 Decision and Information Flows Including Financial Strategy and Management.

in the technologies used by building occupants. The strategy they develop also takes into account corporate policies, guides, standards, and benchmarks, both internal and external.

Within that strategy, it is necessary to develop scenarios for the needed quantities of each functional capability in the portfolio, to specified time horizons. These demand scenarios, expressed in real estate terms, can be matched against what is in the portfolio and projected. Gap analysis will lead to portfolio-level strategies and initiatives, and to project priorities for each budget cycle.

The diagram in Figure 19.19 shows other factors that must be taken into account: the concerns of other stakeholders and investors, outside the enterprise; and compliance with industry and building regulations and codes. Figure 19.19 adds financial strategy and management to the diagram. The chief financial officer typically demands that the real estate group demonstrate, at strategic and operating levels, that is it spending the right amounts of money in the right places on the right things. At the strategic level, the real estate group links the overall real estate strategy to the financial strategy of the enterprise. It plans for the target ratio of ownership to leasing, or synthetic leasing, or other modes of tenure. It sets the strategy and manages migration of the portfolio toward the target financial mix.

The real estate group typically may also have the expertise to lead implementation of the financial migration of the portfolio. The group would conduct the financial side of real estate transactions, including financing of acquisitions, dispositions, and refinancing, and advise on, or directly manage, the financial side of real estate held primarily for investment purposes, rather than as a factor of production.

Figure 19.20 adds enterprise-wide functions to the diagram. It shows the overall enterprise strategy as not only the driver of enterprise operations, with their resultant scenarios and decisions, but also as a direct decision driver of the overall real estate strategy. The overall enterprise strategy and

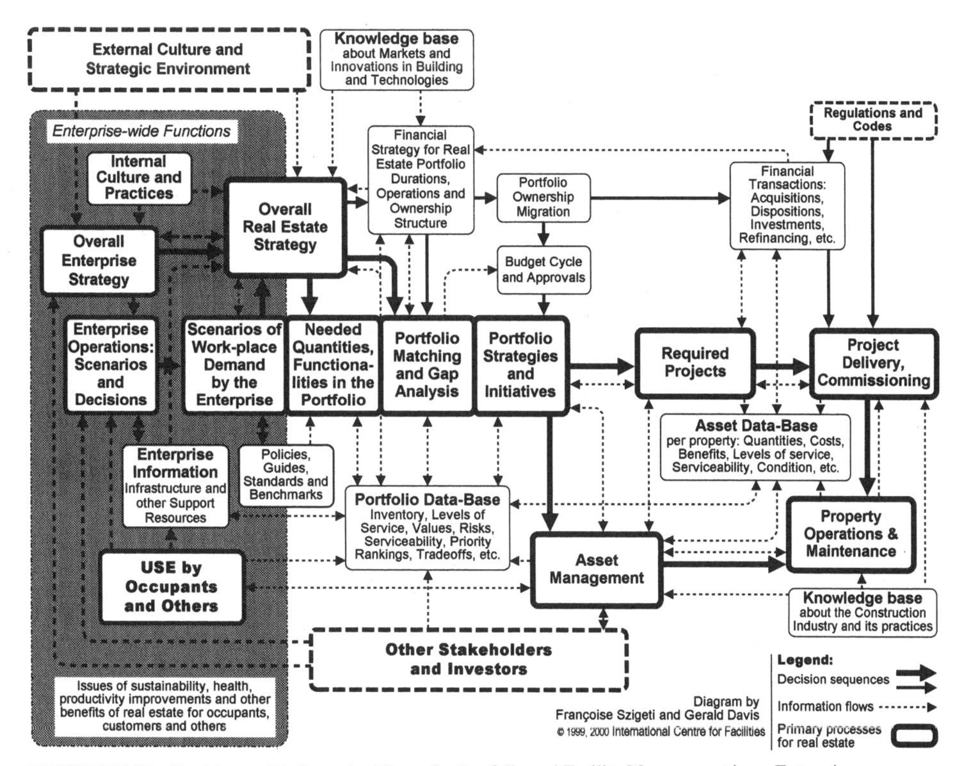

FIGURE 19.20 Decision and Information Flows for Portfolio and Facility Management in an Enterprise.

the overall real estate strategy are mediated by the enterprise's internal culture and practices and by the strategic environment and culture external to the enterprise. The enterprise's information infrastructure and other support resources are used directly in developing the overall real estate strategy, as well as in normal enterprise operations. Information from other stakeholders and investors is also a resource for the enterprise decision makers.

The users of facilities, both occupants and others, including visitors and the general public, are shown at the lower left of the diagram. Information about these groups and their facility needs are basic inputs to the managing of enterprise operations, and of course to the enterprise's information infrastructure. This diagram also lists at the bottom left some other kinds of issues, which are, or should be, taken into account when developing and implementing a real estate management strategy. Among the list are: sustainability, health, productivity improvements, and other benefits of real estate for occupants, customers, and others.

Figure 19.21 has the same boxes and arrows connecting them, in the same places, as in the previous diagram, Figure 19.20. What is different are the line weights. In Figure 19.21, visual priority is given to databases and information flows. The portfolio database contains a copy of information on each facility in the asset database, or has ready access to it. The portfolio database adds roll-ups and cross links from real estate totals to data in the enterprise information infrastructure.

Today, the very best real estate databases are inadequate. Corporate data about real estate is typically held in different formats on disparate systems across many data silos. There are several software packages that do an excellent job for data users for some parts of day-to-day operations or for design and construction. There are a few that meet needs over a fairly wide range of tactical information needs. An analysis by Fransson and Nelson, published in April, 2000,[14] asserts that the level of data integration required for high-level portfolio management decisions simply does not exist.

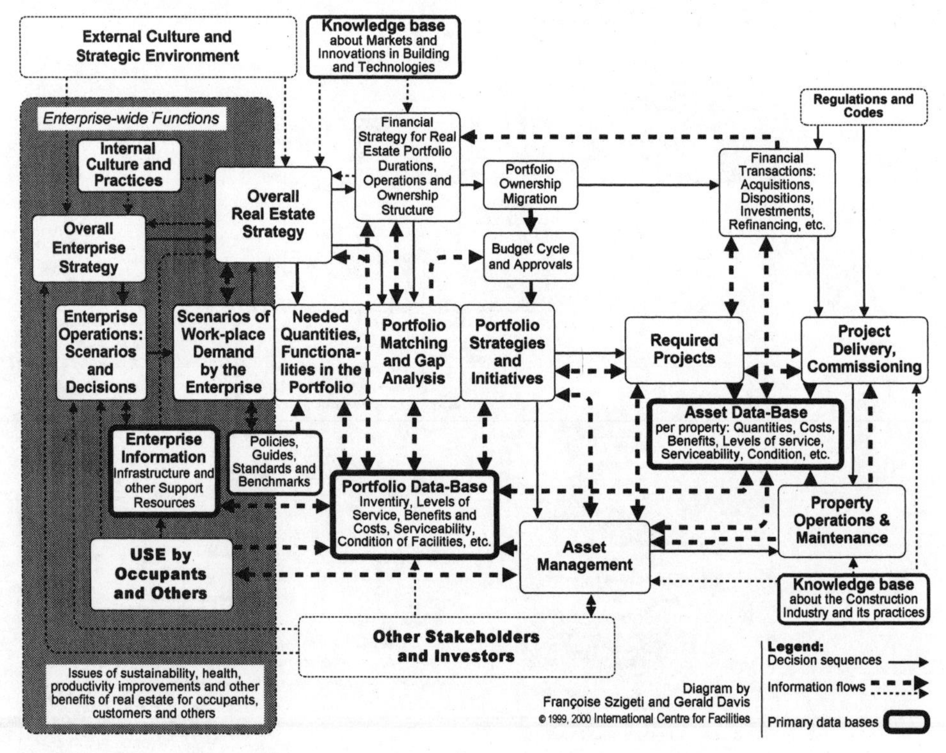

FIGURE 19.21 Information Flows and Databases.

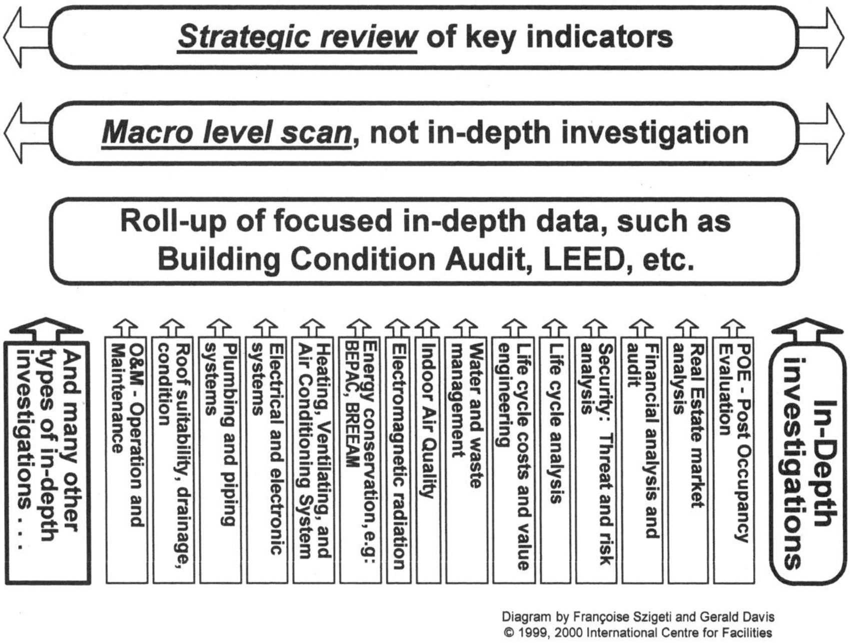

Strategic review **of key indicators**

Macro level scan**, not in-depth investigation**

Roll-up of focused in-depth data, such as Building Condition Audit, LEED, etc.

And many other types of in-depth investigations . . .
O&M - Operation and Maintenance
Roof suitability, drainage, condition
Plumbing and piping systems
Electrical and electronic systems
Heating, Ventilating, and Air Conditioning System
Energy conservation, e.g: BEPAC, BREEAM
Electromagnetic radiation
Indoor Air Quality
Water and waste management
Life cycle costs and value engineering
Life cycle analysis
Security: Threat and risk analysis
Financial analysis and audit
Real Estate market analysis
POE - Post Occupancy Evaluation
In-Depth investigations

Diagram by Françoise Szigeti and Gerald Davis

FIGURE 19.22 Strategic to In-Depth.

The authors report on their findings from the Corporate Real Estate Portfolio Management Alliance, that comprehensive solutions have yet to be created. They compare the situation to that of corporate information systems in general, before the advent of software for enterprise resource planning (ERP).

This section has focused on the portfolio management function of corporate real estate. It has emphasized the need for a strategic view as the primary driver for decisions about corporate real estate. In Figure 19.22, the many types of investigations shown at the bottom of the page provide a foundation of information about individual assets. Roll-ups for a specific site or complex of facilities are used for building condition audit, or in building labeling programs, such as LEED (Leadership in Energy Efficient Design). Portfolio management operates at the two higher levels, with macro-level scans, not in-depth investigations, and with strategic review of key indicators. Earlier in this chapter, the serviceability tools and methods have been discussed as tools for that macro-level scan, and the strategic review.

19.5 FUNCTIONALITY, SERVICEABILITY, AND SUITABILITY TOOLS FOR MEASUREMENT: ASTM STANDARDS ON WHOLE BUILDING FUNCTIONALITY AND SERVICEABILITY

19.5.1 Measurement Matters: If You Can't Measure It, You Can't Manage It!

Gerald de Kerchove[15] points out that managing how well a facility functions as a support for the mission and operations of an enterprise or government organization requires some means of measuring that support. The facility may provide a high, low, or moderate level of service, but without measurement, management is reduced to guesswork. In his work, de Kerchove took as his starting point Joe Ouye's[16] observation that workplace performance can be decomposed into three distinct parts:

- *Strategic Performance.* How is the workplace supporting the mission, goals, and objectives of the [core] business?
- *Worker Performance.* How well are the workers who use the workplace performing their functions?
- *Workplace Effectiveness.* How effectively does the workplace support the performance of the workers?

Across those categories, de Kerchove uses a matrix table to map the headings from the balanced scorecard of Norton and Kaplan: financial, customer, internal processes, and learning and growth.[17] The ASTM standards for whole building functionality and serviceability provide a set of tools for measuring the functionality of the physical setting of work consistently across all required types of workplace capability in this matrix.

Most organizations and work groups have only a few generic types of functions, such as general purpose office functions, or office functions requiring special security provisions, or receiving many visitors, or requiring special high-technology, or a mix of "dry" laboratory functions and offices. For each type of function, the facilities are required to have a different mix of capabilities; that is, more or less capability on each of a variety of topics. Each such mix of functions has a different profile of functional requirements; that is, it requires a different mix of levels of serviceability in its facilities.

In the ASTM standards, the serviceability of a facility, such as a workplace, or a home, is its capability to perform as and when required, to support the activities or functions of users, owners, and facility managers. This section explains the principles and general method of establishing what will be suitable for a mix of functions, and of assessing the serviceability of a facility. To know whether a facility will be suitable for a particular use or function, one needs to know the profile of functional requirements and the profile of serviceability of the facility—then, these profiles can be compared, and the significant differences can be analyzed. This approach is also being used for constructions that are not buildings, such as highways and municipal waste systems. This section deals only with its application to buildings and building-related facilities.

19.5.2 Sample Scales to Illustrate Format and Components

There is one set of functionality and serviceability scales for each topic. Topics may cover a wide range of user requirements. Figure 19.23 contains an example of a list of topics. For each topic, there is a set of one requirement scale stating levels of requirement and one rating scale describing features that indicate levels of serviceability. For each level, if the combination of features described in the rating scale is present in a facility, then that facility is likely to meet the combination of functional requirements described at the same level of the requirement scale.

In Figures 19.24 and 19.25, a single serviceability scale is used to illustrate and explain these principles. In Figure 19.24, a set of matching scales is presented at full size. In Figure 19.25, it is presented at a smaller size and annotated to explain how its parts comply with the principles stated in the following section.

19.5.3 Requirement Scales are Provided to Classify According to Requirement Levels

The ASTM standards for whole building functionality and serviceability provide a tool for setting a level of required functionality on each of a range of topics. The levels typically will vary from one topic to the next, according to need. The standards do not specify what level is correct; instead, they provide a framework, process, and tools by which the requirement profile for a facility can be

A. GROUP AND INDIVIDUAL EFFECTIVENESS

A.1 Support for Office Work (E 1660)

A.1.1 Photocopying
A.1.2 Training rooms, general
A.1.3 Training rooms for computer skills
A.1.4 Interview rooms
A.1.5 Storage and floor loading
A.1.6 Shipping and receiving

A.2 Meetings and Group Effectiveness (E 1661)

A.2.1 Meeting and conference rooms
A.2.2 Informal meetings and interaction
A.2.3 Group layout and territory
A.2.4 Group workrooms

A.3 Sound and Visual Environment (E 1662)

A.3.1 Privacy and speech intelligibility
A.3.2 Distraction and disturbance
A.3.3 Vibration
A.3.4 Lighting and glare
A.3.5 Adjustment of lighting by occupants
A.3.6 Distant and outside views

A.4 Thermal Environment and Indoor Air

A.4.1 Temperature and humidity
A.4.2 Indoor air quality
A.4.3 Ventilation air (supply)
A.4.4 Local adjustment by occupants
A.4.5 System capability and controls

A.5 Typical Office Information Technology (E 1663)

A.5.1 Office computers and related equipment
A.5.2 Power at workplace
A.5.3 Building power
A.5.4 Data and telephone systems
A.5.5 Cable plant
A.5.6 Cooling

A.6 Change and Churn by Occupants (E 1692)

A.6.1 Disruption due to physical change
A.6.2 Illumination, HVAC and sprinklers
A.6.3 Minor changes to layout
A.6.4 Partition wall relocations
A.6.5 Lead time for facilities group

A.7 Layout and Building Features (E 1664)

A.7.1 Influence of HVAC on layout
A.7.2 Influence of sound and visual features on layout
A.7.3 Influence of building loss features on space needs

A.8 Protection of Occupant Assets (E 1693)

A.8.1 Control of access from building public zone to occupant reception zone
A.8.2 Interior zones of security
A.8.3 Vaults and secure rooms
A.8.4 Security of cleaning service systems
A.8.5 Security of maintenance service systems
A.8.6 Security of renovations outside active hours
A.8.7 Systems for secure garbage
A.8.8 Security of key and card control systems

A.9 Facility Protection (E 1665)

A.9.1 Protection around building
A.9.2 Protection from unauthorized access to site and parking
A.9.3 Protective surveillance of site
A.9.4 Perimeter of building
A.9.5 Public zone of building
A.9.6 Facility protection services

A.10 Work Outside Normal Hours or Conditions (E 1666)

A.10.1 Operation outside normal hours
A.10.2 Support after-hours
A.10.3 Temporary loss of external services
A.10.4 Continuity of work (during breakdowns)

A.11 Image to Public and Occupants (E 1667)

A.11.1 Exterior appearance
A.11.2 Public lobby of building
A.11.3 Public spaces within building
A.11.4 Appearance and spaciousness of office spaces
A.11.5 Finishes and materials in office spaces
A.11.6 Identity outside building
A.11.7 Neighborhood and site
A.11.8 Historic significance

A.12 Amenities to Attract and Retain Staff (E 1668)

A.12.1 Food
A.12.2 Shops
A.12.3 Day care
A.12.4 Exercise room
A.12.5 Bicycle racks for staff
A.12.6 Seating away from work areas

A.13 Special Facilities and Technologies (E 1694)

A.13.1 Group or shared conference centre
A.13.2 Video teleconference facilities
A.13.3 Simultaneous translation
A.13.4 Satellite and microwave links
A.13.5 Mainframe computer centre
A.13.6 Telecommunications centre

A.14 Location, Access and Wayfinding (E 1669)

A.14.1 Public transportation (urban sites)
A.14.2 Staff visits to other offices
A.14.3 Vehicular entry and parking
A.14.4 Wayfinding to building and lobby
A.14.5 Capacity of internal movement systems
A.14.6 Public circulation and wayfinding in building

B. THE PROPERTY AND ITS MANAGEMENT

B.1 Structure, Envelope and Grounds (E 1700)

B.1.1 Typical office floors
B.1.2 External walls and projections
B.1.3 External windows and doors
B.1.4 Roof
B.1.5 Basement
B.1.6 Grounds

B.2 Manageability (E 1701)

B.2.1 Reliability of external supply
B.2.2 Anticipated remaining service life
B.2.3 Ease of operation
B.2.4 Ease of maintenance
B.2.5 Ease of cleaning
B.2.6 Janitors' facilities
B.2.7 Energy consumption
B.2.8 Energy management and controls

B.3 Management of Operations and Maintenance (E 1670)

B.3.1 Strategy and program for operations and maintenance
B.3.2 Competences of in-house staff
B.3.3 Occupant satisfaction
B.3.4 Information on unit costs and consumption

B.4 Cleanliness (E 1671)

B.4.1 Exterior and public areas
B.4.2 Office areas (interior)
B.4.3 Toilets and washrooms
B.4.4 Special cleaning
B.4.5 Waste disposal for building

FIGURE 19.23 Topics of the Serviceability Scales (ST&M).

A.11 Image to Public and Occupants

Scale A.11.6. Identity outside building

User Requirement Scale		Facility Rating Scale
9 ❑ ❍ ***Public exposure***: Operations require maximum exposure to the public. ❍ ***Ease of locating and identifying building***: The address, building and signage must be very easy for pedestrians or motorists to find and recognize, even for those unfamiliar with the locality.	9 ❑ 8 ❑	❍ **Identity of building:** The building is a well known landmark. The building and entrance are clearly visible and recognizable. ❍ **Corporate identity and signage:** The organization's identity is clearly recognizable, and readily visible from all directions. Direction signs are placed at main nearby transit stops. ❍ **Quality of external signs:** The building has special custom signage, e.g. stand-alone elements, special lighting, and full information. All signs are in as-new condition.
7 ❑ ❍ ***Public exposure***: Operations require above average exposure to the public. ❍ ***Ease of locating and identifying building***: The address, building and signage must be easy to find and recognize, even for those not very familiar with the locality.	7 ❑ 6 ❑	❍ **Identity of building:** The building and building entry are clearly visible to passing motorists and pedestrians, and recognizable. ❍ **Corporate identity and signage:** The organization is well identified from all directions.; Signage is adequate, and clearly visible on every approach to passing motorists and pedestrians. ❍ **Quality of external signs:** Building signage is appropriate and typical, e.g. street address, building name, principal occupant group(s). Signs have no visible deterioration.
5 ❑ ❍ ***Public exposure***: Operations require average exposure to the public. ❍ ***Ease of locating and identifying building***: The address, building and signage must be easy to find and recognize, for those familiar with the locality.	5 ❑ 4 ❑	❍ **Identity of building:** The building and building entry are visible to passing motorists. The building is identifiable, and not easily confused with its neighbors. ❍ **Corporate identity and signage:** The organization is identified to a minimum level. Signage is generally visible to passing motorists and pedestrians. ❍ **Quality of external signs** Building signage is appropriate and typical, e.g. street address, building name and, if appropriate, principal occupant group(s). Signs have no damage or major deterioration.
3 ❑ ❍ ***Public exposure***: Operations do not require much exposure to the public. ❍ ***Ease of locating and identifying building***: Most visitors are regulars. Corporate image is not a high priority.	3 ❑ 2 ❑	❍ **Identity of building:** The building is obscured by other buildings from some directions, and from people approaching along the street from one direction. The building is very similar and hardly distinguishable from adjacent buildings. ❍ **Corporate identity and signage:** The organization is not clearly identified. Signs are obscured from some directions, or are in poor light. ❍ **Quality of external signs:** Signage is minimal or impaired, e.g. minimal information, weathered surfaces, partly damaged.
1 ❑ ❍ ***Public exposure:*** Operations require that the office is obscure to the public, e.g. for security reasons. ❍ ***Ease of locating and identifying building:*** There is no requirement at this level.	1 ❑	❍ **Identity of building:** The building is obscured by other buildings until viewed from directly in front, or, the building is not distinguishable from adjacent buildings, e.g. facades are almost the same. ❍ **Corporate identity and signage:** There is no evidence of the organization's identity on the exterior of the building. Signs are obscured, e.g. by vehicles or other buildings. Signs are very poorly located or hard to read, e.g. signs are too high on the building, too small, the lettering is too small or low in contrast, or signs are in shadow. ❍ **Quality of external signs:** Signage is minimal or badly damaged, with incomplete information, e.g. no street number or building name.

❑ Exceptionally important. ❑ Important. ❑ Minor importance.	
❑ Mandatory minimum level (threshold/criticality) =	❑ NA ❑ NR ❑ DP ❑ LI

NOTES *Space for handwritten notes on Requirements or Ratings*

FIGURE 19.24 Example of a Pair of ASTM Standards Serviceability Scales.

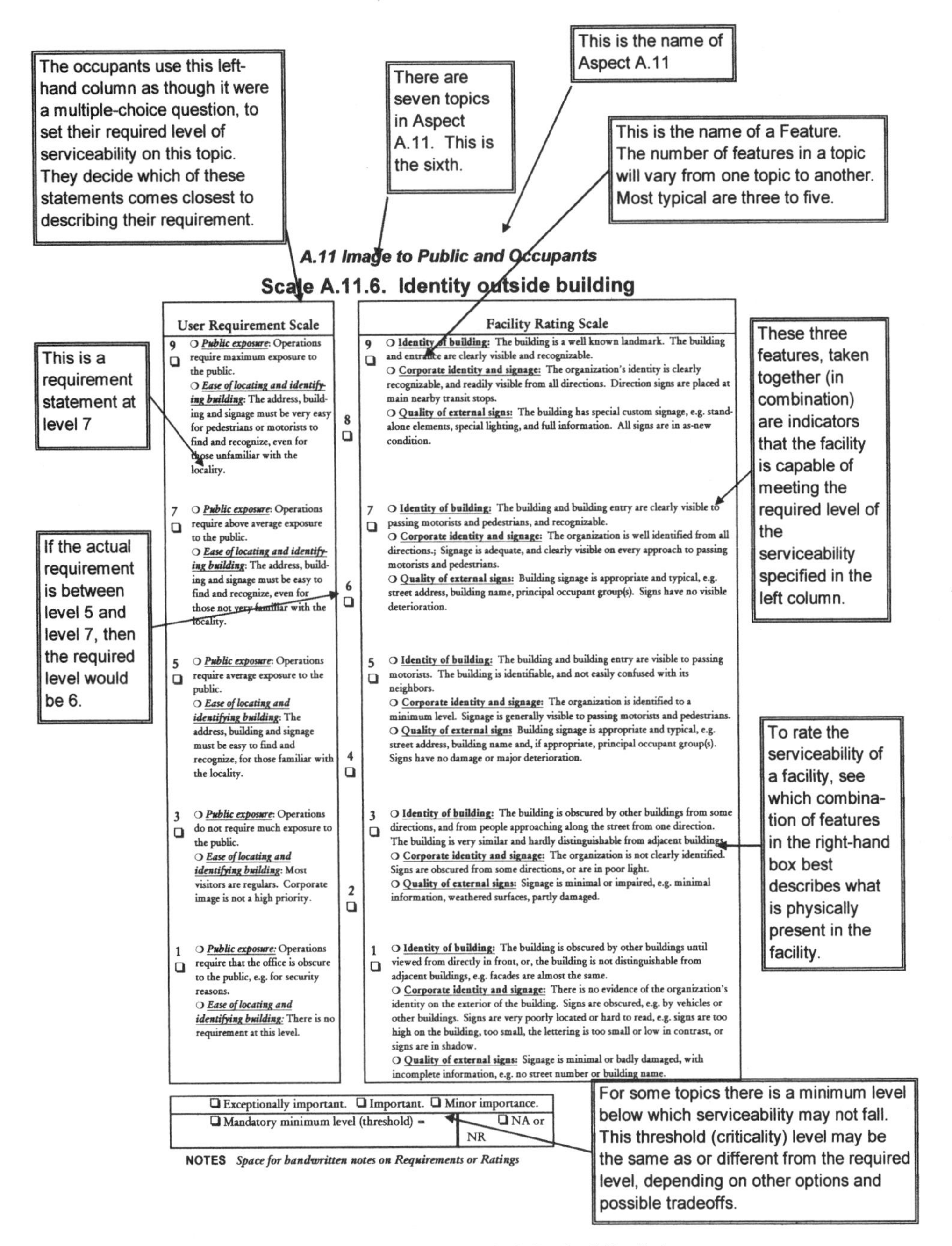

FIGURE 19.25 Explaining the Parts of a Pair of ASTM Standards Serviceability Scales.

tailored to a given situation, and compared to other profiles, using a culture-neutral standard classification approach. This has allowed them to be used satisfactorily in several countries and cultures. In addition to becoming widely used in the private and public sectors in Canada and in the United States, they have been used in the United Kingdom, the Netherlands, and New Zealand, and are being tested in Australia and Japan.

Each requirement scale is used like a multiple-choice questionnaire to select the required level of functionality. Each scale contains five descriptions of requirements for that topic, classified in a range from least to most; for example, from Level 1–9. For clarity, and precision of statement, the description at each level may be subdivided into descriptions of required functions which, taken together, precisely describe the overall functional requirement for one topic. If an actual requirement falls in between one of the five descriptions, then the assigned level may be an in-between level, which would be an even number.

Requirement scales are written in everyday language. They express what occupants need to be able to do, see, feel, hear, or otherwise experience, while in or about a facility. Any occupant or other user should be able to select from the functionality scales the block of text that best describes what is needed without technical knowledge and without interpretation by building professionals. The requirement scales also express what owners and managers need from a facility without using the terms of building professionals.

When comparing the requirement profile prepared by one organization with that prepared by another organization, it is essential that both use the same set of requirement scales. Organizations may use an ASTM standards set of scales; they may create their own; or they may adapt the standards scales for specialized, internal application. In the latter two cases, the organization forgoes the possibility of external comparison. Overall required functionality is displayed as a profile of required levels (i.e., not as a single number) and may be presented as a bar chart.

19.5.4 Facility Rating Scales Classify According to Levels of Serviceability

The ASTM standards also provide the means for assessing the actual serviceability levels of a facility, request for proposals (RFPs), or design for a facility. Such an assessment results in a rating profile of the serviceability of the facility or design. The rating profile of one facility can be compared to that of other facilities (e.g., for benchmarking). The rating profiles of several facilities can be compared to a profile of functional requirements to decide which facility to lease.

Each rating scale is also used like a multiple-choice questionnaire to select the appropriate level of capability. Each scale contains descriptions of certain physical features that would likely be actually present in a facility at each level of serviceability. The rater is asked to identify which of five blocks of text in a rating scale best describes what is physically present in that facility or in the design for a facility. This requires some limited technical knowledge about buildings and building-related facilities. On the other hand, a person using the rating scales to identify levels of capability is not asked to exercise professional judgment about the goodness, technical appropriateness, or quality of a facility. Setting a rating level only involves finding the best match between one of five descriptions, and what is physically present in the facility.

Overall facility capability is displayed as a profile of levels of serviceability (i.e., not as a single number) and may be presented as a bar chart. When comparing the rating profile prepared for one facility with that of another, it is essential that both use the same set of rating scales. Users of these serviceability standards may use an ASTM standards set of scales; they may create their own; or they may adapt the standards scales for specialized, internal application. In the latter two cases, the organization forgoes the possibility of external comparison.

19.5.5 Facility Features Are Considered in Combination, Rather than Separately

More than one building feature normally contributes to the capability of a facility to respond to a single stated requirement. A building is more than the sum of its parts. Until now, however, building elements and systems have been tested and assessed separately from each other. Yet, for example, the effectiveness of the lighting system will depend, in part, on the color of the paint on the walls and the light absorbency of the carpet and furniture.

Each level of the rating scale contains the description of a combination of features that, acting together, indicate that the facility is likely to have the capability to meet the requirement that is stated in the corresponding level of the requirement scale. The descriptions in the rating scale are calibrated to match the corresponding descriptions in the requirement scale, again (e.g., from Level 1–9).

The rating scales do not describe or identify all the features of a facility, which, at a specific level, are likely to contribute to meeting that level of requirement. That would result in an unmanageably long list of features for each topic and for most topics that would include some features that the person conducting the rating typically cannot easily observe directly, nor ascertain from documents.

19.5.6 Comparing Requirement Levels and Rating Levels

For each topic, the requirement scale and the facility rating scale are matched. At each level, each scale is a translation of the other. Since matched scales are used to set requirement levels and to rate facilities, it is easy to ascertain the capability of a facility (or of a facility that is designed but not yet built) to meet occupant or user requirements. On any topic, if the rating level is the same as the requirement level, then the facility is capable of meeting the requirement. If the rating level is higher than the requirement level, then, on that topic, there is a surplus of capability, and similarly, if the rating level is lower than the requirement level on a topic, there is a shortfall of capability.

These matches, surpluses, and shortfalls can be graphed or tabulated for ease of analysis. Significant differences can be identified and presented in summary form for management decision. Note, however, that if the requirement level is zero, signifying "must not have, or must not be present," then any facility rating level greater than zero is undesirable and should be graphed so that the level is not seen as a surplus of capability, but instead is identified as a problem.

19.5.7 Ratings and Comparisons are Expressed as a Profile of Levels and Not a Bottom-Line Score

A serviceability rating profile is far more informative than a single score and far less misleading. Adding up the numbers for all the levels together would not be meaningful and would lump together many more fruits than just apples and oranges. Worse, it would likely mask significant gaps, which would be lost in a single score.

A profile can be scanned quickly and permits "management by exception." One building or facility is not necessarily better or worse than another, although they may have a different mix of levels of quality and capability. A building or facility can be quite appropriate for one occupant group and unsuitable for another. For instance, a building might have a mini-conference facility directly accessible off the main lobby, which could be very functional for a group with many visitors from outside the organization, whereas another group with a high need for confidentiality and few visitors would function better if its meeting rooms were dispersed and directly accessible to the staff areas. Therefore, even though the arithmetic total or mean of rating level numbers might be identical, the building would match the requirement profile of one group, but not the other. Depending on the functional needs of an occupant group, its location requirements, and the financial envelope for the project, trade-offs can be made, based on such functionality and serviceability information.

19.5.8 Levels 1–9 Signify Least to Most, Not Worse to Better

Within the framework of the standards described here, a Level 9 in the scales indicates the "most" demand for a requirement or the highest capability and Level 1 indicates the "least." These levels do not indicate which level is better or worse for a particular occupant, user, or owner. For instance, a retail outlet and a corporate headquarters might each have a commercial need to be highly visible in their town. They might require a Level 9 on the topic of identity outside building. On the other hand, a country's secret service might require a Level 0 for its covert operations, signifying a need to be invisible, or at least as unobtrusive as practicable.

19.5.9 The Scales are Calibrated to the Building Stock, at Levels 5, 9, and 1

The scales are calibrated to the building stock in North America, which is a relatively stable base that evolves slowly over decades. A Level 5 on the rating scales describes what one would expect to find, for each topic, in a town of about 50,000 population, in a commercial building at least 10 years old, that would be classified as Class B according to the scale of the Building Owners and Managers Association International (BOMA). Class B buildings compete for a wide range of users with rents in the average range for the area. Building finishes are fair to good for the area, and systems are adequate, but the building does not compete with Class A at the same price.

As the building stock evolves, the scales will need to be recalibrated, some of them likely each decade. During the year 2000, that recalibration began. For instance, in response to changes in how typical buildings are wired to accommodate computers and telecommunications, the scales for cable plants are being updated.

In each scale, a Level 9 is calibrated to the most that one would expect to generally find for that topic for the type of facility listed in the scope for these standards. Other types of facilities may have more or less capabilities than that described in these scales. For instance, the most impressive image in a building lobby might be found in the corporate headquarters of a corporation of high prestige and national scope. The most security might be found at a facility where highly valued proprietary information is worked on, and the best access to public transportation might be in a government office with very high visitor traffic. On the other hand, these scales would not include enough security for a jail or for the core zone of an embassy, so neither of these facility types is included in the scope section of these ASTM standards classifications. Similarly, a Level 1 was calibrated to the least that one would expect to find for that topic. For instance, an organization with no public visitor traffic, and a high need for information security, might require a Level 1 for identity outside the building and for access to public transportation.

The ASTM standards scales can also be used in any country outside North America and would provide objective comparisons among facilities regardless of location. Requirement profiles can be compared regardless of company, culture, or country. It is unlikely, however, that in countries outside North America, a straight Level 5 in the ASTM standards scales will calibrate to a recognized component of the national building stock. Instead, the mid-level of the older building stock would have a profile that could then be compared to the North American Level 5. Alternatively, the scales might be edited for a particular country or region, for instance, to make a Level 5 reflect the mid-level of that country's older building stock.

19.5.10 A Requirement for All Level 5 is Never the Best Fit

After individual and group interviews with thousands of users, owners, and facility professionals, a straight Level 5 on all topics has never been the optimum requirement profile: it is not the most functional, not the most effective for the organization, nor the cheapest. In the few instances where senior management has directed that requirement at Level 5 be provided across the board, the results were counterproductive, and the directive was eventually abandoned. That is because Level 5 is not calibrated to what is best or average for occupants; instead, it represents what commercial developers expected would be best for them in their particular circumstances at time of construction two, three, or more decades ago.

The generic requirement profile that requires the most topics at a Level 5 is for a general administrative office, based on interviews with thousands of occupants. Yet, a fifth of its topics are not at Level 5: seventeen topics are higher and three are lower than Level 5.

19.5.11 Quality Management and ISO 9000

Quality is described in ISO 9000 as the "totality of features and characteristics of a product or service that bear on its ability to satisfy stated and implied needs." Those who provide a product or service (e.g., a facility, and its management and operation (O&M)) should ascertain the explicit and implicit requirements of the customers (occupants and owner), decide to what level those needs should be met, meet that level consistently, and be able to show that they are in fact meeting those requirements.

The starting point for quality assurance programs is the ability to determine and assess features and characteristics of the product or service, to relate them directly to customers' needs, expectations, and requirements, and to document it all in a systematic, comprehensive, and orderly manner. A quality management system should include the means to monitor the compliance of all production phases and to verify that the final product meets those stated and implied needs of the customer.

The ASTM standards for whole building functionality and serviceability provide that starting point for a quality management system. The standards include the means to monitor and verify compliance, with respect to facilities. These ASTM standards provide explicit, objective, consistent methods and tools applicable to the field of building construction and real property.

19.5.12 Applications in Corporate Real Estate, Portfolio and Facility Management, and User Operations

The ASTM standards have been used to answer such questions as:

- Should we stay in this facility or move?
- Which properties on offer should we decide to negotiate about?
- How do we assess the functional quality of our portfolio?
- How do we assess the functional quality of designs proposed for new or remodeled facilities?
- Which of these designs best match our program requirements?
- What are our typical functional needs? Can we take care of our special requirements?
- Can we prepare corporate norms?
- For our regional offices, what are our typical needs?
- Can we compare the typical needs of similar groups?

19.6 APPLICATIONS OF THE ASTM STANDARDS ON WHOLE BUILDING FUNCTIONALITY AND SERVICEABILITY

The serviceability scales of the ASTM standards can be useful throughout the occupancy or ownership of a building, from when it is first conceived, to when it is considered for demolition.

19.6.1 Prepare the Serviceability Profile of an Existing Facility

Some corporations, large and small, and some units of government require that the asset management plan (AMP) of each major facility include a serviceability rating (with variants as appropriate). As noted earlier, this enables asset managers and project managers to give priority to functional issues that will best contribute to the effectiveness of occupants. It also enables them to ensure that when maintenance and repair projects occur, functional enhancements can occur at the same time, if this will save costs compared to treating functionality as a separate task to be accomplished separately, later. A major benefit is that for projects that disrupt occupants, the elapsed time that they are disrupted can often be reduced to a fraction.

Often, however, using serviceability profiles has been optional and is only now becoming mandatory. For instance, when the serviceability scales were first introduced, Public Works and Government Services Canada (PWGSC) reported publicly each year what proportion of its inventory of hundreds of large office buildings had been rated; but PWGSC did not mandate updating the AMPs, nor inclusion of a serviceability rating with each asset management plan. Now, as PWGSC has contracted for the first updates of its AMPs since 1993, a serviceability rating is required in each new or updated AMP.

Serviceability ratings are also a key data source when portfolio managers build strategy and set priorities for which properties to continue to occupy as-is, which to remodel or refit, and which to dispose of or replace. When the physical workplace is considered as a "factor of production," then how well it matches the serviceability requirement profile of its users is essential information for developing or adjusting the strategy for portfolio management in each scenario for probable corporate futures.

19.6.2 Provide Strategic Advantages in Real Estate Negotiations

If a relocation is planned, developing and applying a specific user profile to prospective sites before entering into lease negotiations provides a strong, strategic advantage. A clear, concise serviceability analysis makes clear the relative business value of each location on offer, with significant strengths and concerns objectively outlined. With such information, the negotiator of a real estate transaction, whether for buyer or seller, landlord or tenant, is far better equipped. For instance, more of the costs associated with paying for any necessary upgrades to support the occupants can be identified in advance, and overpaying for needless amenities can be avoided. It also becomes much easier to rate the capability of the facilities to be operated and maintained economically. Later, when it comes time for a tenant to exercise any lease options or renewals, an updated analysis can be applied to identify deficiencies that will support the need for additional tenant improvement dollars.

19.6.3 Evaluate Existing Sites/Real Estate Portfolios

Businesses with multiple facilities find they must have the ability to shuffle groups around to accommodate operations and/or make effective determinations on the disposition of property. In large organizations, this is a continuous process as department needs change, divisions grow, new groups are added, and real estate portfolios fluctuate, but which locations will really be the best at supporting the needs of the occupants and the organization?

By developing a site profile database, the user requirement profile(s) can be applied to determine the capability of each building within a portfolio to support the organization's core businesses. Similarly, new and updated user profiles can be compared to the serviceability ratings of those buildings with space available and the subsequent report will immediately determine the optimum location for each group. This can eliminate unnecessary upgrade expenses, reduce political issues, improve employee morale, and ensure the best value from real estate investments.

19.6.4 Standardize Functional and Architectural Programming and Evaluate Plans

For many large organizations, a new program is developed for every single project, when in reality, more times than not, there are many basic similarities among all or most occupant groups that are requirements of the core business itself, with each group having just a few specialized functional requirements. With a user requirement profile, the basic employee requirements are established in a concise form for reference by the architect/interior design firm. This reduces up-front fees and helps to avoid the wish list mentality leading to over-design and inflated construction budgets.

Similarly, by comparing the user requirement profile to the architectural concept design, and then to design development drawings prior to the development of construction documents and specifications, any deficiencies and surpluses can be adjusted well in advance of construction. This can lead to huge cost savings by reducing change orders during construction and eliminating retrofits after occupancy.

19.6.5 Make Budgeting Decisions with More Knowledge

The dollars are never available to take care of all of the work that should be done as building condition assessments are developed and maintenance needs are listed and budgeted. It is always necessary for someone to determine how the available funds will be spent. A serviceability analysis becomes an essential tool for making objective budget decisions. Through the application of a user requirement profile(s) to a particular facility, performance deficiencies can be easily identified and the money spent where it will have the most significant impact.

For companies with multiple facilities, each building's individual capability can be rated according to user requirements to determine which best meet the needs of the occupants. Deficiencies can then be compared to make informed determinations on where budget allotments will allow for the greatest improvement in how the buildings perform for the organization and deliver best performance to the core businesses.

19.6.6 Assign Real Dollar Value to Chargebacks

Serviceability can help normalize and adjust chargeback dollars based on how well the physical space is accommodating the users' needs. When space usage changes, the chargeback changes. By differentiating chargebacks through serviceability analyses, appropriate costs are allocated with reference to how well each building's capabilities meet each occupant's requirements. Overall satisfaction with facilities and services increases and the volume of complaints is reduced.

19.6.7 Establish Service Level Set-Points

Service levels for workplaces are usually somewhat vague. Measurements of these service levels can range from using a scale of 1–10, to frequencies of performance of some support function, to response times when repairs are needed. But how to make valid, consistent comparisons over time?

Serviceability is the only performance standard available that can establish service level set-points that, because the scales are an ASTM standard, will not change over time, nor from place to place, nor from one assessor to another. By providing a numerical value based on how good is the match between occupants' needs and the buildings' capabilities, target service levels can be set which can then be tied to budgets. Gap analyses are then practicable that identify specific actions needed from the real estate and facilities group, to fulfill a company's vision or mission statement.

A reference point is needed for setting a performance guideline or target. Serviceability provides just such a reference.

19.6.8 Use at U.S. State Department for the Chanceries

This use of the ASTM standards on current projects for chanceries of the U.S. State Department is a demonstration of how to audit the functionality of a request for proposals (RFPs) and of design–build proposals against an established profile of requirements. In January 1999, this use of the ASTM standards for whole building functionality and serviceability received the support of the assistant deputy secretary responsible for foreign buildings (A/FBO) at the U.S. State Department. The U.S. State Department has also applied the ASTM standards to design–bid–build RFPs and proposals.

Steps in Process to Date. The ASTM standards for whole building functionality and serviceability have been used at the U.S. State Department to:

1. Rate the current "design guide" for several major projects, and the design concepts prepared by architects in response to the design guide.
2. Ascertain the imputed user requirements profile for a U.S. Office Building Chancery (OBC), by "reverse engineering" from the design guide.
3. Prepare a detailed cross-reference table, giving section number and text quotes to link the design guide and the scales of the ASTM standards. This table included the rating levels for each feature and topics covered in the design guide, and identified those topics that were not covered by the design guide, or addressed in the design concept, or both.
4. Identify scales that require editing and topics for which there are no scales at the present time, to respond to the unique requirements of the OBCs.
5. Prepare a "main" requirement profile for the overall base building and seven "variants" for different zones of an OBC. This was done based on over forty group interviews of staff and in-house experts, and data from post-occupancy evaluations (POEs).
6. Rate the RFP for a set of design–build projects.
7. Rate proposals from the respondents to RFPs for OBCs to be procured through the design–build process and compare the ratings of the proposals to the rating of the RFP. Prepare a "gap analysis" of the significant differences.
8. Compare the rating profiles of the RFP and of the proposals against a selection of topics from the profiles of requirements, against each other, and against ICF generic profiles.
9. Prepare a "gap analysis" of the significant differences.

Steps by the U.S. State Department

- Deal with the shortfalls in the rated proposals to-date, and in the generic RFP
- Use rating approach during design reviews
- Integrate findings from the "gap analysis" into the next sets of RFPs

Training. Training sessions are organized to introduce project managers, in-house teams, and external design teams to the ASTM standards and to their use for the U.S. Chanceries. One session is aimed at managers, lasts two days, and focuses on "How to buy serviceability ratings and manage the service providers who will do these ratings." The other session is aimed at professionals, lasts four days, and focuses on "How to do" actual ratings of buildings, RFPs, and design proposals, as well as designs during design reviews.

The first morning is the same for both training sessions: an introductory session aimed also at more senior managers who need to know what the ASTM standards are about. Then, of the next day and a half of the two-day course, one half day is used to review what this approach to functionality and serviceability is about and how it is used, mostly based on the information from a building tour and using slides. Half a day is used applying the lessons learned to a design proposal and a half-day is used on practical tips about how to choose and manage a contractor. The four-day session includes the overview, the building tour on slides, an actual tour of the building, the rating of the real building toured, and half a day of rating an actual set of documents at the design-concept stage.

19.6.9 The Whole Picture

In today's tight employment market, it is more important than ever to attract and retain qualified personnel. Providing suitable workspace will not only help ensure the type of high-performance staff every business wants and needs—it will also increase productivity of those people, to maximize both human resource and facility dollars.

The effective measurement of facility performance relies on more than just one metric. We will always need to know the operating costs per square foot in order to stay competitive, but these costs do not represent the whole picture. If optimal use of a facility is to be achieved, then there must be an understanding of how well the building supports its customers. Without the application of different metrics, including serviceability, chances are any facility will fall short of meeting its customers' requirements and significantly reduce profitability potential.[18]

APPENDIX A FURTHER READING

Amiel, M. S., and Vischer, J. C. 1997. Space design and management for place making. Proceedings of the 28th Annual Conference of the Environmental Design Research Association (EDRA). Edmond, OK.

Aronoff, S., and Kaplan, A. 1995. *Total Workplace Performance, Rethinking the Office Environment*. Ottawa, ON, Canada: WDL Publications.

ASTM (American Society for Testing and Materials). 1999. *ASTM Standards on Building Economics*. West Conshohocken, PA.

ASTM (American Society for Testing and Materials). 1996. *ASTM Standards on Whole Building Functionality and Serviceability*. West Conshohocken, PA.

ASTM (American Society for Testing and Materials). 1987. *ASTM Standards on Computerized Systems*. West Conshohocken, PA.

Baird, G., Gray, J., Isaacs, N., Kernohan, D., and McIndoe, G. 1996. *Building Evaluation Techniques*. Wellington, New Zealand: McGraw-Hill.

Becker, F., and Joroff, M. 1995. *Reinventing the Workplace*. Norcross, GA: Industrial Development Research Foundation.

Becker, F., Joroff, M., and Quinn, K. 1995. *Toolkit: Reinventing the Workplace*. Norcross, GA: Industrial Development Research Foundation.

Becker, F., and Steele, F. 1995. *Workplace by Design, Mapping the High-Performance Workscape*. San Francisco: Jossey-Bass.

Brand, S. 1994. *How buildings Learn, What Happens after They're Built*. New York: Penguin Books.

Cameron, I., and Duckworth, S. 1995. *Decision Support*. Norcross, GA: Industrial Development Research Foundation.

Cotts, D. G., and Lee, M. 1992. *The Facility Management Handbook*. New York: AMACOM, a division of American Management Association.

Eley, J., and Marmot, A. F. 1995. *Understanding Offices, What Every Manager Needs To Know about Office Buildings*. Middlesex, UK: Penguin Books.

ISO (International Organization for Standardization). 1994. *ISO 9000 Compendium—International Standards for Quality Management*, 4th ed. Genève, Switzerland: ISO, Central Secretariat.

Joroff, M., Louargand, M., Lambert, S., and Becker, F. 1993. *Strategic Management of the Fifth Resource: Corporate Real Estate*. Norcross, GA: Industrial Development Research Foundation.

Kozlowski, D. 2000. New measures for building performance—Rating systems offer benchmarks that could rewrite the rules about designing and managing buildings. *Building Operating Management*. March, 59.

Lambert, S., Poteete, J., and Waltch, A. 1995. *Generating high-performance corporate real estate service*. Norcross, GA: Industrial Development Research Foundation.

Loftness, V., et al. 1996. *Re-valuing Buildings: Invest inside Buildings to Support Organizational and Technological Change through Appropriate Spatial, Environmental and Technical Infrastructures*. Grand Rapids, MI: Steelcase.

Lundin, B. L. 1996. Point of Departure—Standard offers a new way to determine if a building measures up. *Building Operating Management*, September, 154.

Lynn, M., and Davis, G. 1998. Building measurement standards and real estate securities. *Real Estate Review*, Fall.

Mohr, A. 2000. The new metrics—Going beyond costs per square foot to measure building performance. *Today's Facility Manager*, February.

O'Mara, M. (Ed.). 2000. Special Issue on Portfolio Management. *Journal of Corporate Real Estate* 2(2).

O'Mara, M. 1999. *Strategy and place: Managing the corporate real estate in a virtual world.* New York: The Free Press.

Pertz, S. 1997. Redefining strategic facilities planning. *Facility Management Journal*, November/December.

Pinkwater, D. M. 1977. *The big orange splot*. New York: Scholastic.

Roodman, D. M., and Lenssen, N. 1995. *A building revolution: How ecology and health concerns are transforming construction*. Washington, DC: WorldWatch Institute.

Rondeau, E. P., Brown, R. K., and Lapides, P. D. 1995. *Facility management.* New York: John Wiley & Sons.

Sims, W., Joroff, M., and Becker, F. 1996. *Managing the reinvented workplace.* Norcross, GA: Industrial Development Research Foundation.

Smith, P., and Kearny, L. 1994. *Creating workplaces where people can think.* San Francisco: Jossey-Bass.

Spillinger, R. S., in conjunction with the FFC (Federal Facilities Council) 1998. *Adding value to the facility acquisition process: Best practices for reviewing facility designs.* Technical Report #139. Washington, DC: National Academy Press.

Steele, F. 1986. *Making and Managing High-Quality Workplaces.* New York. Teacher College Press.

Steele, F. 1973. *Physical Settings and Organizational Development.* Reading, PA: Addison-Wesley Publishing.

Sullivan, E. 1994. Point of departure—ISO 9000: Global benchmark for manufacturers may be future facility challenge. *Building Operating Management,* October, 96.

U.S. General Services Administration. 1998. Office of Government Policy, Office of Real Property. Government-wide Real Property Performance Results. Government-wide baseline. Private Sector Performance. Building Profiles. Washington, DC, December.

U.S. General Services Administration. 1998. Office of Government Policy, Office of Real Property. Government-wide Real Property Performance Measurement Study. Washington, DC, June.

U.S. General Services Administration. 1997. Office of Government Policy, Office of Real Property. Office Space Use Review. Washington, DC, September.

Wineman, J. D. (Ed.) 1986. *Behavioral Issues in Office Design.* New York: Van Nostran Reinhold.

ENDNOTES

1. See Figure 19.12 from the executive report of the Corporate Real Estate Portfolio Management Alliance Research Study, available from The McMahan Group, San Francisco.
2. The authors first used the concept of "demand" and "supply" in this context in 1994 during a demonstration of the ASTM Standards on Whole Building Functionality and Serviceability to the Dutch Government Building Agency. See also Chapter 3 of this Handbook.
3. See Section 5 of this chapter for a more detailed description of the ASTM Standards.
4. The PBS survey was managed by Booz, Allen, and Hamilton, and conducted by the Gallup Organization. It is reported to be the largest survey, and with the most data points, ever done by Gallup.

5. See Bibliography, Szigeti, F. and Davis, G., 1997.
6. Both definitions are from ASTM Standards Facility Management Terminology (E 1480–92), included in ***ASTM Standards on Whole Building Functionality and Serviceability***, 2nd ed. 2000. ASTM stock number WBDG2000.
7. The authors appreciate the support, collaboration, and information that has been provided over the years by many staff members in Public Works and Government Services Canada (formerly Public Works Canada) both at headquarters and in the regions. They have been exceptionally generous in contributions of information, insight, expertise, and hard work. The authors also express their appreciation for the opportunity to develop the serviceability scales and related tools and methods.
8. Based on a survey by the authors of major private and public organizations from several countries, including: Australia, Canada, Finland, Ireland, France, Germany, the Netherlands, Norway, Spain, Sweden, the United Kingdom, and the United States. We gratefully acknowledge the contribution of these organizations to the survey.
9. The series of seven figures are based on diagrams by Frans Evers, then director general of the Dutch Building Agency, of the government of The Netherlands.
10. See Endnote 8.
11. This section draws in part on research conducted during the corporate real estate alliance portfolio management project, led by The McMahan Group, of San Francisco, from September 1998 through April 1999. The project was sponsored by corporate real estate organizations, which included: BellSouth Telecommunications, Boeing Realty Corporation, Fidelity Investments, Florida Power & Light, Microsoft Corporation, Pacific Gas & Electric, State Farm Insurance, Sun Microsystems, U.S. General Services Administration, U.S. West, and Washington Mutual. The researcher members of the Alliance included: Tom Bomba, Andrew Light, and Sven Pole of The McMahan Group; John McMahan of both The McMahan Group and the Fisher Center for Real Estate at the University of California, Berkeley; Wade Fransson of CB Richard Ellis, with David Nelson; Martha A. O'Mara of the Harvard Graduate School of Design; Françoise Szigeti and Gerald Davis of the International Centre for Facilities; Barry Varcoe of Johnson Controls; Kevin Deeble of TriNet Corporate Realty Trust; Joseph Gyourko and Yonheng Deng of Zell/Lurie Real Estate Centre at the Wharton School.

    The work to develop the set of diagrams presented here was done from April 1999 to April 2000 by the authors as part of continuing development undertaken by the International Centre for Facilities (ICF). These diagrams have been tested in discussions and presentations to a number of other major public and private organizations. For instance, the authors are working (Spring 2000) with the real estate division of a large corporation with thousands of employees and about 100 sites. To better respond to changes in overall corporate direction, the head of that division has now directly mapped his organization onto the diagram in Figure 19.19 and has identified present and prospective managers for each of the functions.
12. Bomba, T. 2000. The alliance program: Establishing a new knowledge base on the corporate real estate portfolio management competency. *Journal of Corporate Real Estate* 2(2): 109. ISSN 1463-001X.
13. O'Mara, M. A. 1999. *Strategy and Place: Managing Corporate Real Estate and Facilities for Competitive Advantage*. New York: The Free Press.
14. Fransson, W., and Nelson, D. 2000. Management information systems for corporate real estate. *Journal of Corporate Real Estate* 2(2): 154–169. ISSN 1463-001X.
15. de Kerchove, G. 2000. *High performance workplaces.* Paper presented at Environmental Design Research Association (EDRA) 31/2000 Conference, San Francisco.
16. Ouye, J. A. 1998. *Measuring workplace performance.* American Institute of Architects PIA Conference, Cincinnati, OH, March.
17. Kaplan, R., and Norton, D. 1996. *The balanced scorecard.* Cambridge, MA: Harvard Business School Press.
18. Part of the text in this section was prepared in collaboration with Ann Mohr.

BIBLIOGRAPHY

Davis, G. 1986. *Building Performance: Function, Preservation, and Rehabilitation.* Philadelphia, PA: ASTM.

Davis, G. 1978. A process for adapting existing buildings for new office uses. In W. F. E. Preiser, ed. *Facility Programming.* Stroudsburg, PA: Dowden, Hutchinson and Ross.

Davis, G. 1974. Applying a planned design process and specific research to the planning of offices. In D. H. Carson, ed. *Man–environment interactions: evaluations and applications—The state of the art in environmental design research,* pp 63–89. Stroudsburg, PA: Dowden, Hutchinson and Ross.

Davis, G. 1969. The independent building program consultant: His role and work in planning buildings. In *Building Research,* April/June, pp 16–21.

Davis, G. 1969. What is new about building programming? *Construction Specifier,* October.

Davis, G., et al. 1993. Serviceability tools: Volume 1—Methods for setting occupant requirements and rating buildings; Volume 2—Scales for setting occupant requirements and rating buildings; Volume 4—Requirement scales for office buildings (excerpt from Volume 2); Volume 5—Rating scales for office buildings (excerpt from Volume 2). Ottawa, Canada: The International Centre for Facilities.

Davis, G. et al. 1985. *ORBIT-2 main report and supporting volumes.* Norwalk: Harbinger Group.

Davis, G., and Altman, I. 1976. Territories at the workplace: Theory into design guidelines. In *Man–environment Systems* 6(1), 46 53. Also published, with minor changes, in *Appropriation of space.* 1977. Proceedings of the Third International Architectural Psychology Conference, Louis Pasteur University, edited by P. Korosec-Serfati. Strasbourg, France.

Davis, G., and Ayers, V. 1975. Photographic recording of environmental behavior. In W. Michelson, ed., *Behavioral Research Methods in Environmental Design,* pp 235–279. Stroudsburg, PA: Dowden, Hutchinson and Ross.

Davis, G., Ford, P., Halas, S., Henning, D., and Thatcher, C. 1996. *How Are Your Facilities Measuring Up? Using the New ASTM Serviceability Standards.* Panel discussion in International Facility Management Association (IFMA) proceedings from World Workplace '96, Vol. 2, Houston, TX: IFMA.

Davis, G., Schley, M., and Meyer, W. 1992. How to get the building you need. *ASTM Standardization News,* November.

Davis, G., and Szigeti, F. 1996. Preparing for ISO 9000 for the new ASTM standards for whole building functionality and serviceability. *International Facility Management Journal (IFMA),* November/December.

Davis, G., and Szigeti, F. 1996. Using functionality and serviceability standards for strategic benchmarking. *The Corporate Real Estate Executive (NACORE),* November/December.

Davis, G., and Szigeti, F. 1996. Quality and functionality standards at the core of portfolio analysis. *The Corporate Real Estate Executive (NACORE),* July/August.

Davis, G., and Szigeti, F. 1996. Quality and functionality standards aid in worldwide site selection. *The Corporate Real Estate Executive (NACORE),* June.

Davis, G., and Szigeti, F. 1996. Standards are now available for building quality and serviceability. *The Corporate Real Estate Executive (NACORE),* May.

Davis, G., and Szigeti, F. 1995. Do your facilities measure up? *The Total Quality Review, Measuring Performance,* May/June.

Davis, G., and Szigeti, F. 1985. Using a joint planning process in adaptive reuse. In W. F. E. Preiser, ed., pp 67–81. *Programming the built environment.*

Davis, G., and Szigeti, F. 1983. *Development of a functional program for overall building performance.* In Proceedings for the Conference on People and Physical Environment Research. Wellington, New Zealand: Ministry of Works and Development.

Davis, G., and Szigeti, F. 1982. Programming, space planning and office design. *Environment and Behavior* 14(3), 299–317.

Davis, G., and Szigeti, F. 1979. Functional and technical programming, when the owner/sponsor is a large or complex organization. In *Conflicting experiences of space,* Proceedings of the Fourth International Architectural Psychology Conference, Louvain-la-Neuve.

Davis, G., Szigeti, F., and Atherton, A. 1996. *Why cutting costs isn't enough, building operating management (BOM).* Milwaukee, WI.

Davis, G., and Ventre, F. T. 1990. Performance of buildings and serviceability of facilities. ASTM STP 1029. Philadelphia, PA.

Szigeti, F., and Davis, G. 1997. *Programming and evaluation: Relationship to the design, management and use of facilities.* Invited paper, in Proceedings of EDRA 28/97, Edmond, OK.

Szigeti, F., Davis, G., Atherton, A., and Henning, D. N. 1997. Changes in the relationships between customers and providers of facilities, and usefulness of the new ASTM/ANSI serviceability standards for the procurement of office facilities. In Proceedings, CIB W92 Symposium, Montreal, Canada.

CHAPTER 20
THE DESIGN AND CONSTRUCTION PROCESS

John D. Macomber
President
Collaborative Structures, Inc.
E-mail: macomber@costructures.com

20.1 INTRODUCTION

Facility managers often are called on to create new space, in addition to carrying out their on-going roles managing and operating existing space. In this novel process, the facility manager is being asked to shepherd an idea that originated elsewhere in the organization, give it substance, negotiate its passage through issues of cost, revenue, and politics; and then lead the delivery of the project with a team that may include dozens of outside vendors. This chapter helps the facility manager succeed in that task by offering:

- A context in which to approach the project
- An organizing structure for identifying risks and roles
- Elements of basic theory around key delivery issues

The intent is to equip the facility manager to work intelligently with both his or her own internal constituency, and with outside vendors who come and go on an as-needed basis.

20.2 THE TASK: FROM NEED TO OCCUPANCY

20.2.1 Skills

When the organization considers building new space, the facility manager is called upon to show a completely different set of skills than those which are used in day-to-day management. These skills range from the technical to the financial, and they touch on the aesthetic. Of course, the facility manager must show throughout the process a continuous high degree of political skill.

20.2.2 Circumstances

Understanding the particular circumstances of the project at hand comes even before marshaling the resources, assembling the team, and making decisions about the space. For example, is there no substitute for building new space for this user group, or is new space just one of several possibilities? A pharmaceutical company likely requires new construction for new products, but a financial services company could well reuse existing space, rent space somewhere else, or even stay where they are.

20.2.3 Considerations

The most obvious considerations around a project are, first whether it is all new construction, a major renovation, or simply interior fit out. Second, is the job large or small? Third, are there overarching design criteria including adjacencies, performance of the space, and aesthetics; or is it a simple utilitarian buildout? Finally, and of critical import, who are the internal clients? Is there a small and tightly led user group, or a large, diverse, and powerful collection of important users from throughout the company?

Following these particular characteristics, each project is different, so there can be no one menu. The intent of this chapter is to lay out key elements in the process, the product, and the team, so as to help the facility manager accomplish successful projects of all varieties.

20.3 ASSEMBLING RESOURCES

The facility manager must assemble resources of many kinds in order to carry out a new project. The most obvious resources are land and funding. For many facility managers, new project development consists almost entirely of finding the site or the space, and finding the money. For others, the land and the money are straightforward, but getting a commitment for time and attention of user groups, as well as collecting the resources to staff and monitor a large project which is outside of the organization's normal breadth of work, is the key challenge. Often finding the whole package of in-house resources is harder, and more visible, than finding the design and construction team.

20.4 PHASES OF WORK

The task of the facility manager changes with each phase of the project. There are four principal phases in any project: formulation, design, construction, and on-going operations. In order to both control the project and to use his or her own time to the most advantage, the facility manager must play quite different roles in each phase. Figure 20.1 indicates the key goals that the facility manager needs to achieve in each phase.

20.4.1 Formulation

Formulation is the very first phase, during which time all of the basic decisions are made. Some of these are highly strategic with regard to the organization's plans. Can we control the land? Shall we finance the work with a loan or from our balance sheet? Is there an opportunity to use our credit to occupy on a sale leaseback basis? Who will occupy the building? How will they compensate the facilities function? Typically, in this phase, the process is highly unstructured, the decision steps are hard to lock down, the decision makers are hard to find, yet the clock is running. Facility managers with skill in managing the budgetary uncertainties and political pressures of this phase are particularly adept and particularly valuable.

The other key aspect of the formulation phase is to properly set up the structure of the team that will execute the work. Depending on the particular circumstances of the project, there may or may not be, for example, an early need for a technical consultant. There may be so many uncertainties in

Phase	What the facility manager needs to accomplish in this phase
Formulation	• Determine the use of the facility; who are the occupants, what do they need? • Determine the funding for the facility; directly from tenants or otherwise? • Commit the resources, including land, funding, overhead • Establish the setup of the team, their contracts, and their respective responsibilities
Design	• Draw out, understand, record, and commit the program needs of the users. • Match the needs to the funds and time available • Lead decisions around the right trade-offs between cost, time, quantity, and quality; lead the aesthetic vs. pragmatic trade-off decisions • Make the first cost vs. life-cycle cost investment choices
Construction	• Determine who takes what responsibilities and risks and why; enter into contracts accordingly • Monitor budgets and spending as a fiduciary for your organization • Control changes to the work from all sources—users, designers, tradespeople
Operations	• Serve the users, including collecting rent and maintaining the space, both as cost effectively as feasible • Affirm that the facility receives the appropriate level of investment to avoid deferred maintenance problems down the road • Continually assess the value of owning the facility vs. leasing the space

FIGURE 20.1 Phases of Work.

a complex renovation that it is critical to have a service-oriented trustworthy contractor engaged from day one; or by contrast, the project may be on a clean site, simply defined, and so lend itself to a more commodity oriented delivery team.

During this time of high uncertainty in the formulation phase, feeling the rush to get going, and realizing that for many facility managers design and construction are infrequent tasks, the formulation phase can be hastily concluded without properly setting up the team. This haste universally leads to problems down the road, when parties didn't understand the scope of the work, the contractual obligations, or the risks being entered into. Section 20.5 covers the selection process in more detail.

20.4.2 Design

During the design phase, a new level of granularity enters into the process. The team is no longer deciding who'll be in the building or whether to build at all; rather, issues of structural systems, department adjacencies, and of course aesthetic appeal are decided. During this phase, the core skill of the facility manager evolves into making the right trade-offs. At this stage, resources are becoming constrained, and the facility manager often is balancing between different constituencies of users. With respect to physical choices in the project, a course needs to be set, balancing the tripartite closed system of time vs. cost vs. scope. The facility manager has to be a facilitator and discussion leader; the facility manager has to make major trade-off decisions; and the facility manager probably needs to be an enforcer. The facility manager is in the middle, as discussed in Section 20.4. In the design phase, this middle comes under pressure more than at any other time.

20.4.3 Construction

The construction period involves a third set of facility manager skills. Now, instead of marshaling resources or facilitating cost and design trade-offs in the abstract setting of conference rooms and drawings, the facility manager turns into a monitor making sure that everything has been ordered, the work is coming as expected, and every nickel and dime is accounted for. This is a time of list making and control.

20.4.4 Operations

Finally, in the operations phase, the facility manager re-enters familiar territory. Most buildings are in operation for decades; the design and construction process can be as short as 12 months. Now the facility manager shows yet another set of skills having to do with maximizing the value of the asset to the organization in the big picture, doing the little things like collecting rent and sweeping the floors, and making sure that there's adequate funding to keep the properties up and avoid a painful deferred maintenance makeup expense.

Figure 20.1, Phases of Work, highlights the different skills the facility manager needs to use to make each phase build on the successes of prior phases.

20.5 THE PLAYERS: THE FACILITY MANAGER IN THE MIDDLE

The facility manager is the person in the middle during a design and construction project. Not only must the facility manager manage "down" as he selects, directs, and controls a large array of vendors, including architects, contractors, engineers, consultants, and subcontractors; but the facility manager also must manage up. All of the organization's end-users must be managed by the facility planner too. These can include very influential executives, fund-raisers, researchers, or manufacturing people. Figure 20.2 shows how the facility manager must manage upward to an internal group of users, sideways to peers in the organization, and down to a large collection of vendors.

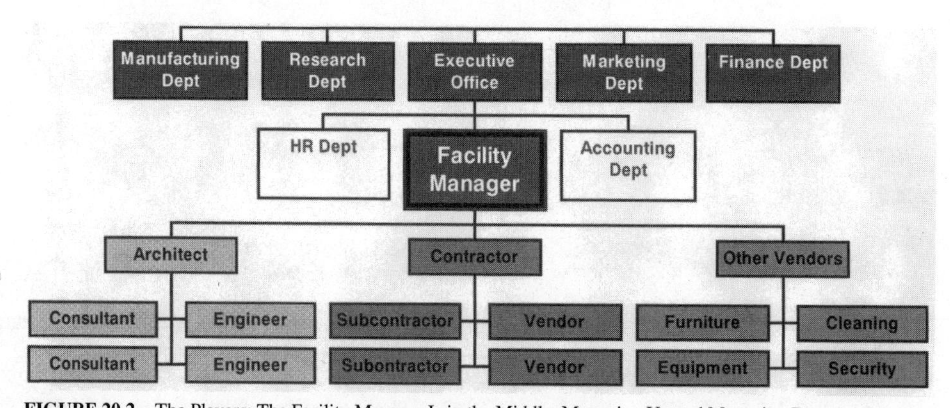

FIGURE 20.2 The Players: The Facility Manager Is in the Middle, Managing Up and Managing Down.

20.5.1 Internal Clients

The architects and contractors understand the facilities design and construction process deeply, since they do it every day. Most of the internal users do not; doctors or physicists or fund managers don't build much, but they are used to getting their way in most areas of life. This can lead to tremendous complexity for the facilities manager in serving these users. Typically, communication skills, clarity, clear expectations, and the loan of a big stick from upstairs are critical in leading the project.

20.5.2 Influence

A variety of skill sets and solutions are seen in practice. Some facilities managers rely on force of personality or tenure with the organization to keep the process on track and make decisions stick. Others rely on organizational structure, requiring the user group to designate a single spokesperson, and channeling all the work through that person. Still others rely on charismatic consultants, usually drawn from the ranks of the architect, an outside program management consultant, or the contractor. Sometimes the value of these third parties is greater in leading the process than it is in their nominal core skill. Smart facilities managers recognize this range of value added as a supplement to their own skill in building a team that's appropriate for the particulars of the project at hand. Figure 20.2 depicts the range of clients that the facility manager must interact with.

20.6 ELEMENTS OF THE PROJECT AND WHAT TO CONSIDER

Every project has the same basic elements: site, foundation, structure, envelope, mechanical systems, finishes, and furniture and equipment. Yet the risks involved in each element are very different, as is the delivery of design and construction work. Subsequently the facility manager's tactics need to be different. Table 20.1 lists key considerations for each element of the project, and also discusses how to buy or contract for components of each element.

20.6.1 Site

Site considerations can be the major factor for many projects. Most obvious is the decision process around acquiring land. Second is being sure that the land is buildable. Some sites have too much liability for hazardous materials; other sites have questionable conditions relating to bearing capacity for foundations, site drainage, or various conditions restricting parking. For many projects, simply getting all the permissions to build is the hardest part of the facility manager's job. Typically, advice in this area is purchased from knowledgeable lawyers and consultants, on an hourly fee basis, with all the risk assumed by the organization.

20.6.2 Foundation and Structure

The foundation, structural, and mechanical systems have the strange characteristic of never being seen again once the building is complete. The foundation and structure will likely never be maintenance issues for the facility manager ever again. Yet the decisions made in the foundation and the structure have two major impacts on the project.

20.6.3 Schedule and Flexibility

First, choices of foundation methods and structural systems can influence the schedule of the project. A deep foundation with an open hole takes quite a bit longer than a shallow foundation with no retaining walls, but the shallow foundation might forever give up valuable space that could have been saved for parking, storage, research, mechanical systems, or other important future building program elements. Similarly, a building with a concrete frame can likely be erected faster since a steel frame has to be fabricated first, and concrete will likely have shorter floor-to-floor heights. However, a concrete frame is not nearly as flexible for cutting holes and moving mechanical systems for future unanticipated uses.

With respect to the structural system, a consulting engineer will have a good handle on costs. During construction, the cost of the structure can typically be part of the general contract. However,

TABLE 20.1 Key Considerations

Element of the project	Key considerations for the facility manager	How to buy it
Site	• Legal control of the site • Physical configuration of the site • Comfort with environmental liabilities • Comfort with soils for foundations	• Advice from legal and geotechnical experts • Typically hourly rates
Foundation	• This will never be seen again and probably will never be a maintenance issue, but . . . • Selection of foundation system impacts usable square footage and purpose • Foundation means and methods for installation impact schedule and first cost and disruption to neighbors	• Part of the design package, typically subcontracted to a structural engineer. • For complex projects, consider hiring the foundation contractor in advance of the overall general contractor so as to a) help with accurate costing and b) remove an area of uncertainty from contractor pricing
Structure	• This will never be seen again and probably will never be a maintenance issue, but . . . • Choice of structural system impacts cost and speed of construction • Choice of structural system impacts future flexibility to change uses • Sometimes choice of system impacts usage by effecting story heights	• Part of the design package • Typically part of the general contracting package, subcontracted out • The design team can probably assess cost trade-offs
Envelope	• What will be the life-cycle costs to maintain roof, windows, skin, over years of sun, rain, wind, and freezing? • How does it look? • What is the first cost?	• Part of the design package. Be sure that one designer has overall responsibility. Emphasize life-cycle cost along with aesthetics. • Part of the general contracting package. Be sure one entity has the overall installation and warranty responsibility.
Mechanical systems	• These will never be seen again but will be the largest source of maintenance expenses and of tenant complaints around heat, humidity, noise, and power • Life-cycle cost including maintenance and energy can be a major consideration compared with first cost to install • Evaluate trade-offs in efficiency, complexity, and reliability	• Part of the design package. Be sure there is one point of responsibility. Be sure the life-cycle costs are fully considered. Consider an outside check engineer. • Part of the construction package. Be sure one entity has responsibility.
Finishes	• Although only 10–20% of the cost of the project, this is what users see and touch • How does the layout work for the users? • How do the carpets, doors, hardware look? • What is the maintenance expectation?	• Emphasize thoughtful consideration of how the users will use the space • These can be bought with the general contracting package or separately as the scope is easier to understand and manage
Furniture, fittings, equipment	• Be sure that these costs are explicitly carried in someone's budget • Be sure that for each item it is clear who specifies, buys, and installs it • Often there are capital vs. expense issues and differences in funding sources	• Keep a detailed table • Consider using a specialist consultant

for foundations, if the situation is very complex, it can pay to engage a contractor from the beginning to work out the details and the costs. If the foundation uncertainty can be contained before the award of the general contract, then that reduction in uncertainty will lead to reduced prices.

20.6.4 Envelope

The building envelope is an area where facility managers need to exert their strongest skills at understanding and making trade-offs during the design phase. Typically, the most aesthetically interesting skin treatments also are those with the most life-cycle problems relating to sun, water, wind, and freezing. Sometimes the most appealing treatments are also the most expensive. Yet, the image of the building, and often the image of the institution, depends to a very large extent on appearance of the skin. Attractive buildings tend to have tangible positive results, including attracting higher caliber students to a university, attracting and satisfying high-quality employees to a corporation, or attracting high-paying tenants for a landlord.

Often, the exact pricing of the building envelope is very hard to determine in advance, since there are so many interrelated systems, including the structure, the curtainwall, the windows, the waterproofing and sealants, the roof, and many items of decoration. There can also be physical problems in installing complex systems involving many skilled craftsmen from different trades who need to coordinate work that is many stories above the ground. For this reason, typically projects with more complex façades are likely to be projects where a contractor is selected earlier in the process.

20.6.5 Mechanical

As indicated in Table 20.1, mechanical systems are a different class to consider. None of these systems will ever be seen again once the building is finished, but they are certain to be the major cost driver in the building life cycle. Replacements of worn-out parts and the energy costs for day-to-day operations are highly dependent on choices made by the facility manager in the design stage. Most good engineers have a good handle on both first cost and life cycle costs, although sometimes a check engineer can be considered.

Mechanical systems will also be a major source of complaints in the building, from users who are too hot, too cold, too humid, are getting water dripped on them, are subject to too much noise, or who get stuck in elevators. The facility planner is well advised to consider the facility operator when major choices are being made in the mechanical area.

20.6.6 Finishes

The building finishes—the partitions, carpets, doors, plumbing fixtures, and the like—represent only a small portion of building cost. Yet these are the major contributors to the effectiveness of the space for the occupants. Effectiveness ranges from how well the floor plan layout works, through the feel and longevity of doorknobs.

The finishes element typically is well understood by designers, so there's relatively little value compared to the other elements in engaging a contractor early on. Similarly, the scope of work in these tangible components is understandable, and problems are easily remedied. Accordingly, some facilities managers retain this work for themselves and their staff even in new construction, and take a still greater role in renovations.

20.6.7 Furniture, Fixtures, Equipment

If the choice of foundation systems is large scale and limited, managing the design, procurement, and installation of furniture, fittings, and equipment is the opposite: small scale and wide open.

Here, the facility manager needs an obsessive approach to list making and to detail. There are dozens of items (in a hospital, thousands of items) where there can be great confusion among the numerous possibilities for these items. Many are difficult to retrofit after the core construction work is done; so some facilities managers reduce risk by hiring special consultants just for this function.

While this section has not covered the engineering details of structures and mechanical systems, we have tried to indicate the major areas of judgment, control, and risk that the facility manager must consider (see Table 20.1, Key Considerations).

20.7 BUILDING THE TEAM: VALUE ADDED, RISKS ACCEPTED

The composition of the project team is critical to the success of the project and to the success of the facility manager. There are many reference sources regarding contract types and contract language, which are too arcane to go into here. However, it is possible to consider the major classes of participants in the project and think hard about three aspects of each role:

1. What is the core value added of the participant?
2. What risks need to be considered with respect to this role?
3. What are the key considerations in selecting players for each role?

A solid understanding of these issues, followed by consideration of the particulars of the specific project, will hold the facility manager in good stead when he or she is ready to look more deeply into specific contract issues and legal terms.

Table 20.2 presents issues around the value added and risks assumed by the key classes of participants. Two classes of participants are not selected by the facility manager: the end users and the peers of the manager within the organization. All of the others are within the array of choices made by the facilities manager.

20.7.1 Project Management

Supplemental project management is a key make-or-buy decision for every project. This can be done in-house, through consultants, or by one of the major vendors, like the architect or contractor. The first consideration is simply capacity; many organizations don't build regularly, so there's not surplus talent waiting on the bench for a new project to come along. The second consideration is expertise. Since the vendors (the architect and contractor) are familiar with the process, it is often useful to have another person on board who is well versed in the process, too.

20.7.2 Designer

Selection of the lead designer, usually an architectural firm, of course has a major impact on the success of the project from both a process and a product point of view. It is critical to consider the specific needs of the project before starting the selection process. No designer can do their best work in all possible circumstances, ranging from high-end new construction to low-end renovation. Similarly, sometimes the process and politics will be so complex that a charismatic personality is more important than a pure gift for design. Sometimes the opposite is true. Consider the needs of the project, check references heavily, and make clear what the consequences are for time and budget problems.

TABLE 20.2 The Value Added by the Participants

Classes of participants	What is the core value added?	What risks to consider?	What are the key selection considerations?
Internal users, tenants, occupants, or customers of the facility	The reason the facility is being built	• Will they occupy? • What is the finance plan? • Is the program need clear? • Who decides and speaks for them?	• No choice for the facility manager
Peers of the facility management function: HR, IT, accounting	Support to the organization in similar staff roles; interface with needs of the facility	• How do they need to be involved before, during, and after the project?	• No choice for the facility manager
Internal project management	Time and expertise if facility manager has other demands or less experience with respect to new projects	• How capable is the consultant? • What risks do they accept regarding time and money?	• Typical make versus buy considerations • What are present skills and resources? • What is experience and cost of consultant/staff?
Prime designer: Architect	Leadership, judgment, experience, aesthetics, management of team, resources to design	• Who is responsible if the design is over budget? • How does responsibility flow for errors and omissions? • Who leads the project team?	• Experience with similar client companies • Experience with similar project types • References from prior project teams • Cost
Consulting engineers	Specialized expertise	• What is the relevant expertise? • How does responsibility flow for errors and omissions?	• Experience with similar projects • Delivery method • Life-cycle cost orientation
Prime constructor (construction manager or general contractor)	Management of many vendors, site coordination, cost of construction, capital base, accounting and payables, scheduling, budgeting	• Is the price guaranteed? How? • Is the capital base adequate? • Is the completion date guaranteed? How? • How do responsibilities flow for lien claims, property damage, personal injury, delay?	• Organizational capacity and fit with size of project and culture of client • Experience with similar projects • Financial capacity • Cost for overhead, administration, profit • Not necessarily cost for subcontractors
Subcontractors	Actual execution of the work, purchasing the materials, managing quality, coordinating with others	• Is the capital base adequate? • Does the facility manager have a say in their selection?	• Manpower capacity • Prior experience • Financial capacity • Scope of work (fit with design) • Price
Other vendors working directly for facilities manager	Needed materials and services	• Who pays them, coordinates them, manages their scope and their schedule?	• Relevant experience • Fit of product lines with needs • Price

20.7.3 Constructor

Selection of the primary constructor, typically a general contractor or construction manager, follows the same general rules. Match the capabilities and experience of the firm to needs of the project. General contracting can be purchased on a highly collaborative, professional services basis where the contractor works very closely in all phases with the owner and architect; and construction can also be purchased on a low-cost, commodity basis, where the general contractor and its subcontractors are arm's-length vendors. The basic philosophical trade-off is between the coordination advantages of the cooperative approach, compared to the price advantages of a competitive approach. The first is suitable for complex situations with unknown conditions; the second is best suited for simple situations with easily described conditions. Table 20. 2 describes other considerations with respect to selection of these and other participants.

20.8 CONCLUSION

In summary, the design and construction process always follows the same four phases: formulation, design, construction, and operations. Further, the facility manager is always in the center of the same array of players: internal users, peers, the design team, and the construction team.

Every project's circumstances are different, so construction risk, process risk, and internal resources vary greatly. Facility managers are well served to:

- First, understand the process
- Second, understand the players
- Third, think about the issues of each element of the project
- Fourth, pick the right team and set them up to accept (or not accept) the right risks

Finally, the core skills and roles of the facility manager evolve drastically as the phases of work progress. The effective facility manager handles unstructured processes and ambiguous decision making in the formulation phase, is a gifted facilitator of trade-off decisions in the design phase, is a list making monitor in the construction phase, and in the operations phase is capable of thinking of everything from asset allocation in the real estate function to low-cost building cleaning, snowplowing, and security in the administrative function. Armed with these references, first-rate communication skills, and commonsense, the facility manager survives and thrives as the indispensable orchestrator: the person in the middle.

CHAPTER 21
SPACE AND ASSET MANAGEMENT

Bill Tracy, AIA
Managing Principal
Building Area Measurement LLC
E-mail: tracywb@mindspring.com

21.1 FINANCIAL VS. FACILITY ASSET MANAGEMENT

Asset management is not a new concept in business; it is a central activity in the management and measurement of investment and businesses health. Business performance ratios, such as return on assets (ROA), and return on investments (ROI), are fundamental yardsticks that influence stock values, cost of capital, and many critical determinants of business success. Even in the public sector, guidelines such as the Government Accounting Standards Board Rule #34 are beginning to recognize the importance of accounting for government assets and infrastructure.

Accounting textbooks usually define an *asset* as any claim that has value to its owner, which includes investments, accounts receivable, inventories, and the like, in addition to physical plant investments. Most management and accounting textbooks approach asset management in this financial/legal sense. This chapter addresses asset management from the standpoint of the facility manager. This is an important distinction for the following reasons:

1. It focuses on a subset of corporate balance sheet assets that are physical in nature, rather than purely financial or legal.
2. It enlarges the scope of "assets" to extend beyond those assets indicated on corporate balance sheets, including innumerable "noncapitalized" assets such as leased space, office equipment, human resources, and the like.
3. It includes operating assets—assets, which, unlike investments, depreciate and wear out over time, and require regular maintenance and repair to maintain their functionality and avoid catastrophic failure.
4. It involves the utilization and deployment of assets in space and over the various parts of an organization.
5. It accounts for the "return" on "investment" (often just expenses, really), which is often in the form of increased productivity of occupants, a very difficult value to quantify and measure.
6. It takes into account that the facility manager may or may not have authority over disposition (purchase, lease, or sell) of all or a portion of the assets that he or she manages.

Let us review these concepts to gain a better understanding of the distinction between the role of the facility manager and the roles of line management in the field of asset management.

21.1.1 Facility Assets Are a Subset of Balance Sheet Assets

The facility manager does not manage balance sheet assets like bank accounts, securities, production inventories, and accounts receivable. Rather, facility managers focus upon balance sheet assets such as "building and improvements" or "fixtures and equipment." It is noted that these assets are usually classified as "noncurrent" assets, meaning that they cannot be converted at their full value to cash within a normal business cycle without affecting the ability of the company to stay in business.

21.1.2 Facility Assets Are More Than Balance Sheet Assets

It is common for companies to focus their capital on production-related assets, such as production equipment and inventories, and avoid tying up expensive capital in facilities. They often accomplish this by leasing office space, equipment, and, sometimes, even furniture. Although this reduces the need for capital and keeps those items off the balance sheet (they are technically not the company's assets if the company leases instead of purchasing them), this does not mean that they are not candidates for a facility manager's asset management efforts.

In fact, managing assets that are owned by a lessor can be more demanding than managing owned assets because those assets have to be accounted for and returned at the end of their lease. Leased office space can represent a significant expense that can be minimized by effective space management.

The working environment is another area that can significantly influence the productivity of staff, the salaries of whom are a significant expense in most companies. As the common saying goes: "People are our biggest asset." Although this has nothing to do with assets in the balance sheet sense, staff productivity is influenced by the working environment and productivity is a central concern of the facility manager (as well as line management); therefore, the working environment is an appropriate focus for asset management efforts.

21.1.3 Facility Assets Are Operating Assets

In managing operating assets, the battle against entropy requires an optimal stream of maintenance expenses to keep them running efficiently. Anticipating and budgeting for these expenses requires not only good recordkeeping, but also a good crystal ball to determine where and when funds may be needed to avoid expensive breakdowns and minimize energy expenses. The most obvious examples of such costs are building systems operation, such as pumps, fans, compressors, roofs, etc. Less obvious is the "cost of churn"; that is, the cost of rearranging facilities to respond to new business demands. In today's fast-moving, customer-driven business environment, churn is becoming an increasingly critical facility management (FM) focus.

21.1.4 Facility Assets Have Locations

Most financial assets reside in "accounts." With a few exceptions, like inventories, physical location is not a key consideration in their management. In contrast, facilities assets all reside in physical locations, which is an important consideration in their management. Space, by definition, incorporates the concept of location. Since space (for reasons to be discussed later) is usually the first facilities asset to be graphically documented and systematized, it provides a context for graphically locating all other physical assets (equipment, furnishings, people, etc.). This attribute of graphic location is often missing from other asset management systems. Graphic location also provides a means of allocating both the cost of space, and the furniture and equipment therein, to portions of organizations (cost centers) that occupy the space. Since the advent of graphic computer systems, this capability, often called "charge-back systems," has become economically more feasible for space, and is another factor that distinguishes facility asset management from financial asset management.

21.1.5 Return on Investment Is Viewed Differently for Facility Assets

In facilities asset management, return on investment, or ROI, can be very different than in financial asset management. In financial asset management, you invest capital with the expectation of receiving an income stream. The income stream, or ROI, may be uncertain, but it is usually measurable, quantifiable, and often predictable within limits. In facility asset management, the investment is often not capitalized as an asset, but is an expense intended to produce an improvement in operating efficiency or in productivity. These operating savings or productivity improvements are returns that are notoriously difficult to measure and quantify, even though they may appear reasonable. For example, how will the productivity of staff change with a 10% change in their workstation sizes? What about a 20% increase in lighting levels or another air-change or two per hour? These "returns" on "investment" in facilities are often impossible to measure. The investments or expenses themselves, however, are eminently measurable.

21.1.6 Disposition of Assets Are Line Management Responsibility

Most companies exist to provide goods and services, not just to occupy facilities. Line management, which means chief operating officers, division or department managers, have the ultimate responsibility for profitable operations of the company. In order to be effective, line management must make decisions on all asset acquisitions, their management, and disposal. The facility manager is not usually line management, but does operate in a support role to line management as a specialist in planning, budgeting, operating, and maintaining facilities necessary to support the business plan. This puts the facility manager in the primary role of putting facility asset information into the hands of those line managers who are responsible for asset management decisions, rather than making those decisions themselves. Exceptions to this are common when line management delegates some of its responsibilities (usually within limits established by periodic budgets) to facility managers. However, it is uncommon for facility managers to have line management authority in the leasing, purchase, or sale of major assets.

21.1.7 Summary

Because facility asset management is different in all these respects from financial asset management, it would perhaps be appropriate to rename "facility asset management" to be called something like "facility resource management." This would eliminate much of the confusion and discussion that exists concerning the facility manager's control over disposition of assets, which is clearly constrained. This renaming is not something we can accomplish within the scope of this chapter, so when the reader encounters the term "asset management" in this chapter, he or she should be careful to think of it from the point-of-view of the facility manager.

21.2 GOALS OF FACILITY ASSET MANAGEMENT

There are three fundamental goals of space and asset management:

1. To increase the productivity of facilities, furniture, fixtures and equipment (FF&E), and, most importantly, staff who work in the facility
2. To minimize the operating and ownership costs of facilities and FF&E

3. To provide an information foundation, historical, current and future estimates, for facilities strategic planning and business planning

Each of these goals deserves some expansion because there are a multitude of ways in which the facility manager can achieve them.

21.2.1 Increasing Productivity of Facilities

There are various measures of facility productivity. One that is often used in an office environment is staff density, or average area per person. Since rentable area per person is a popular benchmark, it is tempting to minimize workstation sizes to minimize occupancy costs until you understand that, by chopping workstation sizes, you may also be negatively impacting productivity of staff. Since salary costs per square foot are in the order of ten times facility costs, this approach can be costly in the long run. A more logical approach is to seek an optimum workstation size and arrangement that maximizes staff productivity at an optimal occupancy cost. Collateral concerns include full utilization of space and equipment to eliminate unused capacity. Space and asset management systems seek to provide information that assists management in avoiding excess capacity and achieving full utilization of facilities resources. This includes planning for and providing adequate facilities to support the business plan of the company.

21.2.2 Minimizing Facilities Costs

The reduction in facilities expenses are usually expected to justify the costs associated with setting up and running an FM system. These savings are realized by a reduction in facilities' area and/or in operating costs. The most popular mechanism for controlling occupied area is the facility occupancy cost charge-back system, which allocates occupancy costs to divisions or departments based upon the area that they occupy. Another approach is to calculate revenue per square foot occupied, staff densities, and similar ratios, and compare this against benchmark data within an industry or business sector. Both approaches can produce a bias on the part of division or department managers toward occupancy of less area. Less area often translates into a savings in occupancy costs, which often more than offsets the cost of implementing and maintaining a space or asset management system.

Another focus is, through better information, to directly lower facilities operation costs by more efficient construction, maintenance, and operation. Efficient facility design and construction is the most significant of these because it can limit the ultimate efficiency of a facility. However, for any given facility, there is an optimal level of maintenance of systems, such as HVAC and roofs, which will result in the longest service life at the lowest life-cycle cost. Facilities systems can assist in these maintenance efforts. Daily operation of facilities generates many opportunities for both waste and efficient operation since the demand for energy (light and heat) varies over space and time, requiring that energy flow be modulated accordingly in order to minimize energy expenses. Since businesses and facilities are rarely static, there are always changes within facilities that have to be implemented as efficiently as possible, which generates more potential for either waste or cost savings.

21.2.3 Providing an Information Foundation for Strategic Planning

In addition to knowing the answer to "Where are we going?" a strategic planner must also know the answer to "Where are we now?" with respect to facilities resources and utilization. Chapter 3 on Strategic Planning describes the management cycle consisting of the four phases of planning, implementation, monitoring, and evaluation. Space and asset management systems provide information useful in the monitoring and evaluation stages of this cycle.

Useful information on facilities resources and utilization can be extracted by physically surveying facilities, analyzing current drawings of facilities, or by generating reports from an up-to-date space and asset management system. Most commonly, a combination of two or all three approaches is utilized. However, the fastest way to provide the answer to "Where are we?" is reporting from a current facility system, and this can also be the cheapest way, depending upon the size and rate of change of the organization.

21.3 SYSTEM PRINCIPLES

21.3.1 Design or Select an Appropriate System

The size, organizational complexity, and dynamic characteristics of an organization dictate the approach to asset and space management. For very small firms with no departments or divisions, low churn rates and no prospects for reorganizations, mergers, or acquisitions, an expensive, formal facility asset management system is not justifiable. This is because the business value of information provided by a facility asset management system would be exceeded by the cost of establishing and maintaining the system. In these cases, informal and manual systems provide the necessary information for FM at lower cost. Manual recordkeeping and budgeting utilize existing accounting records, paper plans, spreadsheets, log books, and human memory.

As companies grow in size and organizational complexity, the need for information to manage facility space and assets increases. An aggravating factor is the dynamic aspect of the company. Companies that reorganize frequently, and which are often involved in merger and acquisition activity, have a much higher need for current information on facilities resources, and can justify higher expenses of setting up and maintaining space and facility asset management systems. At one end of the large/complex/dynamic spectrum, a company can easily justify the cost of fully automated graphic FM systems that are fully integrated with other management systems, such as accounting, human resource, and management information systems. Indeed, the information provided by such a system can significantly leverage the business plan of the company by helping it plan better and become more responsive to change at lower cost.

The transition from small simple companies and facilities systems to large, complex, and dynamic facilities systems is invariably a difficult one for facility managers. An organization that is growing larger, more complex, and more dynamic should be setting up facilities systems that are appropriate for their future, rather than their current needs. Since it is usually difficult to justify the cost of forward-looking systems based only on historical experience, facilities systems always lag behind the need for them. It is, therefore, critical that facilities systems be capable of implementation on an incremental basis.

21.3.2 Start with Highest Measurable Payback Issues

Space and asset management covers many issues, including allocation of space, furniture, fixtures and equipment, building systems, construction, maintenance and repairs, and the accommodation of people in the performance of their activities and transactions. The payback on FM efforts ranges from the measurable and quantitative, such as reduction in rent and operational expenses, to the highly subjective, such as an increase in sales that might result in a better image in retail spaces or higher productivity of office staff that might be achieved by a more attractive and functional office.

In a 1991 roundtable of approximately fifty IFMA facility managers, moderated by the author, a list of space and asset management issues was developed, ranked by the perceived value of incorporating each issue in a FM system. That list is included as Appendix A. The list reflects the perceptions of a wide range of companies and facility managers. It would not reflect the priority for implementation at any given company.

As will be discussed in the section on implementation, key to the success of a space and asset management system is selecting an appropriate scope that will optimize the benefits of the system relative to the cost of implementation and maintenance of the system. A critical first step is for the facility manager to identify all candidates for inclusion into the system and rank them by benefit–cost ratio, similar to Appendix A, but in the context of his or her own organization.

21.3.3 Start Small and Grow Incrementally

The three most devastating mistakes a facility manager can make in implementation of a space and asset management system are:

1. To start too modestly
2. To start too aggressively
3. To use a system that cannot be expanded

Starting Too Modestly. The cost of implementing and maintaining a space and asset management system can be significant in the context of a facility manager's budget (see Appendix B). There is a tendency to "scrimp" by restricting the scope and cost of an asset management system. This can increase costs in the long run since it is generally more expensive to incrementally expand the capabilities of a system than it is to incorporate capabilities (up to some optimal level) initially. This is, however, a cautious approach that lowers the risk of a large failure to produce expected benefits, and can defer some expenses until the application produces proven benefits. This approach is more acceptable than the second approach.

Starting Too Aggressively. Facility managers who do a good job of anticipating the full cost of a space/asset management system seldom run the risk of establishing a scope that requires a level of expenditure that will exceed the benefits. In reality, costs are often underestimated and benefits overestimated, which leads to the common mistake of "biting off more than one can chew." This can lead to unanticipated costs and implementation difficulties that will undermine the success of the system. A company that is growing rapidly may eventually grow into a system that offers more than its current needs. However, the costs of this problem are often greater than those of starting too modestly because the system can lose credibility and fall into disuse, and cost over-runs will not bring the facility manager into good standing with line management. This problem is still preferable to the last critical error: using a system that cannot be expanded.

Using a System that Cannot Be Expanded. It has been said that the only constant is change and nowhere is this more true than with information technology. Whether a company itself is changing rapidly or not, it is absolutely critical that a space and FM system be expandable (scalable) and capable of interfacing flexibly with other IT systems in a company, a feature called "open architecture." The most costly mistake is to implement a system then discard it and start over from scratch because it could not be made to meet increasing or unanticipated demands. Indeed, a step-by-step approach, incorporating asset management issues into a system one at a time, is often the recommended approach because of the lower initial costs and lower risk of this approach. This is possible only with a system that offers an expandable open architecture approach.

21.3.4 The Reality >Data >Information Link

The information produced by an asset management system reflects the reality of the facility only indirectly. The system itself incorporates, via its databases, a model of the facility, and that model has to be kept in close synchronization with reality or the information it produces becomes worthless. The effort of keeping the system and the physical facility in sync is one of the most frequently underestimated costs. However, the cost of *not* keeping the system in sync with reality is potentially far greater since information that is out of date is often useless.

21.3.5 The 80/20 Rule and the Law of Diminishing Returns

In selecting what facility elements to include in a space/asset management system, it is often the case that 80% of the benefits are available from tracking only 20% of the candidate elements. The "law of diminishing returns" suggests that there are always some elements for which lower benefits are produced at higher cost if incorporated in an asset tracking system. Experience indicates that the law of diminishing returns accurately characterizes the fact that it is possible to rank elements in order of benefit-to-cost ratio, taking into account not only the initial implementation costs, but also the on-going cost to keep the facility databases current. However, the 80/20 rule suggests that there may always be 20%, or thereabouts, of facility elements for which it makes sense to track. This is not always the case.

Even when the benefits are hard to measure and quantify, the decision of what to include in an asset management system must always be based upon incremental and current benefit–cost estimates as relates to the particular organization and their specific need for facility information. Results of this approach yield radically different asset management systems between companies and industries and explain why some successful systems include items as small as light bulbs and drawer contents and others seem to max out on benefits with just tracking space occupancy or leaseholds. New technologies are having a dramatic impact on lowering costs, with the result that benefit–cost ratios are changing for everyone. Computing is getting cheaper, and future generations of "smart chips" will dramatically reduce the cost of entering and keeping current the data on assets. This makes it unwise to rely on rules of thumb like the 80/20 rule.

21.3.6 Timing Issues—Info Now or Later

It is useful to consider facility information needs in two fundamental categories: information that is time-critical, and information that is not. This is a concept that distinguishes "management accounting" systems from "financial accounting" systems, and is useful in establishing what facility elements to include in a system and how frequently the updating process needs to be performed. Consider two examples:

The first example—a large architectural engineering firm with many dynamic project teams having important adjacency requirements. It is not uncommon for such organizations to reorganize staff monthly and to grow and contract suddenly as major projects flow through the firm. Occupancy information that is more than a month or two old is useless, so any occupancy tracking systems must be frequently updated, resulting in high data maintenance costs. The accuracy of the occupancy data is not as critical as its currency. This requires a "management accounting" approach to tracking occupancy.

In contrast, consider a landlord who manages an office building that has a few dozen tenants on 5- and 10-year leases. Occupancy here changes much more slowly. Currency is less critical than accuracy since a mistake in figuring rentable area is highly leveraged by the rent rate over the duration of the lease and can cause havoc with annual expense allocation among the tenants. This requires a "financial accounting" approach to tracking occupancy.

Facility space/asset management systems incorporate elements that individually may require either management or financial accounting approaches to data updating and reporting. If this is not recognized in the design of systems to update facilities data, there will be needless expenses. The need for currency and accuracy of data must drive the design of the system and the method of updating its data.

Because information needs vary so much from one company to another, it is difficult to offer guidelines for how much to budget for keeping facility data up-to-date. Some companies find that administration and updating of a system that tracks space occupancy by space type and organizational unit only takes one full-time person per million square feet. However, companies that are less dynamic would require less time than this and companies that track additional data would need more time.

21.3.7 Maintain Data in One Location Only

Computing in many industries has gone through cycles of centralization of data on mainframes and decentralization (or distributed) onto servers or individual PCs. A facility manager contemplating a space/asset system usually finds that portions of the needed data already exist in company databases, sometimes in more than one place. This gives rise to several system design issues. Data that exist primarily in more than one location:

1. Must be updated redundantly, increasing the cost of keeping it current
2. Can get out of sync, reducing reliability and creating reconciliation costs

There are two ways of dealing with these issues and they both start with identification of one location as the primary one for each piece of data. For instance, if an individual were primarily assigned to a certain department in the company payroll system, it would be inappropriate for a facility manager to be able to change this fact by reassigning an individual in the facility human resources tracking module.

The process of identifying a primary location for each piece of facility system data (e.g., defining the data dictionary) can be one of the most time-consuming aspects of designing a space/asset management system. It requires the facility manager to discover where all current asset information is kept, to identify all the stakeholders in that data, and how the data are to be kept up-to-date.

There are two additional concepts that influence the approach to this task:

1. Those who update data should be the ones most capable and knowledgeable to efficiently perform that task
2. Those who are most impacted by certain data should own the responsibility of keeping it current

Resolving these issues can be complex and is made more complex by the internal politics and the often unfortunate existence of incompatible data formats and systems within some organizations. However, if these issues are not resolved and a solid data dictionary developed, an asset/space management system is unlikely to be successful.

A closely related issue is who should be allowed to change data in a space/asset management system. An oft-heard statement is "people who don't know what they're doing have no business messing with my data!" This means that the converse of item number 2 mentioned previously is also true—asset data should be safeguarded from being changed by persons who are not responsible for keeping it current. Controlling access to data is easier if it is located only in one database.

21.3.8 Use Data Exhaust

In setting up and maintaining a space and asset management system, the facility manager has an opportunity to lower costs of entering data by drawing on data maintained by others. For example, there are many firms that utilize CAD to keep floor plans updated. There are three ways of transferring this CAD data into a space/asset management system:

1. Manually reenter the CAD data
2. Import the CAD data
3. Reference the CAD data

Manually reentering CAD data into an FM system is obviously the least cost-effective approach, both initially and in updating facility plans. Where plan information changes frequently, which is common in many companies, the economics of keeping two sets of plans up-to-date become worse and leads to inconsistencies between systems. This approach is the least desirable.

Importing CAD data into an FM system greatly reduces the cost of data entry by using the "data exhaust" of another process. The other process can be the original construction of the facility or the on-going updating of plans in the course of renovation projects by in-house or consulting A/E firms. This is a less expensive and more accurate way of feeding updated plan information into an FM system. However, there are three important caveats. First, the original CAD data must be drawn in strict adherence to CAD standards required by the FM system (data format, layering, line weights and colors, text, and symbology). Second, the importing process must happen on a timely basis or the FM system will be out-of-date. Third, the company must have the contractual right to use data supplied by the consultants. This right must be negotiated with A/E firms, which traditionally retain all rights to drawings that are considered "instruments of service" and not the property of client organizations.

Referencing CAD data means that the FM system has the capability to display CAD data without copying it. The CAD data can reside with the CAD system and is merely referenced by the FM system. This approach has three major benefits. The FM system will always be as up-to-date as the CAD data; there is no cost to keeping FM data in sync with the CAD data; and there is a much lower possibility that the CAD and the FM systems will be out of sync. Caveats are that the CAD data must meet certain standards, as mentioned previously, and, in addition, must reside where the FM system can see it. As the Internet and Intranets proliferate, this is becoming less and less of a problem as software can "see" practically anywhere you need it to. In addition to having the lowest cost for data entry, this approach can avoid some data ownership issues and prevents a user of the FM system from changing plan information that should be updated only from the CAD system. Some who would like to update plan information in the FM system regard this last feature as a disadvantage. However, this author contends that the capability to update plan data should reside in one location only. If that location is on the CAD system, the FM system should not be able to alter plan information. If it is important for plan information to be altered on the FM system, one of the first two approaches (entering the plan information directly or importing it) should be used.

These same principles can be applied to using non-CAD data exhaust. In most companies, data that could be integrated into a space/asset management system may already be maintained by a multitude of parties in many locations on a company system, sometimes redundantly and in a multitude of formats. As an example, if it is desired to track staff and staff phone numbers in the FM system, the primary source for phone numbers is usually the IT/Communications group. Employee names and IDs are usually maintained by the human resources group. So, there are two important existing databases to reference to ensure that any specific staff member is or is not an employee and what is his or her phone number.

If this example seems a challenge to implement successfully, things become far more challenging as additional data is tracked in the FM system. Appendix C presents a list of data that is frequently found on multiple databases maintained by different company departments. Consideration of this list makes it obvious that intelligent implementation of a wide-scale space/asset tracking system will involve most departments of a company because the existing data and stakeholders are widely distributed.

21.3.9 Don't Lose the Big Picture

As previously stated, the primary goals of asset management are to increase productivity and lower management costs. Studies of computerization have shown that attempts to computerize existing systems and procedures usually result in only marginal productivity gains. However, when the new capabilities of computers are merged with the redesign of existing systems that will take advantage of the new capabilities, that is where the real productivity gains occur. To quote Loren Hitt of the University of Pennsylvania's Wharton School, summarizing a study of 599 firms between 1987 and

1994, "Our results suggest that computers increase productivity both directly and by making new types of work structures possible over time."

The implications of this, along with the widespread location of asset data in many companies, strongly suggest that space and asset management systems should impact the basic management systems of a company if it is to realize the most benefits. Multiple distributed databases usually evolved not as a result of any overall planning effort, but as a result of individual efforts to apply personal computers to solve problems. A centralized space/asset management system is often a wonderful opportunity to coordinate and streamline the parochial databases of many separate departments and individuals. This has the potential of increasing productivity beyond the realm of pure FM and should not be overlooked.

There is one caveat in this regard. Enterprise information systems, such as SAP™ and their various integrated application modules, have a wide-ranging impact on data useful for space and asset management. Depending on the FM software selected, interfacing with these systems will be either very easy (since the data will be in a single integrated architecture) or very difficult, since that data architecture is typically very expensive to modify. Special knowledge, skills, and systems are critical to implementing FM systems in organizations that employ enterprise information systems.

21.3.10 Have Implementation Support from the Top

We have seen in the previous discussion that space and asset management systems have widespread impact in a company. They draw on data that may have existed only as databases in parochial management domains. This can result in a sharing of information across department and divisional boundaries that has not occurred in an organization before the system is implemented. This information sharing may not be welcome. Some managers may resist allowing access to "their" databases as well as the time and effort it takes to coordinate their data with the FM system (including making potentially disruptive modifications to a system that may be running smoothly). Exacerbating this condition is the fact that the facility manager is often perceived as a "low-ranking" manager ("weren't they the janitors just a few years ago?") who is using an FM system to "build their empire" at the expense of historically comfortable parochial control.

This concern is not unwarranted. In today's organizations, information is power. Sometimes giving out information results in loss of local control. An example might be a manager "sitting on" some vacant space that he or she might need for his or her group's future expansion when there is an urgent current need for space for a new project team, and the only other alternative is to lease additional space for the team. It might be in the company's best interest to use the vacant space for the team now and lease expansion space later. However, the manager to make that decision is more likely to be more senior, since he or she is responsible for the company's overall performance. It is because space and asset management systems lead to many scenarios like these that their implementation must have support from the senior management of an organization.

21.3.11 Focus Information Support on the Top

It is also precisely because space and asset management systems allow useful facility information to rise to the top of an organization that make them so valuable as a planning tool to be used for the strategic advantage of a company.

In today's corporate environment, reorganizations, mergers, and acquisitions are common occurrences and have tremendous impact on facilities. Nearly every merger or acquisition creates redundant functions and facilities that need to be identified, evaluated, merged, or shed. Reorganization of company structure nearly always requires physical redeployment of staff and functions. "Right-sizing" efforts in particular should be reflected in facilities if benefits are to be fully realized.

The sections in this book on facilities strategic planning discuss these issues in great depth, but a critical task of a space and asset management system is to provide information to top management that will be useful for strategic planning. The benefits of having efficient access to accurate facility information are the greatest at the top level of a company. Also, consider that you must ask top management for their support in the implementation and administration of an FM system, and this information, this power, is the most important *quid pro quo* that a facility manager has to offer in return for the time and effort that top management will expend to support and implement the system.

21.3.12 Leverage Decision Making at All Levels

Appropriate distribution of useful information within an organization can empower good decision making at all levels. A space/asset management system can perform a key role in maintaining and distributing such information. The key words are "appropriate distribution" and "useful information." These are critical in the design of any space/asset management system.

Appropriate distribution of information means that data availability is constrained to those who are authorized and have a need to know it. Many organizations have processes or projects that are company confidential and sensitive for competitive or contractual reasons. To the extent that space or asset management systems reflect information about such processes and projects, this information must be safeguarded by the system and made available only to authorized staff. Space and asset management systems must contain features to control access to data, and the data must be structured in such a way as to allow sensitive data to be segregated from the more generally useful data. Controlling access to data is often done through a system of "permissions" implemented by passwords that the system recognizes and correlates to the database structure so as to make only appropriate data available. Implementation and maintenance of this system of permissions and passwords is a significant task in the initial design and the on-going maintenance of any space and asset management system unless access is constrained to only a very few staff, which would limit the potential usefulness of the facilities information. In some organizations this is a valid trade-off, but in others it is a waste of potential value.

Useful information is that which applies to things over which an individual has control, or things that impact upon an individual. Information that is not useful is worse than irrelevant. It can obfuscate useful information like static on a radio. Also, to be useful, information must be in a form that can be easily comprehended. The graphic capabilities for many space/asset management systems are very useful in visually communicating spatial relationships, but they can also present considerable static if not thoughtfully designed and implemented. Design of the user interface must provide a capability for sorting and filtering information as needed by the user. Systems that lack a flexible approach to viewing and reporting information will provide less useful information than those with strong capabilities in this area.

21.3.13 Monitor and Reevaluate Frequently

Since the only constant is change, space and asset management systems should be constantly monitored and frequently reevaluated. Some of the occurrences that would require modification to a system include:

1. Unanticipated needs, such as information filters, reports, and graphs
2. Scheduled expansions to system capabilities
3. Changing organizational environment
4. Improved technology

Unanticipated needs should be minimized by a thorough evaluation of the requirements of all users and stakeholders in the course of the system's design. Unfortunately, experience reveals that this nearly always constitutes the largest post–implementation effort. This effort can be minimized by building systems that have maximum user flexibility, providing information filters, and generating custom reports. The trade-off here is that these capabilities require additional user training. "Canned" reports are cheaper to implement and simpler to use. They may be appropriate for some applications, but generally, they limit the usefulness of space/asset information and should be employed sparingly. Consider including the ability to export asset data into analytical tools that are already familiar to users, including Microsoft Excel™ or Lotus 1-2-3™. The system designer can build on the users' existing knowledge of these tools to avoid additional training, call-backs for custom reports or graphs, and the like. In addition, the presentation capabilities of these common office tools greatly leverages the ability of the facility manager and other users of asset information to communicate their needs.

It is, perhaps, unfair to imply that unanticipated needs are always the result of sloppy implementation. It is frequently the case that wide availability of new information will cause rethinking of how work is done, giving rise to new work structures. This ability of information to cause change in work process often requires changes to the space/asset management system that could not be anticipated. This will require the facility manager and company line management to work together to promote productive change to work process and the organizational changes they sometimes attend. As was noted earlier, these changes to work structure are often where the biggest payback exists in implementation of information technology, but it also has an impact on the IT system itself that must be understood and implemented.

Scheduled expansions are the result of the incremental implementation process described previously (see Start with Highest Measurable Payback Issues). In the context of an organization that is changing, any system implementation schedule must itself be reevaluated in light of changes, some of which are the result of implementation of earlier phases of the space and asset management system. While it is sometimes possible to predict where new information systems will take an organization, it is more common that work patterns will change unpredictably, causing adjustments to both the planned system and its implementation. As a general rule, expect to reevaluate future plans in step with current systems to best accommodate change.

New alternative work strategies (see Chapter 8) are caused by many other factors in company environments. Some examples include "hoteling," where offices and workstations are shared by multiple individuals, and "home officing," where staff members work at home for some or all of the time. These new work structures, and others that may not be invented yet, can profoundly impact the design of a space and asset management system. Other organizational changes that will change space/asset system requirements may include expansion of a company to multiple sites domestically or internationally, company reorganizations, new products or business lines, a major contract, new laws or statutes regarding safety, codes, taxation, and other regulatory matters. Because a company's environment is in constant flux, its IT systems must be constantly reevaluated and often redesigned.

Finally, in the constant stream of new technology that becomes economically feasible each year, there are capabilities that quickly enable new systems and make existing ones obsolete. Several years ago, the availability of hand-held bar code readers and laser scanners revolutionized the retail industry and led to the ubiquitous UPC (Uniform Product Code) bar codes that appear on all retail products. This had a profound impact on inventory tracking and work structures in many retail organizations. The bar code was adopted for asset management in many organizations, especially for tracking IT assets, because it led to similar economies. This incorporation of retail technology made many asset management systems economically feasible for the first time. At this writing, retail giants Wal-mart and Procter & Gamble are backing a new EPC (Electronic Product Code) which may be in use within 5 years. The EPC codes will be electronically embedded in all products and will cost a penny each. Scanning will be accomplished electronically and "smart shelving" will even be capable of notifying a store manager when shelf inventory is running low.

It doesn't take much imagination to understand how easy it will be to track company assets when this technology becomes available. Items that currently are uneconomical to track because they "grow legs" will be easily locatable in any facility. Even clients who want a system that will help "find their car keys" may get their wish. A new realm of asset management capabilities will be unleashed along with a new set of challenges, ranging from how to filter and present all this data so that it becomes useful information, to how to safeguard data that may be a breach of security, or an invasion of personal privacy. It is not clear at this time how all these issues will shake out. It is clear that space management and asset management systems will be a dynamic and exciting field which will have a major impact on management in the coming years.

APPENDIX A 1991 IFMA FACILITIES EXPO: FACILITY MANAGEMENT ISSUES APPROPRIATE FOR FM SYSTEMS

1. Space allocation for user groups, divisions, and departments
2. Location of existing walls and doors
3. Location of data/communications wiring, outlets, and equipment
4. Facilities/FF&E work-order and maintenance tracking
5. Tracking space types (office, conference, lab, warehouse, etc.)
6. Facilities/building systems and work-order tracking
7. Tracking HVAC systems (fans, ducts, plumbing)
8. Tracking nondata electrical (power and lighting) systems
9. Tracking security and keying systems
10. Hazardous materials tracking and reporting
11. Tracking systems furniture panels
12. Tracking location of staff
13. Tracking location/utilization of major equipment (copiers, etc.)
14. Tracking leaseholds and subleases
15. Parking management
16. Reflected ceiling plans (light fixtures, HVAC, sprinklers, etc.)
17. Type and condition of floor, wall, or ceiling finishes
18. Site and landscaping management
19. Tracking signage
20. Tracking systems furniture nonpanel components
21. Tracking nonsystems desks, chairs, files, etc.

APPENDIX B COST CONSIDERATION FOR SPACE AND ASSET MANAGEMENT SYSTEMS

1. Hardware and software licenses
2. Setting up and configuring hardware and software
3. Integration with company IT/data architecture
4. Integration with company organization/reporting structure
5. Initial data load into system
6. Debugging system
7. Roll-out user training
8. On-going data entry and updating
9. System administration
10. Report generation
11. Capability expansions/extensions
12. Software upgrades and re-training

13. Fixing things that break
14. Cost of downtime

APPENDIX C DISTRIBUTED DATA TO LOOK FOR

1. Floor plans maintained on CAD systems
2. Data/communications data maintained on CAD systems
3. Building system data maintained on CAD systems
4. Building system maintenance record databases
5. Staff data maintained in HR and payroll systems
6. Organization structure/charts in HR systems
7. Phone and e-mail directories
8. Security system databases
9. Work order system data
10. Asset records for FF&E and real estate in accounting systems
11. Lease records for FF&E and real estate in accounting systems
12. Hazardous materials tracking systems
13. Regulatory and contractual reporting databases
14. State property tax records
15. Insurance files

APPENDIX D CANDIDATES FOR SPACE AND ASSET MANAGEMENT SYSTEMS

1. Building areas (gross/rentable/usable/assignable, etc.)
2. Building areas by type of space (office/lab/storage/conference, etc.)
3. Building areas by use (occupied, vacant, under construction, etc.)
4. Building areas by occupancy (departmental, common, etc.)
5. Building areas by lease or sublease
6. Site or property areas, locations, characteristics
7. Facility locations, functions, and capacities (geographically)
8. Building core and shell plans
9. Building interior partition/layout plans
10. Building systems
11. Floor systems (type, materials, load capacities, etc.)
12. Partition systems (type, characteristics, etc.)
13. Ceiling systems (type, materials, heights, etc.)
14. Ceiling features (lighting, HVAC, fire protection, etc.)
15. Finishes (type, colors, condition)

16. Mechanical, electrical, plumbing systems (including controls)
17. Safety systems (fire detection and suppression, etc.)
18. Special process systems (gasses, chemical, exhaust, waste, etc.)
19. Fixed conveyance systems (cranes, lifts, pneumatic tube, conveyors, etc.)
20. Security systems (surveillance, access control, keying, etc.)
21. Signage and wayfinding systems
22. Site systems
23. Boundaries, easements, code or environmental restrictions
24. Roads and walkways
25. Site utilities (drainage, irrigation, lighting, etc.)
26. Landscaping and surface materials
27. Site signage and traffic signaling
28. Parking management (surface and structured)
29. Special uses
30. Security
31. Business assets
32. Furnishings (movable systems, desks, seating, storage, etc.)
33. Fixtures (lab equipment, major tools and machinery, etc.)
34. Equipment (storage and filing, reprographic, A/V, etc.)
35. Data/communications (computers, phones, wiring, etc.)
36. Special systems (robotics, etc.)
37. Interior flora
38. Artwork (paintings, sculpture, etc.)
39. Tools and inventory
40. Hazardous materials tracking and reporting
41. Live assets
42. Employees (payroll)
43. Contractors (nonpayroll)
44. Tasks
45. Maintenance management systems
46. Work orders systems
47. MAC (Move/Add/Change) systems
48. Janitorial and cleaning management systems
49. Occupancy cost chargeback systems

BIBLIOGRAPHY

1. Hamer, J. M. 1988. *Facility Management Systems.* Van Nostrand Reinhold Company, New York.
2. International Facility Manager's Association. 1997. Benchmarks III, Research Report #18.
3. Moffitt, N. 2000. The real story behind the IT revolution. *Wharton Alumni Magazine,* Winter.
4. Nelson, E. 2000. Big retailers try to speed up checkout lines. *Wall Street Journal,* March 13.

CHAPTER 22
OPERATIONS AND MAINTENANCE

Graham Lane Thomas
Vice President
Business Development
Graphic Systems, Inc.
E-mail: gthomas@graphsys.com

22.1 PURPOSE AND GOALS

The focus of this chapter will be on the features of a Computerized Maintenance Management System (CMMS) application to support the Operations and Maintenance (O&M) processes required for a typical facility organization. It will analyze the features of the software available today and their relevance to the business processes for a facility management (FM) department. It will look at the impact a CMMS application can have on the O&M operations and how such a system extends to other areas of not only facilities, but also the entire enterprise. There are many excellent books on plant operations showing how to maintain equipment and this chapter will not duplicate that effort. For a facilities department, there are three essential areas that consume resources:

- *Demand Work:* where the client calls in for service, where breakdowns in equipment require repairs and emergency events that affect the facilities department.
- *Preventive Maintenance Work:* where a scheduled program of work maintains the investment in the physical assets for a corporation. These assets may be equipment assets or facility assets.
- *Project Work:* where changes to the business focus requires a reorientation of space and people or the changes in regulations require upgrades to maintain compliance, such as ADA, EPA, or OSHA.

This chapter addresses the first two categories. Other chapters address project management issues in detail. The recognition of the importance of the maintenance to the corporation requires the facility manager to present the business case for how their unit can contribute to the bottom line.

22.2 THE EVOLUTION AND APPLICATION OF WORK MANAGEMENT

The effective implementation of a CMMS or work management program has become mandatory for facility departments. The implementation of a CMMS application represents one of the greatest challenges that a facility department will face, as the potential for radically changing the character of a facility department lies at the heart of a CMMS application. The first section explains why a facility department must use a CMMS today and outlines potential growth of the role of a CMMS with the evolution of the digital business environment. What follows illustrates why a successfully implemented CMMS can provide the business framework to build a successful FM department.

The CMMS can help increase the quality of customer service and help in managing resources efficiently as an organization maintains its asset base. Meeting these two goals will help the facility department define its role in a corporation and increase the visibility of the department within the organization providing the recognition it deserves given the financial impact of the property. The facility department has stewardship over the second largest expense, the property assets, for many organizations after labor costs.

Maintenance is mandatory. Maintenance management dictates that: "What you do not maintain today you will have to maintain tomorrow, except that the costs will be far greater." The CMMS has evolved from the narrow scope of providing a preventive maintenance schedule to building engineers responsible for the maintenance of mechanical systems. Today, the current leading vendors in the CMMS "information marketplace"[1] have expanded the scope of their applications. Why have these systems evolved? As the market for their products has matured, the facility manager expects work management software to do more, to integrate with other business applications. The typical facility manager has become more sophisticated in the use of work management software. There is a requirement that the software blends in with the business processes. At the same time, the software vendors see an opportunity to acquire more of the market by appealing to a broader range of clients. Figure 22.1 illustrates the spectrum of work management tools relevant to the facility market. They range from pure space management applications, through hybrid products, to the infrastructure software vendors. Most of these players claim to be in either the enterprise asset management market or the infrastructure management market.

At the farthest end of the spectrum are the Enterprise Resource Planning (ERP) systems (see Table 22.1) that often include features found in the "maintenance management marketplace."

There is a merging of spheres of influence. The ERP marketplace is much larger; the vendors larger. The CMMS vendors are small in comparison. There will be significant changes in the features provided: who provides them and how the vendor delivers them to the user. New players are entering the market offering exciting alternatives to the traditional delivery of CMMS services. The Internet changes the equation in how application services are delivered, where the data can reside, and the initial cost of the investment. In the technical section, the range of software and hardware solutions illustrates how the Web will open up access to the information created through the work management systems. The facility manager will exploit the new power provided by control of the information.

22.2.1 What is CMMS?

A fully implemented computerized maintenance management system (CMMS) will track all the work performed by a facilities organization. Figure 22.2 shows the modules of today's enterprise asset management application.

The CMMS can be thought of as a sophisticated "to do list" that allows the FM group to perform three key tasks: the creation of a predefined schedule of tasks known as preventative maintenance; the tracking of requests called in to an FM service center, often referred to as demand work orders; and the tracking of projects. While this may appear to be simple to do, you have to recognize that to list all the information necessary to support these three efforts requires a complex database program that can store all the information. A typical building might have over 1,000 pieces of equipment to maintain with ten tasks per item. It may perform six PMs per item each year. Each piece of equipment may track twenty items related to it, such as the manufacturer and vendor, date of purchase, replacement cost, model, and serial number. A large organization, such as the University of Minnesota, may process over 500,000 work orders a year, each with 100 items of related data requiring the organization to track and store 50 million data fields. The CMMS provides the

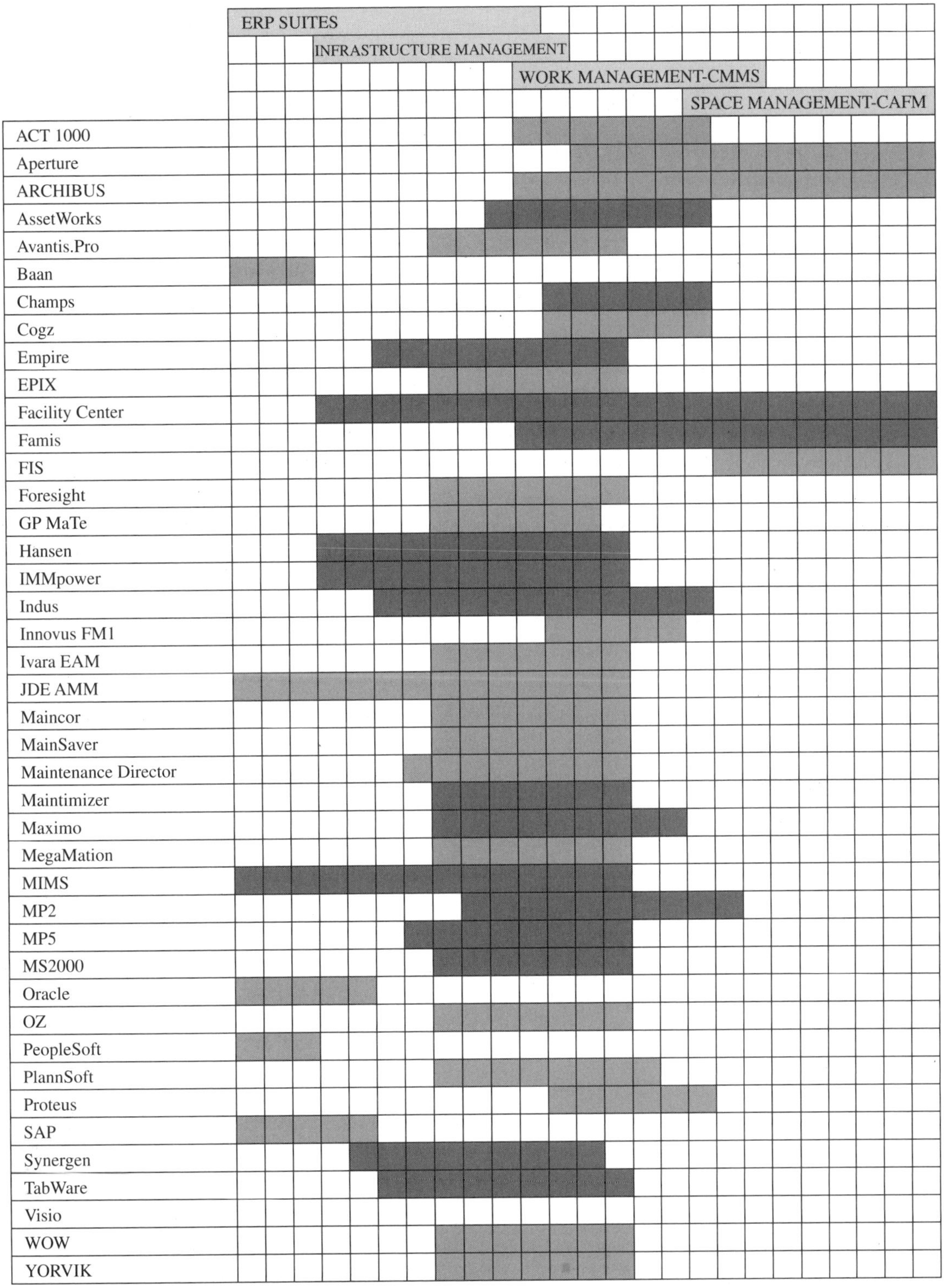

ERP SUITES
INFRASTRUCTURE MANAGEMENT
WORK MANAGEMENT-CMMS
SPACE MANAGEMENT-CAFM

ACT 1000
Aperture
ARCHIBUS
AssetWorks
Avantis.Pro
Baan
Champs
Cogz
Empire
EPIX
Facility Center
Famis
FIS
Foresight
GP MaTe
Hansen
IMMpower
Indus
Innovus FM1
Ivara EAM
JDE AMM
Maincor
MainSaver
Maintenance Director
Maintimizer
Maximo
MegaMation
MIMS
MP2
MP5
MS2000
Oracle
OZ
PeopleSoft
PlannSoft
Proteus
SAP
Synergen
TabWare
Visio
WOW
YORVIK

FIGURE 22.1 Spectrum of Information Tools.

TABLE 22.1 Elements of an ERP System

- Financial suite
  - General accounting
  - Accounts payable
  - Accounts receivable
  - Fixed assets
  - Financial modeling and budgeting
  - Multicurrency processing
  - Cash basis accounting
  - Time accounting
  - Canadian payroll
- Logistics/distribution suite
  - Forecasting
  - Requirements planning
  - Enterprise facilities planning
  - Sales order management
  - Advanced pricing
  - Procurement
  - Work order management
  - Inventory management
  - Bulk stock management
  - Quality management
  - Advanced warehouse management
  - Equipment management
  - Transportation management
  - Job cost
  - Service billing
- Services suite
  - Contract billing
  - Subcontract management
  - Change management
  - Property management
- Manufacturing suite
  - Configuration management
  - Cost management
  - Product data management
  - Capacity planning
  - Shop floor management
  - Advanced maintenance management
- Project management suite
  - Procurement
  - Inventory management
  - Equipment management
  - Job cost
  - Work order management
  - Subcontract management
  - Change management
  - Contract management
  - Contract billing
  - Service billing
  - Property management
- Energy and chemical suite
  - Agreement management
  - Advanced stock valuation
  - Sales order management
  - Bulk stock management
  - Load and delivery management
- Payroll suite
  - Payroll
  - Time accounting
- Human resources suite
  - Human resources
- Customer service management suite
  - Customer service management

data structures to store this information efficiently with the tools to create and modify the data. In the second section, the anatomy of a work order will illustrate how this core document links together any functions performed by an organization. There are many modules to a CMMS application. Some of these modules overlap into other department roles, such as finance, human resources, and information technology. Companies such as Peregrine Systems recognize the overlap and synergy that can be provided by treating the asset as the base unit and linking all the information to allow the tracking of the full life cycle.

22.2.2 What Value Can a CMMS Bring?

The CMMS can bring value at several levels—it provides a vehicle for improving customer service and a roadmap for staff to follow in providing service. It provides benchmarks to show whether the facility manager has met the level of service or has improved service. It provides both the summary and the detailed information to monitor that service standards have improved. Internally, the detail provides the justification for the FM group. It can show management the volume of work performed and the specific tasks completed. It provides the documentation on what went right and what went

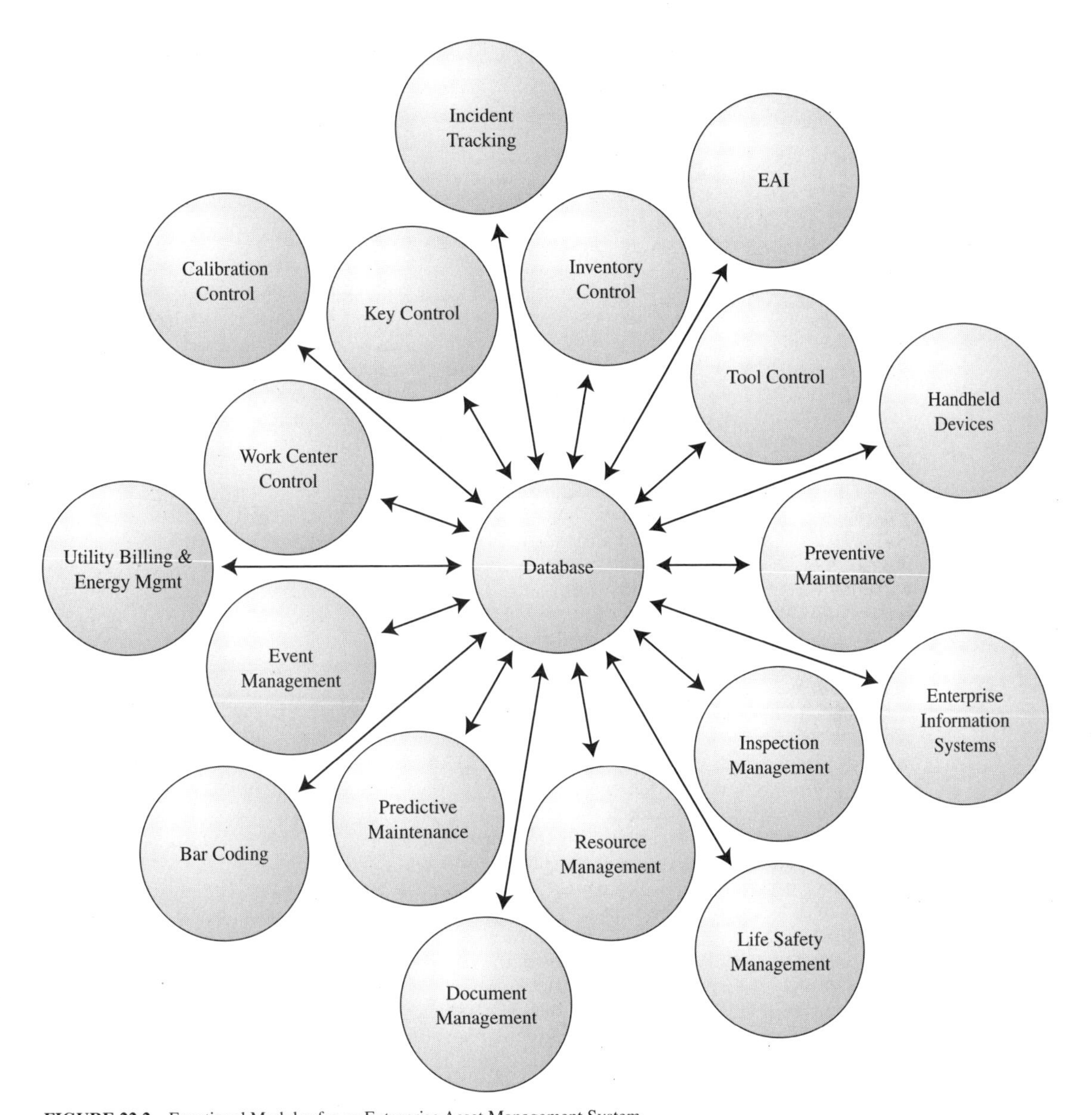

FIGURE 22.2 Functional Modules for an Enterprise Asset Management System.

wrong. It stores how things work, the raw knowledge. In short, it provides a digital memory of past FM events.

In surveys conducted by IFMA, customer service is still the number one concern of facility managers. The integration of the CMMS "call center" component into a central facilities customer service desk will provide the foundation to improving the delivery of services. By knowing what you are providing, where, when and to whom, the facility manager can start to analyze the patterns of service. Using this information wisely, the facility manager can develop a methodology to measure key performance indicators (KPIs) over time to provide benchmark statistics on how the de-

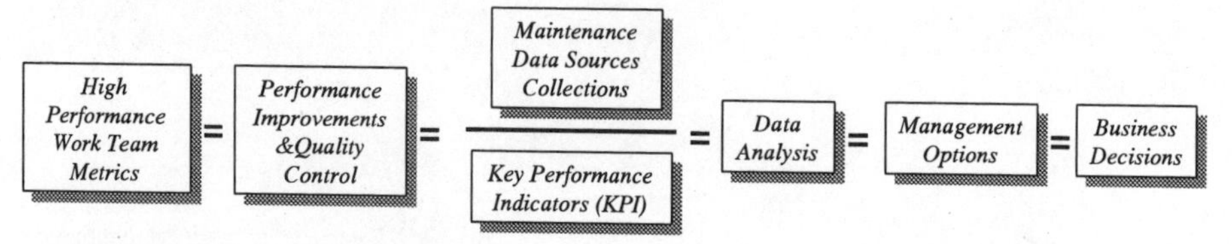

FIGURE 22.3 Basic Cost Equation.

partment is providing service. Gaining customer feedback through the CMMS will allow the facility manager to pinpoint where the department is failing to provide quality service. The facility manager can use this information in two ways, to provide a plan to improve the service and to create a budget justification for costs associated with the improvements.

As part of the implementation of a CMMS, standards are defined as to how the facility manager intends to deliver services or perform maintenance tasks. These standards form the basic template for the department. The facility manager will modify standards over time to meet changing needs, but the CMMS captures the knowledge of the institution on how best to perform maintenance tasks. Often, organizations claim to be unique, but in reality they perform similar tasks. However, each organization may have a different way of solving a specific task, or how to gain access to a certain valve. These lessons are learned by trial and error over the years. When these experienced building engineers leave after 20–30 years of service, their knowledge will leave with them. Therefore, by capturing their knowledge in the CMMS, a road map is provided for new technicians, as well as a history of what has gone before.

By capturing work performed in sufficient detail, it provides a gold mine of data to show how the department is providing service. It is interesting to note that the last part of many of the CMMS products to receive attention has been the report writer. This has changed. PSDI, the vendor for Maximo, has introduced a sophisticated data analysis tool for combining maintenance data into specific data cubes relating to customer service and maintenance tasks. Peregrine Systems has embedded Cognos as part of their report-writing tool in the Facility Center. It is no longer acceptable to use the CMMS as a "tickler" file on what needs to be done, just a calendar of tasks for the year. The introduction of KPIs to measure the performance of the facility department and the use of industry specific benchmarks to set standards provides management with the tools to analyze how well the facility department is performing. The result is preserving the knowledge of the staff and measuring performance results in the development of an institutional memory.

How does a facility department achieve the changes? Once these standards and modules have been established, integrated, and implemented within the CMMS, the FM department gathers KPI data to measure the efficiency and benefits of the CMMS within a department. The FM department collects the KPI data within the CMMS to show the benefits and cost avoidance potential that the CMMS implementation brings to a department.

The definition of the methodology for determining cost avoidance, benefits, and savings from the CMMS system follows a simple formula. Figure 22.3 shows the basic cost equation.

The application of the formula is as follows: it defines the meaning of the formula and shows how a facility department can substitute a range of KPIs into the equation to demonstrate how it is applied.

22.2.3 CMMS Cost Methodology Summary—Bringing It All Together

How will a department make use of the information gathered by the Computer-Aided Facilities Management (CAFM) application and use it to enhance the performance in the delivery of facilities services? The equation defined in the executive summary is at the core of the methodology. Performance improvement is the goal of the department. What does it mean? Figures 22.4 and 22.5 analyze the equation, showing how the components affect the overall performance of an organization.

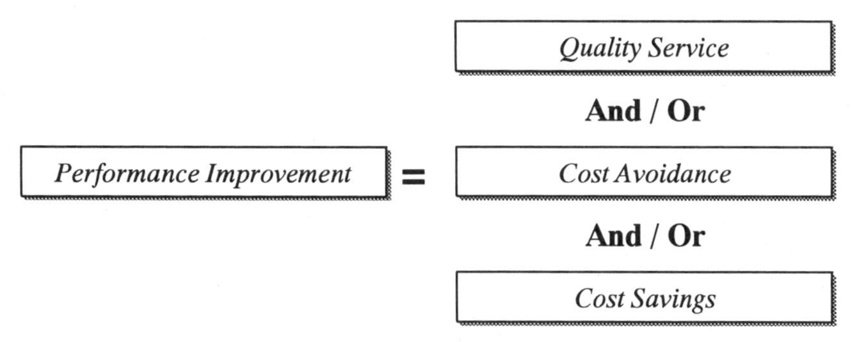

FIGURE 22.4 Derivatives of Performance Improvement.

The document lists much of the data collected for the three primary areas: demand work orders, preventive work orders, and facility projects. This information represents the CMMS data collection. The KPIs are the data blades that will help the department slice and dice the data so that the meaning is revealed when trends are dissected. A select range of KPIs has been included in the document to illustrate their power to analyze the CMMS information. The division of the KPIs into the CAFM data results in data analysis, the examination of data over time. Trends in data, with the use of benchmarks to measure improvement, reveal insights into the operation and function of the department. The result of this analysis provides management a range of options. Options supported by information will result in better business decisions. Organizations are already making use of this methodology. Figure 22.6 shows the use of the equation.

22.2.4 Creating an Institutional Memory

A CMMS will become the institutional memory for a facilities organization. The CMMS has evolved from a building engineer's tool to track mechanical equipment and to schedule preventative maintenance tasks into what CMMS vendors refer to as an enterprise asset management system (EAMS). Analysts have defined the CMMS as a "to do list." Figure 22.7 shows the amount of information that a facility manager can track as part of a CMMS to support an FM operation. Even over a period of several years, massive amounts of data can be stored in the database. The ability to report on that data in an intelligent manner, either recounting past facts and statistics or generating trends for decision making, makes the application of tremendous value, not just to the FM group but to the whole organization.

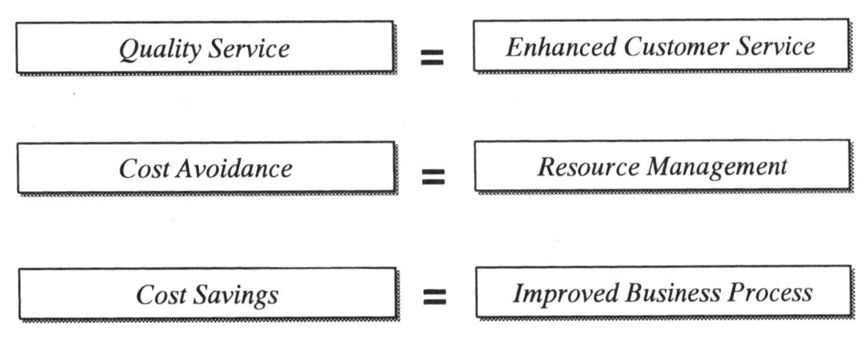

FIGURE 22.5 Outcomes of the Derivatives of Performance Improvements.

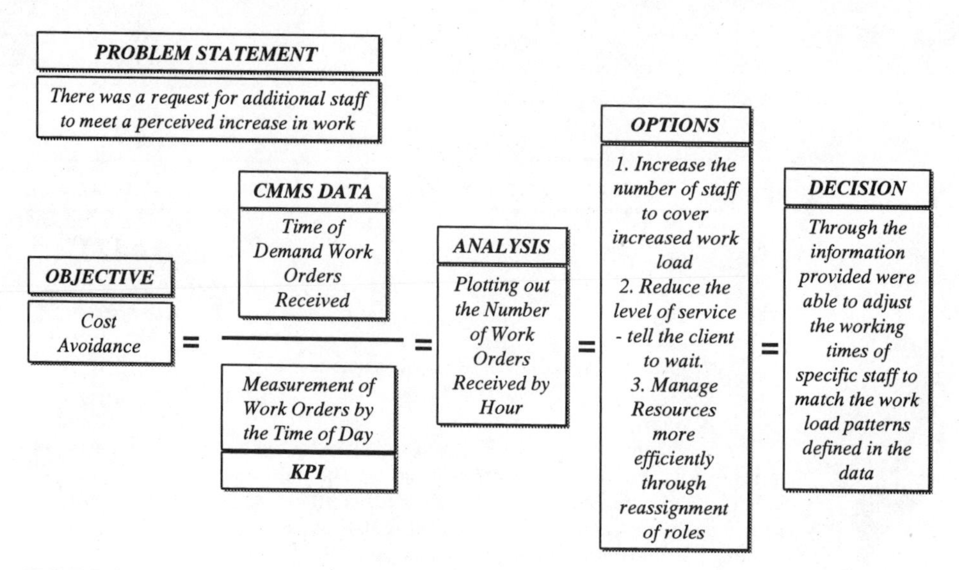

FIGURE 22.6 Cost Equation Applied to a Specific Problem.

22.2.5 Going Digital

For a correct implementation of a CMMS, the whole process should be digital. For technology to work effectively the facility manager must take its implementation to the logical conclusion. In business, this means looking at the process involved and seeing if the technology applied to it improves the process or just moves the problem to another part of the business operation. Implementing a portion of the CMMS, but not completing a logical life cycle for a piece of information, results in frustration by the users who perceive it as a duplication of effort. The digital process must minimize the number of times that information drops out of the digital realm into paper. Figure 22.8 shows the life cycle of a simple work request using two methods: a digital process and a mixed CMMS paper/digital process.

22.2.5.1 Digital Work Reception. The ability of the customer to enter service requests electronically through the Internet opens the work management scope. No longer is there a reliance on the customer service desk. The Internet extends the hours that O&M must support to 24 h a day, 7 days a week. The expansion of telecommuting combined with OSHA extending the employer's responsibility to the home office potentially expands the scope for the facility manager. As the organization becomes "virtual," is there a need for the virtual FM?[2] The ability for customers to e-mail requests into the system, fax requests, and use mobile phones or personal digital assistants (PDAs) to call in problems means that the facility manager has to start to deal with a highly mobile workforce. How will the FM department meet the increases in mobility? The answer is by going digital and mobile themselves.

22.2.5.2 Digital Dispatching. Once the customer service desk has received the request, it is logical to use digital techniques to dispatch resources. Call center software has the ability to e-mail, fax, web, and page, or uses a PDA to send a request to shops or vendors for outsource services. For example, if pest control is performed by a specialist company on call, the company can be digitally notified once the business approval process has been determined. Figure 22.9 shows facility center's call center screen used for dispatching.

22.2.5.3 Digital Scheduling of Resources. When requests are beginning to move in an instant, time is compressed and responses have to be quick and correct. The facility manager will have to

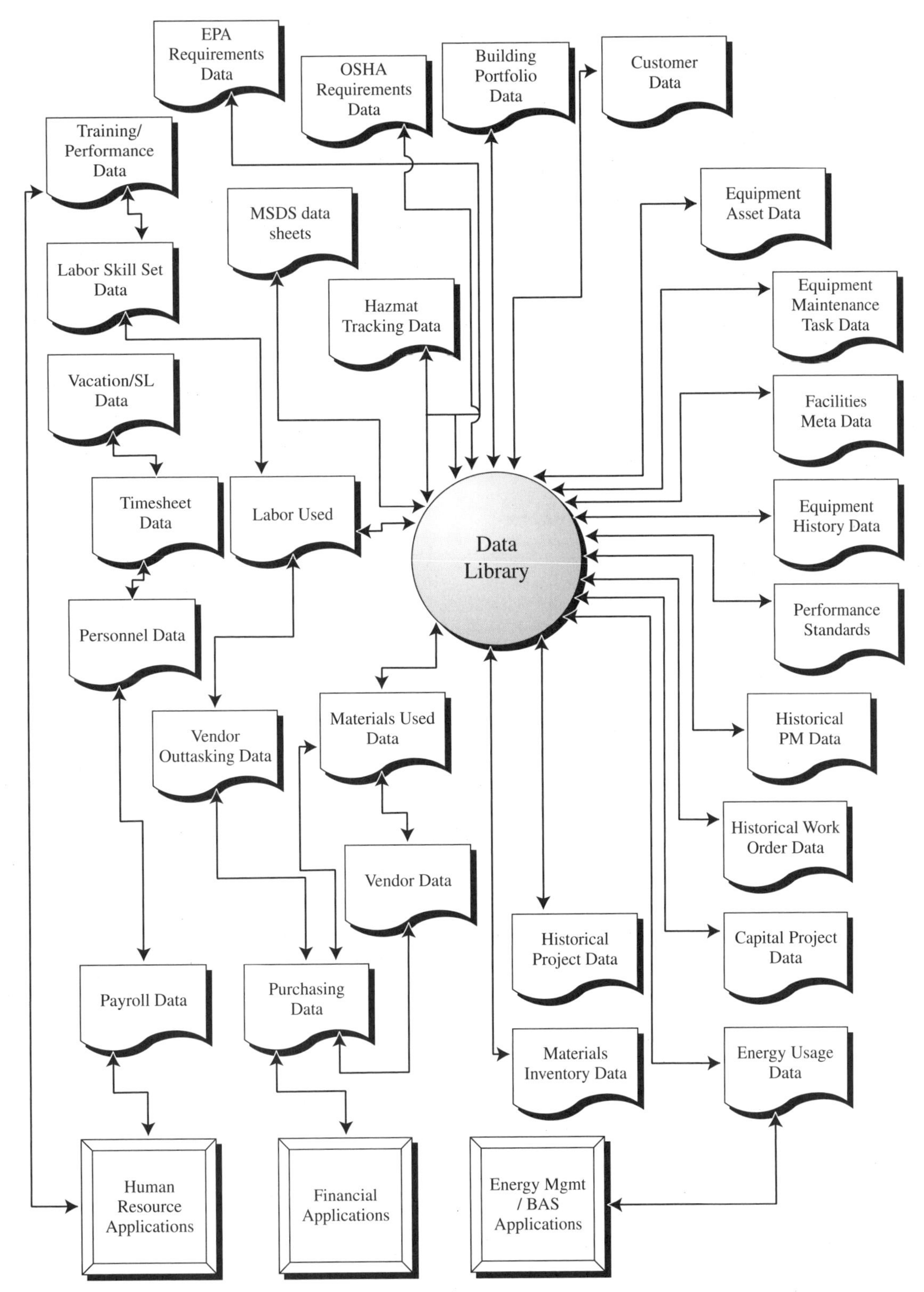

FIGURE 22.7 Knowledge Development Base for a CMMS.

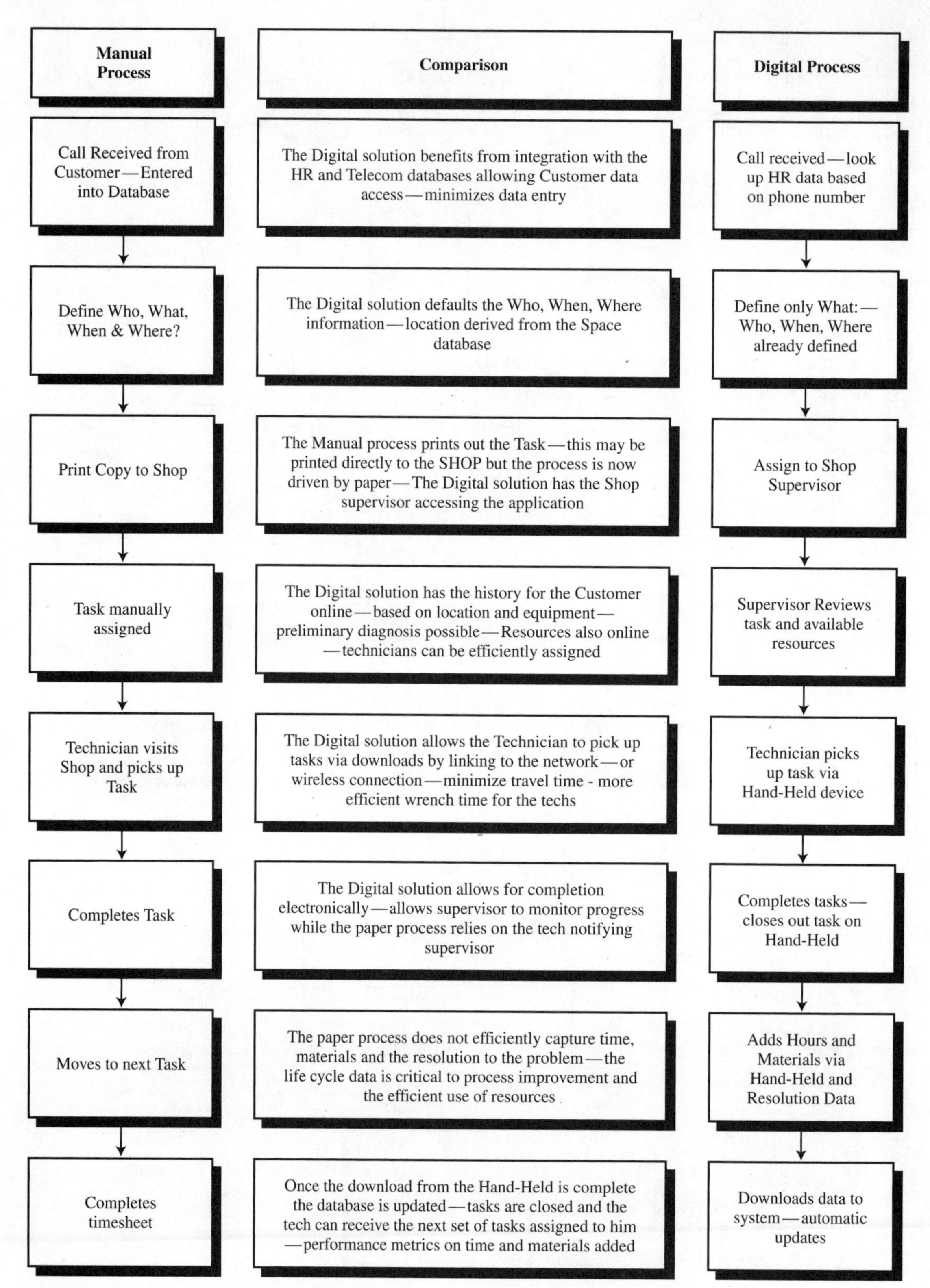

FIGURE 22.8 The Life Cycle of a Work Request Using a Manual Process vs. a Digital Process.

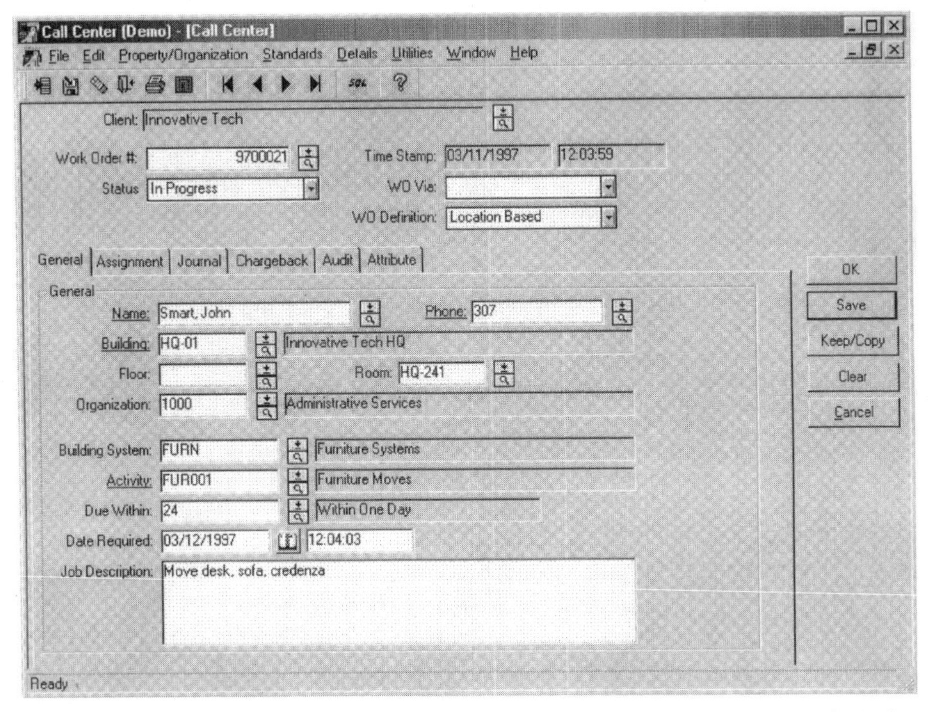

FIGURE 22.9 Facility Center Call Center Screen by Peregrine Systems. Provides a Central Location for the Customer Service Department to Initiate Work Requests and Dispatch Calls.

improve his or her ability to manage resources, specifically labor, space, and materials. In order to manage labor, the CMMS will have to track all the time for employees paid either directly or indirectly through vendor organizations. Getting a full picture of the available labor resources will allow the CMMS applications to assign resources more effectively. It will require mental effort, additional skills, and additional training. Facility managers will have to reassign some employees. However, research statistics suggest that the largest growth areas for PC users are those over sixty, predominantly the retired. Given that "being digital"[3] frees you from geographical constraints, management will demand effective scheduling of labor resources. In a similar vein, the management of warehouse inventory and materials management will demand a "just in time" management style. By scheduling tasks and projects where the engineer specifies the materials, procurement economies can maximize cost savings and minimize warehouse space requirements. Providing credit card purchases for smaller items using e-commerce sites speeds up access to the materials and provides great freedom and responsibility for the facilities employee to make a positive contribution. Space management has always been one of the drivers for CAFM technologies. Efficient use of space is important, but the reduction in space needs to balance the provision of a stimulating environment for the worker. Research has shown that reducing space standards to save on lease costs can have a hidden cost through the attrition of quality employees who do not want to surrender an efficient private office. The cost of replacing a single quality employee outweighs the space savings. Given the dynamic nature of many corporations, space management means effective move, add, and change operations for existing space. Application vendors are adding move management solutions to their work management solutions. Figure 22.10 shows the facility center move management module.

*22.2.5.4 **The Digital Web.*** While the Web can make the entire CMMS application accessible, there are a number of complementary tools that are now available through the Web to assist facility

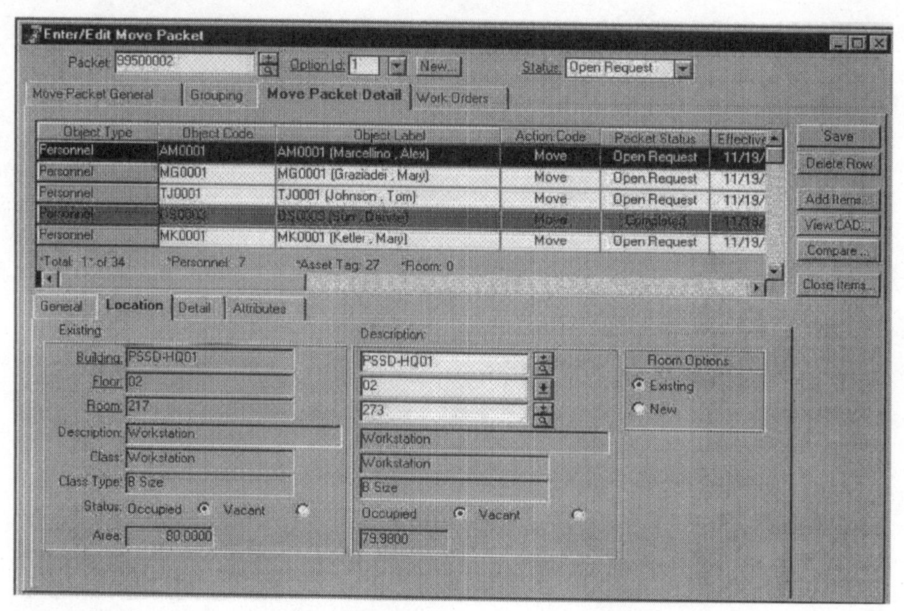

FIGURE 22.10 Facility Center Move Manager.

managers in their digital environments. The collaborative project Web site is creating a new work environment for interfacing with project teams. As the pressure to outsource tasks increases, facility managers are still responsible for ensuring a successful outcome. Therefore, they need a tool that can manage a decentralized workforce. The project Web site provides the environment for sharing information. Events can trigger notifications for a project manager. The client can redline drawings electronically. Staff can post reports and meeting minutes for comment. All these functions minimize the administrative tasks associated with manual transactions of sharing information. The project site can create the audit trail for actions. The RFI function and routing ensures that everyone knows where the buck has stopped.

The ability to access information changes the dynamics of working through more people knowing what is happening. While there is still a tendency to try to hide information, open access to facility data and the metrics associated with the data will help the performance of the facility department. Senior management will insist on open accountability. DynCorp, a multi-million dollar facilities outsourcing operation for military bases works on the open book philosophy, where the Air Force can drill down to each work order to view the labor and materials charged through the Web. The University of Rochester Facilities Department uses the Web to publish the invoices for services rendered for each of the other university departments. Each department, through a secure login, accesses its own set of accounts. It can drill down to the detail at the work order and check on who performed the work and what materials the technician used. Figure 22.11 shows the Prism Web tool.

The publication of reports and analyses of performance shows the volume of the work performed. The publication of the value of the work performed, demonstrating the effort that a facilities department puts in to provide a quality work environment, is a positive marketing tool. Communication is critical to a successful department.

22.2.5.5 Digital Timekeeping. A major expense for a facilities department is labor. A major problem for a facilities department is the shortage of labor. Organizations demand accountability. A CMMS work management tool is perfect for tracking where shops use labor resources. Some orga-

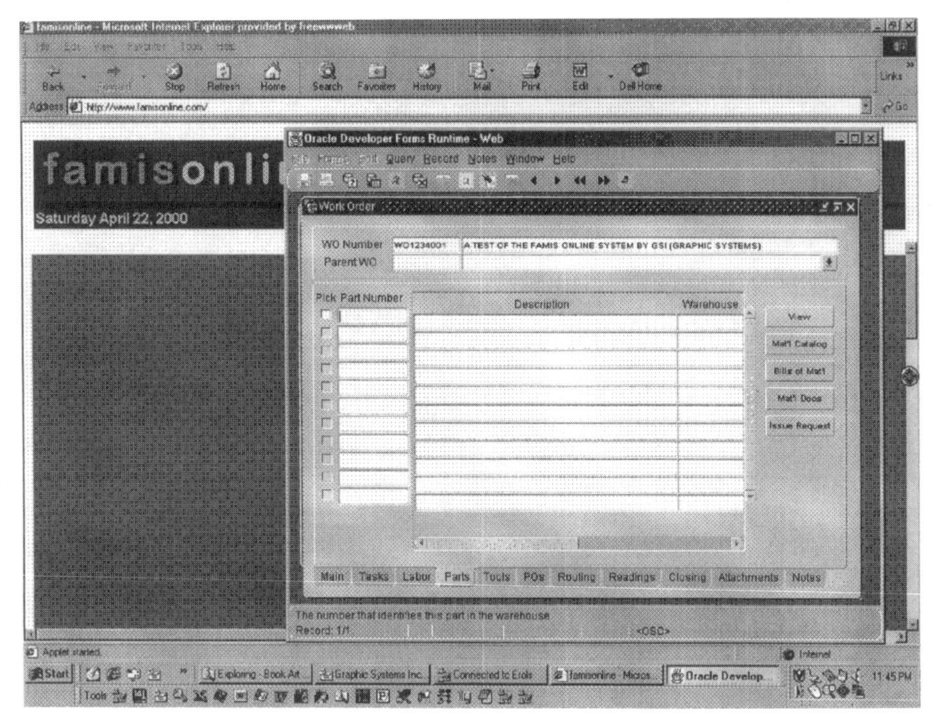

FIGURE 22.11 Prism FAMIS Online—Fully Functional Web-Based ASP Service for Work Management.

nizations, particularly in the education field, pay particular attention to where labor expenses are charged. It is possible for all employees to enter their own time. Applications provide time tracking modules of varying sophistication. The FAMIS[4] application follows the work process through staff to enter time, supervisors to check and approve online, then pass the data to a payroll application for payment. Figure 22.12 shows the labor approval screen. The advent of handheld devices using wireless technology that can receive work orders and close them out are available. Adding the functionality to add hours allows the organization to have a true digital solution.

22.2.5.6 Digital Accounting. As stated, the ability of displaying account information over the Web is one part of the financial process. The integration of debit and credit charges and the use of the financial applications chart of accounts will accelerate the digital flow to create a virtual facilities information system. For work orders to integrate the business process with that of materials management, the inventory needs to be available to the user of the CMMS. When materials are not available, the user will initiate the purchase request process. The merging of the purchase request process on the CMMS side and the purchase order on the financial side requires detailed planning. The matching of the receipts, purchase orders, and invoices requires flexibility in both applications. However, vendors have developed solutions for a number of institutions. With the increase in integration, extending from ERP applications, it will become a standard operation that is transparent to the user of these systems.

22.2.5.7 Digital Meeting Reservations. Many e-mail systems allow the scheduling of meetings. In Novell's GroupWise, it is possible to define meeting rooms as resources so that staff can check the availability. Today, it is possible to integrate the reservation of the room with the request for food and audiovisual services, the viewing of room layout options, and the notification of the par-

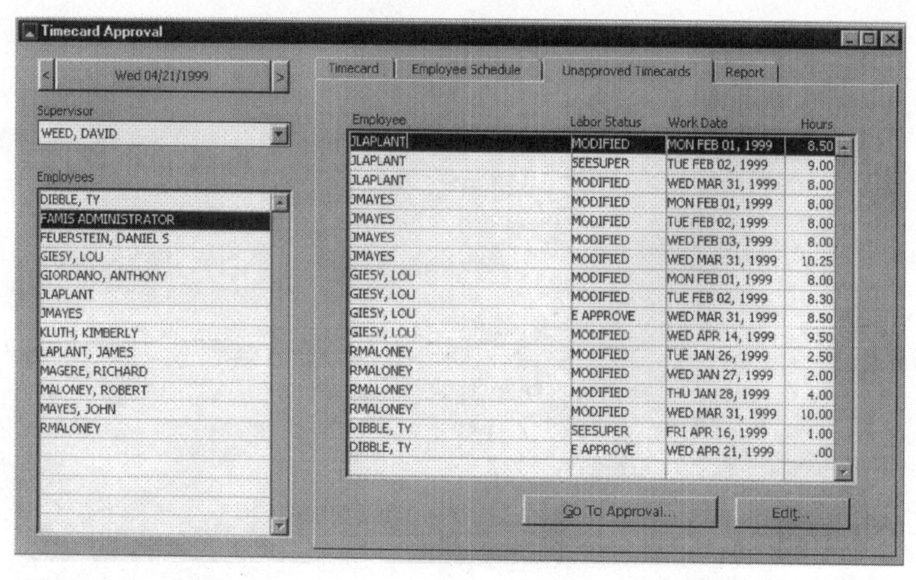

FIGURE 22.12 Prism FAMIS—Labor Approvals for Timecards.

ticipants, all over the Web. Reserve is an application integrated with facility center to provide a complete digital experience for the user. Figure 22.13 shows the reserve application allowing for an on-line solution for conference rooms and meetings.

22.2.5.8 Building Automation Systems. Many of the early BAS applications in the 1970s included rudimentary preventive maintenance applications. Today, these applications will integrate directly with the work management systems. Work orders can be automatically generated based on alarm conditions with the on-call engineer paged with the details, as important run time data on equipment is gathered. Preventive maintenance schedules are generated based on this run time data, thus reducing the number of PMs needed. Figure 22.14 shows the Honeywell system in operation at the Bangkok Bank, Thailand.

22.2.5.9 Digital Utility Billing. For a campus facility, there are multiple meters for a range of utilities such as gas, electricity, water, and sewer charges. The facility manager needs to input and allocate costs accurately. This information comes from information contained on the meter and provided by the supplier. Utility companies have the ability to use wireless technology from the meter to a central collection point. Developments by the CMMS vendors allow the facility manager to gather this data digitally. With deregulation in the utility sector, the client has the ability to make the best buys. To negotiate effectively, the facility manager must have a clear picture of where and when a facility incurs the costs down to an hourly time scale. Electrical companies are beginning to use the power lines themselves as a means of transferring digital information. Facilities organizations will be able to read their meters online.

22.2.5.10 Digital Manufacturer's Documentation. One of the largest costs in implementing a CMMS lies in the loading of the equipment data and the associated preventive maintenance tasks required. A consortium is working to propose electronic standards for manufacturer's documentation on equipment. The objective is to have the designer of the building put together an electronic operations manual that will feed into any CMMS that supports the standard. This will happen as manufacturers start to publish information on the Internet about their product. The printing industry is moving toward a digital process so the manufacturer has to get the data into

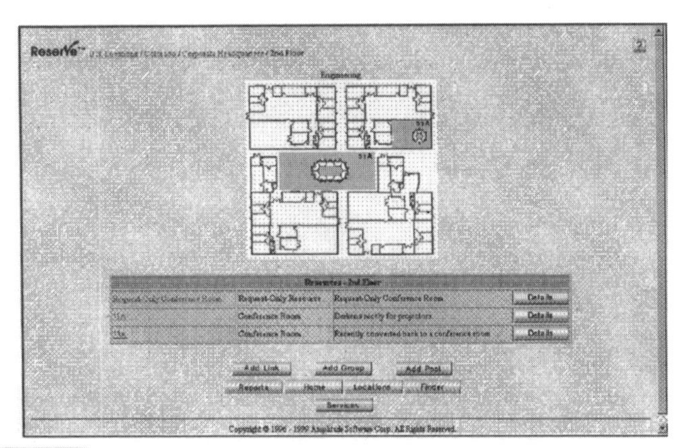

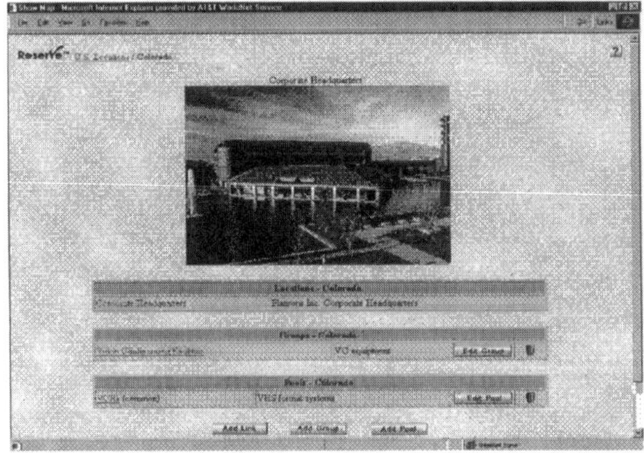

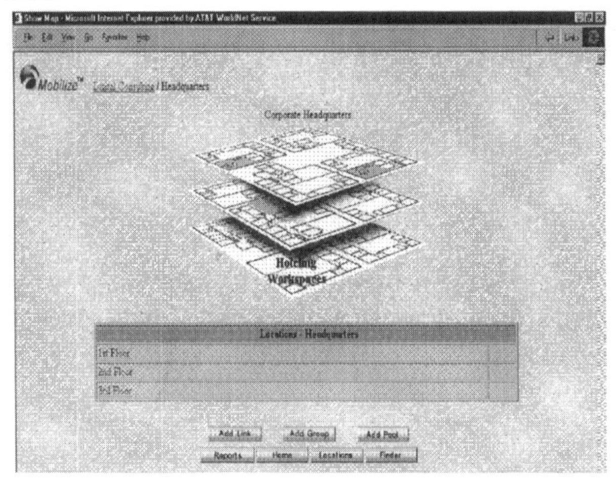

FIGURE 22.13 Reserve—Ability to Reserve Conference Rooms Online along with Additional Resources—Audiovisual Equipment, Catering.

FIGURE 22.14 Honeywell Building Automation and Energy Management Systems—Bangkok Bank, Thailand.

digital format even for paper brochures. The important element is to set a standard that both the A/E profession and the CMMS vendors will adopt. The advantage will be the ability to have a fully configured preventive maintenance schedule on startup for a new building. It will contain all the maintenance tasks, diagrams, images, and drawings necessary and document the warranties and contact data. This effort will have one of the biggest impacts on maintenance management, as few organizations have been able to invest in documentation to this degree. Only large manufacturing facilities that have a critical requirement to keep the production lines running assign the necessary budgets.

22.2.5.11 Personal Digital Assistants (PDAs) and Wireless Nets. Getting the information into the database and then maintaining the information has been a major reason for CMMS systems not achieving their full potential. It takes a considerable degree of commitment to keep the records current. Often the excuse of "I do not have the time," or "We do not have the manpower to fill in the data" are used. In some ways, it is true, as it takes a mental commitment to keep records current. How many of us can keep our home accounts current? In defining business processes, it is important to ensure that the CMMS reduces the number of tasks rather than adding a data entry function. The increased sophistication in PDAs that can receive and send work orders through a wireless net will allow the data to be captured at a logical place in the work order process. Figure 22.15 shows a Palm Pilot of PDA devices. Vendors are bringing to market the tools for downloading work orders, then modifying them based on the tasks completed and uploading the results. Building technicians will be able to add the information about what work was done, how long they took, the materials needed, and the resolution of the problem.

The future of these devices is endless. The Boeing Corporation has developed a headset to allow technicians building the plane to have the drawings and documentation appear in front of them as

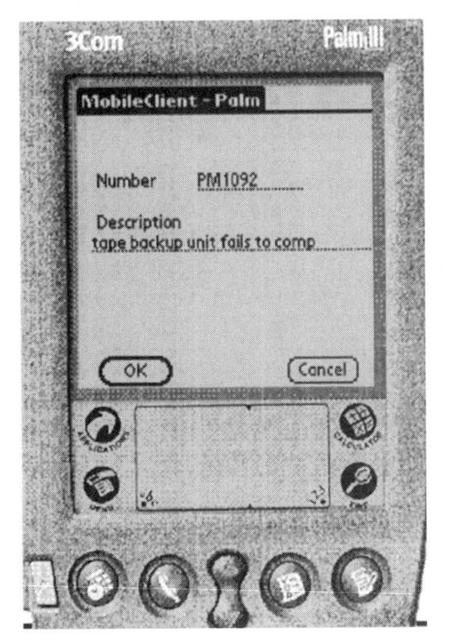

FIGURE 22.15 Palm Pilot Handheld Device for PM Work Orders.

they work with their hands free (see Figure 22.16). Voice recognition is advancing to the point where the computer can be controlled verbally, allowing for the quick input of data. The opportunities and savings that will come from a fully loaded data model of the facility are nearby just waiting for the organization with the commitment to make it happen.

22.2.5.12 ***Barcoding.*** Everyone recognizes barcoding, but few organizations have successfully integrated it into the business process. Major applications for the CMMS relate to materials management and inventory control. Organizations use barcoding to track assets. The U.S. Senate, for example, uses barcoding to track over 100,000 furniture assets. Many pieces have historical significance based on who used the furniture. The Architect of the Capitol case study (see Chapter 28) illustrates the integration of assets with the business process. Iowa State has taken the integration of barcoding and business process a step further. In their key control module, the user can request a key through a Web interface. Designated security staff approve requests online. The work ticket then prints out on a tagged card with a barcode containing the necessary data to make the key. The key is cut and attached to the barcoded tag and sent up to the key control reception desk. When the person comes to collect the key(s), the receptionist scans the request with a light pen and the barcoded key to ensure that the data matches and then electronically issues the key. The entire process is digital.

22.2.5.13 ***Natural Language Voice Recognition.*** The impact of voice recognition in the CMMS process will have two key effects: it will make it easier to enter data and to update information. Voice recognition will also free the hands to manage other tasks. With security access moving to voice profiles in terms of biometrics, think of the customer service desk knowing who is calling based on their voice imprint stored by the corporation. Advantages include:

- The flexibility in defining user access based on voice when entering data into the system
- The ability to dictate the description problem for a work order

FIGURE 22.16 Xybernaut Headset.

- The ability for building engineers to dictate the work performed and the time they took
- To support this level of interaction, the vendors will have to enhance the user interfaces beyond the data centric forms to an event driven model based on the business processes.

22.2.6 Web-Based Solutions

Being digital and making information active means finding new ways of talking to your customer base. The Internet in its various guises has provided an exciting opportunity for customer and service operations to interact. The facility manager can use the Web as a dynamic tool, interacting with the CMMS to provide updated information. The customer must be able to submit requests through the browser, check the status of requests, and review the charges for that request when complete. Using this method, there is little need to print out the work order or to send out invoices. The FM process can play its part of a true digital nervous system for the overall organization. All the major players in the CMMS marketplace are developing Web solutions. Figure 22.17 illustrates a number of vendor approaches. Prism Computer Corporation's product, FAMIS, is a pure Oracle product and the application compels the forms to run using Oracle 8I's Application Web Server. Prism is the first to achieve a production version of the product that purely uses Web technology. Vendors are taking a different path to the same objective, to have a fully functional version of their product available over the Web. Prism hosts the entire FAMIS application as an application service provider (ASP) for a chain of exclusive retirement homes over the Internet.

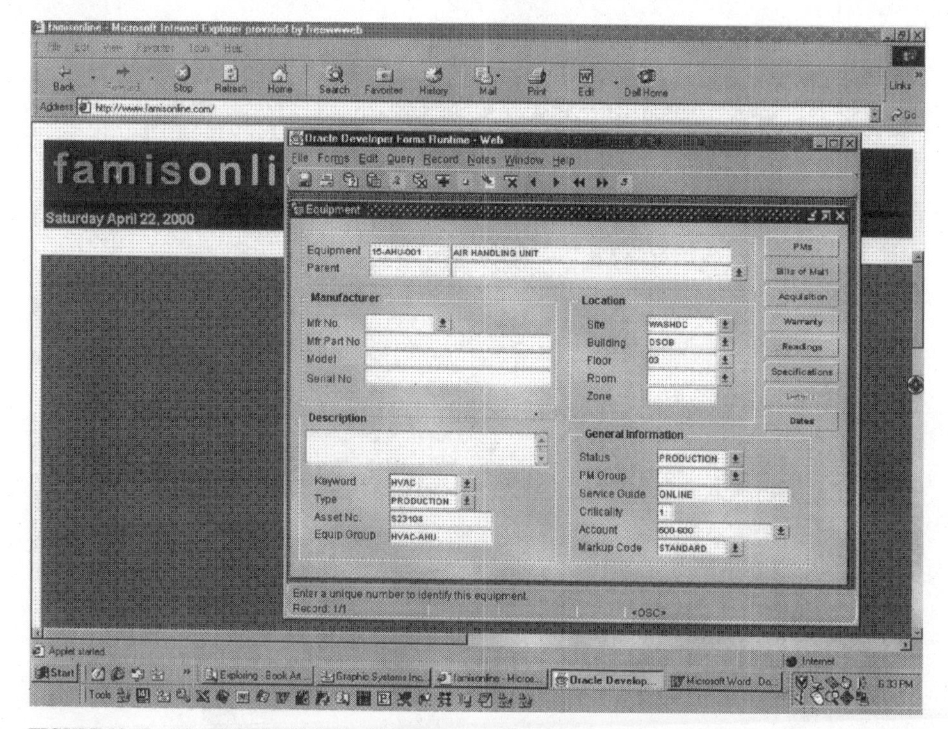

FIGURE 22.17 Prism CC FAMIS Online Fully Functional Web-Based Solution for Work Management Based on Oracle Tools.

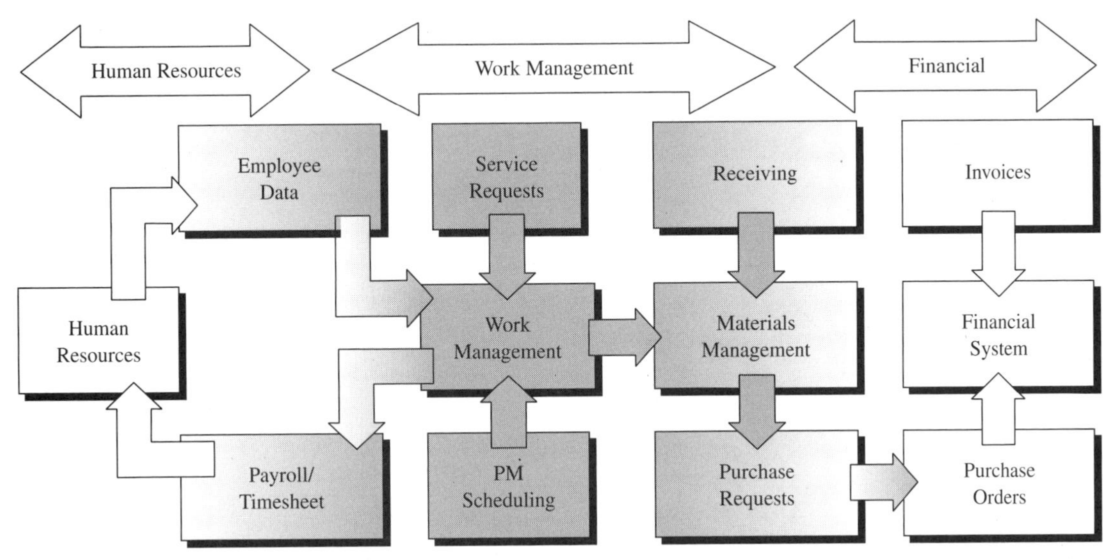

FIGURE 22.18 Relationship of the Applications and How Control Shifts from One Functional Business Area to Another Creating Gray Areas.

22.2.7 What Features Does a CMMS Provide?

Most CMMS applications come in a modular form. Users purchase the core product then add on modules that are needed or that can be phased into the FM group. The CMMS has evolved into an enterprise asset management (EAM) system. To reflect that change the vendors have expanded outside the old focus on maintaining equipment. Figure 22.19 shows the potential model for an EAM application. The spectrum of functions (Figure 22.18) for the leading vendors puts the onus on the facility manager to select the most appropriate package for their needs. In examining what features the vendor provides, the facility manager must also examine the marketplace that the vendor serves and has evolved from in developing the product. Maximo has a strong following in the manufacturing industry where their products have the ability to create intelligent hierarchies out of the assets. Building into interlinked systems provides much of the core requirements for a large manufacturing plant where the interrelationships of the many building systems assets and processes must be understood. Prism and AssetWorks have a strong following in the educational field where their products support the detailed financial tracking and reporting requirements for state and federal agencies.

In examining the features of a CMMS, it quickly becomes apparent that these systems integrate with other information sources within an organization. In configuring the CMMS, data is required relating to people, accounts, and materials. Depending on the overall business processes for an organization, the data may reside in many different locations and have different owners. One of the important tasks during implementation is to ensure that all parties involved accept the information map. This effort moves the CMMS from the technical world into the political world where the facility manager has to reach agreements for real progress to be made.

22.2.8 Cost Savings

The facility manager views the cost savings from a CMMS project at a holistic level. A well-implemented CMMS will save money both in the efficient use of resources, but more critically in the extension of the life of the organization's assets through correct maintenance. Initially, an FM group will not see a reduction in workload, rather they will see an increase. The increase is because the

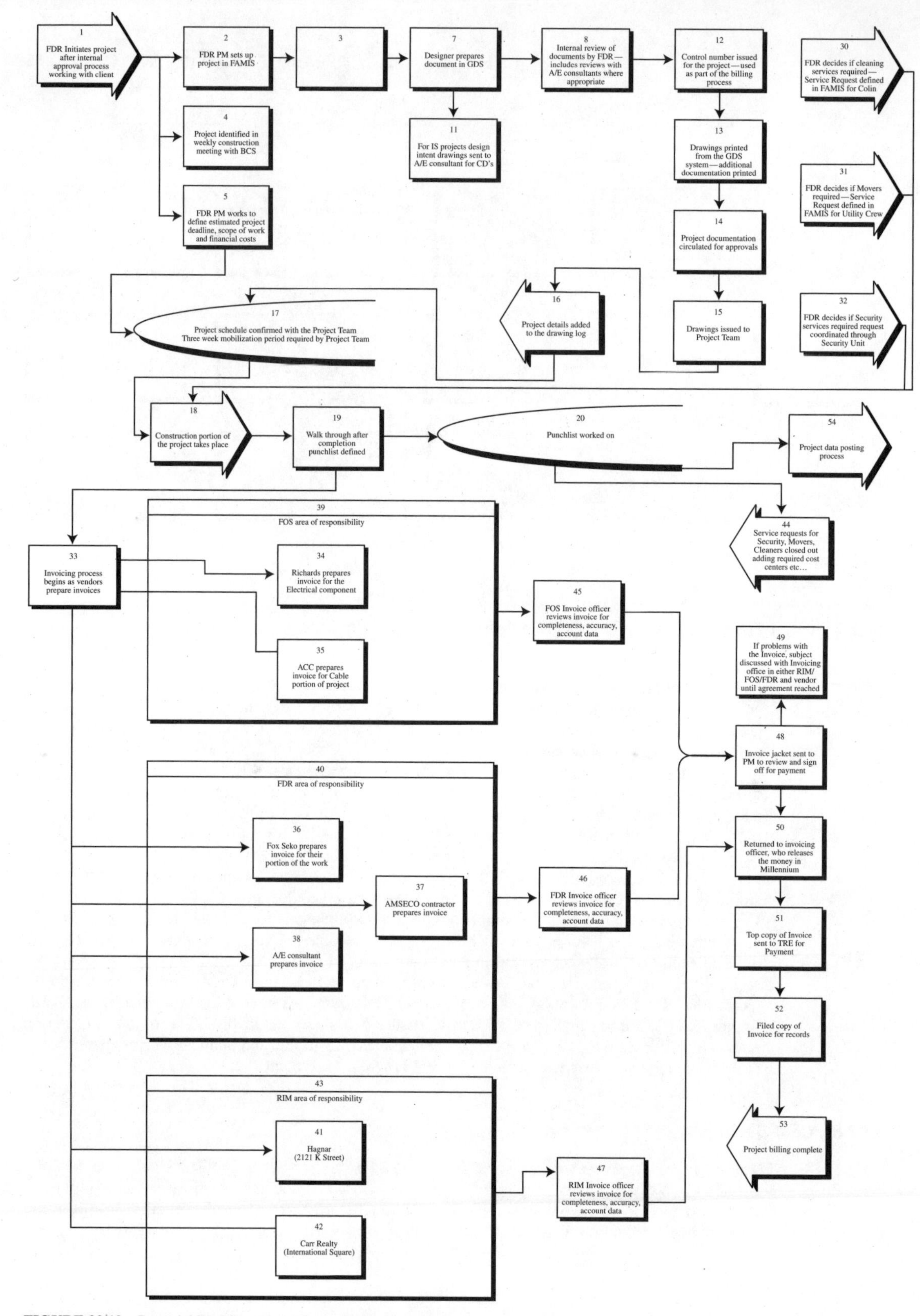

FIGURE 22.19 Potential Workflow Model for an EAM Application.

CMMS acts as the institution's memory and it does not forget tasks. When using a manual system or a partially implemented CMMS, staff can drop older tasks or less visible tasks. Often the less visible tasks relate to the PM of equipment, resulting in an increase in deferred maintenance. Staff does not have the time to perform the work in accordance with the industry standards. How can the facility manager generate cost savings for a CMMS?

The facility manager will have to be careful in managing expectations for "cost savings." The exact amount of savings generated will vary from organization to organization. A poorly managed organization that aggressively implements a CMMS for the first time may see significant savings, while a well-organized manual facility group may see little initial improvement in performance. The major cornerstone for a CMMS is accountability. The facility manager can account for the resources consumed by the department, whether they be people, materials, vendors, or suppliers. Whether the facility manager has outsourced the work or has it performed in-house, if the staff tracks all time and materials within the CMMS for the department, it is possible to account fully for these resources. Minimizing downtime can save time, more efficiently assigning resources to preventive tasks that can prolong the life of equipment rather than emergency work.

Developing a cost methodology for an organization is a critical component of the implementation plan. The cost benefit ROI for initially moving to a CMMS can easily be generated by tracking the savings in time communicating tasks to staff or by building complex maintenance schedules. The use of the CMMS as a tool in the O&M world to help measure performance and provide a tool to define options for management that will drive business decisions. The better the business decisions, the greater the improvement in performance, whether defined by cost savings, cost avoidance, or increased quality of service.

22.2.9 Customer Service

Business has recognized that to survive it must satisfy its customers. FM is customer oriented. To determine whether your FM group is improving, you must define key performance indicators tied to customer service. By developing the CMMS, you build a data warehouse that allows you to mine for patterns of service. What are examples of customer service that the facility manager can measure? How quickly do you respond to a customer's request for service? Do you complete the work on the first visit? How do you set your customer's expectations for the delivery of service? How easy is it for the customer to request service? How easily can the customer monitor the progress of the request? How easy is it for the customer to see the cost of the service? How can the customer provide comments on the performance of the service? The CMMS stores the answers to all of these questions.

AssetWorks and Prism allow for the automatic generation of quality control e-mails. AssetWorks allows the facility departments to custom tailor the form letter to the specific service provided. The Web is changing the way customers interact with a facility department. The facility manager can use the Web to receive requests from users. This has been the first footstep for many vendors, such as Peregrine, AssetWorks, ARCHIBUS, or Prism. Further facility groups are developing sophisticated Web sites for clients to allow for one-stop shopping of services. Universities have developed some of the best sites for facilities organizations.

22.2.10 Integration

One of the important messages in the industry today is integration. The CMMS does not stand alone—it requires information from other applications and can feed information back to other applications. The diagram shows flows of information in the digital nervous system[5] that are related to FM. The process of integrating the whole process for CMMS requires the FM group to coordinate with other groups to ensure a quick clean flow of data. Figure 22.18 shows the applications and who controls them in a typical business. Today's Web technology shows that the clear boundaries of ownership are disappearing.

"Digital information flow makes it possible for a company to create a boundaryless organization, but it takes a new corporate mind-set and culture to turn suppliers (of information) from 'them' into 'us'."[6]

The value from a CMMS comes through its integration with other systems. The object is to create a "value network"[7] within the organization allowing the digital flow of information between facility manager's customers and suppliers. This organization cannot create this network overnight and therefore the planning and implementation for an enterprise CMMS requires up-front planning to develop a detailed implementation plan that spans multiple phases and fiscal years.

22.2.11 Implementation

The effective planning of the CMMS project is the first step in a successful implementation. A checklist of tasks that facility managers need to consider shows the areas that they must address before starting the actual work. Today, the need for a CMMS is mandatory for success. A focused FM group will already have defined its overall strategic goals and having recognized the importance of the CMMS, developed budgets to allow for implementation. One of the first steps in implementation is to develop the project team. Who should be on the project team? What roles should they play? These initial questions go to the heart of a successful implementation.

There are a number of key components that a facility department must face when building a CMMS implementation. Many consulting organizations have developed methodologies around the implementation of technology. Whatever the flavor, they encompass some basic truths. An organization must decide why they are embarking on the exercise that may shake the facility group to its foundations. What will be the vision for the facility department once the CMMS is running? Many books have been written about how to define the vision, but before the vision is reached, it must be documented. Parallel to the vision is the goals for the CMMS. These are more tangible than the "touchy feely" vision. When defining goals, the facility department must know when it has achieved the goals; therefore, a means of knowing when to stop must be included.

In this culture of instant gratification, overlooking the analysis of where an organization stands causes business issues later. Some methodologies even scorn the need to study the past, as their solutions will wipe away all traces so why waste time? Given the record of accomplishment of some of these fast-track approaches, the hare and the tortoise fable is appropriate. How should we document the track record for an existing department? The pulse of an organization is felt through the electrons flowing through its processes. Defining the functions and the processes that carry out the functions is critical to understanding how an operation works. If you are proposing to implement new technology in this department, you must predict how it will change the way you do your work today. If you want to make improvements in how you work and not automate manual processes or poorly conceived automated processes that are no longer relevant, defining the as-is process is critical to obtaining a full understanding of how you operate. With tools to create models of your organization and to allow for the simulation of these processes, it does not have to be a lengthy effort (see Figure 22.19). In looking at the as-is environment you can see what is working and try to preserve this part. You will see where the problems lie in the process and try to use the technology to streamline these bottlenecks. You will also define the obstacles for the implementation, which may be political, financial, or logistical. By knowing your "enemy," you can develop solutions and build alliances to achieve your goals.

Once you have defined your vision, goals, and as-is documents, you will have the basis for your requirement document. With this in hand, you will be prepared to look at the range of systems on the market and select the system that blends in with your organization's culture. You must understand that no system will provide you with 100% of your requirements and that you must be prepared to compromise. All the leading vendors provide many of the features; however, the ability to select a system that matches the culture of your organization is more important than seeing which system has the most features. Many organizations believe themselves to be unique when in fact they are not. Studies performed by McKinsey showed that all organizations perform the same basic functions for their business category, such as manufacturing, service, or financial. The cost of customizing a CMMS to meet quirks in the facility group has both short-term and long-term implications. Many organizations start with the goal of not customizing, but end up doing so as the pressure mounts.

The following rules can be useful in determining whether to customize and how to achieve that goal with the minimum of risk. By understanding how the facility department performs business, the facility group is able to see where processes no longer meet the requirements of the client or where there are more efficient methods of doing work. With that in mind, the first rule, when faced with a conflict between the way the application works and the business process of the organization, is to ask the question: "Can the business process be improved by changing it to support the applications logic?" Often the answer will be "Yes!" If after close examination the application falls short of what is required, look at what will be coming in the next release. The next release should be one that is only a few months away, not vaporware. Most vendors have a 12–18 month development cycle for major releases. Given the implementation schedule may be 9–12 months, the chances of falling in with a major release are possible. In fact, experience has shown that the application version you may start the project with may not be the version you live with!

If the answer to your business process is not in place, the next step will be to discuss your needs with the vendor and consultant. They may have come across your situation and can recommend how other groups have solved the problem. If the feature you are looking for makes good business sense and enhances the product, the vendor may agree to incorporate it in the next release. Many vendors rely on the expertise of their facilities customers to generate new features in their software. If the vendor commits to the applet, then continue to work with the vendor to make sure it is delivered. Remember that you still may have to compromise to make the applet generic and applicable to the entire vendor's customer base. While this step may work, often the vendor will want to be paid for the development work. If that is the case, this is often the decisive test for whether you really need the functionality. You have to present the case to your management to pay more money. If you are prepared to invest in the applet, make sure that the vendor will include it as part of the core product and not drop it at the next release. If all else fails, you may have to decide whether it's worth the risk and the cost of having the applet completed without any assurances from the vendor. The decision for many organizations to select computerised off the shelf (COTS) solutions is to minimize the costs associated with upgrades. Further, once you let the genie out of the bottle for one special interest group, you find it difficult to turn down the next group who wants to have a custom feature. It takes a strong implementation team to hold firm on these requests. It requires that the goals are well defined and adhered to throughout the project.

The establishment of "standards" is a major effort that the FM department must complete successfully when implementing a CMMS project. The application defines standards as the values used in the "look up" or "validation" tables. While this effort may appear straightforward, these standards will embody the business rules of the organization. Take a simple example of the typical work order: There is a field showing the status of the work order—Open, On-hold, Approved, Pending, Complete, or Closed. In establishing the standards, the implementation team must define what each of these statuses means to the business process. What is the difference between "Complete" and "Closed" for a work order? Depending on the facility department, the answer may be that when the building engineer has performed the work successfully, the work may be deemed complete. However, there are a number of closeout tasks that technicians must perform before they can say the work is truly finished. The technician has to charge labor and materials against the work order, and details on the resolution of the work performed must be added to the work order. Where a vendor was involved, the invoice may need to be reconciled against the work order. All of this takes time. In order to monitor performance statistics, there will be several stages of closure to a work order. Once everything has been charged to that work order, the status of "Closed" can prevent any more charges from being made against that work order. Some applications will allow you to embed business rules into the application through the control of these statuses. AssetWorks has built in several layers of sophistication where the facility department can build workflow into their processes. PSDI's Maximo takes the sophistication a step further by having a true workflow tool embedded into their application.

A CMMS will change the character of an FM group from a fragmented collection of independent groups loosely managed by staff who do not have the information to make well-informed strategic decisions and develop adequately supported fiscal budgets. The CMMS will allow for the empowerment of the people doing the work. It will result in direct accountability of staff and decentralize responsibility, but provide sophisticated tools to allow mangers to support the FM efforts. An essential part in the implementation of the CMMS will therefore be "change management."

22.2.12 Change Management

The problem with change management, reengineering, process improvement, and other philosophies is that too many have failed for a variety of reasons. On paper, they look good, bringing a fresh perspective to a problem, but too often they are seen as a silver bullet. You have to recognize that any project will require hard work and the ability to stick to the plan in the face of adversity. "Starting over a decade ago with approaches such as management by objectives (MBO) and quality circles, change management fads have deluged companies until many people are numb to them. Executives try an approach until it doesn't deliver results quickly enough and then switch to the next approach."[8] "Employees take an attitude to 'wait and see' if management is serious this time, and that undermines the credibility of the change effort."[9] "The change survivors: are cynical people who have learned how to beat the system by waiting out the change movement without changing at all."[10]

The right approach to implementation will help avoid the problems of change for the sake of change. The team should involve as many people touched by the project as possible. The organizational chart is based on standard change philosophy focused on maximizing the value of the new technology while recognizing the impact of the change on people and processes. The focus of the process is how to improve the process through the implementation of technology by getting the people who will use the technology to develop the new process and therefore take ownership of the project. They will be accountable for the success of the project and have to overcome the obstacles and put in the effort to learn the new system. A new system has to be learned by staff. There is no magic potion—it takes time and effort. Management must provide the time for the staff to put in the effort.

22.3 CMMS IMPLEMENTATION FOR OPERATIONS AND MAINTENANCE

When implementing a CMMS, two changes will happen: the CMMS introduces new technology to the FM department and this technology will bring new ways for performing the work. The section of this chapter entitled "Going Digital," illustrates how far technology can take the O&M group. The new technology and its additional functionality will bring with it the opportunity for changing the business process. Managing the people in this change process is critical to the success of the system. The implementation of the CMMS requires two approaches that must be blended together to ensure that the technology and the business process support each other. Following on David Cott's (co-author of the *Facilities Management Handbook*) observations that the design of the information management system must support the facility plan and allow staff to gather information around the budget areas, the CMMS must complement this concept. Existing business processes and new opportunities presented by the technology will compete for attention. The facility manager needs to reconcile these two issues. Likewise, in developing an overall strategy for the implementation of the CMMS, the implementation plan has to address these two issues.

The project plan tackles the technology implementation, ensuring that all of the standards are defined, data is migrated or gathered, software and hardware is correctly installed, and training takes place. The project plan provides the overall project management structure, focusing on budget, scope, and schedule. The project methodology tackles the organizational issues of how to get the most from the new technology and enhancing the business process while building the acceptance of the users who may be adverse to change. Merely taking a cookie cutter approach to the implementation of the CMMS may get it installed in record time, but it will not ensure that it successfully meets the business objectives for the FM department. The facilities department tailors the project plan and the project methodology to meet the specifics of each implementation; however, a basic framework is common to all organizations. The involvement of the O&M staff is critical to the success of the organization because the staff has seen many management philosophies trying to improve the way they perform work, from total quality management, business reengineering, or quality circles. The project methodology must gain the "buy in" of all of the O&M staff for the CMMS to work successfully. The following section examines the features of a CMMS application and focuses on the relevance to the O&M team.

22.4 THE ANATOMY OF THE CMMS

This section looks at how a typical CMMS gathers information through the various work management modules to accomplish the business tasks of an organization. At the heart of the program is the work order. Figure 22.8 shows the business life cycle for a work order. The hypothesis presented shows how a CMMS application can provide a management tool that will track all of the work and the resources for a facility department. Many educational organizations are using an "enterprise CMMS" application to track all the hours of their staff and all of their material resources. Every work order has a credit and debit account assigned for labor and materials that feeds back into the financial applications.

22.4.1 The Work Order Business Life Cycle

A CMMS has been described as a simple "to do list." This concept provides the basic anchor for understanding the work order (WO). The WO is at the heart of any system, as it records information about the request made by the customer and provides the history of the request. There are a number of phases that the facility department must address in the WO's life cycle. How well the application meets these requirements, how flexible it is and how flexible the facilities department is, will determine the smooth running of the application.

22.4.1.1 Stage 1: Call Received. For a demand WO, the first phase is the initial request for service (a later section will deal with the preventive WO). This is the first interface between the business process and the application. How many different ways can the user create a request in the application? The latest additions to the list are the Web input and the handheld devices. Does your application support all these sources of input? Does your business process support all these sources of input? During the implementation phase, the project team will address these questions and define the business processes. Figure 22.20 illustrates one type of input form for the user which focuses on collecting data from the customer about their request. In the simplest scenario, the customer will call the facility customer service desk. This call will initiate a request.

Some applications provide for a two-phase system, where phase one is the request for service. The business rule will allow the customer to make any request however outlandish; there is no filtering applied. This minimizes the decision making on the front line; however, this step does not create a work order, merely a request. Nothing appears at the shop level. Although this phase may seem straightforward, there are subtleties that differentiate applications. Often the person calling the service center is not the person who needs the service. For example, on Capitol Hill, rarely will a senator or a congressman call when a light needs replacing. Can the application support the concept of the "contact person" as well as the person who needs the service? Can the application automatically populate the required fields for each of these people? The level of detail captured during the initial call will help determine where the facility department is allocating its resources. It will help determine if there are trends emerging for specific buildings, zones, or departments. Can the

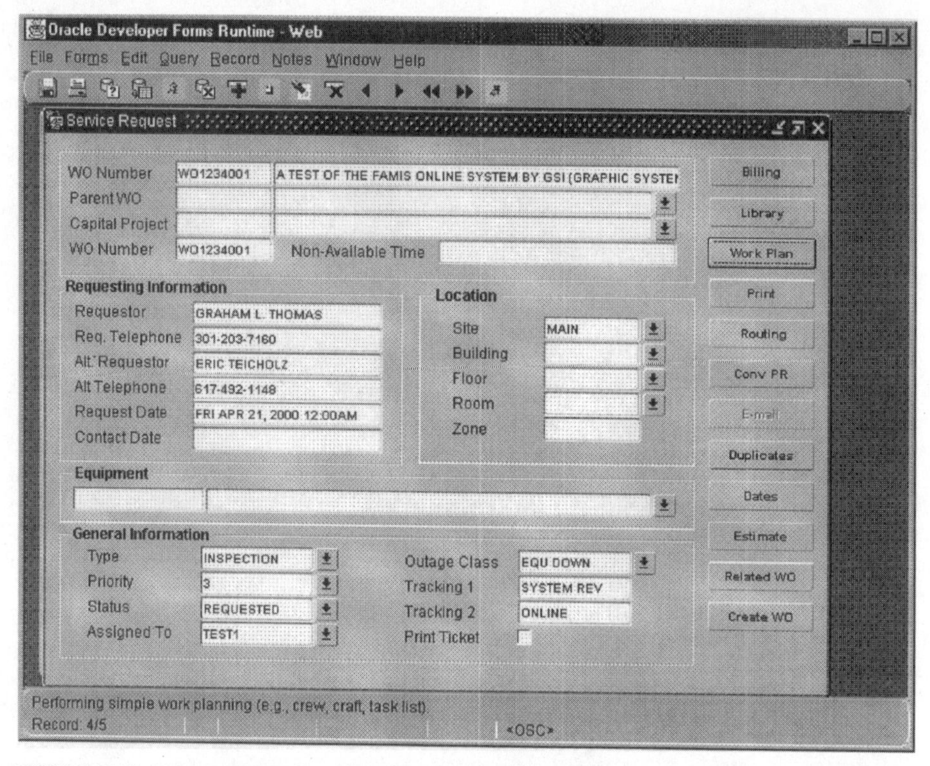

FIGURE 22.20 Prism FAMIS On-Line Web-Based Service Request Form.

application track the organization and contact information for the customer calling? Can it track the location of the customer and the location of the problem? The supporting databases directly relate to the effectiveness of the "call center." Personnel provides organizational data; telecommunications provides telephone, cell, and fax; the IT group provides e-mail; and the space management application can provide the location data. An overall "administrative support information system" strategy for an organization will ensure that these data sources share "common database keys" allowing data to flow freely.

Advances in technology allow data to get into the CMMS through new gateways. Applications have started to embrace the Web as a means to accepting customer requests. Major vendors, such as Peregrine, AssetWorks, PSDI (Maximo), and Prism have deployed Web solutions. Figure 22.21 shows a sample Web screen from FAMIS. The first generation was a simple form to capture data, log it into the database, and track the status of the request. The next generation allows for a fully functional application running through the Web.

The use of a Web requester application will reduce the volume of calls to a customer service desk. If the customer is starting to input requests through the Web, how will the facility department react to the request? Can the application notify the customer service center that Web requests are waiting to be processed? How will the facility department manage the business process work flow? If the facility department rejects the request, how will the facility department transmit that news to the customer? Technology can improve the flow of information and make it less personal; however, if the goal of the facility department is to improve the level of customer service, there has to be the appropriate level of interaction. Talk to your customer!

There are a number of business models for a call center. The simplest model is to take the call, gather the basic contract information, list the problem, and then move on to the next call. This

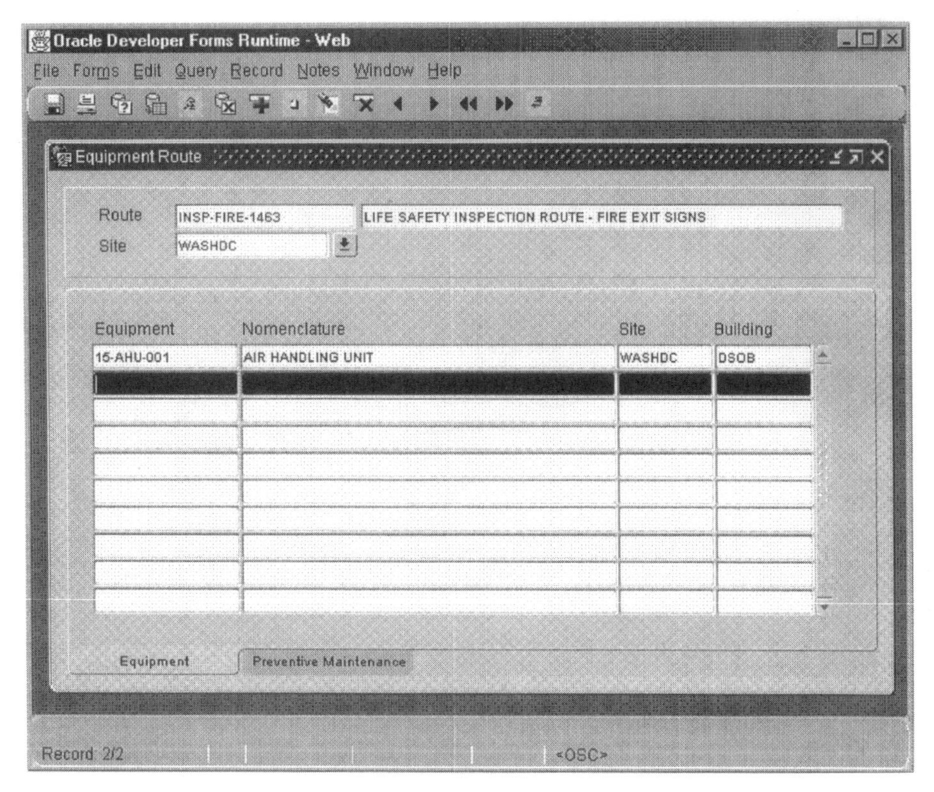

FIGURE 22.21 Prism CC FAMIS Online—Equipment Routes.

model can work well where there is a high volume of calls. The telephone operators need a minimal background in O&M. The facility department can provide a checklist of types of calls. It is possible to outsource this data collection task. The operator makes no decision on the merits of the call. It is important to log all calls, as this will define the total volume of work. Each call takes time. Often saying "No" to a customer takes longer than saying "Yes." In analyzing workload statistics, a department often misses this segment. In some operations, an employee with a facilities background manages the customer service desk. If this is the case, then some of the assignment of the work can be done at the receiving level.

22.4.1.2 Stage 2: Assignment of the Work. The next state in the life of a work order is determining what happens to the request—it may never turn into a work order if the request does not meet any of the business rules established. The customer may be asking for furniture outside of the corporate standard. The request may be a duplicate of an earlier call. Often, if the air conditioning fails for a zone, the service center may receive multiple calls. Can the application recognize duplicate calls and tie them back to a single work order? The data standards for the status field of the work order become critical in managing the work. Embodied in the syntax are the business rules. Defining and understanding what each status means and how it affects the business process is critical to a well-implemented CMMS. The application configuration tables hold the values for the status field. The project team enters these values during the implementation phase for the project. There is a description field in this table. Often the project team leaves the field blank or they repeat the value. If the project team does its job, the information in the description field needs to encapsulate the business rule defined. What does an open work order mean? Has the work order been "assigned," "approved," and "scheduled?" CMMS applications are starting to recognize the importance of work flow. PSDI (Maximo) has embedded a work flow engine into their product to support the enforcement of the business rules.

It takes a certain amount of skill to make the correct assignment of a WO to a shop. In a well-implemented system, the user is presented with a complete list of activity codes. This list represents all the types of work that a facility department can perform. Can the application assign the appropriate shop based on the activity? Can the user override the assignment? The ease of routing of the work order and its support for the business process is important to the effective management of the work of the department. Any request may change and grow in complexity. A simple leak may become a major repair project. How can the application track the history of a work order? The AssetWorks (Figure 22.22) model of multiple phases to a work order provides a sophisticated tool allowing the scope of the work order to grow without the need to create additional "numbers."

FAMIS (Prism) (Figure 22.23) allows secondary labor to be assigned to a work order for additional tasks. Prism also allows you to turn a work order into a project and link work orders together.

Facility center (Peregrine) allows you to create a "facility project" (Figure 22.24) then link a number of existing work orders together to achieve a grouping of tasks.

Facility center's move manager (Figure 22.25) uses a hierarchy concept of parent–child to build complexity into the work order. Different applications use different approaches to achieve a similar result.

Whatever the application or the service center model, the facility department has to take the request and assign it to a shop to either perform or investigate the work. Some requests do not fall into any predefined business process based on the information provided, so more research may be needed. Some requests may need special approvals based on financial limits. Can the application provide for controls on approvals? AssetWorks has developed an elegant approval mechanism tied into the user login and status for the work order. For certain types of approval (user-defined), only the designated user can move the work order forward. Once they do move the work order forward, the work order history shows that they have approved the expenditure of funds or authorized the work.

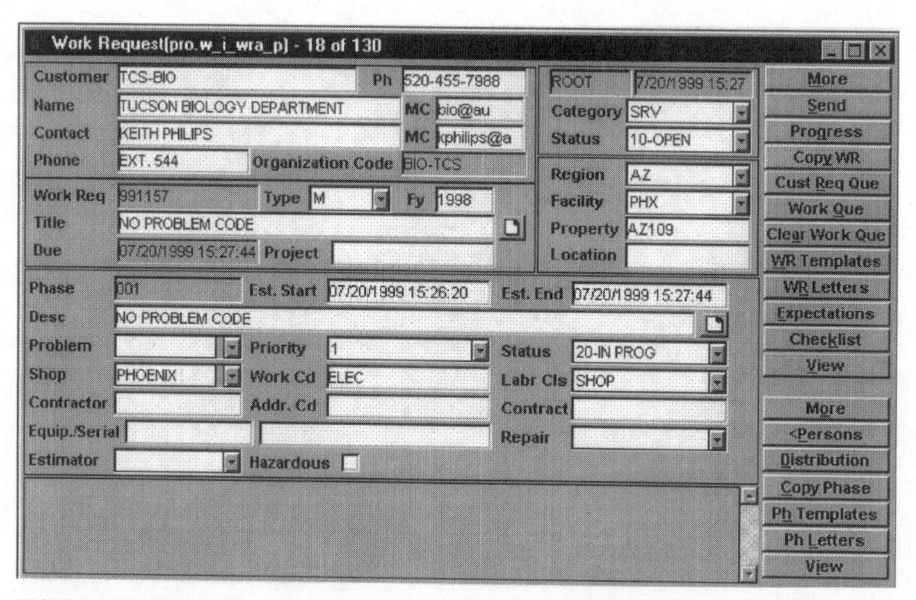

FIGURE 22.22 AssetWorks—Multiple Phases for a Single Work Order—Allowing Facilities Groups to Track Multishop Tasks under One Work Order.

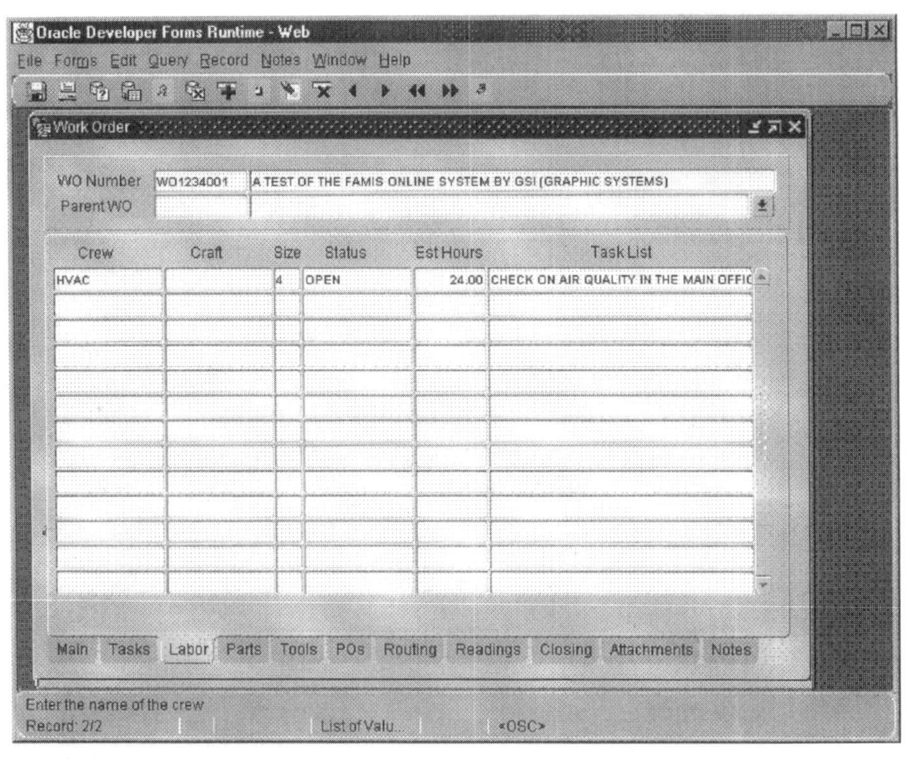

FIGURE 22.23 Prism CC FAMIS On-Line—Secondary Labor Form Provides the Ability to Have a Primary Lead Crew Supported by Additional Crews.

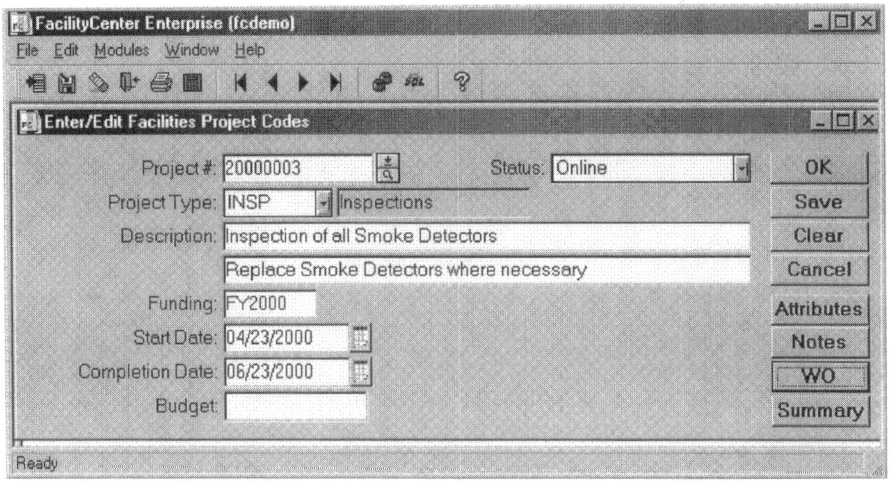

FIGURE 22.24 Facility Center—Facility Projects—A Work Order Linking Number for Tying Together Multiple Work Orders under a Mini Project.

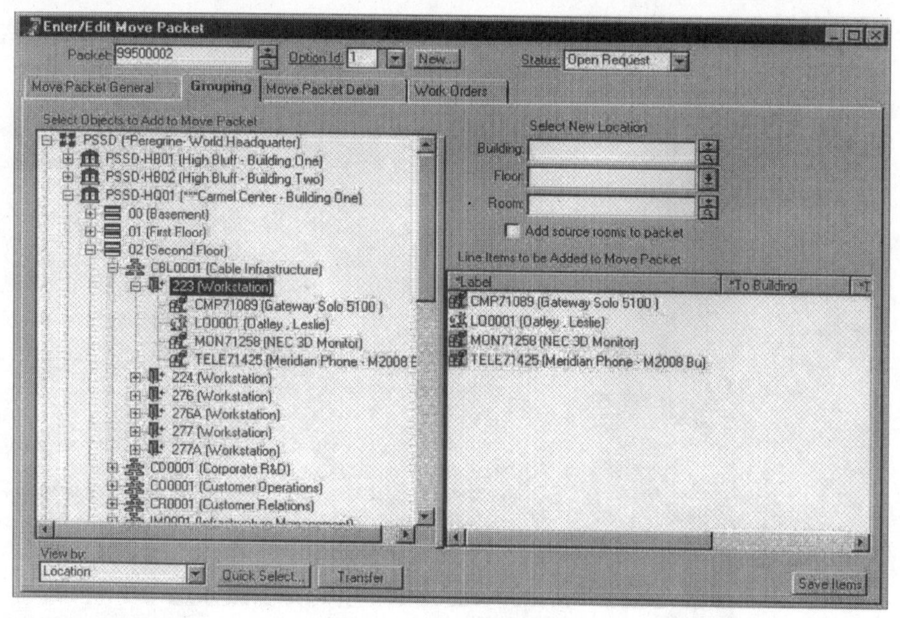

FIGURE 22.25 Facility Center Asset Hierarchy for the Move Manager.

22.4.1.3 Stage 3: Resource Management and Scheduling. The assignment of a task to the shop is relatively easy. The customer service desk will assign the majority of calls if staff performs the service in-house. A "hot/cold" call will go to the HVAC shop, a power problem will go to the electrical shop. However, once a shop has received the request, the supervisors must decide how to schedule the work based on the resources available. Supervisors have to integrate the demand work orders with other work derived from preventive work or project work. They can do this by the priority assigned to the work order. Priority codes are common to most of the CMMS applications. They help trigger certain dates and performance benchmarks. From a business process, there are some predefined priorities. If the call in any way affects the life safety for the facility, staff must attend to it immediately. If the CEO requests help, then the normal priorities often do not apply. Does the application provide the tools to manage work effectively? Most applications are beginning to provide more sophisticated scheduling tools. Facility center has an elegant "drag and drop" feature to assign work orders to specific staff. It then allows staff to generate a daily or weekly task list (see Figure 22.26).

The level of scheduling and resource management depends on the size and type of organization. It also depends on the business process and the experience of those running the shops combined with the IT resources provided to the shops. Putting PCs into the shops is a new concept for many organizations. During a project implementation, there is a need to put in the basic cable infrastructure. As technology changes and new opportunities for interaction occur, all the members of the shops will interact directly with the electronic version of the work order. CMMS vendors are introducing handheld technology with wireless communication. Vendors have already passed through the first generation. Stamford University used Apple Newton's to receive work orders in the field. The second generation of wireless connectivity is here. Facility center allows staff to transmit work orders to digital telephones. All the vendors are working on two-way PDAs to allow for processing of work orders. What does this mean to the resource and scheduling phase? Staff will perform electronically with no paper printed. The old method of using pegboards or pigeonholes for paper versions of the work order will phase out.

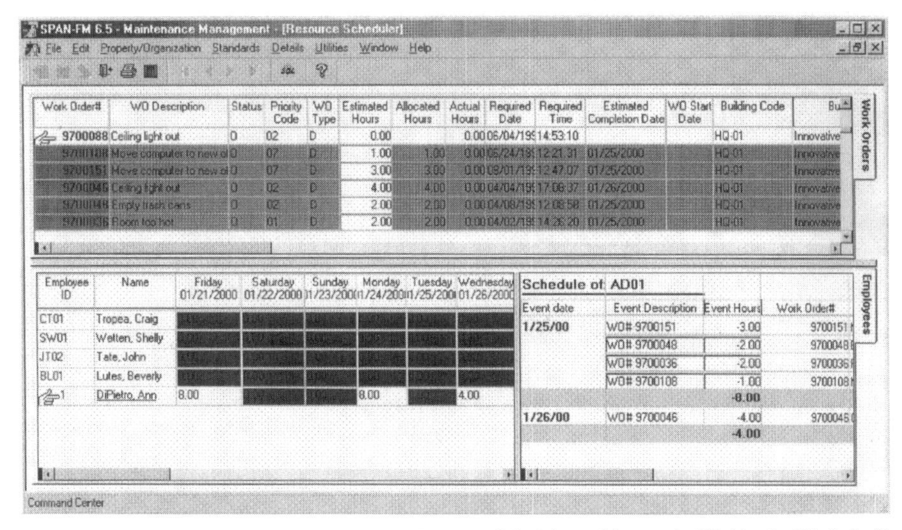

FIGURE 22.26 Shop Supervisors Can Use the Resource Scheduler to Manage the Workload of Their Staff.

22.4.1.4 Stage 4: The Action Phase. Once the supervisor has approved, assigned, and scheduled the work, the building technician will have to work on the task. The first step will be to visit the location of the problem with his or her standard repair and tool kit. If all goes well, he or she may fix the problem in a single visit. Often the problem may be more sophisticated. The work may be for another shop, additional help may be needed, or special materials may be required.

22.4.1.5 Stage 5: The Feedback Loop. Feedback on the resolution of the problem, or feedback on the change in status of the work order is critical to the effective management of the facility department. As previously discussed, the status field defines business rules and also provides feedback to the customer. With the introduction of Web technology, the customer can look into the work order to see what is happening. Changing the status of the work order to reflect what is happening can minimize the need for further calls and help provide quality customer service. While fixing a problem quickly is the best solution, often staff cannot do the work without a special part. A status of "waiting for materials" can help the customer understand why the work is not complete. Other status's can convey important business rules, such as "waiting for approval." Many applications will track the change in status for a work order providing additional history for a problem. When the work goes smoothly, people question why they need to add information, but when there is a problem, this historical information can help solve a problem or prove that the facility department is performing.

An example of this comes from an international financial institution. The organization had a training auditorium. The head of the training department called the head of facilities complaining that she had called many times to have lights fixed, but no one was responding and she was upset at the lack of service. Investigating the problem through the CMMS history for that location brought to light several issues. First, there were 10–15 calls logged for that auditorium. The head of training was correct in that she had called; however, the work orders showed that several building technicians had been out to replace bulbs. The facility department can prove they were doing their job and the head of facilities had the history. On further investigation, the CMMS showed that a relamping project had recently changed all the bulbs in that theatre to energy efficient lights before the problem arose. Further research showed that the manufacturer was having problems with that model of lamp. The institution was able to get all instances of that type of light bulb replaced with a reliable energy efficient light.

TABLE 22.2 Partial Listing of Resolution Codes for Orders

HVAC-AC01	Air compressors—Checked	
HVAC-AC02	Air compressors—Repaired	
HVAC-BG01	Bearings—Lubricated	
HVAC-BG02	Bearings—Replaced	
HVAC-BT01	Belts—Adjusted	
HVAC-BT02	Belts—Replaced	
HVAC-CN01	Controls—Adjusted	Thermostats, pneumatic, elec, DDC
HVAC-CN02	Controls—Calibrated	Thermostats, pneumatic, elec, DDC
HVAC-CN03	Controls—Checked	Thermostats, pneumatic, elec, DDC
HVAC-CN04	Controls—Installed	Thermostats, pneumatic, elec, DDC
HVAC-CN05	Controls—Repaired	Thermostats, pneumatic, elec, DDC
HVAC-CN06	Controls—Replaced	Thermostats, pneumatic, elec, DDC
HVAC-CN07	Controls—Reset	Thermostats, pneumatic, elec, DDC
HVAC-CS01	Coils—Cleaned	
HVAC-CS02	Coils—Repaired	
HVAC-CS03	Coils—Replaced	
HVAC-DD01	DDC Controls—Adjusted	
HVAC-DD02	DDC Controls—Calibrated	
HVAC-DD03	DDC Controls—Checked	
HVAC-DM01	Dampers—Adjusted	
HVAC-DM02	Dampers—Lubricated	
HVAC-DM03	Dampers—Repaired	
HVAC-DM04	Dampers—Replaced	
HVAC-DR01	Drive systems—Calibrated	
HVAC-DR02	Drive systems—Checked	
HVAC-DS01	Dryers—Checked	
HVAC-DS02	Dryers—Repaired	
HVAC-DU01	Ducts & Vents—Checked	
HVAC-DU02	Ducts & Vents—Cleaned	
HVAC-DV01	Devices—Adjusted	dampers, coils
HVAC-DV02	Devices—Checked	dampers, coils
HVAC-DV03	Devices—Installed	dampers, coils
HVAC-DV04	Devices—Repaired	dampers, coils
HVAC-DV05	Devices—Replaced	dampers, coils
HVAC-EQ01	Equipment—Adjusted	Pump, drive, compres'r, refr, dryers
HVAC-EQ02	Equipment—Checked	Pump, drive, compres'r, refr, dryers
HVAC-EQ03	Equipment—Disabled	Pump, drive, compres'r, refr, dryers
HVAC-EQ04	Equipment—Installed	Pump, drive, compres'r, refr, dryers
HVAC-EQ05	Equipment—Repaired	Pump, drive, compres'r, refr, dryers
HVAC-EQ06	Equipment—Replaced	Pump, drive, compres'r, refr, dryers
HVAC-FC01	Fan coils—Cleaned	
HVAC-FC02	Fan coils—Repaired	
HVAC-FS01	Filters—Cleaned	
HVAC-FS02	Filters—Replaced	
HVAC-GS01	Gaskets—Checked	
HVAC-GS02	Gaskets—Replaced	
HVAC-IN01	Insulation—Checked	
HVAC-IN02	Insulation—Installed	
HVAC-IN03	Insulation—Repaired	
HVAC-IN04	Insulation—Replaced	
HVAC-PE01	Pipe/fittings—Checked	
HVAC-PE02	Pipe/fittings—Installed	
HVAC-PE03	Pipe/fittings—Removed	
HVAC-PE04	Pipe/fittings—Repaired	
HVAC-PE05	Pipe/fittings—Replaced	

TABLE 22.2 *(Continued)*

HVAC-PM01	Pumps—Checked
HVAC-PM02	Pumps—Repaired
HVAC-RE01	Refrigeration equipment—Adjusted
HVAC-RE02	Refrigeration equipment—Checked
HVAC-RE03	Refrigeration equipment—Repaired
HVAC-SE01	Sheaves—Adjusted
HVAC-SE02	Sheaves—Replaced
HVAC-SF01	Special fluids—Checked
HVAC-SF02	Special fluids—Installed
HVAC-SF03	Special fluids—Replaced

Even if the work order is not completed on the first try, providing feedback or transferring responsibility to the appropriate shop ensures that the ball is not dropped. The facility department has to realize that the work order is a living document. Once the work is complete, there is still the need for feedback. What was the problem? What piece of equipment did the staff fix? What was wrong that caused the service call? These questions enhance the knowledge base for the facility department. It provides the data for analyzing the performance of the building and allows for effective life-cycle analysis to take place. The more information entered about what was the nature of the problem and how the building technician fixed the problem will result in more efficient business practices. Facility departments do not enforce the effective tracking of the closeout data that can be supported by many applications. In a well thought-out application, the resolution codes can speed up the closeout process. Table 22.2 shows a partial listing of resolution codes used by this author for a client.

22.4.1.6 Stage 6: Accounting Reconciliation. Accountability is a driving force in the implementation of many CMMS applications. In the educational field, for example, there is a need to show where federally funded resources are spent for indirect cost recovery purposes. For large institutions, millions of dollars could be at stake if the information is not present. Medical facilities have to prove how they use the building and maintain it to ensure sources of funding from NIH for research, Medicare for federal reimbursement, and also OSHA and the EPA for compliance purposes. Meeting accreditation requirements drives the need for accurate information. Information flows from many sources and there is an interaction between the financial systems and the CMMS, particularly in the case of materials management and purchase orders.

Linking performance to costs makes the results easy to understand by management. It also provides a solid benchmark for performance improvements. It is recommended that all organizations work toward the goal of tracking all expenses for their facilities department, even if there is no true chargeback mechanism. By charging for the work performed, the facility department gains accurate information on how they measure up to the competition, the drive to outsource services. In many areas, the internal team can provide services cheaper than the "outsource company" as they have a home field advantage.

To close out the work order, it is necessary to assign all the appropriate labor costs. There are several business models for entering time against work orders. In time, all staff will electronically add their own hours into the system. They will account for all their hours, including time not spent on work orders, but in training, meetings, sick leave, vacation, holidays, and personal time. Users will have access to handheld devices similar to PDA, that may incorporate communication devices, both voice and data. While there are initial claims that building technicians are not capable of using computers, those making the claims forget that these engineers often maintain far more sophisticated equipment. The fastest growing segments of computer users are retired workers. Soon the

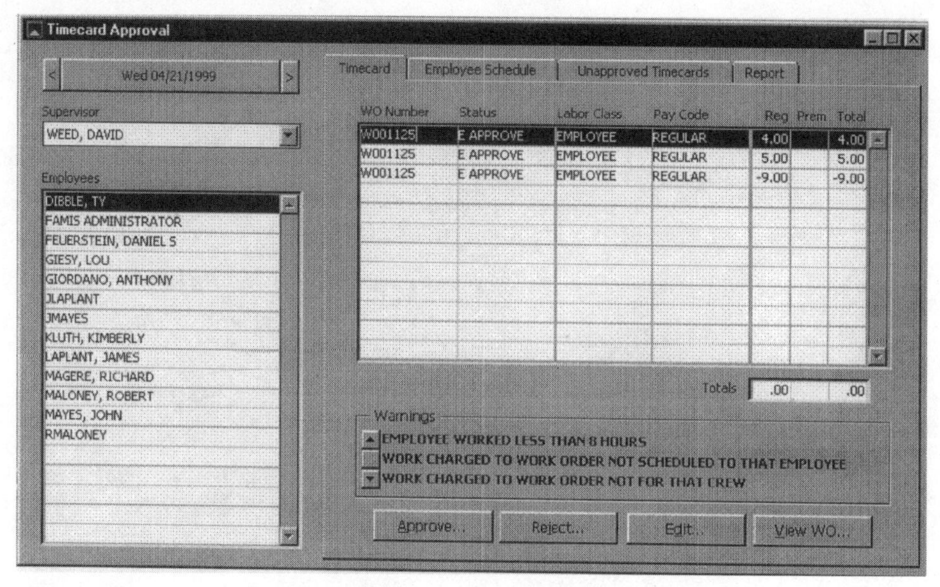

FIGURE 22.27 Prism CC FAMIS Daily Timecard Approval Screen for Employees.

computer illiterate will fall into the same category as the individual who cannot read. All the major CMMS applications have time entry features. Most require that labor be entered against a work order. Standing work orders are created to track hours spent on "nonbillable" tasks, such as training, meetings, or administrative tasks. Figure 22.27 shows typical timecards. The FAMIS application of Prism has thought through the business process for time allocation very completely. A detailed set of supervisor approval screens allows a facility operation to use the timecard process to feed the payroll system.

The next level of complexity lies in ensuring that the correct accounts are used in charging back services. AssetWorks and Prism have powerful account tracking processes. They do not provide an accounting system, but allow the CMMS to gather all the necessary data to charge accounts in the financial system. Prism uses a series of summary tables to gather information, roll up data, and, if necessary, roll it back. The system supports the need to track multiple accounts for tasks, split the costs by percentage, or allocate dollar values for chargeback. It is the most detailed of all the current CMMS applications and allows staff to use unlimited accounts on unlimited phases. The staff handles debit accounts at the work order level and credit accounts at the system configuration level. Each shop can have a credit account and each warehouse can have a credit account, while the facility department may have its own debit account to track costs that are absorbed internally. In dealing with charges to clients, the facility department has to track two types of work, fixed price and time and materials. The CMMS must be able to pass accrued costs to the financial system for projects that may last over several months. The CMMS must be able to encumber accounts on the financial application and then charge against that budget. For fixed cost projects, the CMMS must be able to deal with the overages and underages on these projects.

Materials are part of the costs staff assign to work orders. When staff draws materials from the warehouse, the process of charging back to the work order is straightforward. The shop must be able to look up the availability of the material in the warehouses, to see what may be on order. If the part is not available, then the purchasing process needs to be able to begin. The generation of the purchase request defines the various line items required for the job. The work order will be on hold while the material is on order. AssetWorks is able to define a status that will notify the shop when materials arrive in the warehouse. For certain purchases, the ability to create vouchers, use open purchase orders with local hardware stores for set amounts, paid by credit cards, are needed to allow the CMMS to match the business processes and provide the flexibility for the facility department.

The CMMS must complement the financial process. Several of the CMMS applications interface with the financial systems. Some are scripted based on database triggers to pass data between a financial application, such as PeopleSoft, while Prism allows the user to view the records through FAMIS that reside in PeopleSoft. Different vendors provide solutions to the dividing line between the need to track costs in the CMMS and the need for a single financial accounting system. The definition of each organization's accounting and business practices will require tailoring the CMMS interfaces.

22.4.2 The Anatomy of the Preventive Maintenance Work Order

Preventive maintenance is a core module of a CMMS and is an important responsibility for the facility department. In defining cost benefits of the CMMS, the ability to move resources from demand or emergency work orders to a preventive maintenance program is cited as an essential element in savings. It is more efficient to work against a planned schedule and to be able to deploy your engineers in a proactive program, rather than working in panic mode under crisis management. In IFMA's Benchmark III report, the cost relationship between the repair work orders compared to the preventive work order was $1.20/RSF for demand and $0.84/RSF for PM. Reporting on the correlation of demand to PM is a critical benchmark in defining the value and effectiveness of the CMMS (see Figure 22.28).

22.4.2.1 Stage 1: Gather the Data. Unlike the demand work order, the preventive maintenance process has greater structure and requires up-front planning. The first step on the road to a PM program is to define the scope of the equipment that you intend to maintain. The HVAC building system is a major candidate, given its equipment focus. It is necessary to define the level of detail that staff will need to collect. Equipment consists of hierarchies, with the top level defined as the building system. There is a parent–child relationship for each piece of equipment and it is necessary to understand these relationships. Which motor drives the fan for which AHU? Once defined, which

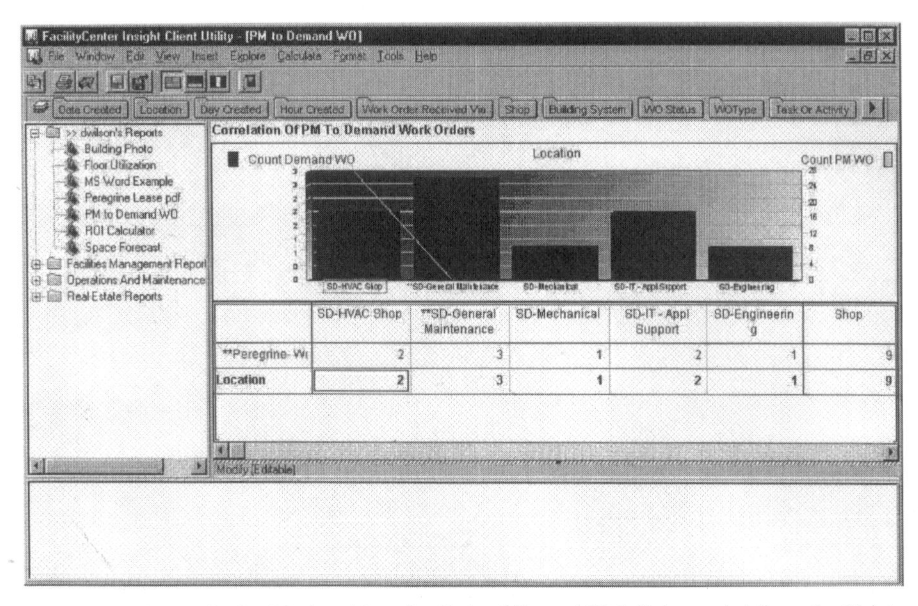

FIGURE 22.28 Application Metrics—Measuring Ratio of Demand Work Orders against Preventive Maintenance Work Orders.

systems will be maintained through the PM program? Information has to be collected about each specific element. In reviewing the diagram, you see the volume of information. As mentioned previously, efforts are underway to have the manufacturer assemble relevant data so that it flows into these fields. Relevant information should also flow from financial systems. How well does the CMMS support the building of the equipment database? How easy is it to build the equipment hierarchies? An effective PM module will support the development of the PM program.

*22.4.2.2 Stage 2: **Define the Maintenance Tasks.*** Once the information about the equipment is gathered, the actual PM tasks can be defined. Based on the experience of the building engineers and the manufacturer's recommendations, it is possible to define the frequency that PMs should take place. These PM tasks will fit into a PM schedule once defined. Recognizing that there will be hundreds, if not thousands, of equipment assets, takes time. Facility departments often hire an experienced engineering firm to assist in documenting the PM tasks. Does the application allow the engineer to easily define the tasks and then relate them to equipment? Can the engineer look at a PM task and see all the equipment that uses that PM? Can the engineer look at a piece of equipment and see all the PMs run for that specific asset? Does the application support the ability to drill down to the detail? Figure 22.29 shows the Peregrine's facility center task form.

*22.4.2.3 Stage 3: **Define the Procedures for Each Task.*** The previous state defines the monthly, quarterly, annual, or seasonal PM tasks. Defining the steps that need to happen for that task to ensure that the work is completed correctly and safely also needs to be documented. Figure 22.30 defines the itemized procedures for a specific task. The systems can provide a checklist to follow. The procedure can also store information about notifications and timing of the work.

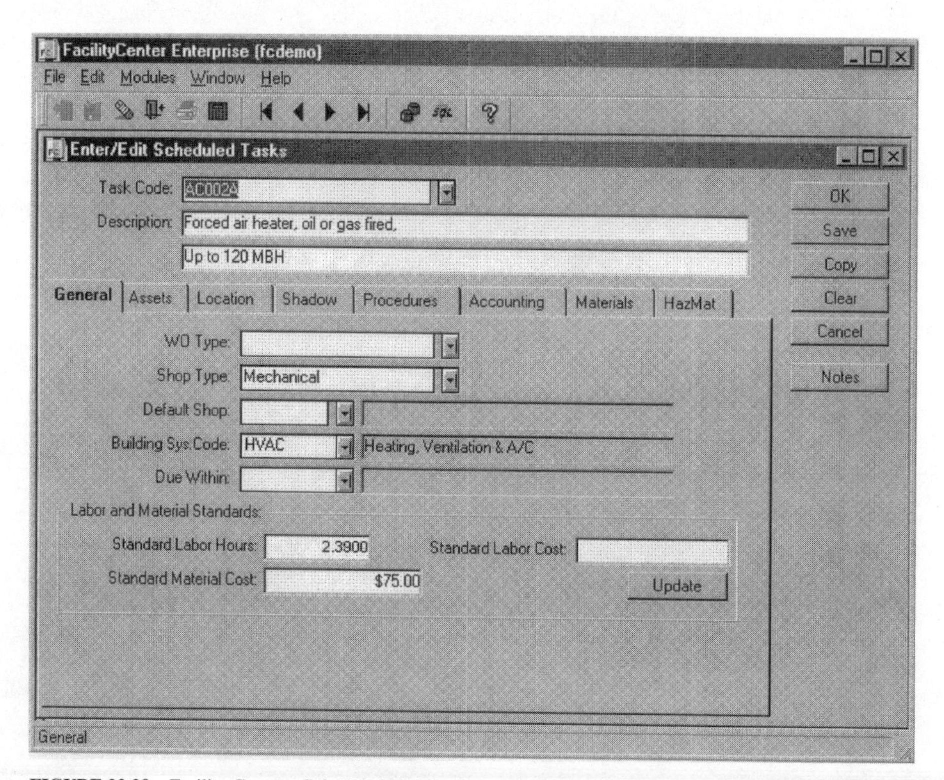

FIGURE 22.29 Facility Center—Scheduled Task for Preventive Maintenance.

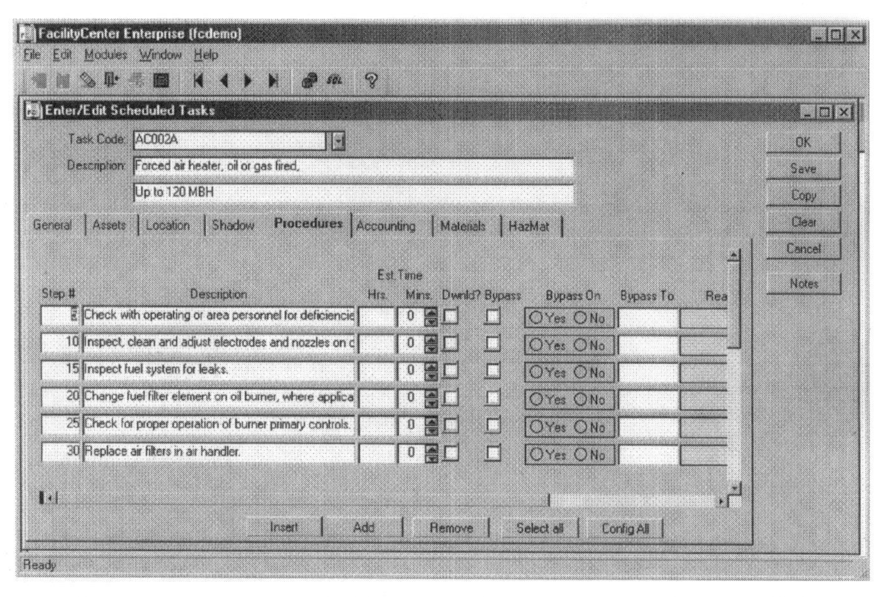

FIGURE 22.30 Facility Center—Detailed Procedures for Completing a Specified Task under a PM Program.

Some tasks can only be performed in off-hours, some need special access, and some need to be performed in confined spaces and have special life-safety requirements. It is worth spending the time to add the anecdotal information as it builds the maintenance knowledge base on how to perform the work. Over the years, the engineers know which tools to use, and where access to a piece of equipment is good or poorly designed. Information such as this can help define the amount of time it takes to perform the work and helps justify resources. Too often facility organizations allow the experience to leave when the senior engineers retire. Can the application allow for the development of a knowledge base? Can the application help build a checklist of items that can be printed on the work order? Can specific health and safety warnings be printed with the work order? How rich is the information that can be associated with the procedure? Does it include drawings, diagrams, images, and manufacturer's manuals? Figure 22.31 shows scheduling time period options. Figure 22.32 depicts shadowing of duplicate PM tasks.

22.4.2.4 Stage 4: Define the Resources for Each Task. In defining the procedure for each PM task, the engineer will define the materials and the labor necessary to perform the work. It will be necessary to factor in both the travel times for the PM and that the material is available. For each task, staff will assign the parts and special tools required to the PM. The staff will define the estimated duration for the task based on benchmark data. The value gained from documenting this information is two-fold. It allows the scheduler to see what labor resources the PM needs and what are the skill levels. It allows the procurement function to become proactive and perform just-in-time delivery or to combine purchasing requests to achieve volume discounts. A well-organized purchase plan can combine these two options by committing to buying the parts in advance and achieving the discounts, but only having them delivered in time, thereby minimizing the storage requirements. How much can staff save? The Boeing Corporation estimates that it was able to save $30 million by restricting its facilities procurement strategy.

22.4.2.5 Stage 5: Define the PM Schedule. While each PM task has a specific period, the schedule will be a combination of time and resources available. There may only be one or two qualified staff to perform the work. It does not make sense to assign all monthly PMs to start on the

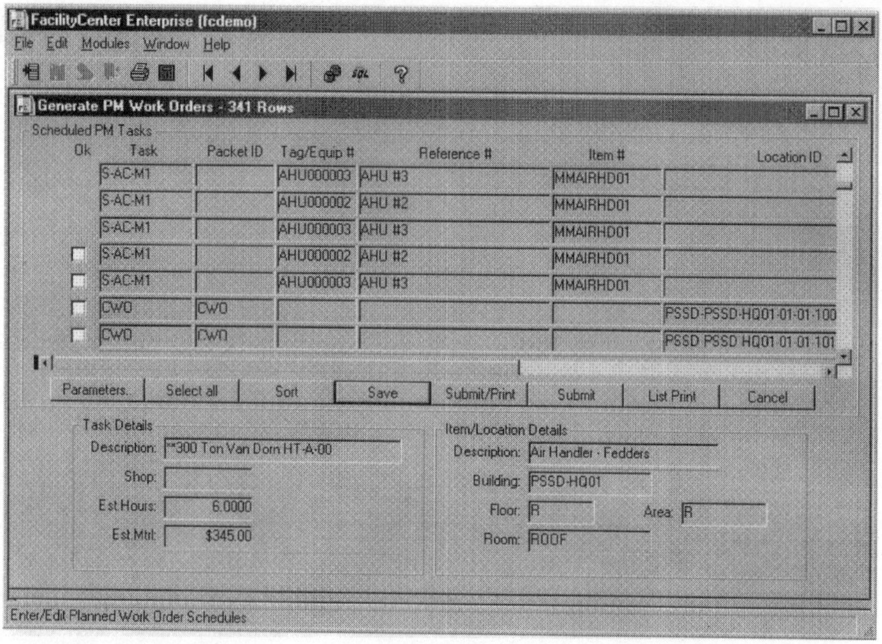

FIGURE 22.31 Facility Center—After Preventive Maintenance Tasks Are Scheduled the Actual Work Orders Can Be Selectively Created Based on Current Resources.

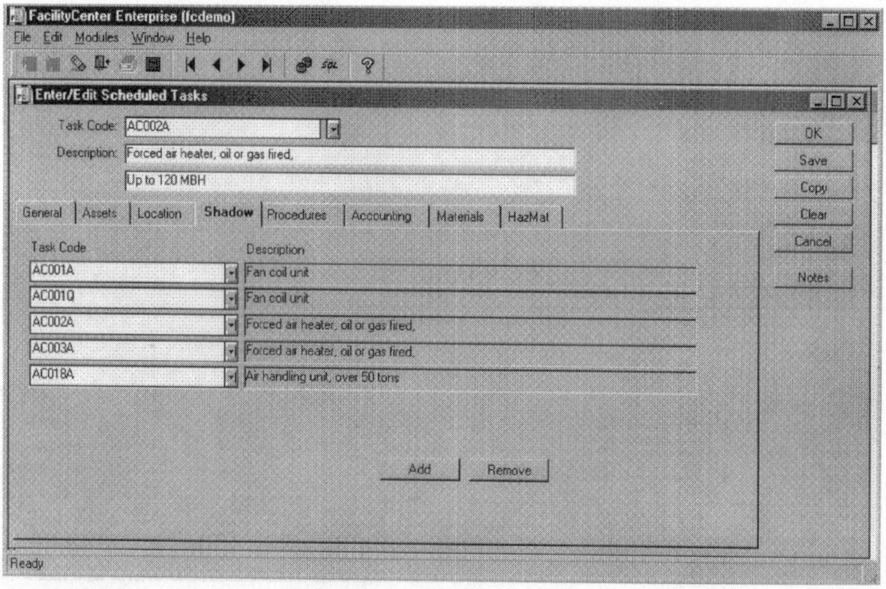

FIGURE 22.32 Facility Center—Shadowing of Overlapping Preventive Maintenance Tasks—Such as the Monthly and Quarterly PMs for a Specific Piece of Equipment.

first of the month. Therefore, the ability to plan out when PMs take place based on resources will help ensure the most efficient use of staff time. While the CMMS tool makes it easy to assign frequencies to PMs, it takes an experienced engineer who knows the building to be able to set the schedule. Each PM is not an isolated event. The development of PM groups, a set of related tasks that can take place together when staff shut down equipment, helps improve efficiencies. The development of PM routes where a similar task needs to take place for many pieces of equipment, such as grease bearings or check fan belts, reduces the amount of travel time for engineers. Does the application interface with the work calendars for staff? Does it interface with the inventory module?

22.4.2.6 Stage 6: Generate the PM Schedule. Up to this point, staff has defined the work effort on the front end. At this point, the supervisor can generate the PM work orders ahead of time. Until staff assigns the work order, they are not active. The supervisor may generate PM WOs for the current month or for the quarter. These tasks sit as open items of the shops work schedule until the shop assigns the work to a daily schedule. Figure 22.33 shows a daily schedule for work assignment. How well does the application relate PM work order with the demand work order? If a service request comes through for a piece of equipment, can the application recognize that a PM is scheduled in a week and that the two efforts should be combined? Depending on how the procedures for the PM tasks have been defined, can the application skip the monthly PM task when the quarterly PM is due to minimize the administrative tasks?

22.4.2.7 Stage 7: Complete the Work. At some point, the actual work has to be performed. The labor and materials have to come together with the instructions to perform the PM task. Does the application make it easy for the building engineer to get at the information related to the PM? The concept of a "workbench" developed by Prism brings together the information needed to perform the work (see Figure 22.34).

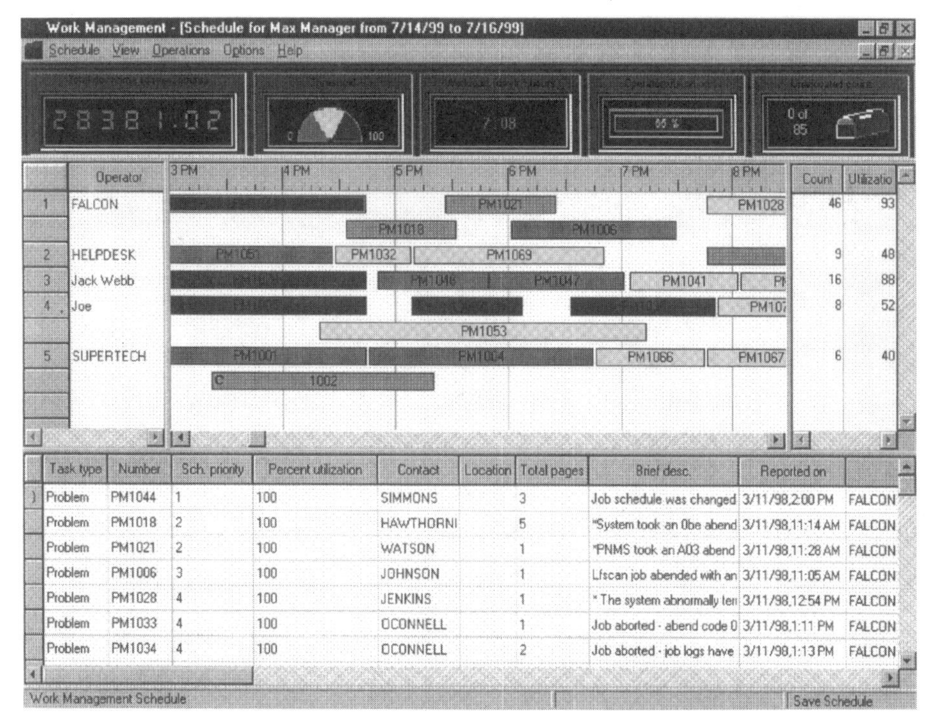

FIGURE 22.33 Work Management and Scheduling PM Resources.

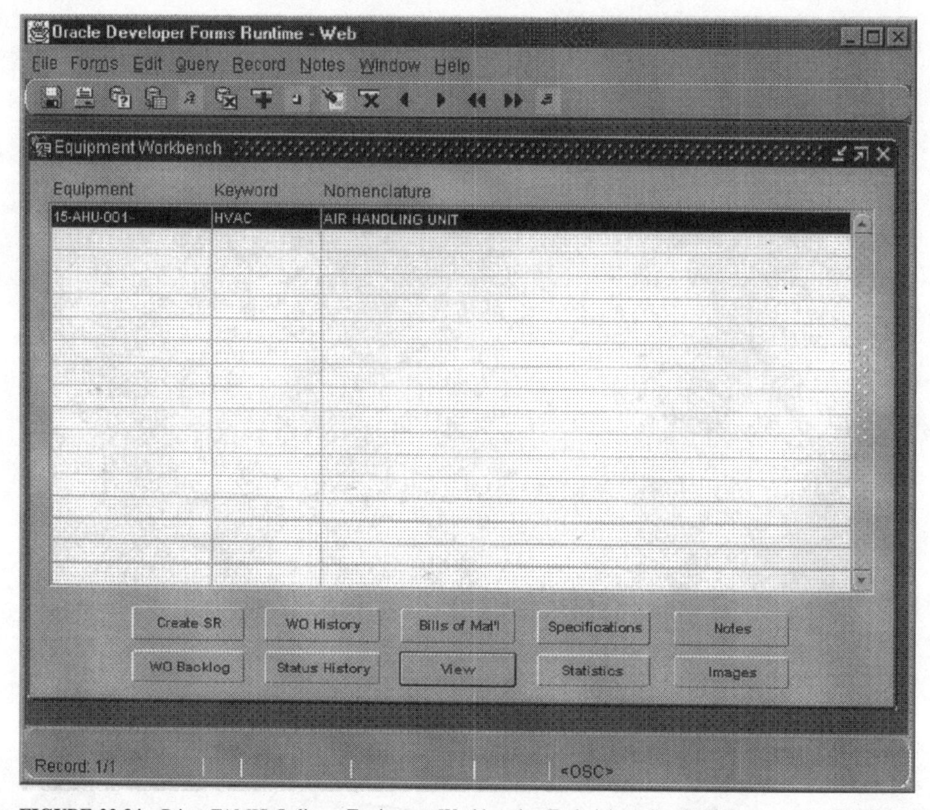

FIGURE 22.34 Prism FAMIS Online—Equipment Workbench—Technicians Can Drill Down to the Full Set of Data through the Workbench Concept.

22.4.2.8 ***Stage 8: Close Out the PM Work Order.*** Getting feedback from the work performed is important. The default response will be to close the work order. Everything was normal when the engineer performed all of the procedures set out in the PM task. For many pieces of equipment, specific readings are required, whether they are run time for motors or for calibration data for specialized pieces of equipment, this information must be fed back. The engineer may be concerned about the performance of the equipment, but does not have time to investigate or feels that when they perform the next major PM the problem can be addressed. This data is needed to build the history, provide accountability, and help determine trends in equipment performance, either degradation or outright failures. We cannot stress enough the importance of gathering data from a preventive maintenance program—it provides the fuel for the analysis of the data. Predictive maintenance feeds off the effort.

Closing out the PM takes a degree of discipline by the engineer and management oversight. It is part of the work of the building engineer and is not an option.

22.4.3 Equipment vs. Facility Preventive Maintenance Tasks

There are two distinct types of preventive maintenance: the traditional preventive maintenance tasks tied to equipment and the category of recurring maintenance that is tied to the facility itself, often defined by a location. The process described previously can work for either facility or equipment related PMs. The facility PM focuses on preserving the fabric of the facility, the cleaning of the facility, and programs that require a set schedule of work that will repeat over a period of time. This period of time may not necessarily be annually. An example may be the need to paint all offices once every 3 years. The advantages of using the PM program to define the work and schedule it provides two key benefits. First is that information about the tasks is preserved for accountability. The work is scheduled and therefore labor and resources can be assigned as part of the total amount of work provided by the facility department. By planning out the effort and capturing the results, it is possible to look at how individual tasks are accomplished and then look to improving that specific process. Secondly, the changes can be monitored and the effectiveness of the change documented. Information derived can be added to budget requests for additional funding or resources.

22.4.4 Inspection Programs

In discussing the types of PMs, the concept of a PM route was included. A specific type of PM relates to inspections. These inspections are important, as they often pertain to life-safety issues. An example of an inspection route is to ensure that all of the fire extinguishers are functioning. Documenting that your facility has complied with federal and state requirements is important. Combining barcoding technology to ensure that the engineer has visited each fire extinguisher on the inspection route provides further accountability with the advantage of minimizing supervision.

22.4.5 Predictive Maintenance

Gathering data on the performance of equipment is part of being proactive with maintenance management. Parallel to a wellness program in medicine, predictive maintenance looks to prevent breakdowns by fine tuning the equipment before it breaks. Predictive maintenance requires a regular feed of data from sensors. These can be supplied by a building automating system, such as Johnson Control's Metasys product. Data relating to temperature, vibration, airflow, and other factors can be sent to the CMMS. If the CMMS supports predictive maintenance, it has the ability to set ranges for each variable fed into the application. If the equipment moves out of that range, then the application generates an alarm with a work order for the engineer.

22.5 CONCLUSIONS

In today's business environment, a CMMS is critical to the operations of an O&M group, whatever the size. This chapter has focused on the O&M aspects of the use of a CMMS. The functionality and capabilities of these systems have expanded the scope of how they are used. Rather than just managing the narrow spectrum of equipment assets, these systems focus on the infrastructure and assets of an enterprise. Corporations recognize that they must keep the fabric of the business tuned, as well as the business operations. Therefore, a CMMS implementation must mirror the business objectives of the facility plan. It must gather the information necessary to monitor compliance with the business plan.

The information contained within the CMMS is precious. It provides an accounting of the work performed. Management holds facilities organizations accountable for how they spend the money allocated. Without unlimited budgets the facilities organization must have the tools to prioritize the projects that they will invest in for the current fiscal year. To prioritize they must have accurate and current information. By tracking all the work performed in the CMMS, and the materials associated with the work, an accurate model is created and maintained, allowing for projections for planning and historical analysis for performance improvement programs based on solid metrics.

The changes in technology, the introduction of wireless devices, and the impact of the Internet on how a facilities organization can operate will make the next 5 years a revolutionary period in business. Those facilities organizations that can measure up to the challenge will enable their businesses to thrive both internally and in competition.

ENDNOTES

1. Dertouzos, M. 1997. *What will be.* Dertouzos invented the term "information marketplace." San Francisco: Harper.
2. Lynch, M. 1998. Facilities Solutions, Inc.
3. Negreponte, N. 1995. *Being digital.* New York: Random House.
4. FAMIS from Prism Computer Corporation.
5. Digital Nervous System. 1999. Bill Gates Business @ The Speed of Thought. New York: Warner Books.
6. Ibid. p. 217.
7. The phrase "value network" indicating a digital flow of information was developed in Tapscott, D. 1993. *Paradigm shift: The new promise of information technology.* New York: McGraw-Hill.
8. Youngblood, M. D. 1994. *Eating the chocolate elephant: Take charge of change.* Richardson, TX: Micrographix Inc.
9. Hamel, G., & Prahalad, C. K. 1989. Strategic intent. *Harvard Business Review,* May–June, 67.
10. Duck, J. D. 1993. Managing change: The art of balancing. *Harvard Business Review,* November–December, 109.

CHAPTER 23
ENERGY MANAGEMENT

Al Ferreira
President
Ferreira Service, Inc.
Specializing in Air Conditioning Maintenance, Repair, and Energy Cost Reduction
E-mail: Al_Ferreira@ferreira.com

23.1 ENERGY MANAGEMENT DEFINED

Because energy is *not* a fixed cost, energy management is important. As a variable, the cost of energy is determined by the amount used and the cost per unit. Energy management is the process of analyzing how efficiently energy is used and how competitive the cost per unit is in the deregulated marketplace. The concept of how energy is used and charged is best understood by focusing on the electric power meter, as shown in Figure 23.1.

The meter measures the amount of power consumed over time. The unit of measurement is watts, determined by multiplying the amperes of current by the voltage, divided by 3 (for three-phase power). Because buildings use a large number of watts, the power measurement is usually referred to as kilowatt (1,000 watts) units. Lights, air conditioning (HVAC), receptacles, elevators, computers, etc., all consume a certain amount of power. If all the power is added up at a given time, the resulting total kilowatt number is referred to as kilowatt demand (kW). The power company must supply enough power to meet that demand after losses of power through the distribution system.

23.2 UTILITIES INVOICE

One component of the energy bill is for kilowatt demand (kW), based on the highest demand required for each billing period (usually one month); another component is kilowatt-hours (kWh). To establish the utility cost, the meter records how many kWh were used and what the highest kW was and these are multiplied by the rate charge. Figure 23.2 depicts what a typical utility invoice might look like.

Until now, rates charged by the utilities for energy have been regulated. Utilities were allowed an investment return on assets and rates were established accordingly. In many states, deregulation has changed this aspect of rate monopoly. Today, most states have legislation either pending or implemented that allows customers to shop for energy pricing. Even so, states like California and New Hampshire, that have already implemented deregulation, are phasing it in over many years. The full effect of deregulation in California, for example, will not be realized until the year 2002.

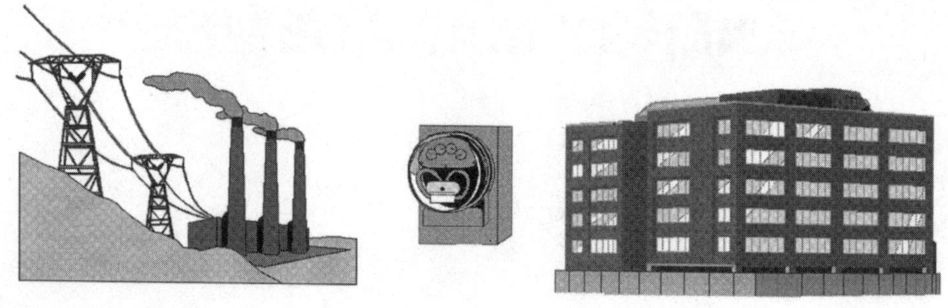

FIGURE 23.1 Concept of How Energy is Used.

During the phase-in process, utilities are allowed to recapture stranded costs, such as power plants built under a regulated environment. The stranded costs appear on utility bills under headings such as competition transition charge (CTC). This charge makes it difficult for consumers to save on energy costs during the deregulation transition period.

While anticipating cost reduction due to deregulation, it is important to understand that only power generation cost will be deregulated; *not* distribution and transmission costs. Power generation cost is usually about \$0.02–\$0.03/kWh, depending on the time of day and time of year produced. Because the power cost is only a small percentage of the total bill, savings, in the early stage of deregulation, are very small in states that have a high CTC.

23.3 DEREGULATION

At the time of this writing, few facilities have seen very significant cost savings from deregulation. Future savings will depend on the cost of oil, public utilities (PUC) rate increase allowances for

Service/Billing Period – 08/06/98 to 09/04/98 (29 days) – Summer Season					
Facilities Rel. Demand	862 kW	x	$6.40000	$	5,516.50
Summer On Peak	862 kW	x	$17.55000		15,128.10
Summer Mid Peak	851 kW	x	$2.80000		2,382.80
Demand Charge Total				$	23,027.70
Summer On Peak	96,523 kWh	x	$0.09485	$	9,155.21
Summer Mid Peak	93,125 kWh	x	$0.05989		5,577.26
Summer Off Peak	120,552 kWh	x	$0.03810		4,593.02
Current Energy Bill				$	19,325.49
Customer Charge					298.65
Summer Pwr. Factor Adj.	620 kVar	x	$ 0.23000		142.60
Long Beach City Tax	$ 42,794.44	x	$10.00000%		4,279.44
StateTax	310,200 kWh	x	$ 0.00020		62.04
Current amount must be paid by 10/01/98				**$**	**47,540.46**

Your Account's Total Balance Due

FIGURE 23.2 Typical Utility Invoice.

transmission and distribution, and demand for power. The immediate consequence of deregulation, however, has been to compel facility managers to focus on energy cost. This is good, as there are huge opportunities to reduce costs on the load or demand side of the power meter. Before focusing on the specifics of these savings, let's discuss some aspects of deregulation that will greatly affect facility managers.

As a result of utilities undergoing tremendous change in how they do business, there is a corresponding change in the marketplace that affects end users. In a word, this change is in the marketing of utilities. Because of deregulation, utilities are now in competition to attract and retain customers. Huge amounts of capital are being invested on the aggressive marketing of utilities. The result is a confused atmosphere for the facility manager. The utilities do not want to sell only power. Power is going to become a commodity and, therefore, will require large volume to make money. As the utilities restructure, what they would ultimately like to be is a one-stop shop for facility managers' total energy requirements. Everyone is claiming to have expertise in energy management. Utility companies are now offering bundled services, including:

- Power
- Metering
- Energy monitoring Web sites
- Energy audits
- Energy project management
- Energy efficiency equipment financing
- Utility bill consolidation
- Performance contracts

Purchasing bundled services from anyone without knowing exactly how competitive each component of the bundle is can be very dangerous and even end up costing managers huge amounts of money in lost opportunity.

Bundled services are usually made attractive to users by offering a small discount on the power component of the utility bill for a set amount of time. Although this may look attractive initially, facility managers should be aware of exactly what terms they are committing to. In most cases, the power is being discounted, but the user must commit to using the power provider for all of the other services listed. The user becomes "locked in" to the power provider long term and can become somewhat of a captive customer. In many cases, the user does not have the right to sign an energy cost reduction contract with another energy service company (ESCO), for example, even if that ESCO could save more energy than the power provider.

The objective of a bundled service contract is to set up the energy requirements of the user as a funnel where all services pass through the provider's company, entitling it to fees on anything used (Figure 23.3).

23.4 BENCHMARKING USAGE

If the power provider provides the absolute best values for the services going into the funnel, plus giving the user a good power rate, there would be nothing wrong with the decision to purchase bundled utility services. This is, however, seldom the case, and facility managers need to be very aware of what constitutes *good values* so they can evaluate the best total decision related to their energy situation.

One of the absolute best ways to evaluate services is to know ahead of time how efficiently your specific facility is using energy. Everything will relate back to this. If you don't know this, you become an easy target for marketing people who will want to keep discussions about the specifics of your power needs and the cost of services very vague, with few hard facts included. Also, by not

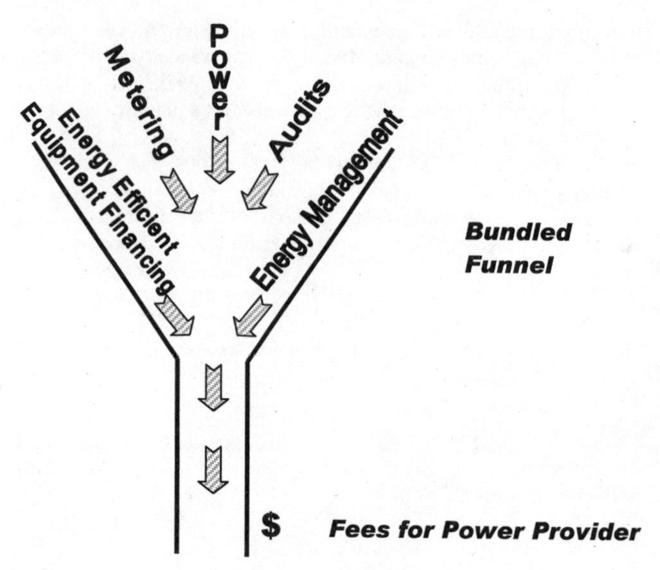

FIGURE 23.3 Bundled Service Contract Model.

knowing how efficiently your building uses energy, a discount in power cost may look attractive to you, when in fact the real opportunity is to reduce the amount of kWh you are using, as well as paying 1/2¢ less for each one.

23.4.1 EPA's Energy Star Buildings

To find out how efficiently your building uses energy, you must benchmark the use against an efficient model. Here we will use the EPA's *Energy Star Building* model because it has been verified against thousands of facilities and is very accurate. The EPA has modeled efficient facility use for over 250 cities around the United States. The model calculates energy requirements in BTUs/ft^2/yr. This can be converted to kWh/ft^2/yr, which is more useful because utility invoices are billed in units of kWh and therms. The therm component is for gas and is usually low enough not to be a factor in energy conservation. The big usage of energy is kWh and converting BTUs to kWh is accomplished by dividing BTUs by 3,413.

The EPA's *Energy Star Building* model breaks down into approximately the following geographical benchmarks:

kWh/ft^2/yr	
West coast	16
East coast	18
Mid West	20
South	22

These benchmarks are a very good tool in evaluating the energy efficiency of your facility—just as you would analyze the fuel usage of your car by comparing miles per gallon. To determine how your facility compares to the EPA benchmark number, total the last 12 months of energy usage in kWh and divide the total kWh by the net usage square footage (exclude warehouse, storage, and parking garages). The number you come up with will be the kWh/ft^2/yr that can be compared with the benchmark number. You should be in the 16–22 kWh/ft²/yr range depending upon your geographic location, and provided your facility is almost 100% standard office use.

As an example, let's assume that you manage a 500,000 ft² facility on the west coast. You do the math described previously and find out that your facility is using 24 kWh/ft²/yr.

24 kWh/ft²/yr	Existing usage
16 kWh/ft²/yr	EPA benchmark
8 kWh/ft²/yr	Difference

The 8 kWh difference is a condition of energy inefficiency, and since West coast energy is about $0.10/kWh:

$$8 \text{ kWh} \times \$0.10/\text{kWh} = \$0.80/\text{ft}^2/\text{yr}$$

$$\$0.80 \times 500{,}000 \text{ ft}^2 = \$450{,}000/\text{year of energy waste}$$

Do not make the expensive mistake of thinking that your building is unique and cannot achieve the EPA standards. Nothing could be further from the truth. The fact is, almost all buildings use energy in a very predictable fashion and in pretty much the same way. Buildings also usually waste energy in the same way, and it is very important to know why and where the energy is being wasted.

Another important point about energy waste is that whenever you identify energy waste you also need to know what the payback is on capital invested to eliminate the waste. This is another area of misconception. Many facility managers assume 7-year paybacks or more are required to justify the correction of energy inefficiency problems. This is simply not so. In reality, most inefficiency problems can be corrected for 2–3-year paybacks or less! The reason for this is that most energy waste comes from the *operation* of the facility, not the design or equipment efficiency.

23.4.2 Energy Usage Components

To understand why most waste comes from operations, you need to know what the components of energy usage are in a facility. There are three basic components:

1. Lights
2. Receptacle load
3. HVAC

Each one of these components uses energy at a certain rate. Lights, for example, consume about 32 watts of power (fluorescent single lamp). The watts consumed times the hours of consumption gives us the watt-hours, which when divided by 1,000 converts the power into kWh. kWh is how the utilities charge for power, so the first component of power consumed is lighting. If you add up all of the light power in watts for the entire facility (space lighting, night lighting, outside lighting, etc.) a typical office building will use between 1–3 watts per square foot. One watt per square foot is very efficient and is probably due to the use of high efficiency lamps with electronic ballasts, reflectors, etc. Three watts per square foot are usually what older lighting systems use if they have not been converted to high efficiency.

Multiplying the watt/ft² usage times the hours used per year gives you the kWh/ft²/yr benchmark component that we will use to compare to the EPA benchmark mentioned earlier. If the lighting is on 3,400 h per year and the average consumption is 1.6 watts/ft²/yr then:

$$3{,}400 \text{ h/yr} \times 1.6 \text{ watts/ft}^2 \times 90\% \text{ (avg. on time)} \times \text{kW}/1{,}000 \text{ watts} = 4.9 \text{ kWh/ft}^2/\text{yr}$$

Each of the three basic components of energy usage in a facility can be calculated in this fashion. Receptacle load usually equals about 4.2 $kWh/ft^2/yr$ for a typical building with PCs on each desktop.

The first two components, lighting and receptacle load, are fairly constant over a typical business day; that is, the lights are turned on in the morning and consume 1.6 $watts/ft^2$ consistently until they are turned off at the end of the day. Receptacle load is also fairly constant, with PCs turned on at the beginning of the workday and staying on until the end of the day. If the lighting component of energy use is 4.9 $kWh/ft^2/yr$ then the receptacle is:

$$2{,}860 \text{ h} \times 1.6 \text{ watts/ft}^2 \times \text{kW}/1{,}000 \text{ watts} = 4.5 \text{ kWh/ft}^2\text{/yr}$$

The lighting load plus the receptacle load equals 9.4 $kWh/ft^2/yr$. Because the average commercial office building consumes about 18 $kWh/ft^2/yr$, then the third component of energy use, HVAC, must consume $(18 - 9.4) = 8.6$ $kWh/ft^2/yr$.

We could further verify that the HVAC uses about 8.6 $kWh/ft^2/yr$ by calculating the energy as follows:

$$\text{Fan energy} = 3{,}400 \text{ h/yr} \times \frac{1 \text{ watt/cfm/ft}^2}{1{,}000 \text{ watts/kW}} = 3.4 \text{ kWh/ft}^2\text{/yr}$$

$$\text{Chiller energy (chillers, towers, and pumps)} = \frac{3{,}400 \text{ h}}{\text{year}} \times 1 \text{ ton}/350 \text{ ft}^2$$

$$\times \frac{1.3 \text{ kW}}{1 \text{ ton}} \times 0.4 \text{ avg.\% full load} = 5 \text{ kWh/ft}^2\text{/yr}$$

$$\text{Hot water pump} = 0.2 \text{ kWh/ft}^2\text{/yr}$$

Fan	= 3.4 $kWh/ft^2/yr$
Chiller	= 5 $kWh/ft^2/yr$
Hot water pump	= 0.2 $kWh/ft^2/yr$
	8.6 $kWh/ft^2/yr$

The previous calculations are accurate for an average system, but it is very possible to have systems much more efficient. In some systems, the total HVAC load can approach 5 $kWh/ft^2/yr$ depending on the geographical location. The example is very simplistic, and in reality, an engineer would project a much more accurate calculation by utilizing actual trend logs of temperatures, pressures, and kW measurements and then adjusting load and run hours by weather data.

The point of all this is simply to demonstrate that building energy use can be very accurately projected or modeled for any type of building. A benchmark can be calculated to show what any building's energy consumption should be.

If, as in the previous example, your building uses 24 $kWh/ft^2/yr$ and the EPA benchmark shows that an efficient building uses 16 $kWh/ft^2/yr$, then what you are looking for is where the 8 $kWh/ft^2/yr$ of waste is, and what the cost to correct the inefficiencies will be. The most accurate way to identify the inefficiencies is to perform a technical audit.

As we have discussed previously, it is possible to break down the three major components of energy use (lights, receptacle load, and HVAC) and measure each through a technical audit.

Lighting. Figure 23.4 is a lighting audit recap showing existing vs. proposed changes. A lighting audit measures the lighting load in each fixture (task lighting, night lighting, outdoor lights, etc.) and multiplies it by the quantity and run hours of all the lights in the facility. Lighting can then be broken out of the total usage on a $kWh/ft^2/yr$ basis. Additionally, if more efficient lighting can be installed, like converting from 40 watt fluorescent lamps to 32 watt lamps, 8 watts per lamp could be saved and a hard calculation could be made to show specific savings by converting to high efficiency lights. Costs for the lighting retrofit could also be calculated and a payback determined.

		EXISTING FIXTURE TYPES				PROPOSED MODIFICATIONS			
#	Area	Fixture Description	Hrs	Qty	Watts	Fixture Desc. & Lamp Type	Hrs	Qty	Watts
1	1ST FLR								
2	Offices	2L 4' T8 SP35K, EB, 1x4 TROFF	5460	359	59.3	1L 4' T8 841K, EB, Refl.	5460	359	31.1
3	"	2L 3' T8 SP35K, EB, 1x3 TROFF	5460	43	45.7	1L 3' T8 841K, EB, Refl.	5460	43	26
4	"	2L 2' T8 SP35K, EB, 1x2 TROFF	5460	12	85	1L 2' T8 841K, EB, Refl.	5460	12	19.7
5	2ND FLR								
6	Offices	2L 4' T8 SP35K, EB, 1x4 TROFF	5460	213	59.3	1L 4' T8 841K, EB, Refl.	5460	213	31.1
7	"	2L 3' T8 SP35K, EB, 1x4 TROFF	5460	35	45.7	1L 3' T8 841K, EB, Refl.	5460	35	26
8	3RD FLR								
9	Offices +	2L T5 BIAX SP35K, 2x2 TROFF	3120	36	72	3L 2' T8 841K, EB	3120	36	47
10	Offices	3L T8 SP35K, EB, 2x4 TROFF	3120	47	88.95	2L 4' T8 841K, EB, Refl.	3120	47	59.3
11	Elec. Room	2L T12 CW ES, 1x4 Strip Refl.	3120	27	72	2L 4' T8 841K, EB	3120	27	59.3
12	Telecom	2L T12 CW ES, 1x4 Strip Refl.	3120	37	72	2L 4' T8 841K, EB	3120	37	59.3
13	Training Rm	3L T8 SP35K, EB, 2x4 TROFF	3120	12	88.95	2L 4' T8 841K, EB, Refl.	3120	12	59.3
14	"	2L T5 BIAX SP35K, 2x2 TROFF	3120	10	72	3L 2' T8 841K, EB	3120	10	47
15	4TH FLR								
16	Offices	3L T8 SP35K, EB, 2x4 TROFF	3120	129	88.95	2L 4' T8 841K, EB, Refl.	3120	129	59.3
17	Offices +	2L T5 BIAX SP35K, 2x2 TROFF	3120	18	72	3L 2' T8 841K, EB	3120	18	47
18	5TH FLR								
19	Offices	3L T8 SP35K, EB, 2x4 TROFF	3120	103	88.95	2L 4' T8 841K, EB, Refl.	3120	103	59.3
20	Offices +	2L T5 BIAX SP35K, 2x2 TROFF	3120	19	72	3L 2' T8 841K, EB	3120	19	47
21	Offices-not	3L T8 SP35K, EB, 2x4 TROFF	3120	26	29.65	2L 4' T8 841K, EB, Refl.	3120	26	29.65
22	6TH FLR								
23	Offices +	2L 4' T8 SP35K, EB, 1x4 TROFF	5460	164	59.3	1L 4' T8 841K, EB, Refl.	5460	164	31.1
24		2L 3' T8 SP35K, EB, 1x3 TROFF	5460	24	45.7	1L 3' T8 841K, EB, Refl.	5460	24	26
25		2L 2' T8 SP35K, EB, 1x2 TROFF	5460	7	85	1L 2' T8 841K, EB, Refl.	5460	7	19.7
26	7TH FLOOR								
27	Offices +	2L 4' T8 SP35K, EB, 1x4 TROFF	5460	321	59.3	1L 4' T8 841K, EB, Refl.	5460	321	31.1
28	"	2L 3' T8 SP35K, EB, 1x3 TROFF	5460	50	45.7	1L 3' T8 841K, EB, Refl.	5460	50	26
29	"	2L 2' T8 SP35K, EB, 1x2 TROFF	5460	13	85	1L 2' T8 841K, EB, Refl.	5460	13	19.7
30									
Existing KW:		109.59	Proposed KW: 63.18				KW Saved:	46.41	
Existing KW/H		511203	Proposed KW/H: 284762				KW/H Saved:	226441	

FIGURE 23.4 Lighting Fixture Inventory and Proposed Modifications.

Lighting payback example

Existing use	511,203 kWh
Proposed use	284,762 kWh
Savings	226,441 kWh/yr

$$226{,}441 \text{ kWh/yr @ \$0.09/kWh} = \$20{,}378$$

$$\text{Cost to achieve} = \$61{,}134$$

$$\text{Payback} = \frac{\$61{,}134}{\$20{,}378} = 3.0 \text{ years}$$

Additional lighting calculations and paybacks can be made for:

- Electronic ballasts
- Night sweep-off systems
- Motion sensors
- Daylight perimeter dimming

After the cost effective[1] retrofit opportunities for lights have all been calculated, the new kWh/ft^2/yr that could be realized is subtracted from the total kWh/ft^2/yr facility usage to see how close to the EPA benchmark this component will get you.

Assuming that we can cost effectively[1] save 2 $kWh/ft^2/yr$ from the facility that uses 24 $kWh/ft^2/yr$—this would mean that we could get down to 22 $kWh/ft^2/yr$ by lighting retrofit.

HVAC. The next step in the audit is to calculate potential energy savings in the HVAC system. If the facility is using 24 $kWh/ft^2/yr$, and we just found 2 $kWh/ft^2/yr$ in lighting savings which will take us to 22 $kWh/ft^2/yr$, then we would expect to find about 6 $kWh/ft^2/yr$ in HVAC savings because the target benchmark is 16 $kWh/ft^2/yr$. The receptacle load usually does not allow for any savings because it is a fairly constant load while the facility occupants are in the building. This means that after the lighting load savings potential is calculated the rest of the potential is usually HVAC.

The HVAC audit is much more complex than the lighting because it is not a constant load. It varies with outside temperature, solar load, and inside load, such as people and lights. Additionally, HVAC provides comfort so the technical audit must take into account system design, comfort requirements, and weather patterns. Fortunately, from an engineering standpoint, all of these items can be projected, measured, and calculated to provide an engineering solution to identify and eliminate energy waste.

One of the first things energy auditors mistakenly focus on when analyzing a facility for energy efficiency is the HVAC equipment itself. The standard audit evaluates, for example, what the effect of installing a more efficient piece of equipment, such as a chiller, will be.

Chillers have become more efficient over the last 10 years, with older chillers (20+ years) utilizing energy at a rate of 0.8 kW/ton. This means the chiller requires 0.8 kW to provide 1 ton of cooling (12,000 BTU/h heat removal). Chillers manufactured in the last 5 years can achieve efficiencies approaching 0.5 kW/ton, so there is definitely some savings by upgrading chillers. The problem, however, is that any equipment replacement, such as chillers, is not going to be cost effective. Chiller replacements average 6–8 year returns on capital invested. Most chief financial officers will not want to wait this long for a rate of return. Most CFOs will require 3-year or less paybacks to provide the capital to fund energy projects; therefore, equipment replacement is probably not where you should focus.

It turns out that to get fast paybacks on HVAC energy efficiency upgrades you again need to focus on the operation (or misoperation) of the HVAC system. This is a very important concept that is a missed opportunity in 90% of all energy audits performed. Engineers and facility managers always assume that the HVAC system is properly maintained, control systems are properly calibrated, and the system was designed to operate efficiently, so it probably does. Nothing could be further from the truth. This concept is so important that I want to repeat it to emphasize the importance. *Most of the following assumptions are almost always wrong!*

1. HVAC systems are properly maintained and operated

2. Control systems are efficiently programmed and calibrated

3. This HVAC system was designed by an engineer to operate efficiently, so it probably does

These three misconceptions cost facility owners billions of dollars per year, and the sad thing is that these three misconceptions have the fastest payback to fix.

Why then, are they almost always missed? Because they are complex to understand and recognize. For example, take something as simple as maintenance. Everyone assumes that it is done. Maintenance personnel will say it is done. We even have sophisticated computerized maintenance systems that spit out reports showing it is done. The problem is that 90% of maintenance functions are not easily verified. Heating and cooling coils inside ductwork can be plugged solid with dirt and you can't see it. Economizer damper actuators can be loose on the drive shaft, and you cannot see this from a report. The outside air damper could be frozen open (or shut), and you cannot see it. Moreover, heating valves could be leaking; heating and cooling could be occurring simultaneously; fan belts could be slipping; air balance dampers could be throttled too far closed; air and water flows could be different from design; start/stop schedules for fans, pumps, and chillers could be bypassed; and you will never see any of it on a maintenance report!

HVAC systems are designed to operate under a wide variance of conditions. Full load design is a rare occurrence. Because these systems are so flexible, they can be severely misoperated, or there can be a large amount of deferred maintenance and the system will still function. It just won't function very efficiently, and may or may not provide comfort, but the energy usage will be abnormally high.

It is difficult to find deferred maintenance, therefore most auditors either do not recognize it when they see it, or they are unfamiliar with just how much inefficiency results from this condition. Additionally, if a computerized maintenance system reports that it has been maintained (even if it has not been), and a maintenance mechanic simply checked off that it was done, an erroneous assumption has been made.

The control system is another area where wrong assumptions cost large dollars in energy waste. Control systems that are not programmed properly are difficult to uncover. First, an engineer would have to be very experienced in control strategies that maximize energy efficiency while keeping building occupants comfortable. Remember that because HVAC systems provide comfort, even when they are functioning inefficiently, you won't know what's really happening unless you analyze the programming code. A simple 100,000 ft^2 building could have 50 pages of programming code that direct a microprocessor-based direct digital control system. There is a tendency, from control companies, to use generic programs that can be loaded quickly or even preprogrammed into a microprocessor. The generic program may not be the most efficient for your specific system, but it is fast to set up, it works, and it gets the control programmer (who may cost $80.00 per hour), off the job at the end of the installation. Because your project was probably awarded to the lowest bidder, the contractor's priority is usually to get the project up and running as quickly as possible.

There is usually absolutely no incentive for a control contractor to save energy, so the system is usually installed with minimal or no energy tuning. This means that an energy auditor must find the control inefficiencies. Reviewing 50 or 60 pages of programming codes is a tedious process, and it is easier to just assume that there is no need for it and to focus on something else, like lights or a more efficient chiller.

Assuming that the HVAC system was designed by an engineer to operate efficiently is another great misconception. It is easy to see how this happens. A great example of this is California, which has a law stipulating, for almost the last 20 years, that new building designs pass an energy efficiency requirement called *Title 24*. This regulation requires a professional engineer to certify that the new mechanical design has the capability, due to the system selection, equipment efficiency, and envelop construction to use a minimum amount of energy. The minimum amount is approximately 50,000 BTU/ft^2/yr, and varies based on facility use. The 50,000 BTU/ft^2/yr equals about 15 kWh/ft^2/yr.

This regulation has been in effect for approximately 20 years, which means for the last 20 years any new building built in California had to have professional engineer certification that it could comply with that criteria.

However, very few buildings in California are using 15 kWh/ft^2/yr of energy. Why? The answer is in the fact that the regulation states that the design would allow them to operate this efficiently, but nothing requires them to operate as designed. As a result, due to misoperation, poor control calibration, deferred maintenance, etc., very few buildings operate in the design efficiency specified by Title 24, and most buildings in the world have the same problem.

Proving this is simple. Benchmark your facility and see how many kWh/ft^2/yr it is using. If you are in California, for example, and the benchmark for your use is 15 kWh/ft^2/yr, you are okay. If not . . . start looking at the three wrong assumptions discussed above. A detailed energy audit is the answer to most facilities' HVAC energy problems. The audit should start with the installation of data loggers on the HVAC equipment to trend temperatures, pressures, events, kW, and comfort levels. These trend logs key the auditor into the opportunity areas for savings. An experienced professional engineer, doing a proper audit, will notice abnormalities in the system operation from an efficiency standpoint. This will allow the engineer to focus on the best approaches to quantify and qualify the energy waste that was identified by subtracting the EPA's *Energy Star Building* guidelines from the facilities actual energy usage in kWh/ft^2/yr.

A sampling of selected examples of a comprehensive HVAC audit follows (Figures 23.5 through 23.9).

The first thing you need to perform an audit is data. Real data, *not assumed data.* You can accomplish this by putting trend loggers on equipment to determine the operating profile of each fan, pump, chiller, coil, motor, control, etc. A trend log for a fan, for example, would continuously record supply air, outside air, return air, and mixed air temperatures to determine whether the economizer cycle is functioning, heating and cooling are occurring at the same time, or if there is deferred maintenance required on the unit.

Each piece of HVAC equipment is trend logged in this fashion over a time frame of 4–6 weeks. An engineer then uses the data to analyze the operating profiles and do calculations to determine energy savings potential. An example of this can be seen in Figure 23.5. Trend logs determine the return air, mixed air, cooling supply, heating supply, average zones, and outside air temperatures.

Current sensors can trend fan power and run hours. From this data a model can be programmed to calculate the total annual gas and electric use for that specific fan system. It is very accurate because it uses real data from that fan, *not assumptions.* In Figure 23.5, we can see that the model calculated the existing fan energy usage to be 11,732 therms and 128,402 kWh.

Figure 23.6 shows a second model, but this time the engineer makes changes to the operating profile and calculates the data between the two models using the 52.5 outside air temperature bin. You will note the following differences between Figures 23.5 and 23.6:

Differences between Figures 23.5 and 23.6

	Existing energy use	Proposed energy use
Operating time	982 h	202 h
Cold deck air flow	4,076 CFM	3,311 CFM
Heating deck air flow	7,824 CFM	4,913 CFM
Total fan energy	6,988 kWh	885 kWh
Return air temperature	68.5°F	72°F
Mixed air temperature	56.5°F	70.1°F
Cooling supply air temperature	50.5°F	70.1°F
Heating supply air temperature	90°F	93°F
Average zone supply air	76.5°F	83.8°F
Gas usage	2,224 therms	197 therms
Chiller usage	1,945 kWh	0
Net cooling (htg.) delivered to zones	−102 kBTU/h	−104 kBTU/h

By modeling a piece of equipment (e.g., the fan, the pump, the chiller, etc.), a very accurate comparison can be shown, that defines exactly what has to be corrected or modified to acquire the projected cost reduction. Additionally, it shows that the cost reduction can be achieved at no change in comfort level. That is why in the previous example, the very bottom row, which compares the net cooling or heating delivered to the zone, had almost no change, even though most of the temperatures, air flows, and total energy use changed dramatically. The model is showing that there is a more efficient way to control the system with no penalty in comfort. The proposed model then becomes the programming strategy that a control engineer would use to set up the control functions in the 52.5 outside air condition. Implementing a control strategy that allows the proposed model to control the equipment in this fashion will deliver the savings projected.

Figures 23.7 and 23.8 are other examples of what an audit will investigate. In these examples, Figure 23.7 shows a chilled water pump's existing operation compared to proposed operation with 258,195 kWh per year in energy savings. Figure 23.8 shows two secondary pumping loops that

Audit Summary Sheet • Existing Energy Use

Outside air temp. 5 degree intrvls.	°F	27.5	32.5	37.5	42.5	47.5	52.5	57.5	62.5	67.5	72.5	77.5	82.5	87.5	92.5	97.5
Annual operate. Time in ea. range	hrs	0	0	34	424	651	982	1290	1547	1414	1047	808	348	143	56	16
Cold deck airflow	cfm	2400	2400	2400	2400	3288	4076	4865	5654	6443	7232	8021	9500	9500	9500	9500
Heating deck airflow	cfm	9500	9500	9500	9500	8613	7824	7035	6246	5457	4668	3879	2400	2400	2400	2400
Total annual fan energy use	kwh	0	0	242	3017	4633	6988	9180	11009	10063	7451	5750	2477	1018	399	114
Return air temperature	°F	67.0	67.0	67.0	67.5	68.0	68.5	69.0	69.5	71.0	71.5	72.0	72.0	72.0	72.0	72.0
Mixed air temperature	°F	37.4	41.1	44.9	48.8	52.6	56.5	60.4	64.3	68.4	72.3	76.1	79.9	83.6	87.4	91.1
Cooling supply air temperature	°F	37.4	41.1	44.9	48.8	50.0	50.5	51.0	51.5	52.0	52.0	52.0	52.0	52.0	52.0	52.0
Heating supply air temperature	°F	90.0	90.0	90.0	90.0	90.0	90.0	90.0	88.0	86.0	84.0	82.0	8.0	83.6	87.4	91.1
Avg. zone supply air temperature	°F	79.4	80.1	80.9	81.7	78.9	76.5	74.1	70.7	67.6	64.6	61.8	57.6	58.4	59.1	59.9
Annual gas usage	therms	0	0	126	1436	1811	2224	2323	1983	1175	496	159	1	0	0	0
Annual chiller energy use	kwh	0	0	0	0	455	1945	4766	9033	12084	12420	12664	7465	3480	1524	482
Net clg. (htg.) delivered to zones	kBTU/h	-159	-169	-179	-182	-141	-102	-65	-15	44	89	131	184	175	165	156

TOTAL ANNUAL GAS USAGE	11,732 therms	TOTAL ANNUAL ELECTRICITY USAGE	128,402 kwh

FIGURE 23.5 Defining Existing Energy Use.

Audit Summary Sheet • Proposed Energy Use

Outside air temp. 5 degree intrvls.	°F	27.5	32.5	37.5	42.5	47.5	52.5	57.5	62.5	67.5	72.5	77.5	82.5	87.5	92.5	97.5
Annual operate. Time in ea. range	hrs	0	0	18	123	132	202	322	520	570	462	417	190	78	31	12
Cold deck airflow	cfm	1200	1200	1200	1200	2256	3311	4367	5422	6478	7533	8589	9644	10700	10700	1070
Heating deck airflow	cfm	7140	7140	7140	6398	5655	4913	4170	3428	2685	1943	1200	1200	1200	1200	1200
Total annual fan energy use	kwh	0	0	77	481	538	885	1416	2370	2690	2254	2102	1061	478	190	114
Return air temperature	°F	70.0	70.5	71.0	72.0	72.0	72.0	72.0	72.0	72.0	72.0	72.0	72.0	72.5	72.5	73.0
Mixed air temperature	°F	65.8	66.7	67.7	69.1	69.6	70.1	70.6	71.1	67.5	72.1	72.6	73.1	74.0	74.5	75.5
Cooling supply air temperature	°F	65.8	66.7	67.7	69.1	69.6	70.1	70.6	71.1	67.5	72.1	72.6	73.1	74.0	74.5	75.5
Heating supply air temperature	°F	92.0	93.0	95.0	99.0	96.0	93.0	91.0	84.0	72.0	72.1	72.6	73.1	74.0	74.5	75.5
Avg. zone supply air temperature	°F	88.2	89.2	91.1	94.3	88.5	83.8	80.5	73.6	67.1	63.3	59.3	56.1	58.3	58.8	59.8
Annual gas usage	therms	0	0	30	204	171	197	237	199	60	0	0	0	0	0	0
Annual chiller energy use	kwh	0	0	0	0	0	0	0	925	748	3115	4366	2828	1183	470	181
Net clg. (htg.) delivered to zones	kBTU/h	-164	-169	-181	-183	-141	-104	-79	-15	49	89	134	186	183	177	170

SUB-TOTAL PROPOSED GAS USE	1,098 therms	ANNUAL GAS SAVINGS	10,635 therms	SUB-TOTAL PROPOSED ELECTRICITY USE	28,402 kwh	ANNUAL ELECTRICITY SAVINGS	100,000 kwh

FIGURE 23.6 Defining Specific Energy Waste.

Audit Calculation Sheet • (1) Pump Operating at 23.5 BHP Peak, on VFD

Prop.d usage = hours x BHP x (kw /HP /motor eff.) x average annual load:
= 8766 hours x 23.5 BHP x (0.746 /0.9) x 0.26
= 44,396 kwh /year

Pump Savings = Existing - Proposed Usage:
= 254,311 - 44,396
= 209,915 kwh /year

Pumping Reheat of Chilled Water:
H = Pumping reheat of chilled water (kwh per kwh of excess usage)
H = Percent of pump power into Ch W x 3413 BTU /hr-kwh x 0.90 kw /T /12,000 BTU /hr-T
= 0.23 kwh per kwh excess pumping

Reheat Savings = 209,915 kwh x 0.23 kwh /kwh = 48,280 per year
Annual Savings = Pump savings + Reheat Savings = **258,195 kwh per year**

TRIM PRIMARY PUMP IMPELLERS:
The 4 primary pump impellars are oversized: flow control valves are throttling GPM, placing unnecessary pressure differential on the pumps.

Excess pumping power = GPM x Excess head / (5308 x motor eff. x pump eff.)
= (1100 GPM avg) (10.5 FT avg.) (10.5 FT avg.) (5308 x .90 x .75)
= 3.22 kw

Annual Pump Savings = Power x Time
= 3.22 kw x 8766 hours
= 28,226 kwh /year

FIGURE 23.7 Engineering for Results.

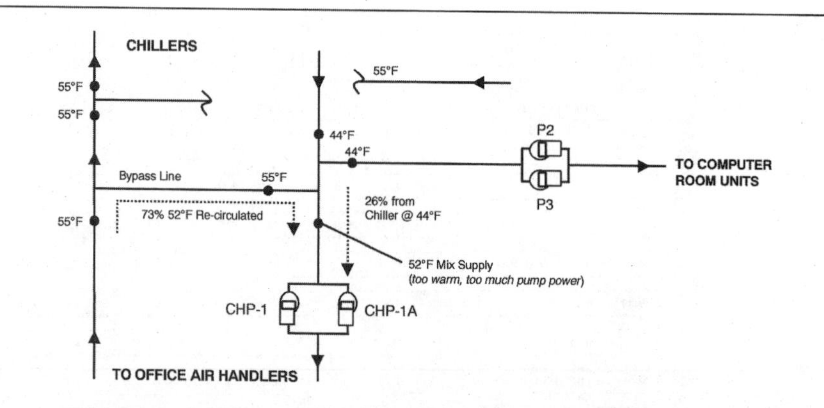

Solution:

Install a variable frequency (VFD on chilled water pump CHP-1A) and reduce flow until there is minimal mixing of primary and secondary loop flow, as shown in the figure below. This should be automated, so that the air handlers can be converted to variable chilled water flow as recommended in phases 3 and 4. This also allows the system's operation to be adjusted over a wide range of flow conditions, to accommodate growth and modification, which are likely given the facilities history to date.

Oversized chilled water pump CHP-1 should be downsized as well. This can be done by increasing the impeller size and changing the motor from the existing 50 HP and 2450 rpm to 15 HP and 1750 rpm (to match the capacity of CHP-1A). It should also be equipped with a VFD, and the two pumps should alternate between being on-line and emergency backup.

FIGURE 23.8 Engineering for Solutions.

Audit Summary Sheet

Summary of Savings

	Annual Energy Savings	Project Cost	Simple Payback (years)	Audit Cost	Payback (including Audit)
HVAC System and Controls Upgrade ▶	$221,271	$292,000	1.32	$32,400	1.47

Energy Conservation Methods

ECM	Project Description	Peak kw Saved	Kwh/Year Saved	therms/Year Saved	$/Year Saved
1	Optimization of AHU 1 & 2	****	227,557	40,019	$38,705
2	Optimization of AHU 3	****	112,655	13,583	$15,624
3	Heating System Recontrol	****	11,530	272	$964
4	Chiller Plant Optimization	****	2,164,858	****	$151,959
5	Recontrol & Upgrade AH 1 & 2	****	177,950	2689	$14,019
Electric Rate: $0.0702/kwh Gas Rate: $0.568/therm	TOTALS ▶	0	2,694,550	56,563	$221,271

© Ferreira Service Inc.

FIGURE 23.9 The Audit Summary.

have a considerable amount of recirculation, which only burns pump horsepower, while providing no increased benefit.

In Figure 23.8, the trend loggers picked up this misoperation on the graphs and the auditing engineer found that if the recirculation was eliminated by tighter control, the existing 50 hp pump motor could be downsized to a 15 hp pump motor.

Figure 23.9 is a summary sheet from an audit showing five energy conservation measures (ECMs) that were found to be cost effective to implement. The summary of savings shows a potential $221,271 per year of savings that can be realized for a cost of $292,000, allowing for a 1.32 year simple payback. Each of the ECMs, 1–5, would have supporting calculations or models based on real trend log data similar to the samples in Figures 23.5 through 23.8.

23.5 ENERGY AUDIT BENEFITS AND COSTS

An audit is an invaluable tool in analyzing energy cost reduction opportunities. It becomes the plan for where you are going and how you are going to get there. The biggest advantages, however, are as follows:

1. An audit shows you exactly how to achieve the energy savings projected by benchmarking your usage against the EPA *Energy Star Building* benchmark.
   - If the EPA benchmark is 16 kWh/ft^2/yr and your facility is using 26 kWh/ft^2/yr, the audit shows you where the 10 kWh/ft^2/yr of waste is.
   - The audit shows you how much of the savings is lighting and how much is HVAC.

- The audit calculates the cost to achieve the savings, so a simple payback can be analyzed and prioritized.

2. The audit lets you know exactly how much potential savings is cost effectively available so that you can have a meaningful discussion with any utility or energy service company about energy projects.

   - Most facility managers do not know where the opportunities are, or what paybacks are realistic, so anything the utility or energy service company (ESCO) tells them is difficult to evaluate.
   - Not knowing where you could be in kWh/ft^2/yr, vs. where you are now, is the biggest mistake you can make when trying to negotiate energy projects.

One more important point about audits, they cost money! Audits for both lighting and HVAC with as much detail as discussed previously, and like examples shown, cost about \$0.10–\$0.20 per ft^2 when performed by a professional engineer. If a professional engineer does not perform the audit it will usually not be very beneficial. You will undoubtedly be offered, at some point, a "free audit" by utilities or ESCOs, as an enticement to get you to do business with them. Many times, to get the "free audit," you will be required to sign a letter of intent that generally requires that you pay for the audit if you do not enter into a contract to perform energy work with the utility or ESCO, based on the findings of the audit. This is a very dangerous situation to get into, because a "free audit" is a marketing strategy to lock you into bundled services. Remember what was discussed earlier in this chapter about bundled services: each service should stand on its own as the best competitive opportunity you can get. Anything that has value is seldom free. "Free" is *always* a marketing enticement and should be viewed with skepticism. The value of a professional audit is that it defines exactly the best value from a capital investment in the form of simple payback analysis. Additionally, an audit defines the scope of work required to achieve specific results that can be competitively procured in an unbundled format.

If, for example, you knew that your existing energy usage was 26 kWh/ft^2/yr, and an audit calculated a 10 kWh/ft^2/yr savings for a 3-year payback, you could then negotiate effectively with a utility or ESCO, your existing HVAC contractor, or lighting contractor, for competitive and unbundled services. If the ESCO was proposing that you sign a letter of intent for 7-year payback projects, undefined in scope until they performed their "free audit," you know not to commit to this because you would know it was not the best payback on your investment.

After you have completed a professional audit and have learned how much energy reduction opportunity is available for a cost-effective payback, the next step is to implement the projects defined. There are many ways to do this, but the most effective way to implement an energy project is through a performance contract.

23.6 PERFORMANCE CONTRACT

A performance contract is defined as a contractual obligation where there is a defined consequence to not performing. Energy performance contracts should be kept very simple. The reason for this is that the energy cost reduction process is complex and difficult to understand, and what you want as a facility manager is the result without putting a lot of time and effort into managing the process. With a performance contract, the responsibility of achieving the result should rest with the contractor.

Energy cost reduction in facilities is not a new concept; it has been around since the 1973 oil embargo. In all that time, however, the industry in general has not produced very impressive results. If it had, there would not be such a large opportunity to reduce usage now. It has not been accomplished very successfully because it is difficult to do. This does not mean that it cannot be done, and there are certainly plenty of success stories to prove it, including the EPA's *Energy Star Building* history, but you must be careful to get the results you want and expect. The way to do this is with the performance contract.

Focus on what you want to achieve and specify a consequence if it is not achieved. For example, if your audit defined a 10 kWh/ft^2/yr energy cost reduction for a 3-year payback, you would expect the following financial results:

- 10 kWh/ft^2/yr @ \$0.10/kWh = \$1.00/ft^2/yr energy savings
- For a 100,000 ft^2 facility the savings would be \$100,000 per year
- If the payback was 3 years, the project would cost \$100,000 × 3 years = \$300,000
- So you invested \$300,000 and you are expecting \$100,000 per year in cost reduction for 3 years

Assume, however, that you did not get the \$100,000 per year savings, but \$50,000 per year instead. That would mean that over the 3 years, you would have a shortfall of \$150,000 for savings not achieved. Under a properly written performance contract it would be the ESCO's or contractor's obligation to refund the \$150,000 to you if all other contractual obligations were met.

A performance contract has two distinct advantages:

- It tends to focus the contractor on achieving results and not promising more than can be delivered.
- It allows the facility manager to be assured that the investment objectives presented to upper management will be achieved one way or another.

A few things to consider when putting a performance contract together are:

1. Make sure energy savings are not achieved at the expense of lighting level reduction or comfort. A good way to achieve this is to specify a minimum lighting lumen level required at workstations, and to use ASHRAE (American Society of Heating Refrigeration and Air Conditioning Engineers) comfort standards of 62–84° at design outside air conditions.
2. Use the kWh and therm savings specified in the audit, at the utility rates applicable at the time of the audit, as the performance goal. Make the monthly utility invoice the vehicle to determine savings. If you do not get the savings on the invoice, it does not do you much good. Sometimes you may need to correct for weather changes, but using 30-year weather averages for the audit will usually get you close enough in the audit calculations that you will not have to correct for weather when making monthly savings comparisons.
3. Be very aware of assumptions made in the audit process, because a performance contract will tie directly to those assumptions. For example, assuming the facility is used between 7:00 A.M. and 6:00 P.M. 5 days per week, and if after the energy work is complete the schedule is changed, a correction will need to be made on the calculations before the monthly savings can be determined.
4. HVAC maintenance will be very important in achieving energy savings. If new equipment on control system strategies and calibration are not maintained it could void the guarantee on the performance contract. Make sure you provide for how this will be handled and monitored.
5. Verify that the performance contractor has experience in this type of contract. Check references carefully, paying close attention to paybacks promised and delivered. Make sure that the contractor has a long history of delivering energy savings promised, and in the range that you are expecting. Some very large ESCOs today have trouble providing good references. If you are looking for a 2- or 3-year payback, make sure you can talk to several referrals where they have done this before. Some ESCOs specialize in government work. Governments typically expect 10- to 15-year paybacks on energy projects. This is usually far too long for nongovernment facilities, so make sure your contractor has experience on fast payback projects.

The listed items should assist you in writing a successful performance contract. Additionally, remember contractors and ESCOs that have lots of experience in reducing energy costs will not hesi-

BUILDING NAME	
ADDRESS	
CITY, STATE ZIP	
TOTAL AIR CONDITIONED	
COMPUTER ROOM SQ. FT.	
WAREHOUSE SQ. FT.	

ENERGY	
UTILITY	
AVERAGE RATE COST	$
	$
	$

1. Determine actual energy cost consumption for the last 12 months for gas (therms), electrical (KWH), and total dollars.

KWH/ MONTH	DOLLARS/ MONTH	THERMS/ MONTH	DOLLARS/ MONTH	TOTAL DOLLARS/MONTH
	$		$	$
	$		$	$
	$		$	$
	$		$	$
	$		$	$
	$		$	$
	$		$	$
	$		$	$
	$		$	$
	$		$	$
	$		$	$
	$		$	$

2. Compute KWH/sq. ft./yr., therms/sq. ft./yr., and total dollars/sq. ft./yr. and compare to benchmark numbers.
3. Check off areas to investigate for energy reduction.

- ☐ Time Scheduling
- ☐ Holidays
- ☐ Fan Speed Control
- ☐ Pump Speed Control
- ☐ Chiller Optimization
- ☐ Boiler Optimization
- ☐ Heating and Cooling Controls
- ☐ Economizer
- ☐ Night Flush Strategy
- ☐ Variable Air Volume Retrofit
- ☐ Static Pressure Control
- ☐ CO_2 Monitoring
- ☐ Duty Cycling
- ☐ High Efficiency Motors
- ☐ Design Modification
- ☐ Energy Management System
- ☐ Deferred Maintenance
- ☐ Load Shed
- ☐ Motion Sensors
- ☐ Delamping
- ☐ Reflector Retrofit
- ☐ After Hour Lighting Monitoring
- ☐ Lighting Control System
- ☐ Incandescent Lamp Retrofit
- ☐ High Pressure Sodium Retrofit
- ☐ High Efficiency Re-lamping
- ☐ Demand Side Management
- ☐ Utility Rebates

4. Comfort Problems:
   - A. Acquire complaint logs from building occupants
   - B. Total complaints per week per HVAC systems
   - C. Construct monthly HVAC repair costs (should not exceed $.20/sq. ft./yr.)
   - D. Construct monthly total HVAC maintenance costs (should not exceed $.10/sq. ft./yr.)
   - E. Check off areas to investigate for comfort improvement:

- ☐ Air Distribution (Air Balance)
- ☐ Zone Selection
- ☐ Original Design Integrity
- ☐ Control Calibration
- ☐ Control Valve Leakage
- ☐ Control Damper Leakage
- ☐ Primary Control Functionability
- ☐ Secondary Control Functionability
- ☐ Scheduling
- ☐ Duct Cleaning
- ☐ Indoor Air Quality
- ☐ ASHRAE Standard 62-89
- ☐ Direct Digital Control Retrofit
- ☐ Building Warm-Up
- ☐ Humidity & Temperature Targets
- ☐ Preventative Maintenance
- ☐ Coil Cleaning
- ☐ Cooling Tower Treatment

FIGURE 23.10 Cost Control Worksheet.

tate to back-up their projects with performance contracts and guarantees. Be very wary of contractors and ESCOs that hesitate.

23.7 CONCLUSION

Energy cost reduction in today's environment of deregulation is very achievable, but it can only be achieved by dealing in facts. This will prove difficult from the standpoint that you will have to sift through lots of marketing and advertising strategies to sort fact from fiction; however, it is well worth the effort. Here is a brief summary of the useful information provided in this chapter:

- The most important thing to know before you talk to anyone about energy savings is your current energy benchmark in $kWh/ft^2/yr$.
- Compare your current energy use to the EPA *Energy Star Building* benchmarks to determine how much energy you could save.
- Realize that 90% of the time you can save the energy difference identified for a 2- or 3-year simple payback.
- If the paybacks being proposed to you are longer than 2 or 3 years, the audit has probably missed a large amount of operating savings and is more focused on large equipment change outs. This is the wrong approach. You want to do the fast payback items first, then look further out.
- Use performance contracts to simplify your efforts and put the responsibility on the people doing the work. Make sure there are real consequences for not performing. You will be glad you did.

The cost control worksheet model (Figure 23.10) will help you get started. The following steps will help you fill out most of the cost control worksheet and give you a place to start your analysis.

1. Take total monthly kWh and total monthly therm usage from your utility invoice and enter where indicated. Do the same for the dollar cost per month and fill in all numbers for the last 12 months.
2. Total the kWh, therms, and dollar cost for each of the last 12 months.
3. Divide the yearly kWh by the usable facility ft^2 (exclude garages, storage, etc.), this will give you your most important benchmark of $kWh/ft^2/yr$.
4. Do the same for the therms. This is not as important, but is useful in determining heating waste. (A typical facility uses 0.2–0.4 $therms/ft^2/yr$.)
5. Divide the total kWh dollar cost by the yearly total of kWhs to get cost per kWh.
6. Do the same for the therms.

If you compare your energy use to EPA's *Energy Star Building* benchmarks listed at the front of this chapter, you will be able to know how much energy savings potential is available in your facility.

The bottom half of the control sheet will give you some direction as to where to look for waste and comfort problems. Interestingly enough, the two usually happen together.

ENDNOTES

1. Cost effective paybacks for most facilities means a 3-year or less return on capital invested.
2. Ibid.

CHAPTER 24
SECURITY

Roy Spillenkothen[1] (deceased)
Architect/Engineer
Federal Government

Ronald J. Massa, Ph.D. (deceased)
Sr. Vice President
Rolf Jensen & Associates, Inc.
Framingham, MA
E-mail: rmass@rjagroup.com

24.1 CONTEMPORARY SECURITY THREATS

The on-going revolution in electronics and computer technology, as well as advances in materials technology, have greatly enhanced the arsenal of effective security tools, both high-tech and low-tech, available to combat historical security threats. The techniques and equipment to thwart ballistic attacks, breaking and entering, theft, and physical assault are addressed in many security texts.

New security threats such as bomb attacks and bomb threats, and chemical and biological weapon attacks are not as thoroughly treated in present texts. This chapter will focus on bomb threats and bomb attacks, treating both historical approaches to bomb defense and new tools for dealing with these threats. The chapter will conclude with a brief introduction to chemical and biological weapons and how they influence present security concerns.

24.2 WEAPONS OF MASS DESTRUCTION

Until the first atomic bomb was dropped on Hiroshima in the closing months of World War II, the concept of a weapon of mass destruction was never demonstrated in warfare. During World War II, the 10,000-lb high explosive bomb, the 16″ artillery shell, the V-2 rocket, and even poison gas demonstrated their capability to cause substantial damage and injury. Individual deployments of these weapons had the potential to kill or injure, at best, a few hundred people—then came the atomic age. The first attack on Hiroshima killed more than 40,000 people and leveled an entire city. The weapon of mass destruction was born and deployed in a declared war.

In the few decades since Hiroshima, we have seen many potential weapons of mass destruction deployed in terrorist attacks as opposed to declared wars. The sarin gas attack in a Tokyo subway and the explosive attacks at the World Trade Center and the Murrah Building are examples of the successful deployment of potential weapons of mass destruction. While these attacks caused only hundreds of casualties, they could easily have been configured to claim thousands of casualties from a population, which was not forewarned, by a state of declared war.

For instance, if the World Trade Center bomb were located a few hundred feet away near a public square filled with lunchtime crowds, rather than two stories below ground, this same weapon might have claimed 10,000 or more casualties. Clearly, those who design or manage facilities, which could be the targets of such attacks, must have an understanding of effective security measures, which can reduce the casualties and property damage weapons of mass destruction are capable of inflicting.

24.3 HISTORICAL DEFENSE MEASURES AGAINST EXPLOSIVE WEAPONS

Prior to the World Trade Center bombing in February 1993, American security interests were not focused on the large explosive weapon as a credible threat to the lives and property they were charged with protecting. Even though attacks in urban centers using such explosive weapons had taken place in Peru, Colombia, the United Kingdom, and elsewhere in the world, the magnitude of this new threat was only appreciated—in the United States—after the Murrah Building bombing in April 1995.

By April 1995, a great deal had been learned about this new threat to our security. Large bombs could be detonated just about anywhere—deep *inside* a building in New York City or just *outside* of a building in Oklahoma City. The perpetrators of such acts were not necessarily foreigners whose culture permitted or encouraged such actions, but could be a "typical" American or Americans. A potential target did not require either high visibility or high value. A federal building in Oklahoma City was hardly the U.S. Capitol. The perpetrators apparently faced little risk of death or injury in handling, transporting, or detonating such weapons. Relatively little technical knowledge was required to successfully mount such an attack.

All of this information posed a dilemma for building security interests. How do you defend against this threat—a threat that progressed from "not considered" to "very serious" in just about 2 years?

Security and law enforcement people reached quickly for the most basic procedural and technology based defense measures that could be put in place almost immediately. The following bomb defense improvements were implemented at many locations within weeks or months.

1. More uniformed security personnel were deployed to the extent budgets permitted.
2. Rudimentary vehicle barriers were installed and vehicle checkpoints were implemented around buildings deemed at risk from explosive attack. Figures 24.1, 24.2, and 24.3 show typical vehi-

FIGURE 24.1 Vehicle Barriers (Bollards); U.S. Capitol Grounds.

FIGURE 24.2 Vehicle Barriers (Planters); U.S. Capitol Grounds.

cle barriers deployed in defense of prominent U.S. government buildings in the Washington, D.C. area.

These low-tech, labor-intensive defense measures are still in place at many locations in the United States. It is instructive to briefly examine the rationale behind the adoption of such security measures and their effectiveness. Uniformed security personnel are used primarily to convey the notion that a particular target is well defended. If they are successful in conveying this target status, ideally they will cause most perpetrators to seek another target. A relatively small number of perpetrators will give up entirely and decide that they do not make good, fearless terrorists. The reader should note that uniformed security guards are an expensive and largely ineffective defense against bomb attack. Studies of many bombings around the world indicate that nearly all had some measure of uniformed security at the time of attack and none of the in-place security was able to prevent or significantly mitigate the effects of the attack. At the bombings of the two U.S. embassies in Africa, the uniformed guards outnumbered the terrorists by more than 10 to 1. Both embassy attacks were successful.

Vehicle barriers are more closely related to the science of bomb attacks. To do a great deal of damage, a large bomb—perhaps containing several hundred or several thousand pounds of explo-

FIGURE 24.3 Vehicle Barriers (Concrete Bollards); The White House.

sive—is required. Each 100 lbs of explosive occupies more than about 1 ft^3. Hence, a 1,000-lb bomb (approximately the size bomb used at the World Trade Center) occupies more than 10 ft^3 and is far too heavy to be brought to the detonation site by hand. Historically, a vehicle—car, van, or truck—is used. It follows that if you can keep a vehicle carrying a bomb away from a building you can keep the bomb away from the building as well. More importantly, the effects of an explosive detonation diminish very quickly with distance from the explosive. Thus, to a point, the scientifically based vehicle barrier is a far better defense measure than a uniformed security guard. That point is frequently reached in urban settings where there simply is not enough distance—or enforced standoff—available to sufficiently reduce the effects of the explosive detonation.

Nonetheless, the vehicle barrier, often cast in the shape of an outdoor planter, was the first attempt to bring science to bomb defense. Taking a closer look at improvised, explosive devices (IEDs) and the building targets they are directed against, we can find many other areas where science can be brought to bomb defense.

24.4 EXPLOSIVE WEAPONS

Explosive weapons, IEDs, generally comprise two components: a quantity of high explosive and a detonation device. The high explosive could be a manufactured solid explosive, such as TNT or C-4, or an improvised explosive, such as ANFO.[2] Table 24.1 shows some important properties of common explosives. The strength of an explosive is determined by its specific energy. This is the energy available, per pound of explosive, to cause damage or injury through blast effects. From the small sample of explosives in Table 24.1, it is apparent that the strengths of high explosives do not vary greatly. The strength differences between the listed explosives are less than 3 to 1. Note that some of the properties of ANFO are not reported in the table. ANFO is most commonly a "homemade" explosive with limited quality control or consistency in mixing. Despite these issues, ANFO remains a powerful explosive capable of inflicting great damage.

Explosive effects are primarily based on the weight of the explosive involved. A 2-lb charge has more damage potential than a 1-lb charge. For those involved in bomb defense, the density of an explosive used as a potential weapon is very important. Table 24.1 indicates that a density of about 100 lbs per cubic foot matches most high explosives. From this average density, we can conclude that a typical videocassette ($5\frac{1}{4}'' \times 1\frac{1}{2}'' \times 9''$) could contain about 4 lbs of high explosive; a package the size of a ream of paper ($9'' \times 12'' \times 2''$), which fits nicely into a briefcase, could contain 12.5 lbs of high explosive; and a typical "suspicious object" ($11'' \times 13'' \times 5''$) might contain as much as 41 lbs of high explosive. All of these packages must be treated as very serious security threats to people and property.

TABLE 24.1 Physical Properties of Explosives

Explosive type	Specific energy (foot-lbs/lb)	TNT equivalent weight (for pressure)	Density (lbs/ft^3)
TNT	1,514,200	1.00	99.8
C-4	1,631,450	1.37	98.6
Dynamite	907,850	0.60	81.1
ANFO	—	0.82	—
Nitroglycerine	2,244,500	1.42	99.2
PETN	1,943,000	1.28	110.5
RDX (cyclonite)	1,795,600	1.14	102.9

24.5 QUANTITATIVE CHARACTERISTICS OF EXPLOSIVE WEAPONS

Explosive materials share a property, which is very useful in analyzing the blast effects resulting from high explosive detonations. Blast effects obey a simple scaling relationship called cube root scaling. First, we shall examine blast effects and the parameters which define them, and then illustrate how cube root scaling facilitates prediction and analysis of these blast parameters.

When a liquid (such as nitroglycerine) or solid (such as C-4) explosive is detonated, the explosive material is very quickly converted to a hot gas. In this conversion process, a considerable amount of energy is liberated. Some of this energy, typically 10–20%, is liberated as heat through the extremely high temperature (several thousand degrees centigrade) of the gas by-product of the explosion. A great deal of the blast energy is liberated in the form of a shock wave that propagates radially away from the detonation site. This shock wave moves outward at a hypersonic velocity, referred to as the *shock front velocity*. For high explosives, the shock front velocity begins at many times the speed of sound and decays rapidly with distance from the detonation site. The shock wave from the detonation is followed by a second energetic phenomenon, an expanding gas cloud moving radially outward at roughly the speed of sound.

If you were to observe a detonation from a radial distance, R, from the site you would instantly see the flash of flame as the detonation begins. It travels toward you at the speed of light. Sometime later you would experience a sharp rise in the ambient pressure followed by a decay back to the ambient pressure. This "overpressure" is referred to as the *incident overpressure* and it is described by the idealized triangular waveform in Figure 24.4. The peak incident overpressure is called P_s, the *time of arrival* at R is called t_a, and the duration of the overpressure pulse, t_+, is referred to as the *positive phase duration*.

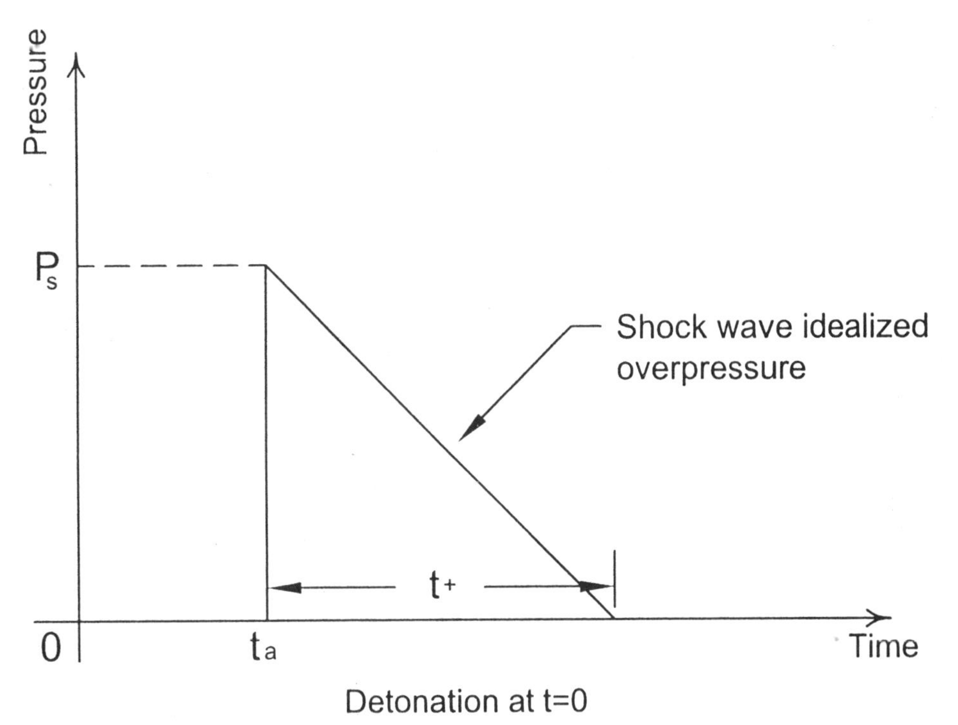

FIGURE 24.4 Typical Idealized Blast Shock-Wave Parameters.

If the incident shock wave were to strike a sizable solid object, such as an exterior wall of a building, it would be reflected from the object. The process of reflection can increase the pressure experienced by the object many times. Typically, *reflected overpressure* at a large surface is 2–10 times the incident overpressure arriving at the surface. This shock wave reflection from a solid surface can result in the transfer of a great deal of shock wave energy to the surface, sometimes causing it to fail.

Because of the cube root scaling of blast parameters, we can use a single chart, such as the scaling chart in Figure 24.5 to compute all of the blast shock wave parameters. For example, what are the blast wave parameters resulting from a 1,000-lb TNT charge at a distance of 100 ft? A 1,000-lb charge at a range of 100 ft defines a scaled distance $Z = 10$. For $Z = 10$, the following blast wave parameters can be read from the graph in Figure 24.5:

- *Incident Overpressure:* $P_s \approx 9.5$ psi
- *Reflected Overpressure:* $P_r \approx 25$ psi
- *Incident Impulse:* $i_s \approx 80$ psi-ms
- *Reflected Impulse:* $i_r \approx 180$ psi-ms
- *Shock Front Velocity:* $U \approx 1400$ ft/s
- *Positive Phase Duration:* $t_+ \approx 26$ ms
- *Time of Arrival:* $t_a \approx 42$ ms

Notice that several of these blast parameters are scaled by the cube root of the explosive weight. Whereas the chart itself depicts blast parameters for a *hemispheric ground burst,* the parameters for a *spherical air burst* can be approximated by reducing the charge weight by 50%. A hemispherical ground burst can often be used to model a mass of explosives placed in a car or truck a foot or two above the ground. A spherical airburst can be used to model an antiaircraft shell exploding in the air high above the ground or any other significant reflector.

This brief discussion of explosive effects is virtually all that is needed to approximate the loads on structures and humans from explosive weapons. Of course, in this treatment, we have made several simplifications, summarized in the following list, which are usually justified in terrorist bombing scenarios:

Cratering: When a bomb is detonated close to the ground it often evacuates a crater in the ground immediately below the charge. Up to 25% of the energy in the detonation (depending on soil materials) can be consumed in digging the crater. This energy loss reduces the detonation energy available to cause other damage or injury.

Fire: The very hot gases immediately following a detonation demonstrate the fact that about 10% or so of a bomb's energy escapes in the form of heat. These hot gases generally do little more than blacken nearby structural components. However, if the detonation occurs in the vicinity of a large number of vehicles in a poorly vented parking garage, these hot gases can ignite automobile tires, gas tanks, and upholstery to create serious fires generating a great deal of acrid smoke.

Vehicle Fragments: When a bomb is detonated in a vehicle, the explosion often tears the vehicle apart, propelling various automotive parts as high velocity shrapnel. These parts are also capable of inflicting damage and causing injury.

Quasi-Static Gas Pressure: If the bomb is detonated in a poorly vented location, such as an underground parking garage, in addition to the shock wave, an expanding quasi-static gas cloud can cause additional damage. For well-vented exterior explosions, this effect can often be ignored.[3]

24.6 BOMB ATTACKS—A HISTORICAL RETROSPECTIVE

Bombs directed against symbols of authority or political adversaries have long been taken as a fact of life in many parts of the world. Oil assets in Colombia have been under attack for decades, as

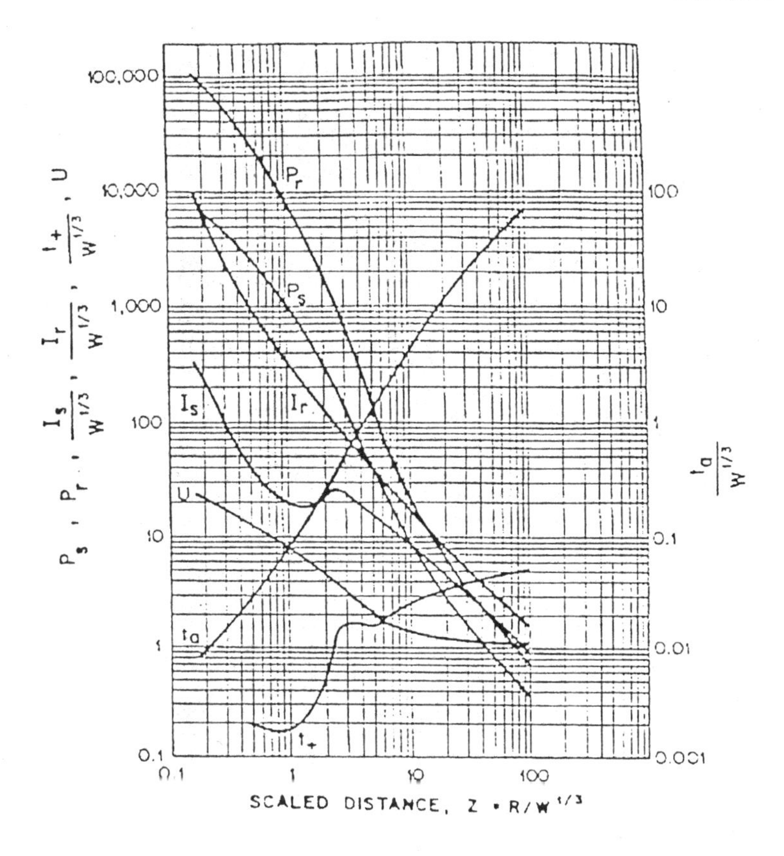

P_s = Peak Positive Incident Pressure (psi)

P_r = Peak Positive Normal Reflected Pressure (psi)

$I_s/W^{1/3}$ = Scaled Positive Incident Impulse (psi-msec/lb$^{1/3}$)

$I_r/W^{1/3}$ = Scaled Positive Normal Reflected Impulse (psi-msec/lb$^{1/3}$)

$t_a/W^{1/3}$ = Scaled time of arrival (msec/lb$^{1/3}$)

$t_+/W^{1/3}$ = Scaled Positive Phase Duration (msec/lb$^{1/3}$)

U = Shock Front Velocity (ft/msec)

W = Charge TNT-Equivalent Weight (lb)

R = Radial Distance from Charge (ft)

Z = Scaled Distance (ft/lb$^{1/3}$)

For spherical air burst use $W = 0.5 \bullet W_{actual}$

FIGURE 24.5 Blast Wave Scaling Chart.

have political targets related to the unrest in Northern Ireland. The United States was dramatically introduced to domestic terror by bomb with the World Trade Center bombing in 1993. Shortly after the World Trade Center bombing slipped from the headlines, the Murrah Federal Building in Oklahoma City was subjected to a major bombing in 1995. We will briefly describe these incidents and two others in different parts of the world to illustrate the actual effects of explosives weapons when

FIGURE 24.6 Bombing at San Isidro, Lima, Peru, 1992.

they are used as weapons of terror. All of these attacks, and many others, share some frightening attributes. Each demonstrates that the terrorist can deliver and detonate a sizeable explosive weapon virtually anywhere.

By selecting the detonation location, the terrorist determines whether a given weapon is a killer or not. The terrorist carries a relatively small quantity of explosives to the scene. These explosives, when detonated, can involve tons of materials in the incident through building collapse and falling and flying glass.

The four incidents depicted in Figures 24.6–24.9 illustrate nearly the full gamut of bomb attack consequences. The San Isidro bombing in a large open square did primarily glass damage. This damage occurred on a score of buildings—some more than 1,000 feet from the detonation. This underscores the importance of glass in urban bombings. Glass is the most common cause of human injury in bomb attacks. Glass damage and glass injury are exacerbated in dense urban environments where there is a great deal of glass to break, fly, and fall.

The World Trade Center is the only example among the four of an interior detonation. Significant structural damage occurred within the building, but very little (comparatively speaking) glass

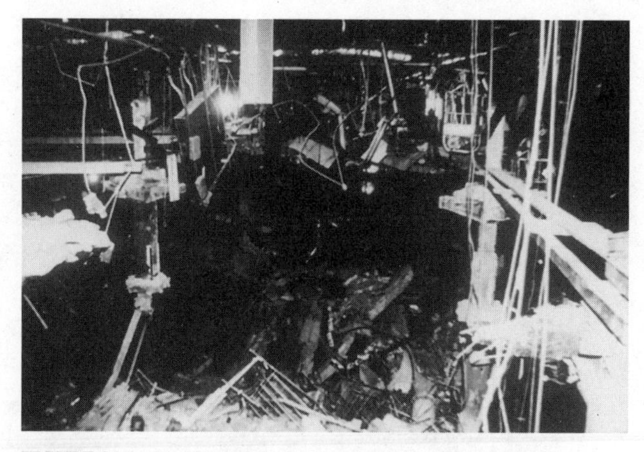

FIGURE 24.7 World Trade Center, New York City, February 1993.

FIGURE 24.8 Bishopsgate Bombing, London, April 1993.

damage occurred. The bulk of the more than 1,000 injuries were caused by smoke from the explosion-induced fires inside the building—specifically the poorly vented parking garage. The tower structure itself was sufficiently robust so that major collapse was not an issue.

The Bishopsgate bombing occurred outside in a densely populated downtown area of London. Fortunately, the area was evacuated because many warnings were issued before the actual attack. More than 500 tons of glass was broken and fell to the streets around and near the building that was the primary target of the attack. Some structural damage occurred to buildings near the detonation. There was no building collapse. There was one death. Ostensibly, the victim ignored the evacuation order.

The Murrah Federal Building bombing involved virtually all of the adverse consequences of urban bomb attacks. The building suffered significant collapse—largely because of the proximity of the bomb. Building collapse is a major source of fatalities in bomb attacks. A total of 168 died in this attack. Glass was broken for a mile or more from the detonation. Scores were injured by flying and falling glass. The proximity of the bomb to a parking lot across the street resulted in an exterior explosion-induced fire that contributed to an environment of dense smoke hampering rescue efforts immediately following the detonation.

FIGURE 24.9 The Murrah Federal Building Bombing, Oklahoma City, April 1995.

24.7 DEFENSE AGAINST BOMB ATTACK

Clearly, the incidents described in the previous section indicate that the terrorist can deliver and detonate large explosive weapons to nearly any public place. What can be done to defend people and property from such explosive attacks? There are three basic components of bomb defense: prevention, mitigation, and recovery. Usually, the best results come from a well thought out combination of all three of these components developed before an incident occurs.

Prevention is primarily a law enforcement and intelligence function that few commercial organizations are prepared to duplicate. However, there are contributions that well-prepared building occupants can make to the successful execution of the attack prevention function. Reporting suspicious persons and/or behavior near or within the facility, screening employees both when hired and when terminated, and maintaining good communications and working relationships with local and federal law enforcement will all assist in incident prevention.

The two areas where a meaningful bomb defense program can pay the greatest dividends are mitigation and recovery. In these areas, proactive management can be very effective in minimizing (or exacerbating) the consequences of both bomb attacks and bomb threats.

A 5-point bomb defense program, requiring minimal expenditures, has proven technically effective. This defense program makes use of detailed technical knowledge involving the facility in question as well as blast effects. Previous experience regarding bomb attacks is also very useful in implementing this program. The technical knowledge required is not difficult to obtain from bomb defense engineering professionals. The five components of the suggested bomb defense program are briefly summarized:

1. *Structural Screening:* Rule out the possibility of building collapse to limit fatalities from credible attacks. The potential for building collapse can be evaluated based on the size of a credible explosive charge at a given location and the structural details of the building. Estimates of structural component failure are readily made and good engineering judgment can often extrapolate between individual structural component failure and partial or progressive collapse of a building. Structural collapse, and its catastrophic consequences, is often relatively easy to prevent. Key structural components can be isolated from the blast or a few additional feet of enforced standoff can be provided.

2. *Analyze Your Glazing:* Many steps can be taken to minimize the consequences of flying and falling glass and limit serious injuries in an actual bomb attack. By virtue of their specific location on a building, some glass lights are less vulnerable than others. Once identified, the lights capable of inflicting the greatest injury become candidates for replacement with glass less likely to shatter, fall, or fly. Either laminated glass or glass lights inherently stronger due to their heat treatment can be used. Glass performance under blast loads is predictable and risk of glass injury can be significantly mitigated.

3. *Analyze Vulnerability to All Credible Threats:* Consider everything from mail bombs in the mailroom to truck bombs at loading docks and develop tactics to minimize loss if an incident occurs. There are several locations in and around a building that offer particular advantages to the terrorist. Executive suites, central computer facilities, uninterruptible power supplies, certain places of assembly, and ceremonial entrances and lobbies all seem to attract the attention of the terrorist. Frequently, simple structural or procedural defense measures can mitigate the effects of attacks directed toward specific components of the building.

4. *Integrated Security Development:* Integrate bomb defense with extant building security. Every security resource should have a specific role in bomb defense. That role should be thoroughly understood by all concerned and even practiced.

5. *Provide Dispassionate and Sober Self-Protection Training to All Personnel:* Even the most rudimentary training has been demonstrated to save lives in a bomb emergency. Even if it is not possible to implement the first four steps of the suggested program, self-protection training should not be overlooked. Such training is the least expensive component of a bomb defense program. Such simple direction as "If you feel uncomfortable or suspicious about anything going on in or around your building, go to the elevator lobby, not the window!" can save lives.

Recovery from a bomb attack usually proceeds in two phases. In the initial, short-term phase, emphasis is properly placed on rescuing the injured and recovering the dead. The subsequent, longer-term phase involves restoration of normal operations. The duration and cost of this phase can be influenced in favorable ways by modest investments made before the attack occurs. Some of these investments include such items as data back-up, planning for clean-up activities, and identification of mission critical activities and development of specific plans for their restoration.

24.8 CHEMICAL AND BACTERIOLOGICAL WEAPONS—A LOOK TO THE FUTURE

As disruptive and heinous as a major bomb attack on an urban complex can be, its ability to kill and injure is dwarfed by the potential damage that could result from the deployment of chemical or bacteriological (C/B) weapons. Defense against a bomb can always be made very simple—keep sufficient distance or standoff between a potential bomb and a potential target. Defense against C/B weapons is often much more complex and difficult, frequently involving sophisticated equipment and techniques.

First, there are a very large number of different chemical and bacteriological agents, each of which exhibits its own characteristic behavior. Sarin gas is a typical potential chemical weapon; anthrax is a typical potential bacteriological weapon.

Fortunately, actual attacks involving C/B weapons in terrorist scenarios are rare. The most notable terrorist attack in the last 10 years involving chemical weapons involved the release of sarin gas in a crowded Tokyo subway station. There are no significant reports of the use of bacteriological weapons during this period.

Chemical weapon attacks usually involve the release of the chemical, which could be a liquid, gas, or solid, in a location from which it can spread to large populations of people. Poisonous gases allowed to filter into an HVAC system will eventually affect virtually every person in the building. Once exposed, these individuals may be fatally injured even though they exhibit no immediate effects.

Bacteriological weapons are even more sinister in their behavior. For example, a biological agent contaminating a water system can fatally infect people for several days before the first individuals show symptoms of the infection.

Defense against chemical and bacteriological weapons involves constant vigilance in monitoring all potential points of entry and/or dispersion in all critical infrastructure components.

ENDNOTES

1. It is with sadness that the editor must report that Roy Spillenkothen passed away while writing this chapter. We are indebted to Ron Massa, his good friend and co-author, for completing his work.
2. A slurry mixture of ammonium nitrate, a common fertilizer, and fuel oil.
3. For a technical discussion about quasi-static gas pressure and how to estimate this load, as well as additional information about blast loads and their effects on structures, see Baker *et al.*, 1983. *Explosion Hazards and Evaluations.* New York: Elsevier.

P • A • R • T • 5

TECHNOLOGY

CHAPTER 25 OVERVIEW AND CURRENT STATE OF FM TECHNOLOGY

25.1 Introduction

This section outlines the various generations of facility management (FM) technology—each generation of which results in a paradigm shift of how computers influence the practice of FM. To understand the current application of technology, it is also necessary to understand infrastructure management, which addresses the relationships between people, assets, and locations from a common technology infrastructure throughout an enterprise.

25.2 Technology Evolution: Generations 1 and 2

Generation 1 of technology was FM on mainframes. These machines were expensive and resulted in centralized, cohesive databases. This changed significantly with the advent of PCs and decentralized computing. FM "islands of automation" brought about widespread use of FM computing for the first time.

25.3 CAFM and CIFM

Generation 3 of FM technology was the integration of applications running off a common database either on a desktop (CAFM) or, using client-server databases, throughout the organization (CIFM). For the first time, it was possible to integrate and share common data between applications.

25.4 Generation 4: The Internet and Infrastructure Management

The Internet is the largest network in the world. At first FM vendors simply used the Internet to publish reports. Increasingly, entire applications moved onto the Internet. A manifestation of this is collaborative project Web sites which offer functions such as work flow, document management, and redlining. Such software, mostly used for construction project management, is changing how clients and design consultants communicate.

E-commerce is another manifestation of software technology migrating to the Internet. The ability to purchase goods and services on the Internet will increase ten-fold between 2000 and 2004.

The final aspect of the Internet discussed in this section is infrastructure management. IM is a centralizing force that manages all FM, real estate, HR, IT, and financial data from a common technology infrastructure.

25.5 Generation 5 and Beyond

Just when we thought that we understood the implications of the Internet, along comes high speed Internet connections, wireless communication, and "smart" assets. This section discusses the implications of these and other Internet devices.

CHAPTER 26 INTEGRATING THE INTERNET INTO FM

26.1 Introduction

Why should the facilities manager be concerned with Internet technology? How do I know what I need to know to successfully leverage Internet technology for facilities management (FM)? How can understanding Internet technology make me a more effective manager? This section provides an introduction to the chapter.

26.2 What Is the Internet and How Does It Work?

This section presents introductory information on the technology behind the Internet. The section discusses the basic infrastructure of internetworking, including how the Internet is structured and how information travels from a remote information server to your desktop. It is necessary to have a basic understanding of how the Internet works to understand how it can be useful to the facilities manager.

26.3 Internets, Intranets, and Extranets

The Internet is composed of three separate and distinct areas of accessibility. These areas or contexts are primarily defined by how they are accessed and who can access them. This section describes the differences between these three areas. The section discusses typical examples of each area and how these areas may be utilized in FM.

26.4 Internet Applications for FM

The Internet presents us with a wealth of rich technology with which we can provide business solutions. This section discusses a wide range of FM and general business applications available to the facilities manager that utilize Internet technologies. E-mail to e-Commerce, the Internet provides almost limitless opportunities for improving business processes through distributed on-line access to information. In this section, we explore Internet applications and implementation strategies for utilizing those applications in FM.

26.5 Internet Technologies

The Internet, by definition, is technology. Discussions of technology are often technical—comprised largely of secret acronyms and buzzwords. Facilities managers implementing Internet solutions can become lost in the terminology associated with presentations made by Internet solution providers. This section includes an Internet terminology primer with descriptions of many of the common terms associated with Internet technology.

26.6 Conclusion

This section provides a brief summary and conclusion to the chapter.

CHAPTER 27 A/E/C INDUSTRY CASE STUDY: THE BIRTH AND DEVELOPMENT OF A COLLABORATIVE EXTRANET, BIDCOM

27.1 Preconception: Before Birth

How did the concept of a collaborative Extranet for the A/E/C industry come about? What industry issues needed to be addressed that created a demand for a collaborative Extranet?

27.2 Birth: You Have to Start Somewhere!

The story of how Swinerton & Walberg Builders met the founders of Bidcom, and how the initial Bidcom product concept was modified to suit the customer's most immediate needs.

27.3 Infancy: The Time to Grow up Is Now

A discussion of the first project owner, Charles Schwab & Co., and the foresight of their management team in agreeing to support the implementation of a brand-new process using the Internet to assist in managing a design and construction project.

27.4 Toddler: Go Quickly from Crawling to Walking to Running

A discussion of how the Bidcom system was developed and deployed on its first project. Some of the issues and challenges of implementation and bringing new features on-line are discussed.

27.5 Childhood: Challenge Doing It the Same Old Way

How can we take our newly developed tool and optimize its ability to assist in improving the efficiency of our business practices? Continuous improvement must be applied in tearing apart all of our processes and mapping them into the project Extranet as each new functional element is added to the system.

27.6 Adolescence: Metrics

What kind of results can we measure in evaluating what kind of efficiencies are gained by using a collaborative Extranet for the A/E/C industry? This section discusses the metrics on the return on investment and the improved profitability of using a collaborative Extranet.

27.7 The "Bottom Line": How Productivity Savings Turn into Increased Profits

An examination of how marginal productivity improvements can drastically effect overall bottom-line profits. This section demonstrates how when fixed overhead costs are covered, small gains in productivity at a job and gross fee level dramatically effect ultimate net fee results.

27.8 Back to Conception: Why Project Management and Administration Tools before Bidding?

Why did Swinerton & Walberg Builders push vendors into working on contract administration improvements using the Internet prior to working on their original concept of on-line bidding? The reasons for replacing a very thoughtful bid analysis and submission process with on-line automation and its potential pitfalls are discussed in some detail.

27.9 Teen Years: Procurement

How will procurement be played out on the Internet? This section includes a speculation of various procurement models and how they will be implemented on the Web.

27.10 Adulthood: Where Can We Go from Here?

Where can Bidcom and other A/E/C Extranet services go from here? This section provides an analysis of the current state of the A/E/C Extranet industry and some thoughts on where the industry might broaden their offerings to continue to work element by element through the design and construction industry to squeeze out efficiency at every step of the process.

CHAPTER 28 TECHNOLOGY AT THE ARCHITECT OF THE CAPITOL: A CASE STUDY

28.1 Who Is the Architect of the Capitol?

This chapter contains a case study on the implementation of CAFM in an extremely challenging environment: Capitol Hill. This first section will introduce the architect of the Capitol, its responsibilities, and other characteristics that will define the requirement for innovation in FM technology.

28.2 Apparent Need

Since 1793, when the agency was formed, demands placed both on the facility and the facility managers have changed greatly. Today, facilities are a vital component of strategic decisions, its issues often being decided in the boardroom. From downsizing to Y2K, converging forces compelled the AOC to change its business culture.

28.3 Strategic Plan

In this section, we will take a look at the steps involved in a typical CAFM assessment and the unique experiences of the AOC. The foundation of an implementation that sticks is in the needs analysis, process definition, and appropriate product selection.

28.4 Implementation

Facility managers know that their departments are complicated. Bringing systems into the department requires close coordination, clear communication of purpose, and often a cultural change. At the AOC, this can be seen through the project management and oversight processes.

28.5 The Road Ahead

Finally, we will take a look at the strategic direction of the CAFM initiative, and how it embraces state of the art technology. The success of the project will be the flexibility of the technology as both FM and CAFM develop at the AOC.

CHAPTER 29 MICHIGAN STATE UNIVERSITY FACILITY MANAGEMENT MASTER'S LEVEL CERTIFICATE PROGRAM: A CASE STUDY

29.1 Introduction

The demand for independent learning is increasing. As organizations adopted the theory of "just in time" processes to improve their effectiveness and efficiency, so are educational institutions developing methods and processes that support "just-in-time" learning. The purpose of this case study is to share the experience and evolution of the Michigan State University (MSU) facility management master's level certificate program, which is delivered entirely over the Internet.

29.2 Program Development

This section traces the evolution of the MSU program from the initial idea, through the identification of educational drivers, to the development of an Internet-based, active learning program. Learn about the new educational paradigm in which the instructor becomes the "guide on the side" and the student is an active participant in the educational process.

29.3 Content Development

The content of the MSU program was developed to cover a key knowledge base needed by facility managers. This section describes the content of the four courses in the program: *Facility Management: Theory and Principles, Information Management for Facility Professionals, Achieving Facility Management Organizational Effectiveness,* and *Facility Real Estate and Building Economics.*

29.4 Course Delivery: A Partnership

Developing and running a master's level certificate program over the Internet requires the expertise and support of many people. This section describes the MSU team, exploring the roles of the instructor, on-campus coordinator, off-campus coordinator, teaching assistants, and the MSU virtual university.

29.5 Design and Structure of the Internet Courses

A consistent organizational structure is used for each of the four courses in the program. This section explains the 3-D matrix design of the course, and describes each component of the course design in detail. Illustrations from one of the courses show you how students navigate through the courses. The communication tools WebTalk and LiveChat are introduced.

29.6 Learning on the Internet

Built into the courses in the MSU program are many methods of keeping the student engaged—from welcome videos to LiveChat discussions to sharing of information. This section shifts away from the physical layout of the course to the more human, interactive aspects.

29.7 Measuring Success

Three critical success factors have been met by the MSU program: to run a pilot project, to meet the demand, and to engage the student. This section summarizes the strategies that the program has used in their successful "high tech, high touch" approach.

CHAPTER 30 GIS CASE STUDY: SPACE MANAGEMENT AT THE UNIVERSITY OF MINNESOTA

30.1 Introduction

This section describes the selection and deployment of GIS (geographic information systems) technology to support the space management needs of the University of Minnesota.

30.2 Background

The University has facility assets valued at over $3 billion, covers 24 million ft^2 of space in 60,000 rooms in over 1,000 buildings spread across five major campuses and more than 20 research sites and experimental stations. At the time this project was deployed, using GIS to manage room level detail represented a new application of this technology.

30.3 Requirements Definition and Strategy Design

This section describes the process that the University followed to select and implement the system. It begins with a discussion of project team roles and the processes that were deployed to develop business and functional requirements for the system.

30.4 Enterprise-Wide Data Sharing Strategy Emerges

This describes the vision of the "enterprise-wide" approach targeted at supporting a broad constituency of planners and administrative staff across the university. As stewards of the university's facilities, FM's role included managing facilities in use across the enterprise. The new space management system would extend that stewardship role to include providing facilities information that would easily integrate with information from other enterprise databases for decision support by academic and administrative groups.

30.5 Technology Evaluation and Selection

This section discusses the decision that led to the choice of a GIS system and other tools that form the basis of the navigation and reporting environment. The ability to connect space management data (*institutional data*) with information in other university and departmental databases for graphical navigation and reporting is described.

30.6 GIS Interface Highlights SPACE System Design Strategy

Describes the design strategy for the new space management system (SPACE) which is centered on a graphical navigation and reporting component built on GIS technology.

30.7 Institutional and User-Defined Data Supports Analysis of Business Processes

Business requirements clearly indicated the need to support integration of data from other systems with information about facility utilization. The concepts of "institutional" vs. "user-defined" data were developed to allow SPACE data to be used in concert with other university or departmental databases.

30.8 Deployment Strategy Designed for Enterprise-Wide Use

This section includes a description of the data warehouse concept that was developed to enable access by university staff outside of the FM department.

30.9 What-If Scenario Planning

Describes the ability to conduct "what if" scenarios for planning and design on their desktop without modifying the institutional data.

30.10 SPACE Data Management

Describes on-going maintenance of databases used by SPACE.

30.11 Implementation Incorporates Pilot Project to Confirm Design Strategy

The discussion of implementation includes an analysis of a pilot project and the basis for the data population and quality control efforts that were deployed leading to a full production implementation of the new system.

30.12 Moving to a Full Production Environment

Activities associated with full integration and deployment of the SPACE system across the university.

30.13 Enterprise Concept Embraced

The chapter concludes with a description of some of system deployment to support several business and planning activities at the university.

30.14 Summary

CHAPTER 31 TEACHING TECHNOLOGY AT THE UNIVERSITY OF NEW SOUTH WALES: A CASE STUDY

31.1 Educational Environment

The design and management of a Master degree course in facility management (FM) is a reflection of the accrediting authority and the client base. Nonattending students place additional demands on decreasing resources. The Internet may be the appropriate delivery vehicle in the future, but there are hidden costs and matters of equity. CD-ROM and DVD together with paper notes offer low-cost guaranteed delivery. Nearly all tuition topics can be demonstrated with commercial software provided at notional or zero cost. This lays the foundation for experiential learning in which the student applies software-based procedures or processes and then becomes acquainted with the underlying theory.

31.2 Facility Planning within a Corporate Planning Context

Facility planning is a component of corporate planning. The changing corporate business objectives define the nature of the organization required to deliver the business outcomes. Both the nature of the organization and its physical accommodation needs define the housing and support services. A regime of intra- and inter-organizational services must be in a state of congruence. If the organization is in a state of constant change, then the regime of services is also changing. If the service regime is changing, then the accommodation requirements are changing.

31.3 Dynamic Simulation

Developing a regime of congruent services merely defines the currently required organizational services. The quantity of these service requirements can be modeled using dynamic simulation. The development of a regime of congruent services followed by a dynamic simulation of the organization's processes can be used to quantify the required physical space.

31.4 Requirements Engineering

Requirements engineering is a well-developed discipline in defense material. It recognizes that the translation of user requirements into a facility is a multi-stage process involving the statement of user needs, a definition of systems capable of supporting those user needs, and the definition of elements and components required to house the selected systems. Requirements engineering provides a bi-directional audit trail throughout the design and construction phases. The impact of a change in user requirements can be traced through the systems and enclosing architecture. Conversely, a change in architecture can be traced through to the interested users.

31.5 Space Economy

Space economy is a gaming process that seeks to model space need with space availability. It is well presented as a pair of hierarchical data sets. The data must be quarantined from the space management data. Space economy seeks to provide a holistic view of the organization's changing space needs and the changing availability of space. These are time-dependent models that can be used to identify milestones in corporate accommodation change.

31.6 Space Location, Allocation, and Optimization

These studies employ the mathematics of optimization to generate a range of possible spatial solutions. The physical constraints of buildings have generated specialized accommodation planning technologies. The results are not necessarily architectural solutions, but rather an indication of appropriate strategic directions. These studies in optimization examine macrogeographic locations and micro-staff locations within a facility. Simulated annealing and genetic algorithms can be employed in the search for optimal solutions.

31.7 Visualization in 3D and 4D

Historically, 3D visualization of a proposed facility tended to be an outcome of the architectural process. Recent low cost technology enables building owners and users to visualize pre-architectural solutions in 3D. If managed effectively, this empowers the building users to define a statement of requirements that can be used to brief the design consultants. 4D technology models in 3D, the assembly sequence of a proposed facility.

31.8 Facility Information Management Systems

Facility information management systems are databases with or without building graphics. The objective is to maintain a register of the current and future state of users, systems, elements, and components. The depth of detail is user defined. The Internet has provided a fundamental change to this technology. Browser, Java, and Windows-based technology impact on the degree of user interaction. The vast array of data maintained in such systems implies an equally large number of interested parties. The Internet can now provide the necessary communications infrastructure.

31.9 Terotechnology

Terotechnology studies the life cycle of physical assets. However, life-cycle costs are a function of higher level policies dealing with building ownership and alternate land use. These in turn impact on the physical quality or performance requirements that underwrite the regime of maintenance expenditures. The nexus between corporate planning and facility maintenance planning is paramount.

31.10 Building Serviceability

Every facility delivers accommodation services. The relevance and value of these services is defined by a taxonomy of user needs. Building functionality and serviceability develops a set of scaled issues that define the performance of a facility. These are compared with an independent set of scaled user requirements. The objective is to establish a "marriage," provided minimum threshold values are honored. Supporting software can undertake an evaluation of the prospective "marriage" or rank prospective facilities.

31.11 Conclusions

CHAPTER 25
OVERVIEW AND CURRENT STATE OF FM TECHNOLOGY

Eric Teicholz
President and Founder
Graphic Systems, Inc.
An independent technology consulting company
E-mail: teicholz@graphsys.com

25.1 INTRODUCTION

This chapter discusses the evolution of technology generations and their effect on facility management (FM)—both in the past and in the near term future. There are at least six generations of technology where there has been a "quantum" (i.e., discontinuous) change between generations. This paper covers a period of almost 50 years, from the early 1960s to approximately 2005—the year it is estimated that the fifth generation of technology (high speed Internet and "smart" assets) will become commonplace within FM. The author makes no attempt to go beyond this timeframe—technology simply changes too rapidly.

To understand FM technology, it is also necessary to understand the current context of that technology. Infrastructure resource planning, or infrastructure management (IM), is an emerging term which describes the trend toward broadening the FM application of computers beyond just FM or real estate. IM addresses the relationships between all people, locations, and assets throughout an enterprise and provides a data automation environment for managing not only computers, but data, telecommunications, applications, and even the work processes associated with changing and maintaining this broad infrastructure. It incorporates not only the FM function, but also real estate, human resources, information technology, and all aspects of finance and procurement. We particularly live in very dynamic times from the perspective of Internet technology and IM and their impact on FM. Understanding how technology, particularly Internet technology, is evolving and the impact of IM is of fundamental importance to facility managers for the effective practice of their professions.

25.2 TECHNOLOGY EVOLUTION: GENERATIONS 1 AND 2

Mainframe computers with applications designed for facility managers existed in the early 1960s, before the term facility manager even existed. There were few vendors at this time, but solutions

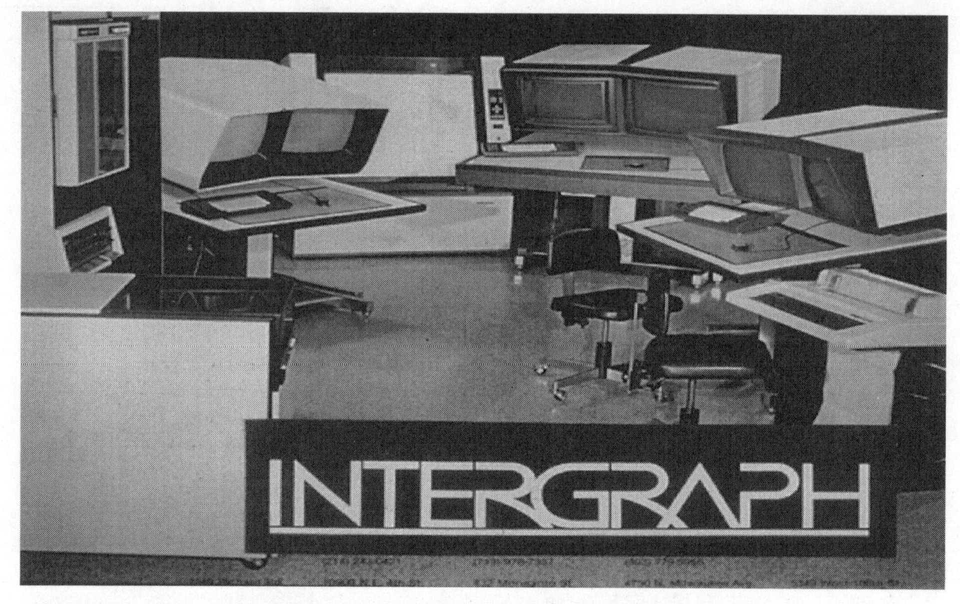

FIGURE 25.1 Intergraph Dual Screen CAD Workstations. *(**Source:** Intergraph Corporation.)*

were quite comprehensive and integrated. Looking at vendor offerings from companies such as IBM and Intergraph at this time, one finds the following applications software:

- Facility inventory management
- Requirements programming
- Computerized maintenance management systems (CMMS)
- Location and layout planning (i.e., stacking and blocking)
- Facility master planning (scenario planning)

Mainframe FM software vendors represented expensive investments for organizations and the integration of graphic and nongraphic databases was complex. Hardware was likewise expensive. Figure 25.1 depicts a CAD graphic workstation from Intergraph Corporation from the late 1970s. These dual screen displays usually cost over $100,000. Usually one screen was used for menuing (option selection) while the other displayed the resultant graphics. Such workstations could only be afforded by the largest of aerospace and oil companies.

Starting in the mid-1980s, this centralized, integrated model for FM software technology became fragmented with the widespread deployment and acceptance of personal computers (PCs). PC applications, representing the second generation of FM technology, were initially built around office automation software. Although some third party software vendors started to emerge for various FM applications, most efforts were the result of someone using a word processing, spreadsheet, or database program for an individual need. No standards existed for graphic or nongraphic databases and these "islands of automation" rarely were designed for integration with other applications.

Figures 25.2 through 25.5 represent the SPACE program, developed by a vendor during the 1980s. SPACE was a space planning and management program, originally written under DOS, for developing stacking and blocking diagrams. Figure 25.2 depicts a stacking plan for a building that

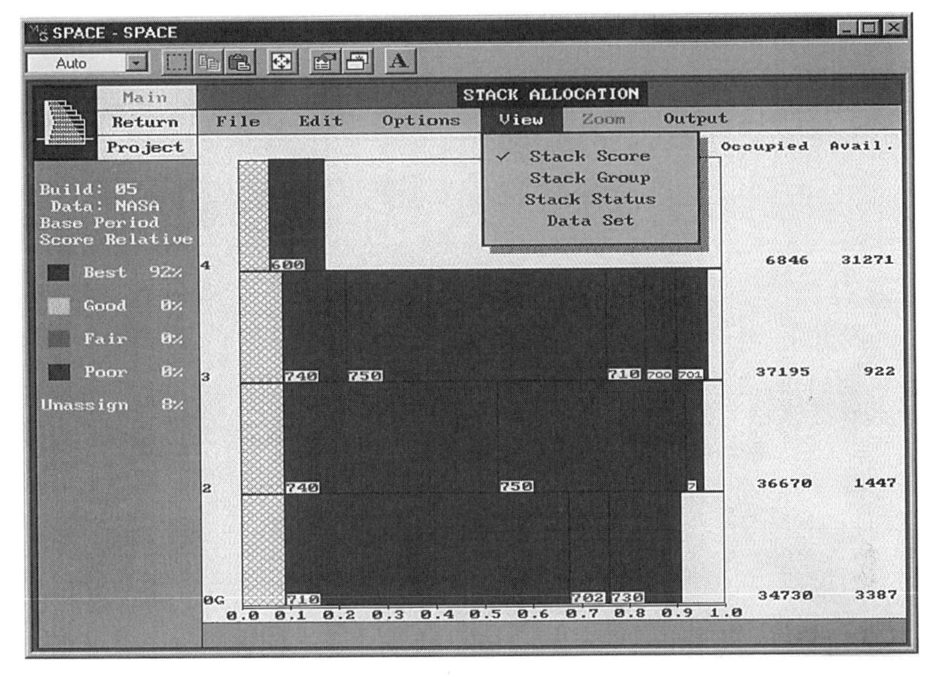

FIGURE 25.2 SPACE Stack. *(**Source:** Graphic Systems, Inc.)*

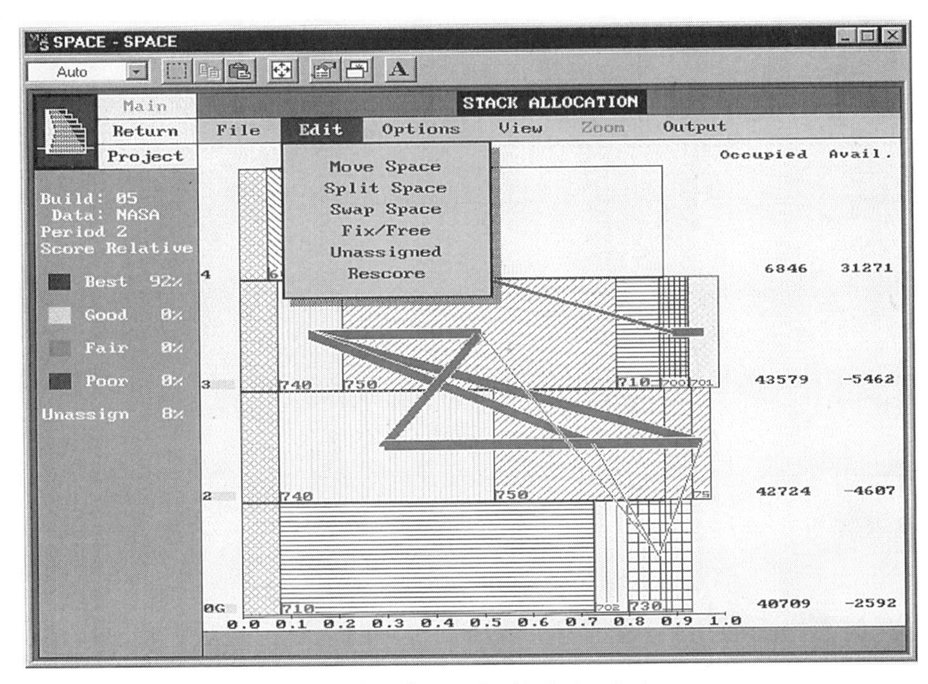

FIGURE 25.3 SPACE Adjacency Analysis. *(**Source:** Graphic Systems, Inc.)*

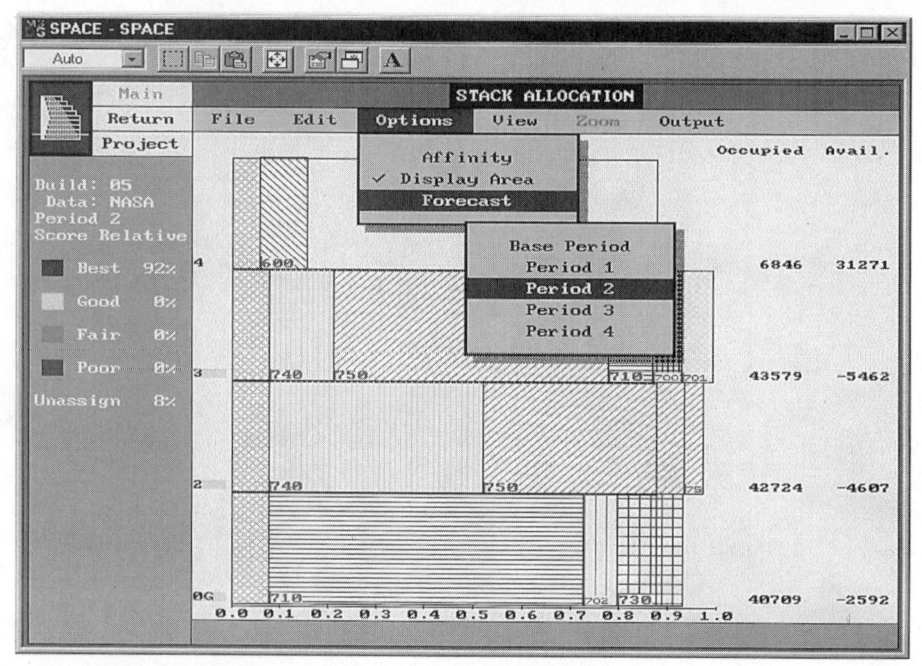

FIGURE 25.4 SPACE Forecasting. ***(Source:** Graphic Systems, Inc.)*

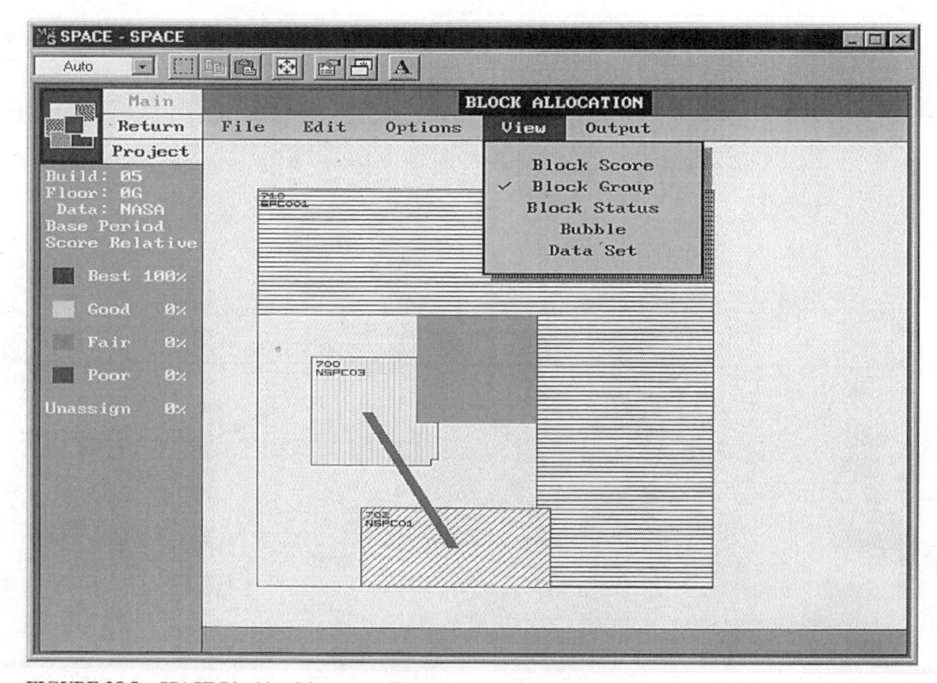

FIGURE 25.5 SPACE Blocking Diagram. ***(Source:** Graphic Systems, Inc.)*

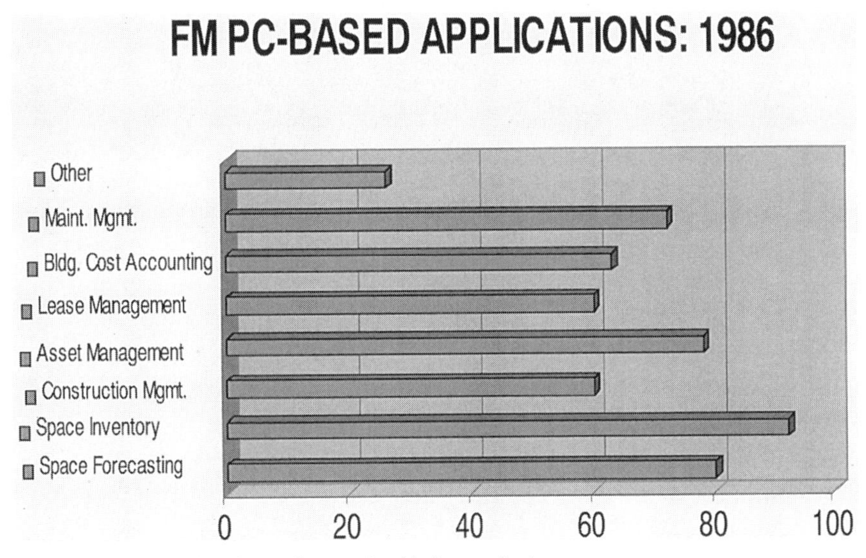

FIGURE 25.6 PC Utilization. *(**Source:** Graphic Systems, Inc.)*

was automatically generated from an adjacency matrix. Such a matrix is input by the user to indicate desired pair-wise relationships between design elements. The lower the number input, the closer the desired adjacency. The software's algorithm uses a heuristic improvement approach to generate a local "optimum" solution. In the SPACE program, color is used as a metric to depict how well the algorithm was able to satisfy the adjacency constraints. Figure 25.3 shows the adjacencies input by the user as lines connecting the two spaces for which there was an adjacency. The thicker the line, the stronger the adjacency. The shading of the rooms depicts the organizational group to which the space belongs. Figure 25.4 illustrates the selection of a new forecasting period for which SPACE will then generate another stack. Thus, users can input various space planning scenarios based on different growth scenarios and determine if the scenario will fit into the building envelope. After a user is satisfied with a particular stack, the resultant spaces are passed to a floorplan for optimization (Figure 25.5).

Not surprisingly, the FM applications that were developed on the PC during this time were significantly different from earlier mainframe systems. An FM technology survey conducted by the author during the mid-1980s revealed again that space inventory (used by over 90% of the respondents) was the primary application of FM PC software. Figure 25.6 depicts the PC-based applications that were popular at this time.

25.3 CAFM AND CIFM

The third generation of FM technology, starting in the early 1990s, was characterized by robust integration between various FM graphic and nongraphic applications, still using the PC as the primary hardware platform. This was an era of enormous market growth and maturing PC technology. The earlier computer-aided facility management (CAFM) systems were desktop solutions whereby data was moved over a local area network (LAN) to the desktop PC. The CAFM software then processed the data and returned the results over the LAN to users. This process tended to be slow (especially for large CAD graphic files) and restricted their use to mostly inventory (property, space, and assets) management and reporting applications. Figure 25.7 depicts the software modules of SPAN FM (now called FacilityCenter from Peregrine Systems). This diagram is meant to

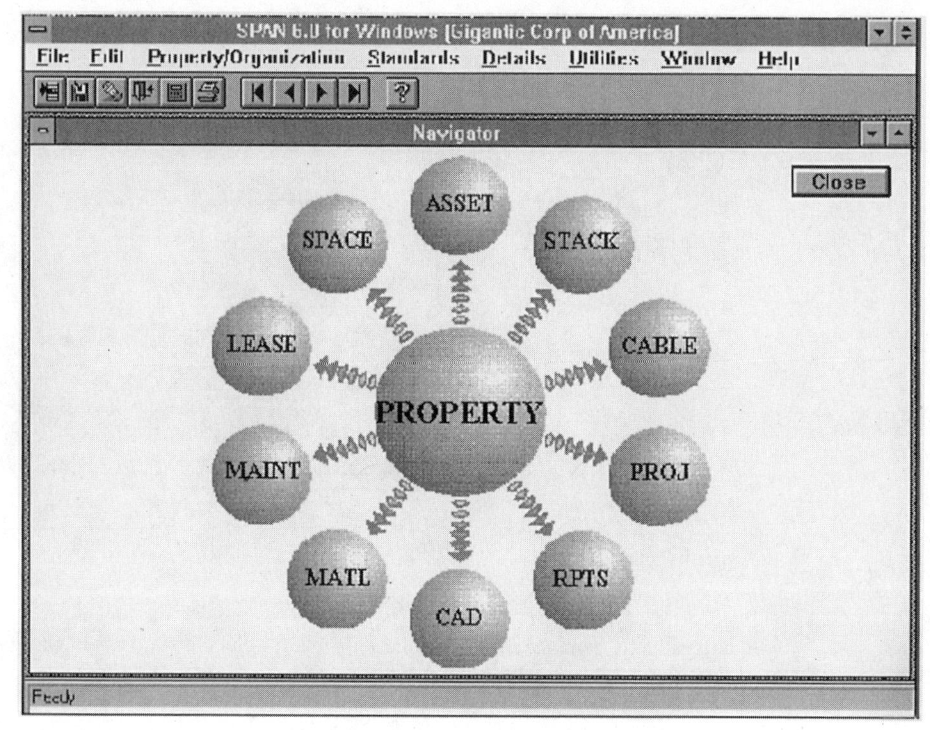

FIGURE 25.7 SPAN Software Modules. ***(Source:** Peregrine Systems.)*

depict the integration of FM modules that shared data between the various application modules. This was in contrast to the previous software generation where various nonintegrated applications often resulted in redundant databases.

The next software improvement came as a result of new database management systems (DBMSs). With the advent of client–server databases from companies such as Oracle and Sybase, enterprise-wide CAFM systems (called computer-integrated facility management or CIFM systems) significantly reduced the data transmission bottleneck and resulted in increased links to external databases managed by non-FM groups such as information technology, finance, and human resources. Client–server databases made it possible to process data remotely and just return the results to the client software. This disbursed processing cut back on the need to load all data on a single machine and, in the process, resulted in significant savings in data transmission time.

Thus, real-time space reports with occupancy information and occupancy cost analyses started to be generated by FM systems. These beginnings of linking FM data with mainstream corporate databases had the result of making CAFM/CIFM more strategic in nature and raised the importance of facility managers and their role in managing spatial and work-related information. Organizations, for example, started to track costs (personnel and assets) associated with space, which in turn led to space chargebacks to the business units.

Figure 25.8 depicts an output from Aperture showing an occupancy plan for a client. Depicted are data associated with room number, person's name, and phone extension (coming perhaps from a human resources' database), and furniture assets. Figure 25.9 depicts the more advanced application of move simulation. This application from FIS has the ability to use scheduling and financial data to generate "what-if" move scenarios for the space planner.

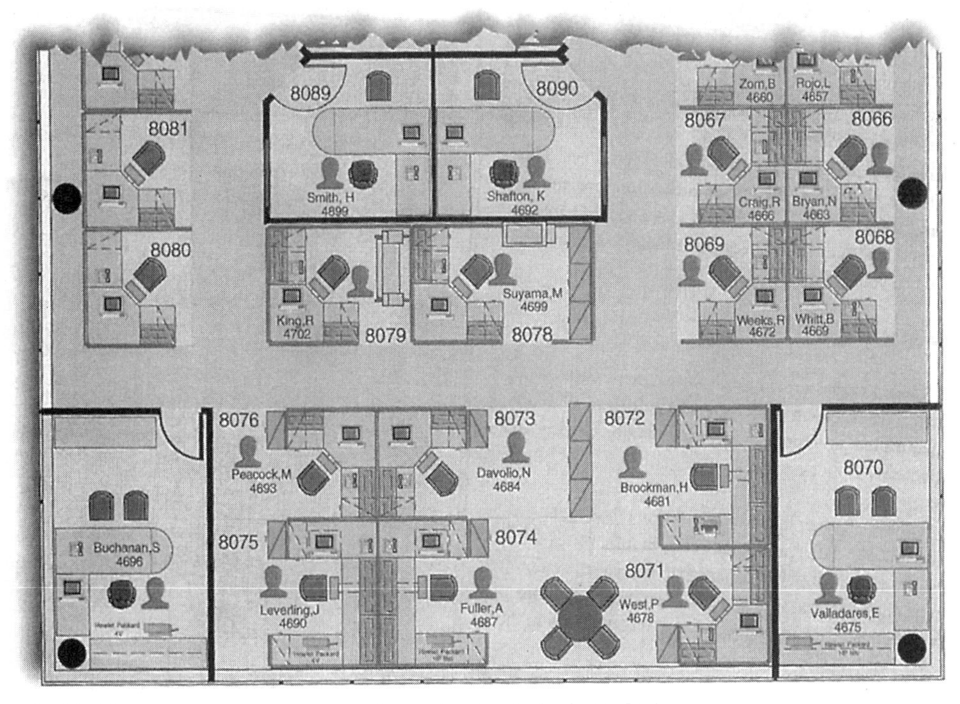

FIGURE 25.8 Aperture Occupancy Plan. *(**Source:** Aperture Technologies, Inc.)*

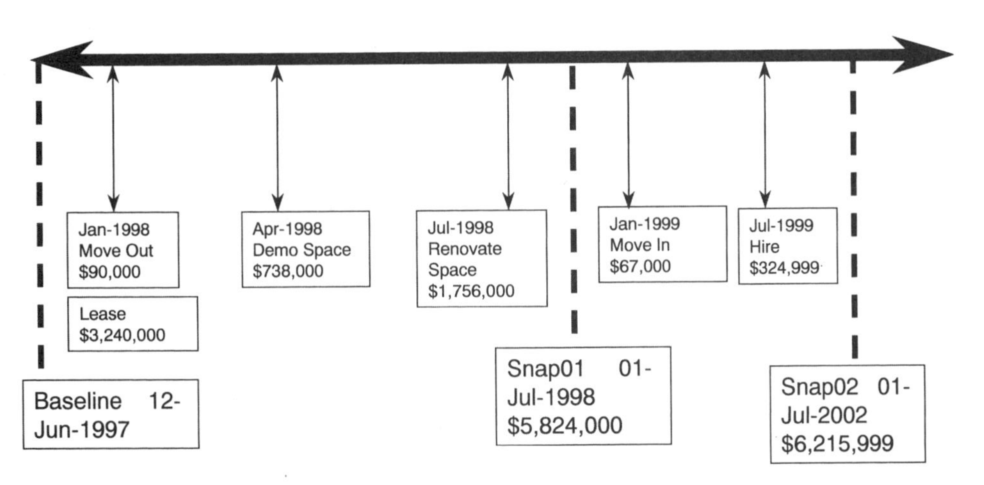

FIGURE 25.9 FIS Move Simulation. *(**Source:** FIS, Inc.)*

25.4 GENERATION 4: THE INTERNET AND INFRASTRUCTURE MANAGEMENT

25.4.1 The Explosion of the Internet

The Internet (discussed in detail in the next chapter) existed over 20 years ago, primarily as an R&D tool between universities and other research organizations. It was not until the early 1990s, however, that someone linked together data stored on various computers by being able to use a mouse to click on graphic or text icons. This rapidly led to the development of the ability to hyperlink various Web addresses (called URLs for uniform resource locators) using a presentation format called HTML for hypertext markup language. HTML is a language for describing how a site appears on a computer display. Simultaneously, companies invented an easy-to-use common access/navigator tool called the browser for moving between various sites that led to the formation of the World Wide Web (WWW).

XML (extensible markup language) is the next generation of Web languages. The big difference between HTML and XML is that XML not only transmits data regarding display parameters, but also interprets what the data means. In 1999, companies started defining an aecXML domain for vendors, contractors, A/E firms, and government agencies in order to establish common protocols that can be used worldwide. At present (mid-2000), there are no fmXML protocols, but it is just a matter of time before fmXML becomes the glue that links and integrates FM data, especially for e-Commerce tools and applications.

Today, the Internet represents the largest telecommunications network that currently exists in the world. Any vendor that does not use this network for all aspects of data communication and reporting will find it increasingly difficult to exist. Initially, FM vendors used the Internet to post static information by publishing HTML-formatted reports from the client–server applications. Next came some transaction server-based Web queries (using Java, Whip!, Active Server Pages, Cold Fusion, and a host of other Web publishing tools) for functions such as entering or querying the status of a work request or requesting a customized graphic or nongraphic report from the software.

Today, CAFM vendors have moved well beyond these simple data collection, querying, and reporting functions. Entire suites of complex integrated applications are moving to the Web using a host of different publishing approaches and tools. Many of the CAFM vendors seem to be writing their own Java code to get their own "look and feel" to the graphic user interface (GUI), but some are using geographic information system (GIS) or other third party publishing tools. Still other vendors are using Web development environments. One potential problem of these diverse approaches is that systems' integrators that need to link and Web-enable diverse Web-based legacy systems (of which FM is just one application) often have a difficult task because of the lack of standards and diverse technologies employed.

The benefits of the Internet (and its associated manifestations in Intranets and Extranets) are clearly discernible and have been well defined and understood. As process-based tools, such as work flow and technical document management are better incorporated and integrated into FM technology, the benefits again increase significantly. The increasing growth and documentation of tangible and intangible benefits of collaborative project Web sites that employ work flow and document management attest to this fact.

25.4.2 Project Web Sites

Collaborative project Web sites basically share information between clients, design consultants, contractors, and anyone else that needs to view and operate on any form of project information (see also Chapter 27). There are currently over one hundred software vendors offering project Web site software—primarily for the purpose of managing construction projects.[1] These sites traditionally offer document management services and track project information such as drawings, revisions, specifications, RFIs, various logs, administrative forms such as purchase orders, and so forth. Most have the ability to "redline" (annotate) drawings on the Web, automatically route data and documents to the appropriate individual, keep certain data secure and, increasingly, link the project Web site software to other applications such as project management, cost estimating, and so forth. The basic difference between products relates to whether a product incorporates, and to what extent, functionality relating to work flow, project management (e-PM), and application knowledge. In general, however, these Web sites make data available to internal/external clients via the Internet or an Intranet. In a facility context, for example, human resources might want to know what staff is affected by a projected move of a corporate unit; finance might want to track project costs for financial control and capital planning purposes; business unit managers might want access to floor plans that affect their staff; and so forth.

Figure 25.10 is an illustration from Framework Technologies' Active-Project server software. The horizontal tabs along the top of the screen can be programmed to link various databases to the graphic user interface. Thus, a user can generate a query for a particular database (e.g., what active work orders are open) and then "drill down" on a campus, building, floor, room, etc.

Collaboration involves individuals working together in teams on joint projects. Collaborative tools involve a document of repository where the team's collective work is stored. In general, the "collaborative" component of project Web sites has been less developed than any other functionality. The ability for multiple users located at various sites to operate on data simultaneously is just beginning to be developed. Thus, for example, two or more individuals would be able to edit the same spreadsheet, drawing, or word document over the Internet. As we move to high-speed telecommunications and the Web better supports audio and video, the "collaborative" function of project Web sites will be widely accepted. Perhaps more than any other application so far, project Web sites will result in a fundamental change in how projects get designed, implemented, and managed—especially when linked to e-Commerce.

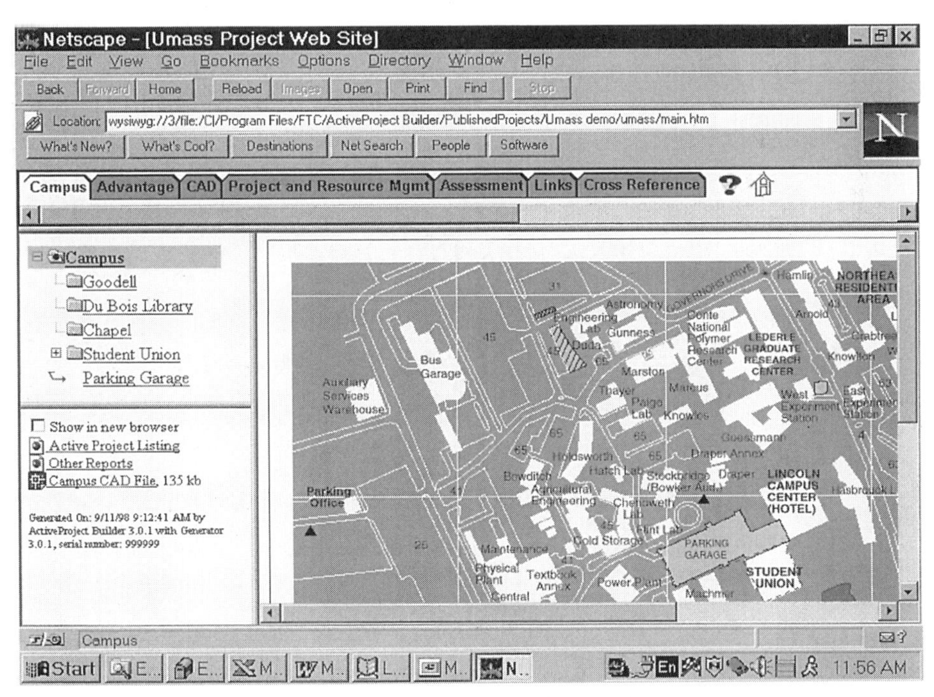

FIGURE 25.10 Project Web Sites. ***(Source:** Framework Technologies, Inc.)*

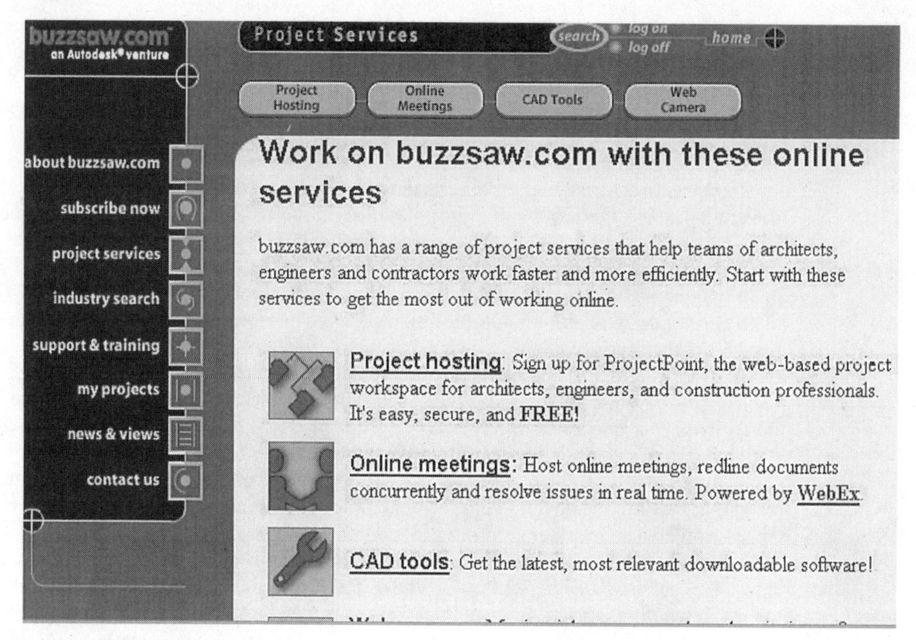

FIGURE 25.11 Buzzsaw.com—Remote Hosting of a Project Web Site. ***(Source:** Autodesk, Inc.)*

Software vendors of project Web site software currently either charge fees for their software or rent functionality while hosting the software on the project Web site vendor's computers. Thus, a user would use a browser to access the vendor's site, set up a project on the vendor's computer, and store the project documents on the vendor's host computer. Because of the potential revenues derived from site advertising, however, some vendors host software free just to get users to come and visit the site. This was dramatically illustrated recently when Autodesk, the developer of AutoCAD, developed a project Web site product called Buzzsaw which users can freely access (Figure 25.11). Autodesk hopes to recoup revenues for creating and maintaining the site by advertising and selling other products and services on this site. Again, the Internet is increasingly playing a role in controlling whom is accessing, using, and managing FM data.

25.4.3 E-Commerce

To use Internet parlance, we are entering a B-B ASP world. That is, financial venture capital funds fueling Internet start-ups (e.g., Softbank, CMGI, and ICGE) are investing heavily on the business-to-business (rather than business-to-consumer or consumer-to-consumer) application service provider (ASP) model. This means that businesses will be hosting e-Commerce Web applications for other businesses. The projected numbers are significant: In 1998, B-B revenues were $12 billion, of which "services" accounted for $1.4 billion (there is no breakdown for A-E-C or FM services); in 2002, there will be $131.2 billion B-B commerce with "services" accounting for $18.4 billion of that figure. This represents over a ten-fold increase in e-Commerce services in 4 years! Project construction Web sites, for example, will not only share documents but construction materials and services will most certainly be purchased on-line with manufacturers literally bidding on-line for the business. It is currently envisioned that almost all FM goods and services will be purchased on the Internet.

To cite a parallel, in a January 2000 edition of the *Wall Street Journal,* General Motors and Ford, "icons of the Old Economy," announced plans to establish on-line bazaars for all goods and services they buy—"everything from paper clips to . . . contract manufacturing." The motivation is to cut costs. It is estimated that billions of dollars will be saved over their current procurement processes for goods and services. All suppliers will have to go through their sites by the end of 2001. In the FM services world, there are literally hundreds of companies being formed that are trying to be the e–marketplace where it is possible to bid on and procure building supplies from manufacturers worldwide.

In an application service provider model, the ASP creates a single vendor solution that replaces all or part of the customer's IT infrastructure and rents the FM application to the customer on the Internet. The appeal is that the customer can interact with a single entity, as opposed to multiple technology and service vendors. The other advantages of the ASP model is that the customer saves the software costs (the application is rented), and saves time (the software expertise resides with the vendor). All the customer requires is a browser to use the application or purchase goods.

In order for the single point solution to occur, the FM ASP must have a great deal of technical expertise. Delivering FM applications as an ASP requires competency in three areas: IT services, the application software, and network infrastructure. At this point, most of the FM ASP vendors are turning to partnerships with third parties to perform at least two of these functions, both to save money and to get to market faster. This makes it difficult for end users to evaluate the ASP since other third parties are involved. For example, an ASP might go to Commerce One or Ariba to provide the portal site and streamline procurement, while the software vendor concentrates on migrating a particular application to the Web.

A Furniture Case Study

Bricks-and-Mortar Furniture. The office furniture market generates at least $32 billion annually and consists of Grade A (large contract), midmarket (small contract), and budget categories. Large corporations make up 25% of the total market. The need for office flexibility has resulted in panel-based systems to replace casegood desks found in fixed wall offices. Panel systems (currently representing over a third of the market) usually require electrical and network wiring and require experienced furniture dealers to perform design, specification, and installation services. Thus, at present, manufacturers rely on the 7,000–10,000 distributors and dealers that make up the current furniture distribution channel. Since 1970, the number of small dealers have continued to decrease and the remaining dealers have formed alliances with manufacturers, distributors, or large buying groups to remain competitive.

Changing Times. Perhaps the most dramatic change in the furniture industry today relates to its sales and fulfillment cycle. In the past, it took several months for a client to work with a dealer doing the design, placing the order, and installing the software. In today's business climate, this time line is unacceptable and is being replaced with a cycle of several weeks. Various "quick ship" programs, whereby manufacturers offer limited options and finishes in order to speed delivery, have been initiated. Dealers need to operate more efficiently and are naturally turning to automation. Additionally, retail stores such as Staples and WB Mason are beginning to offer interior design services so that panel systems, as well as casework, can be sold to businesses.

Furniture as a Dot.Com Industry. The business-to-consumer (B-C) e-retail furniture market has not been very successful on the Internet so far (early 2000). Customers simply did not have the expertise to order expensive and sophisticated furniture—either casework or panel systems. It was estimated that up to 20% of B-C furniture sold on the Internet was returned because people simply changed their minds once the furniture was installed in their homes or offices. Business-to-business (B-B) service models, as mentioned, however, are all the rage with venture capitalists today and a host of B-B businesses are appearing on the Web to sell goods and services.

These businesses are offering "one-stop" shopping for all facets of the life cycle of furniture. Figure 25.12 lists the potential functions included by the .com furniture portal. To be a comprehensive portal, offerings are organized into three categories: content, commerce, and community. Content categories include services such as comparing manufacturers according to design criteria or various types of distribution and installer relationships. Commerce activities include procurement via dealer, manufacturer, or collective buying and disposal through various channels including auctions. Content services add to the value of the portal and include proprietary and nonproprietary data.

Content	Commerce	Community
Workstation comparison engine	Manufacturer fixed price	Office design tips from industry professionals
Room layout technology	Variable regional dealer price	Discussion groups & forums
Dealer partnerships	Collective buying	Branded e-mail
Installer partnerships	Auctions (asset disposal)	Product reviews
Distribution partnerships (MRO)		
Manufacturer partnerships		
Information provider partnerships		
Furniture inventory support (CAFM)		
Move management support		

FIGURE 25.12 Functionality for a Furniture Portal.

Thus, one can conclude that selling furniture on the Web should change the procurement cycle substantially and touch all aspects of the asset's life cycle. The role of the "middleman," in this case the furniture dealer, will change as their role in the procurement cycle diminishes and they take on new client service roles. The success metrics of a successful furniture Internet business will eventually be commerce—how much revenues and profits are generated from the site providing the content, commerce, and community services. Measuring success in the early days of the Internet start-ups (until the Internet stocks crashed during the spring of 2000) had to revolve around the site metrics. Figure 25.13 shows some sample metrics for a furniture .com start-up. Figure 25.13a shows high-level statistics for a 1-month period; Figure 25.13b show visits by hour of day; Figure 25.13c shows the duration of the average visit; Figure 25.13d depicts the number of new users vs. number of first time users of the site; and Figure 25.13e shows the operating system used to access the furniture Web site.

25.4.4 Evolution of Infrastructure Management

IM principles enable the management of strategic and tactical processes related to all corporate assets from a common, integrated environment. It embodies concepts made possible by management's growing realization of the strategic importance of information technology (IT) and that IT can influence *all* aspects of an organization's business. No business unit, FM included, is immune from this centralizing force.

IT already is the largest capital expense in many corporations today (amounting to some 7.0% of revenues and 14.2% of expenses for service industries) and IT investment is growing at a rate of 3% per year.[2] The corporate/organizational IT infrastructure increasingly reaches across IT to all business units. During the first three FM generations of technology, the management of the IT infra-

General activity statistics	
Number of hits	8,898,314
Number of requests	3,196,770
Number of visits	160,016
Average requests per visit	19.98
Average visit duration	00:05:01

FIGURE 25.13a High-Level Statistics for a 1-Month Period.

Visits by hour of the day

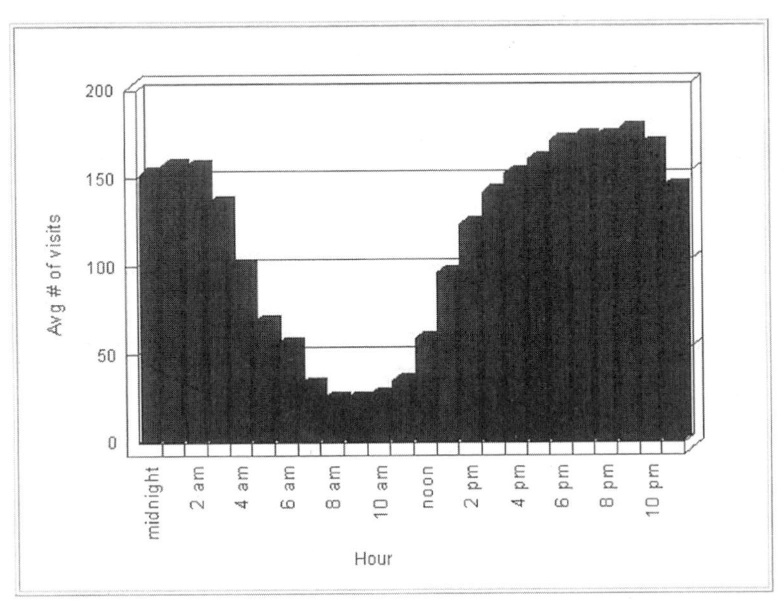

FIGURE 25.13b Visits by Hour of Day.

Visit duration trends

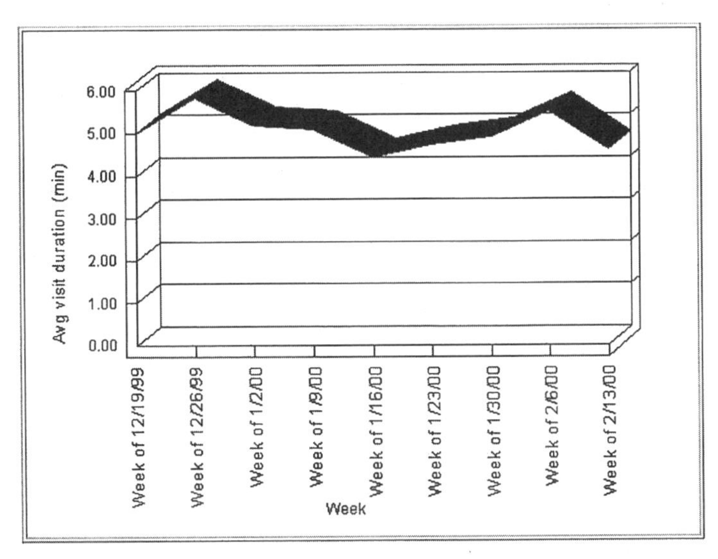

FIGURE 25.13c Duration of the Average Visit.

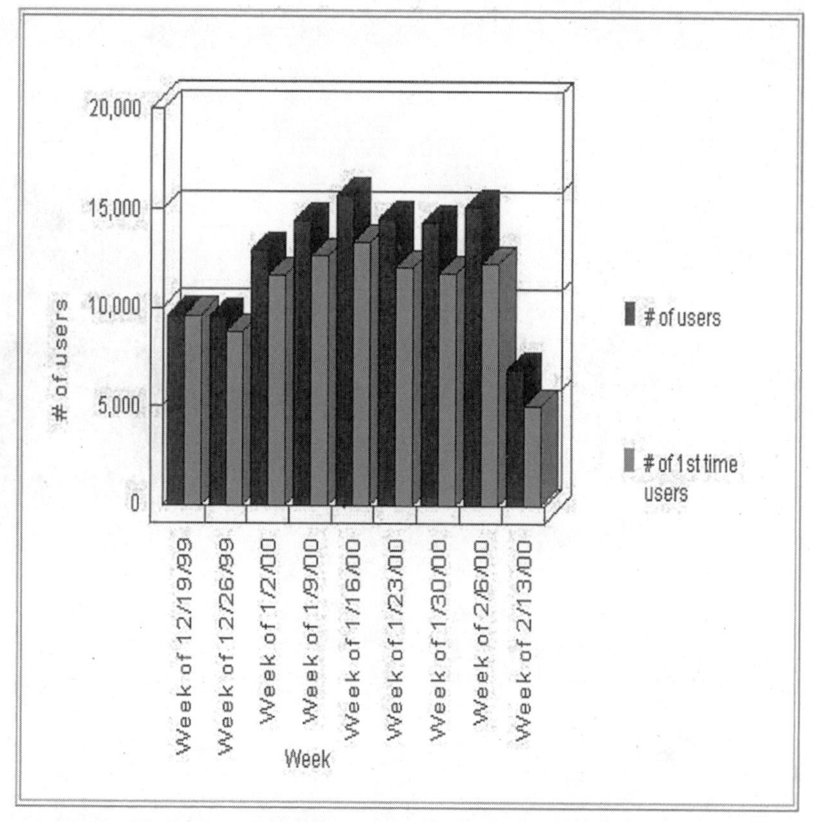

FIGURE 25.13d Shows the Number of First Time Users and Overall Users across the Weeks in the Analysis Period. Larger Gaps between Total and First Time Users Indicate a Growing Level of Repeat, and Presumably Satisfied, Customers.

structure involved in most cases the care of the mainframe computer and associated LANs. Partly because of the potential benefits of IM, however, IT increasingly incorporates:

- PCs and all aspects of computing devices (e.g., laptops and palm tops)
- Communications and telecommunications protocols and vendors
- Vendors and other types of data of corporate-wide interest
- Internal and external Web technology, including inter-/intra- and extranets
- Benchmarking and financial metrics for cost–benefit analyses
- Various aspects of mission-critical work process, such as the development of aspects of procurement and FM

The number of technology-based functions and processes centrally managed is growing and organizations are increasingly spending dollars to understand and manage the risks and returns of such investments.

Benefits and Relationship to Enterprise Resource Planning Systems. Although the cause and effect relationships between IM investment, risks, and rewards are just beginning to be understood

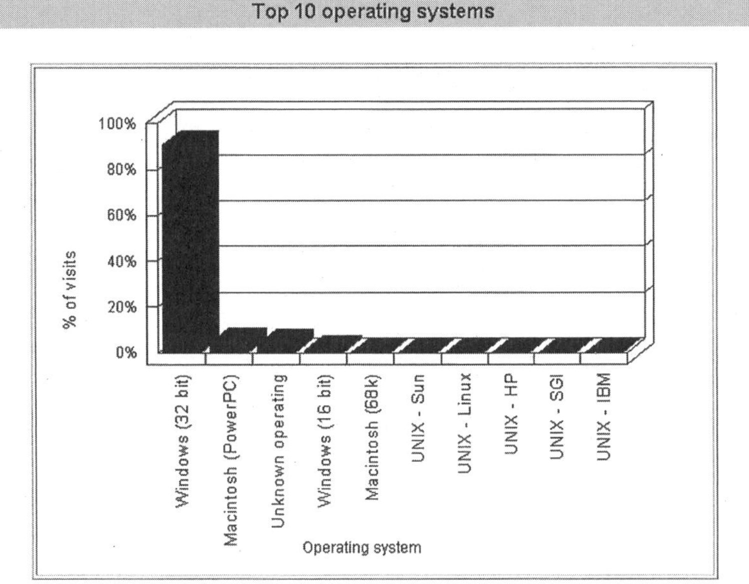

FIGURE 25.13e Operating System Used to Access the Furniture Web Site.

and quantified, there are enough case studies available to reach the conclusion that organizations that understand such relationships and invest heavily in IM receive more business value and return on assets (an increasingly common FM benchmark) than those that do not.

For the facility manager, it means that the role of FM technology will change as IM is better understood and accepted by organizations. It means that an asset (whether it be a chair, a PC, a paper clip, or a building) will be processed by the same technology and with the same benchmarks and work processes that are in place throughout the organization. It means that the support structure required for corporate strategic decisions (e.g., human resources, finances, real estate, IT, and purchasing) will all be using the same technology and business processes and reporting to the same corporate or organization management. Now every resource an organization is responsible for (whether leased or owned) is part of an integrated strategy that affects the entire life cycle of that asset. In the IM environment, understanding the life cycle of assets from purchase to disposal is critical.

As IM increasingly centralizes business functions and processes, it means that CAFM vendors will increasingly be required to integrate with enterprise resource planning (ERP) systems and vendors. Mainstream "mission critical" systems controlling purchasing, financial, human resource, and IT tasks (typically the domain of ERP vendors) need to share information with FM systems to provide the spatial and financial information required by management. Thus, vendors such as Oracle, Peoplesoft, SAP, Bain, etc. are already working with FM vendors to enable such integration. This line that separates FM and ERP technology is still somewhat fuzzy. A number of possible scenarios are possible that would make this line clearer:

1. ERP vendors might focus on mainstream enterprise-wide business processes and FM would manage the environment in which these business processes take place
2. ERP vendors might try increasingly to develop FM and RE functionality within the ERP environment
3. ERP vendors might increasingly partner with FM vendors to provide comprehensive integrated solutions

So far, it looks like the second alternative, ERP development of real estate and facility applications, seems to be the solution of choice for these vendors. Finally, IM has implications on how internal and external customers, for all organizational units, interact with information. The model of having a single point of interface has appeal since it provides a unique location for all organizational data. Thus, the current CMMS Help Desk would be expanded to aid the customers with any problems they are having in utilizing any aspect of the infrastructure. There are also knowledge bases, which are now used to decrease the time spent with a customer while increasing automated problem solving capabilities.

Although the benefits of IM for the facility and real estate manager are just beginning to be understood, some value can be anticipated. Effective IM deployment and management:

- Provides for a single repository of information about all occupants, locations, and assets that an organization owns or leases
- Captures, tracks, and potentially calculates the impact to an asset on an organization, which in turn will minimize downtime and expenses
- Minimizes the cost and time of moving employees, assets, and systems of an organization from one location to another
- Will enable new FM and RE technologies and applications to be deployed throughout an organization
- Will provide much more rapid integration of FM and RE data with HR, IT, and financial systems
- Leverages the large investment made in ERP systems
- Facilitates the definition, collection, and reporting of financial and operational benchmarks that are consistent throughout the organization

25.5 GENERATION 5 AND BEYOND

As we start the next millennium, an additional technology generation is being birthed that, once again, will impact the practice of FM—this is the imminent arrival of high speed Internet computers and networks (both through cables and mobile) and the increasing embedding of microchips into assets.

High speed Internets are already available in universities and most organizations. Using any one of several competing technologies, they will be available to everyone—once again acting as a unifying IM function further linking facility managers with internal and external users and clients. Whether communication will take place on PCs with operating systems (which Microsoft is hoping for) or with network computers (which Sun, Oracle, and IBM would like) is of little significance to the facility manager. Communication will be very high speed, it will be wireless as well as on wires, new data formats such as sound and multimedia video will be included, and there will be very high-speed data networks connected to powerful hub computers. The most immediate impact of these networks should be on collaborative project Web sites. Disbursed communication in all media with integrated Web-based applications running on the Internet and linked to e-Commerce applications will mature very rapidly.

In this post-PC age, assets will also be equipped with embedded chips, likewise connected to the Internet. The chips will have their own operating system (this time pitting Microsoft's CE against Sony's real-time Aperios operating system) and the intelligence will be in the thing (the asset) itself—thereby potentially rendering PCs obsolete. The earliest manifestation of such "information appliances" or "Internet appliances" is specialized computers that only do a few things such as e-mail or Internet browsing. Internet appliances started to appear in 1999. They might automatically download your e-mail every few hours or dial into a fast food restaurant to automatically order a pizza for you. Other Internet appliances include household furnishings such as washing machines that might check the water to regulate temperature or change to the next wash cycle.

When applied to FM, applications of "smart" assets can cover a wide spectrum of uses. For example, a piece of furniture might know where it is and automatically update a CAD drawing and database that is tracking that asset. Assets will be linked to the Internet and perhaps send information back to the vendor so that the vendors will know more about how their products are being used and therefore enable customized education and training. Assets will be able to redefine their functionality on the spot based on how they are being used. Shops will be informed what tools and equipment to bring for demand maintenance work. The list of implications of such technology goes on and on.

There is another aspect to FM Internet appliances that is very powerful but potentially problematic. This is the enabling of a computer to know the location of a person based on their carrying a unique chip with an ID embedded in a piece of equipment, such as a cell telephone or an ID card carried by the person. It has been estimated that workers will be carrying at least three such devices on them by the year 2005. If this does happen, then it will be possible to configure, for example, PC hard drives, telephones, faxes, etc. based on where a person is located at any time—greatly increasing the flexibility of the workplace. Time on task for work management will be automatically calculated. The privacy issues that this raises, however (witness the difficulty Intel and Microsoft are having by enabling the PC or even a document to have a unique ID associated with it), will most certainly slow the deployment of such functionality.

ENDNOTES

1. Dr. Joel Orr publishes a newsletter tracking project extranets. The newsletter, called Extranet News, is available free of charge from Joel's Web site (http://www.extranets.cc).
2. Weill, P., and Broadbent, M. 1998. *Leveraging the new infrastructure: How market leaders capitalize on information technology:* Boston, MA: Harvard Business School Press, p 294.

CHAPTER 26
INTEGRATING THE INTERNET INTO FM

Bruce Cox
Director
Facilities Technology for FMStrategies
A Division of Little & Associates Architects
E-mail: bcox@littlearch.com

26.1 INTRODUCTION

One of the most exciting changes in technology that occurred in the last decade has been the explosion of the Internet. It would be hard to find a business magazine or publication that has not covered the effects of Internet technology on today's business environment. Internet technology has progressed rapidly from a simple communications tool for research collaboration onto our office desktops and into our homes. It has radically changed the way we access and use information and it has even influenced the way we spend our money.

How can a relatively simple technological concept like the Internet so drastically affect the way we do business? What is the impact of Internet technology on the business of facilities management (FM)? What technologies are associated with the Internet and how are they used? What does a facilities manager need to know about Internet technology to make effective decisions about its use? How can a company leverage Internet technology to positively affect its facilities operations? This chapter addresses these questions with the intent of providing the facilities manager with decision-making information on implementing Internet technologies for FM applications.

The Internet almost by definition involves technical concepts and terminology. Changes in Internet technology are so exciting and so dynamic that it is difficult to keep up with all the possible applications. This chapter will explain some of the fundamental technologies that are at work on the Internet today and suggest some possible applications for those technologies in FM.

Internet technology is a vast subject. It will be impossible to cover in detail all the aspects of implementing Internet technology in this chapter. The chapter will instead seek to arm the facilities manager with information necessary to evaluate Internet technology implementation strategies with information technology professionals, technical staff, outsource service providers, and outside technology vendors.

26.2 WHAT IS THE INTERNET AND HOW DOES IT WORK?

Before discussing how the Internet can be used for FM, we must first understand how it works. When thinking about implementing any technology, it is very important to first take the time to research that technology and fully understand how that technology will impact your business. Today, Internet technology is changing very rapidly. Your company may not change or adopt technology at the same rate of speed. It is important to understand your business strategy for adopting Internet technology and to evaluate your company's Internet strategy before moving to implement solutions based on it. The purpose of this section is to give the facilities manager some basic information on how the Internet works so that he or she can make decisions on how it can be used to solve business problems.

26.2.1 Internetworks

The Internet, also known as the World Wide Web (WWW), was born some years ago in government and academia as a simple and failsafe way to foster communications between organizations with dissimilar systems. The Internet, by definition, is a network. In fact, it is a network of networks connecting millions of computers all over the world. This ability to connect dissimilar systems and networks in a global sense is what makes the Internet so important.

The Internet would not be possible without access to a common public infrastructure. The Internet is known as the "Information Superhighway." This highway begins with high-speed optical and satellite networks known as *backbone networks,* which are sponsored and regulated by the government and provided and maintained by the world's largest telecommunications companies. These backbone networks can transmit voice, data, and video information at over 600 Mbps (million bits per second). Tapping into this backbone allows networks to connect to each other and communicate.

Unfortunately, the computer that sits on your desk does not directly connect to the public high-speed backbone. Your information must first travel through a number of intermediate networks and devices before it reaches the public backbone. Speed of transmission and ultimately speed of access to information is one of the most important concepts in understanding Internet functionality and usability.

Information leaving your computer must take several jumps before reaching the public backbone where it can be traded with information from other organizations. If you are working on a computer in a large company your information must first navigate your internal company's local area network (LAN). If your company has multiple locations you may also be connected to those sites via a wide area network (WAN). Large companies will license *dedicated leased lines* from the telecommunications companies to connect to the public infrastructure. A company's access to the Internet may be purchased from an Internet service provider (ISP). An ISP will license large chunks of bandwidth from the telecommunications companies and connect to the Internet through network access points (NAPs) provided by those companies.

Another important concept that impacts the speed of the Internet is simultaneous user access. Bandwidth is a term that represents the amount of information that a given connection, or wire, can handle at one time. Each user on a network shares bandwidth from the Internet connection with every other user attached to that connection. The more users, the less available bandwidth. If you are working from a computer in a large company you are sharing your access to the Internet with your fellow employees. The more users you have, the more bandwidth you need to provide equivalent speed. It's important to understand that the speed you encounter at your desk can be affected by decisions made by your company in how they approach their Internet access.

26.2.2 Client–Server Computing

Although the Internet is really about networking, most of us today think of the Internet in terms of browsing, or surfing for information. This distinction is, and will be, an important one as the technology matures. You may know or recall that many years ago before the invention of the personal computer (PC), most computing was centralized. Large centrally housed mainframe computers served information to terminal workstations, which were known as "dumb" terminals. The terminals themselves had little functionality other than displaying information. This is similar in concept, with some significant differences, to how the Internet works today.

The term server can become confusing in technology discussions because it has several different contexts. Most of us have been, and are on a daily basis, exposed to technology terminology. Most people think of a server as a piece of computer hardware. In one context this is true, but servers can also be software programs or objects that serve up information to other programs. The Internet makes use of both hardware and software type servers. Following are some of the more common server types that are used in Internet computing:

Hardware Servers

- *Internet Server:* Internet servers are computers used to host programs serving information to clients attached to the Internet.
- *Application Server:* Application servers are computers used for hosting programs, objects, or applications.
- *File Server:* File servers are computers used for centrally hosting, storing, and managing information generally in file format.

Software Servers

- *Web Servers:* Web servers provide hypertext documents including graphics to Internet client requesters.
- *FTP Servers:* File transfer protocol (FTP) servers allow users to connect to an Internet server, browse its directory structure, and either upload or download files from that server given the appropriate security access rights.
- *Database Servers:* Database servers host database application software and information.
- *Object Servers:* Object servers host distributed objects for use by distributed applications. Applications known as object request brokers (ORBs) are responsible for controlling the communications between objects on a server and a client application.

There are many other types of servers. These terms are the most common and are the ones we will be discussing in this chapter. It takes a combination of both hardware and software servers to build an Internet application. Servers serve information up to clients.

The term client can also take on many meanings. You are a client, your desktop PC is a client, and the Internet browsing software on your PC is also a client. A client, whether it is a person or a software component, requests information from a server and the server returns information to the client.

The most commonly known client associated with Internet technology is the browser. The browser is a software application that can read and interpret information sent to it by a Web server. Although there are many different browsers available, the two most common and familiar are Microsoft's Internet Explorer and Netscape's Navigator. Browsers are, however, not the only types of Internet clients. Internet clients can also be programs or applications written in almost any programming language that use the Internet's networking infrastructure to communicate with server-hosted programs and information.

Browsers are a very important component of Internet computing. One of the fundamental reasons why the Internet has been so successful is because of its ease of use and ease of access. The browser is the user-visible component that evidences that ease of use. Browsers are typically free and are often included with the purchase of a new PC. They provide an almost universal interface for accessing and displaying information. Browsers can connect with any Web server; it does not matter what operating system or platform that server is hosted on. Browsers require few system resources (memory and disk space) and require little if any maintenance. Browser software can be easily updated by downloading new versions directly from their developers.

Information is transferred back and forth between the client and the server, typically using a network transport protocol known as TCP/IP and a network application protocol known as hypertext transfer protocol (HTTP). If you have used a browser to search for information on the Internet you will notice that the Web address for a particular site, known as the uniform resource locator (URL), is usually proceeded by the HTTP prefix. For example: http://www.fmstrategies.com/.

The URL address represents the user-friendly name for what behind the scenes is simply a number known as an IP address. IP addresses are absolutely unique for each computer connected to the Internet at a given time. The IP address represents the actual address that the networking hardware and software uses to move information from one specific computer to another. An example of an IP address looks like this: 206.74.254.2. Another name for an Internet location, URL, or unique IP address is a domain. There are special software servers that connect the IP address to the user-friendly URL or domain name. These servers are known as domain name servers.

Today, the most common form of information seen on the Internet is viewed through Internet browsers as pages of textual and graphical information. These pages, known as Web pages or Web documents, are composed using a special language called hypertext markup language (HTML) that the browsers can interpret. The browser translates the HTML into formatted user-friendly text and images.

We now have the basic information about how Internet technology works. As shown in Figure 26.1, an end user working on a personal computer using some client software called an Internet browser makes a request to an Internet server. The Internet server hosts some software known as a Web server that responds to the client over the World Wide Web Internet Network using the network transport protocol of TCP/IP and the network application protocol of HTTP. The Web server sends back information to the client PC as HTML code, which in turn is interpreted and displayed by the browser as a Web page, simple right?

The term client–server is used to describe an information system containing at least two components: a client and a server (see Figure 26.2). The amount of work that each component does can vary. In the example mentioned previously of the older technology utilizing dumb terminals, the client end did very little at all other than displaying information. Often the server side will do a lot of the heavy lifting computations and send only the results back to the client. With the introduction of the PC, it allowed the client to do more of the work. Client browsers are beginning to be capable of doing more sophisticated processing. This fact is changing the benefits of Internet technology

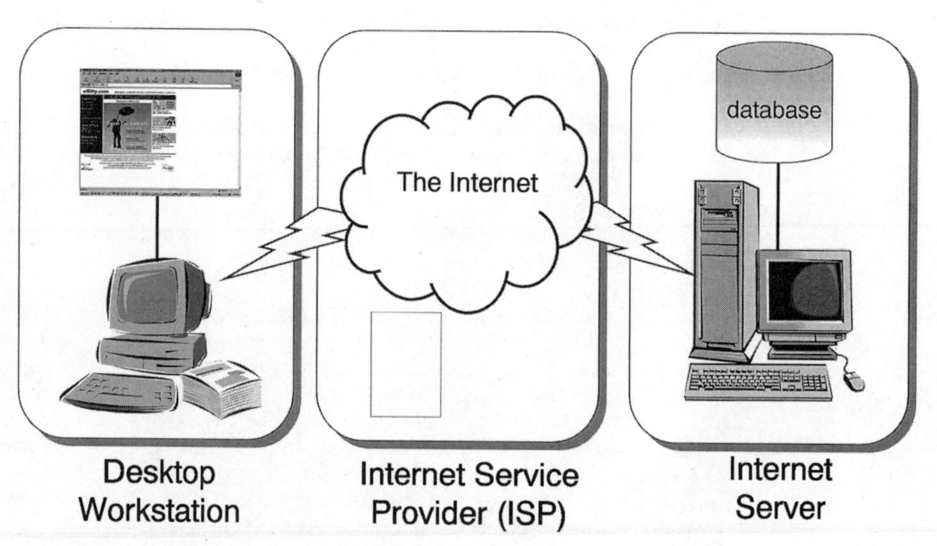

FIGURE 26.1 Basic Internet Technology Model.

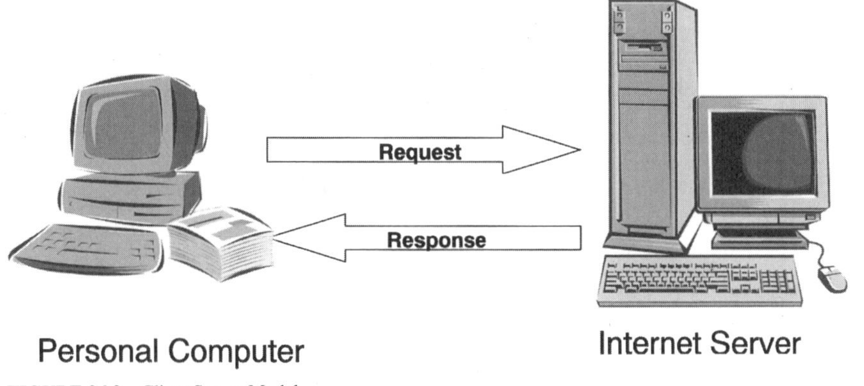

FIGURE 26.2 Client Server Model.

from simply a way of looking at static Web page information to the ability to actually host client–server applications that are capable of doing real work.

This is only an introduction to Internet technology. As we move forward in our discussion of the uses of Internet technology, it will be important to understand that at its heart, the Internet is a network and that with a network you can have access to distributed information and distributed applications. We will explore some potential business uses for the technology and some of the newer enabling technologies that make Internet application development so exciting.

26.3 INTERNETS, INTRANETS, AND EXTRANETS

The Internet proper is divided into three separate application domain types or categories: Internets, Intranets, and Extranets. These domains are distinguished and separated primarily by their need for security. A specific location, address, uniform resource locator (URL), or destination on the Internet is known as a domain or site. The term site can be qualified by adding the application domain type as a prefix. For example: Internet site, Intranet site, and Extranet site are generic references to an Internet domain or location that stores documents in HTML format or that stores Internet-ready applications.

26.3.1 Internet Sites

The most common domain type is simply, but confusingly, known as an "Internet site." The term "Internet" is used in this context to mean public. Internet sites contain information, usually in the form of Internet pages, that is free, open, and accessible to anyone anywhere that has Internet access and some Internet browsing software. It turns out that this is a tremendously useful concept. In fact, today a company can be totally defined by its Internet site.

Internet sites are used primarily for corporate public awareness and advertising as well as a general source of technical and business information. As such, they are not very useful to the facilities manager for business process improvement purposes. They are, however, very useful for business research.

Internet surfing has become a required business skill. There are hundreds of Internet sites that offer valuable information relative to FM. The largest area of use is probably for product information. Manufacturers of building products maintain Internet sites often including information on product specifications. From your desk you can look up specific information on products that is completely up-to-date without the need to lug around a huge paper catalog. In many cases, you can even place orders on-line for these products, improving the order-to-delivery timeframe.

In addition to product information, which is a huge benefit, you can find benchmarking information, process improvement information, business collaboration and networking information, information on industry standards, and information from facilities-related business organizations, including but not limited to organizations like:

International Facilities Management Association (IFMA)	http://www.ifma.org
Building Owners and Managers Association (BOMA)	http://www.boma.org
International Development Research Council (IDRC)	http://www.idrc.org
International Association of Corporate Real Estate Executives (NACOR)	http://www.nacore.org
American Institute of Architects (AIA)	http://www.aiaonlin.com
American Society of Interior Designers (ASID)	http://www.asid.org

There are also sites that have been created to organize building and facilities industry information. Examples include but are not limited to:

FM link	http://www.fmlink.com
Cyber places	http://www.cyberplaces.com

Almost every business publication has an Internet Web site as well. These sites often include information in the form of articles from current and back issues of the publications that you can reference on-line. And, most business consulting companies and facilities-related software solution companies that are dedicated to serving the FM industry have active Web sites that you can reference for information on products and services.

Internet sites offer a tremendous amount of value for research initiatives. The Internet has made information more available than any other time in history—the problem is having time to read and consider all that is available.

26.3.2 Intranet Sites

Intranet sites are internal private sites that are used by a single company or organization to share business information. Intranets use a company's internal network architecture to communicate and do not require an external connection to the Internet. Intranets are potentially the most useful domain types for the facilities manager. An Intranet can be used to share and distribute information within your department and with other departments within your company. It can also be used to improve business processes by distributing business applications.

Intranets are popular for many reasons: they are browser-based, so everyone can be a user; they are easier to develop and are easy to change, update, and customize; they cost less than their traditional development alternatives, such as client–server, and offer higher payback; they are based on standards; they offer good performance; they are more scalable; and, they can be made secure.

A company's Intranet strategy and infrastructure is generally controlled by the internal information services group. In some cases, access to your company's Intranet will be through a centralized server or group of servers maintained by the information services group. In other cases, you may have your own departmental Intranet server that in turn connects to the company's Intranet. You must have easy access to the information stored on the Intranet server so that you can update it.

There are a host of tools available today to make it easy to create and edit HTML or Web page information. In fact, the new breeds of office suite tools that include general business applications like word processors and spreadsheets can now save information directly in HTML format. This makes it extremely easy to leverage Intranet technology for business processes. You no longer need a programmer to realize instant benefits from technology.

The article "Why Build an Intranet?" in the September 1999 issue of *e-Business Advisor* lists typical business and productivity drivers for Intranet solutions as:

- *Employee Productivity:* Dynamic information, shorter cycle times, and increased accuracy.
- *Cost Savings:* On-line information distribution versus paper and lower maintenance costs.
- *Knowledge Management (KM):* KM lets you customize information to meet users' needs.
- *e-Commerce:* e-Commerce typically means customized content, presentation, and transactions.
- *Customer Management Improvements:* Integrating business processes with customer relationship management and client-specific information improves business cycles.

The most important aspects of an Intranet are its cost effectiveness, its ease of use, its ease of implementation, and its distributed nature. In addition, an Intranet system is secure within a company's network firewall. This means that sensitive information stored on an Intranet is not visible or accessible to the outside world of customers and competitors. These factors make Intranets very attractive for business solutions.

26.3.3 Extranet Sites

Extranet sites are external private sites that are primarily used to deliver products and services to paying customers. These sites are usually secure. Users must have authorized passwords to access the information stored and presented at the site, and some form of encryption generally protects the information. These sites can be very useful. A wide range of product information, research information, services, and even products themselves are available through Extranet sites. Extranets can be considered as protected sections of the Internet.

Extranets can be used to provide selective information from a company's Intranet to external users, including its customers and vendors, or an external vendor can use an Extranet site to provide services to its customers. The secret to a successful Extranet is the ability to provide a high level of security for the information it contains. Security is associated with two main concerns: user authentication and data encryption. An Extranet service must be able to limit and control access rights to the site through some form of user authentication or user login. The Extranet service must protect the information it delivers from others who are not authorized to use it or view it.

A common Extranet that you might be familiar with is on-line banking. A bank can provide access to account information and other banking services by giving a user a password and establishing an Extranet site that can validate that password and encrypt the information traveling across the wire to the user. This is a great service for the bank's customers. It allows the customer a convenient method of doing business and can actually save the customer time and money by saving multiple trips to the bank for trivial transactions.

The most interesting types of Extranets for FM includes those that are provided by outsource vendors. There are many potential facilities uses for Extranet services, for example, subscription software services or delivery of outsource service content.

In the first case, subscription software services, a company will provide centrally hosted software that a company can use to service a particular business process. This type of service is now commonly known by the acronym "B2B" or "Business-to-Business." One popular service is project management Extranets. These services provide on-line access to distributed project management applications on either a subscription basis or on a per project basis. The vendor will host all of the data for the project and provide the necessary infrastructure, including backup, disaster recovery, and 24-7 operation. By leveraging the cost of the hardware and development costs among multiple customers, the service can be provided for less cost than if a company had to implement the same service internally. In addition, the service is flexible. You can serve both internal project users and associated external project users for the life of the project. Often there are many outsource consultants involved in a single project. They can all have access to the same information through the Extranet that is actually outsourced to a separate company that specializes in that service. Also, the service is a flexible expense. You can subscribe for a single large project, then cancel the service on completion of the project and capitalize the expense.

Another example of Extranet use is the outsource service provider. For years facilities managers have been making use of outsource services. Typical services include property management, maintenance, security, janitorial services, leasing, asset management, space inventory, and planning services. A large portion of these services is done on site, but in the case of services like space management and planning the services are often performed at the offices of the service provider. Extranets can provide an efficient means of delivering the offsite work, in the form of information over the Internet, to the desktops of the users who need access to that information.

Examples of outsource Extranet services include http://www.bricsnet.com and http://www.citadon.com, project management Extranets specializing in hosting global project information on an outsource basis. Also, http://www.facilitiesmanager.com, a space management outsource Extranet specializing in hosting space and property information for enterprise customers. In addition, traditional services like work order management, timesheet management, invoicing, and project management that involve multiple vendors and employees across multiple sites can benefit greatly from the implementation of an Extranet.

26.4 INTERNET APPLICATIONS FOR FM

The Internet paradigm is so powerful that almost all existing business technologies and solutions will be ultimately affected by it in some way. The technology is relatively easy and cost effective to implement for the benefits that it delivers. The basic paradigm assumes widely distributed, but centrally managed information. Today, many current applications rely on a free browser to access centrally managed information. You can now afford to provide more information to more users. Information is an important element of business. This idea suggests that by involving more users with business information that you may be able to improve work flow and business processes.

In this section, we will explore areas in which Internet technology may be useful in FM. We will look at some beneficial technologies and at how some of the traditional commercial solutions are changing. We will also discuss some of the practical and political aspects of adopting Internet strategies for FM.

26.4.1 Electronic Mail

Arguably one of the most useful and most used features of Internet technology is electronic mail. Electronic mail or "e-mail" has been around almost since the inception of the Internet. Electronic mail is a powerful collaboration tool. Originally used by the Defense Department and universities for project collaboration, it quickly migrated into corporate use. E-mail is the first tool in a facilities manager's Internet toolkit.

For many years companies have been using GroupWare applications like Lotus CC Mail, Lotus Notes, Microsoft Exchange, and Novell GroupWise. These applications run on internal GroupWare servers and offer functionality beyond just mail. The original focus of these applications was internal communications and collaboration. It was not long until these applications took advantage of Internet mail also known as POP Mail. Now within a single desktop application you can communicate with internal resources and external resources, including vendors and suppliers, and, at the same time manage your schedule and your to-do list. Each Internet mail user has an electronic address that must be unique within his or her domain bcox@alltel.net.

These addresses are international in scope. Your e-mail address is now just as important as your telephone number. If you can imagine the size of an international phone directory, the e-mail directory would be similar.

Internet mail can be both a curse and a blessing. You must be careful with your e-mail address if you want it to remain useful. Anyone who has your e-mail address can send you messages. This includes all manners of marketers who would like to sell you something you may or may not want. Like traditional mail, e-mail has its own form of junk mail known as SPAM. If you are not careful, you can easily clog your e-mail system with tons of information you did not want and did not ask for.

Many people in business will receive hundreds of e-mails in a single day. Compare this to the traditional mail you receive at home. Often, over half of the mail you get is junk. Unfortunately, you do not know which messages are junk and which are important until you review them. This process can take up an immense amount of your day. Although e-mail messages generally require a subject header that might allow you to filter or sort the messages into useful categories, there are no standards for what the subject should be, so it is only partially useful for organizing correspondence.

There are things that your network administrator can do to limit the amount of SPAM that reaches your Internet mail box, but once your e-mail address starts to circulate you will find it difficult to filter out all the unwanted information. Good internal e-mail policies can help.

Two of the more useful Internet mail tools are the distribution group and the list server. These concepts are essentially the same. A list of e-mail addresses can be maintained for a given project and a single message can be broadcast to that list. This idea makes it very easy to collaborate and coordinate facilities projects. You can send the same message at the same time to many people on a project team. E-mail can be an extremely efficient means of communication, and when used correctly, it is a vital business tool.

26.4.2 Basic Information Sharing

One of the largest benefits of the Internet has been information sharing. This is the aspect of the Internet that most people are familiar with. An end user can simply sit down at his or her computer, access any one of numerous free on-line high-speed search engines, and ask for some information. Within seconds in most cases, the user will be presented with a listing of topics addressing their needs. By selecting a topic, the information is returned to his or her desktop computer in the form of a Web page or an Internet-hosted document.

The days of the file clerk have been behind us for some time. With the coming of the computer, information could be stored centrally or locally and could be easily accessed by end users. But what does easily mean? Easily in this case is a relative term. Often, access to the information was anything but quick or easy and it often involved a computer professional to mount a disk or run a report or recover some backup information. It was also quite expensive to buy and maintain the software and hardware that the end user needed to access the information, not to mention the specific end-user training that was required for each software application.

The Internet has taken us a large step beyond in our ability to access and organize information. A simple and free Internet browser application installed on a computer connected to the Internet can access volumes of information almost instantly. A user can now easily leverage knowledge stored on computers around the world with great efficiency. One of the difficulties in having so much information available on-line is that finding a specific piece of information can often be time consuming. Numerous companies have been created to address the problem of Internet information mining. Companies like Yahoo, Alta-Vista, Excite, Microsoft, and others have created specialized programs, sometimes known as Spiders or Bots, that surf the World Wide Web automatically and constantly to collect and categorize links to information in large indexed databases. These companies also produce on-line search engines that allow users to search their databases for references to specific information.

Most business have moved rapidly to take advantage of the Internet and have made access to it available at the employee's desktop. There are literally thousands of opportunities for this technology. The opportunities for basic information sharing for facilities managers fall into the three categories discussed previously: Internet, Intranet, and Extranet. Probably the largest use of the Internet in FM is for building product research. Most suppliers, vendors, and service providers now have Web sites where you can look up basic product information and, in many cases, catalog pricing and direct purchasing is available on-line. One of the interesting offshoots of basic information sharing has been the Internet portal. An Internet portal is an Internet site that hosts information from many different sources, organizing that information in one location. These sites can be very useful.

Good examples of an Internet portal serving the building design and construction industry are http://www.aecinfo.com and http://www.cmdg.com. AEC Info and the Construction Market Data Group collect and manage information on the building design and construction industries. Another good example is http://www.fmlink.com. FM Link is a specific facilities management information portal. FM Link organizes information on facilities issues, including benchmarking statistics and information on outsource service providers and CAFM software.

Intranets are now prevalent in many companies. It is easy, cost effective, and very useful to access a wealth of corporate information on-line. We are seeing everything from company marketing, benefits, and profit information to very detailed departmental information on-line inside a company. Basic information sharing of building, property, and lease information on-line is becoming common. Internet technology has simplified document management and document distribution. Facilities real estate and maintenance departments are beginning to share information using their company's Intranet as common ground. Departments are using Intranets to build, share, and maintain processes, policies, and standards through the distributed information media of Intranets. It is extremely easy and cost effective to gain wide benefits in general office administration using Internet technology through basic information sharing.

26.4.3 Applications On-Line

Beyond basic information sharing is on-line application sharing. The idea of renting and not buying software is returning to technology. Interestingly, this is similar to how centralized mainframe software was delivered in earlier days of terminal-based computing. Thanks to Internet technology, you can now buy a subscription to an application. That application is downloaded right to your desktop. The manual is on-line. The software is upgraded automatically when changes occur. There are no boxes or shipping costs. The expense becomes flexible, and the application is always up to date.

There are generally two types of applications that are available. *Local applications* reside and operate on your desktop PC using the resources of that machine. *Network applications* generally split the work between your PC and a remote server. Local applications are downloaded in their entirety to your local PC or workstation over the Internet and are updated by downloading new components when they become available. Network applications are updated centrally and changes to these applications are then broadcast to all users and are available to everyone instantly on access to the application.

There are a number of different technologies used to develop on-line applications. Internet technology is still young, so buyer beware. The largest number of on-line applications available today are known as "thin client" which generally means that they are hosted by the combination of a desktop browser interpreting hypertext markup language (HTML) documents, or Web pages, and a remote server dynamically generating and managing those HTML documents. Keep in mind though, that the Internet is simply a network at heart, and that there are many application development technologies that can exploit its connectivity.

When purchasing an on-line application, there is much to consider. If the application is hosted remotely by a vendor, you must make sure that the content and any local application code will travel from the outside world onto your desktop or server through your company's network security system, including your network firewall. A firewall is a combination of hardware and software that prohibits unauthorized users from accessing your company's network.

HTML is the common language of the Internet and the most prevalent type of on-line application available today. There are several good reasons for this. The HTML language is simple and lightweight. It is a network safe language, meaning that once it is downloaded to your local machine it cannot do harm to your local data or other data on your local network. Applications written using HTML are stateless, meaning that they do not maintain a consistent connection with a backend server. An HTML document is downloaded to your machine and then the network connection is closed. Browsers have been designed to efficiently interpret HTML documents. You can both send and receive information in HTML format.

Although HTML and its newer brother, dynamic HTML (DHTML), are productive, useful, and fast, there are limitations to what you can do with it. Most of the applications work must be done on the server and only the results are returned to the client browser. Because it is stateless, local data validation is difficult. There are other Internet application technologies, like JAVA, that overcome these limitations, but at a cost. The cost is the size of the application code that must be downloaded to the local workstation. The larger the applications the more download time it takes to get the application to your desktop.

JAVA is an interpreted language, meaning that there is a program known as a "virtual machine" that must read precompiled JAVA code and translate it into a form your machine can understand. Currently, each time you visit a Web site that hosts a JAVA application, the code must first be downloaded to your machine, and then it must be loaded and interpreted by the JAVA virtual machine. This can take some time for large complex applications. Once the application is loaded, however, it is very fast and can maintain a consistent connection to a backend server and database so that data can be updated more efficiently.

There are other types of on-line application technologies as well, including hybrid approaches like ActiveX, JAVA Applets, and Netscape Plug-ins, that are hosted inside HTML pages. We are only scratching the surface here. The most important advice for the facilities manager who is contemplating implementation of an on-line system is to carefully explore what technology is driving that application compared to your available network bandwidth. How current is the technology, leading edge, or bleeding edge? How will it work in your technical environment? How will it be updated and maintained? How is it licensed? What is the cost per user? And, of course, what is your return on investment expected to be?

It is clear that application software systems are changing drastically. Our world is driven by technology. It is increasingly important that facilities managers become familiar with these systems, how to evaluate them, how to purchase them, how to manage them, and how to utilize them to their best advantage.

26.4.4 Electronic Commerce, e-Business

Electronic commerce has been around for a long time. As early as the 1960s, industries were beginning to standardize their data formats so that computers could exchange information efficiently. In recent history, the electronic data interchange (EDI) standard for information exchange has governed how companies exchange business information electronically. Using EDI, companies have been doing on-line electronic ordering, purchasing, and invoicing for some time.

By marrying the credit card industry with the Internet, the ability to do on-line purchasing has been drastically expanded. You can now purchase almost anything from anywhere directly from your desktop computer and usually receive it by the next business day. One of the side effects of e-Commerce has been the tremendous increase in distribution and shipping companies. Once this feature became available to individual consumers through inexpensive home computers and dial-up modems, the marketing industry took off. This in turn expanded the availability of on-line product information. Companies are spending millions of dollars to take advantage of the on-line purchasing phenomenon because it is so highly efficient, globally available, and profitable. Facilities managers can now often order products directly on-line and monitor the processing of their orders, including the shipping status of those orders directly without a single phone call to a customer service representative. This ability is continuing to increase and more and more services are available on-line.

For example, FMStrategies, the author's company, provides an ever-increasing range of client services, such as:

- Contract management for contractors/subcontractors
- Management of invoice information for project vendors
- Submission of summarized invoice information to our clients electronically using EDI
- Maintenance of a project Extranet that allows our customers to monitor their capital project costs against projected budgets on-line

Vendors are responsible for updating the status of their work in our on-line application. At any time, a customer can call up a capital project tracking system and see the schedule and budgetary status of a given project. All systems are available to customers, project managers, and to associated vendors from anywhere in the world that has an Internet connection. Internet technology inspires creative thinking. On-line furniture order processing, on-line service requests, outsource collaboration, product purchasing, payroll, system controls monitoring, security, inventory services, you name it, are available as e-Business solutions in some form today.

26.4.5 On-line Collaboration

On-line collaboration has also been around for some time. In addition to e-mail, there have been on-line forums hosted by companies like Compuserve and America Online (now the same company) and hundreds of on-line bulletin boards. These services were originally dial-up services, but have now been replaced by their Internet equivalents. In addition to the commercial services, there are also newsgroups like Usenet for almost any topic you care to discuss.

The commercial services host topic forums that you can subscribe to peppered with a surrounding plethora of additional content and marketing information. Newsgroups are more utilitarian and mainly consist of threaded discussions where one person will post a question and others will respond with answers, opinions, or additional information. These topic forums can be very useful for answering technical questions. They are available globally and are invaluable for things like software support. Topics on architecture and building engineering, as well as hundreds of others, can be found in Newsgroups.

In addition to the generic discussion group collaboration sites, there are the commercially available venues that are primarily geared to facilities and construction project management. The majority of these sites are presented as "on-line applications for rent"; that is, they are sold as subscription services and can be purchased for a given period of time and a given number of users. Examples of these services are included in Table 26.1.

Framework Technologies ActiveProject is sold as an application and can be purchased and hosted on your own hardware. ActiveProject can be modified to suit your individual company and project requirements. EProject is a generic project management site. EProject is currently free, there is no charge for its use. The remaining example project collaboration applications are hosted remotely. You subscribe to, or license these applications, but do not directly purchase the software.

TABLE 26.1 Examples of On-line Collaboration Offering "On-line Applications for Rent"

Company	Product	URL
Bricsnet	ProjectCenter	http://www.bricsnet.com
Citadon, Inc.	ProjectNet	http://www.citadon.com
Framework Technologies	ActiveProject	http://www.frametech.com
Meridian Project Systems	ProjectTalk	http://www.mps.com
eProject.com, Inc.	eProject	http://www.eproject.com

These project collaboration sites include both project collaboration and document management functions. The main idea is that all project participants can log-on to a single centrally hosted site to add to, update, or retrieve project information. Features include RFI management, on-line drawing viewing and redlining, project scheduling, project budget management, project standards management, project resource management, project discussion threading, and project document management, as well as others.

26.4.6 Opportunities for Improving FM Processes and Procedures

Internet technology presents so many opportunities that it will be difficult to find a single business process that would not be benefited in some way by implementing the technology. Again, because of its ease of use, ease of implementation, and cost effectiveness you will often see immediate improvements in business processes upon implementing Internet technology.

Why? Internet technologies' strengths are in its ability to provide better collaboration, communication, organization, and distribution of information. Processes that are shared by more than one person, and there are not many that are not, should be the first targets for implementation of Internet strategy. The more users you have that need to access a specific set of information or specific process the better the benefit. This technology is changing so rapidly that almost anything you can imagine is either already available or is on the drawing boards. Today, you can access the Internet from your network, telephone, cell phone, satellite phone, palmtop, desktop, television, car, and soon probably from your coffee maker.

Through the use of on-line video and Web cams, you can now go to a construction site and make a field observation, in the rain and snow of course, without leaving your desk. You can receive e-mail on your cell phone. You can even receive it on your wristwatch. Through the availability of digital imaging scanners and viewers, CAD vector drawing viewers, 3D virtual world viewers, sound viewers, video viewers, and voice recognition software, much of which is free, there is very little that is done at a desk that cannot be done from anywhere. It is easy to imagine that one could receive a change request, modify and view the changes to a production drawing in 3D, author the change order, and distribute it to every contractor on the planet. Not only that, but one could do it while sitting in a tent at 28,000 ft listening to a downloaded MP3 audio file of Barry White, and sipping some tea on an ascent of Mount Everest.

The list of FM processes that can benefit from Internet technology is truly endless:

- Request management
- Controls monitoring
- Work order management
- Purchasing
- Space and occupancy
- Management budgeting
- Asset management
- Move/change management
- Lease management
- Standards management
- Property management
- Processes management
- Issues management
- Customer relationship management
- Bidding and negotiating
- Resource/vendor management

- Project management
- Survey management
- RFI management
- Contract management
- Benchmarking
- Reporting
- Document management
- Strategic planning
- Design review
- Security

This is a short but incomplete target list to consider when planning your Internet strategy. There is not a single process or service listed that could not benefit from the centralized, collaborative, nature of Internet technology.

FM is a hybrid profession—it includes bits and pieces of many other unique building engineering, construction, and real estate disciplines. To be a facilities manager is to "be all things to all people" or "to know a little about a lot." The advantage of this definition is that as a facilities manager you can take advantage of the detailed research and development of many separate industry focus businesses. The disadvantage is that it will be difficult to find a single solution that encompasses all your needs and responsibilities in a coordinated fashion.

FM is no different than any other business. You must hire, train, manage, and coordinate resources. You must buy and sell services and products. You must report and benchmark your services. You must communicate with your customer. You must create and implement strategic plans. You must manage budgets.

The perfect facilities manager has a degree in mechanical, civil, and electrical engineering, architecture, construction management, computer science, accounting, human resource management, conflict resolution, and, of course, an MBA or two. He or she should also have about 100 years of experience and be a nice guy or gal. The point being made is that when designing your Internet strategy you should first look around. You should spend a great deal of time studying what other people are doing. What is your company doing with its Intranet? What are other associated facilities industries doing? How can you leverage what has already been done? First analyze what is available, then decide whether it fits your needs and whether you need to build, buy, or integrate solutions to improve your business.

Today, in almost any company, you can find numerous on-line projects underway. Check what human resources and finance is doing. Talk with your information services group and find out what their standards are for Internet technology and security before looking for solutions. Look at the commercial products that are currently available. Do these products give you any new ideas?

Before you begin your research, try to generically list the functions and services that you believe can be improved through the use of Internet technology. Prioritize the list. As you do your research, revise the list. As you would with any purchase, ask for references. It may be wise to enlist a consulting firm that specializes in Internet technology and has had experience with real estate and facilities-related systems. It is important that you consider a scalable strategy for leveraging Internet technology as it continues to change rapidly.

26.4.7 Commercially Available Computer-Aided Facilities Management (CAFM) Solutions

It is not surprising that the commercially available computer-aided facilities management (CAFM) software companies have eagerly sought to migrate their products to Internet technology. The CAFM software business really began to evolve during the early 1980s with the advent of PCs and the availability of PC-based database and computer-aided design (CAD) systems. A number of companies have evolved to offer integrated solutions to typical facilities information problems. These companies have now survived Windows, LAN, and client–server phases of technology change and are well positioned to capitalize on Internet technology for their products.

CAFM systems have been designed to help facilities managers organize and maintain facilities information, including applications that help to manage facility business processes. Many companies now offer integrated solutions that address multiple facilities disciplines or multiple business applications. These products are often designed as suites or frameworks of various applications that address facilities information management problems. Examples of the application areas included in currently available CAFM system products are:

- Space management
- Asset management
- Lease and property management
- Maintenance management
- Work request and work order management
- Help desk
- Cable and IT asset management
- CAD drawing and image management
- Project management
- Project budgeting
- Fleet management
- Others

The integrated CAFM systems are designed to work with standard database management systems (DBMS) and generally include some method of displaying and managing associated graphical information that is available from CAD system generated drawings. There are two main areas of functionality that must be addressed when Internet enabling a CAFM solution. The first is the ability to organize, view, display, and report information included in the backend database or application. The second, and more difficult, is the ability to manipulate, manage, maintain, validate, and update the information contained in that backend database system. You can get a great deal of value out of a read-only approach to simply publishing information on the Internet, but greater value can be obtained by being able to actually collect and manage information on-line.

It is important to understand the technology behind a commercial off-the-shelf software (COTS) solution before making a financial commitment to it. This is an increasingly difficult proposition because both the quality and quantity of technologies is expanding rapidly, especially in the world of distributed technologies like the Internet. It is interesting to look at some of the better-known companies offering integrated CAFM solutions to see how they have adapted their products to the Internet. Table 26.2 provides a partial listing of some of the better-known commercially available CAFM solutions.

TABLE 26.2 Partial Listing of Commercially Available CAFM Solutions

Company	Product	URL
Aperture	Aperture Enterprise Solutions	http://www.aperture.com
Archibus, Inc.	ARCHIBUS/FM	http://www.archibus.com
FM:Systems, Inc.	FM:Space	http://www.fmsystems.com
Peregrine, Inc.	FacilitiesCenter	http://www.peregrine.com
Facilities Information Systems, Inc.	FIS/FM	http://www.fisinc.com

TABLE 26.3 Internet Technology Used by CAFM Solution Vendors

Company	Internet implementation
Aperture	Custom CGI Programs
Archibus, Inc.	Allaire, Cold Fusion
FM:Systems, Inc.	Microsoft, Active Server Pages
Peregrine, Inc.	Sybase, PowerBuilder Web
Facilities Information Systems, Inc.	Oracle, Application Server, JAVA

Each of the vendors listed in Table 26.2 has approached its Internet integration utilizing a different Internet implementation technology. You must keep in mind that at the time of this writing these companies are still in the first or near first release of their Internet offerings. It generally takes many years for a software product to mature. The rapid development of Internet technology has caused many companies to move quickly to develop products with Internet capabilities. One expects that these products will continue to grow and change their Internet offerings for some time.

All of the products listed in Table 26.2 have Internet enabled some, if not all, portions of their product. All of the product solutions dynamically generate their Web page content from a backend server-based system and database. But, each of the five example vendors has chosen a different technology to produce their Internet enabled product, as shown in Table 26.3.

When selecting an Internet application there are, like any major purchase, many things you should think about and many questions you should consider before acting. First and foremost, what are your business needs for the system? Does the product solve the business problem you need to solve? What will the maintenance costs for the system be? What will be the actual return on investment (ROI) after system implementation? Answers to these and other questions are both directly and indirectly linked to the underlying technology used to power the system. The author will discuss these and other Internet technologies in the *Internet Technologies* section of this chapter. When fully implemented, these systems can be very powerful and can provide useful information for business management. Often you will be presented with a "build or buy" decision for Internet FM applications. It is fun to build basic Web pages. Most of the basic business software suites that include word processing and spreadsheet functions have direct export capabilities for HTML pages. You can simply "save as" a document to a Web server and it is instantly available to hundreds of people.

Basic value can easily be achieved with simple, static, Internet information developed in-house. This approach is fine for static information that will not be changed often, like office standards, and policy and procedure documentation. If your business need requires adding additional information or constantly changing or updating information, like a work order system, you will need a more sophisticated solution. In most cases, it is more cost effective to buy an already developed solution than to build it yourself. The downside is that in some systems you must adopt a new business process that comes as an integral part of the application. In some situations, this can be undesirable. Flexibility is always a good thing.

These systems are continuing to grow to take advantage of the inexpensive and powerful distribution functionality of the Internet. They can be complicated and involve many subsystems. Larger enterprise systems are moving to adopt multitier object technology, which can require multiple servers, including Web servers, database servers, and object servers to implement.

Enterprise resource planning (ERP) system vendors like SAP are beginning to cross the boundaries of business, finance, and human resources application information management into offering traditional FM systems functions. Facilities are growing. Global companies are being built it seems on a daily basis. These companies must manage their facilities assets across multiple countries and time zones. Internet technology can make enormous information distribution problems conceivable and the commercial vendors can help you deliver systems quickly that will take advantage of Internet technology.

When contemplating adding a new FM system, you should:

- Do a thorough business needs analysis
- Do your homework and research all commercially available solutions
- Work to obtain a basic understanding of the technology
- Consider working with an experienced third party systems integrator that specializes in Internet technology and facilities solutions to assist in preparing your RFI or RFP
- Work with and include your own information services group in your research
- Carefully examine your information security requirements

And finally, think hard about implementing applications that do not include the use of distributed Internet technology as a part of their solution.

26.4.8 Telecommuting

One of the interesting side benefits of Internet technology has been the increased effectiveness of working from your home, or in some cases, from your customer's site. Two factors have made working from home a realistic possibility. The first is the availability of inexpensive PCs configured with dial-up modems. The second is the availability of inexpensive local points of presence (POP) provided by internet service providers (ISPs) and local telephone and cable companies.

For a very small amount of money, you can now work very effectively from your home. This phenomenon has created many new possibilities for staffing and producing work, and many new legal, governmental, and management challenges. At first, users could become very productive simply with a standard word processor and a low-bandwidth e-mail connection to the Internet. With the advent of high-speed cable modems, ADSL connections, and secure virtual private networks (VPNs), there is very little that employees working from their office cannot do at their home, at their client's office, in a hotel room, or even from their car. With a cell or satellite phone connection to the Internet, users can connect to their office literally from anyplace on earth. (The author can remember, in the good old days, when I went into the office at 8:00 A.M. and went home at 5:00 P.M. and got weekends off because work was only done at the office . . .)

With these new high-speed secure and inexpensive Internet connections, employees can indirectly connect to their office networks utilizing the public infrastructure. More and more applications are being rewritten to take advantage of the Internet. Users working from home will not only be able to access office e-mail and files, but they will also be able to call up and use centrally hosted business applications.

Telecommuting is changing the way work is accomplished in many industries and businesses. Although it is popular with many employees because it allows flexibility with work schedules and can provide better work/family balance, there are problems with managing work and with protecting a company's intellectual property. In many cases, the employee's work environment overhead costs can be reduced and in some cases his or her productivity will increase, but there can be a greater risk and liability to the company. Strong telecommuting policies must be considered before adopting and allowing remote working conditions.

26.4.9 Outsource Opportunities

Internet technology has opened the door to new opportunities in outsourcing. Many outsource resources, particularly the ones that rely on information services and information delivery, have been limited by inefficient methods of project coordination and collaboration. One of the interesting anomalies of outsourcing is that you may hire an excellent experienced company with many resources and services that you would like to tap into. Generally, you see a very small portion of what is available from that company because it is only available to you through the people you see on-site and the information that those people can carry with them. These companies often have more resources internally than you can afford to have sitting at your site. Internet technology can help you to get at some of those off-site resources.

Architectural and engineering firms provide design and construction document services for facilities. Typically, architects are project based. They have a contract to design and manage the construction for a given building. In the past, at the end of that project they would provide their client, the facilities manager, with a static copy of the final working drawings for that building. Today, most construction drawings are done using CAD systems. These firms have large, technical, expensive, systems in place to manage and archive construction documents, including drawings and managed construction projects.

Facilities change. The Internet makes it possible to tap those systems remotely without the expense of creating them yourself. In fact, who better to maintain changes to the original work than the entity that prepared it originally? The problem in the past has been that you could not get access to the resource when you needed it. For a fraction of the cost, you should be able to tap those resources directly and leverage their investments in technology. Property management companies like Trammel Crow now have nationally managed on-line help desks that utilize the Internet for distributed communications to both their customers and their field staff. They are able to walk in the door and offer services to their customers virtually the next day. They bring with them a complete business process that is available on-line through their centrally managed and hosted help desk which is accessible from the Internet. What does this mean to the facilities manager?

Typically, if you were to change a business process that involved a technical system such as a help desk, it might be difficult and time consuming to evaluate and justify the expense. Following this, there might be a consultant hired to implement the solution, then perform training and support. If you were lucky, you could get the system really working in about a year. By adopting an outsource service provider and further adopting their turnkey systems available to you directly through the Internet, you have eliminated or greatly reduced the change cost while still maintaining good access to the information. This approach allows you to better concentrate on the process and leave time and money to manage the benefit.

Another interesting example can be found at FMStrategies. Over the past 3 years, they have built a large Extranet system that addresses space and occupancy management and capital project management for its outsource customers. The focus of the site is on change management and internal rent or chargeback management and strategic space and occupancy planning for large multifacility enterprise clients with properties scattered across the country and around the world. A national repository for space and occupancy data is maintained. Information on buildings, properties, building status, business unit occupancy, and capital projects is kept in an Oracle database and is linked to a CAD database that includes thousands of floor plan drawings that accurately depict business unit occupancy, area measurement, and chargeback information. The company maintains the information through a customer relationship management process in association with many other outsource vendors, including architectural, lease, and property management companies across the country. Through a secure on-line Extranet application housed and maintained by FMStrategies, the information is accessed by the customer and all of his or her vendors. The status of capital projects is kept up to date on-line by the various vendors performing the work. Changes in project status indicate changes in occupancy. The occupancy team monitors the project system so that they can make changes to the associated CAD drawings, which reflect the occupancy information. The corporate real estate division can monitor the work from the office or from home and can use it for strategic business planning. All information is housed and maintained centrally, but is shared globally through the Internet.

This outsource approach can be attractive for a number of reasons. The technical resources, including hardware, software, and technical staff that maintain the system are housed off-site and the cost is spread across multiple users. The information is encrypted and secure so that it can be made available both inside the company and outside the company to associated trusted vendors. When a company, for example, acquires another company, its business processes can be provided and implemented almost immediately through a common Internet browser. Both the facilities and the resources that maintain them are widely separated. The Internet makes it quick and easy to distribute management information to anyone almost anywhere and enhances the outsource leveraging benefit.

These are just a few of the examples of how Internet technology can enhance outsourcing; depending on the service, we could find many more examples. By nature, the Internet is an enabling collaborative technology. A facilities manager can benefit by implementing an Internet strategy for use in collaboration with his or her outsource vendors.

26.4.10 Political Realities, Practical Realities

In an article in the March 22, 1999 issue of *Time* magazine written by Bill Gates, the founder and CEO of Microsoft, he stated, "If the 1980s were about quality and the 1990s were about reengineering, then the 2000s will be about velocity." The world is changing—as always this is nothing new—but the rate of change has increased. Business globalization is very much a part of life these days. In Gates's article he lists twelve new rules "For Succeeding in the Digital Age":

1. Insist that communication flow through e-mail
2. Study sales data on-line to share insights easily
3. Shift knowledge workers into high-level thinking
4. Use digital tools to create virtual teams
5. Convert every paper process to a digital process
6. Use digital tools to eliminate single-task jobs
7. Create a digital feedback loop
8. Use digital systems to route customer complaints immediately
9. Use digital communication to redefine the boundaries
10. Transform every business process into just-in-time delivery
11. Use digital delivery to eliminate the middleman
12. Use digital tools to help customers solve problems themselves

This list can be boiled down to three main ideas: collaboration, distribution, and access to information. These ideas are what Internet technology is about. How can, and how should, the facilities manager take advantage of the technology to better his business? No doubt you are probably using some Internet technology in your business today. From time to time, it is good to stand back and review what you are doing. Like any large initiative, you start by stepping back from the day-to-day and try and take a look at the bigger picture. The first words out of high-priced consultants' mouths are always "needs analysis"; in other words, first define the problem you are trying to solve.

This is a critical step, but often organizations will overlook or short-cut it. You need to examine the current sources of your business information, look at your current business processes, review the technology you control, and further review your company's technology direction and compare your approach with your peers' in the industry. Identify the immediate problems you are trying to solve without trying to define solutions. Then, ignore it all, and do some blue-sky thinking or some "wouldn't it be cool if we could" thinking before making any hard decisions about how to proceed.

The facilities manager is, by default, very close to the existing picture, sometimes too close to do a good job of being objective. This is often a good step to outsource to an independent consultant that has an industry-wide view of FM business practices and technology. This consultant may or may not be the entity that actually implements the solution. Following the "needs analysis," it will be time to do some industry research on commercially available tools and solutions. During this phase you should be able to determine the answer to the build vs. buy question. In many cases, it will not be a clear answer. The more likely case is that you will need to buy and integrate, meaning that you will purchase a core solution and then customize it to meet your needs.

You will need to look at and evaluate the return on investment (ROI) and the total cost of ownership (TOC) for any system you propose. This means that you will want to look closely at support and maintenance issues for the system. In addition, you should look at scalability issues considering how you might grow and expand the system in the future, including both users and functionality. You will also need to look at the up-front cost of implementation, customization, and training. In some cases, you will need to add or reorganize staff resources to manage the system. With new technology, you should always do some out-of-the-box thinking.

The Question?	The Answer . . .
What problem am I trying to solve?	Business requirements/needs analysis
What do I have now?	Existing conditions analysis
How do other peer industries solve my problem?	Peer analysis/benchmarking
What is commercially available?	Technical research/market research
What is possible?	Blue-sky forecasting
What is my long-term need?	Big concept creation/strategic plan
What is the value?	ROI/TOC analysis
When can I use it?	Detail implementation plan/phasing plan
Who will use it?	Resource plan
How will I maintain the solution?	Support plan
How will I achieve success?	Training plan
How do I know I achieved success?	CRM plan

Facilities managers must have a diverse set of skills. Two required skills for successful facilities managers are politics and marketing. Expertise with these skills will serve you well when implementing new systems and processes. The key factors that have a bearing on having your proposals approved are the ROI, often stated in terms of increased efficiencies, and the TOC, which has to do with on-going costs of operation for the system. Although it is not always easy to obtain hard answers in these areas, the better homework you do here, the better your chances of having your proposal approved by upper management.

In addition to the normal work associated with preparing a proposal requiring dollars, an Internet technology proposal will most likely require some additional politicking with your internal information services department. Today, now past the Y2K problem, most information services or information technology (IT) groups have their hands full with implementing Internet and Extranet strategies for their companies as a whole. It will be extremely important that you understand their direction for implementing Internet technology and leverage it in your plan if possible.

In addition to normal shared support and infrastructure for Internet systems, the IT group will be very concerned with security and bandwidth requirements for your system. Key questions about how your system will be used and by whom will require specific answers. One of the most important issues will be whether your system will have an Internet, Intranet, or Extranet focus. In other words, will the system require access from outside the company or will it be accessed strictly within the company's firewall? The answer to this question will make a tremendous difference to how quickly and easily your proposal will be approved by corporate IT. This can be a difficult issue. In most FM scenarios, at least some business value is associated with process improvement related to outside resources. You must carefully weigh the value of the need to secure company information with its need for access by your trusted vendors and discuss the idea openly with your internal IT group.

26.5 INTERNET TECHNOLOGIES

Technology is difficult to keep up with. It changes so rapidly that by the time you begin to understand one technology another has replaced it. Internet technology is a particularly difficult topic. The technology is still immature. It continues to evolve at a rapid pace. The object of this section is to familiarize the facilities manager with some of the terms and issues that they should be aware of when selecting and implementing Internet technology. In most cases, the facilities manager is looking for a cost-effective solution to a particular business problem. Internet technology may provide a means to achieving that solution, but not necessarily the end. Generally, the facilities manager will be relying on an internal information services group to assist him or her in selecting technology solutions or, alternatively, will work with an outsource consulting group, or vendor, who specializes in facilities-related solutions.

It is important that the facilities manager have a basic knowledge and understanding of the elements associated with Internet technology so that he or she can make better-informed decisions when selecting solutions. The information contained in Appendix A will identify some of the current buzzwords and hot technologies associated with Internet solutions. Appendix A is not meant to change the facilities managers job title to "Geek," rather it is meant to empower him or her to ask intelligent questions when evaluating new solutions. Appendix B presents some useful Internet links for facilities managers. Technology is a fascinating, often overpowering and confusing, critical factor in business today. Choosing solutions that are cost effective to implement, cost effective to maintain, and actually work and improve your business should be the primary goal of the facilities manager. Appendix A should at least help to identify some of the technology language surrounding the Internet.

26.6 CONCLUSION

It is estimated that as of 1999 the Internet had some 200 million users worldwide. There are over 100 countries that are connected to the Internet and the numbers continue to grow. Although Internet technology continues to grow and change rapidly, it is no longer considered bleeding edge. In fact, more likely it is considered to be a required and often a strategic technology. The facilities manager now has a new tool to assist him or her in solving business problems related to building and property management. Today, it is important for the facilities manager not only to have a strategic space, occupancy, and maintenance plan, he or she must also have a technology plan. That technology plan will most likely be built on Internet technology. There have been hundreds and probably thousands of volumes written about Internet business strategies and Internet technology by now. Interestingly enough, many of them are actually available on the Internet itself. This chapter has presented only an overview of the subject. Its purpose has been to introduce the reader to Internet technology and its potential relationship to the business of managing facilities. It will be incumbent on the astute facilities manager to continue to read and study further developments in Internet technology and its uses in FM.

APPENDIX A INTERNET TERMINOLOGY PRIMER FOR FACILITIES MANAGERS

Definitions in Appendix A have been summarized and/or expanded from more complete definitions available at www.webopedia.com. Further definitions of technology-related terms can be found at www.whatis.com.

Internetworks

Internet. A decentralized global public network that connects millions of computers around the world.

Intranet. A network based on the TCP/IP protocol that is internal or private to a given organization. An Intranet works like the Internet but is fully protected inside a corporation's secure firewall.

Extranet. An Intranet that is partially accessible to authorized outsiders. Extranets are generally used to access commercial services. Requiring a log-in including a unique username and password generally protects Extranets.

WWW (World Wide Web). A series of computers or servers that host documents created in the standard HTML (hypertext markup language) that are connected through the Internet. The HTML document structure allows hyperlinking from one document to the next across the network.

Connectivity

LAN (Local Area Network). A small local network used to connect multiple computers. A LAN usually is contained within the confines of a single building. There are many different types of LANs. LANs are differentiated by their:

- *Topology:* The geometric configuration of a network
- *Protocol:* The rules for encoding the information that is transmitted and received over the network
- *Transmission Media:* The type of wire, cable, or method used to transmit information between devices on a network.

WAN (Wide Area Network). A network that connects two or more local area networks (LANs) across a wide geographical area. WANs often use public networks including the Internet for communication.

ISP (Internet Service Provider or Independent Service Provider). A company that buys direct access to the high-speed public infrastructure backbone of the Internet and then redistributes access in smaller chunks to companies and individuals. ISPs often provide a suite of services including Web site hosting, data storage and retrieval, virtual private networking, and Internet consulting services.

Protocol (Data Transmission Format). An agreed-upon format for transmitting data between two devices on a network.

TCP/IP (Transmission Control Protocol/Internet Protocol). The common networking communications protocol used to connect Internet hosts. The use of the common TCP/IP protocol helps to make the Internet possible.

T-1 Line. Very fast, dedicated telephone connections supporting data rates of 1.544 Mbits per second. Also known as leased lines, businesses will purchase access to the Internet by contracting with a local telecommunications company or ISP for T-1 service. Higher rate transmission lines are also available including T-3 and OC-48 or greater. Large companies will connect to the Internet utilizing one or many T-1 or T-3 connections.

ISDN (Integrated Services Digital Network). A communications standard for transmitting voice, video, and data over digital telephone lines. ISDN lines support rates of 64 Kbps (64,000 bits per second). Telephone companies usually deliver ISDN lines in pairs. One line can be used for voice and the other for data or both lines can be combined to achieve a 128 Kbps transmission rate. ISDN lines have been a popular Internet connection option for small businesses.

DSL, ADSL, SDSL (Digital Subscriber Lines). DSL technologies use sophisticated modulation schemes to pack data onto standard copper telephone wires. DSL technology is a high-data-rate technology, up to 32 Mbps for downstream and up to 1 Mbps for upstream traffic offered by local telephone companies for Internet connections. The technology is not generally available because it requires short-line runs of less than 20,000 ft directly to a local telephone switching station. The technology is generally for home or small business use only.

Broadband Transmission. A type of data transmission in which a single medium (wire) can carry several channels at once. Cable TV, for example, uses broadband transmission. In contrast, baseband transmission allows only one signal at a time. Cable modems connecting to standard coaxial TV cable in your home can offer speeds up to 2 Mbps. Most communications between computers, including the majority of local area networks, use baseband communications.

Hardware

Servers (Centralized Computers). A computer on a network that manages one or many network resources. Servers can be dedicated to hosting a single specialized service. The Internet utilizes many types of servers, including Web servers, database servers, object servers, proxy servers, file servers, and firewalls. Servers can contain one or many CPUs or processors. Servers can be ganged or clustered to scale the services they support.

Routers. A device used to connect any number of LANs. Routers do not generally care about the type of data they handle, they are simply used to forward information from one network to another.

Hubs. Hubs have multiple ports and are used to connect multiple segments of a LAN. Data coming in from one port will be automatically copied to all other ports. There are various types of hubs. Intelligent hubs allow network administrators to monitor network traffic. Switching hubs will read the destination of a data packet and forward it automatically to the correct port.

Firewall. A system designed to prevent unauthorized access to or from a private network. Firewalls can be implemented in both hardware and software, or a combination of both. Firewalls are frequently used to prevent unauthorized Internet users from accessing private networks connected to the Internet, especially Intranets. All messages entering or leaving the Intranet pass through the firewall, which examines each message and blocks those that do not meet the specified security criteria.

Operating Systems

Windows. The term "windows" generally refers to a type of graphical user interface (GUI) that presents information to users in rectangular windows looking structures. Microsoft has developed a series of graphical operating systems known specifically as "Microsoft Windows" that exploits the "windows" metaphor. There are several variants of Microsoft Windows including Windows 95, Windows 98, Windows NT, and now Windows 2000.

The later of these operating systems have fully adopted the Internet in presentation and functionality. The desktop operating systems are shipped complete with an Internet browser and an e-mail client, and the server operating system versions support all types of Internet server technology including Web servers, ftp servers, proxy servers, etc.

Mac OS X. The latest in the series of Apple Macintosh operating systems. Apple has always been known for its attention to the end user with its graphically enabled operating systems. The MAC OS, like most other operating systems, fully supports and integrates the Internet into its desktop interface.

UNIX. UNIX has been around for a long time and is a standard in many corporations for mission critical systems. UNIX can run on many different hardware platforms. There are a number of variants of UNIX supported by companies like IBM (AIX) and Sun Microsystems (Solaris).

The Internet was pioneered on UNIX-based systems. Although the UNIX operating system has not been a favorite for the majority of business users, it is an excellent platform for Internet servers. It is powerful, highly flexible, secure, and maintains a robust reputation.

LINUX. A freely distributed open source version of the UNIX operating system. LINUX has become very popular over the past few years. Major vendors including, but only scratching the surface, like Oracle, Corel, and WordPerfect are now supporting products that can be hosted by LINUX. LINUX can be used as both a desktop and server operating system.

One interesting combination is the open source (free) LINUX operating system hosting the open source (free) APACHE Web server. This is a highly effective Internet solution where both the software and the support, via Internet, can be absolutely free (excluding hardware, setup, and connection services). What is the world coming to?

Internet Servers

Web Server (HTTP Server). A computer running Web server software that serves or delivers Web pages over the Internet to client computers running Internet browser software. There are many commercial and open source Web servers available today. Some of the more popular Web servers include Apache Software Foundation's Apache Web Server, Microsoft Internet Information Server (IIS), Netscape iPlanet Web Server, Lotus Domino, etc.

FTP Server (File Transmission Protocol). Uses the standard file transfer protocol (FTP) to allow files to be downloaded across the Internet. FTP servers can offer both public, often known as "anonymous," or private secure, access to files on a given server.

SMTP Server (Simple Mail Transmission Protocol). An Internet mail protocol used primarily to send mail from a client computer to a receiving mail server. (See POP3.)

POP3 Server (Post Office Protocol). An Internet mail protocol used to retrieve mail from a stored location on an Internet server. POP can be thought of as a "store-and-forward" service. An Internet mail server receives messages for you and holds them until you request them.

NNTP Server (Network News Transfer Protocol). The standard protocol used by computers (servers and clients) for managing the notes posted on Usenet newsgroups. A newsgroup is a discussion about a particular subject consisting of notes written to a central Internet site and redistributed through Usenet, a worldwide network of news discussion groups.

Proxy Server. A proxy server is exactly what its name implies, a substitute for the real server. A proxy server sits between the Internet and a Web server. It intercepts all requests to the real server and tries to fill the request itself. If it cannot fill the request, it forwards the request to the real server.

Proxy servers are most often used to cache frequently used Web pages. This approach can dramatically improve performance for your corporate Internet access. You would not believe the number of similar requests an external Web server gets from the same company. For example, take the day before or after the Super Bowl, you might have a couple hundred internal users looking for the same information. If those pages were cached locally, the time it took to access the pages would be greatly reduced.

Application Server. A relatively new breed of servers used to connect a Web server and the outside world to a company's backend business applications and databases. Application servers can be very sophisticated and can offer a number of services including integrated and dynamic Web page generation. There is a large and growing market for application servers. You will see some kind of application server tier in most large companies in the future.

Internet Browsers

Browsers (Microsoft Internet Explorer, Netscape Navigator). A Web browser is a software program that is capable of receiving, interpreting, and displaying Web pages. At the beginning of the Internet, many different browsers were available, but only two major browsers are prevalent in business today. Microsoft Internet Explorer and Netscape Navigator now dominate the browser market and the corporate desktop.

Unfortunately, although there are numerous international standards for creating Web page content and there are primarily only two browsers to choose from, you still cannot rely on them to interpret Web pages in the same way. This can be a problem when writing content. You must either adopt a single browser, or only use features that can be interpreted in both browsers.

Web Page Authoring/Markup Languages

SGML (Standard Generalized Markup Language). The larger more complex brother of the HTML markup language used today on the Internet. SGML was adopted by the International Organization for Standards (ISO) in 1986 and has been used widely to manage large documents. The SGML standard does not specify formatting, rather it specifies rules for tagging elements.

HTML (Hypertext Markup Language). The primary authoring language used to create documents on the World Wide Web (WWW). HTML uses a system of common imbedded TAGs to define document elements and formatting. HTML includes the ability to embed hyperlink references to associated documents using URLs or uniform resource locators.

DHTML (Dynamic Hypertext Markup Language). A new flavor of standard HTML that includes additional features for dynamic positioning and visibility of document elements without returning to the Web server for additional information.

DHTML is sometimes confused with dynamic page generation technologies like CGI and ASP that will actually dynamically create HTML pages on the server for return to a client based on parameters passed to the server from an initiating Web page.

Web page programming has come a long way. You can now create complete, full featured, thin client applications using a combination of simple markup and scripting languages that look and feel like more traditional and complex compiled applications.

XML (Extensible Markup Language). A pared-down version of SGML, designed especially for Web documents. XML includes the ability for Web designers to separate data and format into different documents. XML also includes the ability to create unique element TAGs and associated schemas that define those TAGs. XML shows great promise in becoming a standard data interchange mechanism both for the Internet and for traditional applications.

VRML (Virtual Reality Modeling Language). A specification for displaying 3D objects on the World Wide Web. You can think of it as the 3D equivalent of HTML. VRML requires a VRML viewer or browser plug-in program to display and view the files. A user viewing a VRML file can move through the 3D environment using the interactive viewer.

VRML is used in preparing conceptual studies of architectural and interiors projects for delivery over the Web. You can easily view the changes to a proposed project and collaborate on the design from multiple locations. Many 3D modeling tools can now export their contents directly to the VRML format. The technology requires high-bandwidth connections to the Internet to be truly effective.

URL (Uniform Resource Locator). Represents the global address of documents and other resources on the World Wide Web. The first part of the address indicates what protocol to use and the second part specifies the IP address or the domain name where the resource is located, for example:

http://www.fmstrategies.com/default.htm	*An HTML web page*
ftp://www.fmstrategies.com/documents/document.txt	*A downloadable file*

Scripting

JavaScript/JSCRIPT (Java Scripting Language). A scripting language developed by Netscape that allows Web page authors to develop interactive pages. JavaScript works in conjunction with HTML and includes interactive capabilities like looping and conditional constructs. Without JavaScript or its cousin VBSCRIPT, Web pages would be completely static. JSCRIPT is Microsoft's version of JavaScript.

VBSCRIPT (Visual Basic Scripting Language). A scripting language similar to JavaScript used for authoring interactive Web pages. VBSCRIPT is based on Microsoft's Visual Basic programming language.

Dynamic Scripting

ASP (Active Server Pages). A Microsoft technology that allows HTML to be dynamically generated by programs running on the Web server. ASP is a powerful technology. It includes features that allow for connections to be made to backend databases and object servers. Much of the work is performed on the server. Complete applications can be developed on a backend server and then served up to a client browser as an HTML Web page.

JSP (JAVA Server Pages). A server-based dynamic page-generation scripting language based on the JAVA programming language. JSP is Sun Microsystems' answer to Microsoft's Active Server Pages. JSP can work with JAVA Servlets, JAVA objects, and Enterprise JAVA Beans to build robust Web-based applications.

Cold Fusion. A server-based dynamic HTML page generation product developed by the Allaire Corporation. Cold Fusion is based on the CFML proprietary markup language. Cold Fusion was one of the first products to make backend database connections easy for developers. The product has matured into a complete Web application development environment.

PERL (Practical Extraction and Report Language). A programming language used primarily on UNIX operating systems to develop dynamic CGI scripts. PERL is known for its strong text-processing abilities. Programs are written on a server that translates their output into HTML on the fly.

In-Line Internet Server Programs

CGI (Common Gateway Interface). A specification for transferring information between a Web server and a server-based CGI program. CGI programs can be written in almost any programming language, including C, Perl, Java, or Visual Basic. CGI programs are the most common method for Web servers to interact dynamically with users. Many HTML forms use a server-side CGI program to process their information.

CGI programs require the Web server to launch a new process every time the CGI program is called. This can be a problem for Web sites that have a lot of traffic or are limited on hardware.

ISAPI (Internet Server Application Interface). Similar to CGI programs, but they are allowed to run inside the Web server's address space. This means that they are generally faster than conventional CGI-based programs. It also means that if they are not written properly, they can crash and bring down the Web server with them.

JAVA Servlets (JAVA-Based S). Special JAVA programs known as applets that are running on the Web server instead of in the context of a browser. Servlets can process HTML forms and return results in dynamically generated HTML. JAVA Servlets are based on the JAVA programming language and are similar to traditional CGI programs. In contrast, Servlets, once started, remain in memory and are thereby faster to access than traditional CGI programs that must load each time they are called. JAVA Servlets are becoming increasingly popular as an alternative to the CGI approach.

Traditional Programming Languages

JAVA. An object-oriented programming language developed by Sun Microsystems. The language's primary claim to fame is its cross-platform capabilities. The language is interpreted, which means that you must have a separate program, known as a JAVA Virtual Machine, running on your computer to interpret the code. JAVA Virtual Machines are available for most operating systems, which means that you can write a program in the Microsoft Windows environment and it will run on a UNIX server.

In addition to its cross-platform talents, JAVA was also designed to take full advantage of the Internet. With JAVA, you can create precompiled programs known as JAVA Applets that will run inside of a Web browser on any platform.

JAVA is a true programming language and is much more powerful than simple markup languages like HTML. Complete interactive programs can be written in JAVA, housed on a server, then downloaded to a client machine from a Web server when it is needed by an end user. This capability presents many interesting ideas about upgrading or actually renting desktop software from a central location.

JAVA is a general purpose programming language with many features. Many large corporations are moving their business logic to JAVA-based systems. JAVA includes a long line of Internet subtechnologies including JAVA Applets, JAVA Server Pages, JAVA Servlets, and Enterprise JAVA Beans to name a few.

C/C++. The C programming language has been around for many years. C++ is the object-oriented version of traditional C. The C/C++ language is one of the most prolific languages in history. You can write almost anything in C. Many operating systems, including Microsoft Windows and UNIX, are mostly written in C (in addition to some assembly and machine language). Even other programming languages are written in C. C/C++ is natively compiled and is known for its exceptional performance characteristics.

C/C++ is a general programming language and does not have specific Internet features. It is primarily used for creating CGI and ISAP programs and backend server, object-based, program layers. Microsoft has created a technology known as ActiveX which builds on its COM and DCOM component object model technology. ActiveX components can be written in C/C++ and hosted inside Web pages as browser plug-ins. This is similar to the JAVA Applet approach to embedding applications in Web pages.

Visual Basic. Developed by Microsoft, this is a general programming language based on the original BASIC (beginner's all-purpose symbolic instruction code) programming language developed at Dartmouth College in the mid-1960s. Visual Basic was one of the first programming languages to incorporate a graphical programming environment. The advances made by Microsoft in visual program development has made Visual Basic, known as "VB," a language of choice for prototyping applications. Visual Basic is a relatively easy programming language to become productive at.

There are no specific Internet features in the Visual Basic language. The language is highly integrated with the Microsoft Windows operating system and with Microsoft active server pages (ASP). Much of the Internet development with Visual Basic has been done in conjunction with ASP by creating backend server-based objects for specific processing tasks.

Visual Basic can be used to create Web page hosted ActiveX plug-ins based on Microsoft's COM/DCOM technology. You can also create server-side CGI programs with Visual Basic.

Object Communication Standards

A program development paradigm shift has occurred over the last 10 years. In the past, most programs were written using procedural or step-by-step methods. Today, most programs are using some kind of object technology that encapsulates data and behavior into one or many logical programming elements known as objects.

Objects are important to the Internet because of its inherent distributed and multitiered structure. Most Internet systems developed today rely on multiple tiers of functionality including browser clients, Web servers, business object servers, and persistent database servers. These tiers or layers are often physically housed on separate computers, which communicate across networks including the Internet.

Object communication standards are required so that objects operating on one machine can communicate with objects operating on the same or physically separate machine.

COM/DCOM (Component Object Model/Distributed Component Object Model). COM is a specification developed by Microsoft that allows binary objects to talk with each other. DCOM is an extension to the COM specification that allows objects to talk with each other over a network. The COM/DCOM standards primarily apply only to objects developed for Microsoft Windows operating systems. COM objects can be written in most languages. Microsoft's OLE and ActiveX technologies are based on COM.

RMI (Remote Method Invocation). A specification for remote object communication. RMI is a relatively simple protocol, but unlike the COM/DCOM and CORBA specifications, RMI is a JAVA-only standard.

CORBA (Common Object Request Broker Architecture). An object communications specification that allows an object written in any language that conforms to the CORBA specification to communicate with any other object conforming to that specification. Objects implementing the CORBA standard can communicate across languages and across operating platforms.

The Internet is a global communications network. The Internet by definition is composed of many different computers operating and communicating through a single enormous network.

CORBA is a powerful specification that allows applications to be completely free to distribute themselves, to collaborate, and to work collectively to solve business problems.

Security

HTTPS/SSL (Secure Sockets Layer). SSL is a protocol developed by Netscape for transmitting private documents via the Internet. SSL works by utilizing a public key/private key, data encryption scheme. SSL creates a secure connection between a given server and a given client for transmitting encrypted information. Web page addresses utilizing SSL begin with HTTPS:// instead of the standard HTTP://.

Web sites use SSL to send and receive confidential information like credit card numbers. Commercial Extranet sites also use SSL to protect customer information that may be sensitive or proprietary in nature.

PGP (Pretty Good Protection). One of the most common ways to protect messages and information sent across the Internet. PGP utilizes a public key/private key method for encrypting information. One key is a public key that you can send to anyone that you wish to receive a message. The second key is a private key used only to decrypt messages you receive. The PGP encryption tool is highly effective and is freely available from the Massachusetts Institute of Technology.

VPN (Virtual Private Network). Secure and encrypted network tunnels from point-to-point through the Internet. VPNs use the public infrastructure to create private networks. VPN technology is often optionally included with commercial firewall systems.

With the advent of high-bandwidth broadband networks, VPNs can create highly effective remote working environments for businesses that need off-site connectivity including work-from-home telecommuting programs.

New Ideas

ASP (Application Service Provider). Third-party entities that manage and distribute software-based services and solutions from a central location to customers across a wide area network. The Internet has made the ASP business an attractive alternative for outsourcing certain aspects of a company's information technology.

According to http://www.aspnews.com, ASPs can be broken down into five subcategories:

- *Enterprise ASPs:* Deliver high-end business applications
- *Local/Regional ASPs:* Supply wide variety of application services for smaller businesses in a local area
- *Specialist ASPs:* Provide applications for a specific need, such as Web site services or human resources
- *Vertical Market ASPs:* Provide support to a specific industry such as healthcare
- *Volume Business ASPs:* Supply general small-/medium-sized businesses with prepackaged application services in volume

ASPs can be commercial ventures that cater to customers, or not-for-profit or government organizations, providing service and support to end users.

VPA (Virtual Private Application). Similar in concept to ASPs. A VPA is licensed or rented from a centralized service provider or vendor and downloaded through the Internet for use on a local computer. Applications can be upgraded automatically and are maintained at the central location. Applications are usually provided on a "per-user" basis. VPAs present a potentially useful concept for solving some business application requirements. The VPA solution could reduce the overall cost of operation for computer applications. The idea provides flexibility and can support on-demand usage. Applications can be used when and where they are needed and disconnected when they are not needed.

APPENDIX B USEFUL LINKS FOR FACILITIES MANAGERS

Consultants

- Adept Facilities & Design http://www.adeptfacilities.com
- Aptek Associates http://www.aptek.net
- Booz-Allen & Hamilton—Management & Technology Consulting http://www.bah.com
- Business Interiors http://www.businessinteriors.com
- CFI: Computerized Facility Integration http://www.cfi-solutions.com
- Collman & Karsky Architects http://www.collman-karsky.com
- Expense Management Solutions http://www.expensemanagement.com
- Facilities First—Comprehensive Services to Meet Your Expansion Needs http://www.facilitiesfirst.com
- Facilities Plus http://www.facilitiesplus.com
- FACS Facility Services Inc.—Worldwide Total Building Solutions Provider http://www.facsfs.com
- Ferreira Service, Inc. http://www.ferreira.com
- FMinternational—Infrastructure Management Solutions http://www.fminternational.com
- FM Strategies—Facility Planning & Space Management http://www.fmstrategies.com
- The Friday Group—Guiding professionals and organizations to find the right strategies and solutions to solve their business challenges http://www.thefridaygroup.com
- Gensler http://www.gensler.com
- Graphic Systems, Inc.—Facility Management Technology: Consultants and Service Bureau http://www.graphsys.com
- Harvey H. Kaiser Associates, Inc. http://home.att.net/~hhkaiser
- Hero, Inc.—Human Environment Research Organization http://www.hero-inc.com
- Indicos, Inc. http://www.indicos.com
- Nelson & Associates—Taking Innovation, Information and Space to New Horizons http://www.nelsononline.com
- The Onyx Group—Tomorrow's Facility Solutions . . . Today http://www.onyxgroup.com

- Port Remote Services, Inc.—Business Solutions for Telecommuting and Mobile Work — http://www.portstrategic.com
- Rolf Jensen & Associates, Inc.—Fire Protection and Life Safety Engineering — http://www.rjagroup.com
- Space, LLP—A Proactive Force in the New Global Business Mode — http://www.workplayce.com
- Space Diagnostics, Inc.—Facility Planning & CAFM — http://www.spacedx.com
- Swinerton & Walberg Builders—Construction Service Provider — http://www.swinerton.com
- Carroll Thatcher Planning Group Inc. — http://www.thatcherplanning.com

CAFM Vendors

- Aperture Technologies, Inc. — http://www.aperture.com
- Archibus, Inc. — http://www.archibus.com
- Bentley — http://www.bentley.com/index.html
- Centerstone Software, Inc. — http://www.centerstonesoft.com
- Drawbase Software — http://www.drawbase.com
- FIS—Facility Information Systems, Inc. — http://www.fisinc.com
- FM:Systems — http://www.fmsystems.com
- OmniLink—Integrated CAFM Solutions — http://www.omnilink.uhd.com
- Vicusoft, Inc. — http://www.vicusoft.com
- Vision Facilities Management Ltd. — http://www.visionfm.com

CMMS Vendors

- AssetWorks, Inc. — http://www.assetworks.com
- Hansen Information Technologies, Inc. — http://www.hansen.com
- Micromain Corporation — http://www.micromain.com
- Peregrine Systems, Inc. — http://www.peregrine.com
- Prism Computer Corporation — http://www.prismcc.com
- PSDI — http://www.psdi.com

Other Vendors

- 20 * 20 Computerized Design (On-line 3-D office design) — http://www.furnitureweb.com
- Bar-Scan (Bar code asset management software and services) — http://www.bar-scan.com
- Best Software, Inc. (Software to manage people, fixed assets, and the planning process) — http://www.bestsoftware.com

- Building Systems Design, Inc. (Software solutions for the A/E/C industry) http://www.bsdsoftlink.com
- CICCorp, Inc. (Software and services for managing the cost of capital equipment) http://www.ciccorp.com
- Datacad LLC (CADD software and services) http://www.datacad.com
- E3 Corporation (Software and services for inventory management) http://www.e3corp.com
- Facet, Inç. (3-D furniture visualization software with space planning and presentation tools) http://www.giza.com
- HID Corporation (Software and services for access control cards and readers) http://www.hidcorp.com
- PenMetrics, Inc. (Software for field data collection) http://www.penmetrics.com
- Primavera (Project management software) http://www.primavera.com
- Spicer Corporation (Web-based viewing and markup software) http://www.spicer.com
- TISCOR (Tomorrow's integrated systems corporation) http://www.tiscor.com
- Vanderweil Facility Advisors (VFA) (Strategic capital planning and management software and services) http://www.vfa.com
- Vianovus (Software and services to manage construction programs and projects) http://www.vianovus.com
- VISCOMM (Web-based document sharing and more) http://www.viscomm.com
- Wind2 Software (Financial management system for Windows) http://www.wind2.com

Professional Organizations

- American Institute of Architects (AIA) http://www.aiaonline.com
- American Real Estate Society (ARES) http://www.aresnet.org
- The American Society of Interior Designers (ASID) http://www.asid.org
- Americans with Disabilities Act Document Center (ADA) http://janweb.icdi.wvu.edu/kinder
- The Association of Higher Education Facilities Officers (APPA) http://www.appa.org
- Building Owners and Managers Associations (BOMA) http://www.boma.org
- Canadian Design-Build Institute (CDBI) http://www.cdbi.org

- Construction Specifications Canada (CSC) http://www.csc-dcc.ca
- The Construction Specifications Institute (CSI) http://www.csinet.org
- Design-Build Institute of America (DBIA) http://www.dbia.org
- The Federal Facilities Council (FFC) http://www2.nas.edu/ffc
- International Council for Research and Innovation in Building and Construction (CIB) http://www.cibworld.nl
- International Development Research Council (IDRC) http://www.idrc.org
- International Facilities Management Association (IFMA) http://www.ifma.org
- International Society of Facilities Executives (ISFE) http://www.isfe.com
- National Association of Corporate Real Estate Executives (NACORE International) http://www.nacore.org
- National Institute of Building Sciences (NIBS) http://www.nibs.org
- Occupational Health and Safety Administration (OSHA) http://www.osha.gov
- Project Management Institute (PMI) http://www.pmi.org
- Standards Council of Canada (SSC) http://www.scc.ca

Project Web Sites

- 4specs (The Internet directory for "specified" construction products) http://www.4specs.com
- Akropolis.net (The Internet community for architecture, design & construction) http://www.akropolis.net/
- Alibre.com (Liberating design teams through Internet connectivity) http://www.alibre.com
- AnyDay.com (Free on-line day planner) http://www.anyday.com
- Aventail Corporation (Extranet service provider (ESP)) http://www.aventail.com
- Binary Tree, Inc. (Groupware/Web solutions) http://www.binarytree.com
- Bricsnet Project Center (On-line services and resources for the building industry) http://www.bricsnet.com
- Builder SupplyNet (Subscription-based Internet marketplace and document management service for the building industry) http://www.buildersupplynet.com

• Building.com (On-line forums and resources for the building industry)	http://www.building.com
• BuildingOnline (On-line forums and resources for the building industry)	http://www.buildingonline.com/
• BuildNet (Business-to-business e-Commerce solution and services for the residential construction industry)	http://www.buildnet.com
• BuildPoint.com (Construction bidding and procurement)	http://www.buildpoint.com
• Bullwhip Extranet (Free Web-based TeamWare)	http://www.bullwhip.com
• Buzzsaw.com (Project management services, building product specs, industry news, bidding and e-Commerce solutions)	http://www.buzzsaw.com
• CADWeb (Comprehensive solution to communicating information about building projects)	http://www.cadweb.co.uk
• Citadon, Inc.	http://www.citadon.com
• Collaborative Structures (Leading host for project communication in the real estate and construction industries)	http://www.costructures.com
• CollabWare.com (Collaborative engineering and design solutions for the WWW)	http://www.collabware.com
• Construction-Zone (Construction materials, education, and more)	http://www.c-z.com
• ConstructWare.com and HealthFLASH (On-line project management software and services for construction and health industries)	http://www.emergingsolutions.com
• ContractorHub.com (Marketplace where general contractors, subcontractors, and suppliers in the construction industry can exchange goods and services)	http://www.contractorhub.com
• Contractors E Source (Auction & RFQ services)	http://www.contractorsesource.com
• Contractors Register, Inc. (On-line bid solicitation and management system and electronic Blue Book)	http://www.thebluebook.com
• Cubus Corporation (The Internet collaboration company)	http://www.cubus.net
• Digital Paper Corporation (Digital document distribution technology and services)	http://www.digitalpaper.com
• Documentum (Content management solutions that accelerate innovation and shorten time-to-revenue)	http://www.documentum.com

• DrawingRoom.net (Free service to share and collaborate with drawings and image documents over the Web)	http://www.drawingroom.net
• e-Builder (Internet-based communication tool)	http://www.e-builder.net
• Edgewater Services Company (Integrated communication and GroupWare application for project managers—software and services)	http://www.projectedge.com
• Engineering Animation, Inc. (EAI) (Service for sharing project information)	http://www.e-vis.com
• eProject.com (Free Web-based project management application)	http://www.eproject.com
• Framework Technologies (Web-based project communications software)	http://www.activeproject.com
• Frontrunner (Project management software solutions and services for the construction industry)	http://www.frontrunner.com
• Glyphics Communications (Conference calling, Web conferencing, global messaging service, etc.)	http://www.glyphics.com
• HotKoko (Web-based communication tool)	http://www.hotkoko.com
• imanage, Inc. (Software and services that provide e-Business and information commerce solutions)	http://www.netright.com
• Informative Graphics (Visualization and redline collaboration software for Intranets, Internets, and Extranets)	http://www.infograph.com
• Instinctive Technology, Inc.—eRoom (Digital workplace for e-Business collaboration)	http://www.instinctive.com
• IntraACTIVE (Web-based collaboration software and services)	http://www.intraactive.com
• Intranets.com (Free, private and secure Intranet, or Extranet, sites)	http://www.intranets.com
• Involv Corporation (Virtual workspace complete with a robust suite of collaborative Web-based GroupWare and information management applications)	http://www.involv.com
• i-scraper (Internet-based collaborative environment for the commercial real estate and construction industry)	http://www.i-scraper.com
• iteamwork.com (Free Web-based project management tool for anyone who needs to manage multiple projects or teams)	http://www.iteamwork.com

• Magically, Inc. (Free suite of fully integrated Web-based messaging, collaboration, and personal information management applications)	http://www.magicaldesk.com
• Marin Research Inc. (Project management software and services)	http://www.marinres.com/
• MarketStreet.com (A business-to-business (B2B) Internet application service provider (ASP) that provides project communication and management software and services)	http://www.marketstreet.com
• Meridian Project Systems (Software and e-Business solutions for multisite, multiproject management, control, and collaboration within the building industry)	http://www.mps.com
• MyEvents.com (Free service for sharing personal and business info such as calendars, to-do lists, invitations, photo albums, and more)	http://www.myevents.com
• Netmosphere, Inc. (Enterprise project management software)	http://www.netmosphere.com
• OfficeCAD.com (Internet-based automated CAD and drawing sharing over the Web)	http://www.officecad.com
• OnlineBuildings.com (Internet solutions for the commercial real estate industry)	http://www.onlinebuildings.com
• Open Text Corporation (Delivers mission critical e-Business applications to organizations)	http://www.opentext.com
• Pacific Edge Software (Software that integrates with MS project and instinctive's e-Room)	http://www.projectoffice.com
• PM Boulevard (Virtual project management office and on-line resources)	http://www.pmblvd.com/
• PrimeContract.com (E-commerce for construction)	http://www.primecontract.com
• Projects On-line (Providing tools and infrastructure for the construction industry from concept to completion)	http://www.projectsonline.com
• ProjectShare.com (Interactive Internet business application that allows you to communicate information to your business partners, vendors, engineers, etc.)	http://www.projectshare.com
• ProjectTalk.com (Where project teams meet)	http://www.projecttalk.com

- Punch Networks (File sharing product and service) http://www.punchnetworks.com/
- PurchasePro.com (Electronic bidding, sourcing, and procurement for the A/E/C industry) http://www.aecconnect.com
- RedLadder.com (Free e-mail and Web sites for contractors) http://www.redladder.com
- SupplierMarket.com (Marketplace for parts and suppliers, on-line bidding) http://www.suppliermarket.com
- The Pigeon Hole (Internet service for the A/E/C industry, drawing management and sharing tools) http://www.thepigeonhole.com
- Visto.com (Free Web-based communications center) http://www.visto.com
- WebProject (Web-based project management software—JAVA-based) http://www.wproj.com
- Yahoo! Calendar (Free on-line calendar product and service) http://calendar.yahoo.com
- ZoomOn (View, mark-up, design, and collaborate on complex and simple graphics on-line) http://www.zoomon.com

Publications and Information Sites

- Architecture http://www.architecturemag.com
- Architectural Record http://www.mcgraw-hill.com
- Bricsnet.com—E-Marketplace for the Building Industry http://www3.bricsnet.com
- Building Operating Management http://www.tradepress.com
- BUILDINGS Magazine http://www.buildings.com
- Business Facilities Magazine http://www.busfac.com
- CADinfo.net—Automated Design, Documentation and Visualization http://www.cadinfo.net
- Construction Market Data http://www.cmdg.com
- Cyberplaces—The Internet Guide for Architects, Engineers and Contractors http://www.cyberplaces.com
- Engineering Automation Report http://www.eareport.com
- Environmental Accounting Project http://www.epa.gov/opptintr/acctg
- FacilitiesNet http://www.facilitiesnet.com
- Facilitilink—North American Alliance of Commercial Furniture Dealers http://facilitilink.com
- Facility Design & Management Magazine http://www.fdm.com
- FMLink—Your Facilities Management Resource on the Internet http://www.fmlink.com

• International Centre for Facilities (ICF)	http://www.icf-cebe.com
• Journal of Corporate Real Estate	http://www.henrystewart.co.uk
• Michigan State University—Human Environment and Design, Master's Certificate in Facility Management	http://www.msu.edu/user/facmgt
• Michigan State University—Virtual University	http://vu.msu.edu/
• Today's Facility Manager	http://www.tfmgr.com
• Wentworth Institute of Technology	http://www.wit.edu/Academics/DF/index.html
• The Wharton School—University of Pennsylvania	http://knowledge.wharton.upenn.edu

ENDNOTES

1. Clements, R. 1999. *IS managers guide to implementing and managing internet technology*. Paramus, NJ: Prentice Hall.
2. Gates, W. 1999. Rules for the digital age. *Time,* 22 March.
3. Sheldon, T. 1996. *The Windows NT web server handbook.* Osborne McGraw-Hill.
4. Cunningham, M. 1999. Why build an intranet. *e-Business Magazine,* September.
5. http://cyberplaces.com. Copyright 1997–2000 RSMeans CMD Group.
6. http://webopedia.internet.com. Copyright 1999–2000 Internet.com Corp.

CHAPTER 27
A/E/C INDUSTRY CASE STUDY: THE BIRTH AND DEVELOPMENT OF A COLLABORATIVE EXTRANET, BIDCOM

Charlie Kuffner
Senior Vice President
Northern California Division Manager
Swinerton & Walberg

27.1 PRECONCEPTION: BEFORE BIRTH

Since its founding in 1888, Swinerton & Walberg Builders has differentiated themselves from their competitors by applying technological innovations to build things faster, better, and cheaper. Innovations include introducing new materials, such as reinforced concrete, to California just after the turn of the nineteenth century; becoming the industry leader in real-time job site cost control and change management during the 1980s; using fiberglass and carbon fiber for concrete column and deck rehabilitation; and using collaborative real-time team communications via the Internet in the 1990s.

Managing Swinerton & Walberg Builders in Northern California, a $300 million/year regional operating group, I have been on the frontlines, communicating with all of the development, design, and construction project participants for the last 18 years. I ask questions, make decisions, distribute answers, and maintain volumes of current documentation. This information is used to order custom materials and build prototypical buildings in the field. All of this documentation ultimately needs to be turned over to the owner at project completion and be archived for their records and our records.

Armed with technical knowledge gained from attending conferences in 1996 and 1997 at the Center for Integrated Facilities Engineering (CIFE), a civil engineering and computer science research group at Stanford University, I was exposed to the concepts of real-time document sharing collaboration using the Internet, which they were experimenting with in their lab courses.

We had worked with construction software tools such as Primavera's Expedition and Meridian's Prolog, and our own in-house systems. We saw a demonstration of real-time drawing collaboration (one of the wide range of Extranet services that many commercial vendors offer, including Bidcom[1]) from Blue-Line On-Line (since then renamed Cephren) at CIFE. At the time, none of these commercial, or custom, software tools offered real-time work flow for contractor administration tools, such as requests for information (RFIs), product submittals and shop drawing approval, meeting minute action items, schedules, or daily job site reports which are mission critical in our business every day on every project in large volumes.

I was looking to extend these collaborative drawing-sharing, real-time Internet technology tools into work process management for our builders and other members of our A/E/C community. Much to my dismay, no one at CIFE or any of their conference attendees was aware of any commercial products that performed these functions. When I informed Stanford University Civil and Environmental Engineering Professor and CIFE Researcher, Dr. Martin Fischer, that I was ready to drive this new bus of Internet tools for construction administration, he informed me there was no place to insert the key. In fact, he informed me that the designs for the foundry to start casting the pieces had not been formulated or designed. So there I was, poised to start pushing for a revolution in the communication and administration of our industry by leveraging the accessibility and transparency of the Internet. I found myself at square one because none of the tools existed yet.

Looking back, I guess I should not have been too surprised at this since the construction industry had not been generally known as one that traditionally invested heavily (if at all!) in R&D. In fact, most of the innovations in the industry have traditionally been developed in a grass roots fashion. Most ideas coming up from the field are based primarily on the necessity of invention and the ingenuity of the builders.

Relative to the Internet and the A/E/C industry in the summer of 1997, I figured that I was where we all too often find ourselves—if you want it done **right** and you want it done **now**, you will have to find a way to do it yourself.

27.2 BIRTH: YOU HAVE TO START SOMEWHERE!

A San Francisco based on-line service provider start-up, Bidcom, came to our offices in September 1997 to pitch and demonstrate their Internet-based general contractor bidding system. Their basic software design concept was to use only an Internet browser (no software required by the user on his or her local computer) to access their proposed Extranet bidding service. Their initial collaborative Web site software was based on the Oracle 8*i* database and had some interesting features. Their concept was to replace a general contractor's bid-day tabulation summary with subcontractor and vendor bidding figures being submitted electronically over the Internet. Completely automated online bidding, although plausible for private works in a controlled environment, posed some fundamental difficulties for lump sum and public works bid construction projects.

I thought about using the Internet to speed up, streamline, and improve contract administration functions crucial to our business success. I asked the Bidcom founders, Darryl Magana and Sal Chavez, with their one employee, Larry Chen, if they had contemplated real-time work flow for construction administration processes, RFIs, submittals, meeting notes, schedules, daily reports, and the like.

I was looking for a repository for digital images for the project site web-cam and wanted the ability to easily playback "movies" of these stored images. I wanted to have all the current drawings and specifications, including change bulletins and other contract document revisions, stored and easily retrievable by all project participants in real-time. When our projects are completed, I wanted the ability to archive all the drawings and documentation in digital format on a CD or other medium without having to physically store the mountains of documentation we have been storing for years. The rent savings from renting less archive storage space is potentially a huge cost savings opportunity for various members of the A/E/C industry.

I encouraged Bidcom to focus their efforts on the area of project process management (e-PM). This is an area of day-to-day time critical processes crucial to the optimization of designing and constructing a custom building project. I told Bidcom if they would let me define and articulate Swinerton & Walberg Builders' business process models, and if they could technically develop these project process management tools on their Extranet, I could find clients that would incorporate them into our projects.

The Bidcom founders told me they had not contemplated these elements, but if we were interested, and we could sell it to an owner, they could develop those capabilities. I told them that software tools are what we needed to speed up and improve the efficiency of the A/E/C industry; thus reducing the cost to provide the high level of service demanded by our clients. Even though Bidcom was disappointed that I did not jump at their original concept and offer to sign a service contract on the spot, they said they could satisfy my requests and go back and work on my concepts.

27.3 INFANCY: THE TIME TO GROW UP IS NOW

Within a matter of weeks of first meeting with Bidcom, Swinerton & Walberg Builders was awarded a fast-track, 440,000-ft^2 18-floor e-Commerce tenant improvement project for a large investment firm (referred to as "ACME Investments"), on Main Street in San Francisco.

I met with the project director from ACME Investments, and told him of the on-going development of the Bidcom project management system in which Swinerton & Walberg Builders was involved. I asked him if we could retain Bidcom to implement real-time work flow on the Internet for our project administration functions with the tenant, design, and construction team. He was intrigued with the concept and, being a seasoned construction professional, was aware of the need. He approved our commitment to utilize Bidcom's e-PM software before he left my office.

The rest of this chapter is about the development and implementation of the first Bidcom Extranet project management system and its evolution into the commercial product it is today.

27.4 TODDLER: GO QUICKLY FROM CRAWLING TO WALKING TO RUNNING

The project, Main Street, was 440,000-ft^2 covering 18 stories in a recently renovated 1960s "black box" high-rise that had been structurally damaged in the 1989 Loma Prieta earthquake and been out of service since then. The landlord abated the asbestos and structurally upgraded the building so the project was similar to a tenant improvement build-out in a "new" building, but with a twist. There was some existing mechanical equipment that was proposed to be reused, and the existing building had no provisions for emergency generator power, uninterrupted power supply (UPS), or sufficient cooling required for ACME Investments' intended hi-tech use.

Swinerton & Walberg Builders' scope of work included swapping out existing air handling units on all 18 floors, adding additional cooling capacity with supplemental mechanical equipment on the roof, adding two emergency generators in the basement with corresponding fuel storage, and installing a substantial UPS system including a new UPS electrical riser from the basement to the roof to service devices on every floor.

The entire design and construction process was fast-tracked in order to move ACME Investments' employees in as soon as possible so they could enhance and grow their very profitable e-business product line. Bidcom had been working on their revised project management product based on my business process model for just 8 weeks when we rolled it out to the Main Street project team.

The Swinerton & Walberg Builders Team worked very closely with Bidcom as each feature was developed, programmed, tested, and then added to the software vendor's site. We held weekly development meetings, sometimes late into the night, strategizing on improvements and new features. As we brought each new feature on-line, we worked through the bugs so everything would work the way we all wanted it to. This was frustrating for some, and at times initial team members were tempted to "pull the plug" and go back to the tried and true methods. But knowing how valuable a completely functional A/E/C Extranet would be for Swinerton & Walberg Builders, and the industry in general, our staff gutted it out and stuck with Bidcom through some difficult early deployment days. Users coming on-line with Bidcom now fortunately do not experience any of the frustration that the first two teams went through as the product was initially being deployed.

One of the beauties of an on-line Extranet service and application provider is, once a new feature is developed, tested, and ready for commercial use, the site can be updated at any time. The next time a user logs in, all of the new features are at the user's fingertips. Gone are the days where software releases have to be copied, manually distributed, and loaded onto each individual user's machine. This is a huge time and cost-saving technique—especially for firms with large numbers of machines that all need to be on the same platform. Upgrades that had taken weeks or even months now happen instantaneously at virtually no cost. Problems during on-going upgrades in which some users are on one version and others are on another are eliminated.

27.5 CHILDHOOD: CHALLENGE DOING IT THE SAME OLD WAY

At ACME Investments' Main Street project, we compiled statistics on the turnaround time for project schedule critical RFIs. We calculated that a "normal" process (see Fig. 27.1) went something like this:

- RFI is handwritten by subsupplier and faxed to general contractor
- General contractor reviews and makes corrections, if necessary, then attaches a transmittal and faxes the revised RFI to the architect to review
- If the RFI requires a subconsultant's review (structural, electrical, mechanical, etc.), then it will be faxed to them
- If the RFI requires sub-subconsultant's input (lighting, acoustical, etc.), then it will be faxed to them

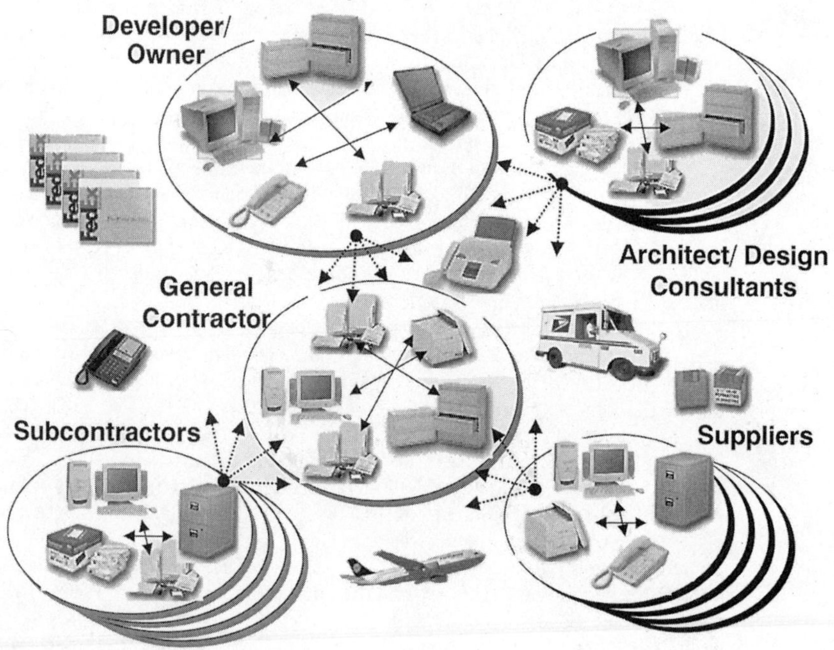

FIGURE 27.1 Industry Processes.

- The RFI is evaluated with multiple parties in-house, they formulate a response, and fax response back to subconsultant
- The subconsultant reviews and approves the RFI, or calls the sub-subconsultant back to modify the document, then faxes it to the architect
- The architect reviews and approves, or calls back the subconsultant to modify the document, then faxes it back to the general contractor and owner
- The general contractor evaluates; if response has a scope change that involves cost and/or schedule changes then general contractor inquires if owner wishes to proceed with the scope change that has a cost and/or schedule impact
- The general contractor receives owner's approval to a) proceed with change and price simultaneously (the most expedient, but also most risky to budget); b) price first, gain approval, then proceed (less budget risky, but more time consuming); or c) reject change, return to architect for different answer (can work, but can also backfire with similar answer many days later and issue is still unresolved and going nowhere)
- Once the disposition status is approved by the owner, the general contractor faxes RFI responses, with defined actions, to all subcontractors and suppliers (from 1–50 entities—usually 6–10 firms depending on the scope of the response)
- Lastly, the general contractor makes sure that the subcontractors and suppliers received the RFI response, both in the field and on the job site at the staff/foreman level, and at the sub/vendor project manager level, in their respective business offices (now that is 2–100 faxes per response)

The "traditional" RFI process averages $1\frac{1}{2}$ weeks (Figure 27.2). Using the e-PM software, our actual turnaround time at Main Street averaged $\frac{1}{2}$ week (Figure 27.3). This is a $\frac{2}{3}$ reduction in

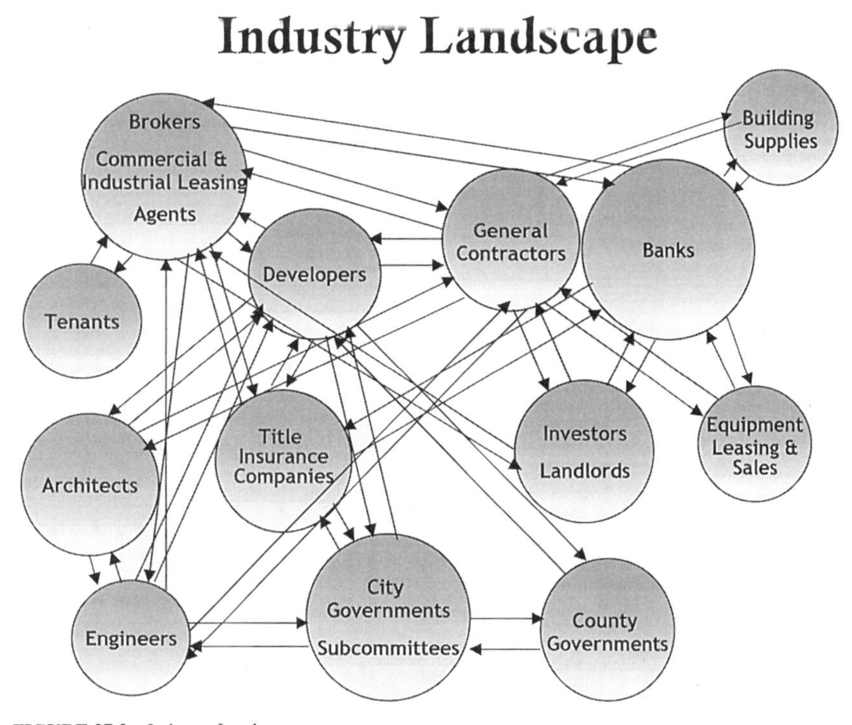

FIGURE 27.2 Industry Landscape.

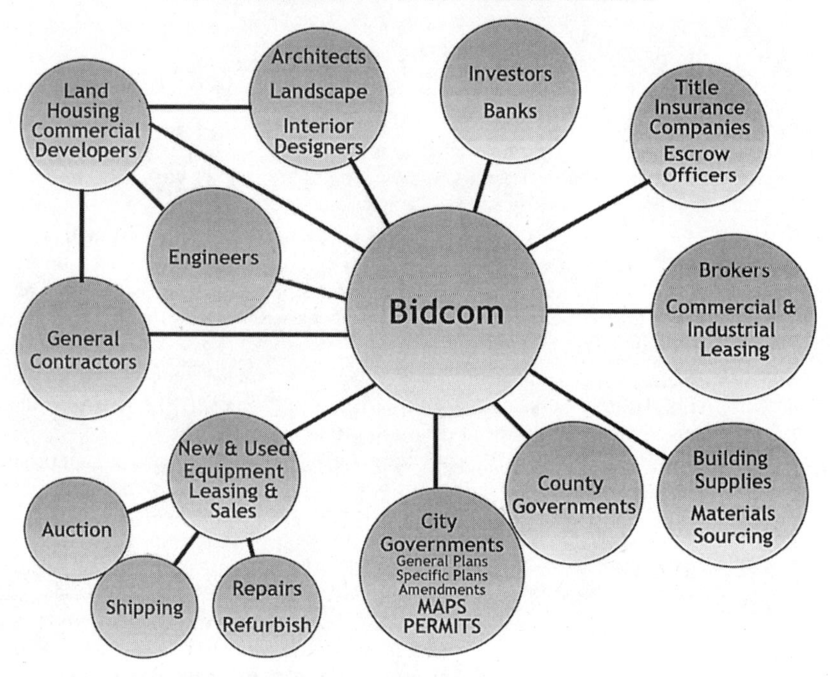

FIGURE 27.3 Central Business Hub.

response time for each issue. On this project, there were 124 RFIs. Any one of these issues could have potentially impacted cost and schedule.

From ACME Investments' standpoint at Main Street, they avoided 2 weeks of potential delays for changes that were incorporated late into the project scope. Swinerton & Walberg Builders was able to complete these changes within the time frame of the original schedule without requiring a schedule extension. These time savings and delay avoidances were achieved by the time and communication efficiencies realized by using the software's project management system.

There are a number of features that make e-PM software powerful, including project workflow, document routing and mark-up, automatic generation of "to-do" lists, scheduling and bidding tools, and so forth. Many of these tools significantly contributed to the time savings in, for example, the turnaround for the RFIs. Instead of calling the architect to find out the status of an RFI that was faxed to a subconsultant engineer, then faxed to one of the sub-subconsultants for response, requiring multiple phone calls to multiple participants that often are at the job site and not in their office to expedite a response to a question; with the software, we could check the status of the RFI and know exactly from whom a response was currently due. If a call is needed to expedite a response, any project participant can tell the exact status of every item and who to call to expedite a response. The Bidcom product also has an audit trail that tells a participant exactly who looked at what document when.

If a client is on the road, or working on multiple projects, he or she can check on the status of their project(s) remotely using any Internet browser. This accessibility and transparency is the tremendous power of a project Extranet viewed via the Internet.

E-PM products often host the project Web site on the vendor's computers, sometimes also known as an application service provider (ASP). E-PM software that incorporates work flow is substantially more powerful than project Extranet Web sites that simply host documents. A project Web site is simply a static repository for information that is only as current as the latest posting. A project Extranet is truly real-time project management, where project Web sites, by the nature of their design, cannot be.

27.6 ADOLESCENCE: METRICS

In terms of rent, the average cost of office space in San Francisco in mid-1998, at the completion of the Main Street project, was \$48/ft^2/yr. This is almost \$1/ft^2/wk. Saving 2 weeks of a 440,000 ft^2 tenant build-out represents \$880,000 of rent exposure saved. The true productive value of that space to ACME Investments for their e-business team to produce quality product out of their new space is some multiple of that savings.

A landlord evaluates time savings similarly to a tenant. For a landlord, the key to his or her success is to have tenants paying rent. Often this is tied to the scheduled completion of the tenant improvements. If construction schedules can be shortened, the landlord can start collecting rent sooner. Savings from using project Web sites will approximate those of the tenant. Another way to look at landlord savings is rent start is also the start of cash flow used to service the debt used to finance the construction of the building and the tenant improvements.

For designers, general contractors, and subcontractors, the sooner a tenant moves in, the sooner they finally complete a project, and the sooner the cost of having that team work on the job can be stopped. These avoided staff labor and job site office costs usually result in dollar for dollar project savings for all project participants. This is a powerful incentive to complete projects as quickly as possible. For a general contractor, these avoided costs can range from \$5,000/mo to \$250,000/mo or more on really large jobs.

For ACME Investments at Main Street, avoided costs for the 2 weeks the schedule was not extended was worth about \$60,000 to Swinerton & Walberg Builders. The architect and subcontractors each have their own calculations of the savings they realized.

A less obvious, but very real, incentive for designers, general contractors, and subcontractors to complete a given project sooner is that a project team that completes a job sooner can go off to work on another revenue and fee-generating project. Depending on the size of the job and the staffing requirements, these gross job profits for a general contractor (before main office overhead and taxes) can range from \$10,000/mo to \$200,000/mo or more on really large jobs.

For ACME Investments at Main Street, opportunity costs that were not lost due to the 2 weeks the schedule was not extended were worth about \$40,000 to Swinerton & Walberg Builders. The architect and subcontractors each have their own calculations of the opportunity cost they did not lose out on due to not extending the schedule.

Another very real area of cost savings that tends to get most of the attention when people talk about metrics, but in the long run is not the largest component of project savings, is the administrative cost savings that are avoided by using project Web sites vs. a traditional paper-based system. Offset by the additional cost of retaining an Extranet project management service, and possibly hooking up a DSL or other high-speed Internet service, these administrative savings can be estimated on a job-by-job basis. The Bidcom software has developed a useful "calculator" to estimate the value of these administrative savings. For Main Street, all of the designers and subcontractors were local to Northern California, and we did not take full advantage of the software's collaborative document management components, so our savings were in the range of \$30,000. Still, even if this were the only savings, it would have paid for the software. All of the other savings that have been discussed would be after the cost of Bidcom was covered by the administrative savings.

In summary, the cumulative savings of most, but not all, of the project participants were as follows:

1. ACME Investments' rent savings by not delaying schedule 2 weeks: \$880,000
2. ACME Investments' value of using the space productively earlier: uncalculated

3. Swinerton & Walberg Builders' avoided site labor/office cost: $60,000
4. Swinerton & Walberg Builders' saved opportunity cost: $40,000
5. Swinerton & Walberg Builders' avoided administrative cost: $30,000
6. Designers avoided and saved opportunity cost: uncalculated
7. Subcontractors avoided and saved opportunity cost: uncalculated

The total estimated calculable savings for tenant and general contractor (no value for designers or subcontractors) was $1,010,000.

1. Bidcom software cost for Main Street: <$30,000>
2. Net calculable savings for tenant and general contractor (no value for designers, landlord, or subcontractors): $980,000
3. Net direct administrative and site labor savings for general contractor (no value for owner, designers, landlord, or subcontractors): $60,000

The Main Street project was approximately a $30,000,000 tenant improvement. The net administrative and site labor savings of $60,000 represents a 200% return on investment of $30,000 for this very first Bidcom project and a 0.2% reduction in job costs.

For the net savings, including ACME Investments' rent savings for not delaying the project 2 weeks, the $980,000 savings is a 3,266% return on investment (ROI) and a 3.27% reduction in job costs. As Bidcom has been scaling up their volume and reducing their prices, the current ROI would be approximately 5,900%, and the job cost reduced by approximately 3.31%. If you factor in current San Francisco leasing rates ($60–70/ft^2), the ROI would be 6,900% and the reduction in job cost 3.91%.

You can quickly see that the cost of the system is dwarfed by the potential savings and, paying a little more to get a lot more with the right A/E/C Extranet for your project, is an excellent return on your investment.

These figures are for the tenant and general contractor savings only, not including the landlord, designers, and subcontractors savings. If all of these savings are added, these ROI and job cost savings numbers for all of the project participants combined increase proportionally.

27.7 THE "BOTTOM LINE": HOW PRODUCTIVITY SAVINGS TURN INTO INCREASED PROFITS

By using efficient methods and e-PM software, a general contractor can lower the fixed portion of main office overhead required for business development, estimating, and senior business management. This results in lowering the main office overhead cost as a percentage of overall volume. This means the net profit that remains from the gross job profit, less main office overhead, can be a larger percentage. This is a good thing, and each firm can put their own numbers to these profit enhancers.

For example, a firm doing $100 million/yr in volume at 3% gross job fee has a revenue of $3 million with a main office overhead of $2 million, or 2%; the net profit (pretax) would be $1 million or 1% of gross revenue. If that group can improve their efficiency by shortening time schedules by 5% (52 weeks/yr = 2.5 weeks shorter schedule for a 52-week job, 1.25 weeks shorter schedule for a 26 week job, or 0.67 weeks shorter schedule for a 13-week job), they could take on 5% more volume a year without increasing their main office overhead. This is an additional $150,000 in gross job revenue, moving their $1 million net to 1.15 million, a 15% increase in net profit (pretax). If you can improve your schedules 10% by using collaborative Internet-based tools, the corresponding numbers are $300,000 in additional gross job fees—a 30% increase. Such figures are compelling reasons to shorten schedules.

27.8 BACK TO CONCEPTION: WHY PROJECT MANAGEMENT AND ADMINISTRATION TOOLS BEFORE BIDDING?

The reasons to use PM and administration tools before bidding are compelling.

- In a lump sum or public works bid project, a general contractor must submit a fixed price for the complete scope of work prior to the specified public opening of the bids. A bid submitted 1 second after the due date and time will be returned unopened, and all of the bidding efforts and pursuit costs expended will be for naught—not to mention the emotional cost to the estimating team.
- Automating the general contractor's lump sum bidding process poses several challenges. One challenge is that general contractors do not know up until bid time which vendors (subcontractors and material suppliers) will submit bids for use in compiling the complete bid. Often a general contractor will receive bids from firms they do not know and must determine if such bidders are qualified to perform the scope of work and if they should be used in the complete bid.
- At the time of the bid, general contractors are often required to list which vendors they plan to use on the project. Once listed, those vendors must be used unless such vendor chooses to withdraw or if a process known as "delisting" is pursued with the public entity. For a general contractor to delist, or substitute one entity for another, would require a substantial irregularity, such as a listed subcontractor not having a valid contractor's license. This is a costly, time consuming, and very often unsuccessful process that contains substantial risk for the general contractor.
- Subcontractors and material suppliers will often take exceptions to plans and specifications that will have to be ferreted out by the general contractor. This process is used to determine if any particular vendor bid is legitimate and can be used to compile the complete bid, can be modified somehow (often by modifying the vendor bid price higher), or if that vendor bid is too risky to be used.
- In order to be listed by a general contractor, vendors often exclude specified items, providing incomplete pricing that gives the appearance of being low. However, with those items added back in as required, their price is actually substantially higher than a competing vendor.
- Almost without exception, bids in compliance with plans and specifications come in with varying amounts of scope that need to be modified on a trade-by-trade basis to be truly complete.
- Many of the vendor bids are received very close to the final bid submission time—often just a few minutes before—and must be adjusted. Vendors then revise their initial bids, sometimes more than once, and occasionally they withdraw all together. Late submission is a form of price protection against being undercut by competitors for the vendors bidding to multiple general contractors. Vendors are also concerned about the security and confidentiality of their bids. Until all of this is "bullet-proof," there will be skepticism and resistance for vendors to submit bids online.
- Vendors do not necessarily submit the same bid price to all contractors proposing on a particular project. Contractors all operate in their own fashion. A bad track record between a vendor and a contractor can drastically affect the ultimate profit or loss when the work is finally completed—months, or even years, later. For example, some general contractors do not supervise projects or manage completion schedules as well as others, others do not pay their vendors in a timely manner, some are more adversarial with vendors than others, and some are more claim-oriented or litigious than others. All of these factors are considered by vendors when they submit their trade bid. Some general contractors are in the market today and gone tomorrow, others are in it for a long time and have more future work potential for that vendor. The general contractor uses these management factors to his or her advantage when the vendors submit their bids. They will request preferential pricing from each vendor in order to optimize their complete bid.

For all of these reasons and more, the public works bid process involves much evaluation, negotiation, and relationship factors that do not lend themselves easily to data automation. The building process is much more collaborative. Two-way communication, relationships, track record, past experience, and negotiation are vital to the mutual success of all parties.

The design and construction process involves specifically skilled labor to build a prototype building from scratch. Everything is custom designed, fabricated, and installed. These labor and performance factors weigh heavily in a subcontract vendor selection.

Imagine if a general contractor received, say, 200 vendor bids electronically 10 min before complete bid compilation and submission and "the system" carries over the "apparent low" bid for each trade, without the benefit of all of the human factors involved. Several problems could exist for the general contractor. He or she must guarantee price and performance of his or her complete bid submission or risk paying a penalty to the public works agency for not honoring the bid. This is usually in the form of 10% of the total bid value, quite a hefty sum in an era of net margins in the $1–1\frac{1}{2}\%$ range.

The contracting business is risky enough without increasing it by trying to automate a very thoughtful human bidding process. An analogy would be trying to automate legal advice or medical services. Besides, the bidding process is only a few days in the life of a building project that may take years to complete, and there is not that much paper involved in comparison to the management and administration of the project itself.

Some elements of Bidcom's initial ideas related to bidding include electronic requests for subcontractor/supplier quotations, electronic acknowledgement if a subcontractor or supplier is planning on submitting a bid, and increased utilization of the software's work flow/automatic notification capabilities.

These elements of Bidcom's original concepts will undoubtedly prove valuable; however, lacking the ability to specifically scope every trade and lock every subcontractor/supplier into complete plans and specifications without qualification, exclusion, or clarification (such as is possible in a controlled private environment), automating a general contractor's bid-day summary for a lump sum or public works bid will require more artificial intelligence than the current unidirectional communication the Internet allows.

For all of these reasons, this particular e-PM vendor has focused its efforts into the area of project process management. This is an area of day-to-day time-critical processes crucial to the optimization of designing and constructing a custom building project.

27.9 TEEN YEARS: PROCUREMENT

Commercially available Internet building materials procurement tools are in their early stages of use as of this writing. By the time this is printed, the industry will have advanced substantially. Most of the currently functional building supply procurement products are basically making pricing requests of sellers by buyers. They are doing a good job of allowing buyers to reach more potential sellers and allowing sellers to identify and be introduced to more potential buyers. Procurement tools will do a better job every day as more firms, both buyers and sellers, get involved.

Buyers are looking forward to the ability to stage reverse auctions for commodity building supplies direct from manufacturers. Sellers want to differentiate themselves from their competitors and abhor the thought of being considered a "commodity." Sellers are working to determine if all of this information and direct access provided by the Internet will help them and not hurt their margins or choke down their volume. Or worse yet, both.

For the same reasons I got involved with CIFE and then with Bidcom in mapping Swinerton & Walberg Builders' business process models for construction administration, I encourage everyone from all segments of our industry to get involved in the on-going efforts in developing and using Internet tools in our industry. All potential buyers and sellers should get involved as soon as they can, or they will be trying to catch up with their competitors later.

I predict that the procurement of building materials, shop, and ultimately field installation labor direct on the Internet (not by making an introduction for a one-on-one negotiation between buyer and seller, which is essentially what most current commercial product offerings allow) will evolve in the following four stages of increasing complexity:

1. Commodity materials (e.g., lumber, fasteners, mastic, hand tools, pipe, sheet metal, and reinforcing steel)
2. Catalog items not inventoried, but built to order for each project (e.g., light fixtures, door hardware, fans, and pumps)
3. Custom materials that require some field measurement or specific project design, but not field installation labor (e.g., custom doors and windows shipped f.o.b., mechanical equipment with site-specific configuration)
4. Custom materials complete with installation labor (e.g., window, wall, and pre-cast exterior panel systems; custom fabricated stone or millwork fully installed in the field)

There is great promise for Internet procurement of building materials. There are theories and hopes that the products will be able to be procured on the Internet for less cost and delivery will be more reliable than exists currently in the marketplace. There is talk of disintermediation of distributors and "middlemen" that will squeeze out savings by cutting out "fat."

Most of the work to date has been focused on the general contractors and subcontractors representing the "buyers." The manufacturers, suppliers, and distributors representing the "sellers" have been underrepresented to date in the development of these Internet procurement tools. These "sellers" will need to get much more intimately involved in order for all of the theorized procurement efficiencies to come to fruition in the near future. Better access to more suppliers creates a more competitive environment for the hugely fragmented building industry, giving more purchasing power to small, local firms that rely on their local or regional lumber yards and hardware stores for a majority of their purchases. In this commonplace local environment, these buyers have very little power and influence to get the most optimum pricing. In global Internet procurement of building materials, theoretically, even the smallest buyer will have access to many sellers and the same purchasing power of much larger buyers.

Setting the cost of procurement aside for the moment, I believe there are great time efficiencies to be gained using these Extranet tools and Internet communications to speed the order placement and supply chain management of delivering custom-built building materials quicker and more reliably to job sites. This is not a trivial issue and is much more complicated in the transitory transactional relationships that exist in the construction business between buyers and sellers as opposed to the relatively steady-state nature of supply chains in most manufacturing businesses. The efforts in developing and implementing these tools will pay handsome dividends in avoiding delays and, ultimately, shortening the lead-time for deliveries. This, in turn, will lead to shortened project completion schedules and, hence, job cost savings.

Everyone knows time is money, and the A/E/C industry is no exception. There are savings to be realized in having real-time procurement and delivery data just an Internet browser away from all of the participants in a particular time critical building project. So even when we put the purchase price component of cost aside, we can see how improved supply chain management will lead to cost savings as it relates to the time, and value of that time, to complete any given project.

I predict that the mapping and tracking of the procurement process will occur with the existing supply chain before any wholesale disintermediation takes place.

27.10 ADULTHOOD: WHERE CAN WE GO FROM HERE?

Where can e-PM Extranet services go from here?

Although I do not believe the public works lump sum bid process will be fully automated, many parts of the process will be—especially for private works in a controlled bidding environment. The solicitation for vendors to be invited to participate in a bid—over the years done by mail, phone, and then fax—is being replaced by Internet requests. The compilation of who is or is not planning to submit a vendor bid will now be electronically tallied instead of the manual method of tallying returned faxes of "yeses" and "nos" and calling to check on the nonrespondents.

For projects that a general contractor is compiling, a bid for the complete bid documents can be loaded onto the software vendor's ASP site. Prospective bidders who currently either come to our office to look at a set of bidding documents, go to a plan room to look at documents, or purchase a complete set of documents to be delivered to their office will now be able to review the documents on-line, decide if that particular job is of interest, and print just the selected drawings and specifications they need for their particular trade—a small subset of the complete bidding documents. This will save vendors substantial time and money and will greatly improve their productivity and, ultimately, profitability. It will also shorten the turnaround time required to compile a bid—an activity that is almost without exception on the critical schedule path. A day saved here is a day saved in the overall completion schedule.

Another area of development for Bidcom, in particular, relates to the submission of monthly progress billings and lien releases. This is an enormous volume of paper from every vendor, both in original submission once a month and lien releases that follow receipt of payments, also once a month from each subcontractor and vendor on each unique project. Since the recent passage of laws recognizing digital signatures, this is a natural extension of their existing work flow technology.

Other opportunities exist in areas such as productivity (work in place per labor and material used expended over time) and scheduling management, and each of these areas will take some development effort to bring these opportunities to fruition. Improved productivity management should lead to more efficient operations and greater ability to manage staff resource levels and job assignments. Not too far in the future we should be able to have field superintendents do daily updates directly into our scheduling software, delivering the most reliable as-built schedule that possibly could be kept. This in itself can be a very powerful and valuable tool.

These are just some of the ideas that have been developed for the use of the Internet in construction in just a few years. Many more will come over time along with the proliferation of these systems.

The Bidcom project management system, and the myriad of competitors in the property development design and construction business play-space (170 firms in the A/E/C play-space with nearly $1 billion in venture capital invested as of the second quarter of 2000), are true applications of the business-to-business use of the Internet that are here today and will be broadening in the future.

No one knows where this evolution of using the Internet is going, the exact path it will take to get there, how long it will take, what technologies and firms will win, and which ones will get passed by. But if you get involved, you can be a part of making it happen.

There are great opportunities for increased productivity using these tools. One of the keys to the success of these tools and to maximize their potential is for everyone in the industry to get on-line and communicate with each other. We will not be talking about the cost of new Internet access hardware and greater bandwidth when doing cost benefit analyses of e-PM systems, because these things will be just a necessity of business just as the telephone, fax machine, computers and CAD software, cell phones, pagers, and e-mail have become over the years. Some of these items were a new "extra" service with "additional fees" just a few years ago.

END NOTE

1. Bidcom merged with Cephren in October 2000 and the combined firms operate as Citadon.

CHAPTER 28
TECHNOLOGY AT THE ARCHITECT OF THE CAPITOL: A CASE STUDY[1]

Jim Barlow
FM Technology Specialist
Graphic Systems, Inc.
E-mail: barlow@graphsys.com

James White
Project Facilitator and Support Specialist
Graphic Systems, Inc.
E-mail: jwhite@graphsys.com

The U.S. Capitol is synonymous with the history of the United States of America. Beneath the corridors of power, a new revolution is taking place. As the architect of the capitol (AOC), Alan Hantman's vision is to make the AOC accountable for all of its work and to enhance a professional facilities organization that can respond to the unique requirements posed by the tenants working on the Hill. The unusual schedules for the House and Senate, the rotating doors and changing seniority of members based on the elections, and the historic fabric of the property portfolio puts tremendous demands on this facility management (FM) organization. Historically, the jurisdictions had operated as independent business units each with their own work processes. Alan Hantman's vision is to enhance the agency and empower his staff with FM technology.

Over the past 2 years, extensive work has been done in the areas of work management, space tracking, and asset accountability. No less than 400 AOC staff members have been trained, and the core business processes radically changed. In this chapter, we will review a case study about a real-life computer-aided facility management (CAFM) implementation at the AOC.

28.1 WHO IS THE ARCHITECT OF THE CAPITOL?

In 1790, the U.S. Congress passed the "Residence Act" which provided that the federal government would be located in a permanent location on the Potomac River by 1800. William Thorton was the first AOC and took office in 1793. Since 1790, only ten individuals have been appointed to this post.

28.1.1 600 CEOs

The architect of the Capitol is both a title and a little-known government agency. Initially, the AOC was one man with a team of assistants responsible for the design and construction of the Capitol. As the history of the Congress has unfolded over the decades, the position has expanded to include oversight of a small, independent city.

The city metaphor becomes clear when one considers that there is in fact not a single CEO or other authority figure whom ultimately governs the corporation as you might find in other FM environments. In this case, there are in excess of 600 figurative CEOs that represent the tenant community. Not only are the many committees, senators, justices of the Supreme Court, and members of Congress that run the business of the United States the customer base of the AOC, they are the creators and enforcers of policies nationwide. Each superintendent within the various jurisdictions recognizes the importance of maintaining acceptable levels of customer service, so that the tenants can focus on the business at hand.

28.1.2 Community of 50,000

The Capitol Hill complex might be best described as a city within a city. This is primarily due to the variety of national organizations, private and public businesses, and state and federal government representatives that work in and around the AOC complex on a daily basis. Recent trends in technology have led some industry analysts to term the D.C. Metro area the Information Valley of the new millennium. This has caused an influx of a new class of technology-driven workers that conduct their business within the boundaries of the AOC. Along with people who work and live around the Capitol complex are three to five million who visit the historic landmark annually. With this mixture of business and tourism, the AOC can have as many as 50,000 people moving about the complex on any given day.

This complicates the balance struck between maintaining the historic nature of the facilities, the need for modern technology and security provisions, high levels of accessibility, and a conducive business environment. Underlying the thousands of daily FM tasks is the requirement to fulfill the operational needs.

28.1.3 Responsibilities

Steeped in historical significance, the AOC's traditional core mission reads like a civics class lesson:

> The traditional core mission of the AOC is to provide for the Congress, on a neutral, bicameral and nonpartisan basis, professional expertise and advice relating to preserving and enhancing the environment of the Capitol Hill complex and operating and maintaining the infrastructure supporting the Congress, other legislative branch support entities, and the Supreme Court.

The AOC has been providing these services for over 200 years. During these years, the Congress has been served by an agency that has responded to constantly changing needs while still maintaining a nonpartisan and continuous role in preserving the Capitol's infrastructure.

The Office of the AOC is responsible for the structural, mechanical, electrical, electronic, fire prevention, safety, custodial care, construction, improvement, and renovation requirements for the various buildings and grounds that comprise the U.S. Capitol Hill complex. Duties and responsibilities the AOC is charged with include space management, planning and move coordination within the Capitol Hill complex, and operation of all Capitol Hill transportation systems.

Since its origin with only the Capitol building, responsibilities have grown to include responsibility for 38 buildings, 14 million ft^2, and 285 acres, including 22.1 acres of historic landscaped grounds. In addition to maintaining interior and exterior areas, the AOC is responsible for facilitating and conducting special events, such as ceremonies held within the buildings or on the grounds. These events range from the Presidential Inauguration, Lying in State, Congressional Hearings, major concerts, and demonstrations. A staff of approximately 2,000 skilled tradesmen and managerial professionals accomplishes this monumental task daily.

28.1.4 All Under the Facilities Mission

The AOC is somewhat unique among its peers—FM is the major purpose of the organization. Typically, facilities takes the back seat relative to departments that are part of the "core mission" of an organization. At the AOC, facilities are not just an overhead function that is left at the bottom of the priority list for funding and other resources.

The AOC employs approximately 2,000 men and women (not including seasonal and short-term contracted tradesmen) dedicated to AOC's mission of maintaining the Capitol Hill complex. Approximately 15% of these individuals make up the staff and management personnel responsible for the execution and administration of day-to-day activities. The remaining 85% of personnel is comprised of front-line shop personnel or service center representatives that in one form or another interact directly with the customer. This interaction ranges from official legislative branch written correspondence to personal walk-ins, electronic communication, and phone calls. AOC management has significantly enhanced the quality of life through complementary customer service and technology initiatives.

28.1.5 Boundaries

The organization has been structured in such a way as to directly meet the unique needs of each of the Capitol Hill tenant organizations. There are seven jurisdictions that service customers within set geographical boundaries (Senate, Capitol, House, Supreme Court, Library of Congress, Capitol Grounds, and Botanic Gardens), as well as several centralized specialty organizations, such as the Electronic Engineering Division, that frequently cross these boundaries (see Figure 28.1).

The AOC is divided into several building systems, whose shops are responsible for maintaining heating, ventilation and air conditioning, electrical, day labor, sheet metal, custodial, masonry, ele-

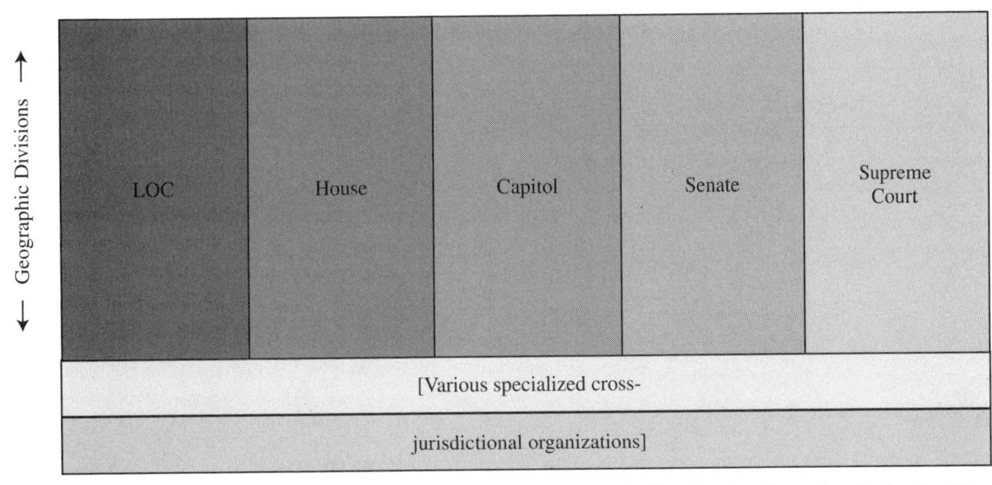

FIGURE 28.1 Most Capitol Hill Jurisdictions Are Broken Down Geographically with a Few Exceptions for Smaller, More Specialized Disciplines.

vator, plumbing, and carpentry requirements. Others, such as subway, cable TV, legislative clock system, grounds, night labor, renovation and projects, sound systems, high voltage, automotive repair, fire electronics, and furniture are also present, but are not represented in each jurisdiction. Combined, all these building systems aid in defining the AOC's core mission.

Senior policymakers within the organization use these tangible boundaries to manage the 14 million ft^2 that make up the Capitol Hill complex. Occasions arise when customer requests require the collaboration of multiple jurisdictions. Some of these specialized events require that senior management coordinate the resources necessary for resolving customer requests with their counterparts in other jurisdictions.

One effort that has worked well in managing all available resources effectively is the Capitol jurisdiction's decision to reorganize. The Capitol has a peculiarity that complicates their mission—its occupants are made up of both House and Senate organizations, at times giving the impression that there are actually two separate Capitol buildings: the House side and the Senate side. The Capitol has reorganized its service center and shops to provide more comprehensive and responsive services to both House and Senate customers.

28.2 APPARENT NEED

Having established a solid foundation for understanding the nature of the AOC, we can now examine the drivers behind the CAFM effort. Government reduction, Y2K, standardization, CAD, space management, and trends in technology are factors that spurred the agency to seek out a robust CAFM solution. Individually, these factors would not merit a heavy or serious investment in a CAFM system that is capable of becoming an enterprise solution; however, once combined, these factors indicate an environment ready to take a calculated step toward embracing FM technology. In this section, we will develop several of the leading factors driving the transition.

28.2.1 Government Reduction

While the legislative branch may find itself exempt from some legislation, government downsizing could not be escaped. Like many others, the budget of the AOC has been constricted, forcing the agency to be more financially responsible and do more with less.

For example, NASA exemplified this principle when its budget was nearly cut in half a few years ago. The agency managed to continue their mission of space exploration with the successful landing of the Pathfinder, which cost less to manufacture and utilize. In brief, better management through decision support allowed tough and thorough strategic analysis. This analysis was performed across the business enterprise, including infrastructure management.

Closer to home, but in a chapter of its history that is not yet complete, the AOC has been able to cut costs while still achieving their mission objective as an FM group. The House Superintendent alone has seen its staff shrink from 700 to approximately 400 within the last 5 years. The scope of services did not shrink during this period, but actually increased, partially due to mandates created by the Congressional Accountability Act of 1995. Managers have accomplished this by downsizing their staff, consolidating shops, and modifying their various business processes.

The CAFM effort aims to further fine tune resources to sustain the quality of service.

28.2.2 Y2K

A large driving force behind the move to modernize at the AOC was Year 2000 compliance. The first phase of the AOC's CAFM effort was part of the $8 billion spent by the government and $100 billion nationally. The legacy work order system had adequately accomplished its goals in its time but was now outmoded, and plagued moderately by Year 2000 issues. The problem was compounded by the fact that there was little documentation because it was written in-house.

In its most basic form, the new CAFM effort was to directly replace the installed system by adopting a commercial off-the-shelf (COTS) software package that met Year 2000 requirements.

28.2.3 Need for Standardization

The CAFM effort is more than just an updated version of the work order system. This system would be used universally, replacing the half dozen disparate systems, and allowing for standardization of processes and data. This fundamental change would enable analytical tools that use maintenance history to perform trend analyses. Additionally, with an enterprise work management system in place, extended functionality is within reach for budgeting, capital planning, and trending. As a by-product, the standardization efforts will improve business processes and customer service.

28.2.4 Investment in CAD

The AOC's CAD staff remembers first encounters with computer-aided design (CAD) systems. CAD experts traveled with suitcase-sized computers barely capable of showing a building footprint. Now, technology offers the ability to place graphics on that building footprint that automatically update based on events such as equipment readings and inventory procedures.

Beginning in the early 1990s, the agency sensed the need for qualitative data in support of the on-going renovation and maintenance of the Capitol Hill facilities. In part, this stemmed from the felt need to maintain as-built plans for the complex in order to preserve the historic components of the complex. A visitor to the Capitol building will quickly see artwork and photographs of the construction of the building. It is not uncommon to hear tales about how many of the original architectural detail drawings were almost lost due to a lack of preservation, something the modern AOC seeks to prevent from happening again. In response, the agency began developing a staff fluent in CAD technology.

The agency standardized on MicroStation (from Bentley Systems, Inc. of Exton, PA), a commercially available CAD package. The AOC began constructing 2D (two-dimensional) and 3D (three-dimensional) plans for all of the buildings it operates. Early on, heavy emphasis was placed on CAD standardization, allowing for multiple designers to manipulate the drawings while maintaining data integrity. The CAD library has matured to over 1,300 drawings. The flexibility and transportability of the CAD drawings has become an integral part of the design process. The files can be sent electronically to construction contractors, as well as in-house staff, thereby reducing design costs to the government and facilitating the as-built drawing process by requiring updated plans.

Within the last few years, as CAD technology has matured, the AOC began to make the drawings "smarter," meaning that they contain nongraphical data and associations to nongraphical data. For example, CAD technicians have built a repository of data that contains information about wall composition in the Capitol, which can be very troublesome in a masonry building with walls upward of eight-feet thick in some cases.

Today, the CAD system maintains primarily site and building plan graphics, which represent actual building elements as well as boundaries of space use and/or ownership, locations of assets, and significant maintenance items. CAD graphics, in a CAFM system, are linked to the space locations and asset information databases to provide highly accurate area measurements and graphic representations of locations, owner or occupant information, and proposed changes. The "smart" drawings that result are perhaps the most expensive for the agency to maintain, but provide the foundation for all other functions (Figure 28.2). The CAD capability can offer some of the best analytical possibilities as problems are displayed graphically, identifying the proximity of resources, equipment failures, and tenant occupancy. The investment in CAD can be leveraged to provide many times greater benefits when used as the foundation for other systems. Integrating the CAFM and CAD systems will allow the AOC to more fully accomplish its care-taking role at the Capitol.

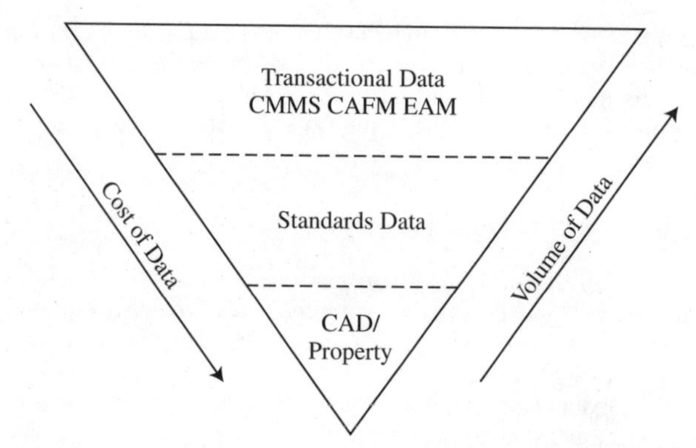

FIGURE 28.2 A Few, Data-Rich "Smart Drawings" Are the Foundation of the AOC's FM System.

28.2.5 Space Management

In conjunction with the maintenance-oriented aspects of CAFM, there is a requirement to improve the space management processes on Capitol Hill. Space design and assignment is done autonomously within the tenant groups in some cases, meaning that a universal approach has not been possible to date. However, the AOC serving as landlord to the Congress now has the capability of inventorying its property portfolio and supporting renovation and redesign. The objective in this area for CAFM is to reduce the amount of time required for space allocation and assist with office design through "what-if" scenarios.

The power of graphics for clearly communicating design options becomes much more valuable when used in conjunction with Internet technology. The AOC has been successful at bringing the tenant into the process by providing floorplans through a Web browser (as shown in Figure 28.3).

28.2.6 Maturation in FM Technology Offerings

CAFM is defined as the application of information technology to the broad field of FM. By design, the CAFM system takes the form of a modern decision support system. Best of breed databases combine with specialized front-end software oriented toward the needs of the facilities engineer of the 1990s. The result is a tool that assists maintenance management, materials handling, asset definition and tracking, space management, project tracking, and cost generation.

As time progresses, technology becomes smaller, faster, and cheaper, which greatly enhances the user's ability to efficiently automate time-consuming tasks. The FM community is a ripe proving ground for the plethora of technological advancements taking place. It is important to note that technology can only be deemed useful if it provides the user some advantage not gained by changing the external tangible process. Technology should provide the user the ability to view the process, offer the possible alternatives, and allow for change. Due to the FM environment's demanding schedule and limited resources, any proposed technology should mirror the organization's business processes. In the event that this is not entirely possible, managers should determine whether there is room for minor technological customization or a modification to their business process. FM technology has in fact matured.

28.2.7 Compelling Force

All of these factors have applied pressure to the AOC of the past and have been the catalysts for transforming the AOC into a modern facilities organization. While the AOC maintains a solid and stable set of business practices, areas abound within these practices that can and should be stream-

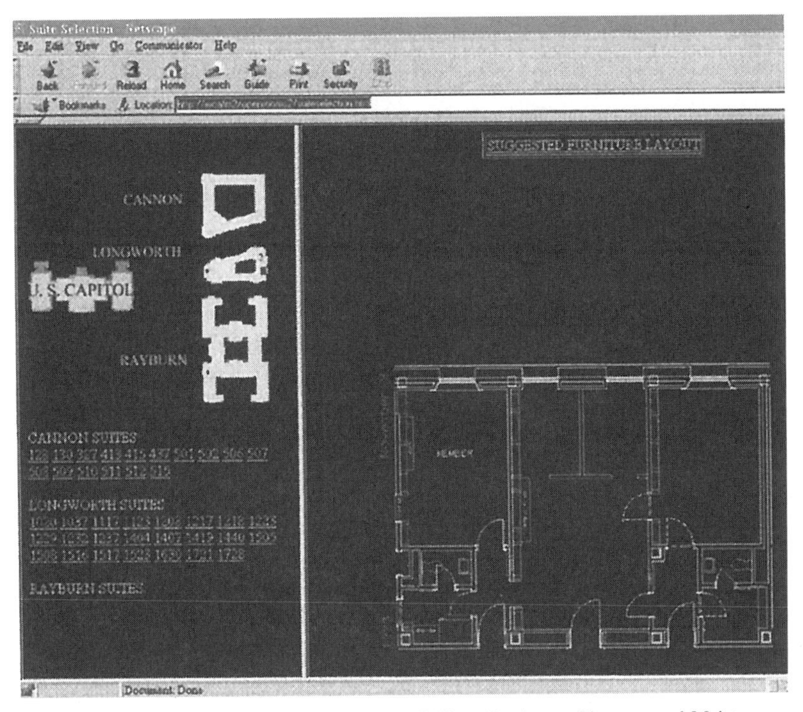

FIGURE 28.3 AOC Floor Plans Are Made Available to Designers, Planners, and Maintenance Staff through a Web Browser, Enhancing the Coordination of FM Tasks.

lined. In some cases, this may be as simple as designing a Web page and removing a phone number from a specific directory; however, the AOC will be examining these alternatives carefully and not automating for automation's sake. This will require thorough analysis, careful planning, and precise execution by senior policymakers in order to provide the agency any tangible value added.

Shortly after his appointment as architect of the Capitol in 1997, Mr. Alan Hantman chartered the AOC Enterprise Steering (AES) Committee, which has jurisdiction over agency-wide initiatives to improve its business processes. The committee identified Computer-Aided Facilities Management (CAFM) as the greatest need of the agency. A study was commissioned with the goal of analyzing the agency's requirements and identifying a suitable course of action.

The study found that there were at least five separate automated systems for demand work orders alone, in addition to numerous paper-based systems. There was little standardization for tasks across the organization. The AOC also had a familiar problem known as knowledge flight. For example, preventive maintenance relied heavily on the head knowledge of the mechanics on the job, a situation demanding attention as downsizing sent staff looking for other jobs. There was no systematic method for inventorying plant assets, albeit some of the maintenance shops had detailed information in disparate systems and paper records. All of these factors combined served as the apparent need which justified the agency's effort to reinvent itself, yet a decisive direction was still missing as to where the agency should go.

28.3 STRATEGIC PLAN

The CAFM initiative was forged from the AOC's vision of streamlining the organization's business processes. This initiative targeted a phased approach for implementing work management, asset management, and space management.

While interest in CAFM has been prevalent within the AOC for years, efforts to bring CAFM to the AOC began in 1996. AOC management commissioned a series of focus groups, a vendor analysis, peer interview, assessment of facility benchmarks, and a formal recommendation. In this section, we will look at a typical assessment approach and compare this with the experiences on Capitol Hill.

28.3.1 Needs Analysis

In November of 1997, the AOC retained a consultant to oversee the selection and implementation of a CAFM system. The consultant served as an impartial third party, independent of vested interests in any particular system. Its tasking included analysis of business processes, deficiencies, and outstanding needs. Many of these needs would be met through the use of FM technology, while others could be addressed through prioritization and procedural changes.

Excerpts from the needs analysis are presented here as documentation of the preliminary stages of implementation undertaken by the AOC.

1. *Survey of Agency:* Any successful business enhancement must be grounded in a solid understanding of requirements. This is especially true when a complicated information technology system is part of the solution. Therefore, significant time was spent in the boardroom and across the complex evaluating the agency's current position and needs.

Summarily, the finding was that the AOC's primary need was to improve the management of their manpower force as they work to maintain and improve their existing facilities. Work orders are the instrument the AOC wishes to use to document a work request, track the progress of work, describe actual work needed and completed, record manpower and materials used, document the technicians completing the work, and indicate supervisory approvals. AOC will implement three principal types of work orders:

- Preventative maintenance (PM) is based on a predefined schedule of maintenance for each end-item of maintenance. A variant of PM work is predictive maintenance, which involves planning maintenance based on the results of equipment instrumentation and/or performance charting. All forms of PM are intended to avoid breakdowns and subsequent service interruptions through proactive actions.
- Demand maintenance (DM) is initiated by a specific request or event that is not proactive in nature. Demand maintenance is grossly lumped into two categories: critical and noncritical. Critical DM includes emergency and immediate service needs due to events that threaten safety, security, or interruptions of essential services. Noncritical DM includes unanticipated requests due to nonessential service interruptions and customer requests for services.
- Custodial maintenance (CM) is routine, daily, scheduled care-taking such as mowing, trash removal, cleaning, and inspection.

2. *Define Current Processes:* Next, a definition phase was executed, recorded existing systems in use, both from a procedural perspective and in definition of data requirements. The AOC is complex in that the organizations have heretofore recorded maintenance activities in separate systems, and to varying degrees of detail.

In each organization, a basic customer service function is performed by a dedicated service center in the larger organizations and by office and administrative staff in the smaller ones as part of other duties. The methods of recording customer requests, dispatching, documentation of work, and customer follow-up varied greatly.

Sample processes in place for customer service range from terminal-based systems on which requests are recorded and through which work orders are remotely printed to maintenance shops, to hardbound logbooks in which calls are transcribed prior to being radioed to maintenance staff in the field.

The management styles in place are generally effective at accomplishing the requested services and have been fine-tuned by each respective manager. For example, the radio dispatch methodology is better suited to a smaller organization. A large organization with many more activities going on at any given time has evolved semi-automated systems for recording requests and quickly dispatching them to the appropriate staff.

The reporting requirements and desire to do larger scale manpower management will require consistent work management processes; however, the standardization must strike a balance with the size and complexity of the organization.

3. *Peer Analysis:* At this point in the process, the agency executed a formal peer analysis of ten other government agencies seen as its peers.

It is important to establish the fact that the AOC is to a certain extent unlike any other government agency. Typically, government is construed as very different from private corporations. Secondly, space type is often considered when identifying a peer from an FM perspective. For example, comparing an executive office complex against a supply center is unlikely to produce valuable management insight. Also, age of a facility affects certain aspects of facilities maintenance. Therefore, a true peer to the AOC would be a government agency with facilities over 200 years old, with a high percentage of executive space, and many elements that must be preserved like a museum. No peer could be found in the United States that could meet these criteria.

Keeping perspective on its findings in mind, the AOC surveyed large government agencies operating significant office buildings. The goal was to elicit information about which FM systems were in use and to what degree they had been integrated to the daily operation—in effect hoping to learn from the mistakes of their peers as it were.

The peer analysis helped to shape the functional areas that should be explored within the AOC community, and eventually that would become the criteria for the system selection.

4. *Focus Groups:* The dominant theme throughout the AOC review was the lack of overall consensus on business procedures and what should change. This is a common problem within a facilities environment. This problem was compounded at the AOC by the multiple independent jurisdictions. Focus groups were established for each of the functional areas in an attempt to better understand the independent issues at hand.

The focus group approach was designed to be democratic in nature to instill all participants with a sense of ownership of the effort. Each group assembled represented the parties that would be affected with the impending system implementation. Using this method of feedback and empirical measurements, with the target goals as guides, this panel can plan specific changes which will improve performance.

The focus groups determined what tasks their new CAFM system should be able to do. For instance, the panel concluded that the work management and space management systems must maintain historical values, as well as contain tools for analyzing and reporting comparative data, estimating the effect of proposed changes, and enforcing uniform standards established by management.

5. *Identify Agency Goals:* The CAFM task force must establish standards and goals for performance of each FM function. Usually, the agency goals are the equivalent of the agency's mission or vision, but at a more detailed level. These goals include targets for ratios of preventative to demand work, estimated ratios of fully documented work vs. partially or undocumented work, targets for space utilization, levels of organizational fragmentation, response times for move requests, the quantity of all manpower and materials consumed, and budget projections. Standardization is the key to improving any in-house process and benchmarking other organizations' best practices because it allows apples-to-apples comparisons across the enterprise or against external organizations.

The following management systems will be used to standardize processes and collect metrics: CMMS, SAMS, and CAD. In each case, goals must be established and interaction to both upstream and downstream systems considered.

Computer-based maintenance management systems (CMMS) automate the work request and work fulfillment functions, data collection, storage, and reporting required during each step in maintenance operations processes. Furthermore, a CMMS is designed to utilize data from and feeds data into financial, personnel, contracting, and material supply systems. Typically, managers in an organization use CMMS data to monitor maintenance performance, improve maintenance functions, and plan organizational strategy and tactics.

Space and asset management systems (SAMS) hold inventories of physical locations and asset items, organizational identification and structure, rules and standards, planning tools, and move transaction tools. Work management uses space and asset information to identify and locate areas of work and items to maintain. Financial reporting uses space owner and occupant information to generate cost allocations. Space planners in the Senate Rules Committee may access space graphics, space occupant information, and organizational attributes for strategic and tactical planning and to transact occupant moves.

Computer-aided drafting (CAD) systems maintain primarily site and building plan graphics which represent actual building elements, as well as boundaries of space use and/or ownership, locations of assets, and significant maintenance items. CAD graphics, in a CIFM system, are linked to the space locations and asset information databases to provide highly accurate area measurements and graphic representations of locations, owner or occupant information, and proposed changes. The agency desires to establish a central system containing all spatial data about the complex.

6. *Vendor Assessment:* After defining the requirements, the AOC commissioned a review of products recognized by the FM industry in the functional areas. The process followed narrowed the field of consideration, saving the detailed analysis and conserving resources for a few products most likely to meet the needs. First, a desk audit of the products was done from marketing literature. A questionnaire was sent to the vendors with an offering that might be a potential fit for the AOC. Next, phone interviews were conducted at each step evaluating the vendor against criteria identified by the AOC. Ultimately, the short list of vendors was invited in to demonstrate their solution in front of an evaluation team.

7. *Recommendation:* The recommendation that follows addresses the AOC's initial infrastructure business needs. First, the recommendations address the AOC's need for a work management system for preventative, demand, and custodial maintenance. Included in these categories are scheduled maintenance, predictive maintenance, emergency and requested service calls, and routine daily maintenance, such as housekeeping, groundskeeping, and refuse collection. Second, the recommendations address the AOC's need for complete and standardized space and equipment inventories and transaction capabilities. Third, this recommendation provides the means by which standardized collection and reporting of expenses for AOC activities can be exchanged with future financial systems and the timekeeping system so that the financial management group can provide improved cost analyses, accounting, and financial management. Finally, this recommendation addresses enterprise integration issues for computer-integrated facility management technologies, of which work management, space and asset management, and CAD integration are the first components. This ensures that the current recommendation not only satisfies AOC's immediate work management needs, but is also compatible with near to mid-range plans being developed.

At this point, a qualitative analysis was done comparing the functional requirements against the product offerings as reviewed in the vendor analysis phase. Eventually, a recommendation was made as to which commercial product offered the best fit based on the agency's needs. The recommendation was ultimately presented to the AOC management for consideration.

28.4 IMPLEMENTATION

The AOC selected a COTS solution that would meet their various requirements for replacing their current [demand] work order system. A consultant was retained to facilitate, coordinate, and ensure the successful integration of the new system with the AOC's business processes. The consultant recommended that portions of the CAFM system be implemented in sections due to the size of the proposed user community, computer experience within the community, and resource requirements.

It is also valuable at this point to represent the agency's position on customized software. Partially due to project failures in the past that relied too heavily on custom solutions, the AOC felt strongly that it needed to use only COTS products. However, as time has progressed, it has been more evident that some degree of modification to the software was required for certain things that make the AOC unique. A concept has developed referred to as tailoring, which differs from customization in that the baseline software is not changed. Tailoring allows for customized configurations, operational use of another application in place of one provided in the COTS package that offers special functionality, and the use of additional business logic.

One of the major requirements of the AOC that has required special attention has been the use of viewing security. This allows for the segmentation of the database so that a user in a given jurisdiction has the perception that only their data is available within the database. Information security for data perceived as sensitive, such as the location and size of offices of members of Congress can be hidden from users in other jurisdictions. At the same time, all of the data kept in the same database offered inherent abilities to keep the data standardized because there is only one copy of common elements. All of this serves to give individuals the support they need while still offering enterprise oversight capabilities to management.

28.4.1 Phased Approach

Figure 28.4 represents the continuum of FM technology. Each of the independent components identified here are interrelated. For example, CMMSs leverage information typically included as part of a CAFM system. Each component is built upon an open database platform. The key element represented in this chart indicated that not all functionality is met by COTS solutions. Typically, the goal is to attain 80–90% of system functions through commercial software and complement them with custom applications that share information and support very site-specialized needs.

It is important to note that each component of the CAFM system can be implemented independently. The AOC was committed to two phases to be performed in the first year of the initiative. However, the requirements analysis took into account other AOC business needs and offered possibilities for future expansion and integration of other CAFM concepts. The phases planned for FY 1998 were the foundational pieces of space management for the Senate and work management for the Capitol, Senate, and House. While these initial phases could stand alone, the processes relied heavily upon data collection and standardization that can be leveraged in the phases to follow.

This methodology leads to improved business processes, which will be outlined in greater detail below. Thus, the AOC's mission objective can be achieved with the explosion of the digital revolution, vastly increasing the quality of modern information systems. These systems complement the AOC's business processes for enhancing operational efficiency through faster and more accurate tracking of building structures, real property, and operations.

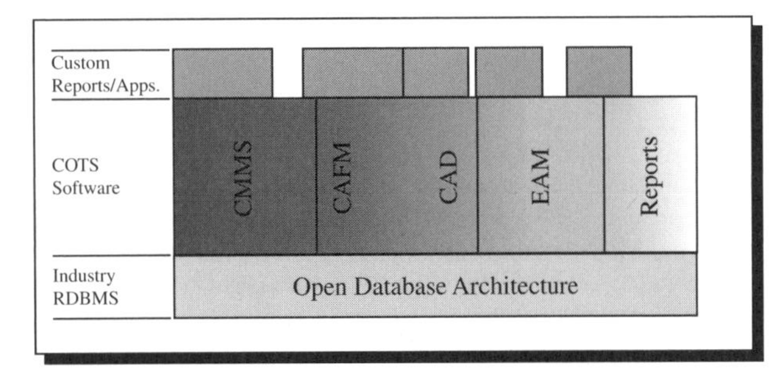

FIGURE 28.4 The FM Technology Architecture Features a Single Data Repository, a Wide-Variety of COTS Software, and a Small Portion of Site-Specific Functionality Custom Applications.

28.4.2 Developing the Strategic Plan

The following yearly tasks were taken from the strategic 5-year plan dated April 1, 1998. These were the initial recommended tasks and dates, but changes have been and will continue to be made as the project continues. Recommendations, by year, of the plan included:

Year 1

- Successful completion of a one-to-one replacement of the extant work order system for demand work orders within the primary jurisdictions
- Resolve Year 2000 compliance issues
- Commence information technology infrastructure upgrades

Year 2

- Expand demand maintenance to the remaining jurisdictions
- Complete the IT infrastructure project
- Further develop the property portfolio

Year 3

- Begin preventive maintenance effort
- Define system integration requirements into newly installed business systems
- Mature reporting capabilities for decision support

Year 4

- Continue preventive maintenance project
- Establish recurring maintenance systems
- Implement automated quality assurance program
- Space planning initiatives
- System maintenance

Year 5

- On-going enhancements
- System maintenance

28.4.3 Constraints

In order to measure the overall progress of the CAFM initiative, the consultant has designated a series of milestones and constraints. The theory behind this logic is that the CAFM system should achieve certain milestones given a list of constraints. Constraints can range from resources to scheduling. Milestones can be defined as major events and AOC business processes that seamlessly merge with the CAFM system. Only at this point can the CAFM system be considered a useful tool for the AOC management.

Y2K. Year 2000 compliance was a primary goal of the CAFM implementation effort. Globally, governments and private industry poured an estimated $100 billion into heading off one of the most perceived technological disasters to hit the twentieth century.

Elections. The three primary jurisdictions are highly sensitive to the election cycles that take place on Capitol Hill. The CAFM system must be able to provide management with a useful tool rather than a loadstone during this very stressful period.

28.4.4 Project Management Approach

With the advent of multiple jurisdictions comprised of varied sizes, responsibilities, and available resources migrating into one enterprise solution, it becomes vital to understand how each group services their customer base. As a whole, the agency must empirically measure the performance of these functions and monitor the relationship between target and actual performance. The work management and space management systems must capture performance measures in sufficient detail and in consistent units of measurement across all departments and functions in order to allow analysis and comparison to target values. Methods of capturing data must be integrated with the work, and to the extent possible, require no additional keying or transcription of values.

1. *Oversight:* It is impossible for the architect himself to attend every meeting and manage every detail concerning the CAFM initiative. Therefore, a management structure has been designated for ensuring that the CAFM application meets or exceeds the architect's vision of a Hill-wide FM solution.

Demographically, the AOC user community is broken into two distinct types of users. Nearly 85% of this group consists of blue-collar workers that have specialized trades and/or skill sets required for the operation and maintenance of the Hill complex. These are the frontline workers, section leaders, supervisors, and foremen. Of this group, approximately 95% either never or rarely used a computer on a regular basis during their tenure at the AOC. The median age for this group is approximately 45 years with over 20 years of service. Quickly educating this group of future CAFM system users in the basics of computer usage proved to be one of the key challenges that faced the CAFM implementation effort. The remaining 15% of CAFM system users utilize computer resources on a daily basis. Service Center personnel, senior management, and various middle management personnel comprise this section of experienced computer users.

The CAFM steering committee comprises senior policymakers from various groups, such as the information resource management group, superintendents of the various jurisdictions, human resources, and financial management group. Designated support staff are also members of this group. This committee is responsible for the direction of the CAFM effort, coordinating necessary resources for continued success, and providing Alan Hantman with progress updates. The CAFM project manager chairs this meeting and executes any direction received from the committee. The project manager is responsible for advancing the agency's overall agenda for CAFM while balancing the significant day-to-day logistical requirements.

The CAFM standards group is responsible for ensuring the details of the CAFM steering committee's orders are carried out. This group is composed of the assistant superintendents and facility managers that are currently using the CAFM system. At times, groups or jurisdictions that have yet to be introduced to the CAFM system are invited to join in the meetings to keep them informed of current events. The CAFM project manager works closely with the members of this group to ensure resolution of any CAFM-related issues that may negatively impact continued success is resolved in a timely manner.

Finally, the CAFM implementation team has experienced a steady growth since the inception of the project. In the beginning, AOC personnel resources were limited, which left the consultant responsible for much of the CAFM implementation. One year later, the AOC was able to hire a CAFM project manager dedicated to the overall initiative. Along with the CAFM project manager, the information resource management group was able to hire an additional technician that could focus primarily on CAFM software and hardware-related issues. Additionally, the Senate, House, and Capitol were able to hire CAFM personnel that would work in conjunction with the CAFM project manager and consultant. Tremendous progress has been made in improving the CAFM implementation team's effectiveness by hiring individuals that are responsible for the CAFM initiative. However, more personnel are required once other modules of the CAFM system are implemented. This type of oversight provides an excellent mechanism for feedback between policymakers, management, and system users.

2. *Viewpoints:* At the writing of this case study, the AOC is approximately $2\frac{1}{2}$ years through its 5-year plan. Any good enhancement initiative must be reevaluated and continually refocused. Significant energies have been applied, and many changes realized. Thus, the AOC must pause briefly at this critical juncture and reevaluate the original strategy set forth.

As the project has gained momentum, foundational tasks have been performed that will allow opportunities never before possible to be realized. For example, the information technology infrastructure has been greatly enhanced. The AOC has over a hundred maintenance units. As with most organizations, the facilities groups are relegated to basements, storerooms, and out-of-the-way spaces. Reaching spaces in older and monumental construction buildings with data network wiring was a challenging effort that the information resource management group eventually succeeded in accomplishing. This task was imperative for the demand maintenance, so that remote dispatching of work orders could be done. This requirement forced the implementation to comply with external schedules, such as cabling efforts. However, now that this infrastructure is in place and fully tested inherently through the demand work-order system, follow-on work can progress much more rapidly.

Likewise, other foundational efforts included the development of a user community through which one maintenance unit assists the others and communication barriers are torn down. Deliberately, the implementation in each jurisdiction included several weeks of user meetings where representatives from each shop and unit worked through process issues together.

The customer service arena has also gotten a facelift. At a time when many other service organizations on Capitol Hill have been going to storefronts, the AOC's review of its processes have allowed for streamlining. On a limited scale, the concept of "one-stop shopping" has begun. Through standardization of work request handling processes and reorganization of service center personnel, fewer calls are turned away to other service units. This process will continue as more of the cross-jurisdictional functions are addressed, helping the bottom line—giving the customer better service.

Other foundational components were general computer training, standardized reporting procedures, and last, but not least, management forums where supervisory staff made joint decisions on strategic issues. All of these issues allow possibilities that were never before available to the AOC for further innovation.

28.5 THE ROAD AHEAD

The road ahead for the AOC will take the CAFM effort in many new directions. First, the CAFM effort will need to continue to expand through each of the functional areas originally identified: from demand maintenance to planned and recurring maintenance, from space inventory to space management, and to an enterprise asset management system. In all implementations, this kind of change does not happen overnight, but rather has to evolve along with the organization. The AOC has not achieved this in the $2\frac{1}{2}$ years spent so far, nor will it in the 5 years covered in the strategic plan. Over years and decades an agency the size of the AOC will mold itself into new ways of doing business. A whole host of tasks outside of the typical CAFM definition await the AOC.

28.5.1 Future Efforts

The resurgence of investment in FM technology in the last few years is partially due to facilities becoming a boardroom subject. Tightening corporate budgets have targeted facilities as the place to increase efficiency; with facilities typically showing up as the second largest capital expenditure, executives look to make it leaner. This makes integration of CAFM to other enterprise systems a natural evolution of the business environment. Enterprise resource planning initiatives are giving way to infrastructure resource planning. Following suit, the AOC will be finalizing its financial management system implementation and commencing its integration to the CAFM system. This effort will yield enhancements in asset accountability, depreciation, and procurement.

Likewise, tighter tie-ins to the human resources system will further enhance boardroom decisions about FM. Building appropriate skill sets in the years to come will be possible through detailed analysis of job requirements. Personnel data will allow for more detailed job costing, such as knowing if a master electrician or a journeyman works on a lighting problem. Integration of a universal time-keeping system will eliminate redundant data entry, as well as more detailed accountability and forecasting.

Capital planning will have a monumental impact on the budgeting process of the AOC. Without detailed information about the life cycle of major building systems, the AOC is forced to be reactive in planning its budget. Through integration of capital planning management systems (CPMS), major expenditures can more adequately be anticipated and scenario evaluation performed, such as investing a smaller amount now to postpone major system replacements.

Frequently, design is far removed from the operations and maintenance efforts within FM. The AOC also has this problem. Design criteria and consideration for asset life cycles often fails to be transmitted from the design teams to the operations and maintenance staff. While there is no prepackaged solution in a box for this problem, integration between the project management system and CAFM will enhance this process. Major system overhaul on a given system (e.g., elevators and conveyors) will eventually feed into the asset repository, providing maintenance guidelines and operating information.

Ultimately, the path taken by the AOC will be different from any plan defined now. In the end, the best value to the government will be the ability to adapt its FM processes to unanticipated needs. Better documentation and management of services today will enable managers to properly position the agency to preserve the U.S. Capitol and adjust to constantly evolving Congressional needs.

ENDNOTE

1. Disclaimer: Information contained in this chapter was neither written nor provided by the AOC. Inclusion of this case study should not be construed as an AOC agency endorsement of any individual, firm, organization, system, product, or policy.

CHAPTER 29
MICHIGAN STATE UNIVERSITY FACILITY MANAGEMENT MASTER'S LEVEL CERTIFICATE PROGRAM: A CASE STUDY

Carroll Thatcher, CFM, IFMA Fellow
Off-Campus Coordinator
Michigan State University Program
Founder of Carroll Thatcher Planning Group
A facility planning and consulting firm
E-mail: cthatch@attglobal.net or cthatch@pilot.msu.edu

Susan Mireley, Ph.D.
On-Campus Coordinator
Michigan State University Program
Faculty member
Department of Human Environment and Design
Michigan State University
E-mail: smireley@pilot.msu.edu

Dana Stewart, Ph.D.
Department Chairperson
Department of Human Environment and Design
Michigan State University
E-mail: dgstewa@pilot.msu.edu

Jean Grant, B.A.
Technical Writer and Editor
Carroll Thatcher Planning Group
E-mail: ctpginc@attglobal.net

29.1 INTRODUCTION

The 1990s environment of dynamic organizations and constant change is continuing into the twenty-first century. Technology and the explosion of information have created an environment in which workers are expected to be knowledgeable and highly skilled, yet are expected to achieve these levels taking minimal time away from the workplace. The "time to knowledge" is critical.

The demand for independent learning is increasing. As organizations adopted the theory of "just-in-time" processes to improve their effectiveness and efficiency, so are educational institutions developing methods and processes that support "just-in-time" learning. The purpose of this case study is to share the experience and evolution of the Michigan State University (MSU) facility management (FM) Master's Level Certificate Program, which is delivered entirely over the Internet.

29.2 PROGRAM DEVELOPMENT

29.2.1 Background of the Michigan State University Program

The Department of Human Environment and Design at Michigan State University has offered a Master of Arts degree in Facility Design and Management since 1987. Numerous practitioners in the southern part of Michigan enroll in the on-campus Master's program. Due to time and work commitments, however, these individuals are often forced to approach their programs in a piecemeal fashion, frequently taking longer than the allowed 5 years to complete the degree.

In the summer of 1997, Dana Stewart, the department chairperson, initiated discussions with faculty and university administrators about expanding the program to meet the needs of the off-campus, professional FM community. Talks centered around making post-graduate level courses available over the Internet. Over time and based upon the findings of a number of research and market studies, the concept solidified.

A Master's Level Certificate Program in Facility Management was developed. The program was designed to:

- Consist of four courses, each worth three graduate level credits
- Be offered over the Internet, enabling students to be involved in the program from any geographic location that has access to the Web—students would not be required to visit the university campus
- Admit students with and without undergraduate degrees, through MSU's Lifelong Learning Program
- Grant a Master's level certificate to students who complete the program courses with an overall grade point average (GPA) of 3.0, the same grade requirement for students enrolled in the on-campus Master's degree
- Allow students with an undergraduate degree and meeting graduate program entrance requirements to transfer certificate credits into the on-campus Human Environment and Design (HED) Facility Management and Design Master's Program, giving students a significant start toward completing the full degree program with these courses as an introduction.

Over the last 3 years, these initial goals have been met. In the summer of 1996, the content of the first course in the program, Facility Management: Theory and Principles, was piloted. It was delivered in a classroom setting at the University of Manitoba, Canada. Then, in the summer of 1998, through MSU, the Internet delivery of the course was piloted. It was 95% Internet-based, with students meeting in person on the first and last weekends. Evaluations of this pilot indicated that it was feasible to deliver the entire program over the Internet. The first course in the series was offered spring semester, 1999, with the fourth offered spring semester, 2000. By that time, well over 100 students had enrolled in one or more of the courses and a core group of students who completed all four courses graduated from the program in May 2000.

During this time, excitement about the program by faculty, administrators, instructors, and students prompted the addition of more components. In May 1999, a Virtual University Tutorial was added to make initial entry into Internet-based learning easier for computer novices. In the summer of 1999, an extensive on-line facility management resource center was added as a core anchor for each of the courses. A frequently asked questions (FAQ) database, focusing on technical issues, was added in June 1999, enabling new students to take advantage of technical explanations given to past students. A Facility Management Certificate Program Web page (www.msu.edu/user/facmgt/index.html), was developed in October 1999 to provide a central point of information for potential students. In March 2000, a 2-week short course was offered that introduced facility managers to working on-line in virtual teams.

Other planned enhancements include a mentoring system, where former students offer assistance to new students, and a database of contact information, containing the names of subject matter experts willing to provide guidance to students in the program. These resources would allow students to expand their network of experienced peers.

Proof of the program's success has come from the students themselves. Those taking the initial round of courses posed questions about the possibility of the program being expanded into a full Master's level degree program. Other students, nearing the completion of the program, bemoaned the fact that the end was near. Still others claimed studying and communicating with fellow facility managers on-line was addictive. After the 1999 Christmas break, some returning students complained that they were bored with the lengthy down time and were relieved to be back in action. Some have used the class to ask fellow facility managers about approaches to solving real, on-the-job problems, prompting a suggestion that a bulletin board center be created that all MSU program participants and graduates could use to communicate with others for ideas and approaches to problems.

This reception has far outpaced the expectations of the program developers. Because of its success, the authors feel background information about the program should be provided to facility managers, to others considering developing Internet-based learning experiences and to technical people. The developers recognize the integral part technicians play in translating subject matter content into an Internet learning environment. Without their design capabilities and the tools they develop to support the learning process, Internet-based learning would not succeed. The following case study, therefore, focuses to a very great degree on sharing information about the types of design and delivery tools that have made the MSU Internet learning experience successful.

29.2.2 Educational Drivers

A combination of market research and focus groups, conducted prior to the development and delivery of the MSU program, showed a number of educational drivers supporting the need for distance learning. The reasons that practitioners gave for wanting to take education courses included:

- The acceleration of new information
- The desire for lifelong learning to expand professional skills and obtain executive level academic credentials
- The desire to master competencies as a basis for professional certification

The drivers that distance learning in particular can address included:

- The need to manage the conflicting and overlapping responsibilities and commitments of family, work, and education
- The inability to leave work to pursue formal education full time
- The distance of students from available educational institutions

This research identified education drivers specific to facility managers, who wanted to be able to:

- Work toward an academic post-graduate level degree, focusing on practitioner needs
- Gain a degree from an accredited university

- Gain access to a program that would be content-based and analytically oriented
- Develop skills in conducting research, writing, and presentation of information
- Have access to educational programs that cover and help facility managers integrate the wide-ranging content with which they deal
- Learn as their time permits, due to the extreme unpredictability of facility managers' work schedules
- Improve their competencies in using the Internet as a learning and communication tool

The majority of facility managers indicated that employers valued and supported formal academic training and that many would reimburse university credit costs. They also felt that formal education would be widely accepted as a precursor to professional advancement. A noted by-product of the development of a university-recognized degree or certification program was the opportunity to further professionalize and advance the credibility of FM as a professional field.

29.2.3 Promise and Peril of On-line Education

Most educational institutions in North America view technology as the wave of the future—for better or for worse. On-line learning is seen as the panacea for education in the new millennium by some, a sham or threat by others. The *Chronicle of Higher Education,* the bastion of education news and views for post-secondary educators, has produced a distance education section on their Internet site that covers 10 years of articles and opinions about distance education. The views expressed are varied and often heated. "Teaching at an Internet Distance," a report by the University of Illinois, offers mixed reviews of on-line education, arguing that it "shows both promise and peril" (*Chronicle of Higher Education,* January 14, 2000).

Foremost in debates on Internet-based programs are questions of quality. Most critics cite reduced faculty control over content, delivery, and evaluation of curricula as reasons for their concerns. The issue of accreditation frequently occurs; critics ask: "Can accreditors actually determine that new, on-line institutions meet the same basic criteria for quality—or, at least, equivalent criteria—that traditional accredited institutions must meet?" (*Chronicle of Higher Education,* October 29, 1999). Supporters argue that institutions will respond to students' demands for quality, particularly when students today are savvy consumers who want value for their tuition dollars.

On-line education offers many advantages for the independent learner—a student who is highly motivated, self-disciplined, and wanting advanced education that is not time or location specific. Foremost among the reasons why potential students consider on-line learning is the need to manage the conflicting and overlapping responsibilities and commitments of family, work, and education. They acknowledge their inability to leave work to pursue formal education, while many students find themselves too far away from educational institutions that offer programs of their choice.

At MSU, Internet delivery was seen as an appropriate vehicle meeting the variety of educational drivers identified in the needs assessment. The MSU Facility Management Certificate Program allows for timely interaction between the student and course content. The program incorporates on-line, active learning. It is holistic and grounded in a "high-tech, high-touch" delivery approach. The World Wide Web and Internet-based multimedia tools serve as the interaction or communication medium binding students, course content, and instructors.

29.2.4 Characteristics of the Internet-Based, Active-Learning Program

Active Learning. Active learning requires student participation in the sharing of experiences, knowledge, and insights, and requires students to challenge each other and the instructor in order to gain a deeper understanding of a subject. The active learning process focuses on a "student interacting with student" model as the learning vehicle, emphasizing maximum student involvement and student contribution. Each student is not only responsible for his or her own learning, but is also responsible for helping educate others in the class.

Holistic educational paradigm	Established educational paradigm
Educator as moderator, the "guide on the side"	Educator as source of all information, the "sage on the stage"
Values: co-operation, share knowledge, and equality	Values: individual power, "own" knowledge, celebrity
Methods: audiovisual, presentations, student is active, does research, and aids others	Methods: rewards and punishment in a selective and competitive system
Student is an active participant in educational process—find, analyze, and present	Student is a receptacle of information—remember and report
Know how to find out	Know-how

Adapted from Weil, P. 1994. *The art of living in peace.* UNESCO: Findhorn Press.

FIGURE 29.1 The Old and New Educational Paradigms.

The virtual classroom supports both passive and active learning. The students are able to read, hear, and see information. They have the opportunity to experiment with the processes demonstrated in the courses. They are expected to talk, lead discussions, write papers, and develop documents and presentations.

Technology supports the asynchronous learning environment, meaning that students can attend lectures in the virtual classroom at a time of their own choosing. They can view videos and other on-line media. They can post questions and receive answers from the instructor and from other students. They can discuss relevant issues with the instructor and each other.

A New Educational Paradigm. To achieve active learning in the MSU program, the instructors use a new educational paradigm. It requires that both students and instructors take on new roles to increase learning (see Figure 29.1).

Location and Time Independence. The entire MSU program is delivered via the Internet. This means that the instructor and students can be located anywhere in the world where Web access is available (see Figure 29.2). Thus far, students have been located in Canada, the United States, Taiwan, Bosnia, and Korea.

Internet courses can be asynchronous; that is, time independent. They can also be location independent, or a combination of both. The MSU program is entirely location independent, and the majority of its components are time independent. For example, the participants read the course content, view videos and PowerPoint audio-synchronized presentations of case studies, do research, complete analyses and writing, and finish self-learning exercises time-independently, at any time of their choosing. In addition, they can post messages to classmates and instructors at any time using WebTalk, an asynchronous bulletin board, useful for communication that does not require instantaneous feedback. The workings and various uses of WebTalk are described in greater detail in section 29.5

While time independent components form the core elements of the courses, there are parts that are time dependent. For example, the courses adhere to the MSU semester calendar, starting and finishing on predetermined dates. The instructors also require that course assignments be submitted on specific dates. At times, synchronous, real-time communication is needed to achieve specific learning tasks. In these instances, LiveChat, a real-time communication tool, is used. Periodically, for example, students are required to participate in discussions concerning current issues in the FM field. LiveChat enables them to post phrases, sentences, and paragraphs, which are viewable instantaneously by other members of the chat group, facilitating an experience not unlike a classroom discussion. LiveChat is also used by students to complete team assignments. Teams schedule LiveChat sessions to discuss their approaches, determine roles to be taken in completing the assignments and discuss content. As writing and editing progress, team members meet to revise and refine the material. LiveChat is described in detail in section 29.5.

Tomorrow the World

FIGURE 29.2 Students from around the World Have Taken the MSU Courses.

29.3 CONTENT DEVELOPMENT

To determine the content of the MSU program, a variety of professional association materials were analyzed. The purpose was to identify a key knowledge base needed by a practicing facility manager to achieve overall professional competence, and to achieve professional recognition by employers and other practitioners in the field. This review was followed by market and focus group research. Final analysis revealed a grouping of content areas and competencies that practitioners feel are mandatory for facility managers to master.

29.3.1 Core Curriculum Needs

The core curriculum needs identified through literature review and market and focus group research included the following performance objectives, allowing students to:

1. Understand the breadth of the field
2. Develop an understanding of the current movement away from viewing FM as a technically oriented field toward a management orientation
3. Develop an understanding of FM as a service-driven profession
4. Increase the ability to gather and manage information
5. Increase the ability to effectively manage complex organizational structures, with an emphasis on the people in that structure
6. Increase an understanding of finance
7. Increase the ability to manage facilities as real estate assets

8. Increase the ability to move from crisis management toward focusing on long-range and master planning, incorporating strategic management to meet goals
9. Increase competence in both written and oral communication
10. Increase competence to seek, identify, analyze, and integrate information to solve problems creatively and effectively
11. Develop networking skills to provide a support system of practitioners to use for help and guidance in meeting future problems

As a result of the analysis, a four-course Master's Level Certificate Program was developed. The eleven curriculum needs were separated into content and process areas, with the content becoming the focus of the specific courses, and the process skills running through each course. The courses build upon one another, although it is not necessary to take them in sequence. One course provides an overview of the field, a second course deals with managing information, a third with organizational behavior and management, while a fourth focuses on finance and real estate. In all courses, students work individually and in teams to conduct research, discuss, prepare, and deliver assignments.

29.3.2 The Content

A brief description of the content of each course follows. More detailed information about the courses can be found at the Facility Management Certificate Program Web page at www.msu.edu/user/facmgt/index.html; the Program page of this Web page contains details for each of the four courses.

1. *Facility Management: Theory and Principles:* This course is viewed as the first in the sequence and is designed to help facility managers understand facilities and their components. This includes the impact of facility form on such factors as individual and organizational effectiveness, construction, operating and maintenance costs, and the environment. By completing the course, students learn about the history and development of FM and the multifaceted role of facility managers in assisting organizations to meet their goals most effectively. The course is designed as a comprehensive initial course for facility managers. It introduces core competencies required for IFMA certification (facility function, communication, finance, human and environmental factors, planning and project management, quality assessment and innovations, and real estate). The course also acts as a base for further in-depth study of individual subjects. Special emphasis is placed on long-range and master planning, including methods and tools to enable facility managers to develop strategic, tactical, and operational plans, space forecasting, planning and management, and project management. The purpose is to help students learn how to plan and manage facilities more effectively.

2. *Information Management for Facility Professionals:* This course focuses on developing an understanding of the types of decisions facility managers must make, the kinds of information needed to make them effective, the cost of information vs. its value, and the management of facility information and knowledge. The course acquaints students with a variety of information technology (IT) software tools and places emphasis on the Internet as an enabling technology. The purpose is to increase student competence to identify, collect, manage, preserve, present, and disseminate information to meet organizational goals. The content focuses on information management techniques, yet provides a broadly based, wide ranging learning experience for facility professionals and for people working toward a better understanding of FM. Successful completion enables students to leave the course with knowledge to more effectively manage information and skills to identify, judge, manage, and present facility-related information more successfully to meet the business needs of an organization.

3. *Achieving Facility Management Organizational Effectiveness:* This course covers organizational effectiveness from the perspective of facility professionals. It focuses on establishing critical links within the FM organizational structure, the delivery system of in-house and external service providers, and the corporate business planning process. A variety of topics are covered, including organizational development and management of the FM function, organization structure, creating an effective work environment, leadership, staff development and empowerment, development of position descriptions, policies and procedures, analytic capability, linkages to business planning, quality FM, benchmarking, marketing, and customer service. Successful completion enables students to leave the course with skill sets to increase their effectiveness in providing FM service and a more comprehensive understanding of the FM profession.

4. *Facility Real Estate and Building Economics:* This course focuses on managing facility real estate as a strategic asset, with successful management being measured by its contribution to stockholder value. The course focuses on five content areas:

1. Strategic business planning and measurements
2. Financial basics
3. Building economics
4. Real estate asset management
5. Exploration of case studies, which results in development of an executive presentation

The purpose of the course is to build a foundation of planning and measurement tools as well as of business analysis and optimization tools that are needed by asset managers. The course culminates in a final case-study-analysis, resulting in the development of a business plan, accompanied by an executive level presentation explaining the plan. One of the instructor's major objectives is to facilitate and draw out the wisdom of the participants for sharing and exchange. The course assumes and encourages participants to undertake independent study beyond the materials presented in the course.

29.3.3 The Content Specialists

The course authors are experienced practicing professionals in various fields related to FM. They are also the primary instructors of their individual courses. While developing their course content, each made use of their own materials, developed through their own experiences. In addition, the authors asked authorities in various content-related fields for permission to synopsize or quote their works, enabling the authors to include the most current or highly recognized material in a specific content area. Finally, the authors relied on practicing professionals to review course content for accuracy and currency. Once draft content material was submitted to the university, it was sent out for a second review to a panel of international reviewers, with comments and recommendations being shared with the authors.

29.4 COURSE DELIVERY: A PARTNERSHIP

29.4.1 The Instructors

The individuals who developed and currently teach the MSU program courses are:

- Carroll Thatcher is the instructor of Facility Management: Theory and Principles
- Guy Thatcher is the instructor of Information Management for Facility Professionals
- Stormy Friday is the instructor of Achieving Facility Management Organizational Effectiveness
- Bernard van der Hoeven, Ph.D., and IFMA Fellow, is the instructor of Facility Real Estate and Building Economics

The biographies of these initial instructors appear in Appendix A.

29.4.2 The On-Campus and Off-Campus Coordinators

Because of very specific needs involved in delivering credit courses to practicing professionals using distance learning tools, responsibility for overall management of the development and delivery of the program rests with two coordinators: an academic program coordinator and an off-campus coordinator.

The academic program coordinator, who is on-campus, is responsible for the academic components of the program and for linking the off-campus practitioners into the university as students. The majority of program participants are not part of the MSU community prior to their first enrollment, and because the program is a certificate rather than a degree program, these individuals do not formally join the MSU community as regular students. Instead, they enroll as life-long education learners. This is the only factor that distinguishes them from traditional MSU students. The students register through the university registrar's office, are provided with university e-mail addresses, and are required to pay tuition fees. Their names appear on formal university class lists, their grades are permanently recorded by the registrar's office, and official university transcripts are sent to them at the end of each semester.

The academic program coordinator is responsible for seeing that all of these procedures take place seamlessly so students experience little or no difficulty taking the courses. The coordinator also acts as somewhat of an academic advisor, student advocate, and friend, answering questions about university policies and procedures, managing records, and answering questions or concerns students might have about the program or interacting with the university.

The off-campus coordinator is one of the instructors, thus a practitioner, and is responsible for working with the other instructors as they develop their course material. The off-campus coordinator ensures that the instructors understand the structure and look of the courses so their material will be easily converted into HTML format for Internet delivery. The off-campus coordinator also works with the academic program coordinator to identify practitioners who might fill various roles in the program, such as content reviewers, student mentors, and teaching assistants.

29.4.3 Teaching Assistants

Teaching assistants (TAs) provide a valuable resource to the instructors in communicating with the students and grading assignments. The instructor is responsible for overall management of the course. Yet, Internet-delivered courses are time-intensive because communication takes place primarily in written rather than verbal form; thus, if enrollments climb above 20, which they regularly do, one teaching assistant is added to work with the additional students, and a course with 40–60 students would have two TAs. The instructor and TAs assume responsibility for specific teams; each monitors their teams' WebTalk conversation areas and their students' WebTalk personal conversation areas, answering questions and providing guidance or redirection if they note problems with content or assignments.

29.4.4 The Virtual University

The Virtual University (VU) is a department in MSU's Computing Libraries and Technology Division. It was established to develop and maintain technologically enhanced educational media and Internet-based courses and instructional program offerings. VU staff are responsible for transferring course content into HTML format for on-line courses. They also work with the instructors, acquainting them with various tools available for inclusion in the courses. VU staff members respond to student questions posted in a conversation area in WebTalk; specifically those questions related to computer problems encountered in taking the courses.

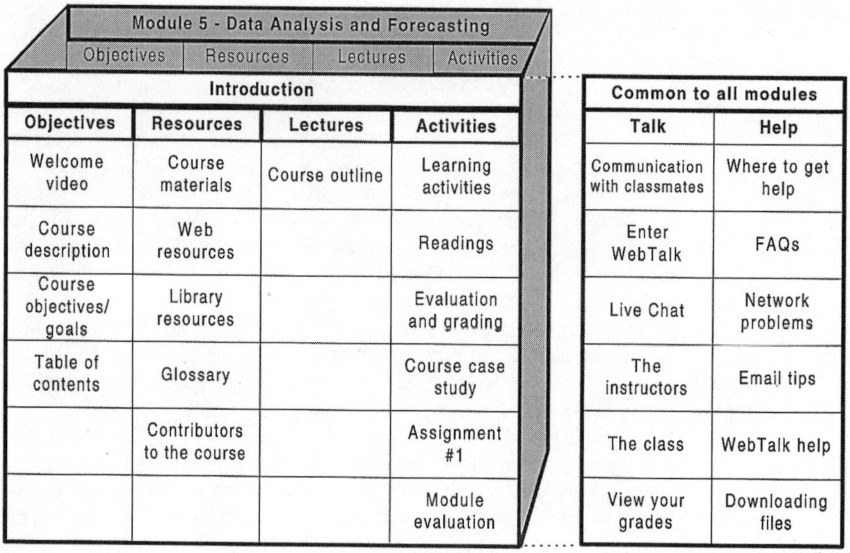

Module 5 - Data Analysis and Forecasting

Objectives | Resources | Lectures | Activities

Introduction			
Objectives	**Resources**	**Lectures**	**Activities**
Welcome video	Course materials	Course outline	Learning activities
Course description	Web resources		Readings
Course objectives/ goals	Library resources		Evaluation and grading
Table of contents	Glossary		Course case study
	Contributors to the course		Assignment #1
			Module evaluation

Common to all modules	
Talk	**Help**
Communication with classmates	Where to get help
Enter WebTalk	FAQs
Live Chat	Network problems
The instructors	Email tips
The class	WebTalk help
View your grades	Downloading files

FIGURE 29.3 3-D Course Matrix—Introduction Module.

29.5 DESIGN AND STRUCTURE OF THE INTERNET COURSES

A consistent organizational structure is used for each of the courses. Each course is structured as a three-dimensional matrix consisting of multiple modules. Each module has the same structure, although the number of modules in each course varies. Each module has the sections "Objectives," "Resources," "Lectures," and "Activities"; these sections are further divided into subsections. The sections and subsections are linked bidirectionally, allowing the students to move around each course in a nonlinear fashion. The sections "Talk" and "Help" are common, meaning that they are the same for all modules.

Figure 29.3 illustrates the matrix, using the first course, Facility Management: Theory and Principles, as the example. The front of the matrix shows the subsections under each section: this example shows the Introduction module. The subsections change within each module.

Figure 29.4 shows how this translates into the actual course.

In Figure 29.4, the menu bar at the top of each site page allows you to navigate through the course. Clicking on the module numbers moves you from one module to another; the number becomes highlighted so you know where you are. Clicking on the colored "tabs" moves you from one section to another. The various sections are color coded so you can tell where you are—the menu bar, sidebar, and graphic in Objectives are blue; in Resources they are green, etc. The example in Figure 29.4 shows the Objectives section of Module 5, specifically the Introduction subsection. The other subsections, Learning Objectives and Table of Contents, are listed at the left.

In the example shown in Figure 29.4, there are a number of choices for moving to another location in the course:

- To another section, for example, if we clicked on Resources, we would be moved to the Resources section of Module 5
- To another subsection, for example, "Learning Objectives"

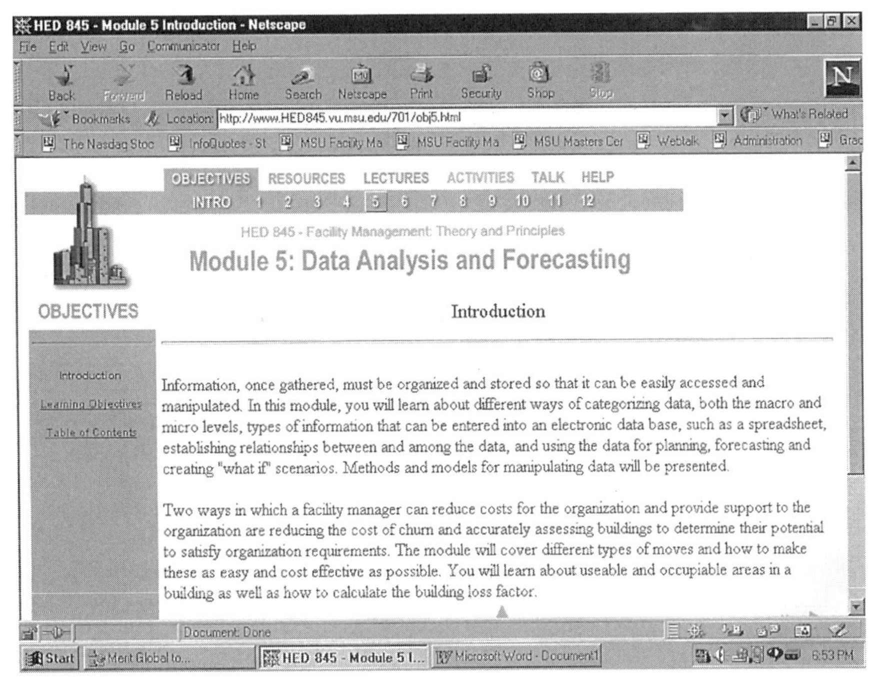

FIGURE 29.4 The Internet Course.

- To another module, for example, if we clicked on the number "3" we would be moved to the first subsection of the Objectives section of Module 3

29.5.1 The Introductory Module

Each course contains an in-depth introductory module, which consists of:

- A brief narrative overview of the content and focus of the course
- A discussion of the instructor's background
- A 5–10 min welcoming streaming video in which the instructor introduces himself/herself, welcomes the student to the course, and discusses the course. This gives a face and a voice to the student, "face time." This was added, at student suggestion, after the first course was run.
- The overall course objectives and goals
- The course syllabus, including a description of assignments and due dates
- A detailed outline of the course content
- A list of required texts, readings, or resources, along with a list of additional suggested resources related to various topics or assignments
- The Talk section, which contains the courses' communications options, tips on communicating, access to WebTalk, LiveChat, the instructors, the class, and the gradebook
- The Help section, which contains support contacts for technical assistance, frequently asked questions (FAQ), network problems, e-mail tips, WebTalk help, downloading files, and tips for AOL users

The Talk and Help sections are described in detail in the Module Structure section that follows.

29.5.2 Module Structure: A Common Format

Each of the modules in a course contains the sections Objectives, Resources, Lectures, Activities, Talk, and Help. Figure 29.5 shows a typical module, in this case Module 5 of Facility Management: Theory and Principles.

Objectives. The Objectives section contains a description of what the module is about, the learning objectives and goals, and a table of contents for the module. At the start of the Objectives section of the Introduction module, students are pointed toward the Tutorial. Prior to beginning each course, students are required to complete this tutorial, which acquaints them with a variety of tools and procedures used in the courses. This tool was developed after the introduction of students into the first course, in January 1999. Over the first 3 weeks of that semester, it became apparent that some students did not have the computer skills necessary to prepare their machines for accessing and moving through the course materials. The Tutorial was designed and prepared by the MSU Computer Center to help students set up the various hardware and software tools needed to successfully complete the courses.

The Tutorial contains written information on hardware and software requirements, explains procedures to navigate through the course sections, and on-line tools that allow students to check their systems for proper configurations to utilize the learning tools included in the courses. It also provides tools to teach students various computer procedures that are used in the courses, such as uploading and downloading files.

Resources. The Resources section (shown in Figure 29.6) points the students to materials that will supplement the course materials contained in the Lectures sections. Resources contains a list of the required readings for the module, as well as a bibliography which lists the research done in the development of the module.

Students are encouraged to use their local libraries and bookstores, and on-line bookstores, to secure the required resource materials. In addition, the university has made arrangements with local East Lansing bookstores to make the materials available. In a number of courses, the instructors have developed sets of required readings; the university uses the services of one local bookstore to

Introduction

Objectives	Resources	Lectures	Activities

Module 5 - Data Analysis and Forecasting

Objectives	Resources	Lectures	Activities
Introduction	Readings	1. Data collation, analysis, ...	Read lectures
Learning Objectives	Bibliography	2. Churn	Exercise 1
Table of contents	Course resources	Calculating area measurem	Exercise 1 answers
	Glossary		Exercise 2
	Contributors to the course		Assignment #5
			Module evaluation

Common to all modules

Talk	Help
Communication with classmates	Where to get help
Enter WebTalk	FAQs
Live Chat	Network problems
The instructors	Email tips
The class	WebTalk help
View your grades	Downloading files

FIGURE 29.5 The 3-D Course Matrix: Module 5 Data Analysis and Forecasting.

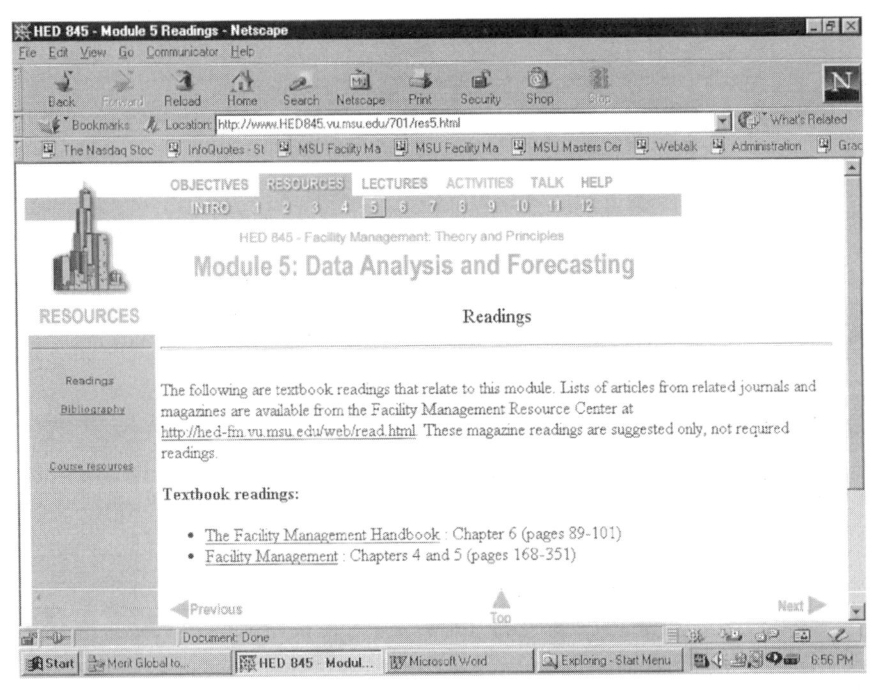

FIGURE 29.6 The Resources Section.

secure copyright permission for the materials and make them available in a copied, bound course package format. The majority of students take advantage of this service, finding it reduces the amount of time required to independently seek out the resource materials. Students taking courses through the Virtual University may also use the resources of Michigan State University libraries. MSU Library Outreach Services is designed to meet the information needs of students registered for off-campus Virtual University courses and provides on-line means for requesting library books, materials, and services. The Resources section also contains a link to the on-line Facility Management Resource Center (see Figure 29.7). The site is available to all students enrolled in the Internet-

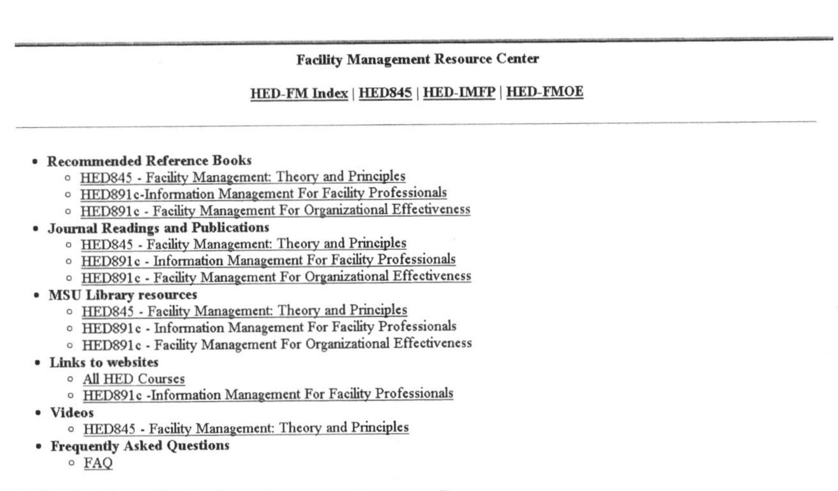

Facility Management Resource Center

HED-FM Index | HED845 | HED-IMFP | HED-FMOE

- **Recommended Reference Books**
  - HED845 - Facility Management: Theory and Principles
  - HED891c-Information Management For Facility Professionals
  - HED891c - Facility Management For Organizational Effectiveness
- **Journal Readings and Publications**
  - HED845 - Facility Management: Theory and Principles
  - HED891c - Information Management For Facility Professionals
  - HED891c - Facility Management For Organizational Effectiveness
- **MSU Library resources**
  - HED845 - Facility Management: Theory and Principles
  - HED891c - Information Management For Facility Professionals
  - HED891c - Facility Management For Organizational Effectiveness
- **Links to websites**
  - All HED Courses
  - HED891c -Information Management For Facility Professionals
- **Videos**
  - HED845 - Facility Management: Theory and Principles
- **Frequently Asked Questions**
  - FAQ

FIGURE 29.7 The Facility Management Resource Center.

based program and in the campus-based Master's program in facility management. It is a core part of each of the MSU courses, acting as a centralized source of information for each.

The Facility Management Resource Center includes:

- A bibliography of highly recommended text and reference books in the field, recognized monographs on particular topics, citations for various professional journals and trade publications, and bibliographic citations for current articles recommended for reading related to a various topic
- A catalogue of the professional journals and publications that the MSU library system maintains Web-based access for, along with an explanation of how students can access them
- Bibliographic citations of each of the various FM-related monographs and publications held in the MSU library collection
- Links to a number of Web sites that contain material of specific interest to facility professionals
- References to various videos

Each of the referenced resources is reviewed prior to inclusion in the site.

Lectures. The Lectures section (shown in Figure 29.8) contains the lecture materials for the selected module; it is where the "content" of the course is found. It is a combination of course on-line textbook, graphics, and links to other Web sites of specific relevance to the topic. Each instructor has developed an extensive, written analysis of the subject matter related to the module focus. Inclusion of these book-chapter-like presentations distinguishes the MSU program courses from many other Internet-based educational courses, the majority of which appear to focus on links to other Web sites for the subject matter content base. Providing the topic discussions in written format allowed each instructor to present the content in the depth and breadth they felt appropriate for the students to comprehend the materials.

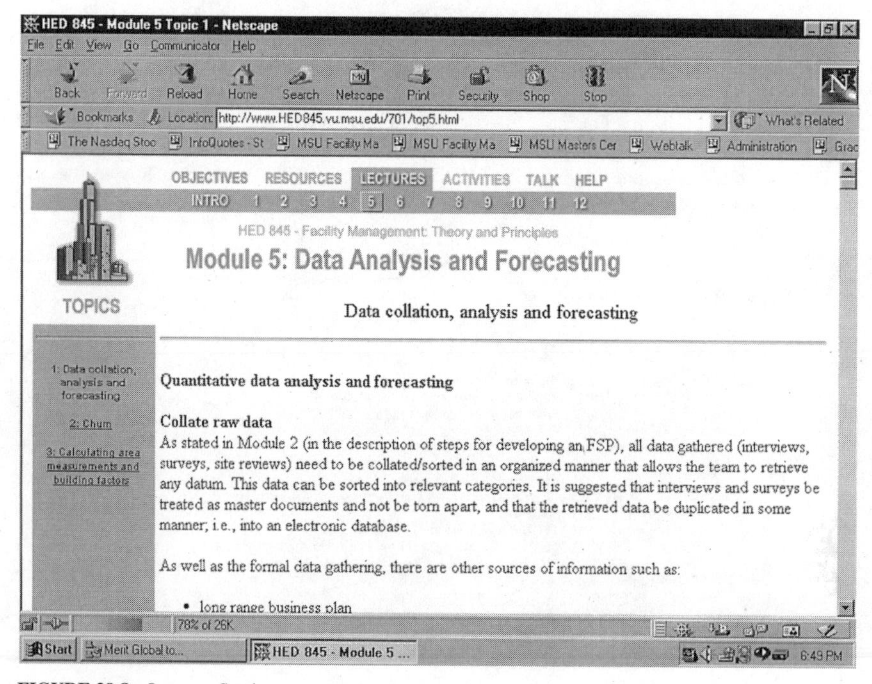

FIGURE 29.8 Lectures Section.

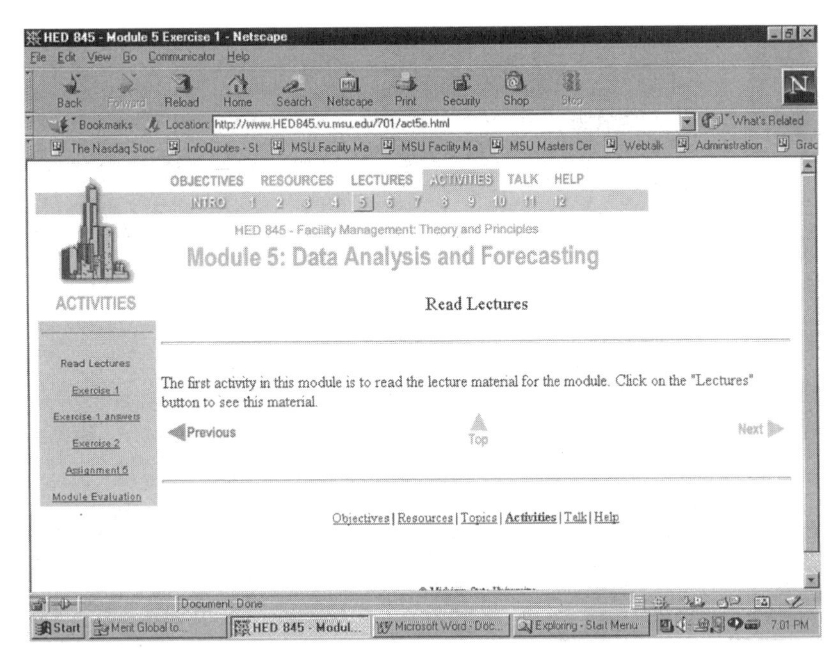

FIGURE 29.9 Activities.

Activities. The Activities section (Figure 29.9) describes the activities and self-learning exercises which the students complete to further their understanding of the subject matter presented in the lectures. As well, it describes assignments that must be submitted to the instructors. The section also contains an evaluation/feedback form that the student completes, anonymously if desired, and uploads to the VU database. Research, analysis, and reporting are possible from this database.

Talk. The Talk section is one of the more heavily accessed sections of the course (Figure 29.10).

The *Communicating with Classmates and the Instructor* subsection has tips on managing teamwork and the communication tools. Because this way of learning and working together over the Internet is new to many students, they are given suggestions to help them and their team members work together and are provided tips on how to best use the communication tools, WebTalk and LiveChat.

The *Enter WebTalk* subsection takes the student to the WebTalk asynchronous communication tool. WebTalk, MSU's Web-based conferencing/bulletin board system, provides a forum to ask questions and have discussions about various issues related to the course content and activities. Students also use WebTalk to upload the completed assignments to the university server. The instructors then download them for review and grading, and upload them once more to the server so the students can review the grades and comments.

It is essential that the student and instructor look in on WebTalk each time he/she logs onto the class, to view updates, announcements, and discussions, as well as communications from team members. In WebTalk, discussion "topics" are created; these are further divided into "conversations." These topics and conversations can be open to all (the classroom), open to some (the team), or private (the student). This means a student can communicate with the instructor only, his/her own team only, or publicly with the whole class.

The topics set up for each class are:

- *Announcements*
- *Help Room:* for posting technical questions

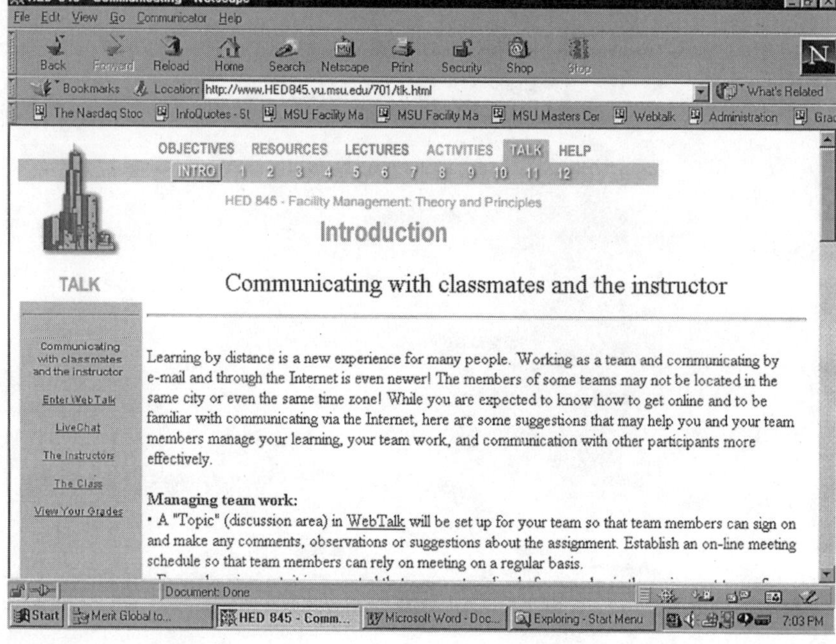

FIGURE 29.10 Talk Section.

- *Classroom:* for posting questions or comments about the course material
- *Workgroup Announcements:* each workgroup (a group of three of four teams) has a common area
- *Team Area:* each team of 4–5 people has its own area to discuss and exchange documents for their team projects
- *Items for General Viewing*
- *Excellent Examples of Assignments*
- *The Virtual Lounge:* for students to post messages about themselves and to indulge in conversations unrelated to the course material
- *Library Resources*
- *Private Conversations:* a conversation for each student, accessible only to that student and the instructor team—used for returning graded assignments or other private communication
- *Instructor's Corner:* for the instructor team to communicate with each other
- *On-line Chat Log:* used to record selected LiveChat conversations

Each topic is then subdivided into conversations. The example in Figure 29.11 shows the conversations in the Announcements topic.

Figure 29.12 shows the conversations for the Workgroup 1–Team 1 topic. This is the area Team 1 would use to discuss each assignment.

The instructor has access to all conversations. This allows the "guide on the side" to watch the discussions and intervene to provide direction, or provide additional information or guidance. Students see only what they have access to; that is, the general topics/conversations, their team area, and their private conversation area.

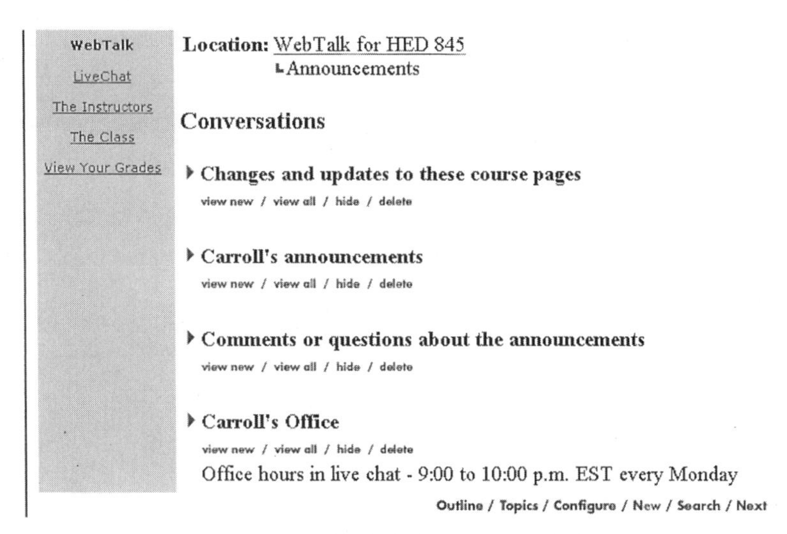

FIGURE 29.11 WebTalk—Conversations in the Announcements Topic.

WebTalk also allows conversation subjects to be threaded. This organizes the comments into logical sequences that can be more easily followed by the reader. The flag "New" indicates that a message has been posted in a conversation in that topic since the last time WebTalk was accessed. The WebTalk conversations are retained and accessible for the duration of the course. At the conclusion of each course, the course material and all recorded communication is burned on a CD. This provides a permanent record of the Web site for that particular delivery.

The *LiveChat* subsection takes the student to the area of the Web site where synchronous, real-time communication takes place (see Figure 29.13). The communication is time dependent in that the students from different time zones wishing to meet on-line must schedule an agreed time.

During the course, LiveChat is used to support a number of different activities:

1. *Impromptu and Scheduled Meetings with the Team Members:* During the course as students work on their assignments there is much communication between team members. The students schedule meetings in LiveChat to discuss their approach to these assignments.

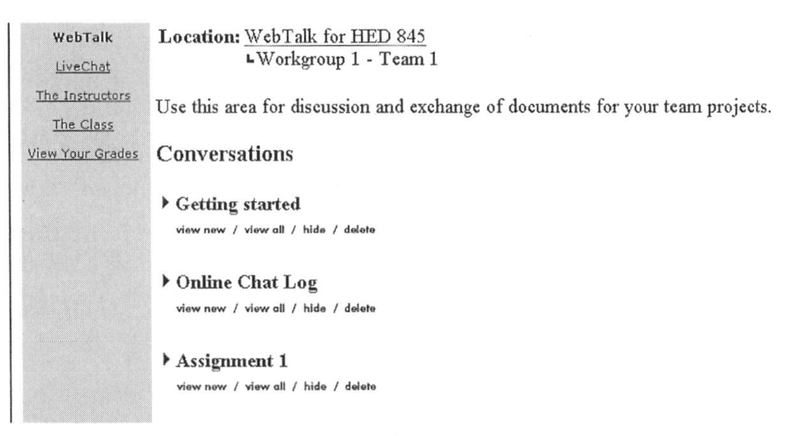

FIGURE 29.12 WebTalk—Conversations in the Workgroup 1–Team 1 Topic.

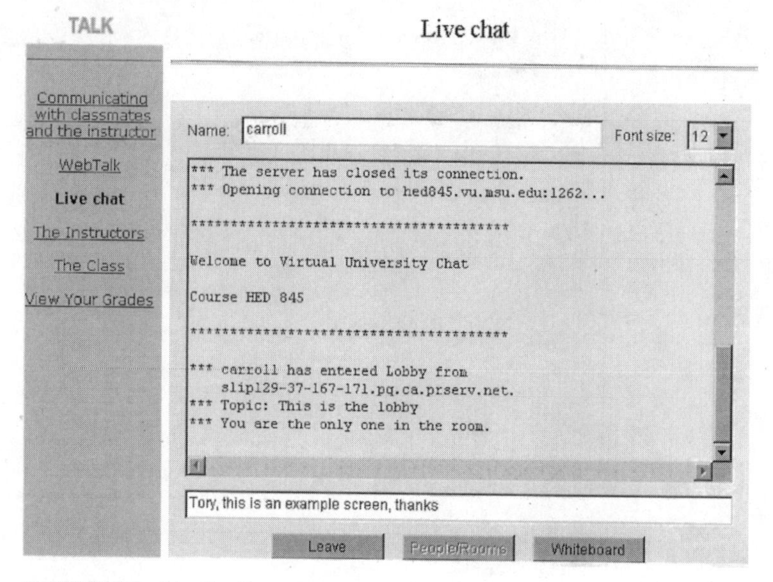

FIGURE 29.13 LiveChat Screen.

2. *Meetings with the Instructor during Office Hours:* Instructors may choose to have regularly scheduled "office hours" in LiveChat, where they remain on-line to answer students' questions.
3. *Class Discussions Scheduled by the Instructor of Each Course:* The dates and times are flexible; this allows the subject matter to be selected in keeping with the class progress. In one of the courses in the MSU program, three scheduled LiveChat discussions are held through the course. The instructor selects the first two topics and then solicits a list of hot topics from the students and selects one for the final LiveChat. Prior to the discussion each student posts a one-to-three paragraph description about his/her knowledge of or experience with the subject matter. Once this has been posted, the student is given permission to read the material posted by the other students.

On the day of the discussion, to make conversations manageable, students are assigned to groups of approximately ten members. The students are selected from various teams for each chat room. In these LiveChat sessions, students from different teams are put together; this allows students to interact with others in the class from outside their own team.

The challenge is to maintain coordinated and coherent conversation flow, which is difficult because of the time needed to type and post complete thoughts. One method that was used was to assign a chat group leader to enforce a round-robin format. The group leader managed the conversation so that it stayed focused on the topic and somewhat sequenced, ensured that all comments were listened to and respected, and made sure that each person had a chance to respond to the question being discussed.

During the discussions the instructor visits each chat room to "listen" to the content. The instructor may inject comments to stimulate discussion and respond to queries, but does not dominate the chat. The content of each of the chat rooms is recorded in WebTalk and is available for all participants to read. The recording of the chat room conversations was added at the students' request, to give them a record of each meeting. Should private chat be desired, rooms may be created that are not recorded.

The *Instructors* subsection takes the student to the area of the Web site where they can access profiles of the instructors. The profiles include contact information and may include a photograph. It is possible to add text that is only visible to members of the instructor team.

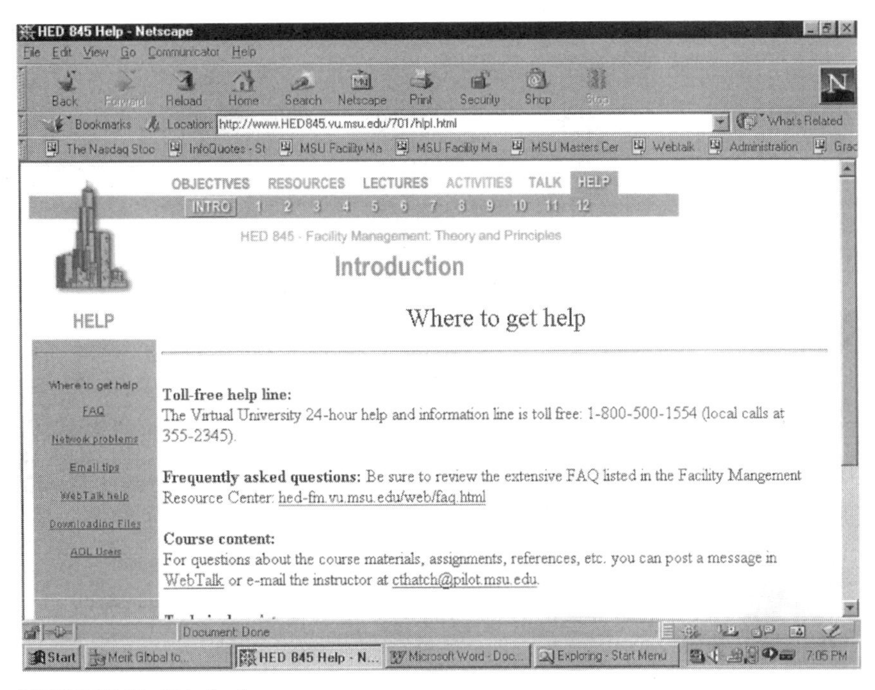

FIGURE 29.14 Help Section.

The *Class* subsection takes the student to the area of the Web site where there are links to students' profiles. These profiles include contact information, such as telephone and fax numbers, mail and e-mail addresses, and a photograph if the student wishes. The class list of students and instructors is provided in a linked format that allows them to send e-mails directly to individuals or to the entire class.

The *View Your Grades* subsection takes the student to the area of the Web site where they may see their own grades and instructor comments on each assignment.

Help. A number of technical aids are also available to students, and the subsections in the Help section link the students to these aids (Figure 29.14). The university's 24-h a day toll-free help line gives students access to computer technicians who can answer questions related to accessing the MSU server, where the courses are stored. When students enroll, they receive a set of instructions on establishing their student computer accounts, which provide them with access to the server. Although the process has been made as painless as possible, difficulties sometimes occur.

Some students find themselves behind firewalls, protective barriers set up by large organizations, corporations, and businesses to protect internal computer networks from hacking problems. Firewalls can cause access problems, as can some on-line service providers. The help line is designed to help students overcome these types of problems.

The Help section also contains a link to a frequently asked questions (FAQ) database. Questions about a variety of course navigation issues were asked during the first course. Anticipating that these issues would surface in future courses, these questions and answers were compiled into a separate help file. FAQ topics include such issues as navigating the course, uploading and downloading files, and procedures to use WebTalk and LiveChat. The FAQ has been continually expanded over time and has helped the students reduce the amount of time needed to acclimate themselves to working in a virtual environment.

Other Help subsections take the students to contact information for the instructor, and sections on network problems, e-mail tips, WebTalk help, downloading files, and using AOL.

29.6 LEARNING ON THE INTERNET

The fact that students meet only in the virtual classroom and not face-to-face means that they start with truly blank slates. No preconceived attitudes based on appearance or first impressions need pigeon-hole people and get in the way of learning or enabling students to contribute fully and equally. To start, the playing field is flat, with sex, age, and socioeconomic status having no impact. Over time, divergence in status occurs, but this stratification is based on behavior and ability rather than preexisting attitudes. The same roles that students play in traditional classroom settings become apparent in Internet courses as well—leaders emerge, as do followers. Over time, exceptional students and poor students are identified.

In Internet courses with a "high-touch" approach used by instructors, the effort expended by each student and the degree of learning each student accomplishes are more easily measured than in the traditional classroom setting. Because, for the most part, communication in Internet courses takes a written form, contributions by each student are apparent. This can enable the instructor to evaluate learning more easily. Instructor monitoring of written conversations also reveals interaction dynamics closely following those exhibited in traditional classrooms. Literature about on-line learning supports what the experience of the MSU program has found to date: there are many learning styles and not all students would thrive in this environment.

The MSU program courses are marked by a degree of complexity and depth in content because they are offered at the graduate level, with high demands placed on the students. They are expected to be motivated self-starters who are both organized and responsible. It is also expected that they will read extensively, seek out resource materials independently, and demonstrate an increasing ability to integrate complex material. Increased proficiency in the communication and presentation of information is also expected.

Students in the program enter as lifelong education students—both practicing professionals with and without undergraduate degrees are accepted into the program. Students with an undergraduate degree are accepted automatically. To ensure that applicants have sufficient background knowledge to function effectively in the courses, the résumés of nondegree students are reviewed prior to admission. If applicants have sufficient and progressively complex work experience, marked by the number of years as a practitioner, and evidence of success with other credit or noncredit education, nondegree applicants are admitted. If not, a suggestion is made that they gain additional work experience before reapplication.

29.6.1 Engaging the Student

As in a regular classroom, it is a challenge to engage a student's interest in the entire course. In an Internet course it is doubly challenging, especially near the end of the course material. The original expectation was that the students would read the material, do the exercises, ask questions about the subject matter, and complete and upload their assignments. The instructor/development team had debates about whether the students needed to be stimulated or whether, as graduate students, the responsibility was theirs to get the most out of the course.

Built into the courses in the MSU program are numerous methods of keeping the student engaged. At the start of the course, in the Introduction module, students can view a 5–10 min welcoming streaming video in which the instructor introduces himself/herself, welcomes the student to the course, and discusses the course. This allows the student to put a face and a voice to the instructor. The videos were added, at student suggestion, after the first course was delivered. Team work, scheduled LiveChat discussions, and WebTalk encourage interaction, as well as exchange of information, knowledge, and experiences. Socializing is also encouraged; students

FIGURE 29.15 Original "Attend Button."

FIGURE 29.16 New "Attend Button."

have a WebTalk topic they can use to talk about their backgrounds and interests and indulge in conversations unrelated to the course material. The first assignment is a team protocol exercise, in which team members discuss and agree on how they will interact as a team throughout the course.

The students may suggest topics for scheduled LiveChat discussions. In one case, a student had a real-life work issue and the scheduled topic was put aside so the other students could share their experiences with similar situations and provide suggestions for solutions. The instructor is available for discussion, support information, and knowledge transfer, both through WebTalk and through regularly scheduled "office hours" in LiveChat. Practicing professionals are guest contributors to the lectures and are available to respond to student queries.

In addition to the assignments that are uploaded to the instructor for comment and grading, there are self-learning exercises that the students take to practice the subject matter being presented. Students are encouraged to research, analyze, and make decisions. There is no right or wrong approach to the assignments; there are different approaches. Grading is based on the logic, thoroughness, content, and presentation of the submission. Superior assignments are posted for the whole class to see. This provides another forum for the sharing and exchange of knowledge. There are feedback forms that provide the student the opportunity to express their opinions on the course content and format as well as on the instructor and other team members. The formal student evaluations that are used for all courses at MSU are now required for Virtual University courses as well.

Even with these built-in methods of keeping the students engaged, it is difficult. For example, in the first course it was apparent from students' comments that some were not viewing the course material. The original name "Topics" was changed to "Lectures" so the students would more readily recognize that these sections contained the course content. To encourage the viewing/reading, an "attend button" was added at the start of each lecture (see Figure 29.15). When students clicked this button, their attendance was registered in a database. This, in conjunction with the instructor checking the database and sending reminders to students, improved attendance, but it was obvious that some students still were not reading the material. The attend button was then moved and changed in appearance—an open book at the end of each topic (see Figure 29.16) and a closed book at the end of each module lecture. In the next version, the book icon will be animated to close when the student clicks on it.

29.6.2 Adapting to the Virtual Classroom Environment

In the recruitment publicity, potential students are told to expect to spend from 10 to 20 h per week working on the course material. In addition, the introductory module of each course is available online in advance and can be previewed free of charge, so potential applicants will understand the requirements involved in the courses. Despite these precautions, course drops occur. Students receive access to the entire course for the first time the day a new university semester begins. For some, this first visit helps them visualize the extensive nature of the experience; a few are overwhelmed and drop the course. Most drops occur within the first week.

When a student drops a class, he or she is contacted by phone to determine the reason and whether university technology and/or teaching staff can help resolve any problems. The majority of students indicate that the reason for dropping is the heavy time commitment involved in interacting with classmates, as well as the time needed to read and analyze written material and do assignments.

There seems to be a common process involved as adult learners acclimate themselves to on-line courses. Not all appear to pass through these stages, or may pass through them less vocally than others. The anonymity which the technology provides may account for some behavior; some students go through a quite vocal stage of commenting on what they perceive to be the difficulties involved in completing the course. These range from the number and difficulty of assignments, the amount and complexity of material to be covered, the newness of it and the time required to process the information, the difficulty of mastering technology new to them, to the requirement that students analyze information and communicate thoughts in written form. Eventually, these students appear to refocus and expend energy on the tasks at hand. Of the initial group exhibiting this behavior, most went on to take more classes or have initiated communications with instructors, indicating they plan to take additional courses. Once students join a second or third class, negative vocalization is seldom exhibited, and particularly not in public.

Contrary to general perceptions about on-line learning, which emphasize solitary study, a surprising degree of social interaction occurs. Once students settle in and become comfortable with the technology, they soon learn to develop connections with many of their classmates. As they interact, listen, think, and learn to absorb or question their classmates' suggestions and observations, they develop similar affiliations as they would in a traditional classroom setting. The MSU program has shown that team members often stay in contact long after their coursework is finished. Some students who have finished one or more courses offer to be mentors to new students because they enjoyed the contact with colleagues and want to share their experiences with on-line communication.

Complaints about the difficulty of post-graduate classes are common. In traditional learning situations, however, the instructor is seldom confronted directly or in public. Internet courses, consisting of adult learners away from traditional learning environments, appear to be different.

29.6.3 Working in Teams

Research has shown that for many students, solitary study does not provide the optimum learning environment. The instructors in the MSU program were committed from the beginning to having the students learn from one another and felt this could be best accomplished by using a team approach. Working closely with fellow professionals with high degrees of specialized knowledge can broaden insights into problem solving. Exposure to practices done by others in the same field is valuable. In addition, exposure to the varied approaches other facility managers use to accomplish the same set of tasks or answer the same set of questions produces an invaluable learning experience. Team learning can accomplish these goals.

Students in each course are assigned to work in teams to complete team assignments. The teams are made up of from four to six students and are designed to be a mix of backgrounds, levels of experience, and academic qualifications. A team can be made up of an architect, an engineer, an administrator, a designer, and a project manager, for example, with each member bringing his or her different FM perspectives to the team experience.

The mix of students in teams has proven particularly positive. The instructor makes a conscious effort before the course begins to choose teams based on a mix of experiences, while respecting the need for similar time and work schedules. When teams break up the workload of team assignments, they soon learn the strengths of their classmates and divide up tasks accordingly. This results in a dialogue about the rationale behind certain approaches, which strengthens the cumulative experiences of the team as a whole.

The team members are location independent, but the teams are made up of individuals from the same time zones, if possible, to make real-time interaction easier. As a first assignment, each team develops a working protocol, which prompts the members to consider together how they will address further group assignments and the task set and roles needed to complete each. For each as-

12) **From**
Date: Sunday September 12, 1999 - 20:25 PM

post
hide
delete

Title: Assignment #2

Hi Team,This has been a really hectic week, so I appologize for not posting this earlier. I think this is a good test for future assignements...so we know how to plan ahead next time! :-)Since we have to turn in Assignement #2 by Monday, tomorrow at noon, EST there is limited time for feedback on this one. Since I am on the west coast I will need to post the assignement by 9AM. Please provide feed back by 6AM, pacific time so I have time to make changes before I post. NOTE: I changed the format to a Word document.

Let's set a time to all get on LiveChat this week to discuss our next assignment and to get to know each other...afternoon/evenings are best for me. Please send out a message with when you can meet.Thanks! :-)

Click here to get the posted file.

FIGURE 29.17 Example of WebTalk Conversation.

signment, the team members accept various roles, such as editor or leader, as the groups research, develop, write, and submit their assignments. The teams also develop communication standards that indicate specifically which communication methods members will use, when they will communicate, and how often.

Over time, as trust develops, it is not uncommon to see group members begin individual assignments in team discussions, working on an understanding of the conceptual basis and learning goals of the assignments before individual work begins. Indeed, levels of trust and respect develop quickly among students. Not only do they value the contributions team members bring to assignments, students begin sharing information about personal work situations and sharing perspectives on how they each address various tasks and responsibilities in the daily work environment. In one case, a course member, faced with a major but unfamiliar project after only days on a new job, discussed his situation with team members during a LiveChat. The planned conversation was quickly put aside as students focused on the new issue. At the end of the project, the student reported back to the group that he found their insights and suggestions invaluable as he moved the project forward.

One of the communication tools which allows this interaction between students is WebTalk. The communication is time independent, allowing the students and instructor time to think in order to respond clearly and in-depth. The example in Figure 29.17 shows the ease of conversation and the way students arrange to meet as a team. It also shows that there is an uploaded file that can be opened for review and comment.

29.6.4 Assignments

The assignments in the courses range from reading both required and suggested materials to developing and submitting written presentations in response to questions or situations posed by the instructors. The core reading in each course is the content written by the instructor—the "lectures" of the course. Instructors may also assign additional reading, often chapters in recognized textbooks or articles from professional journals. The instructors choose material authored by recognized authorities in the FM field, so students purchasing required readings are building their professional libraries as they take the courses. Instructors also provide lists of suggested supplemental readings, consisting of professional journal articles, trade journal articles, written reports, case studies, annual reports, and other resources they feel will provide more detailed information about specific topics. At least two of the instructors require students to develop briefing papers on assigned topics, and the suggested readings provide a starting place to begin the research for these papers.

The number of assignments in the courses ranges from five to seven. Assignment formats are varied as well, from short answers to specific questions, such as finance problems, to extensive, formal presentations. Approximately half of the assignments are completed by students working individually, while the others are completed by the teams. When team assignments are developed, each member is expected to contribute, review, and pass approval upon the final draft of the material before it is uploaded for grading.

When students start in the program, they tend to rely on course content, the study of additional written material, and their own and team members' experience and knowledge to develop answers to questions. Over time, however, the students are required to move away from this more passive approach to developing research tools and conducting field studies on their own. They are also expected to begin identifying their own learning materials, rather than relying solely on suggestions made by the instructors. This helps them develop skills in finding and evaluating sources of information that provide greater depth and detail about specific topics.

In some courses, students are required to develop survey tools and test them, and to conduct interviews on-line with high-level managers of companies and corporations. They are also expected to study the structure and functioning of companies and corporations identified as financially successful or as using best practices or benchmarking examples. This moves the students into studying real world situations and makes them apply their learning by analyzing and understanding best case examples of corporate behavior.

For a few students, problem solving with hypothetical situations is difficult at first. Some complain, for example, that the assignments refer to a type of facility they have no experience managing and feel that their lack of experience with a particular type of organization places them at a disadvantage in answering questions. They also have difficulty in identifying with and seeing the applicability that solving problems related to these organizations has to their own specific situations. As the assignments progress, most students come to understand how the lessons learned apply across organizational types and situations.

All of the assignments are designed to stimulate creativity, critical thinking, and the development of problem-solving skills. Case studies are frequently used as vehicles to help develop these competencies. In one course, a hypothetical case study is presented at the beginning, and the team and individual assignments require analyzing and developing written reports on various aspects of the company's current facility situation and developing approaches to rationalizing their number, location, and operation. The exercises require the student to focus on short-term and long-range planning, and on the development and articulation of management approaches to optimize the company's market niche—with the management of the company's facilities assumed to play a crucial role in determining the company's financial future. The assignments build upon one another, using a developmental approach. The aim is to have students learn processes that can be applied to their personal work situations to help them move from a reactive, crisis management approach to FM toward the use of long-range planning and strategic management to solve facility problems.

To encourage creative thinking, limited detail is provided in the case study. Team members are responsible for developing a scenario by making assumptions, testing them, revising ideas and approaches, and communicating the results. Although students initially find the approach difficult, over time they begin to realize why it is used. One student indicated, for example, that possessing limited knowledge is somewhat like entering a new FM position and beginning to understand procedures, operations, and the goals and functioning of the new employment situation.

Limiting detail also allows students to learn how to structure methods to evaluate current management techniques and introduce changes. One student, for example, transferred the lessons learned in the course to his workplace and used assumptions to structure the identification of a problem and initiate solution approaches. Assumptions were articulated to co-workers, and they responded positively, indicating it allowed them to react to the validity of the assumptions, further clarifying each participant's understanding of the problems faced. After this, they were able to work together more effectively to identify workable solutions.

The final project in each course is a cumulative, integrative experience. The students are required to develop a professional presentation, with the understanding that the material is to be designed as if it were being used as a presentation to senior corporate management to accomplish a hypothetical facility goal, which, the team argues, would further corporate goals in general. The team is expected to develop an integrated, well-reasoned, but succinct argument to make their case. The use of PowerPoint slides and a narrative are required, with the students limited in the number of slides and amount of narrative that can be submitted. Thus, the students must consider not only the quality of the argument, but the quality of the presentation as well. After the presentations are submitted, the instructor, working with input from a group of practicing professionals acting in the role of the corporate team, develops questions about the content. These questions are posted for team members to answer within a set period of time. This corresponds to the question and answer period that would normally take place after an in-person presentation. Conducting these steps in writing over the Internet makes the entire process challenging to both the students' creativity and their problem-solving skills.

Throughout the courses, students who have submitted excellent assignments are asked for permission to post their materials in a WebTalk area created for this purpose, for other students to view. Because assignments are based on practical situations, students studying the examples gain personal insights and can often apply the approaches used by others to solve hypothetical problems in their own work situations. More importantly, however, students can see how other practitioners approach and solve the same problem.

29.6.5 Evaluation and Grading

Evaluating the Students. The students are evaluated in two ways: by the instructor, who evaluates individual and team assignments as well as overall participation, and by their fellow team members, who evaluate their participation in and contribution to the team projects. Grades for the courses are based on participation and written assignments. There is no final exam for any of the courses. Assignments account for a percentage of the final grade. Participation (partly determined by the instructor and partly by fellow team members) accounts for the remaining portion.

Evaluating the Course: Feedback Forms. As well as evaluating their fellow team members, students are asked to evaluate each module of the course. The Activities section of each module contains an evaluation/feedback form that the student completes, anonymously if desired, and uploads to MSU's Virtual University database. Part 1 of the questionnaire focuses on the content of the module, Part 2 focuses on the learning activities and how well these contributed to the student's learning, Part 3 focuses on design aspects of the module, and Part 4 is for an overall rating of the module. Figures 29.18, 29.19, and 29.20 show examples of the feedback forms. Research, analysis,

Instant Message Members WebMail Connections BizJournal SmartUpdate Mktplace RealPlayer

Course: HED845
Title: Facility Management: Theory & Principles

Module # 5
Title: Data collection, analysis and forecasting

Completed by:

Date:

This questionnaire assists you to evaluate your experience with this module and identify some of you own learnings from the module. Your responses will provide direction to the instructor about additional instruction that may be required prior to the end of the course as well as to improve the content and learning activities.

Part 1 focuses on content of the module, **Part 2**, on the learning activities and how well these contributed to your learning, **Part 3**, on design aspects of the module and **Part 4**, on your overall rating of the module

1.

IT IS OK TO INCLUDE MY RESPONSES IN RESEARCH or
PLEASE DO NOT INCLUDE MY RESPONSES IN RESEARCH

Document Done

FIGURE 29.18 Example of Feedback Survey.

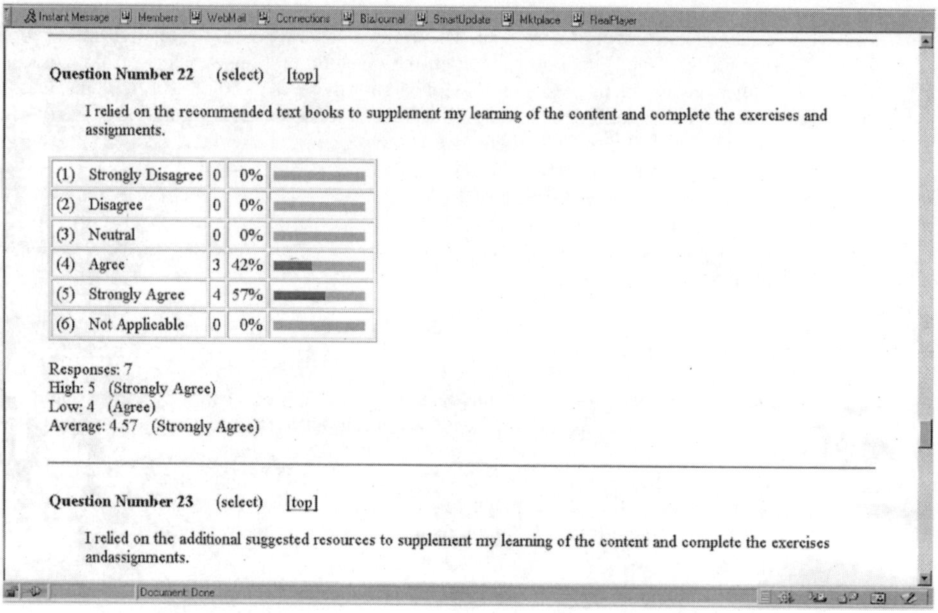

Question Number 22 (select) [top]

I relied on the recommended text books to supplement my learning of the content and complete the exercises and assignments.

(1) Strongly Disagree	0	0%	
(2) Disagree	0	0%	
(3) Neutral	0	0%	
(4) Agree	3	42%	
(5) Strongly Agree	4	57%	
(6) Not Applicable	0	0%	

Responses: 7
High: 5 (Strongly Agree)
Low: 4 (Agree)
Average: 4.57 (Strongly Agree)

Question Number 23 (select) [top]

I relied on the additional suggested resources to supplement my learning of the content and complete the exercises andassignments.

FIGURE 29.19 Example of Response to Numeric Rating Question.

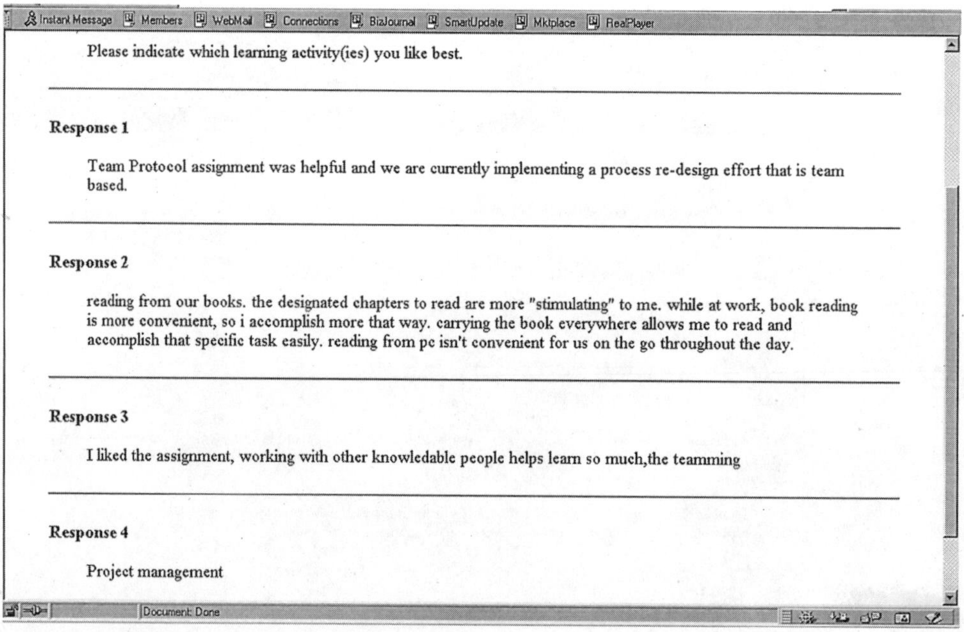

Please indicate which learning activity(ies) you like best.

Response 1

Team Protocol assignment was helpful and we are currently implementing a process re-design effort that is team based.

Response 2

reading from our books. the designated chapters to read are more "stimulating" to me. while at work, book reading is more convenient, so i accomplish more that way. carrying the book everywhere allows me to read and accomplish that specific task easily. reading from pc isn't convenient for us on the go throughout the day.

Response 3

I liked the assignment, working with other knowledable people helps learn so much,the teamming

Response 4

Project management

FIGURE 29.20 Response to a Question Soliciting Verbal Response.

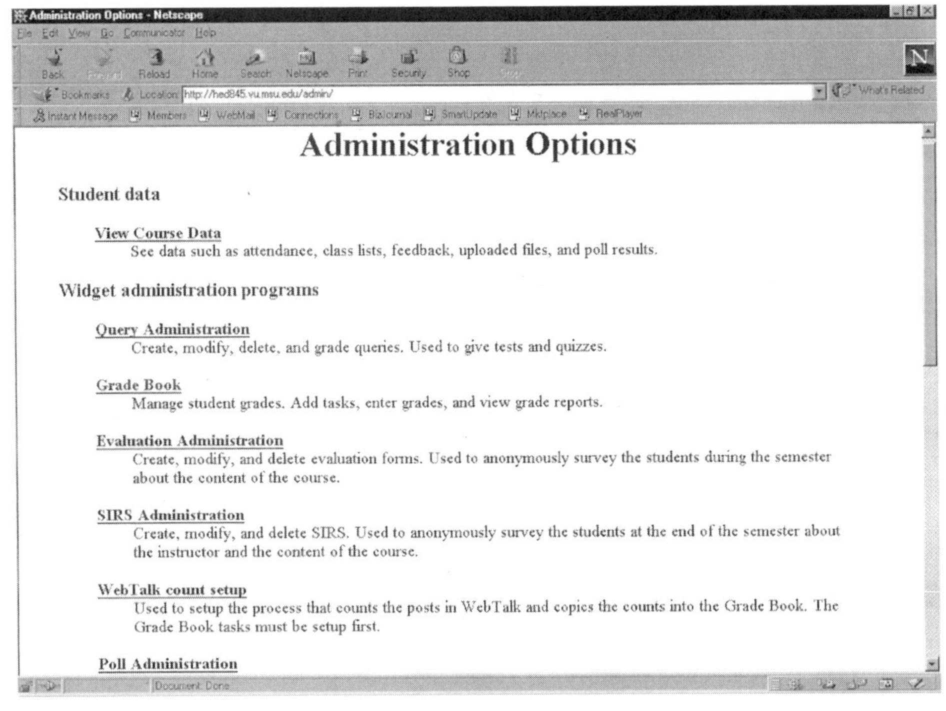

FIGURE 29.21 The Administration Database.

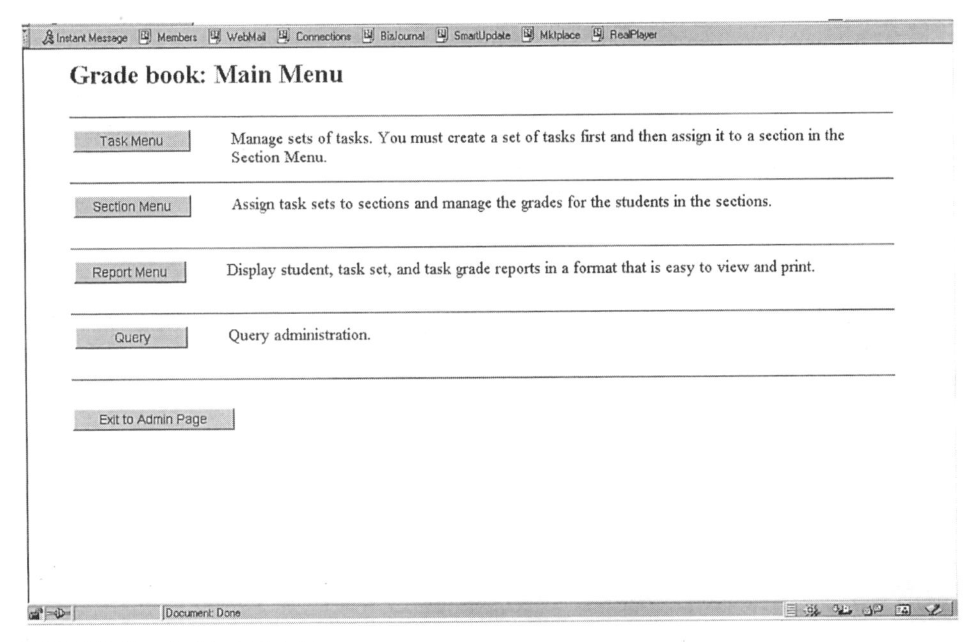

FIGURE 29.22 Gradebook Main Menu.

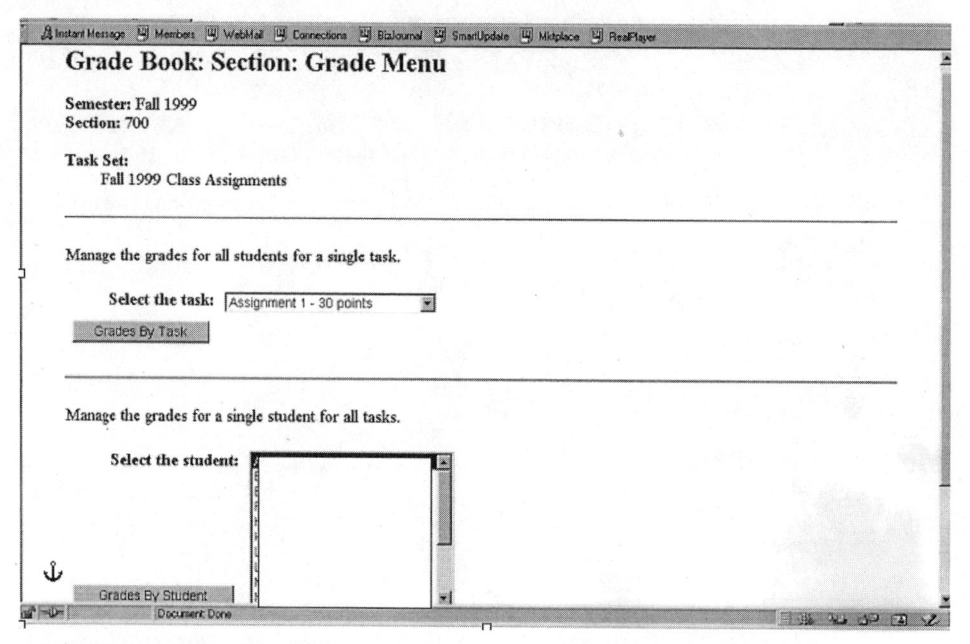

FIGURE 29.23 Gradebook Section Menu.

and reporting are possible from this database, and students are asked if they wish for their comments to be used for research purposes.

The Gradebook. Marks are managed with a tool called the Gradebook, which is available to both instructors and students. From the Talk section of WebTalk, students can click on the "View Your Grades" link to take them to the area of the Web site where they can view their own grades and the instructor's comments on their assignments. Instructors access the Gradebook through an administration database (see Figure 29.21), which also contains the data on the surveys and attendance.

Within the Gradebook, the instructor can manage the students' grades. The following example illustrates the menu where the instructor selects the mode to fill in the grades and the comments. The Gradebook database has a report generation capability and the data can also be exported to other programs, such as Excel. Figures 29.22 and 29.23 illustrate the Gradebook.

29.7 MEASURING SUCCESS

29.7.1 Critical Success Factors

A few key factors can be identified as critical to the success of this experiment in e-learning:

- *Run a Pilot Project:* A pilot course was run in the summer semester of 1998. This small course of six students in two teams showed the feasibility of e-learning to teach FM content.
- *Meet the Demand:* In the fall of 1998 at World Workplace in Chicago, two focus groups were held to determine the needs of facility professionals for graduate-level education. The response from these focus groups articulated and confirmed the approach being taken by MSU.
- *Engage the Students:* The use of a "high-tech, high-touch" approach, using the best of computer communication technology, combined with having students work in small teams and doing

both team and individual assignments, has been highly successful in keeping the students motivated to learn and share their knowledge.

From the time the first course was developed in 1996 and 1997, the course philosophy has always been focused on the student.

In a paper by Bonks and Cummings,[1] the authors list their recommendations for keeping the focus on student-centered learning when offering Internet-based courses. The following list is their twelve recommendations, based on their experience. After each point, we have listed the strategies that have been applied in the MSU program.

1. "Establish a safe environment and a sense of community."
   - An inviting electronic learning community
   - A place for students to introduce themselves
   - A place to socialize, with the instructors encouraging posting about personal activities, such as students' work, personal interests, and family activities
   - An initial assignment to get students to introduce themselves (i.e., team protocol development)
2. "Exploit the potential of the medium for deeper student engagement."
   - Asynchronous communication tools—delayed conferencing to give students and instructors time to think and post reasoned comments and feedback
   - Synchronous communication tools allowing immediate, real-time discussion and feedback
3. "Let there be choice."
   - Choice in LiveChat topics
   - Development of assumptions about the detail of a case study
   - Students encouraged to make decisions
   - Students encouraged to do research, explore the Web, and share findings
4. "Facilitate, don't dictate."
   - Collegial approach, with the instructors acting as "guide on the side"
   - Instructors watching conversations and injecting comments when asked or when students are losing focus
   - Lectures supporting the assignments and vice versa
5. "Use public and private forms of feedback."
   - WebTalk topics of announcements, classroom, team areas, private conversation areas
   - Evaluations requested on course content material
   - Instructor's detailed comments on assignments, in general terms to the class as a whole in the public forum, and to individuals or team privately
   - Excellent examples of work are posted, with permission, for the whole class to see
6. "Vary the forms of electronic mentoring and apprenticeship."
   - Encouraging students to share information with each other in both the team and classroom conversation areas of WebTalk
   - Promoting the idea of a life-long network of contacts
   - Teaching assistants
   - Providing a list of professionals in the field willing to communicate with students
7. "Employ recursive assignments that build from personal knowledge."
   - Course assignments build upon each other
   - Topical subjects for briefing paper assignment and LiveChats
8. "Vary the forms of electronic writing, reflection, and other pedagogical activities."
   - Briefing paper

- Live chat discussions requiring that one to three paragraphs be posted prior to the discussion concerning a student's experience with or knowledge of topic to be discussed
- Responding to classmates' questions and comments
- Developing and responding to questionnaires/surveys
- Assignments requiring students to develop forms, outlines, charters and agreements, project management schedules, and formal presentations

9. "Use student Web explorations to enhance course content."

- Students encouraged to share information found
- Students encouraged to identify useful Web addresses to add to resource center for use by others
- Courses use hyperlinks to encourage Web exploration

10. "Provide clear expectations and prompt task structuring."

- Objectives for each module
- Clear description of assignments and formats to be used
- Assignment due dates given at start of course
- Announcement area in WebTalk where explanations and changes can be provided and feedback received from the students
- Immediate feedback on receipt of uploaded assignment, informing students that their assignments have been received
- General instructor comments to the whole class when assignments are returned

11. "Embed thinking skill and portfolio assessment as an integral part of Web assignments."

- Include Web research in regular assignments
- Half of the required assignments are group assignments
- The instructor's comments and grading both support and encourage different ways of thinking
- Peer evaluations

12. "Look for ways to personalize the Web experience."

- On-line welcome video of each instructor at the beginning of the Web site
- Videotapes used, short ones available on-line using Real Player, longer ones sent to students
- LiveChat
- Photographs posted with profiles
- Instructor photograph attached to every message posted in WebTalk
- Teamwork
- Students know the instructor is watching/listening and willing to react quickly to their comments and questions
- Sharing of ideas and work

In a program that aims to make the student the center of the education strategy, with the instructor acting as the "guide on the side," the best measure of success is the students' response. Feedback from the FM students has been positive.

> I feel fortunate to have "attended" school this semester with a great team. Being able to attend school over the Internet has been an incredible experience. It has afforded me an opportunity to do something I can't imagine having had the time to do in a classroom setting. (*Ellen Harper, Senior Project Manager, Facility Resources, Inc.*)

> Even though some frustrations on my part were occasionally evident in messages, the Instructor Team was always poised, encouraging, and resourceful . . . . I also "met" some great people during the course, especially my teammates. I liked the feeling that you're not alone in the course. (*Chris Kulhanek, Facilities Manager, County of Saginaw, Michigan*)

> Even though I have been in the facilities business for 27 years, I learned a lot! . . . The virtual university technology creates some new challenges of its own, but allows for

course participation that may not be otherwise possible. One benefit that may not be obvious is the network of new friends and future resources that are developed in this course. (*Robert Neitz, Chief—Facilities Management Division, Pennsylvania Department of Transportation*)

You will discover things about yourself and others that you never thought possible, like tenacity, endurance, cooperation, teamwork, trust, abilities, and capacities to learn you thought were lost forever. You will learn all sorts of information that can impact how you work. You will change your paradigm. This is a wonderful opportunity being offered to seasoned veterans and it really opens up new ways of doing and seeing things. . . . The medium once mastered is wonderful. Team members are amazing, the collective knowledge is astonishing. (*Michael White, Director of Facilities, Sylvan Learning Systems*)

29.7.2 Conclusion

MSU has run a total of eight offerings of the courses in their virtual classroom—four of Facility Management: Theory and Principles, two of Achieving Facility Management Organizational Effectiveness, and one each of the others. The evolution of the program has been a steady improvement in the course delivery, processes, and tools. Feedback from the students who have taken these courses indicates that an e-learning approach to FM education is effective and meets the needs of practicing facility professionals.

As the symbiosis of computers and communications continues, it will be an on-going challenge to use the technology wisely to provide graduate-level education to all those who seek it.

APPENDIX A BIOGRAPHIES OF INSTRUCTORS

- Carroll Thatcher is the instructor of *Facility Management: Theory and Principles.* Thatcher is a Principal of Carroll Thatcher Planning Group, in Ottawa, Canada. She is a Certified Facility Manager, an IFMA Fellow, and an internationally recognized strategic space planner and design consultant. Thatcher designed, developed, and taught the first graduate course in FM delivered via the Internet, which is the course included in the MSU Master's Level Certificate Program. She was also the off-campus coordinator for the development of this program. In 1999, Thatcher received the Distinguished Author Award for Instructional Materials from the International Facility Management Association. This award recognizes her development of the material included in this program.
- Guy Thatcher is the instructor of *Information Management for Facility Professionals.* Thatcher is a Principal of Carroll Thatcher Planning Group, in Ottawa, Canada. His long-term interest in information management started when he was a Systems Engineer for IBM in 1965 and 1966. He holds a degree in computing and information science from Queen's University in Kingston, Ontario, Canada, and is a Certified Management Consultant, specializing in information systems. His teaching experience ranges from military topics to more traditional facility content. He speaks extensively on facilities-related topics and has been published in the USA, Canada, and Europe.
- Stormy Friday is the instructor of *Achieving Facility Management Organizational Effectiveness.* Friday is the president and founder of The Friday Group, based in Annapolis, Maryland. She has over 25 years of experience as a manager and consultant for both the private and public sectors. Friday served on the faculty of the University of Manitoba graduate facility management program and has co-authored a book, *Quality Facility Management: A Marketing and Customer Service Approach.*
- Bernard van der Hoeven, Ph.D., and IFMA Fellow, is the instructor of *Facility Real Estate and Building Economics.* van der Hoeven brings 35 years of experience in research and development,

both as a practicing researcher and in a variety of FM roles. He recently retired as the director of facilities, security, and safeguards at the Los Alamos National Laboratory. He has consulted on total quality management and strategic planning in the research environment, working primarily with national laboratories and universities. His facilities organization at Los Alamos was successful in winning two coveted Quality New Mexico Baldrige awards. He is currently teaching, writing, and doing consulting work in the FM field. van der Hoeven is a former member of the board of directors of IFMA and a senior member of IEEE.

ENDNOTE

1. Bonk, C. J., and Cummings, J. A. 1998. A dozen learner-centered recommendations for placing the student at the center of the web-based learning. *Educational Media International* 35(2), 82–89.

CHAPTER 30
GIS CASE STUDY: SPACE MANAGEMENT AT THE UNIVERSITY OF MINNESOTA

David A. Jordani, FAIA
President
Jordani Consulting Group
E-mail: djordani@jordani.com

Debra Gondeck-Becker, Assoc. AIA
Consulting Project Manager
Jordani Consulting Group

30.1 INTRODUCTION

This case study describes the selection and deployment of geographic information systems (GIS) technology to support the space management needs of the University of Minnesota. At the time this project was deployed, using GIS to manage room-level detail represented a new application of this technology. Traditionally, GIS are assigned to the management of land-related databases for information on assets that exist outside "the five-foot line" that surrounds a building. Computer aided facilities management (CAFM) systems are normally used for managing room-level detail, the focus of the effort at the University of Minnesota.

This chapter illustrates, however, that GIS lends itself to the analysis of space as small as a room just as well as for large parcels of land. In its most basic form, a GIS combines the concepts of spatial relationships to analyze and present location based data to support a business process. The detailed analysis of room level facilities data is ideally suited to the spatial analysis and presentation capabilities of GIS. The introduction of GIS by the facility management (FM) department adds a powerful "spatial data analysis" component to the university's technology architecture. In addition to serving FM, it has the potential to integrate facilities data with other location based databases that facilitate the flow of information between business functions and across departmental boundaries.

30.2 BACKGROUND

The University of Minnesota is home to 80,000 students, faculty, and staff on a daily basis. Facility assets valued at over \$3 billion cover 24 million ft^2 of space in 60,000 rooms in over 1,000 buildings spread across five major campuses and more than 20 research sites and experimental stations.

Like other FM groups, the university's FM department faced pressure due to increased workloads, reduced staff levels, and declining budgets, while still challenged to provide the same or even higher level of services.

The mission of the university's FM department is "making facilities work for U." This motto reflects the department's facilities stewardship and service provider roles in support of the university's academic mission. Part of that stewardship role includes providing information about facilities—their function, use, size, and other characteristics—to departments as they evaluate and redesign academic programs and course offerings. A mainframe space management system that served as the repository for information had been in use for almost two decades. As with most legacy systems, information in the existing space management system was difficult to access. Departments and administrative units that needed information about their use of space had to request that information from the FM department. The approach and FM's ability to respond was inconsistent with FM's stewardship role.

Attention turned to the replacement of the legacy space management system. Advances in technology and increased demand for spatial information suggested that customer support could be improved by replacing the existing mainframe space management system. Further, an enterprise-wide approach that would make this information easily available to FM's customers would be more responsive to FM's mission.

Enterprise-wide space management systems have the opportunity to support a broad spectrum of users and business functions. Collaboration and acceptance across departments is crucial for successful deployment. To ensure development met a broad requirements spectrum across the University of Minnesota, yet maintained proper business perspective, a project team with a balance of business and technical viewpoints was established.

To obtain that input and perspective, the project team was organized into three groups: the steering committee, the implementation team, and the user review board. The steering committee included university management to provide direction and assist in coordination with other enterprise data efforts. This group also played an important role in the adoption of common systems' approaches as part of the enterprise-wide deployment. The implementation team included the staff responsible for the delivery of the project. This team comprised staff from the university's planning department (representing the major client for the system), a faculty representative from one of the largest users of space at the institution, FM and university information systems staff, and the consulting team. The implementation team was responsible for all aspects of systems design, development, and deployment. The user review board included academic and administrative units across the university. Their role as part of the project team reflects FM's commitment to deliver technology that would meet the needs of a broad base of users.

The team determined at an early stage that successful deployment of a university-wide system for space management would be best served with a multi-phase project design. The project at the University of Minnesota consisted of five major phases:

Phase 1: Requirements definition and strategy design

Phase 2: Technology evaluation and selection

Phase 3: Detailed design and pilot implementation

Phase 4: Implementation, training, and startup

Phase 5: Production and enterprise-wide deployment

Major milestones included a justification and ROI at the end of the requirements stage, proof of concept at the end of the pilot project stage, and then full deployment target date toward the end of the project.

30.3 REQUIREMENTS DEFINITION AND STRATEGY DESIGN

The project began with a comprehensive analysis of requirements, management procedures, and related business processes to identify space management needs and requirements. Input was obtained from academic, administrative, and operations groups across the university. Existing programs and data were evaluated. Functional and technical requirements were developed through a series of focus groups and facilitated discussion groups. These requirements were documented and presented to the university as the basis of proposal solicitation for a new space management system. Key factors uncovered during the analysis phase included:

- Academic planning and business decisions require access to and synthesis of information from a variety of information systems across the university. The information they need is never in one source. Information about facilities is key to many of these decisions.
- Departments need information about the facilities that house people, equipment, and activities. The information they need includes data about the location, size, condition, accessibility, assignment of space, and other factors.
- Departments need access to accurate, up-to-date drawings of facilities.
- Location, or spatial reference, is an organizing concept for a variety of departmental data, including human resources, security, class scheduling, inventory, and research. These databases include information on topics from staff names and position; to lists of class offerings, meetings, and conferences; AV and technology resources; controlled substances; equipment; and hazardous waste. Associating a building and/or room number to organize departmental or administrative data is a common practice. Departments need to integrate information about facilities with their database values. Many groups use color markers to annotate drawings with these database values.
- Lack of an enterprise-wide consistent spatial model (e.g., consistent site, building, and room numbering scheme) causes other users to develop their own. This leads to multiple databases with inconsistent data and eliminates (or severely complicates) the ability to integrate different databases.
- Departments need robust and flexible data access and reporting tools that incorporate graphical navigation, query, and reporting.
- Accurate space data has a significant economic impact on indirect cost recovery (as related to federally funded research), occupancy, and deferred renewal.
- Legacy space management systems that make data difficult to access frustrate efforts by academic units to make effective programmatic decisions regarding assignment of space. With the existing system, integrity was questionable, data was not current, and some groups did not use the system. Difficulty in use and maintenance of data led to a backlog of uncompleted updates.
- A parallel campus-wide data modeling study concluded that information about facilities is central to the operations of the university. Similar studies in corporate FM and real estate groups substantiate the need for facilities information to support the enterprise.

30.4 ENTERPRISE-WIDE DATA SHARING STRATEGY EMERGES

The concept that emerged from the requirements analysis was that of an enterprise-wide system targeted at supporting a broad constituency of planners and administrative staff across the university. As stewards of the university's facilities, FM's role included managing facilities in use across the enterprise. The new space management system would extend that stewardship role to include providing facilities information that would easily integrate with information from other enterprise databases for decision support by academic and administrative groups. The goal of the project team was to select and deploy a technology that would facilitate the flow of information between business functions and support integration across departmental boundaries. Several things were deemed necessary to achieve this vision:

- The university needed a consistent spatial data model for its facilities, one that would be used by a variety of different database custodians to organize their information to facilitate information exchange. For example, a common spatial reference of "office number" in both human resource data (personnel name, position, and office number) and in inventory management data (equipment name, service call, service date, and office number) allows correlation of personnel to equipment and service calls. This spatial reference provides a consistent vocabulary for locating facility assets and enables business analysis to cross departmental boundaries.

- The team envisioned a database design whereby FM would provide and maintain facilitates information on the occupancy, size, use, programs, accessibility, and other characteristics of space. The technology would provide a platform capable of linking core "institutional" facilities information across departmental boundaries with other "user defined" departmental databases and enterprise information systems.
- To make the system easy to use, the team determined that a graphical interface be deployed as a primary navigation and reporting interface. The concept of using a site plan or floor plan as a navigation device to data implied the ability to graphically select a room or building and dynamically view and report on information from any "connected" facilities, departmental, or enterprise database. The drawings would become the interface for selecting data and the reporting context for displaying their values.
- Processes that would allow people to easily integrate their data with core information about facilities would need to be defined, tested, and deployed. These processes would need to include methods for deploying and maintaining the data.
- An open technology that offered the majority of these features "out of the box," while still allowing customization, especially at the database level, would be a requirement.

The intent was an enterprise-wide automation solution that aids the flow of data between departments. Business processes that would benefit by integration with facilities data included human resources, inventory services, telecommunications, and numerous other processes that use spatial references (building and/or room numbers) as a primary method of organizing data. The team also noted the opportunity to enhance ROI via deployment to support a number of business functions across the enterprise.

30.5 TECHNOLOGY EVALUATION AND SELECTION

Functional and technical requirements were incorporated into a request for information (RFI) for a new space management system. An open (rather than proprietary) technology architecture that would allow the university to achieve the enterprise-wide vision was a key selection criteria. A selection team comprising FM personnel and other university staff was assembled. University policy indicated selection of commercial "off-the-shelf" products, with integration as required, rather than creation of custom software. Vendor financial stability and viability was an important issue.

Responses to the RFI were evaluated and a short list of four vendors was developed. The implementation team prepared a list of performance criteria that were incorporated into a demonstration "script" that was provided along with actual data (drawings and database information) to each of the short-listed vendors. Vendors were given several weeks to prepare for a day-long "performance" of the demonstration script to illustrate the capabilities of their systems to meet the university's functional, technical, and business requirements.

The short list included traditional CAFM and GIS vendors. Space management has traditionally been perceived as a problem best addressed by CAFM software. CAFM's domain includes relational database management and integration with CAD data to manage facilities resources and other assets. Early discussions and research suggested that software from one of the traditional CAFM vendors would likely form the basis of this technology initiative at the University of Minnesota. An early demonstration of GIS arranged through the university's geography department expanded options to include consideration of GIS as a potential technology to replace the legacy space management system.

The project team considered several issues as it evaluated the relative merits of CAFM and GIS technologies:

- Both CAFM and GIS technologies are useful for managing spatially related data. Both technologies pair graphical with tabular data to provide graphical display of query results. Perceptions that GIS is not designed to manage room level detail are due to its ties to land database management and related vendor marketing, or mismarketing for that matter. In fact, GIS can manage the full spectrum of spatial data management issues from the site to a room. While CAFM vendors demonstrate GIS-like interfaces, their data models, at the time, were not geographically continuous.
- The responsibilities of facilities managers are not limited to the space inside of a building, expanding to include a number of business processes such as utility management, capital projects, master planning, inventory services, and infrastructure management services. Though the primary focus of the technology deployment was information about room use, it was reasoned that selecting a system that could deal with a full range of location and spatial relationship concepts could return greater benefits. Though future integration of the university's utility maps was not a high priority item, it seemed foolish to ignore this potential if other requirements were met.
- Initial analysis suggested that since CAFM systems offered a suite of reports and a database design for space management, a CAFM solution would require less customization. In reality, however, none of the vendors, CAFM or GIS, provided a design that would meet the needs of a higher education environment. From the university's perspective, the technologies had an additional factor in common—customization of the database, business rules, and reports would be required. While both technologies were customizable, GIS vendors offered a more open architecture that would better facilitate this requirement.
- The team concluded that GIS tools for display and data analysis exceeded those of the CAFM systems. The concept of "thematic maps" is based on color coding features based on selected data attributes. Within GIS a telecommunications company, for example, may create a thematic map to show area codes assigned to cities by displaying the city boundaries (map feature) and shading each area code (data attribute) with a different color or pattern. The SPACE system extended the concept of thematic maps to include floor plans shading rooms based on function, use, program, departmental assignment, or any other data element that is included in the model.
- GIS also included features that supported analysis of spatial relationships through a consistent, location-based model that can be applied across the enterprise. The selection team perceived additional opportunities with this feature and demonstrations of the spatial analysis capabilities of GIS also influenced the final selection. When used to query data, GIS offers results spatially by displaying the matched attributes across the map. This gives an immediate impression of how the data and location are interrelated. Spatial analysis of the data is also possible. With the attributes displayed against the map, queries can continue with parameters such as "within a distance of" and "intersecting with." In one case, a vendor demonstrated the ability to analyze parking requirements by determining (from a class schedule database) the number of students within a half-mile radius of particular parking facilities at a specific time of day. Another demonstration illustrated the ability for a planner to use information at the site level to identify the impact of underground utility projects on the facilities the utilities support. The planner could then "drill down" to query the room data noting the rooms and their use, pinpointing the need to reschedule classes during the project.

30.6 GIS INTERFACE HIGHLIGHTS SPACE SYSTEM DESIGN STRATEGY

The design strategy for the new space management system (SPACE) centered on a graphical navigation and reporting component built on GIS technology. A customized implementation of ArcView

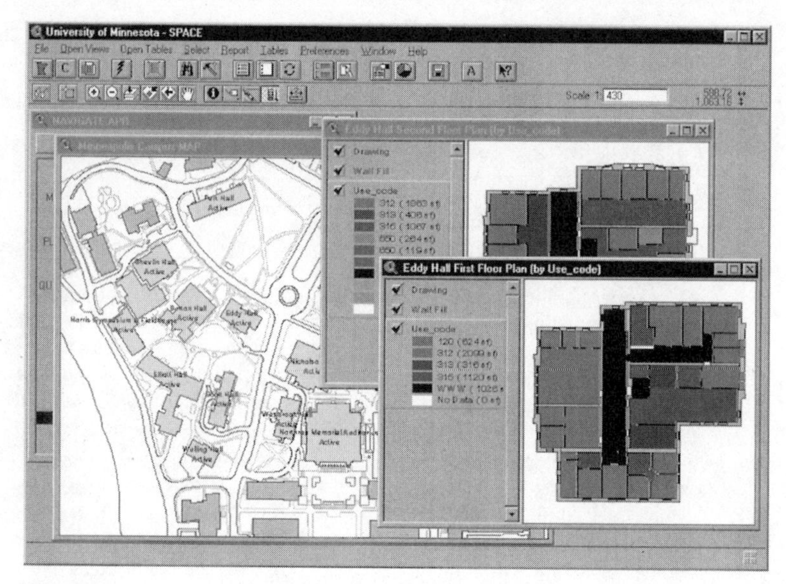

FIGURE 30.1 Campus Map Linked to Color-Coded Building and Floor Plans.

(ESRI, Redlands, CA) provides the graphical navigation and query interface that links 1,400 floor plans and 5 campus map drawings with information about 60,000+, the Oracle database.

Using maps and plan drawings as a navigation method and reporting context, SPACE provides an executive information system (EIS) that can be used to "browse" from a state map, to a campus map, to detailed floor plans. Drawings can be color-coded to reflect various values in the database and incorporated as part of reports, and can also be annotated with any values from the database. Similarly, attribute data can be queried to locate and display specific plans or maps. Campus maps provide information on buildings and floor plans at the scale of a single room (see Figure 30.1). Color-coded plans and charts reflect any value in the database and can be included as part of the reports.

The Navigator software includes a reporting component that produces drawings, charts, and reports at precise scales, including architectural engineering size plots. Selections of rooms from floor plans and tables can also be summarized in standard integrated Crystal Reports from the Navigator. Crystal Reports can also be used as a stand-alone for ad hoc queries and custom reports. A suite of more than twenty standard reports has been developed. Both simple and complex reports can be created with Crystal, incorporating graphics if desired. Since the system is fully Windows compliant, all drawings and text objects can be cut and pasted into a variety of other programs.

Reports are also available via the Web. Nine standard reports (filtered by campus, building, or collegiate unit) are available for display, printing, or download via the data warehouse Web site. Detail data can be downloaded into several formats (Excel, Lotus 123, comma delimited) using custom query tools developed by the university Office of Information Technology (OIT). Information is also accessible using standard query and connectivity tools.

30.7 INSTITUTIONAL AND USER-DEFINED DATA SUPPORTS ANALYSIS OF BUSINESS PROCESSES

Business requirements clearly indicated the need to support integration of data from other systems with information about facility utilization. The concepts of "institutional" vs. "user-defined" data were developed to allow SPACE data to be used in concert with other university systems (such as classroom scheduling, inventory services, human resources, etc.) or departmental databases.

Institutional data provides information about space occupancy, use, and other characteristics of value to a broad spectrum of the university. This data is the core of the SPACE system. A relational data model that met the business requirements of the university was developed along with programs to automatically convert data from the existing legacy system. The Oracle data model employed database triggers and procedures (e.g., research cannot be conducted in certain types of spaces) to ensure data quality compliant with university business rules. The data model incorporated NCES (National Council of Educational Statistics) coding schemes for classification of space.

The SPACE system allows users to connect institutional data with user-defined data. User-defined data is information that departments maintain relative to their unique business requirements, such as instructional activities, research projects, controlled substances, hazardous waste, AV equipment, and other data elements. User-defined data is of primary interest to individual departments for planning purposes and can be analyzed together with institutional data, floor plan drawings, and maps. Responsibility for collecting and updating institutional data resides with FM, while user-defined data maintenance is the sole responsibility of the individual department. In order to connect user-defined to institutional data for analysis, the data sources must have a common key, such as a spatial reference (e.g., building or unique room number) that matches those used in the SPACE system.

Institutional data in the SPACE system is used for strategic planning and programmatic projections, ad hoc reporting, and to support operational and administrative activities. The data is organized in a spatial hierarchy by:

1. Campus
2. Zone (for operations and maintenance purposes) and/or district (for master planning purposes)
3. Building
4. Floor
5. Room
6. Room detail

Most of the information about space is attached to the room and room detail records and is summarized at a building and site level, but it is possible to capture information about an element at any level of the hierarchy. The room record stores all of the information that is valid for the entire room (i.e., square footage, capacity, and ADA accessibility). The room detail record stores information on the activities of a space including occupants (departments), NCES/HEGIS (National Council of Educational Statistics, Higher Education General Information Survey), use and function codes, and grants. Information in the spatial hierarchy will identify the user(s) of a space by department, college, and provost (Chancellor)/vice-president (Vice-Chancellor). Organizational information, which is refreshed from the university's financial system, is linked to information in the spatial hierarchy at the room detail record.

The enterprise-wide spatial data model makes it possible for user-defined data to be connected at any level. Figure 30.2 illustrates a strategy for linking information from human resources and a departmental computer equipment inventory to institutional room level information as part of a strategic planning and move management project. Floor plan drawings provide a powerful visualization tool for navigating and reporting to a variety of data (see Figure 30.3).

30.8 DEPLOYMENT STRATEGY DESIGNED FOR ENTERPRISE-WIDE USE

Deployment was accomplished by defining two primary domains of use: an operational system component to support data maintenance, and a data warehouse implementation to support broad-based access from the university at large.

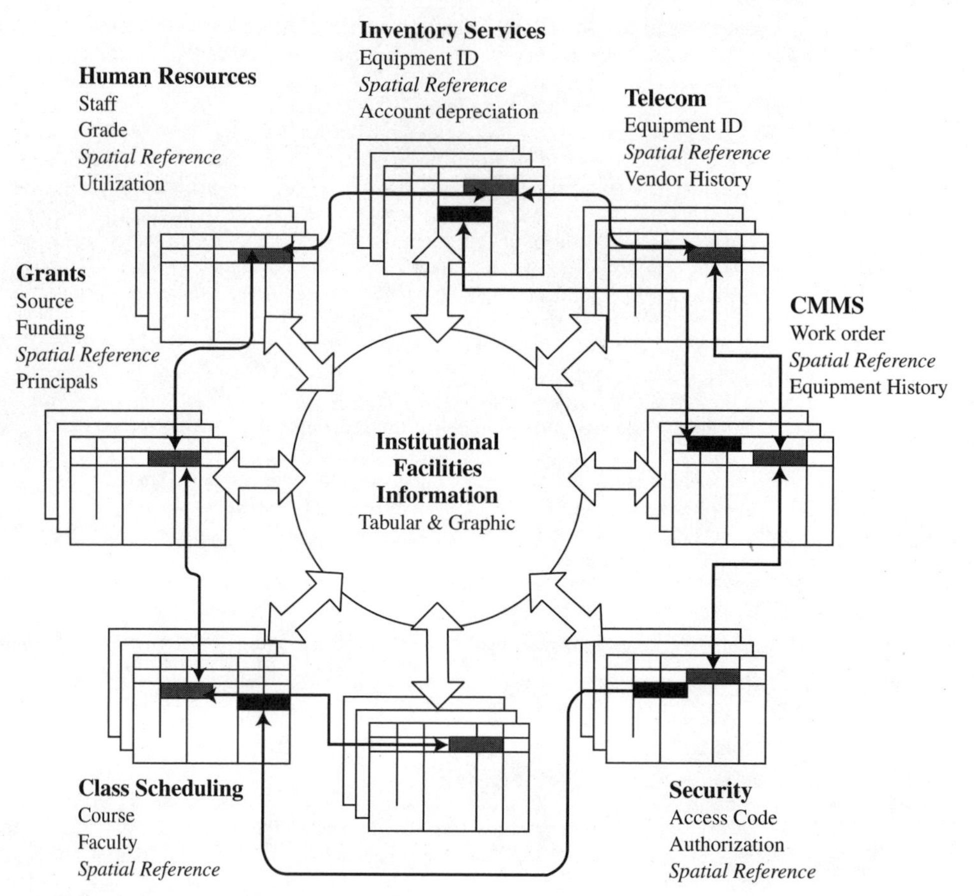

FIGURE 30.2 Enterprise-Wide Spatial Data Model.

The operational system consists of several components maintained on equipment in the university's FM department (see Figure 30.4). The operational system supports accurate and timely maintenance of data by FM staff and departmental space coordinators. Information in this environment is the responsibility of the applications owner (AO), an individual in the planning department who is responsible for data and process integrity and has overall responsibility for the SPACE system. The AO grants authorization to those maintaining the data in response to churn and programmatic changes. The AO also acts as a liaison to other university groups that want to use SPACE to facilitate their planning activities and other business processes. While the operational system also supports navigation, query, and reports, the database schema of this component is optimized for data maintenance. Information from the operational system is uploaded weekly to the university's data warehouse where it is accessible to the entire enterprise.

The data warehouse component is optimized for fast performance during graphical navigation, query, and reports, not data maintenance (see Figure 30.5). The data at the data warehouse represents a weekly "snapshot" of information from the operational system maintained by FM. Data is uploaded from the operational system to the data warehouse each Sunday evening. The data refresh may replace, remove, and add new data to the data warehouse. The upload also provides the opportunity to make minor modifications to the data for security and other business reasons. The data then remains static until the next refresh cycle. Data at the data warehouse is "read-only," no data maintenance can be performed against the data.

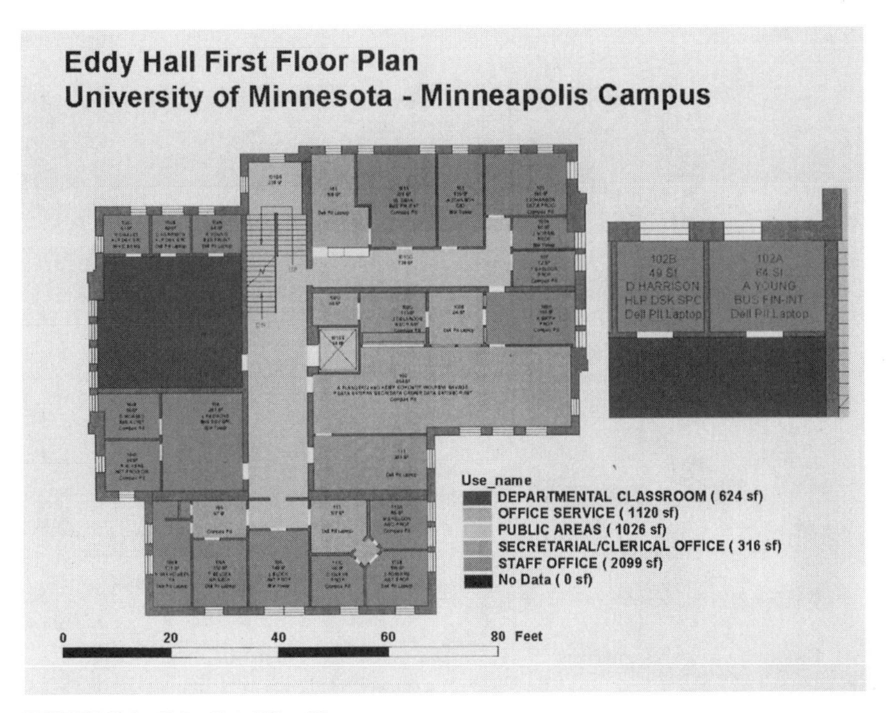

FIGURE 30.3 Color-Coded Floor Plan.

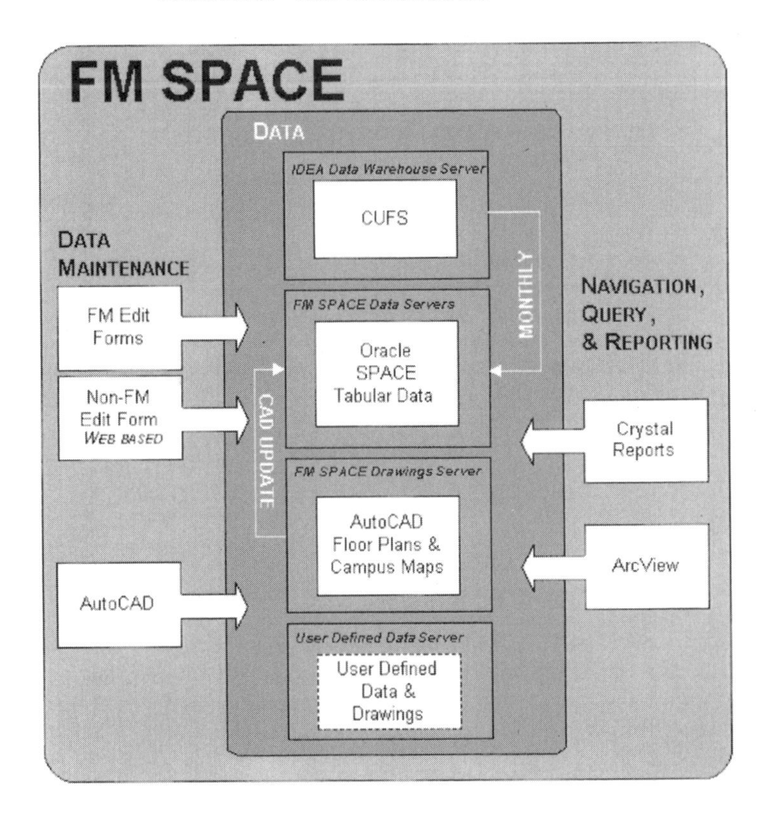

FIGURE 30.4 Operational System Component.

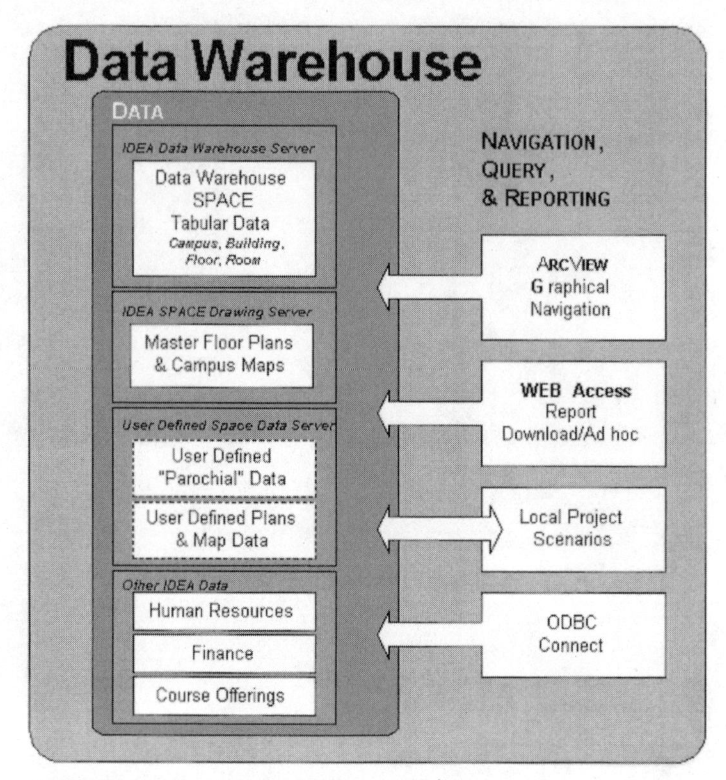

FIGURE 30.5 Data Warehouse Component.

The data warehouse is also the repository for other university information, including financial, human resources, and course offerings. This data, along with other user-defined data that may be stored at the data warehouse, or on user-defined servers, can be connected to the SPACE database for analysis and reporting. For example, a monthly download from the university's financial system (CUFS) provides a record of organizational reporting relationships. Procedures have been implemented to automatically incorporate changes to the university's organizational hierarchy, including a process to resolve any conflicts that may exist as a result of those changes. This approach significantly reduces the data maintenance that results from organizational churn.

The SPACE Navigator software that operates against the Oracle data in the operational system also works against the Sybase database, which is used for the university's data warehouse.

30.9 WHAT-IF SCENARIO PLANNING

The Navigator includes a feature that lets users conduct "what-if" scenarios for planning and design on their desktop without modifying the institutional data. If a user elects to save a project data set to their local hard drive, the data for all of the buildings that are active in the project (including all tabular data, along with the relevant campus maps and floor plans) are copied down to their system. Users can manipulate the data and the institutional data will not be modified. All of the navigation and reporting capabilities are available for use with the local data set. The SPACE Navigator software that operates against the Oracle data in the operational system and the Sybase database at the university's data warehouse also works with the what-if scenario planning model.

30.10 SPACE DATA MANAGEMENT

At any one time, dozens of expansion, renovation, or new construction projects are underway at the university. Departments continue to realign and establish new academic programs and implement alternate uses for existing facilities. Continuous churn in the facilities use required a new approach to eliminate the data maintenance problems and backlog that had plagued the legacy system.

Maintenance of tabular data is accomplished through a set of Oracle Developer 2000 Edit Forms designed for easy and intuitive maintenance of the data. Procedures were implemented as database rules to prevent erroneous data modification. Many of the database rules were embedded in the forms to ensure data quality and consistency. Validation lists enforce quality control and are available as drop-down selection lists for data entry.

Edit forms were Web-enabled to facilitate churn reporting by staff outside of FM. Departmental space coordinators can use a Web form to identify space occupancy and use changes. A limited number of users are authorized to have changes recorded directly in the database. For others, change requests are captured in a separate table for review by FM staff that validates the update requests. Processes define the responsibility of each group, including the method for obtaining access to the data, the security controls that have been implemented, and validation.

Drawing data (floor plans and campus maps) is maintained through the use of AutoCAD. Standards that fully define the information structure of floor plan and campus map data were developed to guide drawing creation and maintenance. Software tools were developed to check the compliance of drawing data with defined standards. Procedures identify the mechanisms for determining when and how drawings should be modified.

Changes to floor plan drawings automatically trigger updates to tabular data as a result of the CAD import procedures that have been developed. Additions and/or deletions of rooms, as well as changes in room sizes (square footage) are detected by the CAD import process, which automatically makes related changes in the Oracle database.

30.11 IMPLEMENTATION INCORPORATES PILOT PROJECT TO CONFIRM DESIGN STRATEGY

Implementation of the SPACE system took place during two phases: pilot project and production. The pilot project was implemented to test the design strategy, during which a detailed evaluation of the implementation tools was also performed. Standards to guide the creation of floor plan drawings and campus maps were developed and tested. The initial version of both the Oracle Forms data maintenance and ArcView navigation tools were implemented and usability testing was performed. Strategies for a number of process issues were developed and tested including the conversion of 60,000 existing database records.

The pilot project tested the full functionality of the system design, limiting the data environment to buildings on one major campus and room level detail for two major facilities. The pilot project concluded with formal presentations and proof of concept. Presentations and hands-on sessions were conducted with users, staff, and faculty to test usability. Detailed project plan and budget estimates were developed for the full implementation phase. Based on response to the design and evaluation of the project plans, a determination was made to proceed with full implementation.

Implementation of the production version of the SPACE system included significant data conversion and creation. Procedures were deployed to automatically convert space management data from the existing mainframe system into the new fully relational data model. More than 1,400 CAD floor plans and five campus maps were developed in accordance with a set of industry compliant standards. Source drawings were reviewed and used when available, occasionally requiring on-site field surveys. Quality control was performed on all data, including drawings that were developed. Data reconciliation was conducted to compare new and old inventory numbers.

Some customization and testing was performed on versions of the navigation, maintenance, and reporting software. All software and data components were integrated. Extensive functional and performance tests were performed on all components of the system throughout this phase.

A number of processes were reengineered to facilitate compatibility with other university business activities. Job descriptions were developed or modified and organizational and reporting relationships were redefined. Full documentation was developed for every aspect of the system. Training was conducted for both system users, as well as those responsible for maintenance of the system.

30.12 MOVING TO A FULL PRODUCTION ENVIRONMENT

The SPACE system was released into full production in two stages. First, the full Oracle database maintenance, query, and reporting capabilities were initiated. Drawing supported graphical navigation to approximately 60% of the inventory at that time. This stage also included access to the university's data warehouse for reports through the university's Intranet.

Full implementation was achieved with the final integration of all drawing data. Activities then turned to integration and deployment of the SPACE system across the university. Enterprise-wide deployment carries with it the responsibility to ensure that all users get the training and technical support they might need. Questions about SPACE data, password access, training, and general use of the system, or integration with other university applications are addressed to the AO. Questions about connectivity to system and infrastructure/telecommunications support issues are the responsibility of the university's Office of Information Technology (OIT), which is responsible for desktop support of other enterprise-wide applications.

The role of the AO emerged as a key factor in achieving broader use. In addition to overall responsibility for the use and quality of the system, the AO acts as liaison to help other university departments implement the SPACE system to address their own business problems.

30.13 ENTERPRISE CONCEPT EMBRACED

Within several months after reaching full production, a number of departments had been trained to use the system. FM's finance department is using SPACE to help analyze the building cost report (BCR) that summarizes utility and other operations costs for university facilities. Once connected to enterprise-wide information, BCR data can be queried to identify buildings, and then the departments and activities in those buildings that have high operating costs. The finance department is using the Navigator to create campus maps that are color coded to reflect the costs of different utilities (see Figure 30.6). They are now able to analyze the information graphically, with more accurate enterprise-wide data. SPACE highlights the managerial financial data that lets them move quickly into action. In the final analysis, this tool is enhancing their operating efficiency.

The Department of Residential Housing merged their database of room assignment characteristics (gender preference, smoking preference, visitation, etc.) with institutional data. The combined data was used to produce drawings used for student registration (see Figure 30.7). Plans call for use of the data to manage custodial contracts, with additional discussion underway to explore the use of SPACE data for Web-based student registration.

In another example of enterprise-wide use, the group responsible for facilities operations and maintenance is integrating the new computerized maintenance management system (CMMS) with the SPACE system. By sharing the Oracle tables that describe the location hierarchy, maintenance planners have access to institutional space data to schedule preventative and unplanned maintenance. This allows a planner to overlay drawings, such as steam tunnel locations, graphically identifying the impact of infrastructure projects on building occupants. For example, the system can locate and highlight building support areas that might be used to stage a steam tunnel repair project (see Figure 30.8).

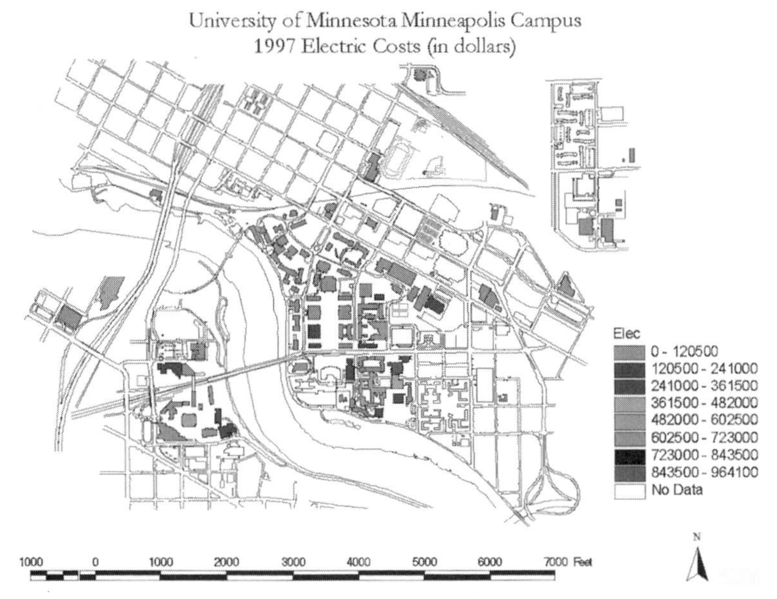

FIGURE 30.6 Color-Coded Campus Map Displaying 1997 Electric Costs.

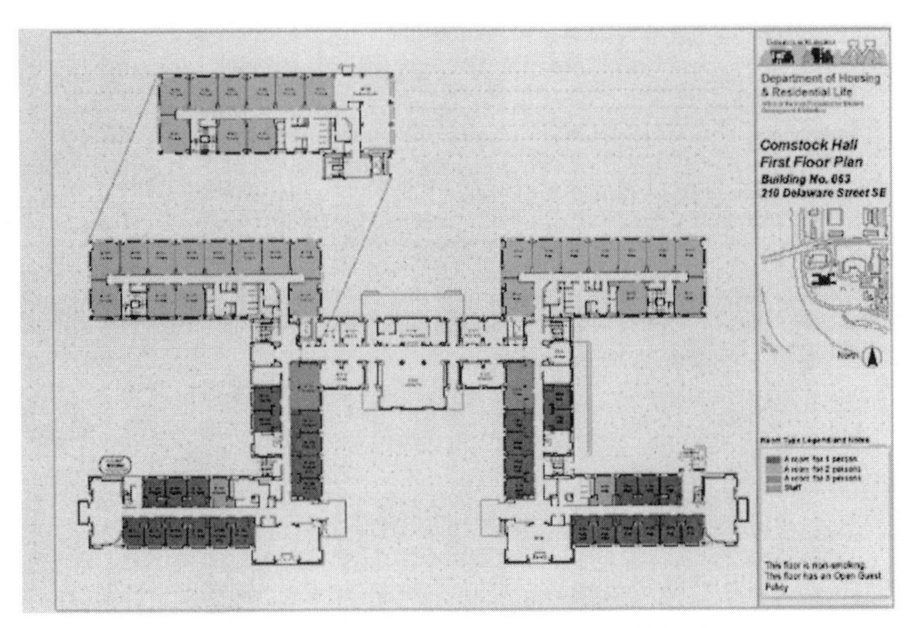

FIGURE 30.7 Example of a Floor Plan Used by the Department of Residential Housing.

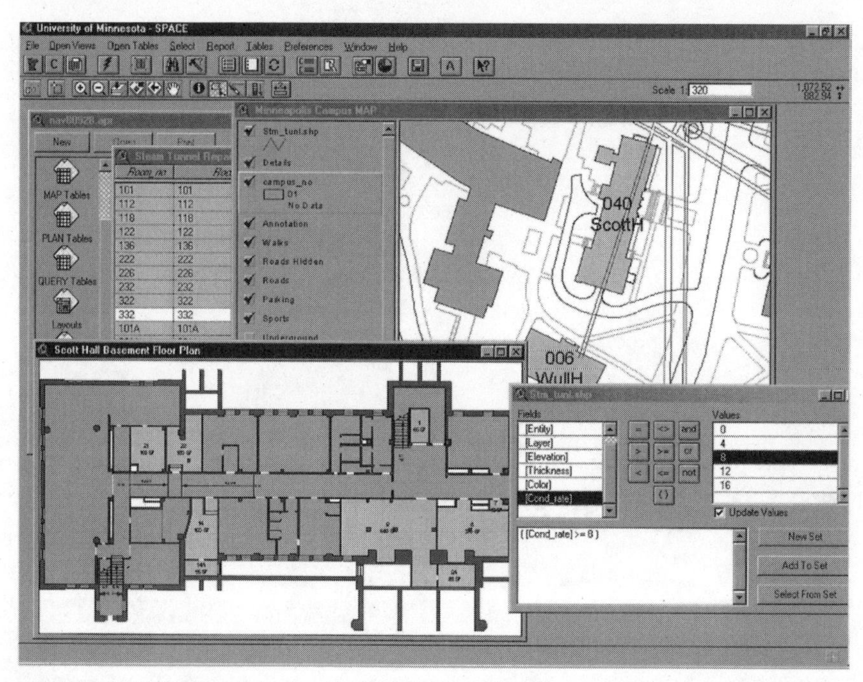

FIGURE 30.8 New CMMS Integrated with the SPACE System.

30.14 SUMMARY

Today's university enterprise requires powerful management solutions to deliver programs that respond to dynamic academic and funding priorities. As stewards of the university's facilities, providing accurate information about those facilities makes FM an active partner in helping the university achieve its academic goals. New information technologies, like the SPACE system, provide a method for assimilating diverse facilities, programmatic, and human resources data into a problem-solving collaboration environment for academic and facility planners. The result is an enterprise-wide solution that facilitates the flow of information between business functions and across departmental boundaries. SPACE is a powerful management and analysis system that is having a much broader impact on the efficiency of facilities utilization and the success of the programs those facilities house.

CHAPTER 31
TEACHING TECHNOLOGY AT THE UNIVERSITY OF NEW SOUTH WALES: A CASE STUDY

Alan White
University of New South Wales
Australian Defence Force Academy
E-mail: a.white@adfa.edu.au

31.1 EDUCATIONAL ENVIRONMENT

The recent composition of facility management (FM) as a body of knowledge finds different expressions across academic environments. With limited and often declining resources, some tertiary educators have resorted to existing courses as the foundation for new programs. FM that is taught within a mechanical engineering environment might display a bias to building systems and environment control. Faculties of design may be inclined to focus on space programming and the human work environment. Most programs display common themes. The differences appear in the areas of emphasis.

Educational paradigms and teaching methods vary across educational levels and societies. Some college level courses address the needs of real estate property management, while tertiary courses might engage the processes of strategic facility planning within a corporate planning context. In this chapter, we will examine a range of teaching resources. In particular, commercial software will be used to demonstrate some fundamental facility planning and FM processes. These products are generally available to educators at a reduced or notional cost, contain multimedia tutorials and additional text-based teaching resources. The Web provides access to software upgrades, multimedia, and supplementary documentation. CD-ROM and DVD provide low-cost media for data distribution, particularly for nonattending students. Current Internet download speeds might make this form of data access inferior to the distribution of disk-based data; however, the near future may significantly change the way students access on-line courseware.

In this chapter, the processes and associated products illustrated are neither exclusive nor conclusive. The primary objective is to give a student experience with a technology through engagement with a software product. The educational process then draws on this personal experience to develop a theoretical context and an insight into commercial issues. The displayed software is a limited sample of the technologies used within the Master's degree course. Not all products are universally available and neither the author nor the University of New South Wales endorses any particular product for commercial application. All products were chosen for their value within a teaching environment. The chapter is loosely structured in a potential chronological sequence of facility-related studies and activities. The initial discussions will examine the design of organizations and the final sections will overview the on-going costs of retaining a building and appraising its serviceability. Each section offers an overview of a facility planning and management issue, software used as a resource to support the educational process, and the educational outcomes sought. Teaching technology is limited to a canvas of decision support methodologies and associated software systems.

31.2 FACILITY PLANNING WITHIN A CORPORATE PLANNING CONTEXT

The evolution of professional facility planning has been a consequence of the corporate planning process. The effective accommodation of the organization's resources is a critical component of strategic business planning, particularly as accommodation related costs are a significant component of the organization's budget. While the direct facility cost is significant, its actual cost may be small compared to the cost of ineffective salaries due to inappropriate accommodation. Historically, major accommodation decisions tend to be infrequent; however, the need for organizations to rapidly adapt to changing market conditions demands continuous assessment of the organization's functional design and its accommodation needs. Strategic facility planning views facilities in terms of (accommodation) services required by the organization. The effectiveness of these services directly contributes to the organization's business outcomes.

31.2.1 Organization Functional Congruence

The history of facility planning has been closely associated with the architectural profession. Within this professional context, there may be a perceived reluctance to engage in the functional design of the client's organization. The facility designer's brief is usually focused on accommodating the organization; however, moving an inefficient organization merely relocates the problem. Every business activity requires individuals and groups to participate in a matrix of functions that provide services. These services are both internal and external to the organization. The matrix of functions is a convenient mechanism to model the static interplay of services. The intersecting cell within the matrix for any two functions can be used to describe the services required and demanded between those two functions. Indeed, at the extreme, there are four service vectors associated with any two functions:

1. Services required by function A from function B
2. Services provided by function A to function B
3. Services required by function B from function A
4. Services provided by function B to function A

An analysis of the congruence of service vectors provides an insight into the mismatch between service provision and service requirement. Modifying the service vectors to achieve service congruence can often lead to a new organization design as redundant services are recognized and new services are defined. This new organization design may generate revised accommodation needs. While this process is not continuous, it may be conducted several times each year. The organization design consequences may not be implemented at the same frequency, but the analysis does provide for an on-going perspective on its effectiveness.

31.2.2 Technology

An obvious choice for modeling the matrix of services from interacting functions is to use a spreadsheet; however, there may be practical limitations on the quantity of textual data. A database can be

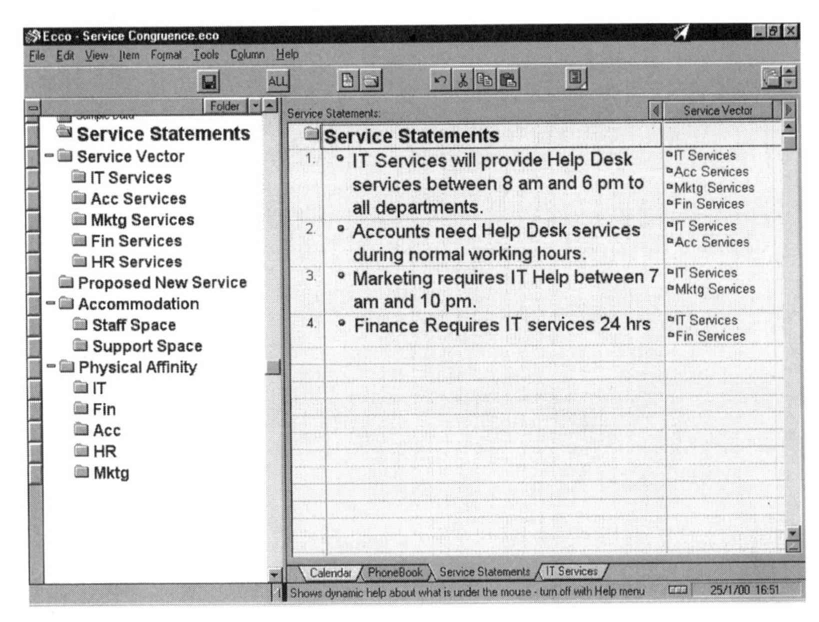

FIGURE 31.1 Organization Functions Presented as a Matrix of Service Vectors. (***Source:*** ECCO *reproduced with the permission of NetManage Inc.*)

designed to fulfill the data capture and reporting function; however, the data management environment should provide a hierarchical structure to the data and offer a roll-up function to higher hierarchical levels. This enables one organization function to consolidate all services demanded by the rest of the organization. Besides using general information management software, some cost estimating software, critical path software, and facility planning software offer a hierarchical structure to textual data. Figure 31.1 illustrates an organization structure with a matrix of mutual service provisions and requirements. Departments, groups, or individuals can assess their service demands and compare these to the current supply of services. The software only provides a capability to record, index, and filter information. Decisions will be required as to which service requirements will be provided from internal and external resources in order to deliver a congruent service delivery regime.

31.2.3 Educational Outcomes Sought

1. A recognition that business outcomes are predicated on a symbiotic organization design, which, in turn, leads to the definition of space need.
2. A recognition that facility planning is a subset of corporate planning.

31.3 DYNAMIC SIMULATION

Dynamic simulation has a strong history in industrial engineering. The availability of PC-based graphical modeling systems has added a powerful tool to the facility planner's portfolio.

Once the organization has developed a congruent service delivery regime capable of underwriting the business objectives, the focus of attention transfers to space type and quantity. Dynamic simulation can provide two powerful insights into the facility planning process. The first is a capacity to optimize the organization's required resources and the second is to quantify the space needed as a consequence of complex interactions over time. Resource optimization computes the resources required within each function in order to deliver the required services. It identifies the under- and over-utilized resources and proposes alternate strategies. Space need computation often models the organization's activities over a typical cycle. An example is an airport with a typical cycle of 7 days and 24 h per day. While there are numerous dynamic simulation software products, it is vital that that the selected product offers an optimization function.

Dynamic simulation software comes with a variety of interfaces; these include 2D flow process visualization using a logic diagram (see Figure 31.2) and 3D flow process using specialized arc and event symbols. In addition, some products provide 2D-flow process with an underlying graphics image, such as a floor plan (see Figure 31.3). There are numerous variants of 2D and 3D graphics. The high-end products use full 3D virtual reality, moving avatars plus user interaction. The mathematical processes may be similar, but the products are differentiated by their graphical presentation and to a lesser degree, interaction.

31.3.1 Technology

The essence of a dynamic simulation is the design of a flow process. This may be in the form of a simple flow chart (Figure 31.2) or a more sophisticated modeling interface; however, in addition to a graphical palette, some products provide a simple scripting language that automatically generates the flowchart. Scripting has the potential to significantly reduce the time required to define the graphical model.

Modeling an existing facility, such as a hospital, can be enhanced with a background floor plan or 3D visualization. The ability to incorporate background images is not available in all products. 3D visualization is available at varying levels of sophistication. Low-level visualization provides a 2D plane with 3D images. Medium-level visualization provides a static isometric image with activity motion in three planes. High-level visualization provides activity motion within a 3D environment together with user interaction (see later at Figure 31.12).

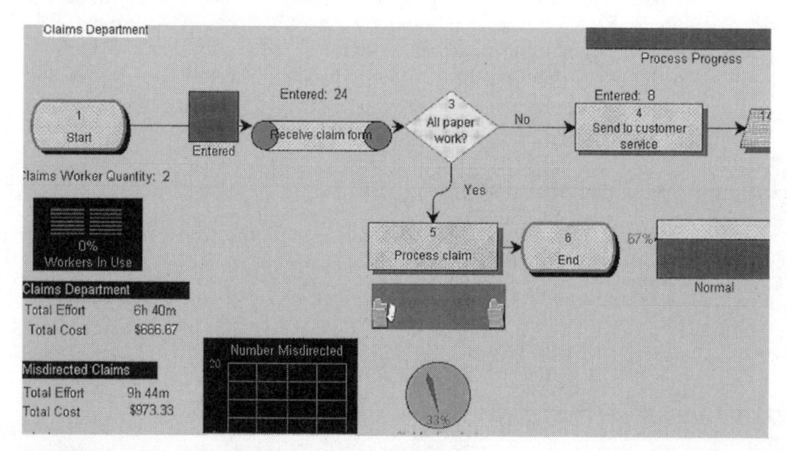

FIGURE 31.2 A Dynamic Simulation Model Flow Chart. (***Source:*** Scitor Process *reproduced with the Permission of Scitor Corporation.*)

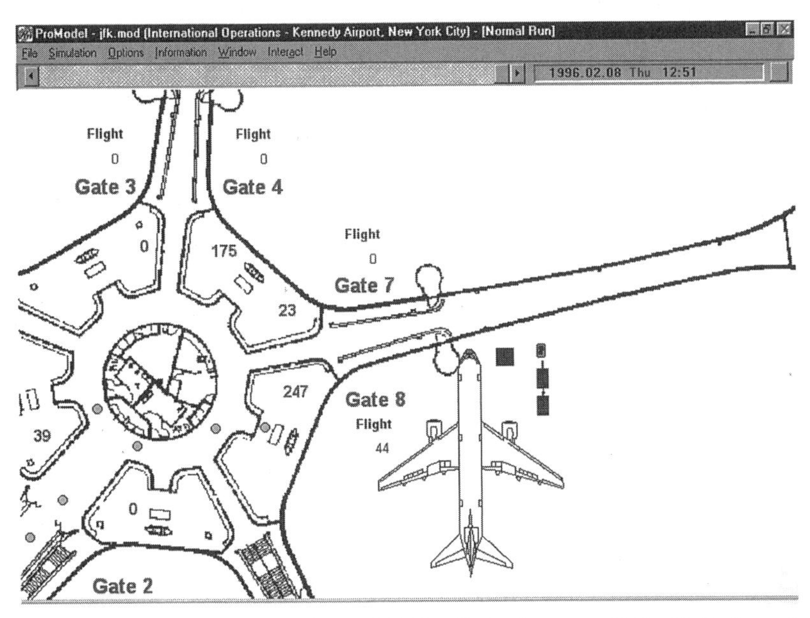

FIGURE 31.3 A Dynamic Simulation Model Using Moving Icons on a 2D Graphic Background. (***Source:*** ProModel *reproduced with the permission of PROMODEL Corporation.*)

31.3.2 Educational Outcomes Sought

Following is a list of the educational outcomes sought through dynamic simulation:

1. To recognize that the computation of space need is an outcome of two processes. The first process models the organization's functional congruence. The second process seeks to optimize sequences and resources. This latter process can be undertaken within dynamic simulation.
2. To recognize that the choice of modeling products is a function of time and presentation needs.
3. To recognize that functional congruence models provide a static model of service supply and demand. The quantification of resources and, hence, space is an outcome of a dynamic simulation modeling the real interactions within the facility.
4. To be aware that all models are constrained by the defined resources. Alternate strategies (not sequences) and resources come from user interaction.

31.4 REQUIREMENTS ENGINEERING

Requirements engineering is a well-established discipline in defense materiel.[1] Its translation to facility procurement promises to add rigor to the facility brief (program) development process. Requirements engineering has a three-phase sequence: user requirements, system requirements (which describe the systems needed to support the user requirements), and system architecture (which describes the physical systems required to deliver all the system requirements). The process should provide a bidirectional audit; that is, user requirements can be traced through to the system architecture selected to satisfy the user needs and vice versa. In addition, multiple system requirements can be grouped to provide a common system architecture. This may be a common solution or a shared solution. All data can originate from disparate sources and every issue will have a unique reference code similar to a work breakdown system code. Change (scope) management can occur at any one of the three phases and the impacts of a proposed change are translated throughout the system.

The conversion of user requirements into a building solution has traditionally been the domain of the architect; however, requirements engineering introduces systems engineering into the facility procurement process. It provides a common data infrastructure for the entire design team and maintains this role during the construction period and through-life management. For example, the interaction between the building structure and the HVAC evolves as a unified specification system within requirements engineering. Equally, the interaction between the hydraulic and electrical plus data cabling systems (of particular importance to communications centers) is analyzed within the composite specification provided by the requirements engineering process. Requirements engineering integrates all specifications (architectural, structural, mechanical, electrical, HVAC, and more) and provides a single interface to all design data.

31.4.1 Technology

A significant number of software systems appear to provide some or most of the functionality for requirements engineering; that is, the provision of bidirectional hyperlinked text with arithmetic functions while being underwritten with a unique code for each requirement. In addition, there are links to all other documentation (objects) and URLs. However, purpose-built requirements engineering software systems are designed for a multi-user environment with associated security, maintain appropriate audit trails, provide a summary of hyperlinked data at the launch point, and provide functionality for the on-going design management of a facility. In addition, there is a formal relationship between the user requirements, systems requirements, and the system architecture. These are presented as matrices for cross-referencing (see Figure 31.4). The mission critical nature of requirements engineering suggests that only high quality fully supported software should be used for this purpose.

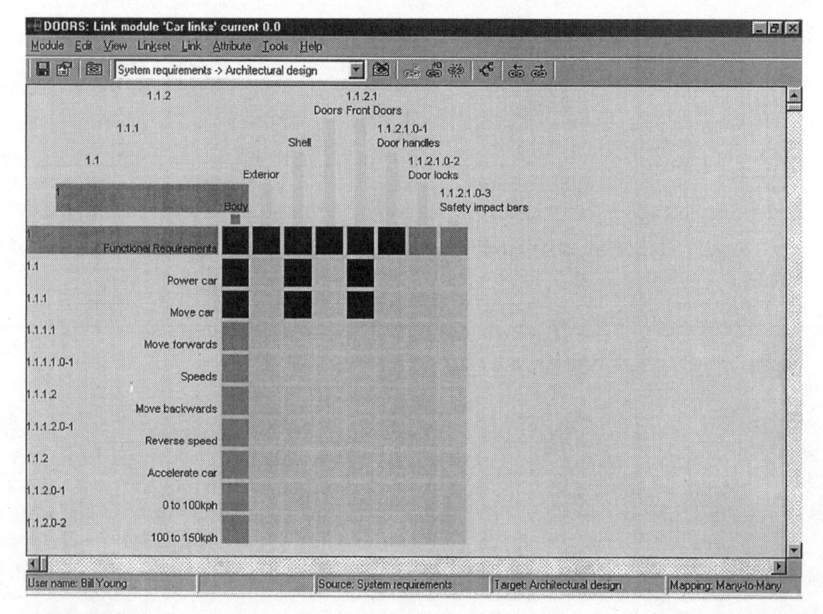

FIGURE 31.4 A Requirements Engineering Matrix Display Relating System Requirements to Architectural Design. (***Source:*** DOORS *reproduced with the permission of QSS, Inc.*)

31.4.2 Educational Outcomes Sought

Following is a list of the educational outcomes sought through requirements engineering:

1. To recognize the relevance of requirements engineering in providing an information core for the design, construction, and on-going management of a facility.
2. To be aware of the political dimensions associated with a concentration of information.
3. To recognize the importance of tracing all architectural components and associated systems back to the original user requirements.
4. To introduce systems engineering into the facility development process.

31.5 SPACE ECONOMY

This discussion on space economy is restricted to the equation between space need and space supply. Typically, this includes space standards and circulation space definitions.[2] However, more recent issues of home-based work and joint venture projects have changed the demand side of the equation and hoteling (amongst others) of office space has changed to supply side. The components of the equation are likely to change in the future; however, the primary data management tool remains the hierarchical database. The notion of a hierarchical database also includes relational and flat-file databases that display data in a hierarchical structure. In fact, two hierarchies work in parallel. One hierarchy models the demand for space and another hierarchy models the supply of space (see Figure 31.5). The demand for space includes primary, secondary, and tertiary (if required) circulation. In addition, space for short-, medium-, and long-term storage is defined.

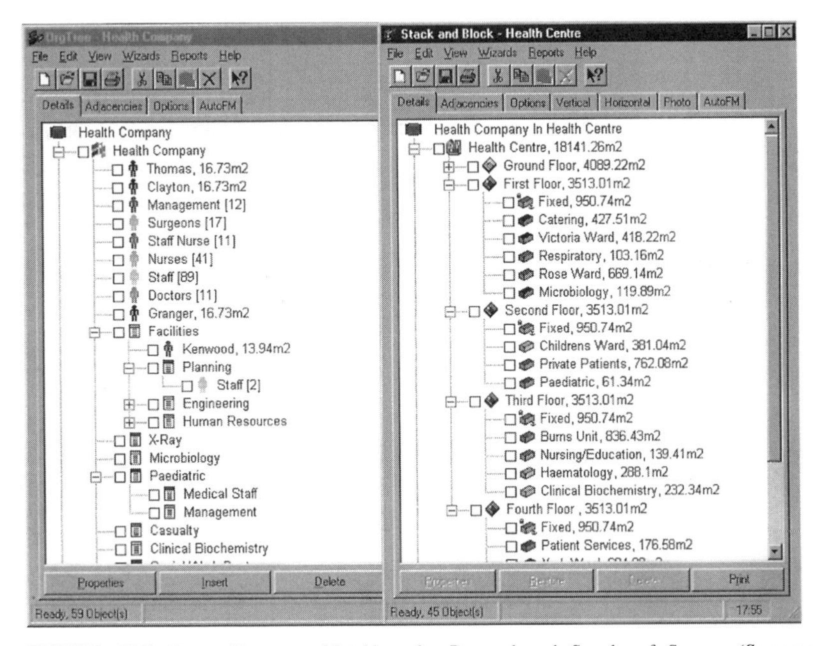

FIGURE 31.5 Space Economy—Matching the Demand and Supply of Space. (***Source:*** OrgTree, Stack and Block *reproduced with the permission of Decision Graphics UK Ltd.*)

Space economy is a gaming process that examines alternate accommodation strategies. Immediate- and short-term accommodation needs tend to be modeled with a static accommodation supply portfolio. Long-term modeling usually includes variability in the accommodation supply portfolio; however, both can be modeled with current software. Physical affinity (see Figure 31.6) can be used to guide placements, but reliance on an automated procedure is problematic. However, physical affinity data is valuable for monitoring the progressive inefficiency of the current accommodation solution. The objective of space economy is to design a holistic view of both the changing organization with its changing space needs plus the changing availability of space. Because it is a gaming process, the data should be quarantined from day-to-day space management data.

The politics of space economy is more evident in organizations yet to embrace accrual accounting where cost or profit centers are responsible for their own accommodation; however, organization data is often located in departmental units (such as human resources) not responsible for space planning. Additionally, some departmental units may prefer the organization use area prediction data based on "hoped for" or current staff levels rather than the space required to satisfy the functionally congruent organization.

31.5.1 Technology

The foundation technology for all space economy is a database displaying data in hierarchical structures. Spreadsheets are in common use, but their limitations become self-evident once the attributes of hierarchical structures are observed. Data roll-up and predefined space standards are key functions. While purpose designed software is preferred, limited functionality can be gained from using the hierarchical structures in some critical-path, cost budgeting, and information management software packages.

The specialized gaming nature of space economy has found few complete implementations in current CAD-based facility information management systems. Most have some, but not all of the

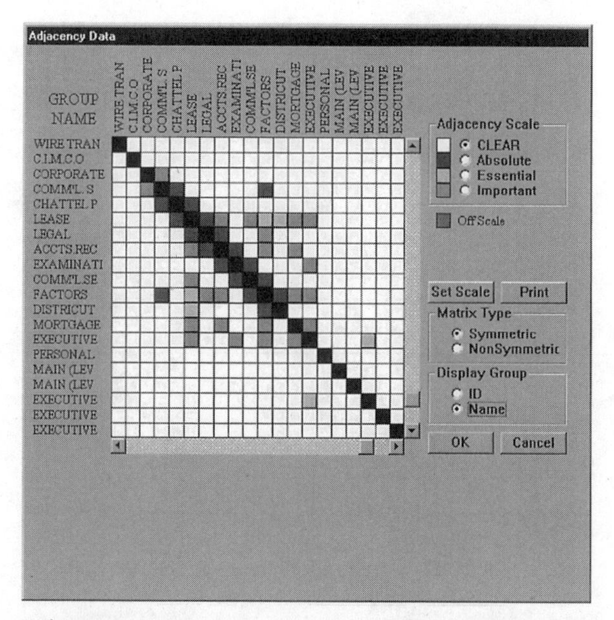

FIGURE 31.6 A Matrix Display of Physical Affinities between Organizational Groups. (***Source:*** WinSABA *reproduced with the permission of SABA Solutions.*)

required functionality. A number of products offer demand analysis. A close to ideal environment comprises two hierarchies (in outline format and in parallel formation), respectively, either representing the demand for or the supply of space (previously displayed in Figure 31.5). The immediate needs of micro-space management while concurrently planning for the near and long-term implies that multiple models will draw from the same base data set. A key software function is the ability to describe the organization in an "outline" hierarchy and to enable structural changes to the organization design by using "drag-and-drop" technology; that is, individuals and groups can be moved around the hierarchy and their impact on space demand is automatically computed.

As individuals and groups are transferred from the organization hierarchy to the accommodation hierarchy, the arithmetic impact on both demand and supply is computed. The software must be able to offer the options of Groups into Buildings, Groups into Floors, and Individuals into Rooms. Additionally, from time-to-time, individuals will be locked into groups or separated from groups and space will be occupied or unoccupied. There is an element of hotel management in this process, as in practice, the organization continues to change during the study period. Sophisticated systems may use allocation optimization algorithms for campus timetable management.

31.5.2 Educational Outcomes Sought

Following is a list of educational outcomes sought through space economy:

1. To recognize that space economy is a gaming activity that uses quarantined data to model the changing supply and demand for space.
2. To recognize that space economy analysis occurs at near-, medium-, and long-term intervals.
3. To be aware that the politics of organization design may make data acquisition problematic.

31.6 SPACE LOCATION, ALLOCATION, AND OPTIMIZATION

There is a clear overlap between this topic and space economy; however, this discussion is restricted to methods for deriving an optimal solution for location and layout studies and the constraints imposed by a fixed geometry, such as a building with multiple floors. The mathematical basis for this overview is described in the affinity or association matrix that identifies the desired physical proximity between any two functions relevant to the organization (previously displayed in Figure 31.6). An organization with η functions has a set of $\eta!$ relationships. A small organization with 20 interrelated functions would generate 20! relationships. The mathematical search for the optimal solution is described as the quadratic assignment problem (QAP).[3] While a method for finding exact solutions to practical QAP formulations is not available, decomposition techniques may be used to solve large scale quasi-QAP formulations. Even the small organization with 20! relationships is beyond the computing power of current desktop systems. The available optimization methods provide very good solutions, particularly, as in practice, the process requires varying degrees of user interaction.

The nature of physical structures has an important impact on the mathematical solution. This can be explained as follows. Consider an organization with only three functions and these functions require equal volumes of space. Assume that these functions have the same mutual physical affinity and that a sphere can represent each function. As all functions are of equal area, the representing spheres are of equal diameter. The optimal geometric layout for this case is triangular; that is, the spheres are in a tangential triangular arrangement. The lines joining all centroids are of equal length because the functions have equal mutual physical affinity. The problem can be extended to four functions in the same context. The spheres in a tangential square formation would not represent the optimal solution, as the diagonal centroid links are longer than the rectilinear links. The optimal solution would be a trapezoidal formation. This is a three-dimensional geometry compared to the three spheres which produced a two-dimensional geometry. The three-sphere solution (triangle) could be housed on one floor of a building, but the four-sphere solution (trapezoid) would require a mezzanine floor. Several issues arise from this exposure. The first is that practical reality would ensure that the four functions (spheres) would be on the same floor even though this solution is not mathematically optimal. The second is that a problem set in which any one function is related to more than two other functions will deliver an unworkable solution in terms of a segmented structure, such as an office building. Software systems seeking to resolve this problem set will introduce user-defined preferences, soft and hard constraints, and perhaps some heuristic algorithms. The results will be good but not perfect solutions. This background understanding of the underlying mathematical processes is important when collecting and moderating actual organization data. Most computer systems enable user interaction, but this only adds value if the software provides immediate feedback on the contributing value of the user inputs.

Problems such as school locations, service delivery points, transport networks, and nodes are a few of the problems resolved through the use of optimal location analysis. Optimal space allocation is a subset of the general location optimization problem but where the hard constraints of building locations and floors must be accounted. While there are many small solving routines, there are only a few software products that provide a CAD interface plus user interaction (see later in Figure 31.11).

31.6.1 Technology

The search for optimal locations can be easily undertaken with a genetic algorithm add-in to an Excel spreadsheet. Genetic algorithms may mimic Darwinian or Lamarckian evolution.[4] Solutions are randomly proposed. Better solutions are given preference and good solutions are "married" to produce offspring with superior "genes." "Mutation" and "cross-over" can be introduced to simulate Darwinian evolution. As with most approximation methods, longer processing time produces better solutions; however, as time progresses, it takes longer and longer to find a better solution (see Figure 31.7).

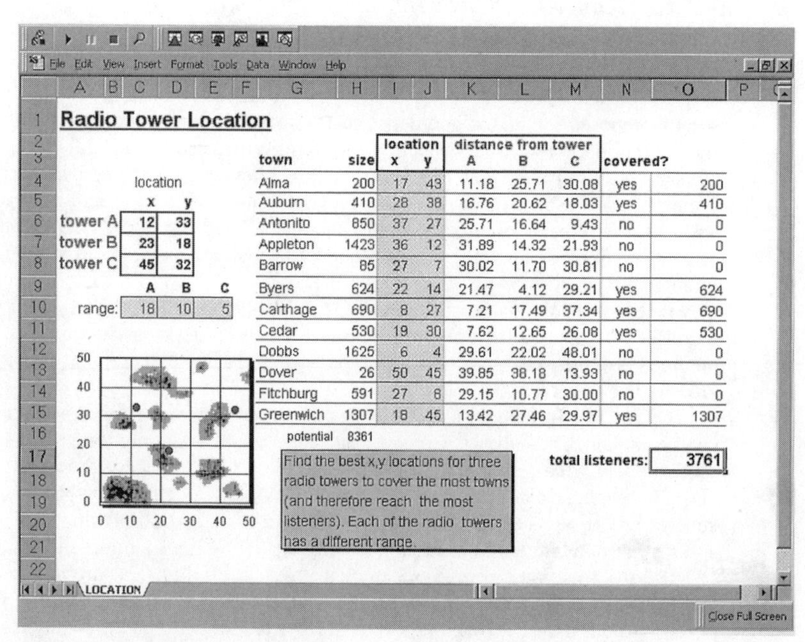

town	size	location x	location y	distance from tower A	B	C	covered?	
Alma	200	17	43	11.18	25.71	30.08	yes	200
Auburn	410	28	38	16.76	20.62	18.03	yes	410
Antonito	850	37	27	25.71	16.64	9.43	no	0
Appleton	1423	36	12	31.89	14.32	21.93	no	0
Barrow	85	27	7	30.02	11.70	30.81	no	0
Byers	624	22	14	21.47	4.12	29.21	yes	624
Carthage	690	8	27	7.21	17.49	37.34	yes	690
Cedar	530	19	30	7.62	12.65	26.08	yes	530
Dobbs	1625	6	4	29.61	22.02	48.01	no	0
Dover	26	50	45	39.85	38.18	13.93	no	0
Fitchburg	591	27	8	29.15	10.77	30.00	no	0
Greenwich	1307	18	45	13.42	27.46	29.97	yes	1307

	location x	location y
tower A	12	33
tower B	23	18
tower C	45	32

	A	B	C
range:	18	10	5

FIGURE 31.7 An Excel™ Spreadsheet Genetic Algorithm Add-In Used for Optimal Location Studies. (*Source:* Evolver *reproduced with the permission of Palisade Corporation.*)

```
CSIRO BCE TOPMET LEVEL 1, TOTCOST= .26509E+07 ESTAB COST= .000      SOLN 14256
I---I---I---I---I---I---I---I---I---I---I---I---I---I---I---I---I---I
I* 1.* 2.* 3.* 4.  5.  6.  7.  8.  9. 10. 11. 12. 13. 14. 15. 16. 17.
IPAR.PAR.SAM.ACC.LEG.CON.EDP.ELC.ELC.CVM.INS.FIN.ASS.   .AMY.CPC.CPP.
I---I---I---I---I---I---I---I---I---I---I---I---I---I---I---I---I---I
I*18.*19.*20. 21. 22. 23. 24. 25. 26. 27. 28. 29. 30. 31. 32. 33. 34.
IPAR.PAR.PAR.LEG.LEG.CON.ELC.ELC.ELC.ELC.   .FIN.CVC.   .AMY.CPC.   .
I---I---I---I---I---I---I---I---I---I---I---I---I---I---I---I---I---I
I*35.*36.*37. 38. 39. 40. 41. 42. 43. 44. 45. 46. 47. 48. 49. 50. 51.
IPAR.PAR.PAR.LEG.LEG.ELM.ELC.ELC.ELC.ELC.COR.FIN.CVC.   .AMY.CPC.GEP.
I---I---I---I---I---I---I---I---I---I---I---I---I---I---I---I---I---I
I*52.*53.*54. 55. 56. 57. 58. 59. 60. 61. 62. 63. 64. 65. 66. 67. 68.
IPAR.PAR.PAR.LEG.LEG.   .AUD.ELC.ELC.PER.COM.CVP.CVP.CVP.AMY.CPM.GEC.
I---I---I---I---I---I---I---I---I---I---I---I---I---I---I---I---I---I
HIGHRISE BUILDING (ON SIDE) TENANCY LAYOUT DEMONSTRATION
  **** BEST SOLUTION ****          INTERACTION   COST = $   2650943.0
                                   ESTABLISHMENT COST = $          .0
                                           TOTAL COST = $   2650943.0
```

FIGURE 31.8 A DOS-Based Implementation of Simulated Annealing. (***Source:*** TOPMET *reproduced with the permission of CSIRO, Australia.*)

Simulated annealing is an alternate to genetic algorithms.[5] Simulated annealing is characterized by proposing a random solution (such as the location of personnel in offices) and then undertaking a series of pair swaps. The objective function seeks to satisfy all the demands of the functional affinity matrix (previously displayed in Figure 31.6). If a pair swap results in a solution closer to the objective function, then it is retained. The slower (hence "annealing") and longer the process, the better the chance of achieving a superior solution (see Figure 31.8). There are other methods for solving the QAP; however, most products remain silent on their underlying methodology.

With so much user interaction required in practice, the focus on mathematical methods may be unproductive. Understanding the underlying mathematical methodology significantly impacts on the practical process of data collection and results analysis. Figures 31.9, 31.10, and 31.11 illustrate technology that focuses the layout optimization problem on buildings with floors and corridors.

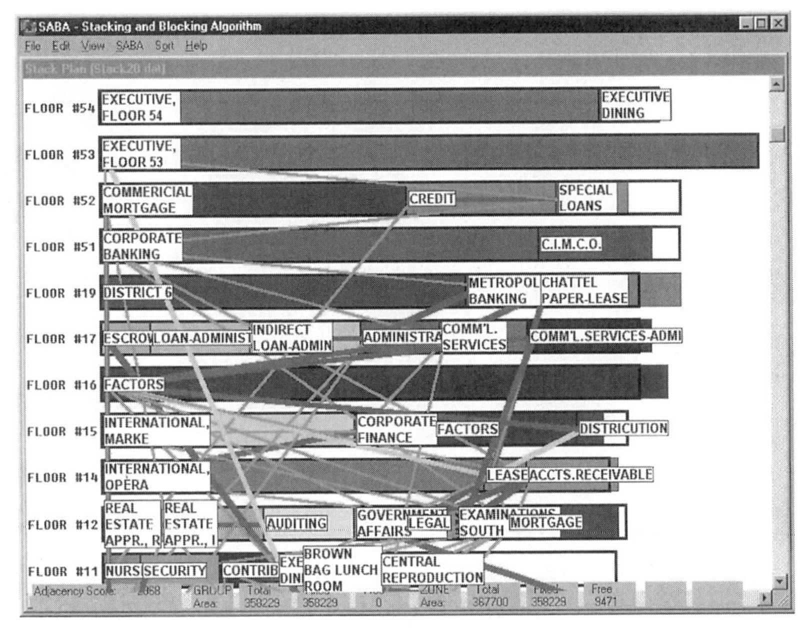

FIGURE 31.9 A Building Stack Model Displaying Group Locations and Their Physical Affinity Network. (***Source:*** WinSABA *reproduced with the permission of SABA Solutions.*)

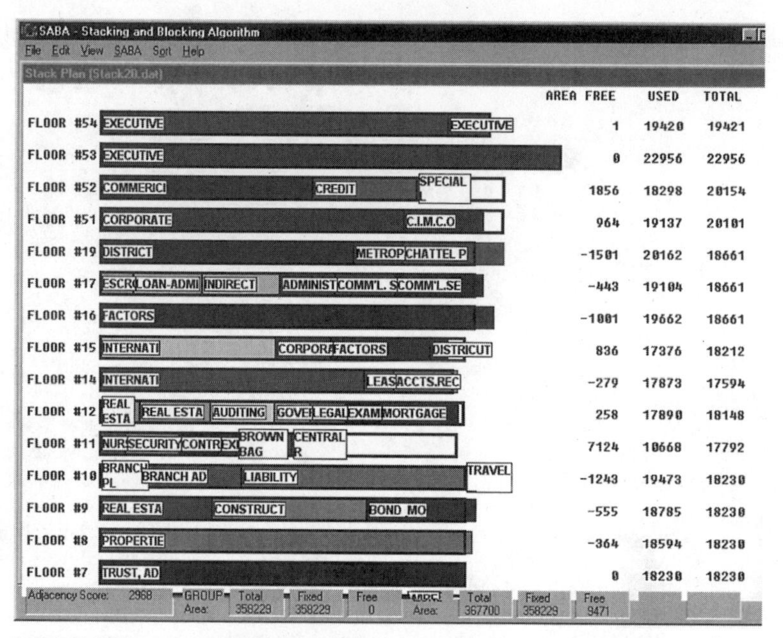

FIGURE 31.10 A Building Stack Model Displaying Group Locations and Space Accounting. (***Source:*** WinSABA *reproduced with the permission of SABA Solutions.*)

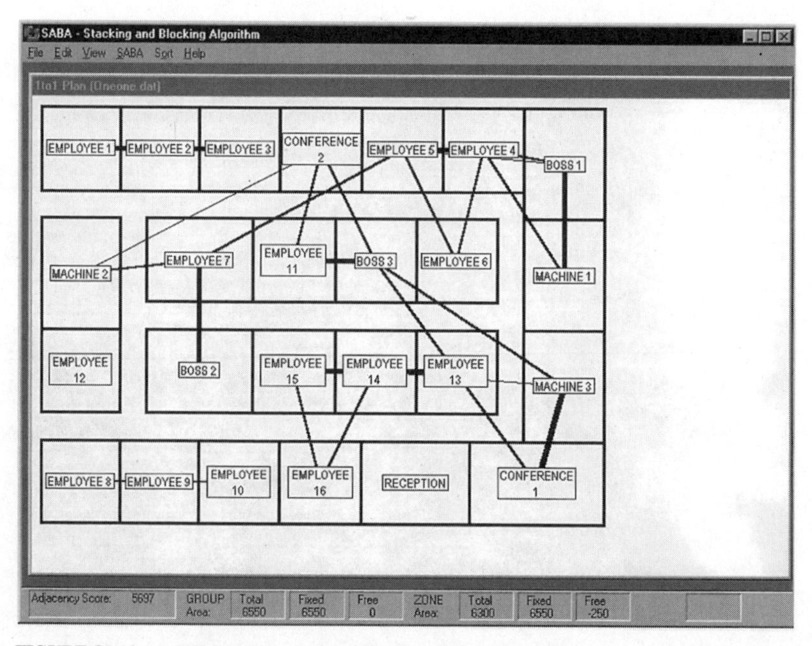

FIGURE 31.11 A Building Floor Block Model Displaying Individual Locations and Their Physical Affinity Network. (***Source:*** WinSABA *reproduced with the permission of SABA Solutions.*)

31.6.2 Educational Outcomes Sought

Following is a list of educational outcomes sought through space location, allocation, and optimization:

1. To be aware of the practical limits to the search for mathematically optimal physical layout solutions.
2. To understand the mathematical processes available to resolve the QAP.
3. To be aware of commercial software employing optimizing techniques and those dedicated to the facility planning process.
4. To be aware of the limitations of the mathematical methods and their impact on data collection specifications.
5. To recognize that these modeling methods merely underwrite an iterative process in which building users appraise the potential impact of optional locations.

31.7 VISUALIZATION IN 3D AND 4D

History is dotted with examples of models and perspective drawings that illustrate a proposed facility.[6] The use of plans, elevations, and sections is a recent development to assist with the scaling and dimensioning of a building; however, few prospective building users can amass all this two-dimensional information and visualize the volume and ambiance of a building interior. The evolution of architectural CAD commenced by replicating the 2D drafting methods employed to document a building; however, it quickly evolved to provide 3D visualization.

3D visualization can profoundly assist the building users to understand the proposed environment (see Figure 31.12). Virtual reality is a term that should be restricted to specialized immersive

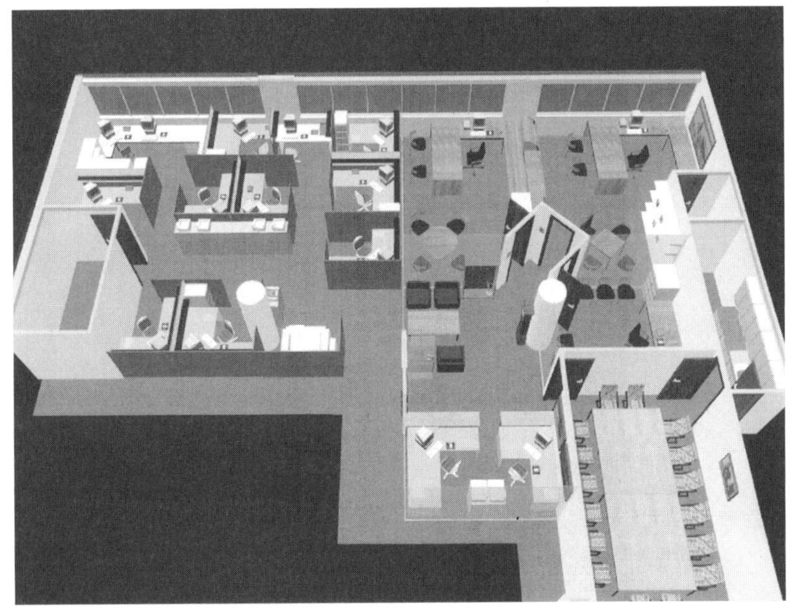

FIGURE 31.12 3D Visualization with User-Controlled Walk-Through Created with 3D CAD. (***Source:*** Concept CAD *reproduced with the permission of the Virtus Corporation.*)

technology where the user is perceptibly immersed in the digital image. 4D adds a dimension of time to the image. This can be an assembly sequence controlled by a critical path or other assembly/production sequence technology and may involve user interaction and animation (see Figure 31.13). Along with these developments in visualization, some CAD systems have evolved into 3D products where every element is presented as a 3D object from the outset. No longer is a pair of lines drawn to represent a wall. The activity of drawing the wall automatically produces a 3D object with height, material properties, and other structural and cost data. Internet-based technologies are enabling remote user interaction to change, for example, surface finishes and fittings in a 3D environment. This higher degree of collaboration and user involvement should produce superior facility outcomes.

The collage of digital effects including 3D visualization, walk-through, orbit, and integrated animation may progressively find its way into facility information management solutions. Computer network management systems appear to be an early implementation of this technology. Users can "fly" through networks and into "boxes" to view a faulty component within the virtual 3D geometry of the computer or similar box. The 3D spatial relationships between service elements in a building are of particular importance for building maintenance. Consider the complexity of service ducts, pipes, and equipment which can be found in cramped ceiling spaces. Walk-through 3D visualization can provide preview access to the desired location, together with animated maintenance sequences and other data. The cost of production is relatively low if the facility documentation was originally generated in 3D.

The 2D facility design process has required a separate coordination process to ensure the alignment of structural elements with architectural outlines, conflicts with service ducts and pipes, and the intersection of utility services with the structure. In addition, all parties to the design process are viewing the evolving facility through 2D images. The introduction of 3D visualization ensures that from the outset, all interested parties have 3D mental images and continue to think within a 3D context.

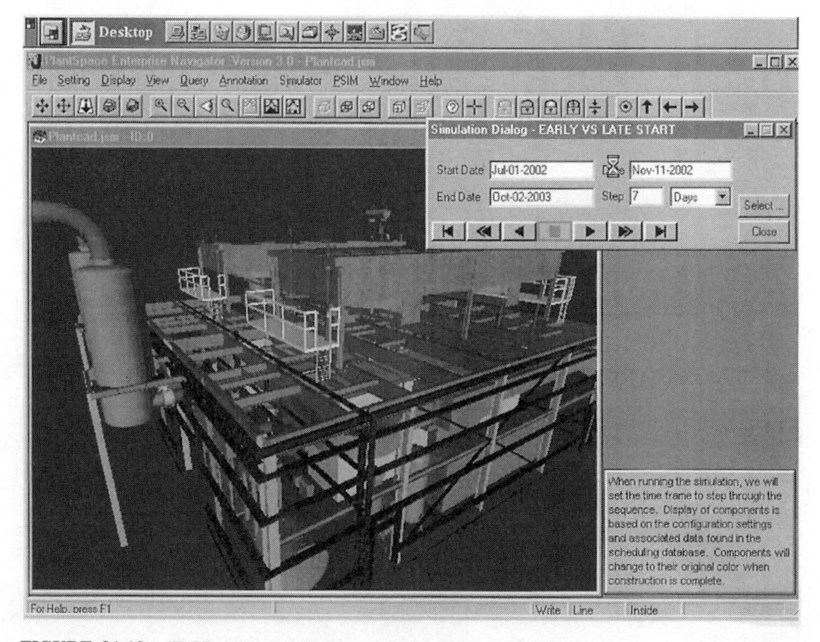

FIGURE 31.13 4D Visualization Displayed as 3D CAD Evolving Over Time. (***Source:*** PlantSpace *reproduced with the permission of Bentley Systems Incorporated.*)

31.7.1 Technology

Consummate implementations of 3D and 4D CAD-based technologies are available in the aerospace industries; however, there appears little evidence that these technologies are being devolved to the built environment. The process plant industries have developed purpose built applications that have yet to be devolved to the general building design process (see previously displayed Figure 31.13). 3D visualization with attendant walk-through and orbit video production is now available in a range of architectural and industrial CAD products. Indeed, some products have reasonable video production facilities; however, the discussion needs to be extended to the kernel of facility information management. We have witnessed the efficiency of "explorer" type interfaces for indexing information. This is a hierarchical index in outline format, but visibility of the entire index is a unique characteristic. However, this outline index is one level of abstraction from the actual information. Just as 3D visualization of a facility removed the abstraction of plans, elevations, and sections, so too does 3D visualization remove the abstraction of the facility index.

The real power of 3D visualization is as an index to all related facility information. The foundation technology is now available as some 3D visualization products enable URLs to be attached to 3D elements or objects. This does not imply that all facility design documentation must be in a 3D format. Access to data is the key issue. Nor does it imply that all design activities must use the same technology; however, it does imply that all design information can be published on the Web. The Internet does not have to be the communication vehicle, and for security reasons, some projects may prefer to distribute Web publishable data on CD-ROM and DVD or similar.

31.7.2 Educational Outcomes Sought

Following is a list of educational outcomes sought through visualization in 3D and 4D:

1. To understand the emerging role of 3D visualization in the design and maintenance of facilities.
2. To recognize that generating the original design documentation in 3D provides the primary tool for on-going building management.
3. To recognize the increasing role of Internet technology to enable collaboration with the evolution of 3D and 4D solutions.

31.8 *FACILITY INFORMATION MANAGEMENT SYSTEMS*

Facility information management systems (FIMS) is a very broad topic and has a number of specialized subsets. Facility information management is essentially a data management task. The availability of low-cost CAD has generated an FIMS design philosophy (there are many FIMS design philosophies) which acknowledged the inherent value of data associated with the CAD objects. A simple example could be the placing of a wall partition in a CAD drawing. The wall partition library object would already have associated data concerning size, supplier, price, and any other relevant data. As each section of wall partition is drawn, additional data is automatically added to the project database. This data could be used for purchasing, depreciation schedules, and maintenance plans. There has also been a long-held unrequited desire to have an automatically produced bill of materials (bill of quantities) and perhaps a building specification as the automated outputs from this technology. STEP and associated activities are designed to further this mission.[7]

From a building owner or manager's perspective, there may be a preference for a unified data environment rather than disparate computer systems. Downside issues focus on system upgrade costs, higher risk, and specialized components that are inferior to their equivalent stand-alone products. There is no denying the feeling of excitement after selecting a CAD object on the screen and viewing associated data either from an associated database or a URL.

The hidden costs in this data design and management philosophy have been the cost of data maintenance, the cost of system upgrade or conversion, and the relevance of this activity with respect to core business activities. The nexus between the CAD file and the data file is a key technical issue that must be understood before committing to a particular product. This technical insight is somewhat unique. A purchaser of a cost accounting system should not need to investigate the internal design of the software. The same generally applies to critical path software; however, the design options available to FIMS significantly impact on the cost of data maintenance. In particular, the link between the CAD file and the database is of critical importance. Before embarking on an FIMS mission, it is important to research the experience of similar organizations and their system maintenance costs. Further investigation should consider system longevity. Some FIMS enable data entry through the browser interface; however, the Internet has resulted in a new generation of FIMS, which maintain the separated CAD and database, and provide a graphical interface for controlled access to all facility data and graphics.

31.8.1 Technology

Simple databases and spreadsheets can be quite effective for basic building information. The simplicity of these systems ensures continuity through staff changes and confident system upgrades. The system design is user defined and can change over time. In a Windows environment, links to CAD files can be achieved through object links. These CAD object links may be lost when the system is upgraded or transferred to another computer; however, the essential data generally remains robust. The following taxonomy of FIMS designs may not apply to any one product. The classification differentiates system design options. A commercial software package may embrace a hybrid system design.

CAD with Integral Data File. The CAD file provides the foundation for this type of system. No data exists without a CAD object. If the CAD object is deleted, then the data is deleted. Additional data entered by the facility manager can only occur if a CAD object is present or inserted. Therefore, most data maintenance activities require a working knowledge of the CAD product. Changing the CAD file (perhaps as a result of layout changes) may be problematic. If the CAD file is deleted, then all associated data is lost. If a new CAD file is inserted prior to deleting the old CAD file, then data must be unlinked from the first CAD file and relinked to the new CAD file or variations of this theme. Data maintenance costs tend to be high. Consequently, there is the danger that the information system becomes progressively out-of-date and ultimately irretrievable.

CAD with Interrelated Data File. With this design philosophy, the data file can be maintained as a separate entity and may never be linked to a CAD file (see Figure 31.14). However, once linked to the CAD file, a number of issues must be considered. The link between the data element and the associated CAD object may display as a small graphic marker on the screen. If this is deleted, then the link to the data is lost, but most products have routines for detecting and replacing lost markers. Data can usually be entered within the stand-alone database and within the CAD interface. From time to time, a reconciliation of both CAD and database files is required. The key advantage of this system is that the database can be updated without any CAD skills, but the database will become incongruent with the CAD file. However, the database will be current.

Inserting a new CAD file can also be problematic. There is a direct link between the old CAD file and the database. These links have to be broken or transferred to the new CAD file. This requires CAD skills at infrequent intervals. The associated risks must be understood before committing to such a system.

Graphical Data Icons Floating above the CAD File. To overcome the problem of linking the CAD file to the database, this design philosophy ensures that the delivered architectural CAD file is agnostic to the data (see Figure 31.15). The CAD file in its full context appears as a background image. Any associated data is defined by selecting a graphic icon (e.g., a workstation) and placing it in the required position on a layer above the background CAD file image. The graphic icon is the link

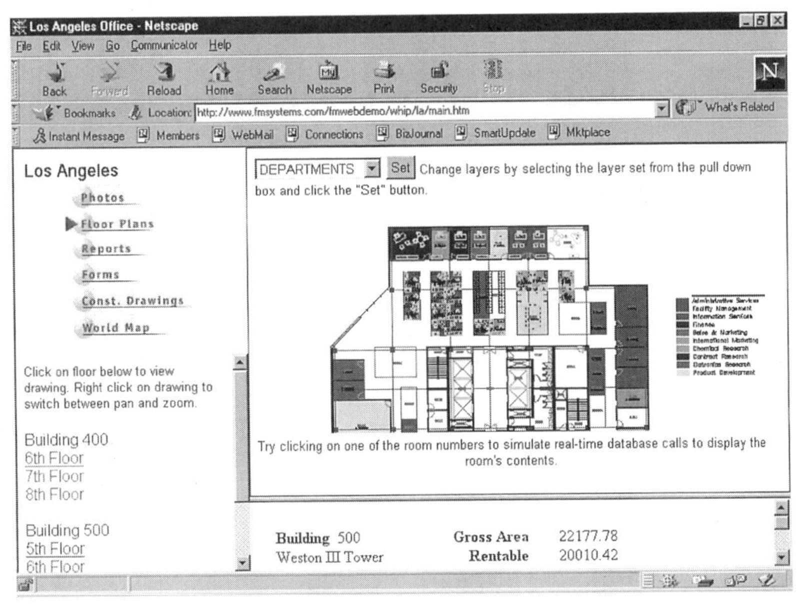

FIGURE 31.14 A Facility Information Management System Using CAD and an Interrelated Database within a Browser Environment. (***Source:*** FMSpace *reproduced with the permission of FM:Systems.*)

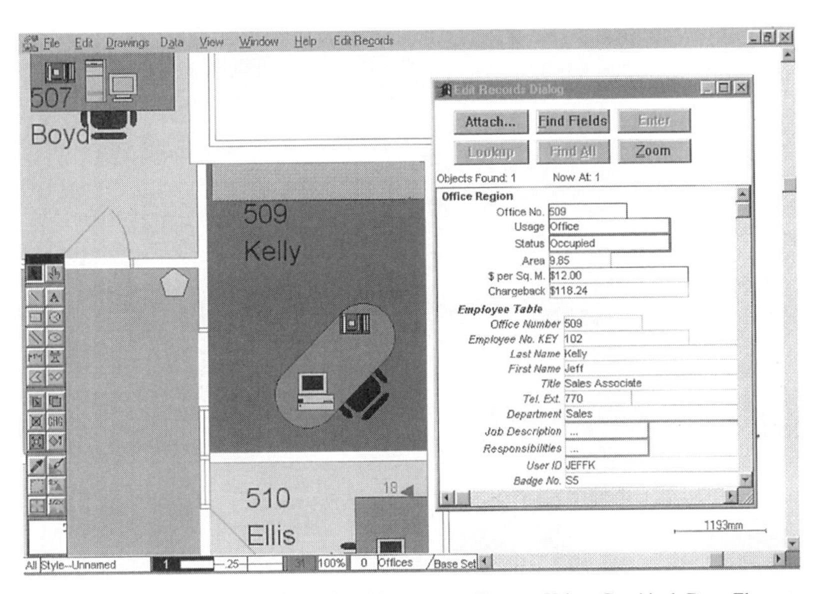

FIGURE 31.15 A Facility Information Management System Using Graphical Data Elements "Floating" above a Background CAD File. (***Source:*** Aperture *reproduced with the permission of Aperture Technologies, Inc.*)

to the internal or external database. Data maintenance is through this graphic interface, which may be more intuitive than an underlying database. The underlying CAD file can be replaced without impacting data integrity, although the data icons may need relocating or amendment. A graphical interface database supporting an underlying CAD file is a superior definition for this system design philosophy.

Logically Located Graphical Data Icons Floating above a Raster Image. This design philosophy further reinforces the supremacy of the database. Indeed, the CAD file is reduced to merely a raster image (see Figure 31.16). Graphical data icons can be manually located above the background image or automatically positioned by the database referencing the X, Y, and Z coordinates of the required location. If the database knows the X, Y, and Z coordinates of the center of a room, then the icon of a workstation can be automatically positioned in that room. Data can be maintained through an "explorer" type hierarchical index or via the graphical data icon floating above the background image. No CAD skills are required for system maintenance. In essence, the database can upgrade the graphic environment. The earlier implementations of these technologies were in data network infrastructure management.

Web Browser Technology. An evolving group of products are focused on the coordination of projects using a predefined interface for promulgating and accessing information. The user can significantly refine the interface design, but in general, no programming is required. Consequently, the information system evolves with the needs of a project.

CAD plug-ins are supplied with these systems along with plug-ins for word-processing, spreadsheets, and numerous other application products. Hyperlinks can be established throughout the system that enables CAD and data to be related. The use of graphical markers "floating" above the CAD is usually employed to avoid a direct connection between the CAD and data. In their full implementation, these systems would coordinate all building design activities plus provide the data infrastructure for the on-going management of the facility (see Figure 31.17).

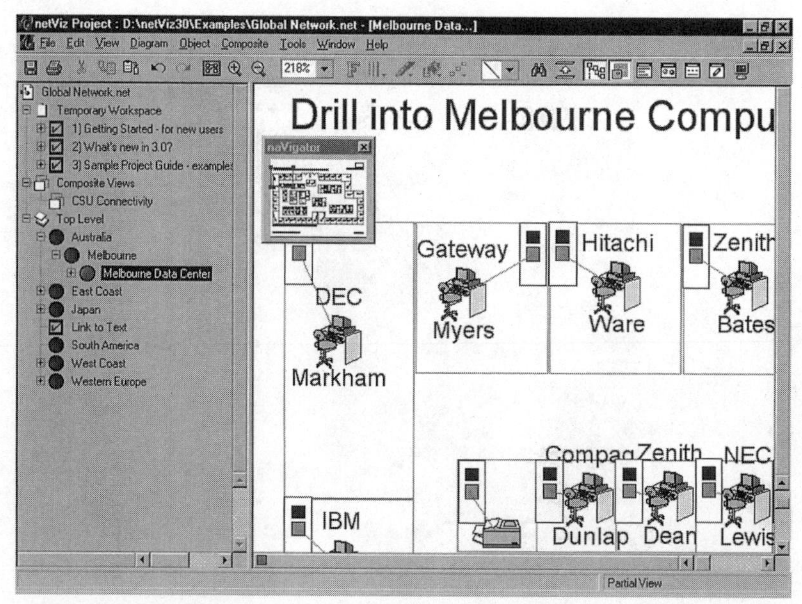

FIGURE 31.16 A Facility Information Management System Using Logically Located Data Icons "Floating" above a Raster Image. (***Source:*** NetViz *reproduced with the permission of NetViz Corporation.*)

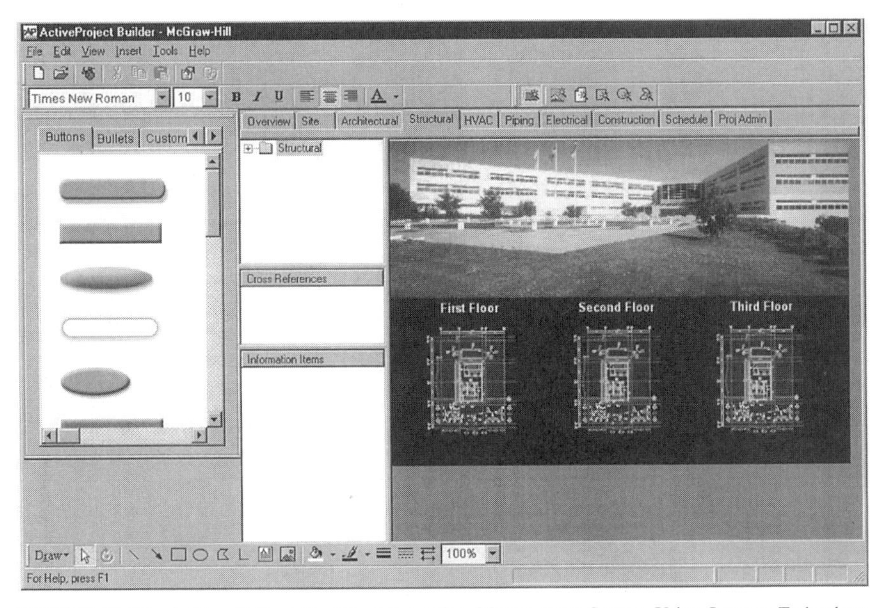

FIGURE 31.17 A User-Defined Facility Information Management System Using Internet Technology. (***Source:*** ActiveProject *reproduced with the permission of Framework Technologies Corporation.*)

31.8.2 Educational Outcomes Sought

Following is a list of educational outcomes sought through facility information management systems:

1. To be aware of the optional technologies that can effect delivery of facility information.
2. To be aware of the data maintenance costs associated with such systems.
3. To recognize the political dimensions of data concentration particularly where ERP systems are being implemented.[8]

31.9 TEROTECHNOLOGY

The maintenance expenditures needed by a facility derive from an equation between the economic utility required from a facility and the cost of achieving that utility. As the required utility will change over time, so too will the maintenance costs and expenditures. The difference between the required costs and actual expenditures can be used as a measure of deferred maintenance for the purpose of a defined facility's economic utility. Measures of current facility condition provide a foundation for estimating the cost to achieve the required facility utility. Simple discounted cash flow (DCF) analysis of investment strategies may be satisfactory for short-term equipment items, such as a chiller with known capital cost, maintenance cycles and costs, and disposal value. Long-term real property assets are not necessarily amenable to DCF procedures, as they are subject to varying future use and hence utility and varying value within a portfolio of facilities. Therefore, the planning of maintenance expenditures is an on-going activity being controlled by short-, medium, and long-term utility values, current physical condition, and available funds. Because maintenance activities are cyclic with different periods, there are times when the majority of cycles peak concurrently. This is better known as the mid-life refurbishment period. Such a high demand on cash can cause a building owner to consider the value of retaining ownership. Knowing the date of the mid-life refurbishment can also impact on perceived utility, building life, and hence the maintenance expenditure policy from now until that period. All these variables contribute to the current expenditure policy. A naive application of cyclic maintenance without reference to the other information inputs can lead to an ineffective maintenance policy. Also, reliance on a fixed percentage of capital costs (maybe 3–5%) for an annual allocation to a maintenance sinking fund may not provide sufficient medium-term cash.

Current software technology appears to segment into technology for managing the planning (see Figure 31.18) and implementation of maintenance activities with associated contact management and cost accounting procedures, and technology for strategic planning and maintenance policy development. Once again, this latter procedure is a gaming process that must use quarantined data. It is unusual to find both capabilities in the one software product.

31.9.1 Technology

Numerous maintenance management software products are available. The reason that so many exist is similar to the large number of critical path packages; that is, each product developer perceives a niche market with unique requirements. Contrast this to the small number of word-processing and spreadsheet packages. The large number of maintenance management systems indicates potential risks when selecting a product. There are different regional accounting and taxation differences, different purchasing conventions, and different service level definitions. In addition, the products offer varying degrees of adaptability to current corporate forms and procedures. Integration with ERP systems adds further complexity.[9] Therefore, maintenance management systems should be appraised and selected from the perspective of a financial accounting system. Ideally, maintenance management systems should be a component of the corporate financial management, contract management, and accounting system.

Facility condition appraisal systems (see Figure 31.19 and Chapter 9) tend to be strategic planning or gaming tools, which evaluate alternate strategies to move from a current facility condition

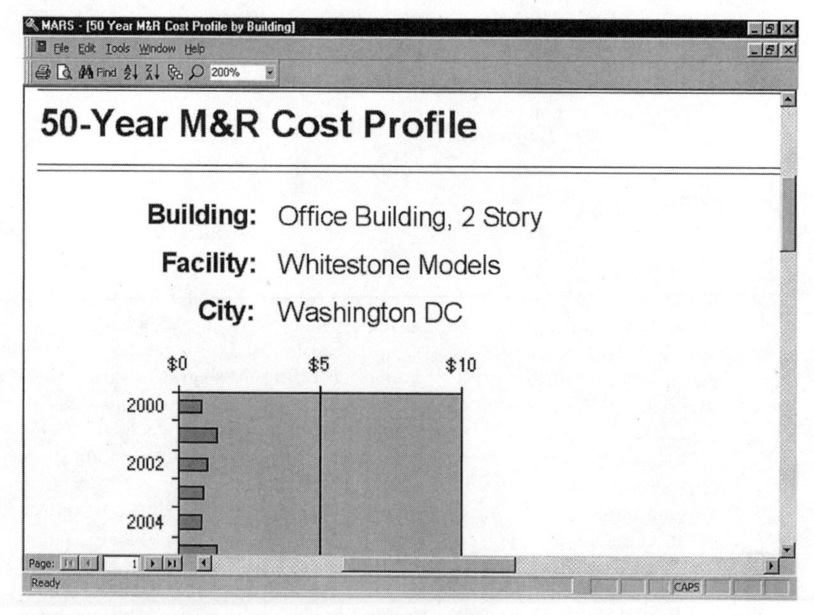

FIGURE 31.18 The Through-Life Cost Prediction Chart for an Example Building. (***Source:*** MARS *reproduced with the permission of Whitestone Research Corporation.*)

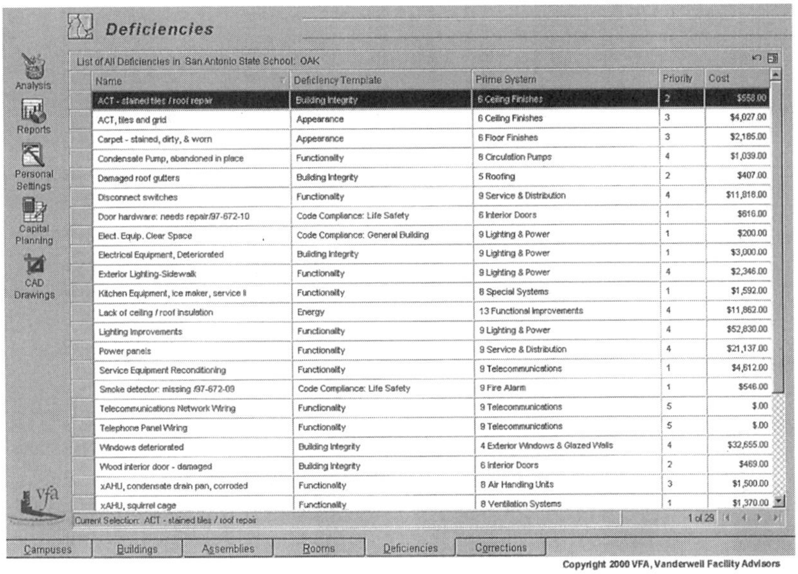

Deficiencies

Analysis
Reports
Personal Settings
Capital Planning
CAD Drawings

List of All Deficiencies in San Antonio State School: OAK

Name	Deficiency Template	Prime System	Priority	Cost
ACT - stained tiles / roof repair	Building Integrity	6 Ceiling Finishes	2	$558.00
ACT, tiles and grid	Appearance	6 Ceiling Finishes	3	$4,027.00
Carpet - stained, dirty, & worn	Appearance	6 Floor Finishes	3	$2,185.00
Condensate Pump, abandoned in place	Functionality	8 Circulation Pumps	4	$1,039.00
Damaged roof gutters	Building Integrity	5 Roofing	2	$407.00
Disconnect switches	Functionality	9 Service & Distribution	4	$11,818.00
Door hardware: needs repair/97-672-10	Code Compliance: Life Safety	6 Interior Doors	1	$616.00
Elect. Equip. Clear Space	Code Compliance: General Building	9 Lighting & Power	1	$200.00
Electrical Equipment, Deteriorated	Building Integrity	9 Lighting & Power	1	$3,000.00
Exterior Lighting-Sidewalk	Functionality	9 Lighting & Power	4	$2,346.00
Kitchen Equipment, ice maker, service II	Functionality	8 Special Systems	1	$1,592.00
Lack of ceiling / roof insulation	Energy	13 Functional Improvements	4	$11,862.00
Lighting Improvements	Functionality	9 Lighting & Power	4	$52,830.00
Power panels	Functionality	9 Service & Distribution	4	$21,137.00
Service Equipment Reconditioning	Functionality	9 Telecommunications	1	$4,612.00
Smoke detector: missing /97-672-09	Code Compliance: Life Safety	9 Fire Alarm	1	$546.00
Telecommunications Network Wiring	Functionality	9 Telecommunications	5	$.00
Telephone Panel Wiring	Functionality	9 Telecommunications	5	$.00
Windows deteriorated	Building Integrity	4 Exterior Windows & Glazed Walls	4	$32,655.00
Wood interior door - damaged	Building Integrity	6 Interior Doors	2	$469.00
xAHU, condensate drain pan, corroded	Functionality	8 Air Handling Units	3	$1,500.00
xAHU, squirrel cage	Functionality	8 Ventilation Systems	1	$1,370.00

Current Selection: ACT - stained tiles / roof repair   1 of 29

Campuses | Buildings | Assemblies | Rooms | Deficiencies | Corrections

FIGURE 31.19 An Example of Measuring the Building's Condition and Defining the Cost of Remedial Strategies. (***Source:*** VFA *reproduced with the permission of Vanderweil Facility Advisers.*)

to a desired condition and utility at a defined date. Funds availability is a key input to the analysis. Condition appraisal systems offer either a condition scale or a condition scale linked to actual cost data. This cost data must be local costs. International products should therefore offer a localization function; however, the description of building systems, elements, and components together with trade practices, varies significantly throughout the world and within countries. Imperial (English) and metric (SI) measures are additional issues. These have all combined to ensure the scarcity of universally adaptive building condition appraisal systems.

31.9.2 Educational Outcomes Sought

Following is a list of educational outcomes sought through terotechnology:

1. To understand the decision inputs to the strategic maintenance planning process.
2. To recognize that maintenance policy planning and maintenance process management will usually require separate software technologies.
3. To be able to develop maintenance management policy within a corporate planning context.

31.10 BUILDING SERVICEABILITY

The architectural profession has developed a well-established procedure for "post occupancy evaluation." Different countries use different terminology, but the essential objective is to evaluate the effectiveness of the completed facility from the users' perspective. The measurement metrics tend to be qualitative with rating scales. This process, which provided a quick evaluation of a building, has now matured into a science using specialized technologies. There are three components: occupant requirement scales, facility rating scales, and multicriteria decision making. Serviceability (see also Chapter 19) takes a longer time perspective than short-term performance measures.[10] Serviceability accounts for current and future needs. It requires quick evaluation methods rather than the detail of performance measurement. The objective is to achieve the highest possible workplace quality and occupants' effectiveness. Above all, it is customer-based.

The occupant requirement scales rely on pedestrian terminology to describe user requirements. The facility rating scales rate what is physically present or what could be present. Threshold values and relative importance weightings moderate the ratings. These ratings have no relation to costs. The scales (occupant requirement and facility rating) are paired for each issue. Building serviceability assessment provides a broad appraisal of user requirements (demand) and the services provided by a facility (supply). The process is readily extended to multiple buildings competing for the users' business. This is akin to a competitive bid (tender) process and adaptations of competitive bid analysis software provide an effective vehicle for processing building serviceability data. The occupant requirement scales provide the specification and the facility rating scales provide the competitive "suppliers" bids.

The design of rating scales for office buildings was developed by the International Centre for Facilities (see Figure 31.20) and adopted as ASTM E 1660.[11] The methodology is equally useful for nonoffice facilities, such as airfields, hospitals, and manufacturing facilities. It differs from requirements engineering, which seeks to manage the design of a new capability. Building serviceability is restricted to the evaluation of existing facilities with respect to user requirements. It is a rapid appraisal methodology.

31.10.1 Technology

ASTM E 1660 is a paper-based (with software support) set of rating scales. The forms are easily translated into screen-based data entry forms. Software for multicriteria decision making has made significant advances in recent time (see Figure 31.21). The limitations of weighted averages, as a method for evaluating competitors is well understood. Pattern matching has been combined with

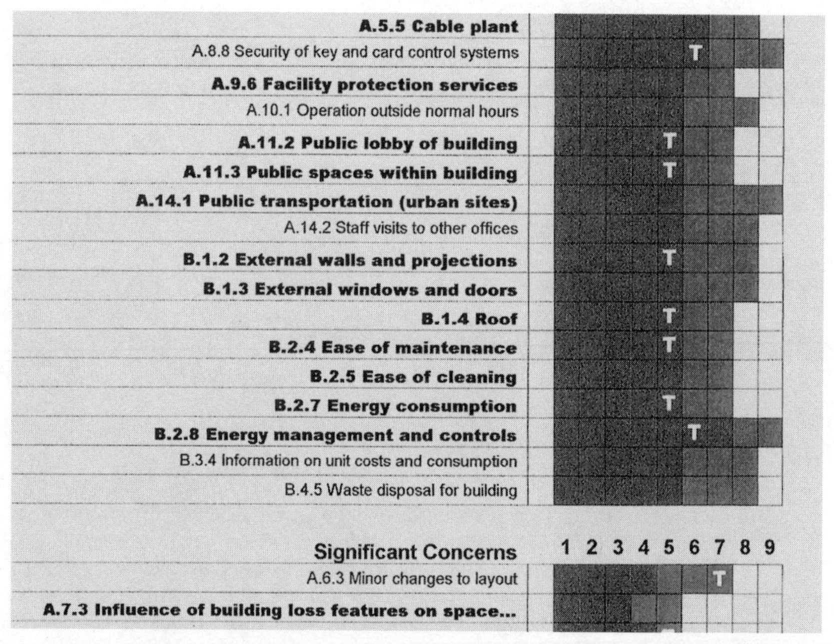

FIGURE 31.20 A Sample Illustration of Facility Rating Scales. (*Source:* Serviceability Tools *reproduced with the permission of the International Centre for Facilities.*)

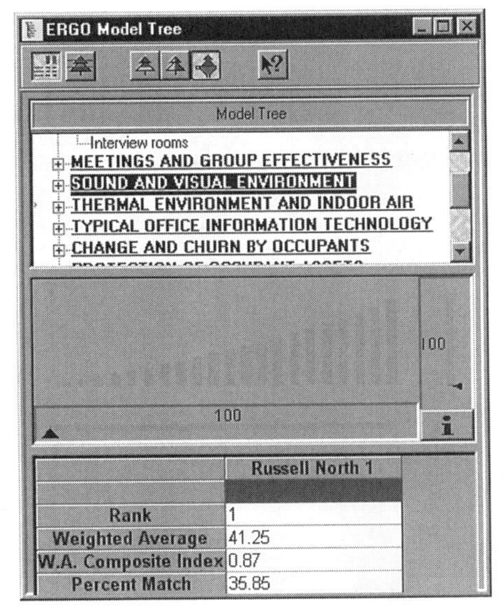

FIGURE 31.21 An Example of Multicriteria Decision Support for Rating Multiple Facilities against Occupant Requirements. (***Source:*** ERGO *reproduced with the permission of Arlington Software Corporation.*)

weighted averages (using local and global weights) to generate a composite index. This composite index is a measure of the composition of the weighted average and pinpoints that combination of factor ratings closest to what is ideal (the ideal being the percentage weight assigned to each factor in the first place). The procedure also accounts for threshold values. The incorporation of sensitivity analysis further enhances the decision making process.

Multicriteria decision-making software is a robust vehicle for analyzing building serviceability data. The concurrent application of a user-controlled methodology to multiple facilities provides an efficient and consistent evaluation process. The inherent audit trail ensures that decisions can be traced back to the underlying analysis.

31.10.2 Educational Outcomes Sought

Following is a list of educational outcomes sought through building serviceability:

1. To understand that building users need to articulate their accommodation needs in non-technical language.
2. To perceive facilities as competing for thc satisfaction of users.
3. To understand the mathematical basis to multicriteria decision making.

31.11 CONCLUSIONS

There are a number of risks associated with this chapter. Only a small number of products have been illustrated to provide visions of concepts. Some of these illustrations may not be current at the time of printing. Some product developers may not agree with the text in proximity to their referenced illustration and some products may be unavailable at the time of printing. The overall intention is not to reference any one product by name. Products evolve over time and any review of functionality must contain a caveat on the use of the review beyond the immediate future. The illustrated products have been used in our Master's degree courses as a vehicle for the education process. The delivered courses use about forty products. There appeared to be a rapid deployment of facility planning and management software in the 1990s. Product evolution appeared to slow in the late 1990s with the incorporation of Internet technology.

Our course is influenced by its military location and this chapter may reflect this bias. The use of commercial software in a teaching environment has elicited varying student reaction. Students with a science and engineering background tend to embrace the acquisition of another software-based skill. The most pleasing responses have come from students directly employed in facility planning and management who comment on their capacity to immediately implement some of the illustrated technologies. The response from nonattending students has been most gratifying. Sincere gratitude is expressed to the product vendors who have provided software to the course and have agreed to the use of their exposure in this chapter.

APPENDIX A SOFTWARE PRODUCTS MENTIONED IN CHAPTER

ECCO
NetManage Inc.
10725 North De Anza Blvd.
Cupertino, CA 95014
www.netmanage.com

Scitor Process
Scitor Corporation
256 Gibralta Dr.
Sunnyvale, CA 94089
www.scitor.com

ProModel
PROMODEL Corporation
1875 S. State St.
Suite 3400
Orem, UT 80097
www.promodel.com

DOORS
Quality Systems and Software
200 Valley Rd.
Suite 306
Mt Arlington, NJ 07856
www.qss.co.uk

OrgTree, Stack and Block
FMx UK Ltd
Westfield House
Bonnetts Ln.
Ifield
West Sussex, RH11 0NY
UK
www.cafmexplorer.com

WinSABA
SABA Solutions
117 9th St.
Manhattan Beach, CA 90266
www.techexpo.com

Evolver
Palisade Corpration
31 Decker Rd.
Newfield, NY 14867
www.palisade.com

TOPMET
CSIRO Building Construction and Engineering
Highett Laboratories
Graham Rd.
Highett, Vic. 3190
AUSTRALIA
www.csiro.au

Concept CAD
Virtus Corporation
114 MacKenan Dr.
Suite 100
Cary, NC 27511
www.virtus.com

PlantSpace
Bentley Systems Incorporated
685 Stockton Dr.
Exton, PA 19341-0678
www.bentley.com

FMSpace
FM:Systems
807 Spring Forest Rd.
Suite 100
Raleigh, NC 27609-9197
www.fmsystems.com

Aperture
Aperture Technologies, Inc.
100-3 Summit Lakes Dr.
Valhalla, NY 10595
www.aperture.com

NetViz
NetViz Corporation
9210 Corporate Blvd.
Suite 150
Rockville, MD 20850
www.netviz.com

ActiveProject
Framework Technologies Corporation
23 Third Ave.
Burlington, MA 01803
www.frametech.com

MARS
Whitestone Research Corporation
610 Anacapa St.
Santa Barbara, CA 93101
www.whitestoneresearch.com

VFA
Vanderweil Facility Advisers
266 Summer St.
Boston, MA 02210-1112
www.vfa.com

ERGO
TEC
Arlington Software Corporation
740 Saint-Maurice
Suite 410
Montreal
Quebec H3C 1L5
CANADA
www.technologyevaluation.com

ENDNOTES

1. Stevens, R., et al. 1998. *Systems engineering—coping with complexity.* Prentice-Hall. Prentice-Hall Europe, Hemel Hempstead.
2. ASTM/ANSI E 1836-98 Standard Classification for Building Floor Area Measurements for Facility Management. Annex A.
3. Sharpe, R., Maersjö, B. S., Mitchell, J. R., and Crawford, J. R. 1985. An interactive model for the layout of buildings. *Applied Mathematical Modelling 9,* 207–214.
4. Holland, J. H. 1975. *Adaptations in natural and artificial systems.* Ann Arbor, MI: University of MI Press.
5. Davis, L. 1987. *Genetic algorithms and simulated annealing.* Palo Alto, CA: Kaufman.
6. Specifying Buildings—A guide to best-practice. NATSPEC Guide 1. See www.cis.asn.au.
7. STEP, the Standard for the Exchange of Product Model Data (ISO 10303). See www.steptools.com.
8. ERP, Enterprise Resource Planning. ERP is designed to model and automate many of the organization's basic processes with the goal of integrating information. See www.cio.com/forums/erp/index.html.
9. Ibid.
10. International Centre for Facilities, 440 Laurier Avenue West, #200, Ottawa, Ontario, CANADA K1R 7X6.
11. ASTM Standards on Whole Building Functionality and Serviceability.

CONTRIBUTOR BIOGRAPHIES

Jeff Austin

Jeff Austin is a strategist on First Union's Corporate Real Estate Team. As such he led the company's efforts to formulate and deploy their Alternative Workplace strategy in the mid-1990's. He is currently working on a cross-divisional team to establish the framework for an integrated workplace strategy and delivery process. Jeff, a registered architect and Malcolm Baldrige advisor, has addressed many Corporate Real Estate executives and industry leaders on a range of strategic issues that CRE groups must address as they realign their divisions to support their company's strategies for the emerging economy.

Azizan Aziz

Azizan Aziz is a Research Associate at the Center for Building Performance and Diagnostics at Carnegie Mellon University. He is actively working on the research and design of commercial buildings based on the principles of the Robert L. Preger Intelligent Workplace. His work includes the masterplan for sustainable development of the US Army Corps of Engineers Research Lab in Urbana-Champaign, the masterplan of the City of Wolfsburg, Germany, and the masterplan of a city block in the Lichtenberg district in Berlin. In addition to the Intelligent Workplace at Carnegie Mellon University, his other building projects include the Adaptable Workplace Laboratory at General Services Administration in Washington DC, the Mercedes Benz Marketing Academy in Stuttgart, Germany, an addition to the Umwelt Campus in Birkenfeld, also in Germany, and the Laboratory of Design for Cognition for the Electricité de France in Paris. Azizan Aziz received his Bachelor of Architecture, and Master of Science in Architecture from Carnegie Mellon University in 1990 and 1991, respectively.

Alan L. Bain

Alan L. Bain, President of World-Wide Business Centers ("WWBC"), Inc. In 1970, Mr. Bain became a naturalized U.S. citizen. He grew up in the U.K., served as an Officer in The Worcestershire Regiment in The British Army, received a BA Jurisprudence and MA from St. Catharine's College, Cambridge was admitted to the English Bar by the Honourable Society of the Inner Temple, where he was an Inner Temple Scholar. Mr. Bain also graduated from Columbia University School of Law in 1964 with an LLB, serving as Class President. Before founding WWBC, Mr. Bain practiced law for six years in New York City.

Mr. Bain, a recognized leader in the executive suite industry and frequent contributor to industry publications and speaker on industry topics, serves on the Board of Directors of ESA, the association for the Office Business Center Industry.

WWBC operates a 40,000 square foot serviced office facility in midtown Manhattan and, through its wholly owned subsidiary, World-Wide Business Centers Network, Inc., operates in over 150 locations in 16 countries on four continents.

Jim Barlow

Jim Barlow serves as FM Technology Specialist for Graphic Systems, Inc. Bringing experience as a system administrator and Oracle DBA, Jim has been instrumental in numerous large-scale implementations of CAFM and related systems over the last 10 years. His assignment for the last two and half years has been to blend process and technology to enhance business at the Architect of the Capitol. Jim has worked extensively in CAD, work management, hand-held devices, and executive reporting.

Stephen Bell

Stephen Bell is a Senior Vice President of Fidelity Corporate Real Estate, the real estate division of Fidelity Investments where he is responsible for the companies strategic real estate planning; acquisitions and development of corporate real estate; and workplace research and development. Since 1996, Stephen has focused on developing methodologies, which identify and link corporate real estate strategies to corporate business strategies. He is also responsible for directing workplace performance research and the development of work environments that enhance corporate performance.

Before joining Fidelity in 1996, Stephen was with Liberty Mutual Insurance Company where he was responsible for all aspects of facilities and real estate for the company's 6 million square feet of corporate space.

Stephen has a Bachelor of Arts degree from the University of Missouri, a Bachelor of Environmental Design degree from the University of Kansas and his MBA from Boston University.

Carolyn Castillo

Ms. Castillo is Program Manager, Metrics, Systems and Process for the Space and Communications Group business segment of the Boeing Company. Prior to joining the Space and Communications Group, Carolyn worked in Facilities Management at Rocketdyne, part of the Boeing Company. While at Rocketdyne, Carolyn initiated efforts to implement a Computer-Aided Facilities Management (CAFM) System. This effort lead to a greater understanding of the work processes involved, organizational change issues and the need for informational sharing within the Facilities organization.

Ms. Castillo holds Bachelor degrees in Psychology and Industrial Technology from California State University Long Beach, and a Masters of Arts in Management (1987) from the University of Redlands.

Bruce Cox

Bruce Cox is Director of Facilities Technology for FMStrategies, a division of Little & Associates Architects based in Charlotte, North Carolina. FMStrategies manages space and property information for over 75,000,000 square feet of real estate for large corporate and institutional customers.

Formerly, Bruce was Vice President of FM:Systems, a well-known Computer Aided Facilities Management software manufacturer in Raleigh, North Carolina. During his six year tenure at FM:Systems, he provided Computer Aided Facilities Management consulting services for Fortune 500, and Fortune 1000 companies across the country.

Bruce is a registered architect and holds a BS in architecture from the Georgia Institute of Technology. For over seventeen years he has been involved with architectural and engineering technology focusing on designing and writing software for Computer Aided Design (CAD), and Computer Aided Facilities Management (CAFM). Bruce currently manages a team of software engineers and programmers specializing in the development of distributed multi-tier Internet systems for hosting and delivering facilities information to FMStrategies customers.

Thomas Davies

Thomas Davies is Assistant Director of Physical Plant–Design and Construction at Amherst College. He has a diverse background, having worked as an engineer, architect, general contractor and program manager. Most recently, Mr. Davies was Executive Vice President at VFA responsible for general management of capital planning assessment projects and business development. With more than 100 million square feet of experience for corporate, government, educational, and healthcare clients, Mr. Davies is well versed in facility planning issues, and building and software systems.

Mr. Davies has written and lectured across the country on the subjects of capital planning management solutions and the use of technology in facility management. As a fellowship recipient at MIT, Mr. Davies researched the practice of capital planning and predicted the growth of data-oriented consulting solutions for strategic facility investment.

Gerald Davis F-ASTM, F-IFMA, AIA, CFM

Gerald Davis is past Chair, ASTM Committee E06 on Performance of Buildings; Chair, ASTM Subcommittee E06.25 on Whole Buildings and Facilities; past Chair, IFMA Standards Committee; and USA (ANSI) voting delegate to ISO Technical Committee 59 on Building Construction.

He led the teams which developed the Serviceability Tools and Methods (ST&M) for defining workplace and facility requirements and for rating existing and proposed facilities, and the ORBIT-2 project, a major North American multi-sponsor study about change in offices, information technology, and the organizations that use them. He is co-author of the first 1987 IFMA Benchmark Report.

He received the Environmental Design Research (EDRA) Lifetime Career Award, 1997, the IFMA Chairman's Citation, 1998, was named an IFMA Fellow in 1999, and "one of 50 most influential people in the construction industry" by the Ottawa Business Journal, also in 1999.

Thomas DeChant

Thomas DeChant is a graduate of the University of Notre Dame and has 20 years of experience in health facility planning and research. He focuses on the financial and functional performance of facility assets, particularly in large hospitals and academic medical centers.

Mr. DeChant has adapted asset management performance measures from other industries to the real estate portfolios of integrated delivery networks, applying these measures to provide strategies for both short and long-term capital facility investment. He also assesses client requirements for computer-integrated facility management (CIFM) systems and provides "second opinions" on alternative facility investment strategies.

Al Ferreira

Mr. Ferreira is President of Ferreira Service Inc., which specializes in air conditioning maintenance, repair, and energy cost reduction. Ferreira Service Inc. currently maintains over 1000 maintenance accounts in Northern and Southern California. Since 1987 the company has reduced the energy costs of its customers by $15,000,000.

In 1991, Mr. Ferreira developed the H.V.A.C. Auditors Training Program for PG&E energy auditors, and for employees of the City of Los Angeles Department of Water and Power. He has provided seminars on energy conservation with Southern California Edison Customer Training Applications Center (CTAC), and is a frequent IFMA presenter.

Mr. Ferreira is a B.S.M.E. graduate of Heald Engineering College, a registered professional Engineer for the State of California, and a licensed Mechanical and Electrical Contractor.

Stormy Friday

Stormy Friday is the Founder and President of The Friday Group, an eleven-year-old firm specializing in strategies and solutions for the facility management profession. Ms. Friday began the firm after twenty years of diversified management and facilities experience, and as a consultant. An internationally recognized speaker in the field of facility management, she has visited twenty-seven countries to consult, train and speak on FM trends; organization development; productivity and motivation; marketing and customer service; strategic planning and outsourcing alternatives.

In 1994, Ms. Friday co-authored a book entitled *Quality Facility Management: A Marketing and Customer Service Approach* published by Wiley and Sons. Formerly on the IFMA Board of Directors, she currently serves as an IFMA trainer, and was named IFMA Fellow in 1999. She is a contributing editor to *The Outsource Report* and writes extensively for other facility management publications.

Loree Goffigon

Loree is Vice President and Strategic Planner at Gensler working on groundbreaking projects for leading professional service firms and Fortune 500 clients. Acting as counselor, guide, and implementation expert, Loree helps complex organizations deal with the havoc change can wreak on a work environment. She helps clients align their workplaces with their business plans and build appropriate strategies to improve performance. Loree brings a modern-day definition of the workplace to her assignments. Her programs are catalysts that help to change the way work is done, making organizations more successful places and the people in them more effective.

Debra Gondeck-Becker

Debra Gondeck-Becker, Associate AIA is a Project Manager for the Jordani Consulting Group (JCG). She has nine years of experience working with technology related to the design field. Prior to joining JCG, Ms. Gonder-Becker worked as a consulting architect and educator.

Her experience includes teaching virtual reality technology courses at the University of Minnesota and Pine Technical College, where she is the Computer and Information Sciences Division Chair. Ms. Gondeck-Becker has also worked as a consultant on technology issues and as an intern architect at a traditional architecture firm, focused on Physical Needs and Handicap Accessibility Studies and housing renovations with the Minnesota Public Housing Authority.

Jean Grant

As Technical Writer and Editor for Carroll Thatcher Planning Group, Jean was involved in all aspects of the development of the courses that make up the Michigan State University Master's Level Certificate in Facility Management, including writing, editing, proofreading, research, and video editing. Jean worked closely with MSU's Virtual University to coordinate the conversion of two of the courses from a text-based format to an electronic format. Jean acted as a Teaching Assistant for the Facility Management: Theory and Principles course in early 1999. Jean has over 15 years experience as a technical writer and editor and has been trained in desktop publishing. Her experience includes writing user guides for computer systems, writing detailed procedures of manual and automated processes, and preparing training materials, reports, and slide presentations.

Stephen R. Hagan, AIA

Stephen Hagan is a program manager overseeing the Project Integration Group within the Capital Development Division at the U.S. General Services Administration's National Capital Region in Washington, D.C. Project integration and information technology tools are being developed and deployed to develop and nurture a community of project managers, design professionals, executives and strategic partners associated with GSA. These tools are also leveraging knowledge and skills to bring enhanced project delivery of programs for GSA's customer agencies. ProjectWeb extranet systems are being deployed to increase project efficiency and communication, and develop vital linkage between project and enterprise-wide information systems.

Hagan is an architect educated at Yale College and the Yale School of Architecture. He has lectured extensively on the subject of the Digital Revolution in Design and Construction and is actively involved in the leadership of the American Institute of Architect's Center for Advanced Technology Facilities Design. Hagan has 20 years of experience in the management of complex, multi-million research facilities, including state of the art laboratories at the National Institute of Health and the National Institute of Standards and Technology.

Volker Hartkopf

Volker Hartkopf has completed research and demonstration projects in Bangladesh, Canada, Germany, Peru, and the U.S. in industrial architecture, housing, commercial buildings, energy conservation, and whole building performance. Since 1988, Dr. Hartkopf has directed the Advanced Building Systems Integration Consortium (ABSIC), an industry-university-government partnership dedicated to improving the quality of the built environment. In 1990, he initiated, conceptualized, and raised the funds on a global basis for the Robert L. Preger Intelligent Workplace; a $4 million living and lived-in laboratory and demonstration facility. Most recently, Professor Hartkopf has been one of the key members instrumental in creating the Global Field Support Consortium (GFS), consisting of seven Global Hardware, Software and Service Providers dedicated to creating the marketing and service environment of a Global 50 Corporation. A frequent keynote speaker worldwide, his work has been cited in the professional literature. He has contributed over 100 technical publications. Volker Hartkopf received his Diplom-Ingenieur from the University of Stuttgart, Germany in 1969, Master of Architecture from the University of Texas at Austin in 1972, and Ph.D. from the University of Stuttgart, Germany in 1989.

Paul Heath

Paul Heath brings an extensive and varied background in strategic management, planning, and design to his position as Principal at SPACE. He is a skilled project consultant and planner with over twenty-five years experience working with organizations and communities to create organizational, facility and development strategies, plans and programs. Most recently Paul has been in charge of strategic planning efforts, workplace planning, and change management at SPACE.

Paul's recent work has focused on the development of alternative workplace/officing strategies, change management support, and strategic planning for high-tech and communications companies. Current/recent projects include: development of alternative officing strategies and change management facilitation for Cisco Systems, Microsoft, AT&T, and Lucent Technologies; programming, planning and work process analysis approaches for US Web; work process analysis for GSA's Portfolio Management Division; and national workplace strategies for the accounting firm of McGladrey & Pullen.

SPACE is an organization that focuses on the design of systems solutions and discoveries that support the continuous change and evolution of the workplace. It offers integrated services, customized to each client's individual and specific needs in the following areas: strategic real estate and facility planning; project design and implementation; facility management; and research.

David A. Jordani

David Jordani is President of Jordani Consulting Group, an independent practice that provides information technology services.

Harvey H. Kaiser

Harvey H. Kaiser is President of Harvey H. Kaiser Associates, Inc., Facilities Management Consultants, and formerly Senior Vice President for Facilities Management, Syracuse University. He has degrees in architecture from Rensselaer Polytechnic Institute and Syracuse University. His firm provides services to higher education for strategic facilities planning, facilities management, campus master planning and architectural programming and design. His clients include U.S. and foreign colleges and universities, governments and agencies, foundations, and corporations.

Dr. Kaiser served for more than 20 years as a Syracuse University Senior Vice President, was the University Architect, and an Associate Professor of Urban and regional Planning in the Maxwell School of Citizenship and Public Affairs. He recently served as the Chair of the SCUP, NACUBO, and APPA Space Management Task Force.

Fred Klammt

Fred Klammt is Principal of Aptek Associates, a quality management consulting firm specializing in improving service quality.

Over the past 20 years, he has worked with Aptek on consulting projects for over 40 Fortune 500 companies including Hewlett Packard, Southern California Edison, Novell, University of Southern California, Paramount Studios, ASU, Sunmaid, and Toyota to mention a few. These projects range from strategic planning, facility relocation evaluations, re-engineering call centers, etc. Overall Aptek focuses on linking facility and front-line operational needs with executive management strategic objectives.

Prior to his current venture, Mr. Klammt was Facility Operations Director for an outsourcer and responsible for over 2.5 million square feet of Toyota Motor Sales facilities nationwide.

Fred has a Bachelor of Science Degree in Electrical Engineering from the University of Wyoming and an Associates in Electromechanical from Southern Illinois University. He also serves on several committees; ASTM TC ISO standards; the California Board of Examiners CQA; 1996–99 Malcolm Baldridge Quality Awards. He is a founder of the Silicon Valley IFMA Chapter; and past chair of The International Program Committee for IFMA. Additionally Fred is a member of the CSU Sacramento and Kufstein, Austria faculty teaching FM and CPM curriculums.

Charlie Kuffner

Charlie Kuffner is Senior Vice President of Swinerton & Walberg Builders' Northern California Division, which includes offices in San Francisco and Santa Clara. A registered Civil Engineer, Mr. Kuffner has been with the firm since 1981. During his tenure at Swinerton & Walberg, Mr. Kuffner has worked as Project Engineer, Cost Engineer, Project Manager, and Operations Manager. Currently, as Division Manager, his responsibilities include office and field construction administration, business and strategic planning, business development and estimating, scheduling, quality control, and professional training of staff.

Mr. Kuffner was instrumental in providing business solutions for the project delivery process to BidCom, enabling them to be mapped into the current commercial In-Site on-line project management product in use today. He also introduced BidCom to the Center for Integrated Facilities Engineering (CIFE) at Stanford University.

Alex K. Lam

Alex K. Lam is President of The OCB Network, a company based in Toronto, Canada specializing in customized corporate workplace learning using team visioning workshops in the areas of creativity, leadership and business transformation. Prior to The OCB Network, he spent 23 years with Bell Canada as General Manager—Facilities where he managed over 2250 facilities with a staff of 379 professionals and an annual operating budget of $70 million.

He serves on the Board of Directors, International Facility Management Association; Board of Trustees, BOMI Institute; Executive Board, International Society of Facilities Executives; and the Editorial Advisory Board of the Canadian Facility Management & Design magazine. He teaches facility management at Ryerson Polytechnic University. He is a frequent speaker at international conferences. He has a Bachelor of Architecture degree from McGill University (1967) and a Master of Theological Studies from Ontario Theological Seminary (1995).

Stephen R. Lee

Stephen Lee is an architect with over 20 years of experience and his research, development and demonstration activities in Carnegie Mellon University's Center for Building Performance and Diagnostics focus on sustainable design and systems integration for commercial and residential architecture. The research involves issues of product, process (including software development for simulation) and field verification of building performance. He continues in his role as Project Manager of the design, construction and operation of the Robert L. Preger Intelligent Workplace. He has completed an experimental house for Armstrong World Industries and The Susquehanna Project and was a team member for the General Services Administration Advanced Workplace Laboratory project in Washington, DC. As a practitioner, Mr. Lee's firm, TAI + LEE, Architects P.C. has been, and continues to be, forerunners in implementing energy effective and sustainable design in their residential and commercial work. Stephen Lee received his Bachelor of Architecture and Master of Architecture from Carnegie Mellon University in 1975 and 1976, respectively.

David Lehrer, AIA

Mr. Lehrer has been with Gensler's San Francisco office for over eight years. In his current role as project architect, he oversees all phases of a project from initial design through construction administration, controlling quality and design intent throughout. He had been involved in a number of project types including office buildings, corporate interiors, retail, and renovation.

David is a leader in Gensler's Sustainable Design Committee that promotes and educates Gensler staff on sustainable design. He is co-chair of the AIA San Francisco Committee on the Environment, and has been invited by a number of groups to speak on sustainable design issues.

Before joining Gensler Mr. Lehrer was employed by UC Berkeley's Center for Environmental Design Research. There he was involved in teaching and research projects focusing on building science issues such as occupant comfort, and sun, wind, and shadow studies for urban development guidelines. Mr. Lehrer holds a Master of Architecture degree from UC Berkeley and a BFA from the University of Arizona.

Vivian Loftness, AIA

Vivian Loftness is Professor and Head of the School of Architecture at Carnegie Mellon University. Extensively published, she has made major contributions to the definition of total building performance and sustainability for a range of building types. In the Center for Building Performance at Carnegie Mellon, Ms. Loftness has been actively researching and designing high performance office environments with DOE, DOD, Department of State, GSA, NSF and major building industries such as EDF, Steelcase, and Johnson Controls. She is a key contributor to the development of the Intelligent Workplace—a living laboratory of commercial building innovations for performance, along with authoring a range of publications on international advances in the workplace. She has served on five National Academy of Science panels as well as being a member of the Academy Board on Infrastructure and the Constructed Environment. Her work has influenced both national policy and building projects, including the recently completed Adaptable Workplace Laboratory at the U.S. General Services Administration. Vivian Loftness received her Bachelor of Science and Master of Architecture from the Massachusetts Institute of Technology in 1974 and 1975, respectively.

John D. Macomber

Mr. Macomber has been in the Architecture/Engineering/Construction (AEC) business for over twenty-five years, with responsibilities progressing from Laborer to Project Manager and culminating in six years as Chief Executive Officer of the George B.H. Macomber Company, a Boston-based General Contracting and Construction Management firm. There, he remains Chairman and principal stockholder. Mr. Macomber is also Chairman of Hamilton Construction Equipment Corporation, a service business that rents construction equipment to a customer list of over five-hundred contractors and subcontractors. He is also part owner of several commercial real estate properties.

Since 1988, Mr. Macomber has been a Lecturer in Civil Engineering at the Massachusetts Institute of Technology, where he developed and teaches a graduate level course in Strategic Management. This course studies the entire value system in the AEC industry, emphasizing the impacts of information technology and industry structure for an audience including engineering, architecture, real estate, and business students. His writings have appeared in *Harvard Business Review, Urban Land,* and other publications.

L. David McDaniel

Dave McDaniel is Chief Scientist for BMS Catastrophe Special Technologies Division. He is responsible for post loss analysis, recommendations for recovery procedures, protocols and remediation. He has been personally involved in most major disasters in the U.S. in the last eight years. He often acts as a consultant to major insurance companies on the effects of fire, water, and acidic contaminants on electronic equipment, documents, industrial equipment, and other contents.

His background includes 11 years as a member of technical staff at Bell Labs Digital Transmission Division, 9 years as Engineering Manager for SIE Geosource and 3 years as Research Director of PCES. He managed and contributed to a wide range of electronic developments from communications to image processing. He has several patents.

He is a current member of ASTM Committee E05 on Fire Sciences, member of the IEEE Computer Society, Faculty Member Zurich American Risk Engineering Management Training, and Research Board member for the Public and Private Business Inc. He maintains close ties with industry laboratories working on corrosivity and other non-thermal damage analysis and restoration techniques. He has lectured widely for Universities and for International Conferences. Articles by Mr. Mc Daniel have appeared in the *Disaster Recovery Journal, IEEE Spectrum, Survive The Business Continuity Magazine, The Bell System Journal,* and *Communications News.*

Dru Meadows

Dru Meadows, AIA, CCS, CSI is a Principal in *theGreenTeam, Inc.,* providing strategic environmental consulting services for the building industry. *theGreenTeam, Inc.* principals have participated in many environmental programs, the majority of which have received significant recognition, including: the White House Closing the Loop Award 1999; the American Institute of Architects Top 10 Environmental Projects 1998; Oklahoma Governor's Award 1996; and the United Nations Earth Summit Global Sustainability Award 1994.

Among her many responsibilities, Ms. Meadows chairs the ASTM Committee on Sustainability; represents the USA on the ISO Committee on Design Life of Building; and is Secretary for the ISO Committee on Sustainability. She has served on the Board of Directors for the Product Emission Testing Laboratory, U.S. Green Building Council, and Oklahoma Recycling Association.

John Messervy

John Messervy is the Director of Capital and Facility Planning for Partners HealthCare System Inc., an affiliation of 4 acute and 4 specialty hospitals with 2400 beds serving eastern Massachusetts. Prior to becoming the PHS Director, he was Vice President for Planning and Construction for Massachusetts General Hospital in Boston.

John was previously the Director of Project Management for the Division of Capital Planning and Operations of the Commonwealth of Massachusetts. He was responsible for $400 million of new construction and was instrumental in introducing alternative project delivery systems into the procurement of public facilities, particularly design/build. John is a registered architect and a graduate of the Massachusetts Institute of Technology with a Masters degree in city planning and advanced studies in architecture.

Susan Mireley

Susan Mireley is a faculty member in the Department of Human Environment and Design (HED) at Michigan State University (MSU). She teaches both undergraduate and graduate courses and guides graduate students. She is also currently the Academic Project Coordinator for the Master's Level Certificate Program in Facilities Management being offered over the Internet by HED. Dr. Mireley received her Ph.D. in Family and Child Ecology at MSU and her Masters Degree in History at Iowa State University. She has been an Extension Specialist with both the Purdue University Cooperative Extension Service and the MSU Extension Service. In these roles, she has developed educational programs for adult education audiences in both states. She has also been responsible for training County Extension Agents to deliver various educational programs in local areas in both states. Dr. Mireley has focused major educational efforts on the areas of rural housing, home ownership education and manufactured housing, developing various educational bulletins, teaching resources, and training practicing professionals in preparation for certification examinations. As an Extension Specialist, she has been recognized for her efforts in securing funding to both develop and deliver programs within her state and she spoken on both manufactured housing and home ownership education across the Midwest and the United States.

Joel Ratekin

Joel Ratekin is the Director of Workplace Strategies for American Express Real Estate. His responsibilities include the creation and implementation of space planning standards, virtual office management, alternative officing strategies and occupancy database networks. A native of Iowa, Mr. Ratekin earned his degree in Design and the Human Environment from the University of Northern Iowa, and has over thirteen years of facility planning experience in the corporate work place.

Ellen M. Reilly

As Founder, President and CEO, Ms. Reilly is responsible for Port's current and future product development, business strategy and management of company operations. Her successful remote work framework and unique telecommuting and mobile work operating and management products have become the foundation for Port Remote Services, Inc.. Port, as a full-service solution provider, has recently expanded to include extensive technology support services. Ms. Reilly holds a MBA with honors from Columbia University's Graduate School of Business and a B.S. from Purdue University. She speaks nationally, has appeared on cable television, occasionally conducts seminars (e.g., the American Management Association) and was the Founder (1996) and past President of the educational, non-profit organization, the International Telework Association Council (ITAC). In addition she has authored articles and is regularly quoted on the subject of telecommuting/remote work in publications such as *The Wall Street Journal, Crains, HR Focus, Alt.Office Journal, United Way Tech News, Computerworld* and *American Banker.* Furthermore, Ms. Reilly has authored numerous financial case studies, of which some were recently published in a ULI book publication earlier this year.

Eric Richert

Eric Richert is the Director of Workplace Operations and Research for Sun Microsystems, Inc., a $12 billion computer company with approximately 30,000 employees worldwide.

Prior to his current role, Mr. Richert was Director of Sun's Workplace Effectiveness Group (now the research arm of Workplace Operations and Research), Director of Design and Construction, and Director of Real Estate and the Workplace for Sun Microsystems Computer company, Sun's largest operating company. He has been with Sun since 1989.

From 1977–1987, Mr. Richert was a partner in a general architectural practice in Palo Alto, California.

Mr. Richert graduated from Syracuse University in 1971 with a Bachelor of Arts and Masters of Architecture degrees, and from the University of California, Berkeley, in 1989 with a Masters of Business Administration degree.

Edmond P. Rondeau, AIA, CFM

Ed Rondeau is the Director of Global Development and Learning for the International Development Research Council (IDRC) which is a global membership association for corporate real estate executives and associated service and development professionals.

He holds a MBA in real estate from Georgia State University and a Bachelor of Architecture from the Georgia Institute of Technology. He is a registered architect, and is a Certified Facility Manager (CFM). Ed served as the 1988 President of the International Facility Management Association (IFMA), and was elected a Fellow of the IFMA in 1992.

Ed has spoken throughout the U.S. and internationally, and is a co-author of a number of books including *Facility Management* for which he received the IFMA Distinguished Author Award in 1997.

Christine Ross

Chris Ross is a Workplace Strategist with Cisco Systems, the world leader in networking for the Internet. She is responsible for developing the strategies, tools, processes and partnerships that support Cisco's innovative, business orientated work environments worldwide.

Prior to Cisco, Ms. Ross was the business unit liaison responsible for the development, implementation and support of the alternative office environment for Tandem Computers Inc.'s U.S. Sales force.

She holds a BS in Organizational Behavior and Industrial Relations from the University of California, Berkeley and an MBA from Santa Clara University.

Jayakrishna Shankavaram

Jayakrishna Shankavaram is a Post-Doctoral Research Fellow in the Center for Building Performance and Diagnostics at Carnegie Mellon University. Dr. Shankavaram has been actively researching the design and management of high performance office environments, along with co-authoring several publications on advanced workplace strategies. He has consulted extensively with leading international corporate and federal organizations on projects ranging from post-occupancy evaluation of building facilities to developing effective facility design and management strategies. His principal research interests include, user-based, flexible building infrastructures, retrofit technologies, economic modeling, developing performance criteria for advanced building technologies and sustainable master-planning of building communities. He received his Bachelor of Architecture degree from the University of Bombay, India in 1990, Master of Science in Architecture and Ph.D. in Building Performance and Diagnostics from Carnegie Mellon University in 1993 and 1999, respectively. He is a recipient of the Rotary Foundation Ambassadorial Scholarship in 1991–92 and has also received the Graduate Service award for outstanding contributions to the Carnegie Mellon community in 1998.

Meredith Spear

Meredith Spear is a graduate of the University of Michigan and Yale University and a former hospital executive. She focuses her consulting practice on facility and operational planning for academic medical centers and integrated delivery systems.

Ms. Spear is nationally recognized for her expertise in the functional and space programming of diagnostic and treatment facilities and for her experience in research space allocation. She is a skilled communicator and is highly regarded by faculty and executives for her broad comprehension of the challenges facing academic medicine, the creativity of her solutions, and her ability to bring groups with diverse or competing interests to consensus.

Timothy Springer

Timothy Springer is a recognized authority on strategy development, decision making, organizational leadership, environment and behavior, ergonomics and productivity in high technology workplaces. He writes extensively on issues affecting organizations, work and workplaces. His comments have appeared in *Business Week, The Wall Street Journal,* and *Consumers Digest* as well as trade and industry publications.

Dr. Springer has served on the faculty at Illinois State University and as Director of the Facilities Management program and the Center for International Facilities Research (CIFR) at Grand Valley State University. He was Chairperson of Human Environment and Design Department at Michigan State University where he developed and led the Facilities Design and Management graduate program.

Dr. Springer earned his Ph.D. from the University of South Dakota.

Gary Steffy

Gary Steffy is the President of Gary Steffy Lighting Design Inc. and is one of the world's preeminent lighting designers. He is a fellow and past president of the International Association of Lighting Designers (IALD), a member of the U.S. National Committee of the International Commission on Illumination (CIE), a member of the Illuminating Engineering Society of North America (IESNA), and a founding director of the National Council on Qualifications for the Lighting Professions (NCQLP). His activities in the IESNA include appointment to its Technical Review Council. He is the author of Lighting the Electronic Office and Architectural Lighting Design. He has written many articles for the lighting magazines Architectural Lighting and Lighting Design + Application. He has taught lighting design at Michigan State University, Pennsylvania State University, University of Michigan and Wayne State University. His firm's work has been featured in *Architectural Record* and *Interior Design* and includes interior and exterior lighting for commercial, hospitality, institutional, municipal and health care clients.

Dana Stewart

Dana Stewart is Chairperson and Professor of the Department of Human Environment and Design in the College of Human Ecology at Michigan State University. Dr. Stewart came to MSU from the University of Manitoba where she was the administrator of the first graduate program in facility management in Canada. Dr. Stewart won the Distinguished Educator of the Year Award from IFMA in 1999 for the development of the Internet-based Master's Level Certificate Program in Facility Management.

Francoise Szigeti

Francoise Szigeti is Vice Chair, ASTM Subcommittee E06.25 on Whole Buildings and Facilities; and past Chair, ISO Technical Subcommittee TC 59/SC 2 on Terminology and harmonization of language.

She is a co-author of the *Serviceability Tools and Methods (ST&M),* including tools for measuring building loss features and calculating the Building Loss Factor (BLF), for determining and specifying occupant requirements, and for rating buildings. Szigeti is also a co-author of the first *1987 IFMA Benchmark Report.* She was a member of the Project team for ORBIT-2. Both were based on her proposal. The ORBIT-2 project (1983–1984) was a major North American multi-sponsor study about change in offices, information technology, and the organizations that use them.

Szigeti received the Environmental Design Research Association (EDRA) Lifetime Career Award, 1997.

Eric Teicholz

Eric Teicholz is President of Graphic Systems, Inc. (GSI), a Cambridge-MA firm specializing in facility management technology consulting and system's integration. Teicholz is a contributing editor for several magazines and the author of nine books, including McGraw-Hill's award winning *CAD/CAM Handbook, Computer-Aided Facility Management,* and *Facility Management Technology: Lessons from the U.S. and Japan* (John Wiley & Sons). Teicholz lectures extensively and has current consulting projects in North America, Asia, Europe and South America.

Teicholz is an architect educated at Harvard University. He remained at Harvard's Graduate School of Design for a number of years as an Associate Professor in Architecture and as Director of Harvard's largest R&D facility, the Laboratory for Computer Graphics and Spatial Analysis, which performed research and software development in the area of CAD and Geographic Information Systems. While at Harvard, Teicholz designed and helped develop the first commercial architectural CAD system. Teicholz occasionally guest lectures at the Massachusetts Institute of Technology on CAD and CAFM technology.

Carroll Thatcher

Carroll Thatcher is the Founder and Expert Advisor of Carroll Thatcher Planning Group Inc., Ottawa, Ontario, recognized internationally as a leading-edge facility planning and consulting firm. Carroll is a lifetime Certified Facility Manager, an experienced strategic space planner and design consultant, with over 25 year's experience.

Carroll is a coordinator for the development of the four courses in a Master's Level Facility Management Certificate Program being developed at Michigan State University (MSU). In 1999, at MSU, Carroll delivered the first course in this certificate program: "Facility Management: Theory and Principles," a distance learning course delivered completely via the Internet.

As a member of the IFMA Standards Committee, she has worked on the development and refinement of advanced methodologies for determining functional requirements and allocating space. She is a co-author of the *Serviceability Tools & Methods* books and a regular contributor to facilities management publications.

Graham Lane Thomas

Mr. Thomas trained as an architect and is registered in the USA and EEC. He has extensive experience in the design and implementation of facilities information systems and integrated architectural and engineering CAD databases. He spent six years leading the evolution of a Facilities Information System for an international financial institution linking space data and drawings to work order applications, network management systems, telecommunication systems, financial systems, HRMS systems, and the building automation systems for energy management, security, and fire alarms. The overall system was nominated for a Smithsonian / Computerworld award for the innovative use of technology in a corporate environment. He lead the project team for an international aerospace corporation to convert 40,000 CAD drawings to AutoCAD from a mainframe CAFM application, train nearly 100 users, and provide a custom interface in AutoCAD utilizing R.14 and Visual Basic for Applications.

William Tracy

William B. Tracy, AIA, is currently proprietor of BAM!, Inc., a new firm focusing on building area measurement and management. Previously, he has been a Senior Associate with the firm of Gensler, Vice President of Cost, Planning and Management, International, and Assistant Manager at Caudill, Rowlett Scott, Architects. He has written and spoken on subjects related to space management to organizations such as IFMA, BOMA and NIAOP.

Bill is an active member in the American Institute of Architects (AIA), and has served as President of the Denver Chapter in 1994. His educational background includes a Bachelor of Architecture from the University of Illinois and a Master of Business Administration from the Wharton School at the University of Pennsylvania.

Alan Stephen White

Alan White is professor at the Australian Defense Force Academy (University of New South Wales), providing a Master degree course in Facility Management. Prior to joining the Australian Defense Force Academy, Alan was the head of the School of Building and Construction Economics at RMIT University. Alan currently holds several facility and construction company directorships.

Alan's current research interests address the applications of Systems Engineering and Systems Theory to the integration of Facility Planning with Corporate Planning.

Mr. White is a graduate of the University of Adelaide and the University of Melbourne. He has thirty years commercial experience in the delivery of facility projects and holds an Unrestricted Class A license for the delivery of infrastructure projects in the Australian Capital Territory.

James A. White

James A. White is a Project Facilitator and Support Specialist working for Graphic Systems, Inc. He has been through various phases of the Congressional Information Facility Management System (CIFMS) implementation effort taking place at the Architect of the Capitol (AOC). Mr. White contains a thorough understanding of the internal business processes and how the system users interact with the work management system.

Current projects include research in the probable merging of Enterprise Energy Management (EEM) systems and Energy Information and Analysis Software (EIAS) with current industry leaders in Computer Aided Facility Management (CAFM) tools and web collaboration for facility managers.

Mr. White holds a Bachelor of Science degree in Management Information Systems and Decision Sciences from George Mason University.

INDEX

C

J

K

L

X

CD-ROM WARRANTY

This software is protected by both United States copyright law and international copyright treaty provision. You must treat this software just like a book. By saying "just like a book," McGraw-Hill means, for example, that this software may be used by any number of people and may be freely moved from one computer location to another, so long as there is no possibility of its being used at one location or on one computer while it also is being used at another. Just as a book cannot be read by two different people in two different places at the same time, neither can the software be used by two different people in two different places at the same time (unless, of course, McGraw-Hill's copyright is being violated).

LIMITED WARRANTY

McGraw-Hill takes great care to provide you with top-quality software, thoroughly checked to prevent virus infections. McGraw-Hill warrants the physical CD-ROM contained herein to be free of defects in materials and workmanship for a period of sixty days from the purchase date. If McGraw-Hill receives written notification within the warranty period of defects in materials or workmanship, and such notification is determined by McGraw-Hill to be correct, McGraw-Hill will replace the defective CD-ROM. Send requests to:

McGraw-Hill
Customer Services
P.O. Box 545
Blacklick, OH 43004-0545

The entire and exclusive liability and remedy for breach of this Limited Warranty shall be limited to replacement of a defective CD-ROM and shall not include or extend to any claim for or right to cover any other damages, including but not limited to, loss of profit, data, or use of the software, or special, incidental, or consequential damages or other similar claims, even if McGraw-Hill has been specifically advised of the possibility of such damages. In no event will McGraw-Hill's liability for any damages to you or any other person ever exceed the lower of suggested list price or actual price paid for the license to use the software, regardless of any form of the claim.

McGRAW-HILL SPECIFICALLY DISCLAIMS ALL OTHER WARRANTIES, EXPRESS OR IMPLIED, INCLUDING, BUT NOT LIMITED TO, ANY IMPLIED WARRANTY OF MERCHANTABILITY OR FITNESS FOR A PARTICULAR PURPOSE.

Specifically, McGraw-Hill makes no representation or warranty that the software is fit for any particular purpose and any implied warranty of merchantability is limited to the sixty-day duration of the Limited Warranty covering the physical CD-ROM only (and not the software) and is otherwise expressly and specifically disclaimed.

This limited warranty gives you specific legal rights; you may have others which may vary from state to state. Some states do not allow the exclusion of incidental or consequential damages, or the limitation on how long an implied warranty lasts, so some of the above may not apply to you.